Electronic Devices and Circuit Fundamentals

RIVER PUBLISHERS SERIES ELECTRONIC MATERIALS, CIRCUITS AND DEVICES

The "River Publishers Series in Electronic Materials, Circuits and Devices" is a series of comprehensive academic and professional books which focus on theory and applications of advanced electronic materials, circuits and devices. This includes analog and digital integrated circuits, memory technologies, system-on-chip and processor design. Also theory and modeling of devices, performance and reliability of electron and ion integrated circuit devices and interconnects, insulators, metals, organic materials, micro-plasmas, semiconductors, quantum-effect structures, vacuum devices, and emerging materials. The series also includes books on electronic design automation and design methodology, as well as computer aided design tools.

Books published in the series include research monographs, edited volumes, handbooks and textbooks. The books provide professionals, researchers, educators, and advanced students in the field with an invaluable insight into the latest research and developments.

Topics covered in this series include:-

- Analog Integrated Circuits
- Data Converters
- Digital Integrated Circuits
- Electronic Design Automation
- Insulators
- Integrated circuit devices
- Interconnects
- Memory Design
- MEMS
- Nanoelectronics
- Organic materials
- Power ICs
- Processor Architectures
- Quantum-effect structures
- Semiconductors
- Sensors and actuators
- System-on-Chip
- Vacuum devices

For a list of other books in this series, visit www.riverpublishers.com

Electronic Devices and Circuit Fundamentals

Dale R. Patrick

USA

Stephen W. Fardo

Eastern Kentucky University, USA

Ray E. Richardson

Eastern Kentucky University, USA

Vigyan (Vigs) Chandra

Eastern Kentucky University, USA

Routledge
Taylor & Francis Group

NEW YORK AND LONDON

Published 2023 by River Publishers
River Publishers
Alsbjergvej 10, 9260 Gistrup, Denmark
www.riverpublishers.com

Distributed exclusively by Routledge
605 Third Avenue, New York, NY 10017, USA
4 Park Square, Milton Park, Abingdon, Oxon OX14 4RN

Electronic Devices and Circuit Fundamentals / by Dale R. Patrick, Stephen W. Fardo, Ray E. Richardson, Vigyan (Vigs) Chandra.

Routledge is an imprint of the Taylor & Francis Group, an informa business

ISBN 978-87-7022-741-4 (print)
ISBN 978-10-0087-974-2 (online)
ISBN 978-1-003-39313-9 (ebook master)

While every effort is made to provide dependable information, the publisher, authors, and editors cannot be held responsible for any errors or omissions.

Contents

Preface

Electronics: Devices and Circuit Fundamentals by Patrick, Fardo, Richardson, and Chandra explores many fundamental topics in a basic and easy-to-understand manner. This book, and the accompanying **DC-AC Electrical Fundamentals** by the same co-authors have been developed using a classic textbook – **Electricity and Electronics: A Survey (5th Edition)** by Patrick and Fardo – as a framework. Both new books have been structured using the same basic sequence and organization of the textbook as previous editions. The previous edition of **Electricity and Electronics: A Survey** contained 18 chapters, 8 in the **Electricity** section and 10 in the **Electronics** section.

    Electronics: Devices and Circuit Fundamentals has been expanded to includes additional chapters, further simplifying content and providing a more comprehensive coverage of fundamental content. Expanded content for this textbook includes 23 chapters. The content has been continually updated and revised through new editions and by external reviewers throughout the years. Additional quality checks to ensure technical accuracy, clarity, and coverage of content have always been an area of focus. Each edition of the text has been improved through the following features:

1. Improved and updated text content
2. Improved usage of illustrations and photos
3. Use of color to add emphasis and clarify content

Organization of the Book

The two separate books, **DC-AC Electrical Fundamentals** and **Electronic Devices and Circuit Fundamentals**, now provide an even better comprehensive reference for the following:

- Survey of Electrical and Electronic Engineering Fundamentals
- Direct Current (DC) Circuit Fundamentals
- Alternating Current (AC) Circuit Fundamentals

- Electronic Device Fundamentals
- Electronic Circuit Fundamentals

Expanded content for **Electronic Devices and Circuit Fundamentals** is a basic introductory text with comprehensive coverage of basic electronic topics. Key concepts in the textbook are presented using the "big picture" or "systems" approach that greatly enhances learning. Many applications, testing procedures, and operational aspects of electronic devices and circuits are discussed through applications and illustrations. The text is divided into two sections: 1) **Electronic Devices** and 2) **Electronic Circuits**. The chapters are organized to include the following:

- Introduction
- Learning Objectives
- Chapter Outline
- Key Terms
- Major Content Discussions
- Self-Examinations
- Chapter Summary
- Formulas/Problems
- Analysis and Troubleshooting
- Glossary of Terms (defined)
- Answers to Self-Examinations

Electronic Devices and Circuit Fundamentals is appropriate for use with the following: Post-secondary technical programs, including colleges, universities, and community colleges, High School (11th and 12th grades) technical programs, business, and industry training programs or for individual self-study to learn about the exciting engineering/technology areas of electronics.

A comprehensive Solutions Manual is available to accompany the textbook. The organization of the Solutions Manual (by Chapter) is–

1. Chapter Outline
2. Learning Objectives
3. Key Terms
4. Figure List
5. Chapter Summary
6. Formulas
7. Answers to Examples / Self-Exams
8. Glossary of Terms (defined).

Acknowledgment

The authors would like to thank the many companies that have provided photographs and technical information for the book.

Dale R. Patrick
Stephen W. Fardo
Ray E. Richardson
Vigyan 'Vigs' Chandra

List of Figures

List of Tables

List of Abbreviations

AP	Access point
AMPS	Advanced mobile phone system
AC	Alternating current
ANSI	American National Standard Institute
AM	Amplitude modulated
ADC	Analog-to-digital converter
AF	Audio frequency
BW	Bandwidth
BFO	Beat frequency oscillator
BCD	Binary-coded decimal
BCH	Binary coded hexadecimal
BCO	Binary coded octal
BJTs	Bipolar junction transistors
CdO	Cadmium oxide
CdS	Cadmium sulfide
CdSe	Cadmium selenide
CdTe	Cadmium tellurid
CRT	Cathode-ray tube
CLCC	Ceramic leaded chip carrier
CCD	Charge coupled device
CB	Citizen's band
CDMA	Code division multiple access
CB	Common base
CC	Common collector
CE	Common emitter
CMRR	Common-mode rejection ratio
CMOS	Complementary metal-oxide semiconductor
CW	Continuous-wave
CDA	Current-differencing amplifier
D-MOSFET	Depletion metal-oxide semiconductor field-effect transistor
DSP	Digital signal processing

DSS	Digital spread spectrum
DTV	Digital television
DAC	Digital-to-analog converter
DVOM	Digital volt-ohm-meter
DC	Direct current
DIP	Dual in-line package
ESN	Electronic serial number
eV	Electron volts
E-MOSFET	Enhancement metal-oxide semiconductor field-effect transistor
FCC	F ederal Communications Commission
FET	Field effect transistor
FF	Flip-Flop
FDMA	Frequency division multiple access
FM	Frequency modulation
GaAs	Gallium arsenide
GaAsP	Gallium arsenide phosphide
GaP	Gallium phosphide
GTO	Gate turn-off
GSM	Global system for mobile communication
HDTV	High definition television
IRED	Infrared emitting diode
IEEE	Institute of Electrical & Electronics Engineer
IGFET	I nsulated gate field-effect transistor
IC	Integrated circuit
IF	Intermediate frequency
JFET	Junction field-effect transistor
LCCC	Leadless ceramic chip carrier
LASCRs	Light-activated silicon controlled rectifiers
LDR	Light-dependent resistor
LED	Light-emitting diode
LCD	Liquid crystal display
LAN	Local area network
LTP	Lower trip point
LF	Low-frequency
MOPA	Master oscillator power amplifier
MF	Medium frequency
MOS	Metal-oxide semiconductor
MOSFET	Metal-oxide semiconductor field-effect transistor

NEMA	National Electrical Manufacturer's Association
NTSC	National Television System Committee
NL	No-load
Op-Amp	Operational amplifier
PRV	Peak reverse voltage
PM	Permanent-magnet
PGA	Pin-grid array
PLCC	Plastic leaded chip carrier
PCB	Printed circuit board
PUT	Programmable unijunction transistor
PRR	Pulse repetition rate
PWM	Pulse-width modulator
RF	Radio frequency
RMS	Root mean square
SSID	Service set identifier
SCR	Silicon controlled rectifier
SCS	Silicon-controlled switch
SPST	Single-pole, single throw
SR	Slew rate
SOIC	Small outline integrated circuit
SST	Solid-state lamp
SDTV	Standard TV
SMPS	Switching mode power supply
SID	System identification code
TCc	Temperature coefficient of capacitance
TDMA	Time division multiple access
TO	Transistor outline
TRF	Tuned radio-frequency
UHF	Ultra-high frequency
UJT	Unijunction transistor
UTP	Upper trip point
VBO	Vertical blocking oscillator
V-MOSFET	Vertical metal-oxide semiconductor field-effect transistor
VHF	Very high frequency
VLSI	Very large scale integration
VOM	Volt-ohm-milliammeter
WAP	Wireless access point
WLAN	Wireless local area network

1

Semiconductor Fundamentals

This chapter deals with the theory of materials that are used in the construction of electronic devices called *solid-state*. **Solid-state devices** are made of semiconductor material. You may recall from your studies of DC and AC electronics that a **semiconductor** is a solid material that has electrical properties that lie somewhere between those of a conductor and an insulator. Semiconductors have made possible such things as radio, television, computers, robots, smart phones, and electronic sound systems.

Semiconductor theory focuses on the flow of current carriers through semiconductor material. A basic understanding of semiconductor theory will enable you to understand the electrical characteristics of solid-state devices and how they operate.

Objectives

After studying this chapter, you will be able to:

1.1 describe the structure and properties of an atom;
1.2 compare and contrast ionic bonding, covalent bonding, and metallic bonding;
1.3 explain the electrical difference between conductors, semiconductors, and insulators;
1.4 explain how current carriers move through semiconductors;
1.5 analyze and troubleshoot semiconductors.

Chapter Outline

1.1 Atomic Theory
1.2 Atom Bonding
1.3 Insulators, Semiconductors, and Conductors
1.4 Semiconductor Materials
1.5 Analysis and Troubleshooting – Semiconductor Materials

Key Terms

atom
atomic numberatomic weight
breakdown voltage
compound
covalent bonding
doping
electron
electron volt
element
extrinsic material
forbidden gap
intrinsic material
ionic bonding
matter
metallic bonding
molecule
neutron
N-type material
nucleus
orbital
proton
P-type material
valence electrons

1.1 Atomic Theory

Nearly every study of electronic devices starts with an investigation of atomic theory. Atomic theory describes the structure of matter and the atom. It is necessary for you to learn this information so that you will be able to understand how semiconductor material responds when energy, such as electricity and heat, is applied.

1.1 Describe the structure and properties of an atom.

In order to achieve objective 1.1, you should be able to:

- describe the characteristics of electrons, protons, and neutrons;
- describe the relationship between the atomic number and atomic weight of an atom;

- explain how electrons are distributed across shells and energy levels;
- define the terms *matter, element, nucleus, electron, proton, neutron, atomic weight, atomic number, shell,* and *energy level.*

Matter, Elements, and Atoms

You should recall from your elementary or high school science classes that anything that occupies space and has weight is called **matter.** Matter can be a solid, liquid, or gas. It is made of one or more types of elements. An **element** is the smallest unit of matter that consists of a single type of atom. An **atom** is the smallest particle to which an element can be reduced and still retain its identity. **Figure 1.1** shows the relationship between matter, elements, and atoms.

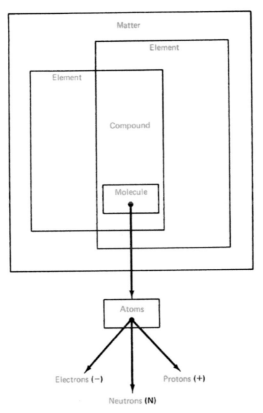

Figure 1.1   The relationship between matter, elements, and atoms. Matter can consist of one or more types of elements. An element consists of the same type of atoms.

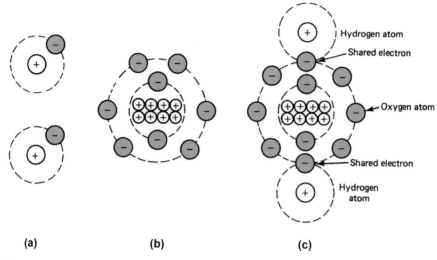

Figure 1.2   Water formed by combining hydrogen and oxygen: (a) hydrogen atoms; (b) oxygen atom; (c) water molecule. Matter is a compound when it consists of two or more types of elements.

Matter can be made of a single type of element, such as copper, or a combination of element types, such as water. Water is a combination of the elements hydrogen and oxygen. Matter that is made of several types of elements is called a compound. A **compound** is two or more elements that have been chemically combined. The smallest particle to which a compound can be reduced before being broken down into its basic elements is called a **molecule (Figure 1.2)**.

Later in this chapter, you will learn about how elements chemically combine. This knowledge will aid your understanding of the electrical differences between conductors, semiconductors, and insulators and how current carriers move through semiconductors.

Atomic Structure

Atoms consist of smaller particles called *electrons*, *neutrons*, and *protons*. **Figure 1.3** shows the particle structure of an **atom** of copper. (This structure is shown in a two-dimensional representation. Keep in mind that an atom is a three-dimensional object.)

Notice that the **nucleus** is the core of an atom. The nucleus of every atom is composed of one or more positively charged particles. Each of these

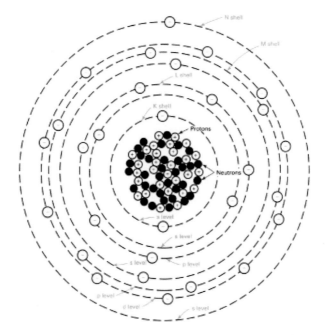

Figure 1.3 Two-dimensional model of a copper atom. The nucleus of an atom contains protons and neutrons. Electrons orbit the nucleus.

particles is called a **proton.** In addition to the protons, in all elements except the lightest, there are one or more neutrons. A **neutron** is a particle that has no electrical charge.

The study of electronic devices is primarily concerned with the electron content of an atom. An **electron** is a negatively charged particle that orbits the nucleus of an atom. An electron is extremely small. It weighs 1/1850 as much as a proton. Therefore, practically, all of the weight of an atom is in the nucleus.

Atomic Weight and Number

The **atomic weight** of an atom is the sum of protons and neutrons in the nucleus. The atomic weight of an *atom* is always a whole number, whereas the atomic weight of an *element* is rarely a whole number because it represents the average value of a large number of atoms. This average value is more accurately referred to as *atomic mass*. The *periodic table* in **Figure 1.4(a)** lists each element's atomic mass. Note the information presented in the periodic table.

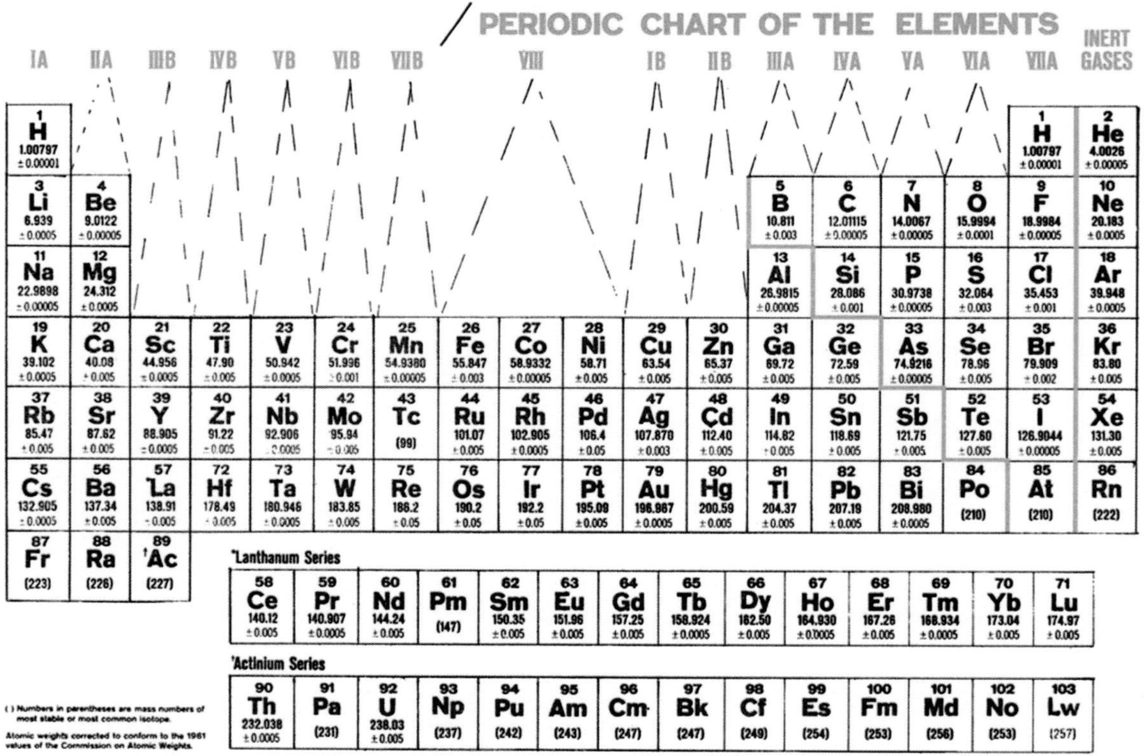

Figure 1.4 **A. (a)** Periodic table of the elements. **(b)** Elements in the periodic table showing groups, symbols, atomic number, atomic weight, and electron arrangement in shells 1−7 (K−Q). Courtesy: Sargent-Welch Scientific Co.

Group	Element	Symbol	Atomic Number	Atomic Weight	Electron Arrangement (shell)						
					K	L	M	N	O	P	Q
IA	Hydrogen	H	1	1.008	1						
	Lithium	Li	3	6.941	2	1					
	Sodium	Na	11	22.99	2	8	1				
	Potassium	K	19	39.10	2	8	8	1			
	Rubidium	Rb	37	85.47	2	8	18	8	1		
	Cesium	Cs	55	132.9	2	8	18	18	8	1	
	Francium	Fr	87	223	2	8	18	32	18	8	1
IIA	Berylium	Be	4	9.012	2	2					
	Magnesium	Mg	12	24.3	2	8	2				
	Calcium	Ca	20	40.08	2	8	8	2			
	Strontium	Sr	38	87.62	2	8	18	8	2		
	Barium	Ba	56	137.3	2	8	18	18	8	2	
	Radium	Ra	88	226.02	2	8	18	32	18	8	2
IIIA	Boron	B	5	10.81	2	3					
	Aluminum	Al	13	26.98	2	8	3				
	Gallium	Ga	31	69.72	2	8	18	3			
	Indium	In	49	114.8	2	8	18	18	3		
	Thallium	Tl	81	204.3	2	8	18	32	18	3	
IVA	Carbon	C	6	12.01	2	4					
	Silicon	Si	14	28.08	2	8	4				
	Germanium	Ge	32	72.5	2	8	18	4			
	Tin	Sn	50	118.6	2	8	18	18	4		
	Lead	Pb	82	207.2	2	8	18	32	18	4	
VA	Nitrogen	N	7	14	2	5					
	Phosphorus	P	15	30.97	2	8	5				
	Arsenic	As	33	74.92	2	8	18	5			
	Antimony	Sb	51	121.7	2	8	18	18	5		
	Bismuth	Bi	83	208.98	2	8	18	32	18	5	
VIA	Oxygen	O	8	15.99	2	6					

Figure 1.4 B.

Elements in the periodic table are listed in the order of the number of protons contained in the nucleus of their atoms. This number is known as the **atomic number.** No two elements have the same atomic number.

For example, hydrogen, which is the simplest atom, has one proton and no neutrons in its nucleus. Any atom that has one proton is always a hydrogen atom. The nucleus of a uranium atom has 92 protons. Any atom that has 92 protons is always a uranium atom. Another example is the element zinc from

	Sulfur	S	16	32.06	2	8	6		
	Selenium	Se	34	78.9	2	8	18	6	
	Tellurium	Te	52	127.6	2	8	18	18	6
	Polonium	Po	84	210	2	8	18	32	18	6
VIIA	Fluorine	F	9	18.99	2	7			
	Chlorine	Cl	17	35.45	2	8	7		
	Bromine	Br	35	79.9	2	8	18	7	
	Iodine	I	53	126.9	2	8	18	18	7
	Astatine	At	85	210	2	8	18	32	18	7
VIIIA	Helium	He	2	4	2				
	Neon	Ne	10	20.17	2	8			
	Argon	Ar	18	39.94	2	8	8		
	Krypton	Kr	36	83.8	2	8	18	8	
	Xenon	Xe	54	131.3	2	8	18	18	8
	Radon	Rn	86	222	2	8	18	32	18	8
IB	Copper	Cu	29	63.54	2	8	18	1	
	Silver	Ag	47	107.87	2	8	18	18	1
	Gold	Au	79	196.97	2	8	18	32	18	1
IIB	Zinc	Zn	30	65.38	2	8	18	2	
	Cadmium	Cd	48	112.4	2	8	18	18	2
	Mercury	Hg	80	200.5	2	8	18	32	18	2

Figure 1.4   Continued.

the periodic table. Note that the atomic number of zinc is 30 and the **atomic weight** is 65.38.

Figure 1.4(b) lists the elements of the periodic table by group. The groups (IA, IIA, IIIA, etc.) are assigned based on the number of electrons in the atom's outer shell. For example, all Group IVA elements have 4 outer **shell** electrons. They are arranged in a vertical column.

Look at the sample "period 2 of the Periodic Table of Elements. Each period is arranged in a horizontal row. The period is determined by the sequence of outer shell electrons in the row (lithium = 1, beryllium = 2, boron = 3, carbon = 4, etc.). Note that neon on the right of period 2 is filled with 8 outer shell electrons. The atomic numbers, shown for each atom, are consecutive (1−8) from left to right. As new elements are discovered, they are entered into the proper position on the lower portion of the periodic table.

The number of neutrons in the atom of an element can be calculated by subtracting the atomic number from the atomic weight. This is represented by the following formula:

number of neutrons = atomic weight − atomic number.

To determine the atomic weight, the atomic mass is rounded to the nearest whole number. This is represented by the following formula:

atomic weight = atomic mass rounded to nearest whole number.

For example, uranium has an atomic mass of 238.03. To determine the atomic mass of uranium, 238.03 is rounded to the nearest whole number. The nearest whole number of 238.03 is 238. The number of neutrons can then be calculated by subtracting the atomic number from the atomic weight.

$$number\ of\ neutrons = atomic\ weight - atomic number$$
$$= 238 - 92$$
$$= 46.$$

Example 1-1:

What is the number of neutrons in hydrogen (H)? (Hydrogen has an atomic mass of 1.00794 and an atomic number of 1.)

Solution:

1. Round the atomic mass to the nearest whole number to determine the atomic weight.

$$atomic\ mass = 1.00794$$
$$atomic\ weight = 1.$$

2. Subtract the atomic number from the atomic weight.

$$number\ of\ neutrons = 1 - 1$$
$$= 0.$$

Related Problem:

What is the number of neutrons in silicon (Si)? (Use the periodic table in **Figure 1.4** to determine the atomic mass and atomic number.)

Shells

The electrons of an atom are not in an equal distance from the nucleus. They orbit the nucleus in shells, or layers. Neils Bohr, a Danish physicist, identified these **shells** by the letters *K*, *L*, *M*, *N*, *O*, and *P*. Take a look at **Figure 1.5.** Note that the K-shell is closest to the nucleus. Each shell or layer has a distinct energy level. Electrons do not exist in the space between these energy levels. The energy levels of a shell are identified by the letters *s*, *p*, *d*, *f*, *g*, and *h*. The s-level is closest to the nucleus. A shell may have from one to six distinct energy levels in its structure.

Shells	Energy Level	Maximum Number of Electrons
K	s	2
L	s	2
	p	6
M	s	2
	p	6
	d	10
N	s	2
	p	6
	d	10
	f	14
O	s	2
	p	6
	d	10
	f	14
	g	18
P	s	2
	p	6
	d	10
	f	14
	g	18
	h	22
Q	s	2

Figure 1.5 Electrons orbit the nucleus in shells. A shell may have from one to six distinct energy levels in its structure.

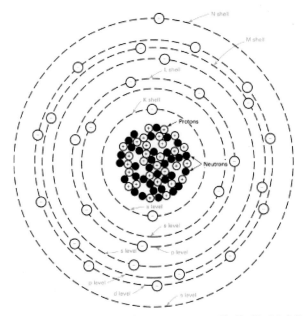

Figure 1.6 Shells and energy levels of a copper atom. Shells K−M follow the typical electron pattern. The N-shell, however, does not contain the full number of electrons for its energy level.

Energy levels represent two things: distance from the nucleus and the amount of energy possessed by an electron. Energy levels closest to the nucleus are of the lowest value while energy levels furthest away are of the highest value. This is because electrons with more energy occupy the energy levels furthest away from the nucleus. This largely depends on the number of electrons in an atom. The number of electrons in each energy level follows a unique pattern. This pattern is 2, 6, 10, 14, 18, and 22. **Figure 1-5** shows the shells, the energy levels found in each shell, and the maximum number of electrons that exist in each energy level.

Note that the P-shell has six energy levels. The maximum number of electrons at each energy level follows the 2, 6, 10, 14, 18, and 22 patterns. This number sequence appears in the other shells but decreases in value with the last level. The complexity of an atom usually dictates its electron shell and energy level assignment. For example, take a look at **Figure 1.6.** Note that shells K−M follow the typical electron pattern but that the N-shell, which is the last shell, contains only one electron in the s-level.

Another example is lithium (see **Figure 1.7**). Lithium has three protons and three neutrons in the structure of its nucleus. It also has three electrons to

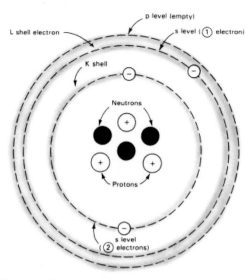

Figure 1.7 Two-dimensional representation of a lithium atom. The K-shell follows the typical electron pattern, but the L-shell does not.

balance the net positive charge of the nucleus. Note that two of these electrons fill the s-level of the K-shell. Since the K-shell only has one energy level, the third electron of the structure must appear in the s-level of the L-shell. The L-shell has s- and p-levels to be filled. The s-level has a capacity of 2, and the p-level has a capacity of 6. The p-level must be filled before the next shell is formed.

The electron in the s-level of the L-shell of lithium is a valence electron. **Valence electrons** are electrons in the outermost shell and represent the highest energy level of an atom. As you will see later in this chapter, valence electrons can effectively enter into some chemical activity with another atom.

Example 1-2:

What is the number of electrons in the energy levels of each shell for aluminum (Al)?

Solution:

1. Determine the number of electrons. Since the number of electrons is equal to the amount of protons, and the amount of protons is equal to the atomic number, the amount of electrons is 13.

electrons = atomic number

electrons = 13.

2. Use the chart in **Figure 1.6** to determine the number of electrons in the energy level of each shell. Electrons are distributed across shells starting with the K-shell.

K-shell = 2s-level
L-shell = 2s-level
= 6 p-level
M-shell = 2s-level
= 1p-level.

Related Problem:

What is the number of electrons in the energy levels of each shell for germanium (Ge)?

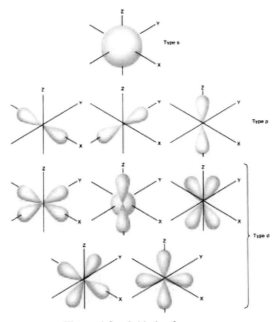

Figure 1.8   Orbitals of an atom.

Orbitals

The exact path an electron follows in the structure of an atom is unknown. However, this path is generally described as an orbital. An **orbital** shows the mathematical probability of where an electron will appear in the structure of an atom. It is believed that there are four types of orbitals in atoms. These are identified by the letters *s*, *p*, *d*, and *f*. **Figure 1.8** shows the representation of s, p, and d orbitals. The f-type of orbital is not shown in this sequence because there is very little evidence that it is as active as the others in atomic bonding, which you will learn about in the following section. This type of orbital is more meaningful in chemical analysis.

Self-Examination

Answer the following questions.

1. The number of protons in the nucleus of an atom is an indication of the _____.
2. The atomic weight of an atom is the sum of the number of _____ and _____ in the nucleus of an atom.
3. The atomic _____ represents the average value of protons and neutrons in a large number of atoms.
4. A negatively charged particle that orbits the nucleus of an atom is a(n) _____.
5. The (K-L-M-N-0-P-Q)-shell of an atom is closest to the nucleus. _____
6. The energy levels of a shell are identified by the letters ___, ___, ___, ___, ___, and ___.
7. The _____ shell of an atom has six energy levels.
8. A(n) _____ shows the mathematical probability of where an electron will appear in the structure of an atom.

1.2 Atom Bonding

In the previous section, you learned that valence electrons represent the highest energy level of the atom and that valence electrons can enter into some chemical activity with another atom. Chemical activity is simply the bonding of one atom with another through the gaining, losing, or sharing of valence electrons. Some chemical activities occur naturally, but most of them are induced through an outside force. Valence electrons of different atoms are frequently bonded together to form stabilized molecules or compounds. This section discusses three types of bonding: ionic bonding, covalent bonding, and metallic bonding.

1.2 Compare and contrast ionic bonding, covalent bonding, and metallic bonding.

In order to achieve objective 1.2, you should be able to:

- describe the characteristics of an atom that is in a stable state and ground state;
- define the terms *ionic bonding*, *covalent bonding*, and *metallic bonding*.

Atomic States

Chemical activity depends on the number of electrons in the outermost energy level of the valence shell. A shell with a full complement of electrons tends to be in a preferred atomic state. A full complement of electrons means that the highest number of electrons is allotted to an energy level. For example, a full complement of electrons for the s-level is 2, and a full complement of electrons for the p-level is 6 (see **Figure 1.6**). An atom with a full complement of electrons in its outermost energy level is in a **stable state** and is generally called a *stabilized atom*. A stabilized atom will not release electrons under normal conditions.

Helium is an example of an element whose atoms are in a stable state. Helium has two protons and two neutrons in its nucleus. Two electrons orbit the nucleus in the s-level of the K-shell. The full complement of electrons for the s-energy level is 2; therefore, helium is in a stable state.

Hydrogen is an atom that contains one proton in its nucleus. This proton, like all protons, possesses a positive charge. Remember that the number of electrons that exist in an atom tends to match the net positive charge of the nucleus. Hydrogen, therefore, has one electron in its structure. This electron orbits the nucleus in the K-shell in an s-type orbital pattern. If no additional energy is added to the electron, the atom is considered to be in its **ground state** or *lowest energy level*. Atoms in the ground state have a stronger tendency to bond in order to reach a stable state. Hydrogen does not have a full complement of electrons in the s-energy level; therefore, hydrogen is in a ground state.

Ionic Bonding

Ionic bonding is the process of bonding atoms through electrostatic force. Electrostatic force occurs through the attraction of opposite net charges between two atoms. This process typically occurs between atoms of different elements. Ionic bonding, therefore, creates a compound. You should recall that a compound is a combination of two or more elements to form an entirely different material.

The atomic structure of lithium lends itself very well to **compound** formation. A compound of lithium and hydrogen is called **lithium hydride**. In this compound, the valence electron of lithium is transferred to the K-shell of a hydrogen atom. The lithium atom takes on a +1 charge because it has lost an electron. It is now considered to be a positive lithium ion.

The lithium atom becomes a **positive ion** and the hydrogen atom becomes a **negative ion**. The net charges of the ionized particles bond the atoms through electrostatic force. An **ion** is an atom that has lost or gained an electron. If an atom loses an electron, it becomes positively charged. Gaining an electron causes an atom to be negatively charged. The hydrogen atom in the example of **Figure 1.9** gains an electron and becomes a negative ion with a charge of -1. Since unlike charged particles have an attracting power, a single molecule of lithium hydride is formed by this combination of atoms. Compounds formed in this manner are described as *electrovalent combinations* or *ionic bonded mixtures*. The net charges of the ionized particles bond the atoms and create a compound.

Table salt is a good example of **ionic bonding** that occurs naturally. **Figure 1.9** shows the ionic bonding between two different atoms. The sodium (Na) atom donates its one valence electron to a chlorine (CI) atom, which has seven valence electrons. On acquiring the extra electron, chlorine becomes a negatively charged ion, while the sodium atom becomes a positively charged ion. Since unlike charges attract, the positive sodium (Na^+) and negative chlorine (CI^{--}) ions are bonded together by an electrostatic force.

Copper oxide (Cu_2O) is also an example of **ionic bonding**. The valence electrons of two atoms of pure copper (Cu) are transferred and combined with one atom of oxygen (O). Copper is a good electrical conductor. Copper has one valence electron in the s-level of the N-shell (see **Figure 1.6**). Oxygen has six valence electrons in the L-shell; two of these are in the s-level and four are in the p-level (see **Figure 1.2**). The combination of two copper atoms and one oxygen atom forms a stable copper oxide compound. The two copper valence electrons combine with four electrons in the p-level of the L-shell of oxygen

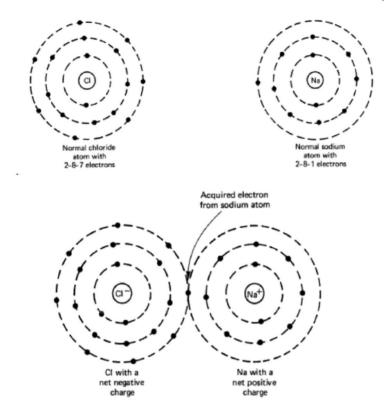

Normal chloride
atom with
2-8-7 electrons

Normal sodium
atom with
2-8-1 electrons

Acquired electron
from sodium atom

Cl with a
net negative
charge

Na with a
net positive
charge

Figure 1.9 Ionic bonding of chlorine (Cl) and sodium (Na) atoms to produce salt. When a valence electron of a chlorine is transferred to the M-shell of a sodium atom, chlorine becomes a negatively charged ion, while the sodium atom becomes a positively charged ion. The positive sodium (Na^+) and negative chlorine (Cl^{--}) ions are bonded together by an electrostatic force.

to form a full complement of six electrons. The stability of this compound makes copper oxide a good insulating material.

Covalent Bonding

Another form of bonding can take place when the outermost shell of an atom is partially filled. The atoms in this form of bonding position themselves so that the energy levels of their valence electrons combine. This form of bonding is called **covalent bonding** or **electron pair bonding.** No ions are formed in covalent bonding because the valence electrons are shared between the atoms. These electrons alternately shift back and forth between each atom.

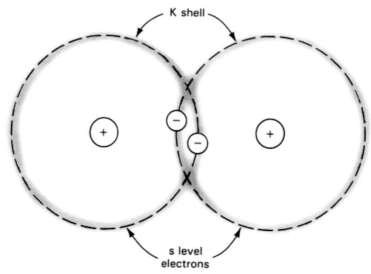

Figure 1.10 Covalent bonding of hydrogen atoms. Electrons are shared in this type of bonding.

A covalent force is, therefore, established between the connecting electrons. The covalent force bonds the atoms in a simulated condition of stability.

A two-dimensional illustration of covalent bonding is shown in **Figure 1.10**. Two hydrogen atoms are bonded in this drawing. Note that the K-shells of each atom overlap so that their electrons can be shared. In doing so, both have the equivalent of a full complement of electrons in their s-level.

The molecules of common gases, such as oxygen and nitrogen, are frequently held together by covalent bonding. Semiconductors such as carbon, silicon, and germanium are also bonded by this process. Covalent bonding is much stronger than ionic bonding.

Metallic Bonding

Metallic bonding is an internal force that holds atoms loosely together in a conductor. It occurs in good electrical conductors. An electrical conductor such as copper has one electron in the s-level of its N-shell. This electron is easily influenced by outside energy and has a tendency to wander around the material between different atoms. On leaving one atom, it immediately enters the orbital of another atom. The process is repeated until the outside energy is removed.

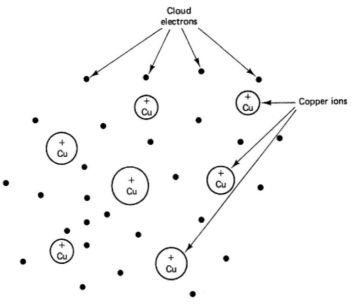

Figure 1.11   Metallic bonding of copper. In this type of bonding, electrons "float" around in a cloud that covers the positive ions. This "floating" cloud bonds the electrons randomly to the ions.

When an electron receives enough energy to leave an atom, it causes the original atom to become a positive ion. In metal, this process occurs at temperatures of 25°C. The process takes place randomly. This means that one electron is always associated with an atom but is not in one particular orbital path. As a result, a large number of the structural atoms of a piece of copper tend to share valence electrons.

In metallic bonding, there is a type of electrostatic force between positive ions and electrons. In a sense, electrons float around in positively charged clouds that surround the positive ions. This floating cloud of electrons tends to bond itself randomly to the positively charged ions. **Figure 1.11** shows an example of the metallic bonding of copper.

Self-Examination

Answer the following questions.

9. Atoms in a(n)_____ state have a stronger tendency to bond in order to reach a stable state.

10. Atoms in a(n) _____ state will not release electrons under normal conditions.
11. When two or more elements are used to form a different material, it is called a(n) _____.
12. The process of bonding atoms through an electrostatic force is called _____ bonding.
13. Table salt and copper oxide are examples of _____ bonding.
14. The condition of stability is simulated when atoms are _____ bonded.
15. Individual atoms of silicon are frequently connected together by _____ bonding.
16. _____ bonding occurs in good electrical conductors.

1.3 Insulators, Semiconductors, and Conductors

Solid materials are used primarily in electronic devices to achieve electrical conductivity. These materials can be divided into insulators, semiconductors, and conductors. As you have learned in your studies of DC and AC electronics, **insulators, semiconductors, and conductors** differ in their potential to achieve electrical conductivity because they vary in resistance. The resistance of these materials is related to the tendency of their electrons to leave the valence band and go into conduction. Materials that make a good conductor are those that are easily influenced by outside energy and, therefore, can go into conduction with a minimal amount of energy. Materials that make a good insulator are not easily influenced by outside energy and, therefore, need a greater amount of energy to go into conduction. The energy needed for a semiconductor falls somewhere in between this range. This section discusses the differences between conductors, semiconductors, and insulators and the energy needed to place them into conduction.

1.3 Explain the electrical difference between conductors, semiconductors, and insulators.

In order to achieve objective 1.3, you should be able to:

- describe the energy bands for conductors, semiconductors, and insulators;
- explain the terms *electron volt*, *forbidden gap*, and *breakdown voltage*.

Assume that we have a cubic centimeter block of an insulator, semiconductor, and conductor. If resistance is measured between opposite faces of each block, some very unique differences will be revealed. The insulator cube will measure several million ohms. The conductor cube, by comparison, will measure only 0.000001 Ω. The semiconductor cube will measure approximately 300 Ω. In effect, this shows a rather wide range of resistance differences in the three materials. All three materials can conduct, but only when a specific amount of energy is applied.

Energy-Level Diagrams

A convenient way to evaluate the relationship of conductors, semiconductors, and insulators is through the use of **energy-level diagrams**. This takes into account the amount of energy needed to cause an electron to leave its valence band and go into conduction. An energy-level diagram represents a composite of all atoms within the material. Energy-level diagrams of insulators, semiconductors, and conductors are shown in **Figure 1.12.** The valence band is located at the bottom and the conduction band is at the top of each diagram. The valence band represents the highest energy level that electrons can attain and still be influenced by the nucleus. Electrons in this band normally combine with valence-band electrons of other atoms to form molecules or compounds. A number of other electrons exist below the valence energy band. As a rule, we are only concerned with the response of valence and conduction band electrons in the operation of semiconductor devices. Electrons in the conduction band are not specifically bound to the nucleus

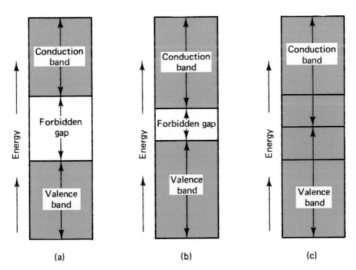

Figure 1.12 Energy-level diagrams for (a) insulators, (b) semiconductors, and (c) conductors.

and are free to move. Conduction band electrons have a higher energy level than the valence band electrons.

The **valence band** represents a composite energy level of the valence electrons of each atom. A specific amount of outside energy must be added to valence electrons to cause them to go into conduction. The area that separates the valence and conduction bands is referred to as the forbidden region. A certain amount of energy is needed to cause valence electrons to cross the forbidden region. If the energy is insufficient, electrons are not released for conduction. They will remain in the valence band.

The width of the **forbidden region** indicates the conduction status of a particular material. In atomic theory, the width of the forbidden gap is expressed in electron volts (eV). An **electron volt** is defined as the amount of energy gained or lost when an electron is subjected to a potential difference of 1 V. The atoms of each element have a specific energy-level value that permits conduction.

Insulators

The **energy-level diagram** of an insulating material has a very wide forbidden gap. A carbon crystal or natural diamond is an excellent insulator. Crystallized carbon has a forbidden gap of approximately 6 eV. The large width of this region keeps valence electrons from crossing the forbidden gap and going into conduction. For valence electrons to travel through the forbidden gap, they must acquire additional energy. The amount of energy needed by good insulators is generally very high. The energy level value that causes an insulator to go into conduction is often called **breakdown voltage.**

Even the best insulators will go into conduction if sufficient energy is applied. Thyrite is an example of this condition. Thyrite is commonly used as a lightning arrester. At normal voltages, it is an ideal insulator. When it is subjected to high voltage, electrons cross the forbidden gap and go into conduction. This shunts the high voltage to ground, thus protecting a device from lightning. When good insulators are operated at high temperatures, the increased heat energy causes valence electrons to go into conduction.

Semiconductors

The forbidden gap of a **semiconductor** is much smaller than that of an insulator. **Silicon**, for example, needs to gain 1.21 eV of energy at absolute zero ($-273°C$) to go into conduction. Energies of this magnitude are not easily acquired. As a result, the valence band remains full, the conduction band remains empty, and these materials respond as insulators. However, an increase in temperature causes the conductivity of this material to change. At normal room temperature or $-25°C$, the valence electrons acquire thermal energy that is greater than the normal eV value. This essentially reduces the width of the forbidden gap and causes a semiconductor to be a conductor. This particular characteristic is extremely important in solid-state electronic devices.

Conductors

The **energy-level diagram** of a conductor is quite unusual compared with other materials. In a sense, the valence band and conduction band are one and the same. Conductivity is explained as having an interaction of different energy levels of the valence band.

Studies show that the atoms of most metals and semiconductors are in the form of a **crystal lattice structure**. A crystal consists of a space array of atoms or molecules built up by regular repetition in a three-dimensional pattern. The energy levels of electrons in the crystal do not respond in the same manner as those of an individual atom. When atoms form crystals, the energy levels of the inner-shell electrons are not affected by the presence of neighboring atoms. However, the valence electrons of individual atoms are often shared by more than one atom.

The new energy level of **valence electrons** is found in a distinct band. The spacing between the energy levels of this band is very small compared with that of isolated atoms. Thus, electrons are free to absorb energy and to move from one point to another, conducting heat and electricity. In good conductors, the energy-level bands of valence electrons tend to overlap. This lowers the energy level of valence electrons and increases the electrical conductivity of the material.

Self-Examination

Answer the following questions.

17. An energy level diagram shows the _____ band at the bottom and the _____ band at the top.
18. The _____ band represents the highest energy level that electrons can attain and still be influenced by the nucleus.
19. The _____ gap or region separates the valence band and conduction bands of an energy-level diagram.

20. The width of the forbidden region of an energy-level diagram is expressed in _____.
21. The energy-level diagram of a(n) _____ shows a wide forbidden gap.
22. The energy-level diagram of a(n) _____ shows a narrow forbidden gap.
23. In the energy-level diagram of a(n) _____, the valence band and conduction band overlap.
24. The energy that causes an insulator to go into conduction is called _____ voltage.

1.4 Semiconductor Materials

1.4 Explain how current carriers move through semiconductors.

In order to achieve objective 1.4, you should be able to:

- describe the characteristics of N-type and P-type semiconductor material;
- define the terms *intrinsic material*, *extrinsic material*, *doping*, and *holes*.

To understand how electronic devices work, you need to have a basic understanding of the structure of atoms and the interaction of atomic particles. This section expands the basic information of the previous section and introduces the **P-N junction**. P-N junctions are the basis of operation for many electronic devices, such as diodes and transistors.

Intrinsic Material

Materials such as **silicon, germanium**, and **carbon** are natural elements found in crystalline form. Instead of being a random mass, these atoms are arranged orderly. A **crystal** of silicon or germanium forms a definite geometric pattern, namely, a cube. **Figure 1.13** illustrates a two-dimensional simplification of the silicon crystal showing only the nucleus and valence electrons. Note that the electrons of individual atoms are covalently bonded together. Germanium has a similar type of crystal structure.

Crystals of silicon and germanium can be manufactured by melting the natural elements. The process is somewhat complex and rather expensive. Manufactured crystals must be made extremely pure to be useable in semiconductor devices. A very pure semiconductor crystal is called an **intrinsic material.** Germanium is considered to be intrinsic when only 1 part of impurity exists in 10^{10} parts of germanium. Silicon is intrinsic when the impurity ratio is $1:10^{13}$. In more sophisticated solid-state devices, the ratio may be even higher.

An intrinsic crystal of silicon would appear as the structure in **Figure 1.15**. Covalent bonding changes the electrical conductivity of the material, causing each group of atoms to have a simulated condition of stability. Thus, only a limited number of free electrons are available for conduction. Therefore, silicon and germanium crystals respond to some extent as insulators.

Intrinsic silicon at $-273°$ C is considered to be a perfect insulator. The valence electrons of each atom are firmly bonded together in perfect covalent bonds. No free electrons are available for conduction. In actual circuit operation, the absolute zero condition is not very meaningful because it cannot readily be attained. Any temperature above $-273°C$ causes silicon to become somewhat conductive. The insulating quality of a semiconductor material is, therefore, dependent on its operating temperature.

At room temperature, which is approximately $25°C$, silicon atoms receive enough energy from heat to break their bonding. A number of free electrons become available for conduction. At room temperature, intrinsic silicon becomes somewhat conductive.

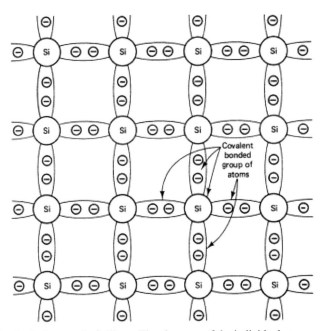

Figure 1.13 Intrinsic crystal of silicon. The electrons of the individual atoms are covalently bonded together.

Holes

An important solid-state event takes place when intrinsic silicon goes into conduction. For every electron that is freed from its covalent bonding, a void is created. This spot, which is normally called a **hole**, represents an electron deficiency. A hole in a covalent bonding group occurs when an electron is released. An increase in the temperature of a piece of silicon causes the number of free electrons and holes to increase.

It should be remembered that when a **neutral atom** loses an electron, it acquires a positive charge. The atom then becomes a positive ion. Since an atom acquires a hole at the same time as the positive charge, the hole bears a positive charge. It should be noted, however, that the crystal remains electrically neutral. For every free electron, there is an equivalent hole. These two balance the overall charge of the crystal.

Hole Flow

When a valence electron leaves its covalent bond to become a **free electron**, a hole appears in its place. This **hole** can then attract a different electron from a nearby bonded group. On leaving its bonded group, the electron creates a new hole. The original hole is then filled and becomes electrically neutral. Each electron that leaves its bonding to fill a hole creates a new hole in its original group. In a sense, this means that electrons move in one direction and holes move in the opposite direction. Since electron movement is considered to be an electric current, holes are also representative of current flow.

Electrons are called **negative current carriers** and holes are **positive current carriers** that move in opposite directions in a semiconductor material. When voltage is applied, holes move toward the negative side of the source. Electron current flows toward the positive side of the source. This condition takes place only in a semiconductor material. In a conductor, such as copper, there is no covalent bonding. The current carriers in metallically bonded materials are electrons.

Figure 1.14 shows how an intrinsic piece of silicon responds at room temperature when voltage is applied. Note that the free electrons move toward the positive terminal of the battery. Electrons leaving the semiconductor flow into the copper connecting wire. Each electron that leaves the material creates a hole in its place. The holes appear to move by jumping between covalent bonded groups. Holes are attracted by the negative terminal connection. For each electron that flows out of the material, a new electron enters at the negative connection point.

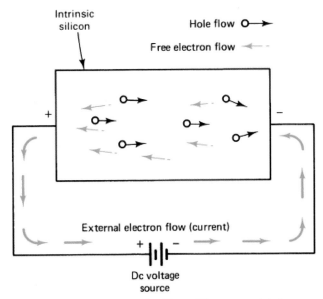

Figure 1.14 Current carrier movement in silicon. Electrons and holes move in opposite directions.

The process of hole flow and electron flow through the material is continuous as long as energy is supplied. Current flow is the resulting carrier movement. An intrinsic semiconductor has an equal number of current carriers moving in each direction. The resulting current flow of a semiconductor is limited primarily to the applied voltage and the operating temperature of the material.

Extrinsic Material

Pure silicon or germanium in its intrinsic state is rarely used as a semiconductor. Useable semiconductors must have controlled amounts of impurities added to them. The added impurities change the conduction capabilities of a semiconductor. The process of adding an impurity to an intrinsic material is called **doping**. The impurity is called a *dopant*. Doping a semiconductor causes it to be an **extrinsic material.** Extrinsic semiconductors are the operational basis of nearly all solid-state devices.

N-Type Material

An **N-type material** is formed when intrinsic silicon is mixed with a Group VB element, such as arsenic (As) and antimony (Sb). When these impurities are added to silicon or germanium, the crystal structure is unaltered.

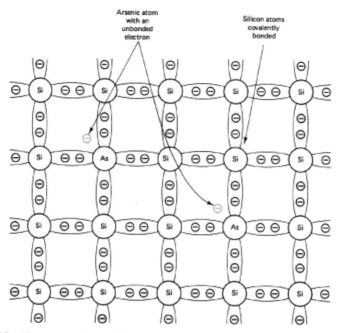

Figure 1.15　N-type crystal material. The extra electron of each impurity atom does not take part in a covalent bonding group and, therefore, does not alter the crystal structure or bonding process.

Atoms of arsenic and antimony have five electrons in their valence band. Adding this type of **impurity** to silicon does not alter the crystal structure or bonding process. Each impurity atom has an extra electron that does not take part in a covalent bonding group. These electrons are loosely held together by their parent atoms. **Figure 1.15** shows how a silicon crystal is altered with the addition of an impurity atom.

When **arsenic** is added to pure silicon, the crystal becomes an N-type material. It has extra electrons or negative (N) charges that do not take part in the covalent bonding process. These electrons are free to move about through the crystal structure. Impurities that add electrons to a crystal are generally called **donor atoms**. An N-type material, therefore, has more extra free electrons than an intrinsic piece of material. A piece of N-material is not negatively charged; rather, its atoms are electrically neutral.

An extrinsic silicon crystal of the N-type will go into conduction with a very small amount of voltage applied. In contrast, an **intrinsic crystal** (pure silicon) requires a rather substantial amount of voltage or energy for

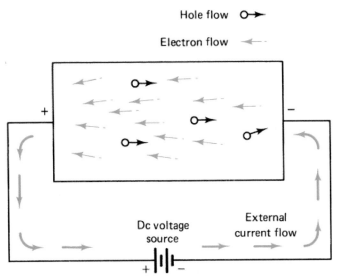

Figure 1.16 Current carriers in an N-type material. Electrons are the majority current carriers and holes are the minority current carriers.

its electrons to go into conduction. Essentially, this means that an N-type material is a fairly good electrical conductor. In this type of crystal, electrons are considered to be the majority current carriers; holes are the minority current carriers. The amount of donor material added to silicon determines the number of majority current carriers in its structure.

The number of electrons in a piece of N-type silicon is a million or more times greater than the number of electron–hole pairs of a piece of intrinsic silicon. At room temperature, there is a decided difference in the electrical conductivity of this material. Extrinsic silicon becomes a rather good electrical conductor because there are a larger number of current carriers to take part in conduction. Current flow is achieved primarily by electrons in this material. **Figure 1.16** shows how the current carriers respond in a piece of N-type material. There are more electrons indicated than holes; so electrons are the majority current carriers and holes are the minority carriers.

If the voltage source of **Figure 1.16** were reversed, the current flow would reverse its direction. This means that N-type silicon conducts equally well in either direction. The flow of current carriers is simply reversed. This is an important consideration in the operation of a device that employs N-type material in its construction. The polarity of the external voltage determines the direction of current flow through the N-material.

P-Type Material

A **P-type material** is formed when intrinsic silicon is mixed with Group IIIA elements, such as **indium** (In) or **gallium** (Ga). Group IIIA elements are often called **acceptors** because they readily seek a fourth electron. This type of dopant material has three valence electrons. Each covalent bond that is formed with an indium atom has an electron deficiency or hole, which represents a positively charged area in the covalent bonding structure. Each hole in the P-type material can be filled with an electron. Electrons from neighboring covalent bond groups require very little energy to move in and fill a hole. Holes in the P-type material tend to wander from one covalent bond group to another.

The ratio of **doping** material to silicon is typically in the range of $1-10^6$ or $1-1$ million. This means that the P-type material has a million times more holes than the heat-generated electron−hole pairs of pure silicon. At room temperature, there is a very decided difference in the electrical conductivity of

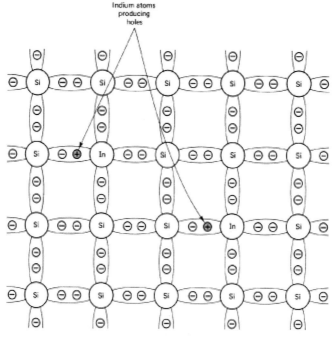

Figure 1.17 P-type crystal material. The atoms of the crystal material form a covalent bond with indium atoms, creating a deficiency or hole in the covalent bonding structure. This hole represents a positively charged area.

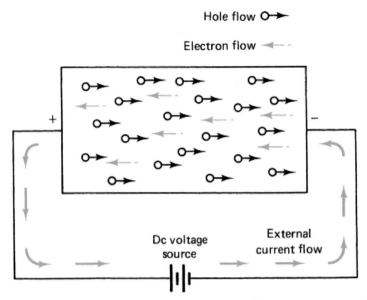

Figure 1.18   Current carriers in a P-type material. Holes are the majority current carriers and electrons are the minority current carriers.

this material. **Figure 1.17** shows how the crystal structure of silicon is altered when doped with an acceptor element. In this case, the dopant is indium.

A P-type material will go into conduction with only a small amount of applied voltage. Intrinsic silicon, by comparison, requires substantially more voltage to produce conduction. Extrinsic silicon, therefore, is considered to be a rather good electrical conductor. In this material, holes are the majority carriers and electrons are the **minority carriers**. The amount of acceptor material added determines the number of **majority current carriers** in its structure.

Figure 1-18 shows how a **P-type crystal** responds when connected to a voltage source. Note that there are more holes than electrons. With voltage applied, electrons are attracted to the positive battery terminal. This is illustrated in **Figure 1.18.** Holes move, in a sense, toward the negative battery terminal. An electron is picked up at this point. The electron immediately fills a hole. At the same time, an electron is pulled from the material by the positive battery terminal, forming a hole. The hole is attracted to the negative battery terminal. Holes, therefore, move toward the negative battery terminal because electrons shift between different bonded groups. With energy applied, hole flow is continuous.

Self-Examination

Answer the following questions.

25. A very pure semiconductor material is called a(n) _____ material.
26. At a temperature of absolute zero, intrinsic silicon is an excellent insulator; at room temperature, it becomes _____.
27. Electrons are called _____ current carriers, and holes are called _____ current carriers.
28. P-type material is formed by mixing a(n) _____ type of element with intrinsic silicon.
29. An N-type material is formed by mixing a(n) _____ type of element with intrinsic silicon.
30. How many electron-pair groups are formed around a single silicon atom in a covalent bonded structure?

  a. One
  b. Two
  c. Three
  d. Four
  e. Five

Summary

- An atom is the smallest particle to which an element can be reduced and still retain its identity.
- Atoms consist of smaller particles called *electrons*, *neutrons*, and *protons*.
- The nucleus (core) of every atom is composed of one or more positively charged particles called *protons* and one or more particles with no electrical charge called *neutrons*.
- For every proton in the nucleus of an atom, there is a negatively charged particle called an *electron* that orbits the nucleus.
- Electrons orbit the nucleus in shells or layers.
- The energy level and location of electrons in the structure of an atom determine its electrical conductivity.
- Valence electrons are electrons in the outermost shell and represent the highest energy level of the atom.
- Stabilized atoms have a full complement of electrons in the valence band.

- A stabilized atom will not release electrons under normal conditions.· Unstable atoms are those that do not have a full complement of electrons in the valence band.
- An unstable atom will try to become stable by drawing electrons away from neighboring atoms.
- Ionic bonding is the process of bonding atoms through electrostatic force.
- Electrostatic force occurs through the attraction of opposite net charges between two atoms.
- In covalent bonding, electrons alternately shift back and forth between each atom and this force bonds individual atoms in a simulated condition of stability.
- In metallic bonding, valence electrons of metal wander between different atoms and form a floating cloud of electrons which permits the material to be a good conductor.
- A specific amount of outside energy must be added to valence electrons to cause them to go into conduction.
- Conductivity is the basis of material classifications such as insulators, semiconductors, and conductors.
- Hole flow and electron flow occur in semiconductor material as long as energy is supplied.
- An N-type material is formed when intrinsic silicon is mixed with a Group VA element, such as arsenic.
- A P-type material is formed when intrinsic silicon is mixed with Group IIIA elements, such as indium or gallium.
- In contrast to intrinsic silicon crystal, extrinsic silicon crystal will go into conduction with a very small amount of applied voltage.
- In N-type material, electrons are the majority current carriers and holes are the minority carriers.
- In P-type material, holes are the majority current carriers and electrons are the minority carriers.

Formulas

(1-1) atomic weight = atomic mass rounded to nearest whole number Atomic weight of an atom.

(1-2) number of neutrons = atomic weight − atomic number Number of neutrons in the nucleus of an atom.

Answers

Examples

1-1. 14
1-2. K-shell = 2 s-level
      L-shell = 2 s-level
            = 6 p-level
      M-shell = 2 s-level
            = 6 p-level
            = 10 d-level
      N-shell = 2 s-level
            = 2 p-level

Self-Examination

1.1

1. atomic number
2. protons, neutrons
3. mass
4. electron
5. K
6. s, p, d, e, f, and g
7. P
8. orbital

1.2

9. ground
10. stable
11. compound
12. ionic
13. ionic
14. covalently
15. covalent
16. Metallic

1.3

17. valence, conduction
18. valence
19. forbidden

20. electron volts
21. insulator
22. semiconductor
23. conductor
24. breakdown

1.4

25. intrinsic
26. conductive or a conductor
27. negative, positive
28. acceptor
29. donor
30. d. Four

Terms

Matter

Anything that occupies space and has weight. It can be a solid, a liquid, or a gas.

Element

The basic materials that make up all other materials. They exist by themselves, such as copper, hydrogen, and carbon or in combination with other elements, such as water, a combination of the elements hydrogen and oxygen.

Atom

The smallest particle to which an element can be reduced and still retain its identity.

Compound

Two or more elements that have been chemically combined.

Molecule

The smallest particle to which a compound can be reduced before being broken down into its basic elements.

Nucleus

The core or center part of an atom, which contains protons having a positive charge and neutrons having no electrical charge.

Proton

A particle at the center of an atom that has a positive (+) electrical charge.

Neutron

A particle in the nucleus (center) of an atom that has no electrical charge.

Electron

A negatively charged particle that orbits the nucleus of an atom.

Atomic weight

The sum of protons and neutrons in the nucleus.

Atomic number

The number of protons contained in the nucleus of an atom.

Valence electrons

Electrons in the outer shell of an atom.

Orbital

The mathematical probability of where an electron will appear in the structure of an atom.

Ionic bonding

Bonding that occurs from the attraction of opposite net charges (electrostatic force) of two atoms.

Covalent bonding

Bonding that occurs by atoms positioning themselves so that the energy levels of their valence electrons interact. The valance electrons are, in essence, shared between atoms.

Metallic bonding

Bonding that occurs due to a floating cloud of ions that hold atoms loosely together in a conductor.

Forbidden gap

Separates the valence and conduction bands. The width of the forbidden gap indicates the conduction status the material it represents. In atomic theory, the width of the forbidden gap is expressed in electron volts (eV).

Electron volt

The amount of energy gained or lost when an electron is subjected to a potential difference of 1 V.

Breakdown voltage

The energy level value that causes an insulator to go into conduction.

Intrinsic material

A very pure semiconductor crystal.

Doping

The process of adding an impurity to an intrinsic material.

Extrinsic material

An intrinsic material that has been doped.

N-type material

A semiconductor material that is formed when intrinsic silicon is mixed with a Group VB element, such as arsenic (As) and antimony (Sb).

P-type material

A semiconductor material that is formed when intrinsic silicon is mixed with Group IIIB elements, such as indium (In) or gallium (Ga)

2

P–N Junction Diodes

P–N junction diodes are used rather extensively in the field of electronics. Radio, television, industrial control, computers, home entertainment equipment, and electrical appliances are a few of the applications. Diodes are probably the simplest of all electronic devices because only two leads, or electrodes, are used in their construction.

In general, diodes are used to control the conduction of electric current. Functionally, diodes are unidirectional, which means that they conduct well in only one direction. Diodes, therefore, are designed to block or pass current according to the polarity of voltage applied (biasing). This permits the diode to be used as a switch. Diodes can also be used to alter the direction of current passing through other electronic parts. In this application, the diode is used to change AC to DC, a process called rectification. **Rectification** is one of the most common applications of the diode and is covered in Chapter 4 – Power Supply Circuits. Other functions of the diode are the operational basis of more complex electronic devices.

Transistors, **integrated circuits**, and other **solid-state devices** are constructed by the same techniques used to produce diodes. A person working in electronics must, therefore, be familiar with the characteristic operation of this device. P–N junction diode theory is the subject of this chapter. Learning this material is an essential first step in understanding how semiconductor devices operate.

Objectives

After studying this chapter, you will be able to:
2.1 describe the physical characteristics of a P–N junction diode;
2.2 distinguish between the forward and reverse characteristics of a diode;
2.3 interpret the I-V characteristics curve of a solid-state diode;
2.4 interpret a manufacturer's data sheet for a P–N junction diode;
2.5 evaluate the condition of a diode as good, shorted, or open.

Chapter Outline

Key Terms

anode
barrier potential
bias voltage
cathode
depletion zone
diffusion
forward biasing
junction
junction capacitance
knee voltage
leakage current
reverse biasing
switching time
Zener breakdown

2.1 P–N Junction Diode Construction

In the previous chapter, you learned about the construction of P-type and N-type semiconductor materials and how these materials respond when they are coerced into conduction. In this chapter, you will learn about the P–N junction diode. The P–N junction diode is a two-terminal, semiconductor device made of P-type and N-type material. One terminal is attached to the P-type material and the other is attached to the N-type material. The current carriers in a P–N junction diode respond differently than those in pure P-type or N-type semiconductor material. In this section, you will learn how the current carriers respond immediately after the diode is formed.

2.1 Describe the physical characteristics of P–N junction diode.

In order to achieve objective 2.1, you should be able to:

• explain how a P–N junction diode is constructed;

- explain the interaction of current carriers when a P–N junction is formed;
- define the terms *junction, diffusion, depletion zone,* and *barrier potential.*

Figure 2.1 shows the crystal structure of a **P–N junction diode.** The crystal structure is continuous from one end to the other. The point where the N-type and P-type materials in a P–N junction diode are joined is called a **junction.** The junction serves only as a dividing line that marks the ending of one material and the beginning of the other. This type of structure permits electrons to move readily through the entire structure.

Figure 2.2 shows two pieces of semiconductor material before they are formed into a **P–N junction.** As indicated, each piece of material has majority

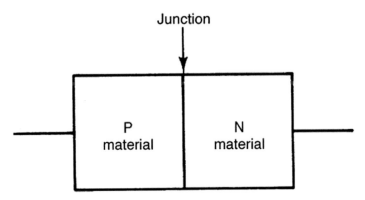

Figure 2.1 Crystal structure of a junction diode. The crystal structure is continuous, allowing electrons to move readily through the entire structure.

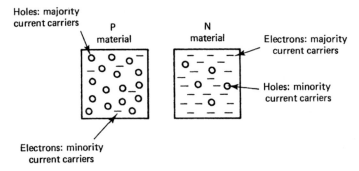

Figure 2.2 Semiconductor materials. Holes are the majority carriers and electrons are the minority carriers in the P-type material. Electrons are the majority carriers and holes are the minority carriers in the N-type material.

and minority current carriers. The number of carrier symbols shown in each material indicates the minority or majority function. Note that electrons are the majority carriers in the N-type material and the minority carriers in the P-type material. Holes are the majority carriers in the P-type material and the minority carriers in the N- type material. Both holes and electrons are free to move about in their respective materials.

Depletion Zone

When a junction diode is first formed, there is a unique interaction between current carriers. Electrons from the N-type material move readily across the junction to fill holes in the P-type material. This action is commonly called **diffusion.** Diffusion is the result of a high concentration of carriers in one material and a lower concentration in the other. Only those current carriers near the junction take part in the diffusion process.

The **diffusion** of current carriers across the junction of a diode causes a change in the diode's structure. **Figure 2.3** shows that electrons leaving the N-type material cause positive ions to be generated in their place. On entering

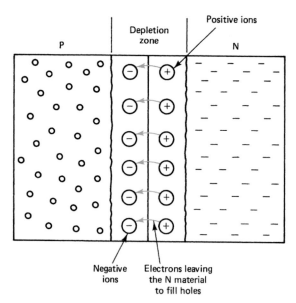

Figure 2.3   Depletion zone formation. The depletion zone is created from electrons leaving the N-type material and entering the P-type material to fill holes.

the P-type material to fill holes, these same electrons create negative ions. The areas on both sides of the junction then contain a large number of positive and negative ions. The number of holes and electrons in this area becomes depleted. The term **depletion zone** is used to describe this area. It represents an area that is void of majority current carriers. All P–N junctions develop a depletion zone when they are formed.

Barrier Potential

Before N-type and P-type materials are joined together, they are considered to be electrically neutral. After they are joined, however, diffusion takes place immediately. The creation of negative and positive charges on the P and N sides of the junction, respectively, tends to drive the remaining electrons and holes away from the junction. This action makes it somewhat difficult for additional charge carriers to diffuse across the junction. The end result is a charge buildup or **barrier potential** across the junction.

Figure 2.4 shows the resulting barrier potential as a small battery connected across the P–N junction. Barrier potential is a small voltage developed across the P–N junction due to diffusion of holes and electrons.

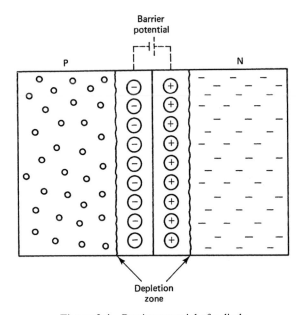

Figure 2.4   Barrier potential of a diode.

Note the polarity of this potential with respect to the P-type and N-type materials. This voltage exists even when the crystal is not connected to an outside source of energy. The barrier potential is approximately 0.3 V for germanium and 0.7 V for silicon; however, these voltage values cannot be measured directly. They appear only across the depletion zone of the junction. The barrier potential of a P–N junction must be overcome by an outside voltage source to produce current conduction.

Self-Examination

Answer the following questions.

1. A(n) _____ is a two-terminal electronic device.
2. The point where N-type and P-type materials are joined together in a diode is called a(n) _____.
3. A diode is formed by joining a piece of _____ material to a piece of _____ material at a common junction.
4. The majority current carriers of a piece of N-type material are _____.
5. The majority current carriers of a piece of P-type material are _____.
6. The minority current carriers of a piece of N-type material are _____.
7. The minority current carriers of a piece of P-type material are _____.
8. The process of electrons moving from N-type material across a junction to fill holes in P-type material is called _____.
9. Electrons leaving the N-type material to fill holes in the P-type material cause a(n) _____ to appear across the junction.
10. The barrier potential is approximately _____ for germanium and _____ V for silicon.
11. An area near the junction of a diode that is void of current carriers is called the _____.

2.2 Junction Biasing

In the previous section, you learned that a barrier potential exists in the junction of a P–N junction diode and that it must be overcome by an outside voltage source to produce current conduction. In this section, you will learn how majority and minority carriers react when voltage is applied to the diode and how the barrier potential is affected. You will see that a P–N junction diode permits current carriers to flow readily in one direction and blocks the flow of current in the opposite direction.

2.2 Distinguish between the forward and reverse characteristics of a diode.

In order to achieve objective 2.2, you should be able to:

- connect a diode so that it is forward or reverse biased;
- explain how majority and minority carriers react when a diode is reverse biased;
- explain how majority and minority carriers react when a diode is forward biased;
- define the terms *bias voltage, forward biasing, reverse biasing,* and *leakage current.*

The barrier potential of a P–N junction can be increased or reduced, depending on how the P–N junction diode is connected or biased in the circuit. A P–N junction diode whose P-type material is connected to the negative terminal and N-type material connected to the positive terminal of the battery is **reverse biased**. A P–N junction diode whose P-type material is connected to the positive terminal and N-type material connected to the negative terminal of the battery is *forward biased.* Any external source of energy applied to a P–N junction is called a *bias voltage* or, simply, a *bias.*

Reverse Biasing

Reverse biasing adds external voltage of the same polarity to the barrier potential and causes an increase in the width of the depletion zone, which hinders current carriers from entering the depletion zone. Take a look at **Figure 2.5.**

Note that the negative terminal of the battery is connected to the P-type material and the positive terminal is connected to the N-type material. This connection causes the battery polarity to oppose the material polarity of the diode. Since unlike charges attract, the majority charge carriers of each material are pulled away from the junction. Reverse biasing of a diode normally causes it to be nonconductive.

Depletion Zone

Figure 2.6 shows how the majority current carriers are rearranged in a reverse-biased diode. As shown, electrons of the N-type material are pulled toward the positive battery terminal. Each electron that moves or leaves the diode causes a positive ion to appear in its place. This causes a corresponding increase in the width of the **depletion zone** on the N side of the junction.

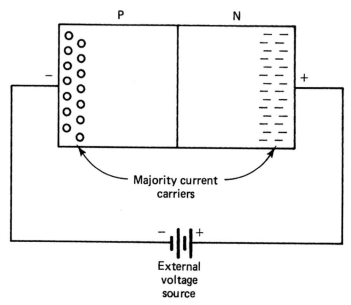

Figure 2.5 Reverse-biased diode. The negative terminal of the battery is connected to the P-type material, and the positive terminal is connected to the N-type material.

The reaction on the P side of the diode is very similar. In this case, a number of electrons leave the negative battery terminal and enter the P-type material. These electrons immediately move in and fill a number of holes. Each filled hole becomes a **negative ion**. These ions are then repelled by the negative battery terminal and driven toward the junction. As a result, the width of the depletion zone increases on the P side of the junction.

The depletion zone width of a reverse-biased diode is directly dependent on the value of the supply voltage. With a wide depletion zone, a diode cannot effectively support current flow. The charge buildup across the junction increases until the barrier voltage equals the external bias voltage. When this occurs, a diode effectively becomes a nonconductor.

Leakage Current

The wider **depletion zone** of a reverse-biased diode represents an area that is almost void of majority current carriers and, thus, responds as an insulator. Ideally, current carriers do not pass through an insulator. However, in a reverse-biased diode, some current actually flows through the depletion zone.

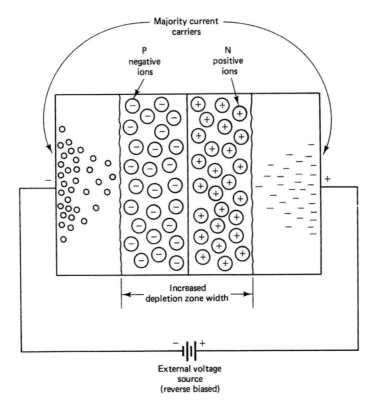

Figure 2.6 Depletion zone creation in a reverse-biased diode. In the N-type material, electrons are pulled toward the positive battery terminal, leaving positive ions in their place. Electrons leave the negative battery terminal and enter the P-type material, filling holes and creating positive ions. The end result is a wider depletion zone.

This is called **leakage current**. Leakage current is dependent on **minority current carriers**.

Remember that minority carriers are electrons in the P-type material and holes in the N-type material. **Figure 2.7** shows how these carriers respond when a diode is reverse biased. It should be noted that the minority carriers of each material are pushed through the depletion zone to the junction, which causes a very small amount of leakage current. Normally, leakage current is so small that it is often considered negligible.

The number of minority current carriers in a semiconductor is primarily dependent on temperature. At normal room temperatures of 25°C (77°F), there are a rather limited number of minority carriers present in a

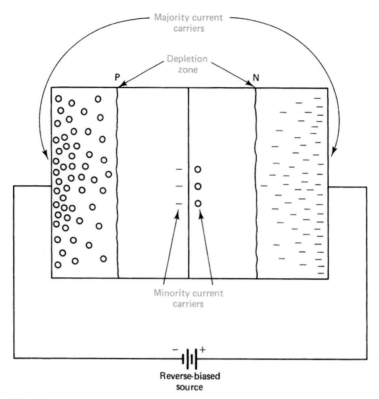

Figure 2.7 Minority current carriers in a reverse-biased diode. Minority carriers of each material are pushed through the depletion zone to the junction, causing a very small amount of leakage current.

semiconductor. However, when the surrounding temperature rises, it causes a considerable increase in minority carrier production, which creates a corresponding increase in leakage current.

Leakage current occurs, to some extent, in all reverse-biased diodes. In germanium diodes, leakage current is only a few microamperes. Silicon diodes normally have fewer minority current carriers; so they have less leakage current. Typical leakage current values for silicon are a few nanoamperes. The construction material of a diode is, thus, an important consideration. Germanium is much more sensitive to temperature than silicon; so germanium has a higher level of leakage current. This factor is largely responsible for the widespread use of silicon in modern semiconductor devices.

Forward Biasing

Forward biasing reduces the barrier potential and causes current carriers to return to the depletion zone. Take a look at **Figure 2.8.** Note that the positive battery terminal is connected to the P-type material and the negative terminal is connected to the N-type material. This voltage repels the majority current carriers of each material. A large number of holes and electrons, therefore, appear at the junction. On the N side of the junction, electrons move in to neutralize the positive ions in the depletion zone. In the P-type material, electrons are pulled from **negative ions**, which cause them to become neutral again. This means that forward biasing causes the depletion zone to collapse and the barrier potential to be removed. The P–N junction, therefore, supports a continuous current flow when it is forward biased.

Figure 2.9 shows how the current carriers of a forward-biased diode respond. Since the diode is connected to an external voltage source, it has a constant supply of electrons. Large arrows are used in the diagram to show the direction of current flow outside the diode. Inside the diode, smaller arrows show the movement of majority current carriers. Remember that **electron flow** and *current* are synonymous.

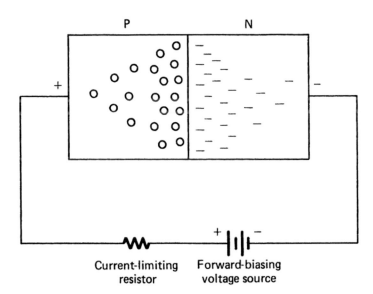

Figure 2.8 Forward-biased diode. The positive battery terminal is connected to the P-type material, and the negative battery terminal is connected to the N-type material.

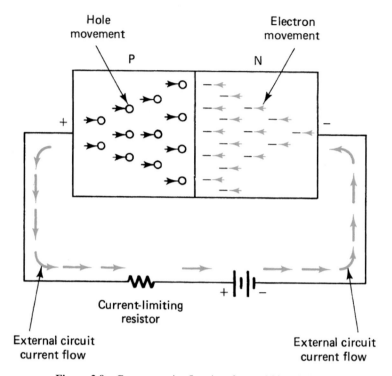

Figure 2.9   Current carrier flow in a forward-biased diode.

   Starting at the negative battery terminal, assume that electrons are flowing through a wire to the N-type material. On entering this material, they flow immediately to the junction. At the same time, an equal number of electrons are removed from the P-type material and are returned through a resistor to the positive battery terminal. This action generates new holes and causes them to move toward the junction. When these holes and electrons reach the junction, they combine and effectively disappear. At the same time, new holes and electrons appear at the outer ends of the diode. These majority carriers are generated on a continuous basis. This process continues as long as the external voltage source is applied.

   It is important to realize that electrons flow through the entire diode when it is **forward biased**. In the N-type material, this is quite obvious. In the P-type material, however, holes are the moving current carriers. Remember that hole movement in one direction must be initiated by electron movement in the opposite direction. Therefore, the combined flow of holes and electrons through a diode equals the total current flow.

The concept of **hole movement** in a P- or N-material may be visualized as a sequence of movement of a holes developed during **electron-pair bonding**.

The **current-limiting resistor**, R_{limit}, of **Figure 2.9** is essential in a forward-biasing diode circuit. This resistor is needed to keep the current flow at a safe operating level. The maximum current (I_{max}) rating of a diode represents this value. As long as diode current does not exceed this value, it can operate satisfactorily.

Self-Examination

Answer the following questions.

12. A(n) _____ permits current carriers to flow readily in one direction and blocks the flow of current in the other direction.
13. The process of connecting a voltage source across the P–N junction of a diode is called _____.
14. When the negative side of the voltage source is connected to the P-type material and the positive side of the source is connected to the N-type material, _____ biasing occurs.
15. When the positive side of the source is connected to the P-type material and the negative side of the source is connected to the N-type material, _____ biasing occurs.
16. Since unlike charges attract, the _____ current carriers of a reversed-biased diode are pulled away from the junction.
17. A reverse-biased diode is considered to be _____.
18. Reverse biasing of a diode causes the width of the depletion zone to _____.
19. The minority current carrier content of a diode determines _____ current.
20. Minority current carrier content of a diode is _____ dependent.
21. Leakage current is greatest in diodes made of _____.
22. _____ biasing of a diode causes the depletion zone to collapse.
23. Current-limiting resistors are used in forward-biased diode circuits to keep the _____ at a safe operating level.

2.3 Diode Characteristics

Now that you have seen how a junction diode operates, it is time to examine some of its electrical characteristics, such as voltage, current, and temperature. Since these characteristics vary greatly in an operating circuit, it is best to look at them graphically. This makes it possible to see how the device responds under different operating conditions.

2.3 Identify the *I-V* characteristics curve of a solid-state diode.

In order to achieve objective 2.3, you should be able to:

- describe the relationship between forward voltage and forward current;
- describe the relationship between reverse voltage and reverse current;
- describe the layout of an *I-V* chart;
- define the terms *knee voltage*, *Zener breakdown*, and *avalanche break-down*.

Forward Characteristics

When a diode is connected in the **forward-bias** direction, it conducts forward current (I_F). The value of forward current is directly dependent on the amount of forward voltage (V_F); see **Figure 2.10.** The relationship between forward voltage and forward current is called the ampere-volt, or **I–V, *character-istic*** of a diode. A typical diode forward *I–V* characteristic is shown in **Figure 2.10(a)**, with its test circuit in **Figure 2.10(b)**. Note, in particular, that V_F is measured across the diode and that I_F is a measure of what flows through the diode. The value of the source voltage (V_S) does not necessarily compare in value with V_F.

When V_F equals 0 V, I_F equals 0 mA. This value starts at the origin (0) of the graph. If V_F is gradually increased in 0.1-V steps, I_F begins to increase. When the value of V_F is great enough to overcome the barrier potential of the P–N junction, a substantial increase in I_F occurs. The point at which this occurs is often called the **knee voltage**. Knee voltage (V_K) is approximately 0.3 V for germanium diodes and 0.7 V for silicon.

If the value of V_F increases much beyond V_K, the forward current becomes quite large. This, in effect, causes heat to develop across the junction. Excessive junction heat can destroy a diode. To prevent this from happening, a protective resistor, called a limiting resistor, is connected in series with the diode. This resistor limits I_F to some point below the maximum current value of the diode. Diodes should not be operated in the forward direction without a current-limiting resistor.

Figure 2.10(b) shows a **forward-biased** silicon diode that has its current limited by a limiting resistor (R_{limit}). Assume that the limiting resistor has a value of 100 Ω. When the diode is connected in this manner, the forward current depends on the source voltage and the value of the limiting resistor.

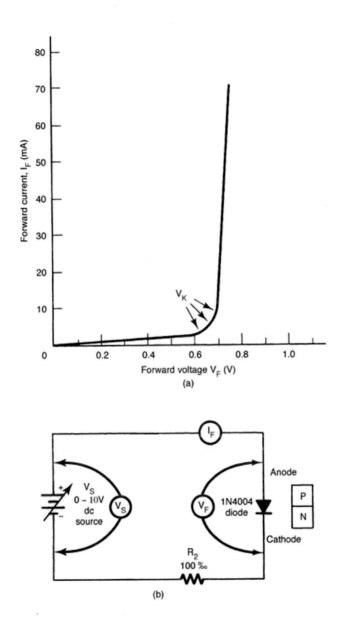

Figure 2.10 (a) I–V characteristics of a diode. (b) Diode characteristic test circuit. Silicon diode connected in forward conduction. Forward current depends on the source voltage and the value of the current-limiting resistor.

Ohm's law can determine the resulting forward current. Forward current (I_F), therefore, is determined by the following expression:

$$I_F = V_S - V_K / R_{\text{limit}}. \tag{2.1}$$

Note that this formula takes into account the voltage drop across the diode when it is forward biased. For a silicon diode, this is 0.7 V. The voltage across the limiting resistor is the source voltage (V_S) minus the knee voltage (V_K) across the diode. Therefore,

$$I_F = V_S - V_K / R_{\text{limit}}$$
$$I_F = 10 \text{ V} - 0.7 \text{ V}/100 \text{ } \Omega$$
$$= 9.3 \text{ V}/100 \text{ } \Omega$$
$$= 0.093 \text{ A, or 93 mA.}$$

Example 2-1:

What is the forward current of this circuit?

Solution

The source voltage (V_S) for this circuit is 10 V and the limiting resistor (R_{limit}) is 200 Ω. The knee voltage (V_K) for a germanium diode is 0.3 V. Therefore,

$$I_F = V_S - V_K / R_{\text{limit}}$$
$$I_F = 10 \text{ V} - 0.3 \text{ V}/200 \text{ } \Omega$$
$$= 9.7 \text{ V}/200 \text{ } \Omega$$
$$= 0.0485 \text{ A or 48.5 mA.}$$

Related Problem

What is the forward current of this example if the diode were silicon?

Reverse Characteristics

When a diode is connected in the reverse-bias direction, it has an $I_R - V_R$ characteristic. **Figure 2.11** shows the reverse *I–V* characteristic of a diode and its test circuit. This characteristic has different values of I_R and V_R. Reverse current is usually quite small. The vertical *I–V* line in this graph has

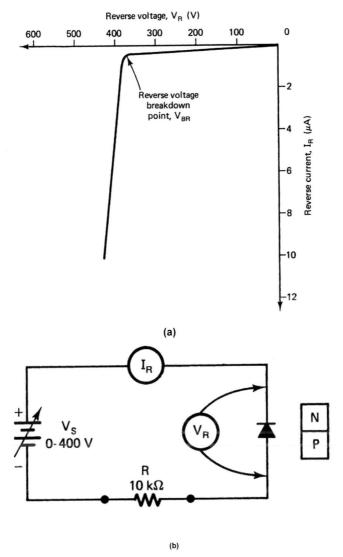

(a)

(b)

Figure 2.11 (a) Reverse *I–V* characteristics of a diode. (b) Reverse diode characteristic test circuit.

current values graduated in microamperes. The number of minority current carriers that take part is quite small. In general, this means that I_R remains rather constant over a large part of V_R. Note also that V_R is graduated in 100-V increments. Starting at zero when the **reverse voltage** of a diode is increased, there is only a slight change in I_R. At the voltage breakdown V_R point, current increases rapidly. The voltage across the diode remains fairly constant at this time. This constant-voltage characteristic leads to a number of reverse-biased diode applications. Normally, diodes are used in applications where the V_R is not reached.

The physical processes responsible for current conduction in a reverse-biased diode are called **Zener breakdown** and **avalanche** *breakdown*. **Zener breakdown** occurs when electrons are pulled from their covalent bonds in a strong electric field. This occurs at a rather high value of V_R. When large numbers of covalent bonds are broken at the same time, there is a sudden increase in I_R.

Avalanche breakdown is an energy-related condition of reverse biasing. At high values of V_R, minority carriers gain a great deal of energy. This gain may be great enough to drive electrons out of their covalent bonding, which creates new electron–hole pairs. These carriers then move across the junction and produce other ionizing collisions and additional electrons. The process continues to build until an avalanche of current carriers is produced, at which point the process is irreversible.

Combined I–V Characteristics

The forward and reverse I–V characteristics of a diode are generally combined on a single characteristic curve. **Figure 2.12** shows a rather standard method of displaying this curve. **Forward-bias** and **reverse-bias voltages**, V_F and V_R, are usually plotted on the horizontal axis of the graph. V_F extends to the right and V_R to the left. The point of origin, or zero value, is at the center of the horizontal line.

Forward and reverse current values are shown vertically on the graph. I_F extends above the horizontal axis, with I_R extending downward. The origin serves as a zero indication for all four values. This means that combined V_F and I_F values are located in the upper-right part of the graph, and V_R and I_R are located in the lower-left corner. Different scales are normally used to display forward and reverse values.

A rather interesting comparison of silicon and germanium **characteristic curves** is shown in **Figure 2.13.** A careful examination shows that germanium requires less forward voltage to go into conduction than silicon. This characteristic is a distinct advantage in low-voltage circuits. Note also that a germanium diode requires less voltage drop across it for different values of current. This means that germanium has a lower resistance to **forward**

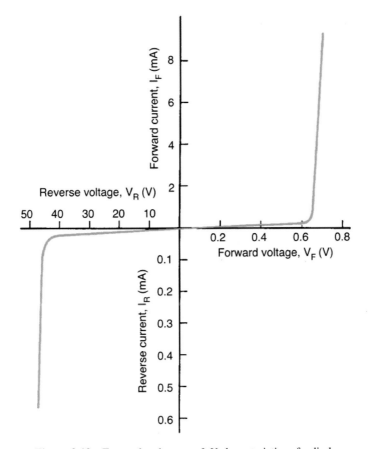

Figure 2.12 Forward and reverse *I–V* characteristics of a diode.

current flow. Germanium, therefore, appears to be a better conductor than silicon. Silicon is more widely used, however, because of its low leakage current and lower production cost.

The reverse-bias characteristics of silicon and germanium diodes can also be compared in **Figure2.13**. The I_R of a silicon diode is very small compared with that of a germanium diode.

Reverse current is determined primarily by the minority current content of the material, a condition influenced primarily by temperature. For **germanium diodes**, I_R doubles for each 10∘C rise in temperature. In a **silicon diode**, the change in I_R is practically negligible for the same rise in temperature. As a result, silicon diodes are preferred over germanium diodes

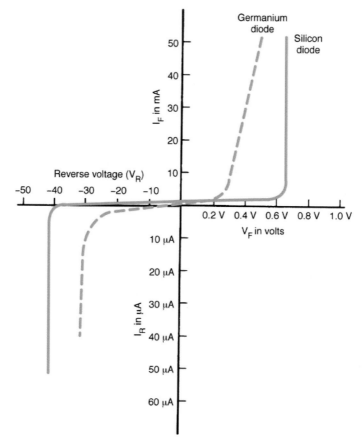

Figure 2.13   Silicon and germanium diode characteristics. The germanium diode requires less forward voltage to go into conduction. The silicon diode remains in a stable state in reverse bias until the breakdown voltage is reached.

in applications where large changes in temperature occur. Comparisons of this type are quite obvious through the study of characteristic curves.

Self-Examination

Answer the following questions.

24. In a diode, forward current (I_F) is _____ related to the value of forward voltage (V_F).

25. Forward voltage (V_F) is measured _____ a diode.

26. Forward current (I_F) is a measure of the current passing _____ a diode.
27. When V_F overcomes the barrier potential of a P–N junction, there is a large increase in _____.
28. The knee voltage (V_K) is 0.3 V for _____ and 0.7 V for _____.
29. When the electrons of a diode are pulled from their covalent bonding in a strong reverse-biased electric field, _____ breakdown occurs.
30. The _____ breakdown of a diode is due to an energy-related reverse-biased condition.

2.4 Diode Specifications

Selection of a diode for a specific application requires knowledge of its specifications. This type of information is usually made available through the manufacturer. Diode specifications generally include absolute maximum ratings, typical operating conditions, mechanical data, lead identification, mounting procedures, and characteristic curves. This section provides you with the knowledge needed to interpret a diode data sheet.

2.4 Interpret a manufacturer's data sheet for a P–N junction diode.
    In order to achieve objective 2.4, you should be able to:

- describe common data sheet specifications for a diode;
- explain how voltage affects junction capacitance;
- explain how temperature affects diode operation;
- define the terms *junction capacitance* and *switching time*.

Diode Data Sheet

Data sheets for semiconductor devices will be placed at the **END** of the chapter. Take a look at the end of this chapter to review representative diode data sheets. Some of the important rating and operating condition specifications are explained in the following:

1. **Maximum reverse voltage**, V_{RM}: The absolute maximum or peak reverse-bias voltage that can be applied to a diode. This may also be called the peak inverse voltage (P_{IV}) or peak reverse voltage (P_{RV}).
2. **Reverse breakdown voltage**, V_{BR} (V_R): The minimum steady-state reverse voltage at which breakdown occurs.
3. **Maximum forward current**, I_F: The absolute maximum repetitive forward current that can pass through a diode at 25°C (77°F). This is reduced for operation at higher temperatures.

4. **Maximum forward surge current**, I_{FM} (surge): The maximum current that can be tolerated for a short interval of time. This current value is much greater than I_{FM}. This represents the increase in current that occurs when a circuit is first turned on.
5. **Maximum reverse current**, I_{RM} (I_R): The absolute maximum reverse current that can be tolerated at device operating temperature.
6. **Forward voltage**, V_{FM} (V_F): Maximum forward voltage drop for a given forward current at device operating temperature.

Some other specifications that may be found in a diode data sheet include the following:

1. **Mechanical data** – this refers to the type of material used to construct the diode (usually molded plastic), maximum temperature values, and dimensions of the device.
2. **Lead identification** – shows the diode symbol or a band on the cathode side to specify polarity.
3. **Mounting procedures** – any special instructions in terms of temperature and dimensions/size restrictions are included.
4. **Characteristic curves** – **Figure 2.14** includes a forward current derating curve that plots maximum average forward current (I_F) at various ambient temperatures.

As you can see, the diode data sheet has a great deal of specific information. This information is simplified and applied in this chapter.

Other values that might be included on a data sheet include the following:

1. **Power dissipation**, P_D: The maximum power that the device can safely absorb on a continuous basis in free air at 25°C (77°F). This may not appear on all data sheets. It can be calculated as $P_D = I_{FM} \times V_{FM}$.
2. **Reverse recovery time**, T_{rr}: The maximum time it takes the device to switch from its on state to its off state. This is not identified on the data sheet.

Diode Temperature

The operation of a diode is directly related to temperature. All semiconductor materials are similar in this respect. The primary effect of temperature is the generation of additional electron–hole pairs because more valence electrons are thermally excited and gain enough energy to move into the conduction band. These additional current carriers cause a decided change in the *I–V* characteristics of the diode.

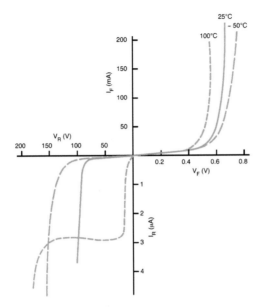

Figure 2.14 I–V characteristics of a silicon diode at 100°C, 50°C, and 25°C. In the forward bias, less voltage is needed to produce conduction at higher temperatures. In the reverse bias, there is a great deal more leakage current at higher temperatures. Lower temperatures produce higher breakdown voltages in forward and reverse biases.

Figure 2-14 shows how the **I–V characteristics** of a silicon diode change with temperature. These operating temperatures are indicated in degrees Celsius. Three main differences should be noted. The first of these is in the **forward-bias** region. This shows that less voltage is needed to produce conduction at higher temperatures. The other differences are in the **reverse-bias** area. One of these shows that there is a great deal more leakage current at higher temperatures. The third difference is in the breakdown voltage. This indicates that lower temperatures produce higher breakdown voltages. These considerations must all be taken into account when selecting a diode for a circuit application.

Electronic circuits that employ diodes are called on to operate in a rather wide range of temperatures. Consumer-grade diodes are usually rated for a range of −50°C to +100°C. The extremes of this range are more meaningful if related to the boiling point of water (100°C or 212°F) and the freezing point of mercury (−39°C or −40°F). Military-grade diodes are rated at −65°C to 125°C. This type of diode is much more expensive than consumer-grade devices.

Junction Capacitance

A **capacitor** is defined as two or more conductive plates separated by an insulating material, called a **dielectric**. Remember that a capacitor develops an electrostatic charge between two conductive plates. The strength of the charge depends on the applied voltage, the size of the conductive plates, and the dielectric constant of the insulating material. The closer the plates are, the more charge may be set up between them.

A reverse-biased diode has a structure that responds as a capacitor. The two independent crystal materials serve as conductor plates, with the depletion zone acting as a **dielectric material**. The term **junction** *capacitance* is used to describe this effect. The value of a capacitor is determined by the thickness of the dielectric material and the area of the two conducting plates. The value of diode junction capacitance depends on the thickness of its depletion zone. **Figure 2.15** shows reverse-biased P–N junction diode. A decrease in **bias voltage** causes a decrease in the depletion zone and

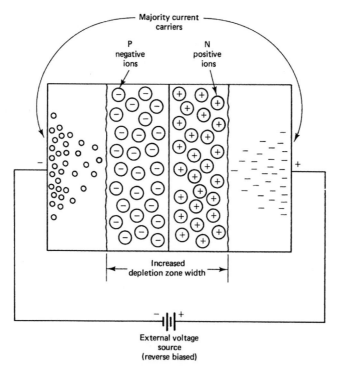

Figure 2.15 Junction capacitance in a reverse-biased P–N junction diode.

an increase in junction capacitance. An increase in bias voltage causes an increase in the depletion zone and a decrease in junction capacitance.

Junction capacitance is a variable value that is dependent on bias voltage (**Figure 2.15**). Small amounts of reverse bias, and, in some cases, forward bias, produce the largest values of junction capacitance because the width of the depletion zone is reduced (**Figure 2.15(a)**). A decrease in the depletion zone or dielectric thickness causes a corresponding increase in capacitance. When the reverse-bias voltage of a diode is increased, there is a decrease in junction capacitance because the width of the depletion zone is increased (**Figure 2.15(b)**). With a thicker dielectric material, there is a smaller junction capacitance.

The **dielectric constant** of a silicon depletion zone is approximately 12. This means that the depletion zone is 12 times better than air as a dielectric. As a result, some rather significant capacitance values appear across a junction diode. This value is extremely important in high-frequency circuit applications. A reverse-biased diode can, therefore, respond as a voltage-controlled capacitor. Special devices known as **varicap diodes** are designed to perform this function. Varicap diodes are discussed in Chapter 5.

Switching Time

One rather important characteristic for some diodes is **switching time**, which refers to the time it takes for a diode to switch from one state to the other. This condition is important in computer applications that involve rapid turn-on and turn-off times.

When a junction diode is forward biased, its depletion zone begins to fill with current carriers. This action produces a large number of electron–hole **recombinations** near the junction. If the bias voltage is suddenly removed, the recombination process does not stop instantly. Current carriers have a type of inertia that causes them to continue moving once they are placed in motion. The nonconduction state cannot be reached until all the current carriers have cleared the depletion zone. This means that the depletion zone has a current-carrier storage function.

Changing state from reverse bias to forward bias is also time dependent in a diode. The delay involved in this characteristic is due to junction capacitance. This capacitance tends to absorb the initial forward-bias current when it is being charged. After the charging operation has been completed, current carriers become available for normal conduction. Special switching diodes are now available for most high-speed control operations.

Diode Packaging

Diodes are manufactured in a wide range of case styles and packages. A person working with these devices must be familiar with some of the common methods of packaging. Element identification and lead marking techniques are essential for proper installation and testing. A diode improperly connected in a circuit may be damaged or may cause damage to other circuit parts. **Figure 2.16** shows the general classes of diodes according to

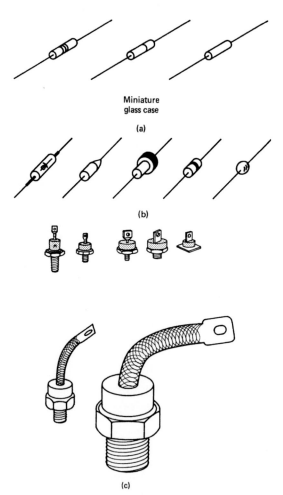

Figure 2.16 Diode packages. (a) Plastic and metal diode outlines ("DO" packages). (b) Low-power diodes. (c) Medium-power diodes.

their current-handling capability. These represent some of the more popular package styles.

The **low-current diodes** of **Figure 2.16(a)** are the smallest of all packages. The body length of these packages rarely exceeds 0.3 cm. The cathode is usually denoted by a painted color band. Glass-packaged diodes often have two or three color bands to indicate specific number types. This group of diodes is generally capable of passing forward current values of approximately 100 mA. The peak reverse voltage rating rarely exceeds 100 V. Low reverse current values for these devices are typically 5 mA at 25∘C.

The **medium-current diodes** of **Figure 2.16(b)** are slightly larger in size than the low-current devices. Body size is approximately 0.5 cm with larger connecting leads. Diodes in this group can pass forward current values up to 5 A. Peak reverse voltage ratings generally do not exceed 1000 V. The anode (+) and cathode (−) terminals may be identified by a diode symbol on the body of the device. A color band near one end of the case is also used to identify the cathode. Low- and medium-current diodes are usually mounted by soldering. Any heat generated during operation is carried away by air or lead conduction.

High-current, or power, diodes are the largest of all diode types shown in **Figure 2.20(c)** and **(d)**. These devices normally generate a great deal of heat. Air convection of heat is generally not adequate for most installations. These devices are designed to be mounted on metal heat sinks, which conduct heat away from the diode. Diodes of this classification can pass hundreds of amperes of forward current. Peak reverse voltage ratings are in the 1000-V range. Numerous packaging types and styles are used to house power diodes. As a rule, the rating of the device and method of installation usually dictate its package type.

Self-Examination

Answer the following questions.

31. The absolute maximum reverse-bias voltage that can be applied to a diode is called the _____.
32. I_{FM} refers to the absolute maximum _____ that can pass through a diode.
33. The *I–V* characteristics of a silicon diode change with _____.
34. The crystal structure of a reverse-biased diode responds as a(n) _____.
35. The depletion zone of a reverse-biased diode serves as the _____ material of a capacitor.

36. A diode responds as a voltage-controlled _____ capacitor.
37. The ability of a diode to change from one state to another is called _____.
38. Diodes are generally packaged according to their _____ rating.

2.5 Troubleshooting Diodes

Diode operation is determined by its connection in a circuit. Current conduction is directly dependent on the polarity of the device. Heavy conduction occurs when it is forward biased, and very little conduction occurs when it is reverse biased. Connecting a diode backward in a circuit reverses its conduction. This can destroy the diode and possibly damage several other circuit parts. A person working with diodes must be absolutely certain that a diode is connected properly in a circuit. They must also be able to test the condition of a diode.

2.5 Evaluate the condition of a diode as good, shorted, or open.
   In order to achieve objective 2.5, you should be able to:

   • use an ohmmeter to test the condition of a diode;
   • identify the electrodes of a diode with an ohmmeter.

2.6 Diode Schematic Symbol

A **schematic symbol** of a diode is commonly used to identify its polarity within a circuit. **Figure 2.17** shows a diode symbol, element names, and the polarity of the crystal material. The term **anode** is commonly used to describe the electrode that attracts or gathers in electrons. The P-type material of the crystal serves as the anode in a diode. The term **cathode** is used to denote the electrode that gives off or emits electrons. The N-type material of the crystal serves as the cathode. Note that forward current passes through a diode from the cathode to the anode. A diode symbol indicates this movement from the bar to the point of the arrow.

   Because diodes are available in many different package types and styles, there is often some confusion about lead identification. A person working with diodes is frequently called on to identify leads. The ohmmeter function of a multimeter is used to perform this operation.

   The **polarity** of the ohmmeter voltage source must be noted for this test to work. Ohmmeters typically have a black, or common lead, as negative and the red lead is positive. A diode connected to the ohmmeter will show a relatively low resistance when forward biased and infinite resistance when

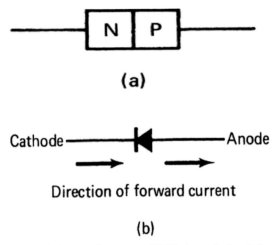

(a)

Direction of forward current

(b)

Figure 2.17 (a) Diode crystal structure. (b) Diode symbol and element names.

reverse biased. Forward biasing occurs when the polarity of the ohmmeter matches the crystal polarity of the diode. The red lead is, therefore, connected to the anode and the black lead to the cathode. Lead identification is simply a matter of determining the forward-bias direction of a diode and noting the polarity of the ohmmeter leads.

Testing Diodes

Figure 2.18 shows an **ohmmeter** being used to test a diode. In preparation for this test, the meter should be placed in the $R \times 100$ or $R \times 1$-K Ohm range. The meter leads are then connected to the diode. Exact resistance readings are not particularly important. Only high-resistance or low-resistance indications are used. **Figure 2.18(a)** shows how an ohmmeter is connected to **forward bias** a diode. This test normally causes the meter to show a relatively low resistance. **Reverse biasing** of a diode is shown in **Figure 2.18(b).** An infinite resistance is indicated by the ohmmeter during this test. A good diode normally shows low resistance when forward biased and high resistance when reverse biased. If a diode does not respond in this manner, it is probably defective. In many digital meters, a diode symbol may appear on the ohmmeter's switching range. This symbol indicates that the ohmmeter will supply the necessary voltage needed to cause the diode to respond when forward biased. Other ranges may not provide the voltage needed to produce conduction when forward biased. When a digital meter is used, look for the diode symbol and place the function switch in this position when evaluating a diode.

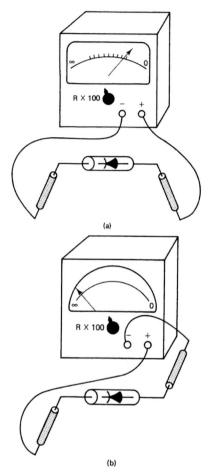

Figure 2.18   Diode testing with an ohmmeter. (a) Forward-bias test connection. (b) Reverse-bias test connection.

A diode that shows low resistance in both directions is considered to be **shorted**. Shorting usually occurs when the maximum current or voltage ratings of the device are exceeded. Shorted diodes often cause damage to other circuit components. Blown fuses, tripped circuit breakers, or overheated resistors are generally good indicators of a shorted diode. It is a good practice to test other circuit components before replacing a shorted diode.

When a diode shows **infinite resistance** in both directions, it is considered to be **open.** Open diodes generally cause a circuit to be nonconductive. This condition rarely causes damage to other circuit parts. Operation simply stops, and there is no current flow through the diode. Open diodes do not occur very often in electronic circuits. A quick ohmmeter check will readily detect this type of fault should it occur.

The value of an ohmmeter's voltage supply is an important consideration when testing diodes. If the supply voltage is less than 0.2 V, it will not be large enough to forward bias all diodes. This could cause a good diode to appear to be **open**. The ohmmeter would, therefore, show an infinite resistance in both directions. You may recall that it takes at least 0.3 V to cause conduction in germanium diodes and 0.7 V in silicon. Many electronic multimeters, especially digital meters, respond in this way. The voltage supply of this type of instrument must be evaluated to see if it can be used to test a diode.

The supply voltage of some ohmmeters may also be extremely large. This type of meter could actually develop enough current or voltage to damage certain diodes. As a rule, high-frequency detection diodes are very susceptible to being damaged in this way. To avoid this kind of problem, do not attempt to test high-frequency diodes with high-voltage ohmmeters.

Self-Examination

Answer the following questions.

39. When a diode is tested with an ohmmeter, _____ conduction occurs when it is forward biased.
40. A good diode normally shows _____ resistance when forward biased and _____ resistance when reverse biased.
41. A diode that shows low resistance in either direction of biasing is _____.
42. A diode that shows infinite resistance in either direction of biasing is considered to be _____.
43. The _____ of the voltage source of an ohmmeter must be known to properly identify the leads of a diode.

Summary

- The term *P–N junction diode* is used to describe the crystal structure of a two-element electronic device made of N-type and P-type materials.
- The *P–N* junction formed when these materials are joined responds as a continuous crystal.
- Current carriers diffuse across the junction when a P–N junction diode is formed.

- The diffusion process establishes a barrier potential across the junction.
- For germanium, the barrier potential is approximately 0.3 V, and for silicon, it is 0.7 V.
- When an external source of energy is applied to the P–N material of a diode, it either adds to or reduces the barrier potential of the junction.
- Forward biasing reduces the barrier potential of a junction and causes the diode to be conductive.
- Forward biasing is achieved by connecting the positive side of the source to the P-type material and the negative side of the source to the N-type material.
- Reverse biasing adds to the barrier potential of a diode and causes it to be nonconductive.
- Reverse biasing occurs when the negative side of the source is connected to the P-type material and the positive side of the source is connected to the N-type material.
- The *I–V* characteristics of a diode show how it responds when connected in a circuit that is forward biased or reverse biased.
- In the forward-biased direction, conduction occurs at a few tenths of a volt. This causes a very rapid increase in forward current.
- In the reverse-biased direction, there is very little current with an extremely large value of reverse voltage.
- Reverse current is generally temperature dependent.
- When selecting a diode for an application, one must consider its specifications, which include maximum reverse voltage, reverse breakdown voltage, forward current, surge current, maximum reverse current, power dissipation, and reverse recovery time.
- Diode operation is related to temperature.
- Junction capacitance is a characteristic that changes with bias voltage.
- Switching time is a variable that refers to the time it takes a diode to change from one state to another.
- Diodes can be tested and the leads can be identified with an ohmmeter.
- A diode's leads are identified by matching the polarity of the ohmmeter leads with the P-type and N-type materials of the diode.
- When the positive lead of an ohmmeter is connected to the P-type material and the negative lead to the N-type material, a good diode will show a low resistance reading; reversing the ohmmeter leads will show a high resistance reading.
- A shorted diode will indicate low resistance in both directions.
- An open diode will show infinite resistance in both directions.

Formulas

(2-1) $I_F = V_S - V_K/R_{limit}$ Forward current of a series circuit with a diode.

Review Questions

Answer the following questions.

1. The construction of a junction diode consists of:

   a. A piece of N-material
   b. A piece of P-material
   c. A combination of N- and P-materials
   d. Electrically conductive materials

2. When electrons leave the N-material to fill holes in the P-material the process is called.

   a. Depletion
   b. Diffusion
   c. Mixing
   d. Doping

3. Diffusion current in a diode is caused by:

   a. Chemical energy
   b. Heat energy
   c. Voltage
   d. Crystal formation

4. The depletion zone or region of a diode is:

   a. Due to reverse biasing
   b. Due to forward biasing
   c. An area created by crystal doping
   d. An area void of current carriers

5. When the P–N materials of a junction diode are formed, diffusion current causes:

   a. A barrier potential to be formed across the junction
   b. Covalent boding
   c. The mixing of current carriers
   d. Forward biasing
   e. Reverse biasing

6. The forward voltage (V_F) of a fully conductive silicon diode is:

   a. Approximately 0.3 V
   b. Approximately 0.7 V
   c. Equal to the source voltage (V_S)
   e. Equal to the voltage drop across the series resistor

7. When a diode is externally forward biased:

   a. The width of its depletion zone increases.
   b. The width of its depletion zone decreases.
   c. Its barrier potential increases.
   d. Its depletion zone or region increases.

8. When a diode is externally reverse biased:

   a. The width of the depletion zone increases.
   b. The width of the depletion zone decreases.
   c. The barrier potential of the junction is reduced.
   d. The depletion zone or region of the junction is reduced.

9. The leakage current of a diode is caused by:

   a. Heat energy
   b. Chemical energy
   c. Internal barrier voltage
   d. The crystal doping material

10. The arrowhead or pointed end of a diode symbol represents the:

    a. Anode
    b. Cathode
    c. N-material
    d. Direction of electron flow

11. A shorted diode tested with an ohmmeter will show:

    a. Low resistance in either directions
    b. High resistance in only one direction
    c. Low resistance in only one direction
    d. High resistance in both directions

12. An open diode tested with an ohmmeter will show:

    a. Low resistance in both directions
    b. High resistance in only the forward direction
    c. High resistance in both directions
    d. Low resistance in one direction and high resistance in the other

Problems

Answer the following questions.

1. The diode of **Figure 2.25** is:

    a. Forward biased
    b. Reverse biased
    c. Shorted
    d. Open

2. The V_F meter of **Figure F2.25** will show a voltage value in the range of:

    a. 0.0–0.4 V
    b. 0.41–0.7 V
    c. 0.71–1.0 V
    d. 1.01–10 V

3. The I_F meter of **Figure 2.25** will show the current to be in the range of:

    a. 0.0–5.0 mA
    b. 5.1–10 mA
    c. 10.1–100 mA
    d. 101 mA to 10 A

4. The V_{R2} meter of **Figure 2.25** will show a voltage value in the range of:

    a. 0.0–0.4 V
    b. 0.41–0.7 V
    c. 0.71–1.0 V
    d. 1.1–10 V

5. If the diode of **Figure 2.25** is reversed in the circuit, the V_F meter will show a voltage in the range of:

    a. 0.0–0.4 V
    b. 0.41–0.7 V
    c. 0.71–1.0 V
    d. 1.1–10 V

6. If the diode of **Figure 2.25** is reversed in the circuit, the V_{R2} meter will show a voltage in the range of:

    a. 0.0–0.4 V
    b. 0.41–0.7 V

     c. 0.71–1.0 V

     d. 1.1–10 V

7. If the diode of **Figure 2.25** is reversed in the circuit, the I_F will show current in the range of:

     a. 0.0–5.0 mA

     b. 5.1–10. mA

     c. 10.1–100 mA

     d. 101 mA to 10 A

Analysis and Troubleshooting – P–N Junction Diodes

In order to perform a test on any electronic device, you must understand the characteristics of the device and be proficient in the use of the test equipment. **P–N junction diodes** can be tested using analog meters or digital multimeters, as shown in **Figure 2.18**. A practical consideration is that some meters provide an output voltage on the ohmmeter range that is less than the diode forward. These meters will give an open indication for good diodes. Analog meters are actually better for this test. When connected as shown in **Figure 2.26(a)**, the diode is forward biased by the ohmmeter. The resistance reading should be low, typically less than 1 kΩ. When connected as shown in **Figure 2.26(b)**, the diode resistance should be extremely high, typically high MΩ. Most ohmmeters give an "out-of-range" indication, showing high diode reverse resistance. If a diode has a high forward resistance or low reverse resistance, the diode is faulty. A defective **open diode** will show an extremely high resistance for both forward and reverse biases. A defective **shorted diode** will show zero or a very low resistance for both forward and reverse biases.

    On low-resistance scales, some ohmmeter sections of a multimeter have enough current output to destroy low-current diodes. It is best to use only the resistance scales rated in the low kΩ or higher. Some multimeter ohmmeter sections supply current from the common (negative) lead, while others supply it from the positive (ohms) lead. Check the (negative) meter to be sure that you know the lead polarities before testing any diodes. Some **multimeters** have a "diode test" function.

    The ohmmeter function of a multimeter can also be used to identify the anode and the cathode of an unmarked diode. Diodes are usually marked to indicate the end that is the cathode. When an ohmmeter is connected to the diode so that it is forward biased, the positive meter lead is the anode terminal, and the negative lead is the cathode of the device.

Data Sheet Analysis – P–N Junction Diodes

The type of information found on **data sheets** for electronic devices varies among manufacturers and the specific type of device. Located at the *end* of this chapter are typical data sheets for P–N junction diodes. We will use the data sheets to analyze typical information available for P–N junction diodes.

Locate the **data sheet** for an **IN4153** small signal diode. This is a relatively low-voltage, low-current device. Use this data sheet to answer the following for the IN4153 diode:

1) Maximum reverse voltage (V_{RM}) = _____ V
2) Average forward current $(I_{F(AV)})$ = _____ mA
3) Power dissipation (P_D) = _____ W
4) Breakdown voltage (V_R) = _____ V
5) Forward voltage (V_F) range at I_F = 1.0 mA = _____ V to _____ V

Refer to the data sheet showing the physical data of the **IN4153** diode package. Note, on the previous data sheet, that "DO-35" is written below the diode's picture. "DO" represents a "**diode outline**." Use this data sheet to answer the following:

6) Overall length of the diode body = _____ mm
7) Length of each wire connector = _____ mm
8) The type of material used for manufacturing is _____

Now, refer to the data sheet showing the characteristics of a general purpose rectifier group **IN5391** through **IN5399**. Each of these nine diodes has similar characteristics except peak reverse voltage (V_{RM}).

Use this data sheet to answer the following:

9) V_{RM} for **IN5392** = _____ V
10) V_{RM} for **IN5397** = _____ V
11) Average forward current (I_F) = _____ A
12) Junction temperature range = _____ °C to _____ °C
13) Power dissipation (P_D) = _____ W
14) Forward voltage (V_F) = _____ V
15) Forward current at V_F = 0.8 V = _____ A (see **Figure 2** of data sheet)
16) Total junction capacitance (C_T) at V_R = 40 V = _____ pF (see **Figure 4** of data sheet)
17) Forward current (I_F) at 130°C = _____ A (see **Figure 1** of data sheet)

Answers

Examples

2-1. 0.0485 A or 48.5 mA

Self-Examination

2.1
   1. diode
   2. junction
   3. N-type, P-type (any order)
   4. electrons
   5. holes
   6. holes
   7. electrons
   8. diffusion
   9. barrier potential
   10. 0.3, 0.7
   11. depletion zone
2.2
   12. diode
   13. biasing
   14. reverse
   15. forward
   16. majority
   17. nonconductive
   18. increase
   19. leakage
   20. temperature
   21. germanium
   22. forward
   23. current *or* I_{max}
2.3
   24. directly
   25. across
   26. through
   27. forward current *or* I_F
   28. germanium, silicon

29. Zener
30. avalanche
2.4
31. peak inverse voltage
32. forward current
33. temperature
34. capacitor
35. dielectric
36. varicap
37. switching time
38. power
2.5
39. heavy
40. low, infinite
41. shorted
42. open
43. polarity

Glossary

Junction

The point where P-type and N-type semiconductor materials are joined.

Diffusion

A process in which electrons from the N-type material of a P–N junction diode move readily across the junction to fill holes in the P-type material.

Depletion zone

An area near the P–N junction that is void of current carriers.

Barrier potential

The voltage that is developed across a P–N junction due to the diffusion of holes and electrons.

Bias voltage

An external source of energy applied to a P–N junction.

Reverse biasing

Adding external voltage of the same polarity to the barrier potential to increase in the width of the depletion zone and, thus, hinder current carriers from entering the depletion zone.

Forward biasing

Adding an external voltage of the opposite polarity to the barrier potential to reduce the barrier potential and, thus, cause current carriers to return to the depletion zone.

Leakage current

A small amount of current that flows through the depletion zone when a diode is reverse biased.

Knee voltage

The point at which the value of forward voltage is great enough to overcome the barrier potential of the P–N junction.

Zener breakdown

A physical process that occurs when electrons are pulled from their covalent bonds in a strong electric field.

Junction capacitance

The capacitive effect that occurs from the two independent crystal materials of a diode serving as conductor plates and the depletion zone acting as a dielectric material.

Switching time

The time it takes to switch from one state to the other (forward to reverse or vice versa).

Anode

The electrode that attracts or gathers in electrons. The P-type material of the crystal serves as the anode in a diode.

Cathode

The electrode that gives off or emits electrons. The N-type material of the crystal servers as the cathode in a diode.

1N4001/L - 1N4007/L

1.0A RECTIFIER

Features

- Diffused Junction
- High Current Capability and Low Forward Voltage Drop
- Surge Overload Rating to 30A Peak
- Low Reverse Leakage Current
- **Lead Free Finish, RoHS Compliant (Note 4)**

Mechanical Data

- Case: DO-41, A-405
- Case Material: Molded Plastic. UL Flammability Classification Rating 94V-0
- Moisture Sensitivity: Level 1 per J-STD-020C
- Terminals: Finish - Bright Tin. Plated Leads Solderable per MIL-STD-202, Method 208
- Polarity: Cathode Band
- Mounting Position: Any
- Ordering Information: See Last Page
- Marking: Type Number
- Weight: DO-41 0.30 grams (approximate)
  A-405 0.20 grams (approximate)

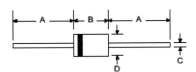

Dim	DO-41 Plastic		A-405	
	Min	Max	Min	Max
A	25.40	—	25.40	—
B	4.06	5.21	4.10	5.20
C	0.71	0.864	0.53	0.64
D	2.00	2.72	2.00	2.70
All Dimensions in mm				

"L" Suffix Designates A-405 Package
No Suffix Designates DO-41 Package

Maximum Ratings and Electrical Characteristics @ T_A = 25°C unless otherwise specified

Single phase, half wave, 60Hz, resistive or inductive load.
For capacitive load, derate current by 20%.

Characteristic	Symbol	1N 4001/L	1N 4002/L	1N 4003/L	1N 4004/L	1N 4005/L	1N 4006/L	1N 4007/L	Unit
Peak Repetitive Reverse Voltage Working Peak Reverse Voltage DC Blocking Voltage	V_{RRM} V_{RWM} V_R	50	100	200	400	600	800	1000	V
RMS Reverse Voltage	$V_{R(RMS)}$	35	70	140	280	420	560	700	V
Average Rectified Output Current (Note 1) @ T_A = 75°C	I_O	1.0							A
Non-Repetitive Peak Forward Surge Current 8.3ms single half sine-wave superimposed on rated load (JEDEC Method)	I_{FSM}	30							A
Forward Voltage @ I_F = 1.0A	V_{FM}	1.0							V
Peak Reverse Current @ T_A = 25°C at Rated DC Blocking Voltage @ T_A = 100°C	I_{RM}	5.0 50							μA
Typical Junction Capacitance (Note 2)	C_J	15				8			pF
Typical Thermal Resistance Junction to Ambient	$R_{\theta JA}$	100							K/W
Maximum DC Blocking Voltage Temperature	T_A	+150							°C
Operating and Storage Temperature Range (Note 3)	T_J, T_{STG}	-65 to +150							°C

Notes: 1. Leads maintained at ambient temperature at a distance of 9.5mm from the case.
    2. Measured at 1. MHz and applied reverse voltage of 4.0V DC.
    3. JEDEC Value.
    4. RoHS revision 13.2.2003. Glass and High Temperature Solder Exemptions Applied, see *EU Directive Annex Notes 5 and 7.*

TYPES 1N4001 THROUGH 1N4007
DIFFUSED-JUNCTION SILICON RECTIFIERS

*electrical characteristics at specified ambient† temperature

PARAMETER		TEST CONDITIONS		MAX	UNIT
I_R	Static Reverse Current	V_R = Rated V_R,	T_A = 25°C	10	μa
		V_R = Rated V_R,	T_A = 100°C	50	μa
$I_{R(av)}$	Average Reverse Current	V_{RM} = Rated V_{RM}, f = 60 cps,	I_O = 1 a, T_A = 75°C	30	μa
V_F	Static Forward Voltage	I_F = 1 a,	T_A = 25°C to 75°C	1.1	v
$V_{F(av)}$	Average Forward Voltage	V_{RM} = Rated V_{RM}, f = 60 cps,	I_O = 1 a, T_A = 25°C to 75°C	0.8	v
V_{FM}	Peak Forward Voltage	V_{RM} = Rated V_{RM}, f = 60 cps,	I_O = 1 a, T_A = 25°C to 75°C	1.6	v

*Indicates JEDEC registered data.

THERMAL INFORMATION

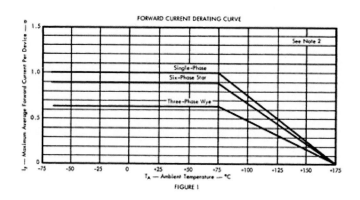

FORWARD CURRENT DERATING CURVE

FIGURE 1

† The ambient temperature is measured at a point 2 inches below the device. Natural air cooling shall be used.

NOTE 2: This rectifier is a lead-conduction-cooled device. At (or above) ambient temperatures of 75°C, the lead temperature ⅜ inch from case must be no higher than 5°C above the ambient temperature for these ratings to apply.

TYPES 1N4001 THROUGH 1N4007
DIFFUSED-JUNCTION SILICON RECTIFIERS

TYPES 1N4001 THROUGH 1N4007
BULLETIN NO. DL-S 63734A, FEBRUARY 1965

50-1000 VOLTS • 1 AMP AVG

- ● **MINIATURE MOLDED PACKAGE**

- ● **INSULATED CASE**

- ● **IDEAL FOR HIGH-DENSITY CIRCUITRY**

*mechanical data

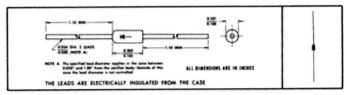

ALL DIMENSIONS ARE IN INCHES

NOTE A. The specified lead diameter applies in the zone between 0.050" and 1.00" from the rectifier body. Outside of this zone the lead diameter is not controlled.

THE LEADS ARE ELECTRICALLY INSULATED FROM THE CASE

*absolute maximum ratings at specified ambient† temperature

		1N4001	1N4002	1N4003	1N4004	1N4005	1N4006	1N4007	UNIT
V_{RM}	Peak Reverse Voltage from −65°C to +175°C (See Note 1)	50	100	200	400	600	800	1000	V
V_R	Steady State Reverse Voltage from 25°C to 75°C	50	100	200	400	600	800	1000	V
I_O	Average Rectified Forward Current from 25°C to 75°C (See Notes 1 and 2)	1							a
$I_{FM(rep)}$	Repetitive Peak Forward Current, 10 cycles, at (or below) 75°C (See Note 3)	10							a
$I_{FM(surge)}$	Peak Surge Current, One Cycle, at (or below) 75°C (See Note 3)	30							a
$T_{A(opr)}$	Operating Ambient Temperature Range	−65 to +175							°C
T_{stg}	Storage Temperature Range	−65 to +200							°C
	Lead Temperature ⅜ inch from Case for 10 Seconds	350							°C

NOTES 1. These values may be applied continuously under single-phase, 60-cps, half-sine-wave operation with resistive load. Above 75°C derate I_O according to Figure 1.

2. This rectifier is a lead-conduction-cooled device. At (or above) ambient temperatures of 75°C, the lead temperature ⅜ inch from case must be no higher than 5°C above the ambient temperature for these ratings to apply.

3. These values apply for 60-cps half sine waves when the device is operating at (or below) rated values of peak reverse voltage and average rectified forward current. Surge may be repeated after the device has returned to original thermal equilibrium.

* Indicates JEDEC registered data.

† The ambient temperature is measured at a point 2 inches below the device. Natural air cooling shall be used.

TEXAS INSTRUMENTS
INCORPORATED
POST OFFICE BOX 5012 • DALLAS, TEXAS 75222

1N5391 - 1N5399

Features

* 1.5 ampere operation at $T_A = 70°C$ with no thermal runaway.

* High current capability.

* Low leakage.

DO-15
COLOR BAND DENOTES CATHODE

General Purpose Rectifiers

Absolute Maximum Ratings* $T_A = 25°C$ unless otherwise noted

Symbol	Parameter	Value									Units
		5391	5392	5393	5394	5395	5396	5397	5398	5399	
V_{RRM}	Peak Repetitive Reverse Voltage	50	100	200	300	400	500	600	800	1000	V
$I_{F(AV)}$	Average Rectified Forward Current, .375 " lead length @ $T_A = 75°C$	1.5									A
I_{FSM}	Non-repetitive Peak Forward Surge Current 8.3 ms Single Half-Sine-Wave	50									A
T_{stg}	Storage Temperature Range	-55 to +150									°C
T_J	Operating Junction Temperature	-55 to +150									°C

*These ratings are limiting values above which the serviceability of any semiconductor device may be impaired.

Thermal Characteristics

Symbol	Parameter	Value	Units
P_D	Power Dissipation	4.8	W
$R_{\theta JA}$	Thermal Resistance, Junction to Ambient	26	°C/W

Electrical Characteristics $T_A = 25°C$ unless otherwise noted

Symbol	Parameter	Device									Units
		5391	5392	5393	5394	5395	5396	5397	5398	5399	
V_F	Forward Voltage @ 1.5 A	1.4									V
I_R	Reverse Current @ rated V_R $T_A = 25°C$ $T_A = 100°C$	5.0 300									$\propto$A $\propto$A
C_T	Total Capacitance $V_R = 4.0$ V, f = 1.0 MHz	25									pF

General Purpose Rectifiers
(continued)

Typical Characteristics

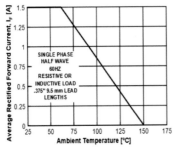

Figure 1. Forward Current Derating Curve

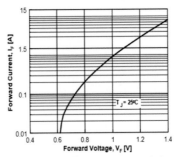

Figure 2. Forward Voltage Characteristics

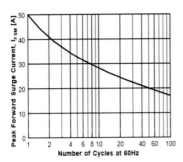

Figure 3. Non-Repetitive Surge Current

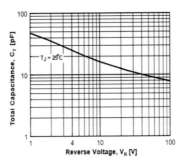

Figure 4. Total Capacitance

3

Zener Diodes

The previous chapter focused on the diode as a forward conducting device. You learned that, as a rule, reverse biasing of a diode does not produce conduction unless the peak reverse voltage (P_{RV}) is exceeded. When this occurs, a P–N junction diode is usually destroyed. Operation of a P–N junction diode is generally related to forward conduction applications.

In this chapter, you will be introduced to a group of diodes referred to as **avalanche breakdown** or **Zener diodes**. This type of diode operates in the reverse direction. The reverse breakdown voltage characteristic of this device is used to a unique advantage: voltage can be regulated. As you will see in this chapter, the reverse breakdown of a semiconductor is an important solid-state operational characteristic.

Objectives

After studying this chapter, you will be able to:

3.1 explain the reverse-bias operation of a Zener diode;
3.2 understand the I–V characteristic curve of a Zener diode;
3.3 analyze the operating characteristics of a Zener diode;
3.4 analyze and test a Zener diode.

Chapter Outline

3.1 Crystal Structure and Symbol
3.2 Zener Characteristics
3.3 Zener Current Ratings
3.4 Analysis and Troubleshooting – Zener Diodes

Key Terms

avalanche breakdown
derating

power dissipation
power dissipation rating
Zener breakdown

3.1 Crystal Structure and Symbol

The Zener diode is a P–N junction device that is different from the P–N junction diode studied in Chapter 2. Zener diodes are designed by modifying their crystal structure during manufacture. The breakdown voltage of a Zener diode is set by carefully controlling its doping level. The symbol of a Zener diode is slightly different from a P–N junction diode.

3.1 Explain the reverse-bias operation of a Zener diode.
In order to achieve objective 3.1, you should be able to:

- summarize the characteristics of avalanche breakdown;
- summarize the characteristics of Zener breakdown;
- recall *avalanche breakdown* and *Zener breakdown*.

The symbol and crystal structure of a **Zener diode** are shown in **Figure 3.1.** Note that the crystal structure is similar to that of a P–N junction diode. The Zener diode symbol, however, is uniquely different. On the Zener diode symbol, the cathode is usually drawn with a bent line in the form of a Z to distinguish it from the P–N junction diode symbol. It only appears as a Z, however, when the symbol is oriented horizontally in a diagram.

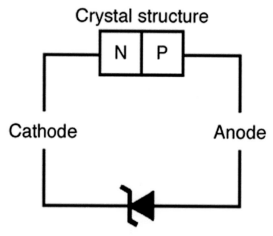

Figure 3.1   Zener diode symbol and crystal structure.

Figure 3.2 shows some Zener **diode packages**. Note that their physical appearance is similar to that of P–N junction diodes. However, the diode symbol is rarely printed on the case of a Zener diode as it is for a P–N junction diode. If a diode symbol is printed on the case of a diode, the diode is ordinarily a P–N junction diode.

Zener diodes are normally connected in a circuit in the reverse-bias direction (see **Figure 3.3**).

This means that the anode must be connected to the negative side of the voltage source and the cathode to the positive side of the voltage source. This action causes the **depletion zone** of the junction to widen (see **Figure 3.4**).

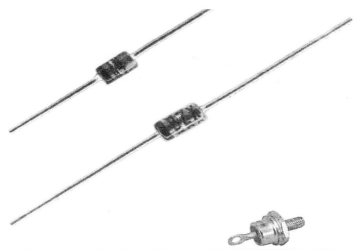

Figure 3.2   Zener diodes are similar in appearance to P–N junction diodes.

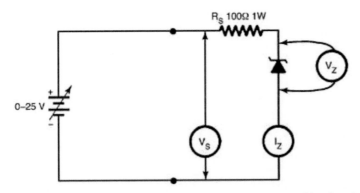

Figure 3.3   A Zener diode is connected in a circuit in the reverse-bias direction.

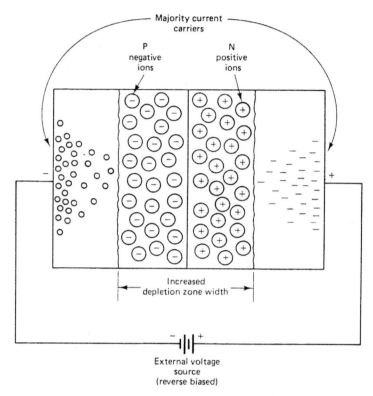

Figure 3.4   Connecting a Zener diode in reverse bias causes the depletion zone to widen.

Under normal circumstances, we would expect a reverse-biased junction to be nonconductive. In a Zener diode, two conditions are responsible for causing this device to become conductive in the reverse-biased direction: **avalanche breakdown** and **Zener breakdown**.

Avalanche breakdown occurs when thermally generated holes and electrons gain enough energy from the reverse-biased source to produce new carriers. These current carriers are the result of removing valence electrons from their normal bonding. These new carriers, in turn, produce additional carriers that disrupt the bonds of other atoms. The process continues with the number of carriers building up in increasing numbers. The end result is high current flow through the reverse-biased junction. Avalanche breakdown is primarily responsible for reverse current conduction above 5 V. Diodes of this type were once called **avalanche diodes**.

Zener breakdown is the result of a barrier potential that appears across the P–N junction. This field causes covalent bonds near the junction to break apart. As a result, a large quantity of new holes and electrons are produced. These newly generated holes and electrons represent a substantial increase in current. Zener breakdown is primarily responsible for conduction of a reverse-biased junction below 5 V. Diodes that conduct in this voltage range are called **Zener diodes**. Zener diodes are heavily doped so that they can begin conduction at relatively low reverse-bias voltages. Today, the term **Zener diode** is primarily used to describe all diodes that operate in the reverse-bias direction.

Self-Examination

Answer the following questions.

1. The _____ of a Zener diode symbol is drawn as a bent line or bar.
2. A Zener diode is normally connected in the _____-bias direction.
3. When a Zener diode is reverse biased, it will go into _____ at a prescribed voltage value.
4. _____ breakdown is the result of thermally generated holes and electrons gaining enough energy from the source to move valence electrons out of their bonding.
5. _____ breakdown is the result of covalent bonds breaking up due to a strong electric field across a reverse-biased P–N junction.

3.2 Zener Characteristics

The characteristics of a P–N junction diode show that it is designed for operation in the forward direction. Forward biasing produces a large value of forward current (I_F) for a rather small value of forward voltage (V_F). Reverse biasing generally does not cause current conduction until higher values of reverse voltage are reached. If reverse voltage (V_R) is great enough, however, breakdown occurs and causes a reverse current flow. P–N junction diodes are usually damaged when this occurs. Zener diodes, however, are designed to operate in the reverse direction without being damaged. In this section, you will learn how current and voltage respond when the breakdown voltage of a Zener diode is reached.

3.2 Understand the *I–V* characteristics of a Zener diode.

In order to achieve objective 3.2, you should be able to:

- apply the proper equation to find the Zener breakdown (V_Z) range of a Zener diode;

- explain how Zener current and Zener voltage respond when the breakdown voltage of a Zener diode is reached;
- explain the relationship of V_Z knee to the tolerance rating of a Zener diode.

Figure 3.5 shows **(a)** the *I–V* characteristics and **(b)** a test circuit for evaluating the operation of a Zener diode under **reverse-bias** conditions. As the value of the applied reverse voltage (V_R) is gradually increased, the amount

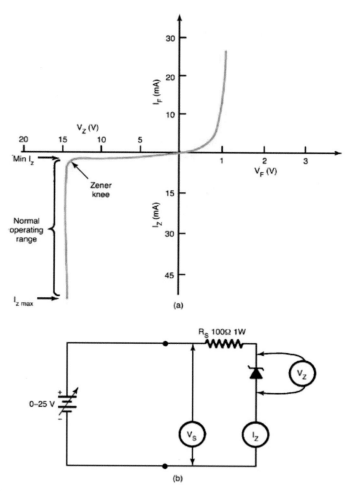

Figure 3.5   **(a)** *I–V* characteristics and **(b)** a test circuit for evaluating the operation of a Zener diode under **reverse-bias** conditions.

of current (I_R) flowing in the circuit is measured. The voltage across the Zener diode will remain constant after the Zener breakdown voltage (V_Z) has been reached. The value of the voltage across the Zener diode and the current flowing through it can be graphed. **Figure 3.5(a)** shows the $I-V$ characteristic of a typical Zener diode.

In the reverse-bias direction, there is practically no reverse current (labeled I_Z) until the breakdown voltage (labeled V_{ZK}) is reached near the "knee" of the curve. When this occurs, a very large change in Zener current (I_Z) is accompanied by only a small change in Zener voltage (V_Z). This means that varying amounts of I_Z can pass through a Zener diode without causing damage. The reverse Zener voltage (V_Z) across the diode remains relatively constant even when the source voltage is increased.

As you will see in Section 3.3, the Zener current has a safe operating limit or maximum value that must be observed. It is determined by the wattage rating of the diode. Nominal Zener voltage (V_{ZT}) is specified on datasheets at a value of reverse current called *Zener test current* (I_{ZT}).

When operated in the forward-bias condition, the Zener diode behaves like an ordinary silicon diode. A large value of forward current (I_F) is produced for a rather small value of forward voltage (V_F).

Zener Breakdown Voltage

Zener diodes are available in a wide range of breakdown voltages. The Zener breakdown voltage (V_{BR}), or simply Zener voltage, is usually identified on datasheets by the letters V_Z. Typical V_Z values range from 1.4 to 200 V. Some low-voltage values shown in the Zener diode **data sheers** at the *end* of this chapter for 1N746A through 1N759A are 3.3, 3.6, 3.9, 4.3, 4.7, 5.1, 5.6, 6.2, 6.8, 7.5, 8.2, 9.1, 10, and 12. A number of other V_Z values are also available. These values vary, to some extent, among different manufacturers.

Zener Tolerance Ratings

The accuracy of the Zener voltage of a diode is an important selection consideration and is largely responsible for the cost of a Zener diode. The accuracy or **tolerance rating** of a Zener diode is a range of values within which the actual V_Z will appear. If the V_Z region of **Figure 3.5(a)** is examined carefully, it shows that Zener breakdown does not occur at a precise location. The curve is actually rounded in this area. This is called the V_Z or the **knee** of the curve. Ideally, the knee should have a sharp edge. When it is rounded,

V_Z occurs more gradually. Zener diodes with a low tolerance rating generally have a sharp V_Z knee.

Some of the more popular tolerance ratings are 20%, 10%, 5%, and 1%. Low-cost Zener diodes have the highest tolerance percentage, while 1% or less tolerance devices are generally the most expensive. A Zener diode rated at 9.1 V with a ±10% tolerance will have a V_Z value that is ±0.91 V of 9.1 V. Therefore, its actual Zener breakdown value could be somewhere between 8.19 and 10.01 V. As a rule, diodes with a 10% tolerance rating will work for most circuit applications. Zener diodes with a 1% tolerance are used only in special circuit applications that require critical voltage values.

Example 3-1:

Calculate the Zener breakdown (V_Z) range for the 1N754A Zener diode. Refer to the end of the chapter for V_Z and the tolerance rating.

Solution

The V_Z of a 1N754A Zener diode is 6.8 V.
The tolerance rating is 5%.

$$V_Z \text{ range} = 6.8 \times 0.05$$
$$= \pm 0.34$$
$$= 6.46 \text{ to } 7.14.$$

Related Problem

Calculate the Zener breakdown (V_Z) range for the 1N759A Zener diode. Refer to the end of the chapter for V_Z and the tolerance rating.

Zener voltages are generally identified at a minimum Zener current value. **Figure 3.5(a)** shows that a certain amount of I_Z is needed to enter the knee of the I_Z curve. In this case, approximately 3 mA of I_Z is needed to produce conduction. After this value has been reached, V_Z remains fairly constant over a wide range of I_Z. The normal operating range of I_Z in this region is determined by the **power dissipation rating**, as you will see in the following section on Zener current ratings.

Self-Examination

Answer the following questions.

6. The characteristics of a Zener diode show that after a minimum value of I_Z is reached, the _____ remains at a constant value.

7. The Zener voltage accuracy of a Zener diode is referred to as _____.
8. The actual Zener voltage of a 6.8-V diode with 10% tolerance is somewhere between _____ V and _____ V.
9. A sharp V_Z knee is generally indicative of a Zener diode with a(n) _____ tolerance rating.

3.3 Zener Current Ratings

In the previous section, you learned that once Zener breakdown voltage is reached, a very large change in Zener current is accompanied by a small change in Zener voltage. Zener current, however, has a safe operating limit or maximum value that must be observed. This value is determined by the wattage, or power dissipation, rating of the diode. This section takes a closer look at the safe operating range of a Zener diode as it covers power dissipation and Zener impedance.

3.3 Analyze the operating characteristics of a Zener diode.

In order to achieve objective 3.3, you should be able to:

- apply the proper equation to find the maximum Zener current that a diode can safely conduct;
- apply the proper equation to find the safe operating range of a Zener diode;
- use a datasheet to determine the operating characteristics of a Zener diode;
- explain the relationship between impedance and the slope of the Zener knee area;
- explain the relationship between the power dissipation rating and maximum Zener current;
- recall *power dissipation*, *power dissipation rating*, and *derating*.

Power Dissipation Rating

Manufacturers rate Zener diodes according to their Zener voltage (V_Z) value and maximum power dissipation (P_D) at 25°C. **Power dissipation** refers to the ability of the Zener diode's junction to give off heat. A principal cause of solid-state device destruction is heat. When power is applied to a device, the device heats up. If too much power is applied, the device will overheat and be destroyed. The **power dissipation rating** of a Zener diode is an indication of how much heat the device can give off or dissipate. It is rated in watts at a certain temperature. Typically, this rating is expressed as 1 W at 25°C (77°F).

The temperature value is an indication of device operation at **ambient temperature**. *Ambient* refers to something that goes around or surrounds an object. In this case, *ambient* refers to the air surrounding the operating device. A temperature of 25°C (77°F) is considered to be the temperature of an inside room. Some manufacturers indicate the power dissipation rating at temperatures such as 50°C (122°F) and 75°C (169°F). These are generally considered to be working temperatures. The actual power that can be dissipated by a Zener diode decreases as the temperature increases and increases when the temperature decreases. In other words, the power dissipation (P_D) rating of a Zener diode has a negative temperature coefficient.

The P_D rating of a Zener diode can be altered, to some extent, by its lead length and its mounting in a circuit. The P_D value can be derated when these factors are considered. **Derating** permits a device to handle more power than its normal P_D value indicates. This is achieved by reducing lead length or mounting the device on a piece of metal, which serves as a heat sink. Manufacturers usually provide data about device derating in a chart that correlates lead length and the area of the heat sink. Many Zener diodes are housed in packages that permit them to be attached to a **heat sink**. In some installations, the metal chassis that houses a circuit may be used as the heat sink. Derating is an important consideration when a Zener diode operates close to its maximum power rating. A derating entry on the datasheet for Zeners **1N4370A−IN4372A** can be referred to at the end of the chapter.

Maximum Zener Current

In the process of dissipating heat, a certain amount of power is used or consumed by the Zener diode. This characteristic gives an indication of the **maximum Zener current (I_Z)** that a diode can safely conduct. A 1-W, 15-V_Z Zener diode, for example, can conduct an I_Z of 0.066 A or 66 mA. This is determined by the basic power equation:

$$P = I \times V \tag{3.1}$$

by solving for *I*. We do, however, substitute P_D for *P*, I_{Zmax} for *I*, and V_Z for *V* (using the typical or nominal voltage value of the Zener from the datasheet) in the formula. The modified formula then becomes

$$I_{Zmax} = [P_D]/[V_Z]. \tag{3.2}$$

Therefore,

$$I_{Z\max} = [1\text{W}]/[15\text{ V}]$$
$$= 0.066 \text{ A, or } 66 \text{ mA.}$$

The $I_Z - V_Z$ **characteristics** of an IN1775 1-W Zener diode are shown in **Figure 3.6.** This Zener diode has a maximum Zener current rating, $I_{Z\max}$, of 66 mA. The minimum Zener current, $I_{Z\min}$, is also indicated by the curve. This value is approximately 5 mA. The region between these two extreme conditions is labeled as the safe operating range. The Zener can operate anywhere between these two points without being damaged. If the current exceeds the maximum I_Z value, additional heat will develop. This heat may become so intense that it can permanently damage the junction. As a rule,

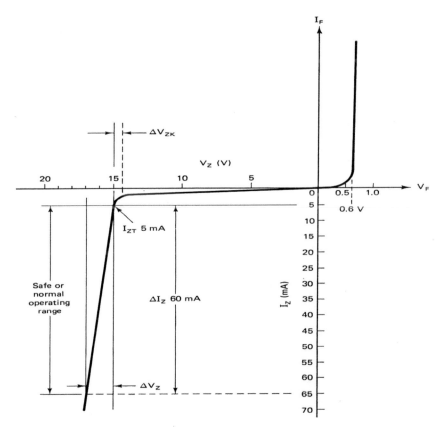

Figure 3.6 $I_Z - V_Z$ characteristic of a 1N1775 Zener diode.

the area beyond the maximum I_Z point should be avoided in normal diode applications.

The safe operating range of a Zener diode is from the minimum Zener current (I_{Zmin}) to the maximum current rating, I_{Zmax}.

$$\text{Safe operating range} = I_{Zmax} - I_{Zmin}. \tag{3.3}$$

Example 3-2:

Calculate the safe operating range for a 1N754A Zener diode. Refer to the data sheets at the end of the chapter for P_D, V_Z, and I_Z values.
Solution

1. Determine the maximum current rating.

$$\begin{aligned} I_{Zmax} &= [P_D]/[V_Z] \\ &= [500\text{mW}]/[6.8\text{V}] \\ &= [0.5\text{W}]/[6.8\text{V}] \\ &= 73.5\text{mA}. \end{aligned}$$

2. Then, determine the safe operating range. The safe operating range is from the minimum Zener current (I_{Zmin}), at which the typical Zener voltage is listed in the datasheet, to the maximum current rating, I_{Zmax}. safe operating range = 73.5 mA − 20 mA.

Related Problem

Calculate the safe operating range for a 1N759A Zener diode. Refer to the data sheets at the end of the chapter for P_D, V_Z, and I_Z values.

Zener Impedance

Another matter of importance in Zener diode selection is Zener impedance (Z_Z). This characteristic deals with the slope of the I_Z-V_Z curve after conduction has been achieved. A Zener diode actually shows a slight increase in the value of V_Z when I_Z increases. This increase is very noticeable because it means that the reverse characteristic is not entirely vertical. Zener impedance is responsible for this characteristic and is determined by the expression:

$$Z_Z = [\Delta V_Z]/[\Delta I_Z] \tag{3.4}$$

where the Greek letter delta (Δ) denotes a change in value.

Figure 3.6 shows an $I_Z - V_Z$ curve that indicates the location of the items discussed. The ΔI_Z value is derived from the safe or normal operating range of the Zener diode. Some manufacturers select the range of values from a prescribed test area near the middle of the reverse current curve. The range of V_Z can be used to determine corresponding values of I_Z. Note the location of the ΔV_Z and ΔI_Z values on the chart. The Z_Z for this range of operation is, therefore,

$$
\begin{aligned}
Z_Z &= [\Delta V_Z]/[\Delta I_Z] \\
&= [17 \text{ V} - 15 \text{ V}]/[66 \text{ mA} - 5 \text{ mA}] \\
&= [2 \text{ V}]/[61 \text{ mA}] \\
&= [2 \text{ V}]/[0.061 \text{ A}] \\
&= 32.79 \ \Omega.
\end{aligned}
$$

The reverse knee impedance (Z_{ZK}) of a Zener diode can be determined by the same procedure as determining Zener impedance. It is represented by the following formula:

$$
Z_{ZK} = [\Delta V_{ZK}]/[\Delta I_{ZK}]. \tag{3.5}
$$

Zener knee impedance shows the impedance of the device near the breakdown point and is a good indication of the slope or sharpness of the knee area of the curve. Look again at **Figure 3.6**. Note that the ΔV_{ZK} value is the difference between the voltage at the start of the Zener breakdown point and the beginning of the linear portion of the curve. The values of I_{ZK} are the corresponding current values on the curve.

$$
\begin{aligned}
Z_{ZK} &= [\Delta V_{ZK}]/[\Delta I_{ZK}] \\
&= [15 \text{ V} - 14 \text{ V}]/[5 \text{ mA} - 1 \text{ mA}] \\
&= [1 \text{ V}]/[4 \text{ mA}] \\
&= [1 \text{ V}]/[0.004 \text{ A}] \\
&= 250 \ \Omega.
\end{aligned}
$$

In practice, a Zener diode with a sharp knee will produce a lower value of Zener impedance than a rounded knee. Ideally, the Zener knee should be very sharp. Manufacturers usually include knee impedance and Zener impedance in their specifications of a device.

Self-Examination

Answer the following questions.

11. Zener diodes are rated according to their _____ and _____.
12. A Zener diode rated at 1 W, 12 V_Z has a maximum current, or I_Z, capability of _____A or _____mA.
13. The space between the minimum and maximum I_Z values of a Zener diode is called the _____ operating range.
14. When the maximum I_Z value of a Zener diode is exceeded, damage will likely occur due to excessive _____.
15. Power dissipation of a Zener diode refers to the rated _____ at a particular _____.
16. When the junction temperature of a Zener diode increases, the P_D rating decreases _____.
17. Zener _____ is a characteristic based on the value of $\Delta V_Z / \Delta I_Z$.
18. A low value of Zener impedance is generally a good indication of the sharpness of the Zener _____.

3.4 Analysis and Troubleshooting – Zener Diodes

A Zener diode cannot be tested in the same way as a P–N junction diode. The Zener diode is designed to break down and conduct in the reverse direction when the reverse voltage equals V_Z. One way to test a Zener diode is to measure the voltage across its terminals while it is functioning in a circuit. If the voltage across the Zener is within tolerance, it is good.

3.4 Analyze and test a Zener diode

In order to achieve objective 3.4, you should be able to:

- use a datasheet to analyze typical information available for a Zener diode;
- use a test circuit to check the condition of a diode.

Zener Diode Datasheet Analysis

The type of information found on Zener diode datasheets provides important specifications for the device. In appearance, Zener diodes are similar to P–N junction diodes. At the end of this chapter, a typical datasheet for a Zener diode group: **1N4370A−1N4372A and 1N746A−1N759A**. A "**DO 35**" **package** is used for these Zener diodes. Use the Zener diode datasheets at the *end* of the chapter to find the following information.

1. Power dissipation (P_D) = _____ mW
2. Maximum Zener voltage (V_{ZM}) for a 1N750A Zener diode at I_Z = 20 mA = _____ V
3. Maximum Zener voltage (V_{ZM}) for 1N758A at I_Z = 20 mA = _____ V
4. Typical Zener voltage (V_Z) for a 1N755A Zener diode at I_Z = 20 mA = _____ V
5. Typical Zener voltage (V_Z) for 1N759A Zener diode at I_Z = 20 mA = _____ V
6. Zener impedance (Z_Z) for a 1N746A Zener diode at I_Z = 20 mA = _____ Ohms
7. Zener impedance (Z_Z) for 1N757A Zener diode at I_Z = 20 mA = _____ Ohms
8. Maximum Zener current (I_{ZM}) for a 1N751A Zener diode = _____ mA
9. Maximum Zener current (I_{ZM}) for a 1N753A Zener diode = _____ mA
10. What are the second and third line markings on the body of a 1N4372A diode in this group?

_____

11. What are the second and third line markings on the body of a 1N750A diode in this group? _____

1. The operating temperature range for a 1N4370A Zener diode is from _____ °C to _____ °C

Zener Diode Troubleshooting

A Zener diode is designed to function as a voltage regulator when it is reverse biased. In this state, it will have a minimum amount of current flowing through it. To test a Zener diode, you can measure the voltage across its terminals while it is functioning. As reverse voltage applied across the Zener diode is varied, the current flowing through it should change. If the current does not change, even when the rated V_Z voltage is applied, the Zener diode terminals may be open. On the other hand, if the Zener is shorted, no voltage will be developed across the diode, and a significant amount of current will flow, even under reverse-bias conditions.

Summary

- Zener diodes are normally connected in a circuit in the reverse-bias direction.
- When power is applied to a reverse-biased Zener diode, the depletion zone of the diode becomes wider.
- Zener voltage (V_Z) is the value of reverse-bias voltage applied to a Zener diode.
- The junction of a Zener diode becomes conductive at a specific Zener voltage (V_Z) value and Zener knee current (I_{ZK}).
- The Zener voltage (V_Z) of the diode remains fairly constant over a large range of Zener current (I_Z).
- Voltage regulation is an important application of the Zener diode.
- Zener voltage (V_Z) has a tolerance rating that indicates the value of its accuracy.
- Low-tolerance V_Z diodes are more expensive than high-tolerance devices.
- A low-tolerance Zener diode has a sharp knee and a very small change in its Zener voltage (V_Z).
- Power dissipation (P_D) refers to the ability of a Zener diode to dissipate heat.
- The power dissipation (P_D) rating is used to determine the maximum Zener current (I_{Zmax}) that a diode can safely conduct.
- The maximum value of Zener current that a Zener diode can safely conduct can be determined by the formula $I_{Zmax} = P_D/V_Z$.
- Zener impedance (Z_Z) is related to the slight increase in the value of Zener voltage (V_Z) when the Zener current (I_Z) of an operating Zener diode increases.
- Zener impedance can be determined through the formula $Z_Z = \Delta V_Z/\Delta I_Z$.

Formulas

(3-1) $P = I \times V$ *Power formula.*
(3-2) $I_{Zmax} = [P_D]/[V_Z]$ *Maximum Zener current.*
(3-3) Safe operating range $= I_{Zmax} - I_{Zmin}$ *Safe operating range of a Zener diode.*
(3-4) $Z_Z = [\Delta V_Z]/[\Delta I_Z]$ *Zener impedance.*
(3-5) $Z_{ZK} = [\Delta V_{ZK}]/[\Delta I_{ZK}]$ *Zener knee impedance.*

Review Questions

Answer the following questions.

1. A Zener diode achieves voltage regulation when it is _____.

    a. forward biased across R_L
    b. connected in series with R_S
    c. reverse biased across R_L
    d. connected in parallel with R_S

2. When thermally generated holes and electrons gain enough energy from the reverse-biased source of a Zener diode to produce new current carriers, it is called _____.

    a. avalanche breakdown
    b. reverse breakdown
    c. Zener breakdown
    d. low-voltage breakdown

3. A strong electric field across a P–N junction that causes covalent bonds to break apart _____.

    a. is called avalanche breakdown
    b. is called reverse breakdown
    c. is called Zener breakdown
    d. results in Zener breakdown above 8 V

4. The reverse current that flows through a Zener diode is called _____.

    a. I_T
    b. I_L
    c. I_Z
    d. I_F

5. The voltage across a Zener diode that is responsible for its ability to achieve regulation is called _____.

    a. V_{RS}
    b. V_L
    c. V_Z
    d. V_{in}

6. The percentage of accuracy that a Zener diode has when it goes into conduction is called _____.

    a. tolerance
    b. knee voltage

    c. minimum V_Z

    d. V_Z safe operating range

7. The P_D rating of a zener diode refers to its ability to _____.

    a. regulate voltage

    b. dissipate heat

    c. control I_Z

    d. change impedance

8. In a Zener diode voltage regulator, a small increase in input voltage (V_{in}) will cause _____.

    a. an increase in V_Z

    b. a decrease in V_{RS}

    c. a decrease in I_Z

    d. an increase in I_Z

9. In a Zener diode regulator, reducing the value of R_L will cause _____.

    a. an increase in V_Z

    b. a decrease in V_Z

    c. an increase in I_Z

    d. a decrease in I_Z

10. Applied input voltage beyond the V_Z point of a Zener diode causes _____.

    a. an increase in I_Z

    b. a decrease in I_Z

    c. the I_Z to level off

    d. the I_Z to cut off

    e. only a trace of I_Z

3.5 Problems

Answer the following questions.

1. A 1N1775A Zener diode is rated 15 V at 1 W. In what range does the maximum I_Z of this device fall?

    a. 0–0.05 A

    b. 1–50 mA

    c. 51–75 mA

    d. 76–100 mA

    e. 101 mA or above

2. A Zener diode is rated at 9.1 V and has a 10% tolerance. In what V_Z operating range will this device fall?

    a. 1–9.1 V
    b. 9.1–20 V
    c. 9.1–10.01 V
    d. 8.19–9.1 V
    e. 8.19–10.01 V

3. In a Zener diode voltage regulator, the value of V_Z is determined by _____.

    a. $V_{in} - V_{RS}$
    b. $V_{RS} - V_{in}$
    c. $V_L - V_{RS}$
    d. $V_{RS} + V_{in}$

4. Design of a Zener diode voltage regulator calls for a 10-V output with a fixed load of 50 mA. The unregulated input voltage varies from 12 to 15 V. What is a suitable value of series resistor for the circuit? Indicate the resistance range where this value will fall.

    a. 1–50 Ω
    b. 51–75 Ω
    c. 76–100 Ω
    d. 101 Ω or higher

5. What would be a suitable wattage rating for the series resistor selected for problem 4?

    a. 1/8 W
    b. 1/4 W
    c. 1/2 W
    d. 1 W
    e. 2 W

6. A 12-V Zener diode has a P_D rating of 5 W. What is the value of I_{ZM}?
7. Refer to the data sheets at the end of the chapter. If a 10-V Zener diode is needed for a voltage regulator circuit, which one should be used?
8. Refer to the data sheets at the end of the chapter. If a Zener diode is needed in circuit with a rating in excess of 105 mA, which one should be used?

Glossary

Avalanche breakdown

P–N junction breakdown that results when thermally generated holes and electrons gain enough energy from a reverse-biased source to produce new current carriers.

Zener breakdown

Current carrier production due to the influence of a strong electric field that causes large numbers of electrons in the depletion region to form covalent bonds across a reverse-biased P–N junction.

Power dissipation

The ability of a device to change energy from one form to another, give off heat, or use power.

Power dissipation rating

An indication of how much heat a device, such as a Zener diode, can give off or dissipate.

Derating

Making physical changes to a device, such as reducing lead length or mounting the device on a piece of metal, to permit it to handle more power than its normal power dissipation value indicates.

FAIRCHILD
SEMICONDUCTOR®

Zeners
1N4370A - 1N4372A
1N746A - 1N759A

Absolute Maximum Ratings * $T_A = 25°C$ unless otherwise noted

Tolerance = 5%

Symbol	Parameter	Value	Units
P_D	Power Dissipation @ TL ≤ 75°C, Lead Length = 3/8"	500	mW
	Derate above 75°C	4.0	mW/°C
T_J, T_{STG}	Operating and Storage Temperature Range	-65 to +200	°C

* These ratings are limiting values above which the serviceability of the diode may be impaired.

DO-35 Glass case
COLOR BAND DENOTES CATHODE

Electrical Characteristics $T_A = 25°C$ unless otherwise noted

Device	V_Z (V) @ I_Z = 20mA (Note 1)			Z_Z (Ω) @ I_Z = 20mA	I_{ZM} (mA) (Note 2)	I_R (μA) @ V_R = 1V	
	Min.	Typ.	Max.			Ta = 25°C	Ta = 125°C
1N4370A	2.28	2.4	2.52	30	150	100	200
1N4371A	2.57	2.7	2.84	30	135	75	150
1N4372A	2.85	3.0	3.15	29	120	50	100
1N746A	3.14	3.3	3.47	28	110	10	30
1N747A	3.42	3.6	3.78	24	100	10	30
1N748A	3.71	3.9	4.10	23	95	10	30
1N749A	4.09	4.3	4.52	22	85	2	30
1N750A	4.47	4.7	4.94	19	75	2	30
1N751A	4.85	5.1	5.36	17	70	1	20
1N752A	5.32	5.6	5.88	11	65	1	20
1N753A	5.89	6.2	6.51	7	60	0.1	20
1N754A	6.46	6.8	7.14	5	55	0.1	20
1N755A	7.13	7.5	7.88	6	50	0.1	20
1N756A	7.79	8.2	8.61	8	45	0.1	20
1N757A	8.65	9.1	9.56	10	40	0.1	20
1N758A	9.50	10	10.5	17	35	0.1	20
1N759A	11.40	12	12.6	30	30	0.1	20

V_F Forward Voltage = 1.5V Max @ I_F = 200mA

Notes:
1. Zener Voltage (V_Z)
   The zener voltage is measured with the device junction in the thermal equilibrium at the lead temperature (T_L) at 30°C ± 1°C and 3/8" lead length.
2. Maximum Zener Current Ratings (I_{ZM})
   The maximum current handling capability on a worst case basis is limited by the actual zener voltage at the operation point and the power derating curve.

Zeners
1N746A - 1N759A

Absolute Maximum Ratings* <small>T_A = 25°C unless otherwise noted</small>

Tolerance: A = 5%

Symbol	Parameter	Value	Units
P_D	Power Dissipation	500	mW
T_{STG}	Storage Temperature Range	-65 to +200	°C
T_J	Operating Junction Temperature	+ 175	°C
	Lead Temperature (1/16" from case for 10 seconds)	+ 230	°C

*These ratings are limiting values above which the serviceability of the diode may be impaired.

NOTES:
1) These ratings are based on a maximum junction temperature of 200 degrees C.
2) These are steady state limits. The factory should be consulted on applications involving pulsed or low duty cycle operations.

DO-35
COLOR BAND DENOTES CATHODE

Electrical Characteristics <small>T_A = 25°C unless otherwise noted</small>

Device	V_Z (V)	$Z_Z(\Omega)$	@ I_Z(mA)	$I_{R1}(\mu A)$	@ V_R(V)	$I_{R2}(\mu A)$ @	V_R(V) T_A=150°C	T_C (%/°C)	I_{ZRM}*(mA)
1N746A	3.3	28	20	10	1.0	30	1.0	- 0.070	110
1N747A	3.6	24	20	10	1.0	30	1.0	- 0.065	100
1N748A	3.9	23	20	10	1.0	30	1.0	- 0.060	95
1N749A	4.3	22	20	2.0	1.0	30	1.0	+/- 0.055	85
1N750A	4.7	19	20	2.0	1.0	30	1.0	+/- 0.030	75
1N751A	5.1	17	20	1.0	1.0	20	1.0	+/- 0.030	70
1N752A	5.6	11	20	1.0	1.0	20	1.0	+ 0.038	65
1N753A	6.2	7.0	20	0.1	1.0	20	1.0	+ 0.045	60
1N754A	6.8	5.0	20	0.1	1.0	20	1.0	+ 0.050	55
1N755A	7.5	6.0	20	0.1	1.0	20	1.0	+ 0.058	50
1N756A	8.2	8.0	20	0.1	1.0	20	1.0	+ 0.062	45
1N757A	9.1	10	20	0.1	1.0	20	1.0	+ 0.068	40
1N758A	10	17	20	0.1	1.0	20	1.0	+ 0.075	35
1N759A	12	30	20	0.1	1.0	20	1.0	+ 0.077	38

*I_{ZRM} (Maximum Zener Current Rating) Values shown are based on the JEDEC rating of 400 milliwatts. Where the actual zener voltage (VZ) is known at the operating point, the maximum zener current may be increased and is limited by the derating curve.

4

Power Supply Circuits

All electronic systems require certain voltage and current values for operation. The energy source of the system is primarily responsible for this function. As a general rule, the source is more commonly called a **power supply**. In some systems, such as a flashlight or other portable device, the power supply may be multiple batteries or a single dry cell. Cellular phones, televisions, personal computers, stereo systems, and electronic instruments usually derive their energy from an AC power line. This part of the system is called an electronic power supply. Electronic devices are used in the power supply to develop the required output energy.

All **electronic power supplies** have a number of functions that must be performed in order to operate. These functions are transformer, rectifier, filter, regulator, and load. Some of these functions are achieved by all power supplies. Others are somewhat optional and are dependent primarily on the parts that are supplied power. This chapter discusses the functions of an electronic power supply.

Objectives

After studying this chapter, you will be able to:

4.1 calculate the output voltage and current capability of a transformer;

4.2 analyze half-wave, full-wave, and bridge-rectifier circuits;

4.3 select the appropriate filter for a given application;

4.4 analyze a Zener regulator circuit under various input and load conditions;

4.5 diagram the current flow of a dual power supply;

4.6 describe various types and the operation of clipper, clamper, and voltage multiplier circuits;

4.7 troubleshoot a power supply.

Chapter Outline

4.1 Transformer
4.2 Rectifier
4.3 Filter
4.4 Regulator
4.5 Dual Power Supplies
4.6 Clippers, Clampers, and Voltage Multipliers
4.7 Troubleshooting Power Supplies

Key Terms

choke
C-input filter
clamper
clipper
dual power supply
filter
L-input filter
pi filter
pulsating DC
rectification
rectifier
ripple frequency
turns ratio
voltage regulator
clipper
clamper
limiter
voltage multiplier

4.1 Transformers

The energy source of an electronic power supply is alternating current. In most systems, the AC power line supplies this. Most small electronic systems use 120-V, single-phase, 60-Hz AC as their energy source. This energy is readily available in homes, buildings, and industrial facilities. Electronic power supplies rarely operate with AC obtained directly from the power line. A *transformer* is commonly used to step the line voltage up or down to a desired value. You should already be familiar with the transformer from your studies of DC/AC electronics. This section reviews the relationship between turns ratio, voltage ratio, and current ratio.

4.1 Calculate the output voltage and current capability of a transformer.
In order to achieve objective 4.1, you should be able to:

- describe the relationship between the turns ratio, voltage ratio, and current ratio;
- identify the primary and secondary windings of a transformer;
- define *turns ratio*.

The **transformer** represents the first function of a power supply, as shown in a **block diagram** of a general-purpose electronic power supply (see **Figure 4.1**).

A **schematic** symbol and a simplification of a power supply transformer are shown in **Figure 4.2.**

The coil on the left of the symbol is called the **primary winding**. The AC input is applied to this winding. The coil on the right side is the **secondary winding**. The secondary winding develops the output of the transformer. The parallel lines near the center of the symbol represent the **core**. In a transformer of this type, the core is usually made of laminated soft steel. Both windings are placed on the same core. Power supply transformers may have more than one primary and secondary winding on the same core, which permits them to accommodate different line voltages and to develop alternate output values.

The **output voltage** of a power supply transformer may have the same **polarity** as that of the input voltage, or it may be reversed. This depends on the winding direction of the coils. A dot is often used on the transformer symbol to indicate the polarity of a winding. The polarity of the input and

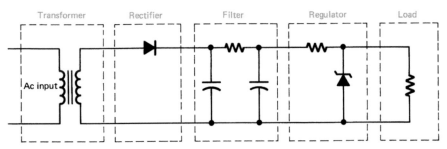

Figure 4.1 Functional block diagram of an electronic power supply. The transformer is the first function of this power supply.

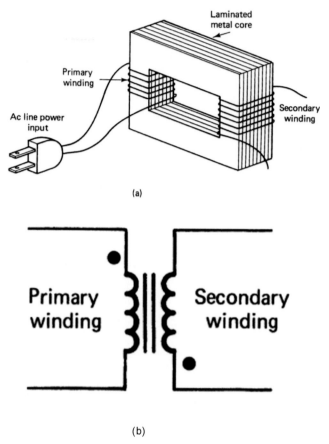

(a)

(b)

Figure 4.2   (a) Power supply transformer. (b) Schematic symbol.

output is the same when the dots are in adjacent locations on the transformer symbol. The polarity of the input and output is opposite when the dots are in opposite corners on the transformer symbol.

The output voltage developed by a transformer primarily depends on the turns ratio of its windings. The **turns ratio** is the ratio of the number of turns in the primary winding to the number of turns in the secondary winding. It is represented by the following formula:

$$N_{\text{pri}}/N_{\text{sec}}. \tag{4.1}$$

The relationship of the turns ratio to voltage is

$$N_{\text{pri}}/N_{\text{sec}} = V_{\text{pri}}/V_{\text{sec}}. \tag{4.2}$$

For example, if the primary winding of a transformer has 500 turns of wire and the secondary has 1000 turns, the transformer has a 1:2 turns ratio.

$$N_{pri}/N_{sec}$$
$$500/1000$$
$$1:2.$$

This type of transformer steps up the input voltage by a factor of 2. With 120-V RMS applied from the power line, the output voltage is approximately two times the input voltage (2 × 120 V).

$$N_{pri}/N_{sec} = V_{pri}/V_{sec}$$
$$1/2 = 120 \text{ V}/V_{sec}$$
$$1 \times V_{sec} = 2 \times 120 \text{ V}$$
$$V_{sec} = 240 \text{ V}.$$

Example 4-1:

If the primary winding has 1000 turns of wire, and the secondary has 50 turns of wire, what is the turns ratio of the transformer?

Solution

$$N_{pri}/N_{sec}$$
$$1000/50$$
$$20:1.$$

Related Problem

What is the turns ratio of a transformer if the primary winding has 200 turns of wire and the secondary has 1000 turns of wire?

Example 4-2:

If a transformer has a turns ratio of 20:1, and 120 V is applied from the power line, what is the output voltage?

Solution

$$N_{pri}/N_{sec} = V_{pri}/V_{sec}$$

$$20/1 = 120 \text{ V}/V_{\text{sec}}$$
$$20\, V_{\text{sec}} = 120 \text{ V}$$
$$20\, V_{\text{sec}}/20 = 120 \text{ V}/20$$
$$V_{\text{sec}} = 6 \text{ V}.$$

Related Problem

What is the output voltage of the transformer with a turns ratio of 1:5 and 120 V applied from the power line?

It is interesting to note that the current capability of a transformer is the reverse of its turns ratio. This is represented by the following formula:

$$V_{\text{pri}}/V_{\text{sec}} = I_{\text{sec}}/I_{\text{pri}}. \tag{4.3}$$

A 1:2 turns ratio steps up the voltage and steps down the current capability. Therefore, 1 A of primary current is capable of producing only 0.5 A of secondary current:

$$N_{\text{pri}}/N_{\text{sec}} = I_{\text{sec}}/I_{\text{pri}}$$
$$1/2 = I_{\text{sec}}/1 \text{ A}$$
$$2 \times I_{\text{sec}} = 1 \text{ A}$$
$$2 \times I_{\text{sec}}/2 = 1/2$$
$$I_{\text{sec}} = 0.5 \text{ A}.$$

The resistance of the wire is largely responsible for this condition. A 1:2 turns ratio has twice as much resistance in the secondary as it has in the primary winding. An increase in resistance, therefore, causes a decrease in current. This is all based on the wire size of the two windings being the same.

Example 4-3:

If a transformer has a turns ratio of 5:1 and 2 A is applied to the primary, what is the secondary current?

Solution

$$5/1 = I_{\text{sec}}/2 \text{ A}$$
$$I_{\text{sec}} = 2 \text{ A} \times 5$$
$$= 10\text{A}.$$

Related Problem

What is the secondary current if a transformer has a turns ratio of 1:5 and 6 A is applied to the primary?

Power supplies for solid-state systems generally develop low-voltage values. Transformers for this type of power supply are designed to step down the line voltage. A turns ratio of 10:1 is very common. With 120 V applied to the input, the output will be approximately 12 V (120 V/10). The **output current** capabilities of this transformer are increased by a ratio of 1:10. This means that a step-down transformer has a low-voltage output with a high-current capacity. Supplies of this type are well suited for solid-state applications.

Self-Examination

Answer the following questions.

1. The _____ winding develops the output of the transformer.
2. The _____ winding develops the input of the transformer.
3. The number of turns in the primary winding to the number of turns in the secondary winding is called the _____.
4. A 1:10 turns ratio steps up the _____ and steps down the _____ capability.
5. A 10:1 turns ratio steps up the _____ and steps down the _____ capability.
6. The output voltage of the transformer with a turns ratio of 10:1 and 120 V applied from the power line is _____ V.
7. The secondary current of a transformer that has a turns ratio of 10:1 and 6 A applied to the primary is _____ A.

4.2 Rectifier

As you have learned in the previous section, most power supplies are initially energized by AC. A transformer is used to step the line voltage up or down to a desired value. However, the primary function of a power supply is to develop DC for the operation of electronic devices. This energy must be changed into DC before it can be used. This process is performed by the rectifier.

4.2 Analyze half-wave, full-wave, and bridge-rectifier circuits.

In order to achieve objective 4.2, you should be able to:

- calculate the composite DC output of half-wave, full-wave, and bridge-rectifier circuits without filtering;

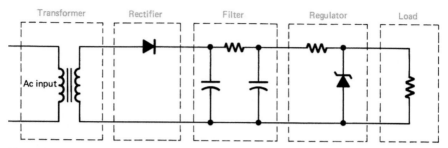

Figure 4.3   A rectifier is the second function of an electronic power supply. It changes AC to DC.

- explain the operation of half-wave, full-wave, and bridge-rectifier circuits;
- explain the characteristics of half-wave, full-wave, and bridge-rectifier circuits;
- identify half-wave, full-wave, and bridge-rectifier circuits.

The process of changing AC into DC is called **rectification.** All electronic power supplies have a rectification function, or **rectifier.** It is the second function of the power supply as shown in **Figure 4.3.**

Rectification can be either half-wave or full-wave. Take a look at **Figure 4.4**, which shows a graphical representation of the rectification function. Note that **half-wave rectification** uses one alternation of the AC input, whereas **full-wave** uses both alternations.

Half-wave and full-wave rectification is determined by the number and configuration of diodes in the rectifier circuit. There are three types of rectifiers: half-wave, full-wave, and bridge. The **half-wave rectifier** uses one diode and produces a half-wave output. Two diodes are used in the **full-wave rectifier**. The full-wave rectifier produces a full-wave output. A **bridge rectifier** uses four diodes and also produces a full-wave output.

Half-Wave Rectifier

In a half-wave rectifier, only one alternation of the AC input appears in the output (see **Figure 4.5**). The resulting output of this circuit is called **pulsating DC.** Note that a half-wave rectifier can remove either the positive or negative alternation, depending on its polarity in the circuit.

When the output has a **positive alternation**, the resulting DC is positive with respect to the common point or ground (see **Figure 4.5(a)**). A negative

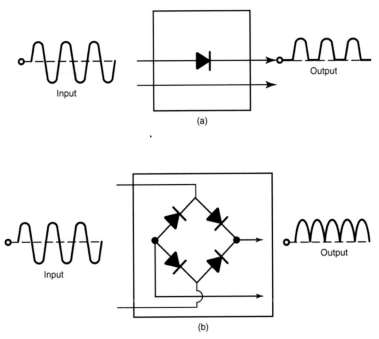

Figure 4.4 Graphical representations of a half-wave and full-wave rectifier. (a) The half-wave rectifier uses one alternation of the AC input. (b) The full-wave rectifier uses both alternations of the AC input.

output occurs when the positive alternation is removed. **Figure 4.5(b)** shows an example of a rectifier with a **negative output**.

The rectification function of a modern power supply is performed by diodes. You should recall that this type of device has two electrodes known as the **anode** and **cathode**. Operation is based on its ability to conduct easily in one direction but not in the other direction. As a result, only one alternation of the applied AC appears in the output. This occurs only when the diode is forward biased. To be **forward biased**, the anode must be positive and the cathode negative. **Reverse biasing** does not permit conduction through a diode.

A simplification of the rectification process is shown in **Figure 4.6**. **Figure 4.6(a)** shows how the diode responds during the positive alternation. This alternation causes the diode to be forward biased. Note that the power supplied to the anode is positive and the power supplied to the cathode is negative. Arrows show the resulting current flow. The output of the rectifier appears across the *l*oad **resistor, R**$_L$. The top of R_L is positive and the bottom of R_L is negative.

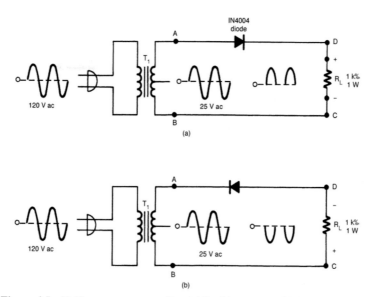

Figure 4.5   Half-wave power supplies. (a) Positive output. (b) Negative output.

Figure 4.6(b) shows how the diode responds during the **negative alternation**. This alternation causes the diode to be reverse biased. The power supplied to the anode is now negative and the power supplied to the cathode is positive. No current flows during this alternation. The output across R_L is zero.

The DC output of a half-wave rectifier appears as a series of pulses across R_L (see **Figure 4.7**). A DC voltmeter or ammeter would, therefore, indicate the average value of these pulses over a period of time. **Figure 4.7(a)** shows an example of the pulsating DC output. The average voltage value (V_{avg}) of one pulse is 0.637 of the peak voltage value (V_p):

$$V_{avg} = 0.637 \times V_p. \tag{4.4}$$

Since the second alternation of the output is zero, the composite average value must take into account the time of both alternations (see **Figure 4.7(b)**). The **composite average value** is, therefore, 0.637/2 or 0.318 of the peak value:

$$\text{composite average value} = 0.318 \times V_P. \tag{4.5}$$

Figure 4.7(b) shows a graphical example of this value.

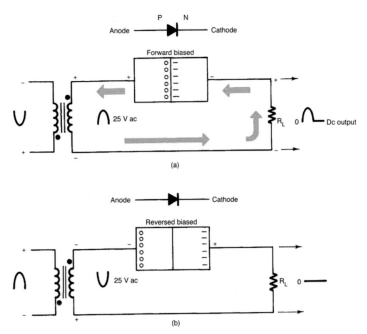

Figure 4.6 Simplification of rectifications. (a) The positive alternation causes the diode to be forward biased. (b) The negative alternation causes the diode to be reverse biased.

Example 4-4:

If the peak value of one pulse of DC at the output is 35 V, what is the composite average value?

Solution

$$\text{composite average value} = 0.318 \times V_P$$
$$= 0.318 \times 35$$
$$= 11.13 \text{ V}.$$

Related Problem

What is the composite average value if the peak value of one pulse of DC at the output is 20 V?

The composite average value of the DC output of a half-wave rectifier without filtering can be calculated when the value of the AC input is known. First, determine the **peak value** of one AC alternation at the input of the rectifier. Do this by multiplying the RMS value by 1.414:

$$V_p = V_{rms} \times 1.414. \qquad (4.6)$$

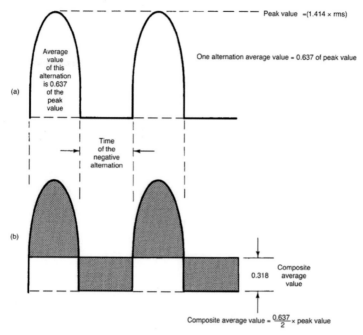

Peak value =(1.414 × rms)

One alternation average value = 0.637 of peak value

Figure 4.7   Half-wave rectifier output. (a) Pulsating DC output. (b) Composite output.

Then, determine the composite average value of the DC output by multiplying the peak value by 0.318:

composite average value = $0.318 \times V_p$.

Example 4-5:

If one AC alternation at the input of the rectifier is 25 V RMS, what is the composite average value of the DC output of a half-wave rectifier without filtering?

Solution

1. Determine the peak value of one AC alternation at the input of the rectifier.

$$V_p = V_{rms} \times 1.414$$
$$= 25 \text{ V} \times 1.414$$
$$= 35.35 \text{ V}.$$

2. Determine the composite average value by multiplying the peak value by 0.318.

$$\text{composite average value} = 0.318 \times V_p$$
$$= 0.318 \times 35.35$$
$$= 11.24 \text{ V}.$$

Related Problem

What is the composite average value of the DC output of a half-wave rectifier without filtering if one AC alternation at the input of the rectifier is 40 V RMS?

A simplified method of calculating the **composite average value** of the DC output of a half-wave rectifier is to combine different values. In this case, the 1.414 value used to calculate peak voltage is combined with 0.637/2 used to calculate the composite average value of the DC output:

$$\text{composite average value} = 1.414 \times [0.637/2]$$
$$= 0.45.$$

The composite average value of the DC output of a half-wave rectifier is, therefore, 45%, or 0.45, of the AC input voltage.

In an actual circuit, we must also take into account the **voltage drop** across the diode. When a silicon diode is used, this would be 0.6 V. The composite average value of DC of an unfiltered half-wave rectifier would, therefore, be expressed by the following formula:

$$\text{composite average value} = (V_{\text{rms}} \times 0.45) - 0.6\text{V}. \qquad (4.7)$$

Using this procedure, you can determine the equivalent DC voltage of the half-wave rectifier.

Example 4-6:

If the RMS input voltage of a half-wave rectifier is 50 V, and the rectifier uses a silicon diode, what is the composite average value of the DC output?

Solution

$$\text{composite average value} = (V_{\text{rms}} \times 0.45) - 0.6 \text{ V}$$
$$= (50 \text{ V} \times 0.45) - 0.6 \text{ V})$$
$$= 21.9 \text{ V}.$$

Related Problem

What is the composite average value of the DC output if the RMS input voltage of a half-wave rectifier is 33 V, and the rectifier uses a silicon diode?

The frequencies of the output pulses of a half-wave rectifier occur at the same rate as the applied AC input. With 60-Hz input, the pulse rate, or **ripple frequency**, of a half-wave rectifier is 60 Hz. Only one pulse appears in the output for each complete sine-wave input. Potentially, this means that only 45% of the AC input is transformed into a usable DC output. Half-wave rectification is, therefore, only 45% efficient. Due to its low-efficiency rating, half-wave rectification is not widely used.

Full-Wave Rectifier

A **full-wave rectifier** responds as two half-wave rectifiers that conduct on opposite alternations of the input. Both alternations of the input are changed into pulsating DC output. Full-wave rectification can be achieved by two diodes and a center-tapped transformer or by four diodes in a bridge circuit. Both types of rectifiers are widely used in solid-state power supplies.

A **schematic** of a two-diode **full-wave rectifier** is shown in **Figure 4.8.** Note that the anodes of each diode are connected to opposite ends of the transformer's secondary winding. The cathode of each diode is then connected to form a common positive output. The load of the power supply (R_L) is connected between the common cathode point and the center-tapped connection of the transformer. A complete path for current is formed by the transformer, two diodes, and load resistor.

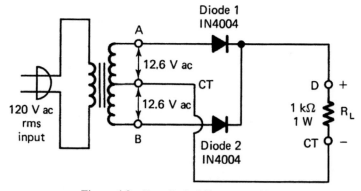

Figure 4.8   Two-diode full-wave rectifier.

When AC is applied to the primary winding of the transformer, it steps the voltage down in the secondary winding. The **center tap** serves as the electrical neutral or center of the secondary winding. Half of the secondary voltage appears between points CT and A, and the other half appears between points CT and B. These two voltage values are equal and are always 180° out of phase with respect to point CT.

Assume that the 60-Hz input is slowed so that we can see how one complete cycle responds. In **Figure 4.9**, note the **polarity** of the secondary winding voltage. For the first alternation, point A is positive and point B is negative with respect to CT. The second alternation causes point A to be negative and point B to be positive with respect to CT. The center tap, therefore, will always be negative with respect to the positive end of the winding. This point then serves as the negative output of the power supply.

Conduction of a specific diode in a full-wave rectifier is based on the polarity of the applied voltage during each alternation. In **Figure 4.9**, conduction is made with respect to point CT. For the first alternation, point A is positive and B is negative. This polarity causes D_1 to be **forward biased** and D_2 to be **reverse biased**. Current flow for this alternation is shown by the solid arrows. Starting at CT, electrons flow through a conductor to R_L, through D_1, and return to point A. This current flow causes a pulse of DC to appear across R_L for the first alternation.

For the second alternation, point A is negative and point B is positive. This polarity forward biases D_2 and reverse biases D_1. Current flow for this alternation is shown by the dashed arrows. Starting at point CT, electrons flow

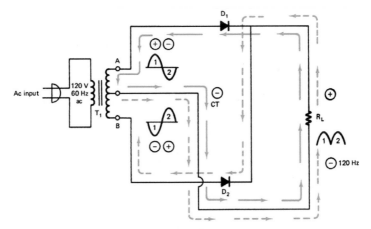

Figure 4.9   Conduction of a full-wave rectifier.

through a conductor to R_L, through D_2, and return to point B. This causes a pulse of DC to appear across R_L for the negative alternation.

It should be obvious at this point that current flow through R_L is in the same direction for each alternation of the input. This means that each alternation is transposed into a pulsating DC output. **Full-wave rectification**, therefore, changes the entire AC input into DC output. The resulting DC output of a full-wave rectifier is 90%, or 0.9, of the AC voltage between the center tap and the outer ends of the transformer. This voltage value is determined by calculating the peak value of the RMS input voltage and then multiplying it by the average value. Since 1.414×0.637 equals 0.9, 90% of the RMS value is the equivalent DC output. The output of a full-wave rectifier is 50% more efficient than that of an equivalent half-wave rectifier.

The actual DC output of a full-wave rectifier is slightly less than 0.9 of the RMS input just described. Each diode, for example, reduces the DC voltage by 0.6 V. A more practical calculation is

$$\text{composite average value} = (V_{\text{rms}} \times 0.9) - 0.6 \text{ V}. \qquad (4.8)$$

In a low-voltage power supply, a reduction of the DC by 0.6 V may be quite significant.

Example 4-7:

If the RMS input voltage of a full-wave rectifier is 50 V, and the rectifier uses a silicon diode, what is the composite DC output voltage?

Solution

$$\begin{aligned} \text{composite DC output} &= (V_{\text{rms}} \times 0.9) - 0.6 \text{ V} \\ &= (50 \text{ V} \times 0.9) - 0.6 \text{ V} \\ &= 44.4 \text{ V}. \end{aligned}$$

Related Problem

What is the composite DC output voltage if the RMS input voltage of a full-wave rectifier is 33 V and the rectifier uses a silicon diode?

The **ripple frequency** of a full-wave rectifier is somewhat different from that of a half-wave rectifier. Each alternation of the input, for example, produces a pulse of output current. The ripple frequency is, therefore, twice the alternation frequency. For a 60-Hz input, the ripple frequency of a full-wave rectifier is 120 Hz. As a general rule, a 120-Hz ripple frequency is easier to filter than the 60 Hz of a half-wave rectifier.

Full-Wave Bridge Rectifier

A **bridge rectifier** requires four diodes to achieve full-wave rectification. In this configuration, two diodes conduct during the positive alternation and two conduct during the negative alternation. A center-tapped transformer is not required to achieve full-wave rectification with a bridge circuit.

The component parts of a **full-wave bridge rectifier** are shown in **Figure 4.10.** In this circuit, one side of the AC input is applied to the junction of diodes D_1, and D_2. The alternate side of the AC input is applied to diodes D_3 and D_4. The diodes at each input are reversed with respect to each other. Output of the bridge occurs at the other two junctions. The cathodes of D_1 and D_3 serve as the positive output. Negative output appears at the anode junction of D_2 and D_4. The load of the power supply is connected across the common anode and common cathode connection points.

When AC is applied to the **primary winding** of a power supply trans-former, it can be either stepped up or down, depending on the desired DC output. In **Figure 4.10,** the 120-V input is stepped down to 25-V RMS. In normal operation, one alternation causes the top of the transformer to be positive and the bottom to be negative. The next alternation causes the bottom to be positive and the top to be negative. Opposite ends of the secondary winding are always 180° out of phase with each other.

Assume that the **AC input** causes point A to be positive and point B to be negative for one alternation. The schematic diagram of a bridge in **Figure 4.11** shows this condition of operation. With the indicated polarity, diode D_1 is **forward biased** and D_2 is reverse biased at the top junction. At

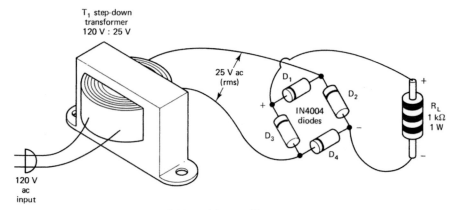

Figure 4.10   Bridge-rectifier components.

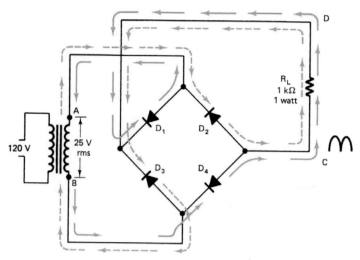

Figure 4.11   Bridge-rectifier conduction.

the bottom junction, D_4 is forward biased and D_3 is **reverse biased**. When this occurs, electrons flow from point B through D_4, up through R_L, through D_1, and return to point A. Solid arrows are used to show this current path. For this alternation, one pulse of DC falls across the load resistor.

With the next alternation, point A of the schematic diagram becomes negative and point B becomes positive. When this occurs, the bottom diode junction becomes positive, and the top junction goes negative. This condition forward biases diodes D_2 and D_3, while reverse biasing D_1 and D_4. The resulting current flow is indicated by the dashed arrows starting at point A. Electrons flow from point A through D_2, up through R_L, through D_3, and return to the transformer at point B. Dashed arrows are used to show this current path. One pulse of DC also falls across the load resistor for this alternation.

The output of the bridge rectifier has current flow through R_L in the same direction for each alternation of the input. AC is, therefore, changed into a **pulsating DC output** by conduction through a bridge network of diodes. The DC output, in this case, has a ripple frequency of 120 Hz. Each alternation produces a resulting output pulse. With 60 Hz applied, the output is 2×60 Hz, or 120 Hz.

The DC output voltage of a bridge circuit is slightly less than 90% of the RMS input. Each diode, for example, reduces the output by 0.6 V. With two diodes conducting during each alternation, the DC output voltage is reduced by 0.6×2 V, or 1.2 V, as shown in the following formula:

$$V_{dc \ (out)} = (V_{rms} \times 0.9) - 1.2 \ V. \tag{4.9}$$

In the circuit of **Figure 4.11**, the DC output is, therefore,

$$V_{dc \ (out)} = (25V \times 0.9) - 1.2 \ V$$
$$= 21.3 \ V.$$

The output of a bridge circuit is slightly less than that of an equivalent two-diode full-wave rectifier. The voltage drop of the second diode accounts for this difference.

Example 4-8:

If the RMS input voltage of a bridge rectifier is 50 V, and the rectifier uses silicon diodes, what is the composite DC output voltage?

Solution

$$\text{composite DC output} = (V_{rms} \times 0.9) - 1 \ V$$
$$= (50 \ V \times 0.9) - 1.2 \ V$$
$$= 43.8 \ V.$$

Related Problem

What is the composite DC output voltage if the RMS input voltage of a bridge rectifier is 33 V and the rectifier uses silicon diodes?

The four diodes of a bridge rectifier can be obtained in a single package. **Figure 4.12** shows several different package types. As a general rule, there are two AC input terminals and two DC output terminals. These devices are generally rated according to their **current-handling capability** and **peak**

Figure 4.12 Packaged bridge-rectifier assemblies. (a) Dual-in-line package. (b) Sink-mount package. (c) Epoxy package. (d) Tab-pack enclosure. (e) Thermo-tab package.

reverse voltage (P$_{RV}$) rating. Typical current ratings are from 0.5 to 50 A in single-phase AC units. P-R-V ratings are 50, 200, 400, 800, and 1000 V. A bridge package usually takes less space than four single diodes. Nearly all bridge power supplies employ the single package assembly.

Self-Examination

Answer the following questions.

8. A half-wave rectifier uses _____ diode(s).
9. A full-wave rectifier uses _____ diode(s).
10. A bridge rectifier uses _____ diode(s).
11. A diode is forward biased when a(n) _____ polarity is applied to its anode.
12. A diode is reverse biased when a(n) _____ polarity is applied to its cathode.
13. The AC ripple frequency of a half-wave rectifier is _____ Hz.
14. The AC ripple frequency of a full-wave rectifier is _____ Hz.
15. The AC ripple frequency of a bridge rectifier is _____ Hz.
16. With 10-V RMS input to a half-wave rectifier, the composite DC output is _____ V.
17. With 30-V RMS input to a bridge rectifier, the composite DC output is _____ V.

4.3 Filters

The output of a half-wave or full-wave rectifier is pulsating direct current. This type of output is generally not usable for most electronic circuits. A rather pure form of DC is usually required. This form of DC is achieved through a filter. This section discusses four types of filters: *C*-input, inductance, pi, and RC.

4.3 Select the appropriate filter for a given application.

In order to achieve objective 4.3, you should be able to:

- describe the main purpose of a filter;
- describe the characteristics of *C*-input, *L*-input, LC, pi, and RC filters;
- identify *C*-input, *L*-input, LC, pi, and RC filters.

The **filter** of a power supply is designed to change pulsating DC into a rather pure form of DC. Filtering takes place between the output of the rectifier and

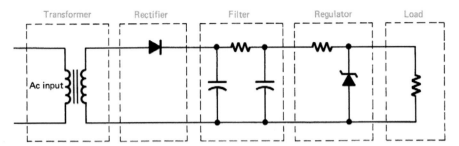

Figure 4.13 The filter is the third function of an electronic power supply. It minimizes ripple from the pulsating DC waveform produced by the rectifier.

the input to the load device (see **Figure 4.13**). Power supplies discussed up to this point have not employed a filter circuit.

The **pulsating DC output** of a rectifier contains two components. One of these deals with the DC part of the output. This component is based on the combined average value of each pulse. The second part of the output refers to its AC component. Pulsating DC, for example, occurs at 60 or 120 Hz, depending on the rectifier employed. This part of the output has a definite **ripple frequency**. Ripple must be minimized before the output of a power supply can be used by most electronic devices.

Power supply filters fall into two general classes according to the type of component used in its input. If filtering is first achieved by a capacitor, it is classified as a **C-input filter**, or **capacitor filter**. When a coil of wire or inductor is used as the first component, it is classified as **an L-input filter**, or **inductive filter.** C-input filters develop a higher value of DC output voltage than L-input filters. The output voltage of a C-input filter usually drops in value when the load increases. L-input filters, by comparison, tend to keep the output voltage at a rather constant value. This is particularly important when large changes in the load occur. However, the output voltage of an L-input filter is somewhat lower than that of a C-input filter. **Figure 4.14** shows a graphic comparison of output voltage and load current for C-input and L-input filters.

C-Input Filter

The AC component of a power supply can be effectively reduced by a **C-input filter**. A single capacitor is simply placed across the load resistor, as shown in **Figure 4.15.** For alternation 1 of the circuit, **Figure 4.15(a)**,

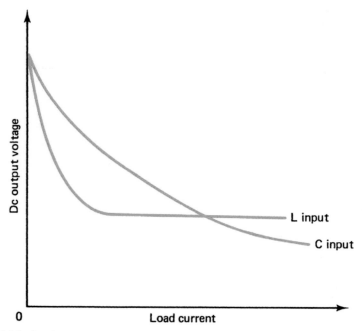

Figure 4.14   Load current versus output voltage comparisons for *L*-input and *C*-input filters.

the diode is forward biased. Current flows according to the arrows of the schematic. The capacitor (C) charges quickly to the peak voltage value of the first pulse. At the same point in time, current is also supplied to R_L. The initial surge of current through a diode is usually quite large. This current is used to charge C and supply R_L at the same time. A large capacitor, however, responds somewhat like low resistance when it is first being charged. Note the amplitude of the I_D waveform during alternation 1.

When alternation 2 of the input occurs, the diode is reverse biased. **Figure 4.15(b)** shows how the circuit responds for this alternation. Note that there is no current flow from the source through the diode. The charge acquired by C during the first alternation now finds an easy discharge path through R_L. The resulting discharge current flow is indicated by the arrows between C and R_L. In effect, R_L has supplied current even when the diode is not conducting. The voltage across R_L is, therefore, maintained at a much higher value. See the V_{RL} waveform for alternation 2 in **Figure 4.15(c)**.

Discharge of C continues for the full time of alternation 2. Near the end of the alternation, there is somewhat of a drop in the value of V_{RL}. This is due, primarily, to a depletion of capacitor charge current. At the end of this

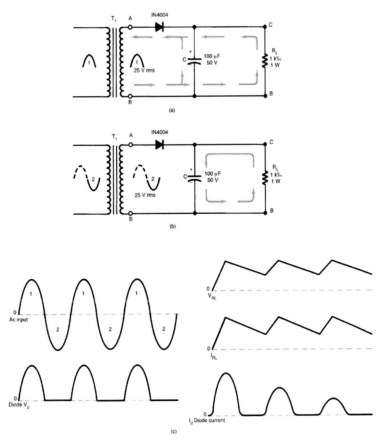

Figure 4.15 *C*-input filter action. (a) Alternation one. (b) Alternation two. (c) Waveforms for AC input, diode voltage (V_D) and current (I_D), and load resistor voltage (V_{RL}) and current (I_{RL}).

time, the next positive alternation occurs. The diode is again forward biased. The capacitor and R_L receive current from the source at this time. With C still partially charged from the first alternation, less diode current is needed to recharge C. In **Figure 4.15(c)**, note the amplitude change in the second 1, pulse. The process from this point on is a repeat of alternations 1 and 2.

The effectiveness of a capacitor as a filter device is based on a number of factors. Three very important considerations are:

- the size of the capacitor;
- the value of the load resistor, R_L;
- the time duration of a given DC pulse.

These three factors are related to one another by the following formula:

$$T = R \times C \qquad (4.10)$$

where T is the time in seconds, R is the resistance in Ohms, and C is the capacitance in farads. The product of RC is an expression of the filter circuit's **time constant**. RC is a measure of how rapidly the voltage and current of the filter respond to changes in input voltage. A capacitor will charge to 63.2% of the applied voltage in one time constant. A discharging capacitor will have a 63.2% drop of its original value in one time constant. It takes five time constants to charge or discharge a capacitor fully.

The **filter capacitor** of **Figure 4.15** charges quickly during the first positive alternation. Essentially, there is very little resistance for the RC time constant during this period. The discharge of C, however, is through R_L. If R_L is small, C will discharge very quickly. A large value of R_L will cause C to discharge rather slowly. For good filtering action, C must not discharge very rapidly during the time of one alternation. When this occurs, there is very little change in the value of V_{RL}. A C-input filter works very well when the value of R_L is relatively large. If the value of R_L is small, as in a heavy load, more ripple will appear in the output.

A rather interesting comparison of filtering occurs between half-wave and full-wave rectifier power supplies. The time between reoccurring peaks is twice as long in a half-wave circuit as it is for full-wave. The capacitor of a half-wave circuit, therefore, has more time to discharge through R_L. The ripple of a half-wave filter will be much greater than that of a full-wave circuit. In general, it is easier to filter the output of a full-wave rectifier. **Figure 4.16** compares half-wave and full-wave rectifier filtering with a single capacitor.

The DC output voltage of a filtered power supply is usually a great deal higher than that of an unfiltered power supply. In **Figure 4.16**, the waveforms show an obvious difference between outputs. In the filtered outputs, the capacitor charges to the peak value of the rectified output. The amount of discharge action that takes place is based on the resistance of R_L. For a light load or high resistance R_L, the filtered output remains charged to the peak value. For example, the peak value of 10-V RMS is 14.14 V. For an unfiltered full-wave rectifier, the DC output is approximately 0.9 × RMS, or 9.0 V. Comparing 9 V to 14 V shows a rather decided difference in the two outputs. For the half-wave rectifier, the output difference is even greater. The unfiltered half-wave output is approximately 45% of 10 V RMS, or 4.5 V. The filtered output is 14 V less 20% for the added ripple, or approximately 11.2 V. The

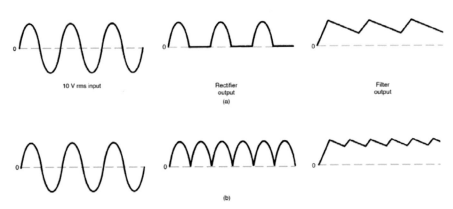

Figure 4.16 Comparison of *C*-input filtering with half-wave and full-wave rectifiers. (a) Half-wave. (b) Full-wave.

filtered output of a half-wave rectifier is nearly 2.5 times more than that of the unfiltered output. These values are only rough generalizations of power supply output with a light load.

Inductance Filter

An **inductor** is a device that has the ability to store and release electrical energy. It does this by taking some of the applied current and changing it into a magnetic field. An increase in current through an inductor causes the magnetic field to expand. A decrease in current causes the field to collapse and release its stored energy.

The ability of an inductor to store and release energy can be used to achieve **filtering**. An increase in the current passing through an inductor causes a corresponding increase in its magnetic field. Voltage induced in the inductor due to a change in its field opposes a change in the current passing through it. A decrease in current flow causes a similar reaction. A drop in current causes its magnetic field to collapse. This action also induces voltage in the inductor. The induced voltage in this case causes a continuation of the current flow. As a result, the added current tends to raise its decreasing value. An inductor opposes any change in its current flow. Inductive filtering is well suited for power supplies with a large load current.

An **L-input filter** is shown in **Figure 4.17**. The inductor is simply placed in series with the rectifier and the load. All the current supplied to the load must pass through the inductor (L). The filtering action of L does not let a pronounced change in current to take place. This prevents the output voltage

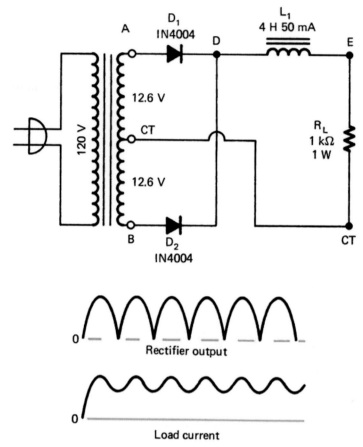

Figure 4.17   Inductive filtering. Since inductors tend to oppose any change in current flow, a larger load current can be drawn from an inductive filter without causing a decrease in its output voltage.

from reaching an extreme peak or valley. **Inductive filtering**, in general, does not produce as high an output voltage as capacitive filtering. Inductors tend to maintain the current at an average value. A larger load current can be drawn from an inductive filter without causing a decrease in its output voltage.

When an inductor is used as the primary filtering element, it is commonly called a choke. The term **choke** refers to the ability of an inductor to reduce ripple voltage and current. In electronic power supplies, choke filters are rarely used as a single filtering element. A combination inductor–capacitor or **LC filter** is more widely used. This filter has a series inductor with a capacitor

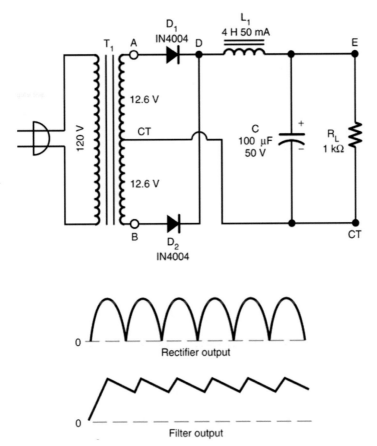

Figure 4.18 LC filter. Voltage and current are controlled in this filter, thus creating a purer form of DC.

connected in parallel with the load. The inductor controls large changes in load current. The capacitor, which follows the inductor, is used to maintain the load voltage at a constant voltage value. The combined filtering action of the inductor and capacitor produces a rather pure value of load voltage. **Figure 4.18** shows a representative LC filter and its output.

Pi Filter

When a capacitor is placed in front of the inductor of an LC filter, it is called a **pi filter.** In effect, this circuit becomes a *CLC* filter. The two capacitors are

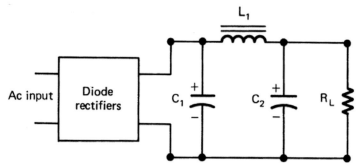

Figure 4.19   Pi filter. A pi filter has very low ripple content only when used with a light load.

in parallel with R_L and the inductor is in series. Component placement of this circuit in a schematic resembles the capital Greek letter pi (π), which accounts for the name of the filter. See the schematic of a pi filter in **Figure 4.19**.

The operation of a **pi filter** can best be understood by considering L_1 and C_2 as an LC filter. This part of the circuit acts on the output voltage developed by the capacitor-input filter C_1. C_1 charges to the peak value of the rectifier input. It has a ripple content that is very similar to that of the C-input filter of **Figure 4.20**. This voltage is then applied to C_2 through inductor L_1. C_1 charges C_2 through L_1. C_2 then holds its charge for a time interval determined by the time constant of C_2 and R_L. As a result, there is additional filtering by L_1 and C_2. The ripple content of this filter is much lower than that of a single C-input filter. There is, however, a slight reduction in the DC supply voltage to R_L due to the voltage drop across L_1.

A **pi filter** has very low ripple content when used with a light load. An increase in load, however, tends to lower its output voltage. This condition tends to limit the number of applications of pi filters. Pi filters have been used to supply radio circuits, stereo amplifiers, and TV receivers. A full-wave rectifier nearly always supplies the input to this filter.

RC Filter

In applications where less filtering can be tolerated, an **RC filter** can be used in place of a pi filter. As shown in **Figure 4.20**, the inductor is replaced with a resistor. An inductor is rather expensive, quite large physically, and weighs a great deal more than a resistor. The performance of the RC filter is not quite as good as that of a pi filter. There is usually a reduction in DC output voltage and increased ripple.

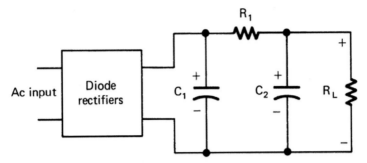

Figure 4.20 RC filter. This filter is similar to a pi filter; however, it uses a resistor in place of an inductor and produces a lower DC output voltage and more ripple.

When operating, C_1 charges to the peak value of the rectifier input. A drop in rectifier input voltage causes C_1 to discharge through R_1 and R_L. The voltage drop across R_1 lowers the DC output to some extent. C_2 charges to the peak value of the R_L voltage. The DC output of the filter is dependent on the load current. High values of load current cause more voltage drop across R_1. This, in turn, lowers the DC output. Low values of load current have less voltage drop across R_1. The output voltage, therefore, increases with a light load. In practice, RC filters can be used in power supplies that have 100 mA or less of load current. The primary advantage of RC filters is reduced cost.

Self-Examination

Answer the following questions.

18. The purpose of a filter circuit is to increase _____ and decrease _____.
19. Two types of inductive filter circuits are _____ and _____.
20. When a capacitor is attached to the front of the inductor of an LC filter, it is called a(n) _____ filter.
21. A(n) _____ filter places an inductor in series with the rectifier and the load.
22. A(n) _____ filter places a single capacitor in parallel with the load.
23. The purpose of a(n) _____ connected across the output of a rectifier circuit is to regulate voltage.
24. The purpose of a(n) _____ placed in series with the load is to regulate current flow.
25. A(n) _____ filter is typically used to supply radio circuits, stereo amplifiers, and TV receivers.

26. The _____ filter is suited for power supplies in which the load requires a high level of current and a pure value of load voltage.

4.4 Voltage Regulators

The DC output of an unregulated power supply has a tendency to change value under normal operating conditions. Changes in AC input voltage and variations in the load are primarily responsible for these fluctuations. In some power supply applications, voltage changes do not represent a serious problem. In many electronic circuits, voltage changes may cause improper operation. When a stable DC voltage is required, the power supply must employ voltage regulation. This section focuses on the Zener diode voltage regulator.

4.4 Calculate percent regulation for a power supply.
   In order to achieve objective 4.4, you should be able to:

- describe the characteristics of a Zener regulator circuit;
- describe the load and no-load operations of a Zener regulator circuit;
- describe how changes in input voltage affect a Zener regulator circuit;
- explain simple transistor and op-amp regulator circuits;
- explain the operation of series and shunt transistor regulators.

A stable DC voltage is achieved through a **voltage regulator**. The voltage regulator is the fourth function in the power supply block diagram in **Figure 4.21**.

A number of voltage regulator circuits have been developed for use in power supplies. One very common method of regulation employs a **Zener diode**. **Figure 4.22** shows this type of regulator located between the filter and the load. The Zener diode is connected in parallel, or shunt, with R_L.

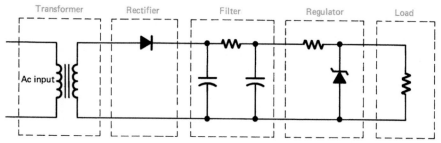

Figure 4.21   The regulator function of a power supply maintains a stable DC output voltage.

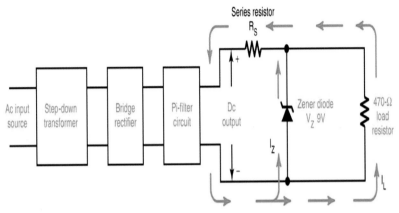

Figure 4.22   Zener diode voltage regulator.

This regulator requires only a Zener diode (D_Z) and a **series resistor (R_S)**. Note that D_Z is placed across the filter circuit in the reverse-bias direction. Connected in this way, the diode goes into conduction only when it reaches the Zener breakdown voltage V_Z. This voltage then remains constant for a large range of Zener current (I_Z). Regulation is achieved by altering the conduction of I_Z through the Zener diode. The combined I_Z and load current (I_L) must pass through the series resistor. This current value then determines the amount of voltage drop across R_S. Variations in current through R_S are used to keep the output voltage at a constant value.

No-Load Operation

A **schematic** of a 9-V **regulated power supply** is shown in **Figure 4.23**. This circuit derives its input voltage from a 12.6-V transformer. Rectification is achieved by a self-contained bridge-rectifier assembly. Filtering is accomplished by an RC filter. The series resistor of this circuit has two functions: it couples capacitors C_1 and C_2 in the filter circuit, and it serves as the series resistor (R_S) for the regulator circuit. Diode D_Z is a 9-V 1-W Zener diode.

Operation of the **regulated power supply** is similar to that of the bridge circuit discussed earlier. Full-wave DC output from the rectifier is applied to C_1 of the filter circuit. C_1 then charges to the peak value of the RMS input less the voltage drop across two silicon diodes (1.2 V). This represents a DC input value ($V_{dc\ (in)}$) of 16.6 V:

$$V_{dc\ (in)} = (V_{sec} \times 1.414) - 1.2\ V$$

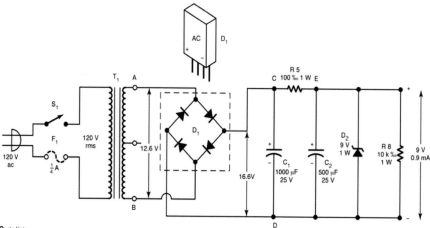

Parts list:

T$_1$: 12.6-V 2-A Stancor P-8130, or Triad F-44X

D$_1$: silicon bridge rectifier 2 A 50 V; General Instrument Co., KBF-005 or Motorola MDA-200

C$_1$: 1000-μF 25-V electrolytic capacitor

C$_2$: 500-μF 25-V electrolytic capacitor

R$_5$: 100-Ω 1-W resistor

R$_8$: 10-kΩ $\frac{1}{2}$-W resistor

D$_Z$: 9-V 1-W zener diode IN5346A or HEP Z-2513

S$_1$: SPST toggle switch

F$_1$: Fuse, $\frac{1}{4}$ A 250 V

Misc. : metal chassis, PC board, line cord, fuse holder, grommets, solder, cabinet

Figure 4.23   A 9-V regulated power supply.

$$= (12.6 \text{ V} \times 1.414) - 1.2 \text{ V}$$
$$= 16.6 \text{ V}.$$

R_{S}, therefore, has a voltage drop of 7.6 V:

$$V_{\text{RS}} = V_{\text{dc (in)}} - V_{\text{DZ}}$$
$$= 16.6 \text{ V} - 9 \text{ V}$$
$$= 7.6 \text{ V}.$$

This represents a total current flow passing through R_{S} of 0.076 A, or 76 mA:

$$I_{\text{RS}} = V_{\text{RS}}/R_{\text{S}}$$
$$= 7.6 \text{ V}/100 \text{ }\Omega$$
$$= 0.076 \text{ A or 76 mA}.$$

With the **bleeder resistor**, R_L, serving as a fixed load, there is 0.9 mA, or 0.0009 A, of load current (I_L):

$$I_L = V_{DZ}/R_L$$
$$= 9 \text{ V}/10 \text{ k}$$
$$= 0.9 \text{ mA or } 0.0009 \text{ A}.$$

The current passing through the Zener diode can be calculated by subtracting load current (I_L) from the series resistor current (I_{RS}). Current passing through the Zener diode is, therefore, 0.0751 A, or 75.1 mA:

$$I_Z = I_{RS} - I_L$$
$$= 0.076 \text{ A} - 0.0009 \text{ A}$$
$$= 0.0751 \text{ A or } 75.1 \text{ mA}.$$

This current must pass through the Zener diode when the circuit is in operation. Nine volts DC will then appear at the two output terminals of the power supply. This represents the no-load (NL) condition of operation.

Example 4-9:

What is the value of current passing through the Zener diode under no-load conditions in **Figure 4.23?**

Solution

To calculate the current passing through the Zener diode, perform the following steps:

1. Calculate the DC output from the rectifier, which is applied to C_1.

$$V_{dc \text{ (in)}} = (V_{sec} \times 1.414) - 1.2 \text{ V}$$
$$= (24 \text{ V} \times 1.414) - 1.2 \text{ V}$$
$$= 32.74 \text{ V}.$$

2. Determine the voltage drop across R_S.

$$V_{RS} = V_{dc \text{ (in)}} - V_{DZ}$$
$$= 32.74 \text{ V} - 9 \text{ V}$$
$$= 23.74 \text{ V}.$$

3. Calculate the current passing through R_S.

$$I_{RS} = V_{RS}/R_S$$
$$= 23.74 \text{ V}/1000\Omega$$
$$= 23.74 \text{ mA}.$$

4. Determine the current passing through R_L.

$$I_L = V_{DZ}/R_L$$
$$= 9 \text{ V}/10 \text{ k}\Omega$$
$$= 0.9 \text{ mA}.$$

5. Determine the current passing through the Zener diode.

$$I_Z = I_{RS} - I_L$$
$$= 23.74 \text{ mA} - 0.9 \text{ mA}$$
$$= 22.84 \text{ mA}.$$

The total current flowing through the Zener diode under no-load conditions is 22.84 mA.

Related Problem

What is the value of current passing through the Zener diode under no-load conditions in **Figure 4.26** if the RMS voltage at the secondary of the transformer is 6 V and the Zener diode rated voltage value is 3 V?

External Load Operation

When the power supply is connected to an **external load**, more current is required. Ideally, this current should be available with 9 V of output. Assume that the power supply is connected to a 270-Ω external load. The total resistance that the power supply sees at its output is 10 kΩ in parallel with 270 Ω. This represents a load resistance of 263 Ω:

$$R_T = [R_L \times R_{ext}]/[R_L + R_{ext}]$$
$$= [10 \text{ k }\Omega \times 270 \text{ }\Omega]/[10 \text{ k }\Omega + 270 \text{ }\Omega]$$
$$= 263 \text{ }\Omega.$$

With 9 V applied, the total load current is 0.0342, or 34.2 mA:

$$I_L = V_{DZ}/R_T$$

$$= 9 \text{ V}/263 \, \Omega$$
$$= 0.0342 \text{ A or } 34.2 \text{ mA}.$$

The total output current available for the power supply is that which passes through R_S. This was calculated to be 0.076 A, or 76 mA. With the 270-Ω load, I_L is 34.2 mA with an I_Z of 41.8 mA:

$$I_Z = I_{RS} - I_L$$
$$= 76 \text{ mA} - 34.2 \text{ mA}$$
$$= 41.8 \text{ mA}.$$

The **output voltage**, therefore, remains at 9 V with the increased load. With a slightly smaller external load, the power supply would go out of its regulation range. Should this occur, no I_Z would flow, and the value of I_L alone would determine the current passing through R_S. An I_{RS} value in excess of 76 mA would cause a greater voltage drop across R_S. This, in effect, would cause the output voltage to be less than 9 V. All voltage regulator circuits of this type have a **maximum load** limitation. Exceeding this limit causes the output voltage to go out of regulation.

If the resistance of the external load is increased, it causes a reduction in load current. This condition reverses the operation of the regulator. With a larger R_L, there is less I_L. As a result, the Zener diode conducts more heavily. The maximum current handling rating of D_Z determines this condition. An infinite load would, for example, demand no I_L. D_Z would, therefore, conduct the full current passing through R_S. This value is calculated to be 76 mA for our circuit. For a 1-W, 9-V Zener diode, the maximum current rating is 0.111 A, or 111 mA:

$$I_{Z \text{ (max)}} = 1 \text{ W}/9 \text{ V}$$
$$= 0.111 \text{ A or } 111 \text{ mA}.$$

In this case, the **Zener diode** is capable of handling the maximum possible I_L that will occur for an infinite load. In the actual circuit, the bleeder resistor, R_S, demands 0.9 mA of current when the load is infinite. The current through the Zener diode (I_Z) will, therefore, not exceed 75.1 mA, which is well below its maximum rating. In effect, the regulating range of this power supply is from an infinite value to approximately 120 Ω. The output voltage will remain at 9 V over this entire range of load values.

Input Changes

A regulator must also be responsive to changes in **input voltage**. If the input voltage, for example, were to increase by 10%, it would cause the supply to see 13.86 V instead of 12.6 V. The peak value of 13.86 V would then be 19.6–1.2 V, or 18.4 V, instead of 16.6 V. The voltage drop across R_S would, therefore, increase from 7.6 to 9.4 V. This, in turn, would cause an increase in total current through R_L of 94 mA. The current flow of the Zener diode would increase to 94.0 − 0.9 mA = 93.1 mA with no external load. Since the Zener diode is capable of handling up to 111 mA, it could respond to this change to maintain the output voltage at 9 V.

A power supply's response to a decrease in input voltage must also be taken into account. A 10% decrease in input voltage would cause the power supply to see only 11.34 V instead of 12.6 V. The peak value of 11.34 V is 16 V. Capacitor C_1 would then charge to 16.0 − 1.2 V, or 14.8 V. The 1.2 V is the voltage drop across two silicon diodes in the bridge. The voltage drop across R_S would be 5.8 V. With this reduced voltage, the total current would drop to 5.8 V/100 Ω = 0.058 A, or 58 mA. This value can certainly be handled by the Zener diode. The maximum low-resistance value of the load would increase somewhat. Under this condition, the load may drop to only 160 Ω before going out of regulation.

Zener Diode Regulator Design

Now, let us apply what you have just learned to the design of a **Zener diode voltage regulator**. The components selected for the regulator consist of a Zener diode and a series resistor. The Zener diode must have a V_Z value and a **power dissipation rating** that will accommodate the circuit ratings. The series resistor must be selected to accommodate the range of voltages used. It must also have a wattage rating that will accommodate the total current. Component selection uses some of the same procedures outlined in the voltage regulator section of this chapter.

Regulator design requires some knowledge of the circuit being controlled and the values of voltage and current being accommodated. Let us assume that the unregulated output of a power supply ranges from 20 to 15 V. The desired regulated voltage is 12 V, and the load current will range from 0 to 40 mA. What values of series resistor and Zener diode are needed to achieve regulation in a circuit attached to the power supply?

Twelve volts is a standard voltage value for a Zener diode. This particular device can be selected from a tolerance range of 1%–20%. The power dissipation rating of the Zener diode will be made after the value of the series resistor (R_S) has been determined by calculation.

A **series resistor** must be selected that will permit the Zener diode to operate within its prescribed range. The value of R_S is equal to the minimum value of the input voltage, $V_{in(min)}$, minus the Zener voltage (V_Z), divided by the maximum value of the load current, $I_{L(max)}$. The voltage difference between V_{in} and V_Z represents the voltage that will appear across the series resistor. This voltage value divided by the maximum load current, $I_{L(max)}$, equals the value of R_S. For our circuit, we, therefore, find the value of R_S to be

$$R_S = [V_{in\ (min)} - V_Z]/I_{L\ (max)}$$
$$= [15\ V - 12\ V]/0.04\ A$$
$$= 75\ \Omega.$$

The **wattage rating** to the resistor must also be determined from the basic power formula, $P = IV$. For this particular application, the wattage rating of the series resistor is determined from the values of the maximum load current, $I_{L(max)}$, and the series resistor voltage (V_{RS}):

$$P = I_{L\ (max)} \times V_{RS}$$
$$= 0.04 \times 3\ V$$
$$= 0.012\ W.$$

Since resistors do not have a standard wattage value of 0.012 W, a value of 0.25 W, or 1/4 W, will be selected for use in this circuit. This resistor can withstand significantly more heat than the calculated value.

The **power dissipation rating**, or P_D, of the Zener diode is a calculated value. In this case, power dissipation is determined by the basic $P_D = I_Z \times V_Z$ formula. The maximum I_Z value of the formula is first determined by finding the total current (I_T) passing through R_S and then subtracting the minimum load current $I_{L\ (min)}$ from I_T. The power dissipation rating of the Zener diode is thus

$$P_D = V_Z \times [V_{in\ (max)} - V_Z/R_S] - I_{L\ (min)}]$$
$$= 12\ V \times [20\ V - 12\ V/75\ \Omega] - 0\ A$$
$$= 12\ V \times [8\ V/75\ \Omega] - 0\ A$$
$$= 12\ V \times 0.1067A - 0\ A$$

$$= 12 \text{ V} \times 0.1067\text{A}$$
$$= 1.28 \text{ W}.$$

Zener diodes are not available with a P_D rating of 1.28 W. Therefore, a 5-W Zener would be selected for this application. This particular Zener could withstand a great deal more current than the calculated value.

Our calculations of the Zener diode and series resistor of the regulator are based on the assumption that the value of V_Z remains fairly constant under normal circuit applications. This is done to simplify the calculations. As a rule, this is a good practical assumption. If the actual value of V_Z does change in the operation of a regulator, the Zener impedance (Z_Z) of the diode could be used to determine changes in I_Z values. Z_Z is generally given in the manufacturer's specifications of a diode.

Example 4-10:

Design a Zener diode voltage regulator that calls for a 10-V regulated output and for load current that will range from 0 to 30 mA. The unregulated input varies from 15 to 20 V DC.

Solution

1. Determine the value of the series resistor (R_S).

$$R_S = [V_{\text{in (min)}} - V_Z]/I_{\text{L (max)}}$$
$$= [15 \text{ V} - 10 \text{ V}]/0.03 \text{ A}$$
$$= 167 \ \Omega.$$

2. Determine wattage rating needed for the series resistor.

$$P = I_{\text{L (max)}} \times V_{RS}$$
$$= 0.03 \times 5 \text{ V}$$
$$= 0.015 \text{ W}.$$

Since resistors do not have a standard wattage value of 0.015 W, a value of 0.25 W, or 1/4 W, will be selected for use in this circuit.

3. Select a Zener diode that will regulate voltage at 10 V. Since 10 V is a standard Zener voltage rating, we will select a 10-V Zener diode for this application.

4. Determine the power dissipation (P_D) rating for the Zener diode.

$$P_D = V_Z \times [V_{\text{in (max)}} - V_Z/R_S] - I_{L \text{ (min)}}]$$
$$= 10 \text{ V} \times [20 \text{ V} - 15 \text{ V}/167 \text{ }\Omega] - 0 \text{ A}$$
$$= 10 \text{ V} \times [5 \text{ V}/167 \text{ }\Omega] - 0 \text{ A}$$
$$= 10 \text{ V} \times 0.030 \text{ A} - 0 \text{ A}$$
$$= 10 \text{ V} \times 0.030 \text{ A}$$
$$= 0.3 \text{ W}.$$

Since 0.3 W, or 300 mW, is not a standard P_D rating, we will select a 10-V Zener diode with a 500-mW P_D rating.

Therefore, a 167-Ω 1/4-W resistor and 10-V, 500-mW Zener diode will be used for the voltage regulator circuit.

Related Problem

Design a Zener diode voltage regulator that calls for a 30-V regulated output and for load current that will range from 0 to 50 mA. The unregulated input varies from 12 to 20 V DC.

Self-Examination

Answer the following questions.

27. The purpose of a Zener diode in a power supply circuit is _____ regulation.
28. A Zener diode voltage regulator consists of a(n) _____ and a series _____.
29. The Zener diode in a power supply is placed across the filter circuit in the _____ bias direction.
30. A Zener diode goes into conduction when it reaches the _____ voltage.
31. Zener current and load current pass through the _____ resistor.
32. The total current passing through the Zener diode is the value of _____ resistor current minus _____ resistor current.

4.5 Dual Power Supplies

The **dual power supply**, or **split power supply**, has been developed as a voltage source for integrated circuits. This supply has both negative and positive output with respect to ground. The secondary winding of the transformer is

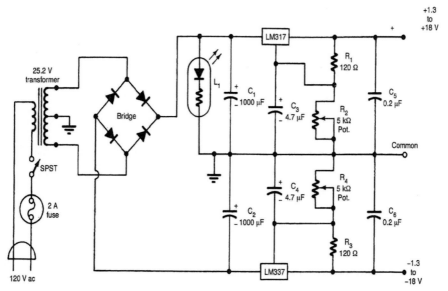

Figure 4.24   Dual power supply conduction.

divided into two parts. The center tap or neutral serves as a common ground connection for the two outside windings. Each half of the winding connects to a full-wave rectifier.

A **dual power supply** with a full-wave rectifier is shown in **Figure 4.26**. Note that the output of this supply is +10.7 and −10.7 V. Diodes D_1 and D_2 are rectifiers for the positive supply. They are connected to opposite ends of the transformer (T_1). Diodes D_3 and D_4 are rectifiers for the negative supply. They are connected in a reverse direction to the opposite ends of the transformer. The positive and negative output is with respect to the center tap of the transformer.

For one alternation, the top of the transformer is positive and the bottom is negative. Current flows through D_4, R_{L2}, R_{L1}, and D_1 and returns to the top of the transformer. The top of R_{L1} becomes positive and the bottom of R_{L2} becomes negative. Solid arrows show the path of current flow for this alternation.

The next alternation makes the top of the transformer negative and the bottom positive. Current flows out through D_3, R_{L2}, R_{L1}, and D_2 and returns to the bottom of the transformer. The top of R_{L1} continues to be positive, whereas the bottom of R_{L2} is negative. The direction of current flow through

R_{L1} and R_{L2} is the same for each alternation. The resulting output voltage of the supply appears at the top and the bottom of R_{L1} and R_{L2}.

Self-Examination

Answer the following questions.

33. A power supply with a negative and a positive output is called a(n) _____ power supply.
34. When the top of the transformer is positive and the bottom is negative, current flows out of point B through D_____, R_L_____, R_L _____, and D_____ and returns to terminal A of the transformer.
35. When the top of the transformer is positive and the bottom is negative, the top of R_{L1} is _____, and the bottom of R_{L2} is _____.
36. When the top of the transformer is negative and the bottom is positive, current flows out of point A through D_____, R_L_____, R_L _____, and D_____ and returns to terminal B of the transformer.
37. When the top of the transformer is negative and the bottom is positive, the top of R_{L1} is _____, and the bottom of R_{L2} is _____.

4.6 Clippers, Clampers, and Voltage Multipliers

P-N junction diodes were introduced in Chapter 2. A primary use is in power supplies as described in this chapter. However, there are many other common applications of diodes. In this section, other diode circuits and applications will be studied. These applications include clippers, clampers, and voltage multipliers.

One type of circuit that uses diodes is the **clipper** or **limiter** circuit. A clipper is a diode circuit that is used to eliminate a portion of a waveform. There are several types of clippers. A **clamper** or **DC restorer** is a diode circuit used to set or restore the DC reference voltage of a waveform. Waveforms are frequently centered around a specific DC reference voltage. For example, a 6-V p-p sine wave might vary equally above and below +5 V_{dc} reference. A clamper circuit can be used to set the +5 V DC reference.

Another diode application is called a **voltage multiplier** circuit. Voltage multipliers produce a DC output voltage that is a multiple of the peak AC input voltage.

Several types of clippers, clampers, and voltage multipliers are described in this chapter.

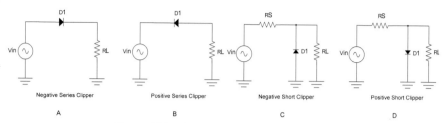

Figure 4.25   Series and shunt clipper circuits.

Clipper Circuits

The basic operating principle of a **clipper** circuit is similar to a half-wave rectifier since a clipper eliminates one of the alternations of an AC input signal at the output of the circuit.

There are four types of clipper circuits, as shown in **Figure 4.25**, which are **negative series clipper**, **positive series clipper**, **negative shunt clipper**, and **positive shunt clipper**. Each series clipper contains a diode that is in series with the load, while each shunt clipper contains a diode that is in parallel with the load.

Note, in **Figure 4.26(a)**, that when the diode in a **negative series clipper** is forward biased by the input signal, it will conduct, and a voltage will appear across R_L. When the diode in the negative series clipper is reverse biased by the input signal, it does not conduct. Therefore, no voltage will appear across the load resistor.

A **positive series clipper** and its associated waveforms are shown in **Figure 4.26(b)**. The positive series clipper operates in a similar manner. Note that, compared to series clippers, the output voltage polarities are reversed. Also, the current directions through the circuit are reversed.

Clippers are used for various functions including altering waveshape, providing circuit transient protection (reducing voltage spikes) and as amplitude modulated (AM) communications receiver detection circuits discussed in a later chapter.

Shunt Clippers

The operation of the shunt clipper is exactly opposite to that of the series clipper. The series clipper provides an output when the diode is forward biased and no output when the diode is reverse biased. The **shunt clipper** provides an output when the diode is reverse biased and shorts the output signal to

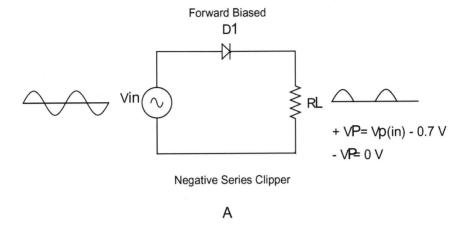

Negative Series Clipper

A

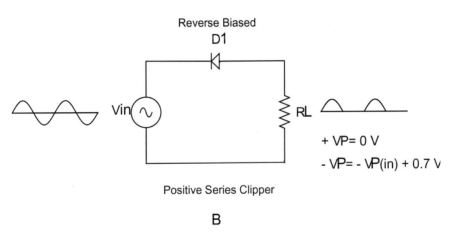

Positive Series Clipper

B

Figure 4.26 Input and output waveforms for series clippers. (a) Negative series clipper. (b) Positive series clipper.

ground when the diode is forward biased. This operation is illustrated in **Figure 4.27**.

The circuit shown in **Figure 4.27(a)** is a **negative shunt clipper**. When the diode in the negative shunt clipper is reverse biased, it is effectively removed from the circuit. With the diode reversed biased, the resistors form a voltage divider and the load voltage can be found by the formula shown in **Figure 4.27(a)**. The output signal is similar to the positive alternation of the input. The peak output voltage is slightly less than the peak input voltage.

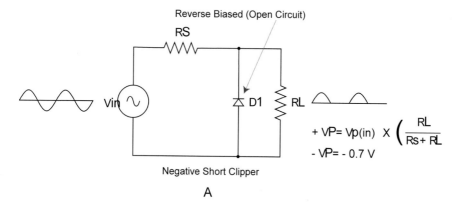

Negative Short Clipper

A

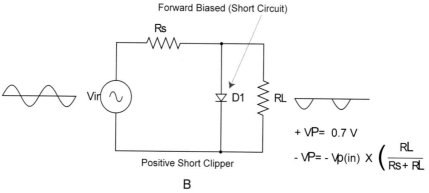

Positive Short Clipper

B

Figure 4.27   Input and output waveforms for shunt clippers. (a) Negative shunt clipper. (b) Positive shunt clipper.

The output of the circuit is approximately −0.7 V when the diode is forward biased.

The **positive shunt clipper**, shown in **Figure 4.27(b)**, is similar in operation to the negative shunt clipper. When the input to the circuit is negative, the diode is reverse biased and, for all practical purposes, removed from the circuit. Note that R_S is added in the shunt clipper as a current-limiting resistor. If the input signal forward biased the diode and R_S was not in the circuit, D_1 would short the source. The signal would be shorted to ground during the positive alternation of the input signal. This could result in the diode being destroyed by excessive forward current. Also, the signal source could be damaged by the excessive current demand of the conducting diode. The value of R_S is much lower than the value of the load resistor (R_L). When the diode is reverse biased, the voltage across the load resistor is approximately equal to the value of V_{in}.

Biased Clippers

Biased clippers have a DC bias source to set the limit of the output voltage of the circuit. **Figure 4.28** shows positive-biased and negative-biased clipper circuits. These circuits are designed to clip input waveforms at values other than the diode V_F (0.7 V). In both circuits, the bias voltage (V_B) is in series with the shunt diode. The diode conducts and clips the input waveform as shown. A **positive-biased clipper** (see **Figure 4.28(a)**) is designed to clip the input signal at $V_B + 0.7$ V. The bias voltage determines the value at which the input signal is clipped. The **negative-biased clipper** of **Figure 4.28(b)** is similar.

A **variable DC source** could be used to provide an adjustable value of V_B. The bias voltage (V_B) would allow an adjustment in this circuit to provide the desired clipping limit. By reversing the direction of the diode and the polarity of the variable DC voltage in both circuits of **Figure 4.28**, the circuit would function as a negative-biased clipper.

Self-Examination

Answer the following questions.

   38. A clipper is used to _____.
   39. Another name for a clipper is _____.
   40. The difference between a series and shunt clipper is _____.
   41. The purpose of R_S in a shunt clipper is _____.
   42. A biased clipper uses a _____ to establish a reference voltage.

Clamper Circuits

A **clamper** or **DC restorer** circuit is designed to shift a waveform either above or below a pre-determined reference voltage without distorting or changing the shape of the input waveform. A positive clamper and a negative clamper circuit are shown in **Figure 4.29**. A **positive clamper** shifts the input waveform so that the negative peak of the waveform is approximately equal to the DC reference voltage of the clamper. A **negative clamper** shifts its

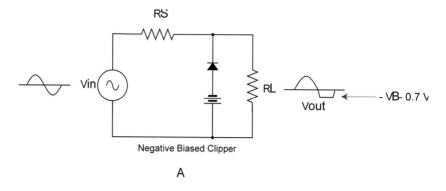

Negative Biased Clipper

A

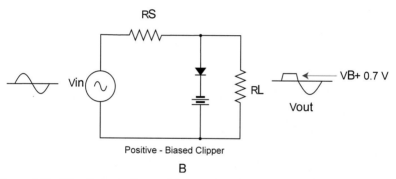

Positive - Biased Clipper

B

Figure 4.28   Biased shunt clipper circuits. (a) Positive-biased clipper. (b) Negative-biased clipper.

input waveform so that the positive peak of the waveform is approximately equal to the clamper circuit's DC reference voltage.

Basically, a clamper adds a **DC reference level** to an AC voltage. To understand the operation of a clamper, refer to the positive clamper of **Figure 4.29(a)**. Starting with the first negative half-cycle of the AC input, the diode is forward biased. The capacitor will charge to near the peak AC input voltage. After the negative peak, the diode is reverse biased, causing the voltage across D_1 to be nearly equal to the peak voltage input. The capacitor can only discharge through the relatively high resistance of R_L; so most of its charge is retained. The capacitor voltage basically acts as a voltage in series with the input voltage. During the positive half-cycle of the input, the input voltage adds to the input voltage value to produce a voltage output as shown in **Figure 4.29(a)**.

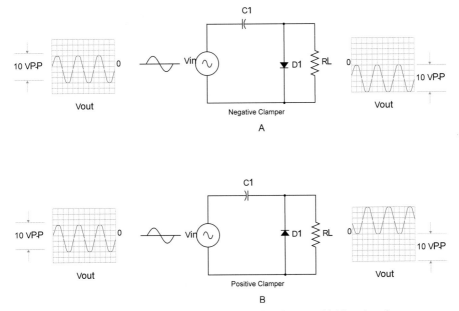

Figure 4.29 Diode clamper circuits. (a) Positive clamper. (b) Negative clamper.

The **negative clamper** of **Figure 4.29(b)** is similar in operation, except that the diode and the capacitor are reversed, clamping the output voltage in the negative direction. Clamping circuits are commonly used in television systems as a DC restorer.

Biased Clampers

Biased clampers, shown in **Figure 4.30**, provide a waveform that is shifted above or below a **DC reference voltage** value. The circuit of **Figure 4.30(a)** is a **biased negative clamper**. The DC reference voltage for the circuit depends on the value of the DC bias source (V_B) value and the setting of the potentiometer (R_1). The **biased positive clamper** is similar except for the polarity of the variable DC voltage negative and the diode. The DC reference voltage of the circuit can be adjusted to any voltage between V_B and 0 V.

Zener Clampers

The circuits shown in **Figure 4.31 Zener clampers**. The Zener diodes are used to set the DC reference voltages of the circuits. When a 2-V

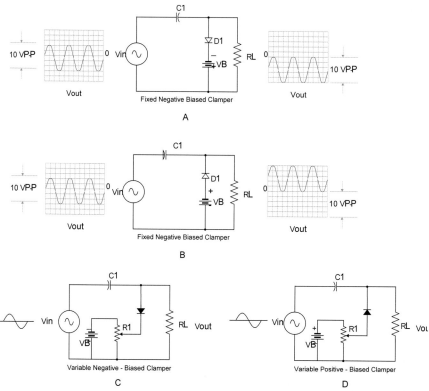

Figure 4.30 Biased clamper circuits. (a) Fixed negative-biased clamper. (b) Fixed positive-biased clamper. (c) Variable negative-biased clamper. (d) Variable positive-biased clamper.

Zener diode is used in a Zener clamper, the DC reference voltage is clamped at +2 V. The P-N junction diodes are necessary to provide one-way current flow. Remember that the Zener diode conducts in both the reverse and forward operating regions. During the positive alternation of the input, the diodes will conduct. The positive peak clamps the peak at $V_Z = +0.7$ V.

During the negative alternation of the input, D_1 is reversed biased and there will be no negligible current flow through the diodes. A Zener clamper may be either a **negative clamper** with a positive DC reference voltage (see **Figure 4.31(a)**) or a **positive clamper** with a negative DC reference voltage (see **Figure 4.31(b)**).

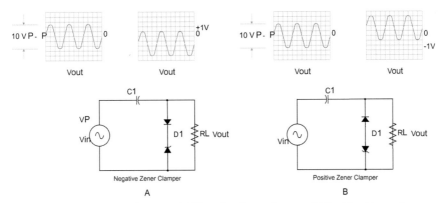

Figure 4.31 Zener diode clampers. (a) Negative Zener clamper. (b) Positive Zener clamper.

Self-Examination

Answer the following questions.

43. A clamper is used to _____
44. Another name for a clamper is _____.
45. Biased clampers provide a waveform shifted above or below a

    _____.

46. A positive clamper shifts a(n) _____ above a DC reference
    voltage.
47. Zener clampers use a Zener diode in series with a(n) _____
    to establish a DC reference voltage.

Voltage Multiplier Circuits

The purpose of a **voltage multiplier** circuit is to increase the value of a DC
voltage output applied to a load. Voltage multipliers produce a DC output that
is a multiple of the peak input voltage. A **voltage doubler** circuit is used to
increase the peak AC input value by a factor of 2. Applications of voltage
multiplier circuits are primarily used for high-voltage DC power supplies.
A **dual-voltage power supply** may also use voltage multiplier circuits for
increased DC voltage output.

The basic voltage multiplier circuits are the voltage doublers shown in
Figure 4.32. In the **half-wave voltage doubler** circuit (see **Figure 4.32(a)**),
when point A has a positive AC input applied, diode D_1 will conduct and
diode D_2 is reverse biased. Capacitor C_1 will charge to the peak AC input

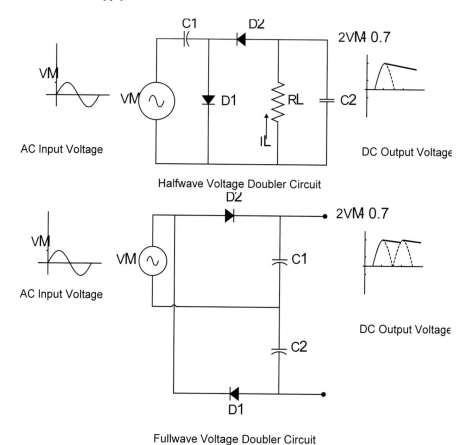

Figure 4.32   Voltage doubler circuits. (a) Half-wave doubler. (b) Full-wave doubler.

voltage. The right side of capacitor C_1 will retain this peak positive charge. During the next half-cycle of AC input, the instantaneous polarities will be reversed. This places point B at a positive potential. Diode D_2 will now conduct, while D_1 is reverse biased. This action places a charge on capacitor C_2 equal to the peak AC input. During the next half-cycle of AC input, the charge on capacitor C_1 will accumulate on capacitor C_2 by discharging through diode D_2. Also, during this half-cycle, capacitor C_2 will discharge through the load resistance (R_L). The sum of the voltages accumulated on C_2 will then approximately equal $2 \times V_P(\text{in})$. The output is similar to a half-wave rectified waveform that has a simple capacitor filter (see Chapter 4). The diode PRV rating must be at least $2 \times V_P(\text{in})$.

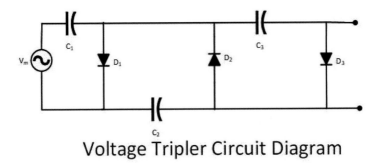

Voltage Tripler Circuit Diagram

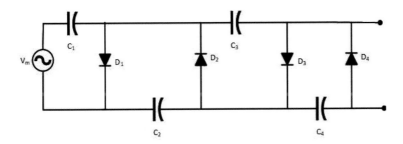

Voltage Quadrupler Circuit Diagram

Figure 4.33 (a) Voltage tripler and (b) voltage quadrupler.

The operation of the **full-wave voltage doubler** of **Figure 4.32(b)** is similar to the half-wave doubler. When point A has a positive potential, diode D_1 will conduct, allowing capacitor C_1 to charge to the value of V_P. The next half-cycle of AC input will place an instantaneous positive charge on point B. Diode D_2 will then conduct, charging capacitor C_2. With no load connected, the DC voltage at the output terminals is approximately $2 \times V_P(\text{in})$. The PRV ratings of the diodes must also be at least $2 \times V_P(\text{in})$. More sections of these circuits can be connected together to form **voltage tripler** and **voltage quadrupler** circuits. Extremely high DC voltages can be produced with these circuits.

Remember that the output power delivered can never be greater than input power. Increased voltage results in a corresponding decrease in current value. A voltage tripler is similar to a half-wave voltage doubler. Note in **Figure 4.33(a)** that diodes D_1 and D_2 and capacitors C_1 and C_2 are the same as the half-wave doubler of **Figure 4.33(b)**. With diode D_3 and capacitors C_3, and C_4 added, the voltage tripler circuit is formed. The half-wave voltage doubler functions to double the DC voltage. Diode D_3 conducts when D_1 is conducting. During the negative input alternation, D_1 and D_3 conduct. This causes C_1 and C_3 to charge to $V_P(\text{in})$. During the positive input, diode D_2 conducts causing C_2 to charge to $2 \times V_P(\text{in})$. The sum of the voltages across C_2 and C_3 now equals $3 \times V_P(\text{in})$. Capacitor C_4 accumulates the combined charge voltages of C_2 and C_3 and also acts as a filter capacitor to reduce the AC ripple of the DC output.

A voltage quadrupler, shown in **Figure 4.33(b)**, produces a DC output that is approximately four times the peak AC input voltage. This circuit is basically a combination of two half-wave voltage doublers (see **Figure 4.32**). The accumulated charge across capacitor C_5 is the sum of the voltages across capacitors C_2 and C_4. Note that they are connected in series so that their voltages will add together.

Self-Examination

Answer the following questions.

48. A voltage multiplier is used to _____.
49. A half-wave voltage doubler uses _____ diodes and _____ capacitors.
50. A full-wave voltage doubler uses _____ diodes and _____ capacitors.
51. A voltage tripler uses _____ diodes and _____ capacitors.
52. A voltage quadrupler uses _____ diodes and _____ capacitors.

Analysis and Troubleshooting

Analysis and troubleshooting for power supply circuits involves similar principles and procedures as **P-N junction diode** circuits described in Chapter 2.

In order to analyze and troubleshoot power supply circuits, you should:

- review analysis and troubleshooting for P-N junction diodes (Chapter 2);
- recall basic analysis and troubleshooting techniques for diode circuits.

In order to improve your analysis and troubleshooting expertise, you should be able to recall basic techniques and common faults that might occur in diode circuits. You should begin by reviewing the "Analysis and Troubleshooting – P-N Junction Diodes" section of Chapter 2.

Basic Troubleshooting Techniques

Always remember "Safety First" – make sure power source is turned **off**.
Identify the symptoms and determine the cause.
Perform a power check – make sure that power is being applied to the circuit.
Make a sensory check – use all of your senses.
Replace damaged components.
Trace signals or voltages (for an active circuit) using an oscilloscope and/or multimeter.
Recognize common faults (for diode circuits).
Power source problem – open fuse or circuit breaker, faulty transformer, etc.

Open diode
Shorted diode
Open resistor
Shorted resistor
Changed-value resistor
Open capacitor
Shorted capacitor
Leaky capacitor

Summary

- The functions of a typical power supply are transformer, rectifier, filter, and regulator.
- The transformer of a power supply is typically used to step down the input voltage.
- The relationship of turns ratio to voltage is $N_{pri}/N_{sec} = V_{pri}/V_{sec}$.
- The relationship of primary and secondary voltages to current is $V_{pri}/V_{sec} = I_{sec}/I_{pri}$.
- A rectifier changes AC to pulsating DC.
- The three types of rectifiers are half-wave, full-wave, and bridge.

- A half-wave rectifier produces a half-wave output.
- Full-wave and bridge rectifiers produce a full-wave output.
- The composite average value of DC output from a half-wave rectifier is 45% of the AC input voltage, minus the voltage drop of 0.6 V across the silicon diode.
- The composite average value of DC output from a full-wave rectifier is 90% of the AC input voltage, minus the voltage drop of 0.6 V across the silicon diode.
- The composite average value of DC output from a full-wave rectifier is 90% of the AC input voltage, minus the voltage drop of 1.2 V across the two silicon diodes.
- Ripple frequency in a half-wave rectifier is equal to the input frequency.
- Ripple frequency of a full-wave and bridge rectifier is twice the input frequency.
- Filters increase voltage and decrease ripple.
- A *C*-input filter consists of a capacitor in parallel with the load.
- An *L*-input filter consists of an inductor in series with the load.
- An LC filter consists of an inductor in series with the load and a capacitor in parallel with the load.
- A pi filter consists of a capacitor in parallel with the load followed by an LC filter.
- An RC filter is similar to a pi filter, except that it uses a resistor instead of an inductor.
- A voltage regulator is used to achieve a steady output voltage from a power supply.
- A dual power supply is commonly used with integrated circuits.
- A dual power supply provides both negative and positive voltages with respect to ground.
- Clippers or limiters are diode application circuits used to remove part of an AC input signal.
- Clampers or DC restorer circuits are used to establish a DC voltage reference for an AC signal.
- Voltage multipliers are diode application circuits used to produce a DC output voltage that is a multiple of the peak AC input voltage applied to the circuit.
- Series clipper circuits have one diode that is in series with an AC source and a load resistance.
- Negative series clipper circuits remove the negative alternation of an AC input signal.

- Positive series clipper circuits remove the positive alternation of an AC input signal.
- Shunt clipper circuits have a diode in parallel with a load resistance.
- A series current-limiting resistor is used to prevent the diode of a shunt clipper from shorting the signal source to ground when the diode is forward biased.
- A biased clipper circuit is a shunt clipper that has a DC voltage source used to bias the diode.
- Positive clamper circuits shift an AC input signal above a DC reference voltage.
- A negative clamper shifts an AC input signal below a DC reference voltage.
- A biased clamper may be used to provide an output waveform above or below a DC reference above (or below) a DC reference voltage.
- Zener clampers use Zener diode in series with a P-N junction diode to establish a DC reference voltage.
- Negative clampers produce positive DC reference voltage and positive clampers produce a negative DC reference voltage.
- Voltage multipliers produce a DC output voltage that is some multiple of the peak AC input voltage applied to the circuit.
- Half-wave voltage doublers produce a DC output voltage that is approximately twice the peak AC input voltage.
- Full-wave voltage doublers produce an output that is approximately twice the peak AC input voltage applied to the circuit.
- Voltage triplers produce a DC output voltage that is approximately three times the peak AC input voltage applied to the circuit.
- Voltage quadruplers produce a DC output voltage that is approximately four times the peak AC input voltage applied to the circuit.

Answers

Examples

4-1. 1:5

4-2. 600 V

4-3. 1.2 A

4-4. 6.36 V

4-5. 18 V

4-6. 14.25 V

4-7. 29.1 V

4-8. 28.5 V
4-9. 4 mA

Self-Examination

4.1

1. secondary
2. primary
3. turns ratio
4. voltage, current
5. current, voltage
6. 12
7. 60

   4.2

8. one
9. two
10. four
11. positive
12. positive
13. 60
14. 120
15. 120
16. (10 V RMS × 1.414 × 0.318) − 0.6V = 3.89
    or
    (10 V RMS × 0.45) − 0.6V = 3.9
17. (30 V RMS × 1.414 × 0.636) − 1.2 = 25.78
    or
    (30 V RMS × 0.9) − 1.2 = 25.8

    4.3

18. voltage, ripple
19. *L*-input, LC (any order)
20. pi
21. *L*-input
22. *C*-input
23. capacitor
24. inductor
25. pi
26. LC

4.4
27. voltage
28. Zener diode, resistor
29. reverse
30. Zener breakdown
31. series *or* bleeder
32. series *or* bleeder, load

4.5
33. dual *or* split
34. 4, 2, 1, 1
35. positive, negative
36. 3, 2, 1, 2
37. positive, negative

4.6
38. Eliminate one AC alternation at the output
39. Limiter
40. Location of diode relative to R_L
41. Avoid having a short-circuited power source
42. Shunt diode
43. Establish a DC reference voltage
44. DC restorer
45. DC reference voltage
46. AC voltage
47. P-N junction diode
48. Increase DC voltage output
49. two, two
50. two, two
51. three, four
52. four, five

Formulas

(4-1)–(4-3) $N_{pri}/N_{sec} = V_{pri}/V_{sec} = I_{sec}/I_{pri}$ Relationship between the turns ratio, voltage ratio, and current ratio.

(4-4) $V_{avg} = 0.637 \times V_P$ The average voltage value of one pulse of DC voltage.

(4-5) composite average value $= 0.318 \times V_P$ Average voltage value of one pulse of DC, taking into account the time of the negative alteration.

(4-6) $V_P = V_{rms} \times 1.414$ The peak value of one AC alternation at the input of a rectifier.

(4-7) composite average value = $(V_{rms} \times 0.45) - 0.6$ V The DC output of an unfiltered half-wave rectifier, minus the voltage drop across a silicon diode.

(4-8) composite average value = $(V_{rms} \times 0.9) - 0.6$ V The DC output of a full-wave rectifier, minus the voltage drop across a silicon diode.

(4-9) $V_{dc\ (out)} = (V_{rms} \times 0.9) - 1.2$ V The DC output of a full-wave rectifier, minus the voltage drop across the silicon diodes.

(4-10) $T = R \times C$ RC time constant.

Glossary

Turns ratio

The ratio of the number of turns in the primary winding to the number of turns in the secondary winding.

Rectification

The process of changing AC into DC.

Rectifier

The function of a power supply that converts AC to DC.

Half-wave rectifier

A rectifier in which one alternation of the AC input appears in the output.

Pulsating DC

A voltage or current value that rises and falls at a rate or frequency with current flow always in the same direction.

Ripple frequency

A circuit that changes pulsating DC into a rather pure form of DC.

C-input filter

A filter circuit employing a capacitor as the first component of its input.

L-input filter

A filter circuit employing an inductor as the first component of its input.

Choke

An inductor that is used as a primary filtering element. Refers to the ability of an inductor to reduce ripple voltage and current.

Pi filter

A filter with an input capacitor connected to an inductor—capacitor filter, forming the shape of the Greek letter pi.

Voltage regulator

A circuit that ensures a stable DC voltage.

Dual power supplies

A power supply that has both negative and positive output with respect to ground.

Clipper

A diode circuit that is used to eliminate a portion of a waveform.

Clamper

Also called a DC restorer, a diode circuit used to set or restore the DC reference voltage of a waveform.

Voltage multiplier

A diode circuit used to produce a DC output voltage that is a multiple of the peak AC input voltage.

5

Special Semiconductor Diodes

In the previous chapters, we were concerned with **P–N junction** and **Zener diodes** and their applications. In this chapter, we examine some semiconductor diodes that have unique characteristics that distinguish them from other diodes. Applications for these diodes are rather specialized and, in some cases, may represent a very small part of diode technology. As a rule, these devices may require some unusual construction technique or a specialized circuit application.

This chapter is included in the study of semiconductor devices to expand your knowledge of diode technology. Through it, you will become familiar with some of the unusual characteristics of a diode. This will provide a better understanding of the diode and its applications. The diode is indeed the building block of all semiconductor devices.

Objectives

After studying this chapter, you will be able to:

5.1 describe the characteristics and applications of tunnel diodes;
5.2 describe the characteristics and applications of varactor diodes;
5.3 describe the distinguishing characteristics of Schottky, PIN, IMPATT, and Gunn-effect diodes.

Chapter Outline

5.1 Tunnel Diodes
5.2 Varactor Diodes
5.3 Miscellaneous Diodes

Key Terms

bistable
Gunn-effect diode
hot carriers
IMPATT diode
negative resistance
PIN diode
quality factor
Schottky-barrier diode
tunnel diode

5.1 Tunnel Diodes

As you have learned, adding a controlled amount of impurities can change the conduction capabilities of a semiconductor. The conduction capabilities of a tunnel diode are changed in a unique way by adding a higher concentration of impurities than a standard diode. In this section, you will learn about the unique conduction capability of a tunnel diode and its *I–V* characteristics, specifications, and applications.

5.1 Describe the characteristics and applications of tunnel diodes.
In order to achieve objective 5.1, you should be able to:

- describe the peak current, valley current, peak voltage, and valley voltage of a tunnel diode;
- locate peak current, valley current, peak voltage, and valley voltage on a tunnel diode *I–V* curve;
- define *tunnel diode*, *negative resistance*, and *bistable*.

A **tunnel diode** is a two-element semiconductor device with construction similar to that of a conventional silicon diode. This device has an anode connected to the P-type material and a cathode connected to the N-type material. The two materials are joined at a common point or junction. The tunnel diode is, however, quite different beyond this point. The P-type and N-type materials, for example, are heavily doped. This means that the material has a rather high concentration of impurities. Typical doping levels may be one hundred to several thousand times that of a conventional silicon diode.

Heavily **doping** the semiconductor material of a tunnel diode causes the N-type material to have a large number of free electrons and the P-type material to have a large number of holes. The high current carrier content of the crystal material causes the width of the depletion region to be very thin. In general, depletion zone width is only 1/100 of that of a regular diode. This permits current carriers to "tunnel" through the barrier rather than move across it when bias voltage is reduced.

Normally, electrons must have sufficient external energy to cross the surface barrier of a junction, as described in Chapter 2 for P–N junction diodes. Due to the increased number of current carriers and the thin barrier that exists in tunnel diodes, electrons tend to pass through the barrier with an extremely small amount of energy. In many cases, ambient temperature may be sufficient to cause some conduction. This can occur in some cases when very little bias voltage has been applied. Current carriers tend to move across the barrier as if it did not exist. **Figure 5.1** shows an example of electrons **tunneling** through the barrier of a P–N junction with a small amount of external voltage applied.

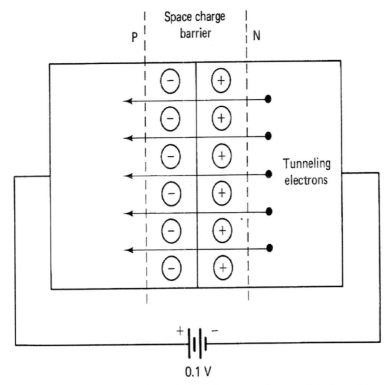

Figure 5.1 Tunneling electrons. The thin depletion zone of a tunnel diode enables current carriers to pass through the barrier with a small amount of bias voltage.

Tunnel Diode I–V Characteristics

The **I–V characteristic** of a tunnel diode is quite different from that of other diodes. The tunnel diode has three distinct characteristics that make it electrically different. Each characteristic depends on a value of bias voltage applied to the device. Within a certain range of bias voltage, a tunnel diode is conductive in both directions. See the *I–V* characteristics of **Figure 5.2** (right and left of the peak voltage point) and the area on each side of the valley voltage. In the second range of bias voltage, the tunnel diode has a negative resistance characteristic. This is located between the forward voltage peak (V_P) and the valley voltage (V_V) point. **Negative resistance** refers to an area where an increase in voltage causes a decrease in current. The third characteristic of bias voltage deals with an area beyond the valley voltage

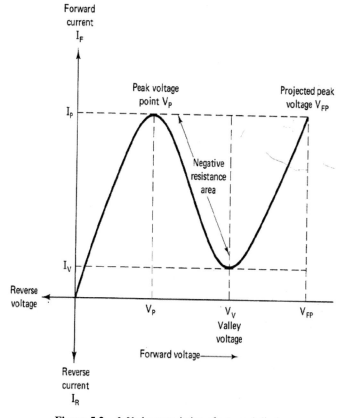

Figure 5.2   *I–V* characteristics of a tunnel diode.

point. Increasing the forward voltage beyond the V_V point of a tunnel diode causes it to respond as a conventional silicon diode. This part of the curve shows that an increase in forward voltage causes a corresponding increase in forward current.

The operation of a tunnel diode can be changed according to the value and polarity of the applied bias voltage. The three bias conditions discussed occur within a few millivolts. The general area of operation is somewhere between 0 and 600 mV. The exact voltage for each bias condition is dependent on the material used in the construction of the diode. **Germanium** or **gallium arsenide** can be used in the construction of most tunnel diodes. The **doping** level of the P-type and N-type materials has a great deal to do with other characteristics. The peak current (I_P), for example, can vary from a few microamperes to 100 A. The peak voltage (V_P), however, is limited to a maximum value in the range of 600 mV. For this reason, a tunnel diode can be easily damaged. A multimeter with a 1.5-V cell energizing the ohmmeter could damage a tunnel diode by simply measuring its resistance. When using the tunnel diode, one must use extreme care in connecting it into a circuit and altering the circuit so that it will produce a desired operating condition.

Tunnel Diode Specifications

Three common symbols for the **tunnel diode** are shown in **Figure 5.3.** Two of these symbols resemble the conventional silicon diode symbol with a slight modification. The third symbol, which is widely used, consists of a line and a half-circle. Note the parts of the symbol that are used to identify the anode and cathode. For the standard diode symbol, the anode and cathode remain unchanged. For the line-half-circle symbol, the anode is represented as a line, and the half-circle denotes the cathode. These symbols may all be drawn within a circle or with the circle omitted.

Tunnel diodes are usually packaged in a special housing. **Figure 5.4** shows a representative package. Note the dimensions of this device. It is extremely small in comparison with a conventional silicon diode. As a rule, the enclosure is usually metal. The **anode** is insulated from the metal housing, and the **cathode** is attached to it. Since tunnel diodes are used primarily in high-frequency signal applications, metal enclosures are purposely used to isolate the internal diode structure from stray electromagnetic fields.

The data from a tunnel diode **specification sheet** is shown in **Figure 5.5.** The tunnel diode represented in the sheet is a low-power device. Note that the data is listed for three different conditions of operation: minimum, typical,

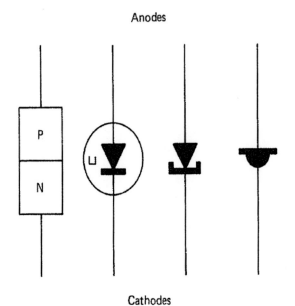

Figure 5.3   Tunnel diode symbols and crystal structure.

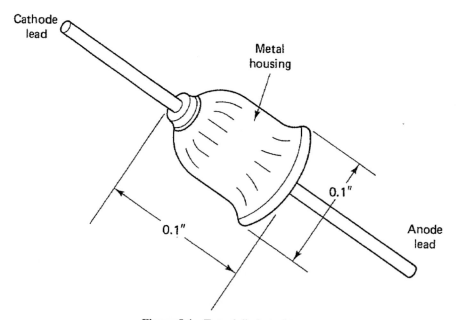

Figure 5.4   Tunnel diode package.

Specifications: Ge 1N2939

ABSOLUTE MAXIMUM RATINGS (25°C)

Current

Forward (−55 to +100°C)	5 mA
Reverse (−55 to +100°C)	10 mA

ELECTRICAL CHARACTERISTICS (25°C)(1/8 in. Leads)

	Min.	Typ.	Max.	
I_p	0.9	1.0	1.1	mA
I_v		0.1	0.14	mA
V_p	50	60	65	rnV
V_v		350		mV

	Min.	Typ.	Max.	
REVERSE VOLTAGE (I_R = 1.0 mA)			30	mV
FORWARD PEAK POINT CURRENT VOLTAGE V_{fp}	450	500	600	mV
I_p/I_v		10		
−R		−152		Ω
C		5	15	pF
L_s		6		nH
R_s		1.5	4.0	Ω

Figure 5.5 Tunnel diode data.

and maximum. This particular diode is used for oscillators, high-frequency amplification, and high-speed switching applications. The specification sheet generally lists a number of characteristics that influence the high-frequency response of the diode, such as capacitance of the junction (C), terminal lead inductance (L_S), negative resistance (−R), and lead resistance (R_S).

Tunnel Diode Applications

When the **tunneling** principle of a **highly doped** P–N junction was first discovered, the tunnel diode was recognized as an important high-speed switching device. State switching could essentially take place at the speed of light. The **response time** of the switching action is primarily limited by diode capacitance. As a rule, this is in the order of $1-10$ pF. This means that switching can occur from the zero point to the peak point with a very short rise time.

The **negative resistance** characteristic of a tunnel diode has also opened the door for a number of other applications. High-frequency signal generation or oscillation can now be achieved by using this characteristic. The negative resistance characteristic also permits the tunnel diode to be used as a high-frequency amplifying device. High-frequency signal control is an application of the tunnel diode.

High-Frequency Oscillator

The **negative resistance** characteristic of a tunnel diode can be used to a unique advantage when it is connected across an LC tank circuit. An inductor and capacitor connected in parallel form an **LC tank circuit. Figure 5.6** shows a **tank circuit** connected to a DC source. When the switch of the circuit is closed momentarily, it causes the tank circuit to be "shocked" into **oscillation**. The resulting **damped** oscillatory waveform is shown in **Figure 5.6**. Note that the amplitude of the wave decreases with each succeeding wave. This is caused by the effective resistance of the tank circuit. The resistance essentially dissipates the power so that it causes the waves to die out after a few cycles of operation. If the tank circuit had only pure inductance and capacitance, the oscillations would be continuous after the initial voltage was applied. In practice, the circuit will always have some resistance. This means that the resulting oscillations will die out after a few cycles of operation.

The **negative resistance** effect of a tunnel diode can be used to cancel or overcome the effective resistance of a tank circuit. This permits the tank circuit to behave as if it has no resistance. As a result, when the tank is shock-excited with voltage, oscillations will occur without a change in amplitude. The frequency of the oscillating wave is determined by the capacitance and inductance of the tank circuit. This oscillator is very simple in operational theory and construction.

To use a tunnel diode for the tank circuit of an oscillator, the diode must be forward biased. The amount of forward-bias voltage needed to reach the negative resistance area is very critical. Ideally, this **operating point** is located between the **peak** and **valley voltage** points. This part of the characteristic

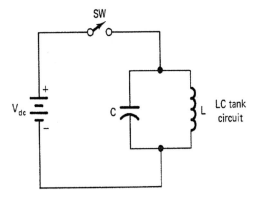

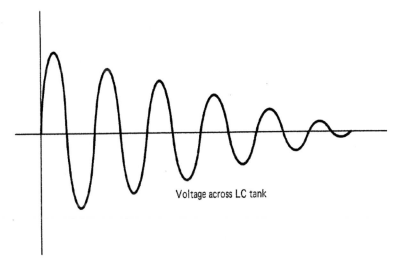

Figure 5.6 LC tank circuit and damped oscillatory waveform.

curve is in the order of 60−350 mV. A representative operating voltage would be somewhere near the center of this voltage range or approximately 145 mV.

Figure 5.7 shows a **tunnel diode oscillator** with a variable capacitor that permits frequency changes. Tunnel diode biasing is achieved by a voltage-divider network consisting of resistors R_1 and R_2. The generated output signal developed by the circuit appears across the tank. An oscillator of this type has good stability and can generate signals in the microwave range. It is, however, rather sensitive to changes in temperature and bias voltage.

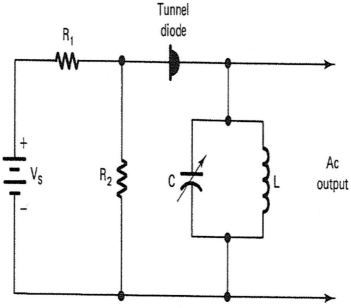

Figure 5.7 A tunnel diode oscillator.

Switching

A very common application of the tunnel diode is in high-speed **switching circuits**. A tunnel diode can perform **logic functions** and **memory**. The tunnel diode offers the following advantages:

- Small size
- Low operating power
- High speed
- Low cost
- High reliability

It is possible to form a simple two-state, or **bistable**, switching circuit by connecting a tunnel diode in series with a voltage source and a single resistor. **Figure 5.8** shows this type of circuit and an *I–V* curve for the diode. To see how the circuit responds as a switch, the operating range of the circuit must be plotted. A solid line extending across the *I–V* curve shows the range of the anticipated operation. If a source voltage of 0.5 V is applied to a 100-Ω resistor, it will cause a forward current of 5 mA. Note the location of these two points (**source voltage** and **forward current**) on the solid line of the *I–V* curve which represent the end points of the line. Note also that the

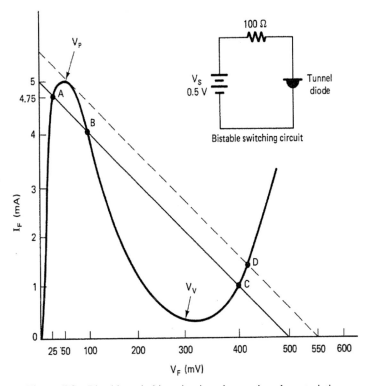

Figure 5.8 Bistable switching circuit and operating characteristics.

operating line crosses the *I–V* curve at three places: point A, point B, and point C. Point A occurs near the peak current point or I_P. Point B occurs in the **negative resistance** region of the curve. This is considered to be an unstable operating point in a switching circuit. Point C is located slightly above the **valley voltage** point, V_V. Points A and C are considered to be the stable operating points of the switching circuit.

To see how a tunnel diode is used as a **bistable switch**, we must consider some specific operating conditions for the circuit in **Figure 5.8**. Point A is located at an I_F of 4.75 mA and a V_F of 25 mV. Point C is located at an I_F of 1.0 mA and a V_F of 400 mV. The circuit can be switched to either of these positions, and it will remain there until a state change occurs. The operational state of the circuit is determined by the value of the source voltage. Before the circuit is energized, both I_F and V_F are zero. When the circuit is energized by the 500-mV source, its operation will change to point A. For the circuit to operate at point C, the source voltage must be increased to 550 mV. This causes the operation of the circuit to shift to the dotted line where it intersects the *I–V* curve at stable point D. The circuit will remain at this operating point as long as the source remains at 550 mV. However, reducing the source to a value of 500 mV causes the circuit to remain in its stable condition by moving to point C on the curve. To change the state of operation back to that of point A, the source voltage would have to drop in value below 25 mV. When the source voltage is changed back to 500 mV, the circuit will remain in its stable state at point A. Thus, the circuit has two stable states of operation depending on the value of the source voltage. This type of circuit is called a **bistable voltage switch.**

Self-Examination

Answer the following questions.

1. Heavily doping the P-type and N-type materials of a tunnel diode causes the depletion zone to be very _____.
2. The _____ effect is due to the heavy doping of P-type and N-type materials and a thin depletion zone.
3. _____ refers to an area of a tunnel diode where an increase in voltage causes a decrease in current.
4. An increase in forward voltage beyond the _____ point causes a tunnel diode to respond as a regular diode.
5. When a tunnel diode is used as a switch, it has _____ stabilized conditions of operation.
6. When a tunnel diode is used as an oscillator, it is biased to operate in the _____ region.
7. A(n) _____ will conduct in either the forward or reverse direction when properly biased.
8. A tunnel diode generally operates in the (forward, reverse) direction for most of its applications.
9. When a tunnel diode is used as an oscillator, the _____ cancels the effective resistance of a tank circuit.
10. The two stable operating points of a bistable tunnel diode switch are to the left of the _____ point and to the right of the _____ point.

5.2 Varactor Diodes

By definition, a capacitor is two or more conductors separated by an insulating material. Since the depletion region of a diode serves as a dielectric medium or insulator and the two independent crystal materials serve as conductor plates, a conventional silicon diode can respond as a capacitor. When a diode has bias voltage applied, its depletion region changes in width. In a sense, this means that a diode responds as a voltage-variable capacitor. Some diodes are specially designed to respond to the capacitance effect. This section discusses these capacitors, which are called *varactors*, *varicap diodes*, or *voltage-variable capacitors*.

5.2 Describe the characteristics and applications of a varactor diode.

In order to achieve objective 5.2, you should be able to:

- describe the electrical characteristics of a varicap diode;
- describe how the internal capacitance of a P–N junction is changed with bias voltage;
- describe how capacitance is produced across the junction of a varactor diode;
- define *quality factor*.

A **varactor diode** is a specially manufactured P–N junction with a variable concentration of impurities in its P-type and N-type materials. In a conventional silicon diode, **doping** impurities are usually distributed equally throughout the material. Varactors have a very light dose of impurities near the junction, but away from the junction, the impurity level increases. This type of construction produces a much steeper voltage−capacitance relationship. **Figure 5.9** shows a comparison of the capacitance between a conventional **silicon diode** and a **varactor diode**. It can be seen that an ordinary diode possesses only a small value of internal capacitance. In general, this capacitance is too small to be of practical value.

Varactor Diode Characteristics

Varactor diodes are normally operated in the **reverse-bias** direction. With an increase in reverse biasing, the depletion region increases its width. This means less resulting capacitance. A decrease in reverse-bias voltage causes a corresponding increase in capacitance. In effect, the capacitance of a diode varies inversely with its bias voltage. This relationship, however, does not change linearly. Note the nonlinear area between 0 and 0.2 V of the varactor diode in **Figure 5.9**. The value of capacitance increases rather significantly when the reverse-bias voltage decreases. The capacitance continues to increase even when the diode is forward biased. In fact, the greatest capacitance is produced just before the forward-bias barrier voltage is

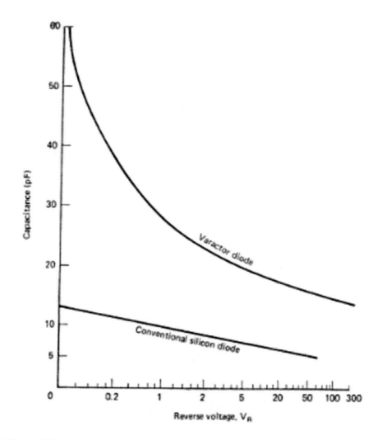

Figure 5.9   Diode capacitance values with ambient temperature, T_A, at 25°C.

reached. This barrier voltage value is approximately 0.6 V for a silicon diode. However, when the **barrier** is reached, the diode becomes conductive. This causes the diode to respond as a shorted capacitor. Because of this condition, varactor diodes should not be used in the forward-bias region. Their normal range of operation is between zero and the **reverse breakdown** voltage.

The **internal capacitance** of a varactor diode changes, to some extent, with temperature. Manufacturers usually take this into account by rating the operational characteristics of the device at some temperature value. The capacitance−voltage (C-V) characteristic of the diode in **Figure 5.9** is rated at 25°C. This shows that these C-V characteristics occur only when the device is operating at or near its rated temperature value. The amount of capacitance change that takes place for a given change in temperature is usually expressed as the temperature coefficient of capacitance. The letters TCc denote this characteristic of the diode. As a rule, the TCc value of a varactor diode has a positive temperature characteristic. This shows that as the temperature increases, the depletion region decreases in width. A decrease in the depletion region causes a corresponding increase in capacitance. A positive TCc value, therefore, shows that an increase in temperature causes a corresponding increase in capacitance.

The electrical characteristics of a representative varicap diode, BB139, are shown at the *end* of the chapter. Note the absolute maximum ratings, a housing outline, and the electrical characteristics for an operation at 25°C. The data listed are for minimum, typical, and maximum operations. Most of these characteristics have some influence on the operational frequency of the device. They are tested at a representative operating frequency. Note, in particular, that this diode has a series resonant frequency (f_r) of 1.4 GHz (1 GHz is 1,000,000,000 Hz). The varactor diodes shown are used in very high frequency (VHF) and frequency modulation (FM) applications. These applications will be discussed in later chapters.

Varactor Diode Operating Efficiency

When the **varactor diode** is used as a capacitor in a resonant circuit, its operational efficiency becomes very important. The efficiency rating of a capacitor is a ratio of the amount of energy stored compared with the actual amount of energy used by the capacitor in the storing process. This relationship is called the **quality factor**, or **Q**, of a capacitor. A varactor diode has a **Q rating** very similar to that of a capacitor. The Q of a varactor diode is an expression of the **capacitive reactance** (X_C) divided by the series resistance (R_S)

$$Q = X_C/R_S. \tag{5.1}$$

Since the X_C of this expression is frequency dependent, the Q will also change with the frequency. The X_C of a capacitor is $1/2\pi fC$. X_C is measured in Ohms. For a varactor diode, the X_C and R_S values of the Q expression are combined into a more workable formula. This is expressed as

$$Q = 1/2\pi fCR_S. \tag{5.2}$$

Note, in this formula, that Q is inversely related to the values of f, C, and R_S. The Q of a varactor diode takes into account such things as internal capacitance, series resistance, and frequency. The series resistance is generally due to lead length and the bulk resistance of the semiconductor material.

In conventional operation, the reverse-biased voltage of the diode blocks DC from passing through the device. The **internal resistance** of a diode is a form of opposition to an AC signal. Since AC passes through the capacitance of a reverse-biased diode, its value is influenced by the R_S that it sees. As a rule, the R_S of a varactor diode is quite small at high frequencies. The data sheet at the *end* of the chapter shows R_S to be 0.35 Ω at a frequency of 600 MHz. This means that a varactor diode offers some measurable amount of opposition to an AC signal. This opposition is frequency dependent. The Q of a varactor diode is often expressed as a number value at an operational frequency. On the **data sheet**, the Q is 150 at 100 MHz.

Typically, these devices have a rather high value of Q. In general, the Q increases at lower frequencies and decreases in value at higher frequencies. This means that the device works very efficiently at some designated range of AC frequency. If the varactor is subjected to frequencies higher than the designated operating frequency, the Q will drop in value. If the normal operating frequency is exceeded, the Q can be reduced to a value of 1. The frequency at which this occurs is called the **cutoff frequency** (f_c). This operating condition is an important selection characteristic for the varactor diode. The cutoff frequency of a varactor diode can be calculated by the following formula:

$$f_c = 1/2\pi C R_S. \tag{5.3}$$

Varactor Diode Construction

Varactor diodes are a rather broad general class of semiconductor devices that are designed for a variety of different applications. As might be expected, the construction of the device has a great deal to do with its use in different applications. The material used in its construction, for example, dictates the **frequency response**. **Silicon** is used largely for devices that operate on frequencies up to 1 GHz. **Gallium arsenide** is generally used for devices that operate on frequencies in excess of 1 GHz. The type of housing or packaging of a device is another important selection factor. A variety of different packages are available.

The DO-35 **package** shown on the data sheet at the *end* of the chapter is preferred for most low-power and low-frequency applications. This type of package is glass enclosed with axial leads that can be soldered into a circuit. The power rating of the device is in the range of 500 mW. **Power rating** is another important selection consideration. Power ratings range from 500 mW to 50 W for stud-mounted packages. The power rating refers to the ability

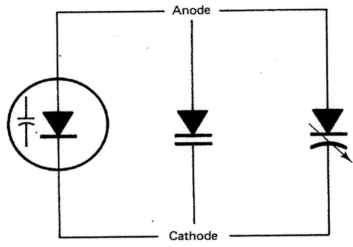

Figure 5.10 Varactor diode symbols.

of a varactor diode to dissipate heat. Generally, this is based on the series resistance (R_S) value of the device. The power dissipated is in the form of AC energy that is being manipulated. AC can pass through a reverse-biased diode that responds as a capacitor.

A number of **schematic** symbols are used to represent the varactor diode. **Figure 5.10** shows three popular methods of representation. The symbol on the left has a small capacitor inside the circle surrounding the diode. This symbol tends to be used more frequently than the other two. Industrial electronic circuits and specialized engineering schematics can use the other two symbols.

Self-Examination

Answer the following questions.

11. The depletion region of a varactor diode serves as the _____ material of a capacitor.
12. The P-type and N-type materials of a varactor diode serve as the _____ of a capacitor.
13. The _____ of a varactor diode is changed by altering the reverse-bias voltage of the junction.
14. The junction of a varactor diode is normally _____ biased when operated as a voltage-variable capacitor.

15. An increase in the reverse-bias voltage of a varactor diode causes the junction capacitance to (increase, decrease).

16. A decrease in the reverse bias of a varactor diode causes the junction capacitance to (increase, decrease).

17. The C-V characteristics of a varactor diode change to some extent with _____.

18. _____ is commonly used in the construction of a varactor diode that will control frequencies below 1 GHz.

19. The _____ rating of a varactor diode refers to its ability to give off or dissipate heat.

20. Varactor diodes are commonly used as the _____ component of an LC resonant circuit.

21. The Q factor of a varactor diode is an expression of _____.

22. When the Q of a varactor diode drops to a value of 1, it is called the _____ frequency.

23. Varactor diodes generally have a rather (high, low) Q rating.

5.3 Miscellaneous Diodes

There are a number of two-terminal devices that have a single P–N junction or its equivalent that respond in some unconventional way. These devices have unique characteristics that distinguish them from conventional silicon diodes. As a rule, these devices may have a different method of operation, terminals, or construction. High-frequency oscillators and amplifiers, power control, and switching are typical applications for this group of diodes. Some of the devices found in this classification are Schottky, IMPATT, Gunn-effect, PIN, and switching diodes.

5.3 Describe the distinguishing characteristics of Schottky, PIN, IMPATT, and Gunn-effect diodes.

   In order to achieve objective 5.3, you should be able to:

   • define *Schottky-barrier diode, PIN diode, Gunn-effect diode, IMPATT diode*, and *hot carrier*.

Schottky-Barrier (Hot-Carrier) Diodes

The **Schottky-barrier**, **surface-barrier**, or **hot-carrier diode** is a two-terminal device that has a number of important applications in electronics. Its areas of application were first limited to the **very high frequency (VHF)**

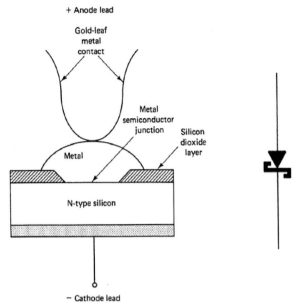

Figure 5.11 Schottky-barrier diode construction and symbol.

range of operation. Now, this device is found in low-voltage/high-current switching power supplies. It is also used in computer integrated circuits, radar systems, communication circuits, and instrumentation.

The construction of a **Schottky-barrier diode** is very different from that of the conventional silicon diode. The junction formed by this device is considered to be a metal-semiconductor structure. **Figure 5.11** shows the structure of a Schottky diode and its **schematic** symbol. The semiconductor material is a piece of N-type silicon. The metal can be a variety of different materials, such as gold, silver, platinum, chrome, or tungsten. The construction techniques and materials of this device each result in a different frequency response and biasing voltage. In general, these characteristics are very similar in many respects. The device can be manufactured so that it will fit into a variety of different applications.

In the two materials of a **Schottky-barrier diode**, electrons are considered to be the majority current carriers. Metal naturally contains an abundance of electrons. The N-type of semiconductor material is purposely doped so that it has a large number of electrons that do not take part in the **covalent bonding** process. When the two materials are joined during the forming process, electrons from the N-type material immediately flow into the adjoining metal.

These electrons possess a rather high level of energy compared with those of the metal piece. The injected electrons are commonly called **hot carriers**. In a conventional silicon diode, the current carriers are represented as electrons and holes. The energy level of the two current carriers is primarily the same. Schottky diodes are unique in that conduction is entirely by majority carriers. The heavy flow of electrons from N-type material to metal causes a depletion of current carriers near the metal junction. This **depletion area** is similar to that of a conventional silicon diode.

The absence of minority current carriers in a Schottky diode makes it an attractive **high-speed switching** device. The minority current carrier content of a diode normally slows the reverse recovery time of a solid-state device. With the minority carrier content at a minimum, the Schottky diode can be used to change states very quickly. **Switching frequencies** approaching 20 GHz are very common for the Schottky diode.

PIN Diodes

Another type of diode is the **PIN diode**. The abbreviation PIN refers to the structure of the semiconductor material, which is quite different from that of a conventional silicon diode. The letter *I* denotes a layer of undoped, or **intrinsic**, semiconductor material between layers of **heavily doped** P-type and N-type materials. **Figure 5.12** shows the structure of a PIN diode and some of the schematic symbols that represent this device.

Probably the most important feature of a **PIN diode** is its ability to respond as almost a pure resistor at high radio frequencies. Resistance can be varied from 10,000 Q to less than 1 Q by the control of current passing through the diode. Most diodes have this characteristic to some degree. The PIN diode, however, is designed to achieve a relatively wide range of resistance with good linearity, low distortion, and low current drive. The PIN diode is widely used in high-speed switching applications, high-frequency control, and microwave circuits.

At **radio frequencies**, a forward-biased PIN diode behaves as a pure resistance. The resistance of a PIN diode is determined by the following:

- Bias voltage
- Thickness of the intrinsic layer
- Properties of the current carriers

The **resistance** is inversely proportional to the **forward-bias current**. Typically, only the on and off resistances are of major concern in the operation of this device as a switch.

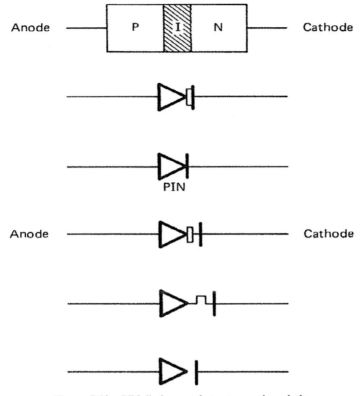

Figure 5.12   PIN diode crystal structure and symbols.

The housing, or **packaging**, of a PIN diode is similar in many respects to that of other high-frequency diodes. **Figure 5.13** shows some representative **high-frequency diode** packages. The application of the device, its operating frequency range, power dissipation, and chip structure are some of the factors that dictate the packaging for a diode.

Gunn-Effect Diodes

Another semiconductor device that is used to control high-frequency AC is the **Gunn-effect diode**. This particular device is primarily designed to operate in the **microwave** region. Typical operating frequencies are in the range of 5–100 GHz. This device is capable of controlling high-frequency AC with a minimum of parts and a low-voltage de-energizing source.

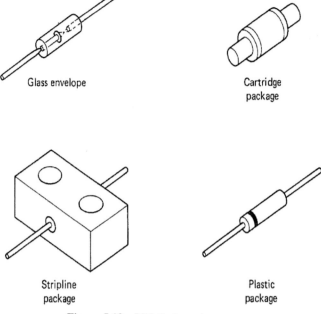

Figure 5.13 PIN diode package types.

The **Gunn-effect diode** differs from other semiconductor devices in its construction. It, for example, does not have a distinct P–N junction. Like the Schottky diode, the Gunn-effect device has a piece of semiconductor material connected between two metal connections. This material is unevenly doped; so the crystal will break into different conduction regions with fields of different intensities across them. With the application of a specific DC voltage value, the device will go into conduction and have a **negative resistance** effect. This effect can be used to generate or amplify RF signals. Operation depends on the amount or bulk of material involved in the structure. **Gallium arsenide** is used to form the semiconductor material of the device.

The **Gunn-effect diode** must have a specific voltage polarity applied to its material to produce the negative resistance effect — that is, it must be properly biased to produce this characteristic. Most devices will be damaged if the polarity of the source voltage is reversed. Typical Gunn-effect devices respond to voltages of $8-12$ V$_{dc}$. The operational current for a 10-mW device is $500-850$ mA. In most microwave applications, the generated signal is radiated directly from the semiconductor material. An application of this device is in portable **radar systems**.

IMPATT Diodes

The **IMPATT diode** is similar in many respects to a conventional silicon diode in crystal construction and operation. The term IMPATT stands for IMPact Avalanche and Transit Time. The term **avalanche** indicates the area of operation, which is the reverse-bias area near the avalanche region of conduction. A small change in reverse voltage causes the diode to produce a **negative resistance** characteristic. This effect can be used to amplify or generate high-frequency AC signals. RF signal control is in the range of 2–10 GHz. The operational efficiency of this device is rather low. At 5 GHz, a working device would have an efficiency rating of 10%. **Efficiency**, in this case, is the relationship of power output to power input. Although an efficiency rating of 10% seems to be exceedingly low, in practice, this figure represents one of the more efficient methods of generating microwave power.

The housing or **packaging** of an IMPATT diode is very similar to that of other **high-frequency diodes**. Most of the packages used for the PIN diode could be used for the IMPATT diode. One very popular package not shown in **Figure 5.13** has the diode mounted inside a small threaded bolt structure. This type of package permits the diode to be placed in a metal cavity that responds to microwave signals and also permits heat to be conducted away from the small P–N junction. Packaging is primarily dictated by the application of the device and the frequency being controlled. The IMPATT diode is very similar in operation to the Gunn-effect diode, but the IMPATT diode has a P–N junction, while the Gunn-effect diode has a metal N-type material structure.

Self-Examination

Answer the following questions.

24. A(n) _____ diode has a metal-semiconductor type of construction.
25. In a Schottky diode, _____ are the majority current carriers.
26. The absence of minority current carriers in a Schottky diode improves its ability to perform high-speed _____ operations.
27. The high-energy electrons of a Schottky diode are generally called _____.
28. A device that has a piece of intrinsic material separating the N-type and P-type semiconductors is called a(n) _____ diode.
29. An important feature of the PIN diode is that it responds as a(n) _____ at high radio frequencies.

30. A Gunn-effect diode has a piece of _____ material connected between two metal terminals.
31. An IMPATT diode operates in reverse bias near the _____ region.

Summary

- A tunnel diode is a two-element semiconductor with construction similar to that of a conventional silicon diode.
- The P–N materials of a tunnel diode are rather heavily doped, which causes the depletion zone to be very thin in comparison to that of a conventional diode.
- Due to the increased number of current carriers and the thin barrier that exists in a tunnel diode, electrons tend to pass through the barrier with an extremely small amount of energy.
- Applications of the tunnel diode are primarily restricted to high-frequency signal control and switching.
- A varactor diode is manufactured with a very light dose of impurities near the junction and an increased number of impurities away from the junction.
- The depletion zone serves as the dielectric material, and the semiconductors respond as the plates of a capacitor; the value of capacitance can be changed according to the bias voltage.
- The depletion zone of a varactor diode serves as the dielectric material, and the semiconductors respond as the plates of a capacitor.
- The value of capacitance in a varactor diode can be changed according to the bias voltage; thus, this device can respond as a voltage-variable capacitor.
- Due to its construction, a varactor diode experiences a higher value of internal capacitance than a conventional silicon diode.
- Applications of the varactor diode are primarily restricted to frequency control.
- Television, FM radio, and automatic frequency control circuits use varactor diodes.
- Schottky diodes are two-terminal devices that are constructed of metal and a piece of semiconductor material.
- The electrons of the N-type material in a Schottky diode possess a rather high level of energy compared with the electrons of the metal.
- When the metal and N-type semiconductor material of a Schottky diode are joined during the forming process, the injected electrons from the N-type material become hot carriers.

- The construction of a Schottky diode causes it to respond entirely to majority current carriers.
- Switching frequencies approaching 20 GHz are common with a Schottky diode.
- A PIN diode is constructed of an intrinsic layer of semiconductor material placed between the P-type and N-type materials of the junction.
- The resistance of a PIN diode can be varied from 10,000 Q to less than 1 Q by control of the current passing through the device.
- PIN diodes have the ability to respond as resistors to high radio frequencies.
- PIN diodes are widely used as high-speed switching devices in microwave control circuits.
- The Gunn-effect diode does not have a distinct P–N junction in its construction; it has a piece of semiconductor material connected between two metal terminals.
- When a particular DC voltage is applied to a Gunn-effect diode, the diode responds by producing a negative resistance.
- The negative resistance produced by a Gunn-effect diode can be used to amplify or generate microwave signals.
- The operating frequencies of a Gunn-effect diode are in the range of 5–100 GHz.
- IMPATT diodes are high-frequency devices.
- The term *IMPATT* refers to impact avalanche and transit time.
- IMPATT diodes operate in the reverse-bias region near the avalanche point of conduction.
- A small change in reverse voltage causes an IMPATT diode to have a negative resistance characteristic.
- The negative resistance produced by an IMPATT diode can be used to control RF signals in the 2–10 GHz range.
- Although these IMPATT diodes have a rather low-efficiency rating, under normal circumstances, this efficiency rating is much better than that of other high-frequency devices.

Formulas

(5-1) $Q = X_C/R_S$ Quality factor of a varactor diode.
(5-2) $Q = 1/2\pi f C R_S$ Quality factor of a varactor diode.
(5-3) $f_c = 1/2\pi C R_S$ Cutoff frequency of a varactor diode.

Review Questions

Answer the following questions.

1. A semiconductor device that resembles a voltage-variable capacitor is the:

   a. Tunnel diode
   b. Varactor diode
   c. Schottky diode
   d. IMPATT diode
   e. PIN diode

2. A semiconductor device that has high-energy electrons or hot carriers is the:

   a. Tunnel diode
   b. Varactor diode
   c. Schottky diode
   d. IMPATT diode
   e. PIN diode

3. A diode that has an intrinsic layer of semiconductor material separating the materials of its P–N junction is the:

   a. Tunnel diode
   b. Varactor diode
   c. Schottky diode
   d. IMPATT diode
   e. PIN diode

4. A diode that has a negative resistance region when forward biased is the:

   a. Tunnel diode
   b. Varactor diode
   c. Schottky diode
   d. IMPATT diode
   e. PIN diode

5. A semiconductor device that operates in the avalanche region is the:

   a. Tunnel diode
   b. Varactor diode
   c. Schottky diode
   d. IMPATT diode
   e. Gunn-effect diode

6. A semiconductor device that has heavily doped P- and N-materials, which produce a thin depletion region is the:

   a. Tunnel diode
   b. Varactor diode
   c. Schottky diode
   d. IMPATT diode
   e. PIN diode

7. When the reverse-bias voltage of a varactor diode increases, the:

   a. Leakage current increases
   b. Capacitance decreases
   c. Depletion zone decreases
   d. Majority current carrier content increases
   e. Negative resistance increases

8. When the Q factor of a varactor diode drops to a value of 1, it shows that the device has reached:

   a. The cutoff frequency
   b. Avalanche breakdown
   c. Full conduction
   d. Heavy leakage current
   e. Thermal breakdown

9. A semiconductor device that operates with a forward-biased metal-semiconductor junction is a:

   a. Tunnel diode
   b. Varactor diode
   c. Schottky diode
   d. IMPATT diode
   e. PIN diode

10. A device that responds as a pure resistance to RF signals is a:

   a. Tunnel diode
   b. Varactor diode
   c. Schottky diode
   d. IMPATT diode
   e. PIN diode

11. The operational efficiency of a varactor diode is based on:

   a. Capacitance
   b. Capacitance and series resistance

    c. Capacitance and inductance
    d. Capacitance and temperature
    e. Leakage current

12. A Schottky diode has the same type of construction as a:

    a. Tunnel diode
    b. Varactor diode
    c. IMPATT diode
    d. Gunn-effect diode
    e. PIN diode

Answers

Self-Examination

5.1

1. thin
2. tunneling
3. Negative resistance
4. valley voltage or V_V
5. two
6. negative resistance
7. tunnel diode
8. forward
9. negative resistance
10. Voltage peak *or* V_P, valley voltage *or* V_V

5.2

11. dielectric
12. conductor plates
13. capacitance
14. reverse
15. decrease
16. increase
17. temperature
18. Silicon
19. power
20. tuning
21. operating efficiency

22. cutoff
23. high
5.3
24. Schottky
25. electrons
26. switching
27. hot carriers
28. PIN
29. resistor
30. semiconductor
31. avalanche

Glossary

Tunnel diode

A two-element semiconductor device similar in construction to the conventional silicon diode, but which contains a high concentration of impurities.

Negative resistance

An electronic condition in which an increase in voltage across a resistive element causes a reduction in current and vice versa. Certain semiconductors have a negative resistance region in their operation.

Bistable

Two stable or stationary operating conditions of an electronic switching circuit.

Quality factor

A ratio of the amount of energy stored compared with the actual amount of energy used by the capacitor in the storing process.

Schottky-barrier diode

A semiconductor device in which a barrier is formed between metal and a semiconductor material. This barrier achieves rectification but avoids slowing down the current carriers, so that high-frequency AC can be rectified.

Hot carriers

The electrons that are injected into a Schottky-barrier diode.

PIN diode

A semiconductor device that contains a layer of undoped, or intrinsic, semiconductor material between layers of heavily doped P-type and N-type materials.

Gunn-effect diode

A semiconductor device that is used to control high-frequency AC. It is primarily designed to operate in the microwave region.

IMPATT diode

An avalanche diode used as a high-frequency oscillator or amplifier. The negative resistance of this device depends on the transient time of current carriers through the depletion layer.

BB139

VHF/FM VARACTOR DIODE

DIFFUSED SILICON PLANAR

- C_3/C_{25} ... 5.0-6.5
- MATCHED SETS (Note 2)

ABSOLUTE MAXIMUM RATINGS (Note 1)

Temperatures
Storage Temperature Range	−55°C to +150°C
Maximum Junction Operating Temperature	+150°C
Lead Temperature	+260°C

Maximum Voltage
WIV	Working Inverse Voltage	30 V

DO-35 OUTLINE

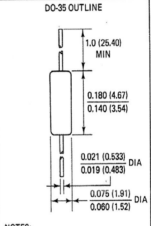

1.0 (25.40) MIN

0.180 (4.67) / 0.140 (3.54)

0.021 (0.533) / 0.019 (0.483) DIA

0.075 (1.91) / 0.060 (1.52) DIA

NOTES:
Copper clad steel leads, tin plated
Gold plated leads available
Hermetically sealed glass package
Package weight is 0.14 gram

ELECTRICAL CHARACTERISTICS (25°C Ambient Temperature unless otherwise noted)

SYMBOL	CHARACTERISTIC	MIN	TYP	MAX	UNITS	TEST CONDITIONS
BV	Breakdown Voltage	30			V	$I_R = 100\,\mu A$
I_R	Reverse Current		10	50	nA	$V_R = 28$ V
			0.1	0.5	μA	$V_R = 28$ V, $T_A = 60°C$
C	Capacitance		29		pF	$V_R = 3.0$ V, f = 1 MHz
		4.3	5.1	6.0	pF	$V_R = 25$ V, f = 1 MHz
C_3/C_{25}	Capacitance Ratio	5.0	5.7	6.5		$V_R = 3$ V/25 V, f = 1 MHz
Q	Figure of Merit		150			$V_R = 3.0$ V, f = 100 MHz
R_S	Series Resistance		0.35		Ω	C = 10 pF, f = 600 MHz
L_S	Series Inductance		2.5		nH	1.5 mm from case
f_o	Series Resonant Frequency		1.4		GHz	$V_R = 25$ V

NOTES
1. These ratings are limiting values above which the serviceability of the diode may be impaired
2. The capacitance difference between any two diodes in one set is less than 3% over the reverse voltage range of 0.5 V to 28 V

6

Bipolar Junction Transistors (BJTs)

The invention of the **transistor** in 1947 signaled the start of a new era in electronics. Circuits could be built exceedingly small, operate without heating power, amplify signals with a low-voltage source, and be extremely rugged. **Integrated circuits**, **microprocessors**, pocket **calculators**, **personal computers**, and communication electronics have all been made possible through the development of the transistor. The impact of this device on the electronics field has still not been fully realized.

A **transistor** is a semiconductor device that is used to control the flow of current. Transistors are used for switching, amplification, and signal generation. There are two major types of transistor: **bipolar junction** and **field effect**. In this chapter, you will study the **bipolar junction transistor (BJT)**. You will become familiar with basic transistor principles and BJT operation, test procedures, and lead identification. A person working with these devices must be knowledgeable in these areas. In a later chapter, you will study the field-effect transistor (FET).

Objectives

After studying this chapter, you will be able to:

6.1 describe the physical construction of NPN and PNP bipolar junction transistors;

6.2 explain the fundamental operation of a bipolar junction transistor;

6.3 predict how a bipolar junction transistor will respond in different regions of operation;

6.4 evaluate the condition of a bipolar junction transistor;

6.5 analyze and troubleshoot bipolar junction transistors.

Chapter Outline

Terms

active region
base
beta
bipolar
collector
cutoff region
emitter
epitaxial growth
gain
mesa transistor
NPN transistor
planar transistor
PNP transistor
saturation region

6.1 BJT Construction

In a previous chapter, you studied the diode – a two-element device with one P–N junction. Diodes are designed to block or pass current according to their biasing and, therefore, are used in switching and rectification applications. A *bipolar junction transistor (BJT)*, however, is a three-element device with two P–N junctions. Altering the voltage applied to the two junctions controls current flow. Through this procedure, it is possible for the device to achieve amplification, switching, or signal generation. The BJT is probably used more than any other single electronic device. Some very sophisticated integrated circuits, for example, may employ thousands of these devices in their operation. Each device responds in the same basic manner. This section discusses the two types of BJTs (NPN and PNP), construction methods, and packaging.

6.1 Describe the physical construction of NPN and PNP bipolar junction transistors.

In order to achieve objective 6.1, you should be able to:

- describe how alloy-junction transistors and diffusion transistors are created;
- identify various bipolar junction transistor packages;
- define *NPN transistor, emitter, base, collector, PNP transistor, bipolar, epitaxial growth, planar transistor,* and *mesa transistor.*

BJT Types

NPN and **PNP** are two types of bipolar junction transistor (BJT). The letters NPN and PNP are used to denote the **polarity** of the semiconductor material used in the transistor's construction. The type of impurity added to **silicon** or **germanium** when the transistor is manufactured determines the polarity of the semiconductor material.

PNP and NPN transistors are **bipolar devices**. The term bipolar refers to conduction by both holes and electrons. Electrons are the majority current carriers in the N-type material, and holes are the majority current carriers in the P-type material. Current conduction in the transistor is based on the movement of both current carriers. The material used in the transistor's construction determines the polarity of the source voltage applied to each transistor element.

A **PNP transistor** has a thin layer of N-type material placed between two pieces of P-type material. The schematic symbol of the PNP transistor and its crystal structure are shown in **Figure 6.1.**

Leads attached to each piece of material are identified as the **emitter, base,** and **collector**. Note that in the PNP transistor symbol, the arrowhead of the emitter lead "Points iN" toward the base. This indicates that the device is of the PNP type. Lead identification is extremely important when working with transistors.

An **NPN transistor** has a thin layer of P-type material placed between two pieces of N-type material. The schematic symbol, crystal structure, and lead designations of the transistor are shown in **Figure 6-2. The schematic** symbol of this device is the same as the PNP symbol except for the direction of the arrow. An NPN transistor has the arrow "Not Pointing iN" toward the base.

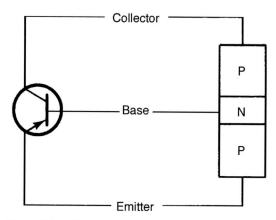

Figure 6.1   PNP transistor symbol and crystal structure.

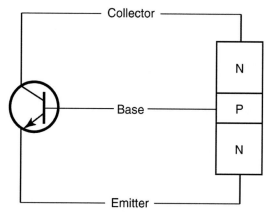

Figure 6.2   NPN transistor symbol and crystal structure.

Construction Methods

Since the invention of the first transistor, a variety of construction methods have been used to manufacture this device. The first transistors manufactured were of the **point-contact** type. In this method, semiconductor materials are connected together by pointed wires that are fused to the material. For many years transistors constructed by this method were primarily used in **high-frequency** applications. Other manufacturing methods are **alloy-junction** and **diffusion**. This section briefly discusses the alloy-junction and diffusion methods. The intention of this discussion is to familiarize you with basic fabrication terms.

Alloy-Junction Transistors

The **alloy-junction** method of bipolar transistor construction is an outgrowth of a procedure that is used to form diodes. In transistor construction, this procedure is achieved by attaching two small pieces of metal on opposite sides of a thin piece of semiconductor material. The metal pieces serve as an **impurity** or **dopant** for the semiconductor. **Figure 6.3** shows a structure with this formation. The entire structure is then heated until the impurity melts into the semiconductor material. The melting process causes an alloy of a different type to be formed on each side of the semiconductor. In this case, a PNP-type of device is formed by the alloying procedure.

The three semiconductor materials formed by the alloying process represent the **emitter, base**, and **collector** of a BJT. Leads are attached to each material. The entire structure is then placed in a housing with the three leads serving as external connection points to the respective material.

Note in **Figure 6.3** that the collector–base area of the transistor is made larger than the emitter–base area. This is purposely done to permit the collector to have greater power dissipation. The alloy-junction technique of transistor fabrication has been used in the manufacture of high-power transistors.

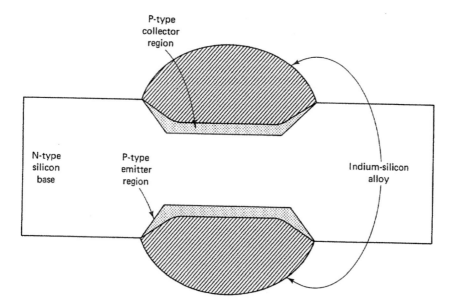

Figure 6.3   PNP alloy-junction transistor formation.

Diffusion Transistors

Bipolar junction transistors are largely manufactured by the diffusion technique. This process involves the movement of N-type or P-type impurity atoms into a piece of silicon. To achieve this movement, it is necessary to heat a piece of silicon to 1250°C in the presence of a controlled impurity vapor. The movement process is very slow. Typical formation rates are 0.00001 inch per hour.

The **diffusion** process is carried out in two steps. The first step consists of heating a piece of silicon in an impurity dopant vapor to form a high concentration of the dopant on its surface. This is called the **deposition** *step*. The silicon piece then has its temperature elevated. This causes the dopant atoms to be absorbed or diffused into the silicon. This is called the ***diffusion step***. Manufacturing involves repeated deposition and diffusion steps.

Figure 6.4 shows transistor formation by the **diffusion** process. The first step involves an N-type semiconductor that has a P-type vapor deposited on its surface. In the second step, increased heat causes the P-type material to diffuse into the N-type material. This forms an N-type collector and a P-type base. The third step is a repeat of the initial process. In it, however, the dopant vapor is a pentavalent material. The deposition area is on top of the base. This causes an area of N-type material to be formed on the base. Diffusion causes this material to move into the base. The structure now has an N-type emitter. The completed transistor is an NPN device.

A refined version of the diffusion process is called **epitaxial growth**. In this fabrication technique, N-type or P-type materials are grown or formed on the surface of another material. The general shape of specific transistor elements can be altered through this procedure. The term *epitaxial* is derived from the Greek words *epi*, meaning *on*, and *taxi*, meaning *arrange*. This describes the process in which atoms are formed on a surface so that they are an extension of the original crystal structure. When the epitaxial growth process is used and the completed structure has a flat top or level plane surface, the transistor is called a **planar transistor**. When the completed structure rises above the primary surface, forming a plateau, it is called a **mesa transistor**. Mesa construction is used to expose the emitter and base regions so that electrical connections can easily be made.

Packaging

Once a transistor is constructed, the entire crystal assembly must be placed in an enclosure or package. The unit is then sealed to protect it from dust,

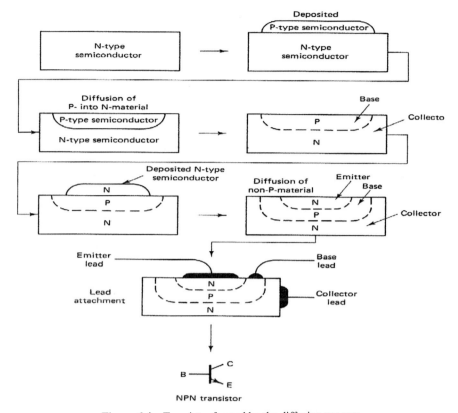

Figure 6.4 Transistor formed by the diffusion process.

moisture, or outside contaminants. Electrical connections are made to each element through leads attached to the package. When **metal packages** are used, the housing also serves as a heat sink, which carries heat away from the crystal. A common practice with this kind of unit is to attach the collector to the metal case. This helps the collector dissipate heat. Packages of this type should not be permitted to touch other circuit components when they are in operation.

Ceramic- and **epoxy-packaged** transistors are also available. As a rule, this type of packaging is somewhat less expensive than metal packaging to produce. However, these devices are usually not as rugged as the metal-packaged devices. Most **small-signal transistors** are housed in epoxy packages. The outside case of this device is insulated from all transistor elements. It is specifically designed for **printed circuit board** construction and close placement with other circuit components.

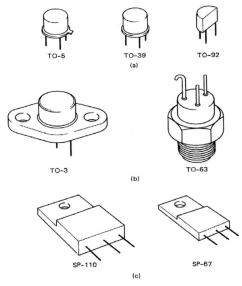

Figure 6.5   Typical transistor packages. (a) Small-signal. (b) Large-signal or power. (c) Epoxy.

Several **transistor package** styles are shown in **Figure 6.5**. **Figure 6.5(a)** shows devices designed for small-signal applications. The letter designation "TO" stands for transistor outline. These same packages are also used to house other solid-state devices. **Figure 6.5(b)** shows some of the packages used to house large-signal or power transistors. Heavy metal cases of this type are used to dissipate heat generated by the transistor. The collector is usually attached to the metal case. **Figure 6.5(c)** shows a few of the common epoxy packages. The letter designation "SP" stands for semiconductor package. This type of package is used to house a number of other semiconductor devices.

With the wide range of package styles available, it is difficult to have a standard method of identifying transistor leads. Each package generally has a unique lead location arrangement. It is usually a good idea to consult the manufacturer's specifications for proper lead identification.

Self-Examination

Answer the following questions.

1. A bipolar transistor has _____ junctions.

2. The centerpiece of semiconductor material of a transistor serves as the _____.

3. The term _____ refers to the conduction of holes and electrons in the operation of a transistor.

4. The arrowhead of a transistor symbol refers to the _____.

5. When the arrowhead of a transistor symbol points toward the base, it shows that the device is (NPN, PNP).

6. _____ transistors are formed by melting two small metal pieces into opposite sides of a thin piece of semiconductor material.

7. _____ takes place when a thin layer of dopant material is formed on the surface of silicon.

8. _____ occurs when a concentration of dopant material is absorbed into a piece of silicon.

9. A fabrication procedure that causes a flat surface structure to be formed produces a _____ transistor.

10. A fabrication procedure that causes a plateau area to rise on the surface of a structure produces a(n) _____ transistor.

11. The letter designation "TO" stands for _____.

12. The letter designation "SP" stands for _____.

6.2 BJT Operation

A P–N junction diode has one junction, which must be forward biased to function properly. A Zener diode also has one junction, but it must be reverse biased to function properly. Bipolar junction transistors have two junctions. For the BJT to operate properly, one junction must be forward biased and the other must be reverse biased. BJT operation is an important electronic principle that is used in the operation of many analog and digital devices. This section discusses the proper biasing of bipolar junction transistors and their operation. It begins with a review of forward and reverse biasing.

6.2 Explain the fundamental operation of a bipolar junction transistor.
  In order to achieve objective 6.2, you should be able to:

  • explain how the emitter, base, and collector of a bipolar junction transistor are biased to make the device operational.
  • describe the relationship between emitter, base, and collector current flow in the operation of a bipolar junction transistor.
  • define *alpha* (α) and *beta* (β).

Forward and Reverse Biasing

In an earlier chapter, you learned that forward biasing of a **P–N junction** is achieved when the polarity of the source is positive to the P-type material and negative to the N-type material. The width of the depletion zone is reduced by this action, and majority current carriers move toward the junction. This condition causes current flow.

   Reverse biasing of a P–N junction is achieved when the polarity of the source is negative to the P-type material and positive to the N-type material. The width of the **depletion zone** is increased by this action, and minority current carriers move toward the junction. This condition does not ordinarily permit current flow. In normal circuit operation, however, reverse-bias current is extremely small.

NPN Transistor Biasing

Consider the biasing of the NPN transistor of **Figure 6.6**. In this diagram, an external voltage source has been applied only to the emitter–base regions of the transistor. In an actual circuit, this source is called the **emitter–base voltage**, or V_{BE}. Note that the polarity of V_{BE} causes the emitter–base junction to be **forward biased**. This condition forces **majority current carriers** together at the emitter–base junction. The resulting current flow in this case would be quite large. In our example, note that the same amount of current flows into the emitter and out of the base. **Emitter current** (I_E) and **base current** (I_B) are used to denote these values. Under normal circuit conditions, a V_{BE} source would not be used independently.

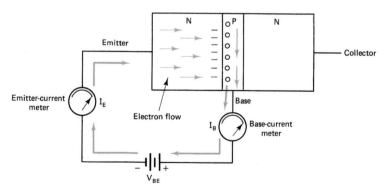

Figure 6.6   Emitter–base biasing.

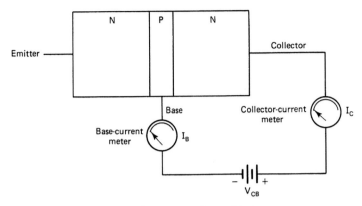

Figure 6.7 Base–collector biasing.

Figure 6.7 shows an external voltage source connected across the base–collector junction of an NPN transistor. The **base–*collector voltage***, or V_{CB}, is used to reverse bias the base–collector junction. In this case, there is no indication of **base current** or **collector current** (I_C). In an actual reverse-biased junction, there could be a very minute amount of current, which is supported primarily by minority current carriers. As a general rule, this is called **leakage current**. In a silicon transistor, leakage current is usually considered to be negligible. In an actual circuit, V_{CB} is not normally applied to a transistor without the V_{BE} voltage source.

For a transistor to function properly, the emitter–base junction must be **forward biased** and the base–collector junction **reverse biased**. Both junctions must have bias voltage applied at the same time. In some circuits, this voltage may be achieved by separate V_{BE} and V_{CB} sources. Other circuits may use a single battery with specially connected bias resistors. In either case, the transistor responds differently when all its terminals are biased.

Consider the action of a properly biased NPN transistor. **Figure 6.8** shows separate V_{BE} and V_{CB} sources connected to the transistor. The V_{BE} source provides **forward bias** for the emitter–base junction, whereas the V_{CB} source **reverse biases** the collector junction. Connected in this manner, the two junctions do not respond as independent diodes.

Figure 6.9 shows how **current carriers** pass through a properly biased NPN transistor. The forward-biased emitter–base junction causes a large amount of I_E to move into the emitter–base junction.

On arriving at the junction, a large number of the electrons do not effectively combine with holes in the base. The base is usually made very thin

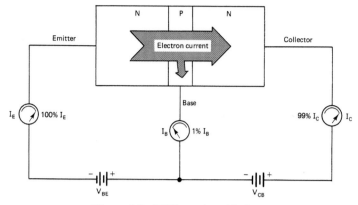

Figure 6.8   NPN transistor biasing.

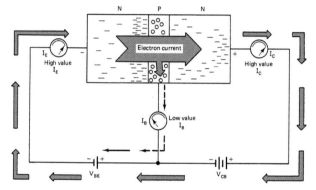

Figure 6.9   Current carriers passing through an NPN transistor.

(0.0025 cm or 0.001"), and it is **lightly doped**. This means that the majority current carriers of the emitter exceed the majority carriers of the base. Most of the electrons that cross the junction do not combine with holes. They are, however, immediately influenced by the positive V_{CB} voltage applied to the collector. A very high percentage of the original emitter current enters the collector. Typically, 95%–99% of emitter current flows into the collector junction and becomes collector current. After passing through the collector region, the collector current, I_C, combines with the base current, I_B, to ultimately form the emitter current, I_E. A large arrow indicates the current flow inside the transistor. Outside the transistor, small arrows indicate current.

The difference between the amount of emitter current and collector current of **Figure 6.9** is equal to the amount of **base current**. Essentially, base current is due to the combining of a small number of electrons and holes in the emitter–base junction. In a typical circuit, base current is approximately 1%–5% of emitter current. With the base region being very narrow, it cannot support a large number of current carriers. A small amount of base current is needed, however, to make the transistor operational. Note the direction of base current, I_B, and its flow path in the diagram. The relationship of **emitter current**, **base current**, and **collector current** in a transistor is expressed by the following equation:

$$I_E = I_B + I_C. \tag{6.1}$$

The largest current flow in a transistor takes place in the emitter. Collector current is slightly less than emitter current. This means that the current flow through a **reverse-biased**, collector–base junction is nearly equal to that of the forward-biased emitter–base junction. The difference in I_E and I_C is I_B.

If the base of a transistor were not made extremely thin and **low doped**, it would not respond as just described. The thin base region makes it possible for large amounts of emitter current to pass through the base and into the collector region. A thicker base would cause more emitter current to combine with holes in the base. Low doping of the base means that the majority current carriers of the emitter cannot effectively combine in the base. Instead of combining in the base, they pass through it and enter the collector. This improves the effectiveness of the base as a control element.

The amount of base current that flows in a transistor is very small but extremely important. Suppose, for example, that the base lead of the transistor in **Figure 6.9** is momentarily disconnected. With this element open, it should be obvious that there will be no base current. Closer examination also shows that V_{BE} and V_{CB} are now connected in series. This means that their voltages are added together. As a result, the collector becomes more positive and the emitter more negative. One would immediately think that this condition would cause the transistor to conduct very heavily. However, it does not permit any conduction at all. In effect, this means that base current has a direct influence on emitter and collector current. The base–emitter junction of a transistor must be **forward biased** for it to produce collector current.

PNP Transistor Biasing

Biasing of a **PNP transistor** is very similar to that of the NPN transistor. The emitter–base junction must, for example, be **forward biased** and the collector–base **reverse biased**. The **polarity** of the bias voltage, however, is

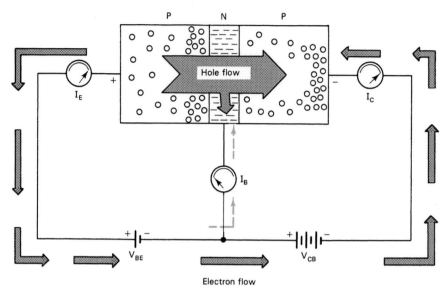

Figure 6.10   Current carriers of a PNP transistor.

reversed for each transistor. The majority current carriers of a PNP transistor are holes instead of electrons as in the NPN device. Except for these two differences, operation is primarily the same.

Figure 6.10 shows how current carriers pass through a properly biased **PNP transistor**. Forward biasing the emitter–base junction causes a large number of holes to move through the junction. On arriving in the base region, a very small number of these holes combine with electrons. This is representative of the base current. Ninety-five percent to ninety-nine percent of the holes move through the base region and enter the collector junction. These holes do not find electrons to combine with in the thin base region. They are, however, immediately influenced by the negative V_{CB} voltage of the collector. This action causes the hole current to flow through the reverse-biased, base–collector junction. The end result is I_C passing through the collector region. It then leaves the collector and flows into V_{CB}, V_{BE}, and returns to the emitter. The large arrow inside the transistor indicates hole current flow. Outside the transistor, electrons achieve current flow. Smaller arrows in the diagram show current outside the transistor.

Compare the transistors of **Figures 6.9** and **6.10** and note the **polarity** of V_{BE} and V_{CB}. The polarity shows the primary difference in NPN and PNP transistor biasing. Note also the type of **majority current flow** through each transistor. This is represented by the large arrow of each diagram. It is different for each device. This means that the two transistor types are not directly interchangeable. Substituting one for the other would necessitate a complete reversal of bias voltage.

PNP transistors are not as widely used as NPN transistors. Electron current flow of the NPN device has much better mobility than the **hole flow** of a PNP device. This means that electrons have a tendency to move more quickly through the crystal material than holes. Because of this characteristic, NPN transistors tend to respond better at high frequencies. In general, this means that the NPN device has a wider range of applications. Manufacturers usually have a larger selection of these transistors in their lines. With a better selection of transistors available, circuit designers find it easier to select devices of a desired characteristic. As a result, the NPN device is much more popular than its PNP counterpart.

Alpha (α) and Beta (β)

The term **alpha** is the ratio of the collector current to the emitter current in a bipolar junction transistor, i.e.,

$$\alpha = I_C/I_E. \tag{6.2}$$

Since $I_E > I_C$, α will always be less than 1. Typical values of α is 0.9 or higher. Since α is a ratio between two current values it does not have any units.

The term **beta** is commonly used to express current gain. Beta can be determined by dividing the collector current by the base current and is expressed as a whole number value. This relationship is expressed by the following equation:

$$\beta = I_C/I_B. \tag{6.3}$$

As you have learned, a small change in transistor base current is capable of producing a very large change in collector current. This, in effect, shows that a transistor is capable of achieving current gain.

Example 6-1:

What is the beta of a transistor circuit with a base current of 20 μA and collector current of 6 mA?

Solution

$$\beta = I_C/I_B$$

$$= 6\,mA/20\,\mu A$$
$$= (6 \times 10^{-3})/(20 \times 10^{-6})$$
$$\beta = 300.$$

Related Problem

What is the beta of a transistor circuit with a base current of 30 μA and collector current of 2 mA?

Transistors are manufactured with a wide range of beta capabilities. **Power transistors**, which respond to large current values, usually have a rather low beta. Typical values are in the range of 20. **Small-signal transistors** may be capable of beta in the range of 400. A good average beta for all transistors is in the range of 100. Beta is a very important transistor characteristic. In manufacturer **datasheets**, the term h$_{FE}$ is frequently used for expressing the current gain instead of the term β. Transistor specifications provided in manufacturer datasheets generally use h$_{FE}$ (or β), rather than α.

Self-Examination

Answer the following questions.

13. Transistor operation depends on the emitter–base region being (forward, reverse) _____ biased and the base–collector region being (forward, reverse) _____ biased.
14. When the emitter–base junction of a bipolar transistor is biased as in **Figure 6.6**, the resulting I$_E$ and I$_B$ are _____.
15. When the emitter–base junction of a bipolar transistor is forward biased, there is a (large, small) _____ value of I$_E$ and I$_B$.
16. A reverse-biased N-P or P–N junction normally causes (zero, medium, maximum) current flow.
17. When the base–collector junction of a bipolar transistor is biased as in **Figure 6.7**, the resulting current flow is (zero, medium, maximum) _____.
18. When the emitter–base junction of a transistor is forward biased and the base–collector junction is reverse biased, the values of _____ and _____ are nearly the same value.
19. The largest value of current that flows in a bipolar transistor is (I$_E$, I$_B$, I$_C$).
20. The difference between I$_E$ and I$_C$ is _____.
21. The term _____ is used to express current gain.

22. The _____ current of a bipolar transistor is usually only 1%−5% of the emitter current.
23. Current gain or _____ of a bipolar transistor is expressed as I_C/I_B.
24. An I_C of 10 mA and an I_B of 10 μA would produce a beta of _____.
25. An I_C that changes from 10 to 5 mA for an I_B of 100 to 50 μA would cause an AC beta of _____.

6.3 BJT Characteristics

When a transistor is manufactured, it is designed with characteristic data that distinguishes it from other devices. This data is extremely important in predicting how a transistor will perform in a circuit. One very common way of showing this information is through the use of a graph called a *collector family of characteristic curves*. In this section, you will learn how to interpret a collector family of characteristic curves graph to predict how a transistor will perform.

6.3 Predict how a bipolar junction transistor will respond in different regions of operation.

In order to achieve objective 6.3, you should be able to:

* interpret the characteristic curves of a bipolar junction transistor;
* define *active region*, *cutoff*, and *saturation region*.

Collector Family of Characteristic Curves

Figure 6.11 shows a **collector family of characteristic curves** for an NPN transistor. The vertical part of this graph shows different **collector current** values in milliamperes. The horizontal part shows the **collector−emitter voltage**. The collector–emitter voltage is the voltage value from the collector to the emitter of the transistor. The value of V_{CE} varies according to the conduction of the transistor. Individual lines of the graph represent different values of base current. The zero **base current** line is normally omitted from the graph. This condition shows when the transistor is not conducting.

A **collector family of characteristic curves** tells a great deal about the operation of a transistor. Take, for example, point A on the 60-μA base-current line. If the transistor has 60 μA of I_B and a V_{CE} of 6 V, there will be 3 mA of I_C. This is determined by projecting a line to the left of the intersection of 60 μA and 6 V. Any combination of I_C, I_B, and V_{CE} can be quickly determined from the display of these curves.

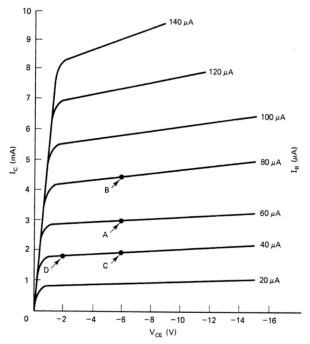

Figure 6.11   Collector family of characteristic curves for an NPN transistor.

Predicting Emitter Current

If the values of I_B and I_C can be determined from a collector family of characteristic curves, they can be used to predict other values. The emitter current (I_E) of a transistor can be determined from the values of I_B and I_C. Remember that

$$I_E = I_B + I_C.$$

Therefore, for point A,

$$I_E = 60\ \mu A \ + 3\ mA$$
$$= 0.00006\ A + 0.003\ A$$
$$= 3.06\ mA.$$

Example 6-2:

Determine the emitter current for point C of the collector family of characteristic curves in **Figure 6.11**.

Solution

$$I_E = I_B + I_C$$
$$= 40\ \mu A + 2\ mA$$
$$= 0.00004\ A + 0.002\ A$$
$$= 2.04\ mA.$$

Related Problem

Determine the emitter current for point D of the collector family of characteristic curves in **Figure 6.11**.

Predicting DC Beta

The current gain or **beta** of a transistor can also be determined from a collector family of characteristic curves. Since beta is I_C/I_B, it is possible to determine this condition of operation. For point A, I_C is 3 mA and I_B is 60 μA. The current gain of a transistor operating at this point would be determined by the following equation:

$$\beta = I_C/I_B$$
$$= 3\ mA/60\ \mu A$$
$$= 0.003\ A/0.00006\ A$$
$$= 50.$$

This condition is called the *direct-current beta*, or **DC beta**, and is represented by the symbol β_{dc}. It shows how the transistor responds when DC voltage is applied. DC beta is an important consideration when a transistor is operated as a **DC amplifier**.

Example 6-3:

Determine the DC beta for point C of the collector family of characteristic curves in **Figure 6.11**.
Solution

$$\beta_{dc} = I_C/I_B$$
$$= 2\ mA/40\ \mu A$$
$$= 0.0002\ A/0.00004\ A$$
$$= 50.$$

Related Problem

Determine the DC beta for point D of the collector family of characteristic curves in **Figure 6.11**.

Predicting AC beta

In many applications, a transistor must amplify AC signals. DC voltage values are applied to the device to make it operational. An **AC signal voltage** applied to the base would then cause a change in the value of I_B. In **Figure 6.11**, assume that the input voltage causes I_B to change between points B and C. Point A is considered to be the operating point. The change in I_B is from 80 to 40 μA. This is normally called ΔI_B. The Greek letter **delta (Δ)** denotes a changing value in I_B. A corresponding change in I_C is determined by projecting a line to the left of each I_B value. The ΔI_C is, therefore, 4.5 to 2 mA. The **AC beta** of a transistor is determined by dividing a change in collector current by a change is base current. It is expressed by the following formula:

$$\beta_{ac} = \Delta I_C/\Delta I_B. \tag{6.4}$$

For points B and C, this is

$$\begin{aligned}
\beta_{ac} &= \Delta I_C/\Delta I_B \\
&= 4.5 - 2\ mA/80 - 40\ \mu A \\
&= 2.5\ mA/40\ \mu A \\
&= 0.0025\ A/0.00004\ A \\
&= 62.5.
\end{aligned}$$

The **AC beta** in this case is somewhat different from that of the **DC beta**. As a general rule, AC beta is larger than DC beta. To amplify an AC voltage, a transistor must operate over a range of different values. This usually accounts for the difference in beta values.

The same **collector family of characteristic curves** can also be used to show the effectiveness of a transistor in controlling collector current. Note points D and C on the 40-μA curve. A change in V_{CE} from 2 to 6 V occurs between these two points. Projecting points D and C to the left indicates a change in I_C from 1.8 to 2 mA. A 4-V change in V_{CE}, therefore, causes only a 0.2-mA change in I_C. Using the AC beta for comparison, a 40-μA change in I_B causes a 2.5 mA change in I_C. This indicates that I_C is more effectively controlled by I_B changes than by V_{CE} changes.

Example 6-4:

Determine the AC beta for points A and B of the collector family of characteristic curves in **Figure 6.11**.

Solution

$$\beta_{ac} = \Delta I_C / \Delta I_B$$
$$= 8 - 6 \; mA/40 - 30 \; \mu A$$
$$= 2 \; mA/10 \; \mu A$$
$$= 0.002 \; A/0.000010$$
$$\beta_{ac} = 200.$$

Related Problem

Determine the AC beta for points B and D of the collector family of characteristic curves in **Figure 6.11**.

Operating Regions

A **collector family of characteristic curves** is also used to show the desirable **operating regions** of a transistor. The center area is called the **active region**. This area is located anywhere between the two shaded areas of **Figure 6-12**. Amplification is achieved when the transistor operates in this region.

A second possible region of operation is called the **cutoff region**. On the family of curves, this is where $I_C = 0$ mA. When a transistor is cut off, no I_B or I_C flows. Any current flow that occurs in this region is due to leakage current. As a general rule, a transistor operating in the cutoff region responds as a circuit with an open switch. When a transistor is cut off, there is infinite resistance between the emitter and collector.

The shaded area on the left side of the family of curves is called the **saturation region**. This area of operation is where maximum I_C flows. A transistor operating in this region responds as a closed switch. A transistor is considered to be fully conductive in the saturation region. The resistance between emitter and collector is extremely small when a transistor is saturated.

Figure 6.13 shows how a transistor responds in the three regions of operation. In **Figure 6.13(a)**, the transistor is operating in the **active** region. The value of V_{CE} is somewhat less than that of the source voltage V_{CC}. Collector current through the transistor causes a voltage drop across R_C. The base current of a transistor operating in this region would be some moderate

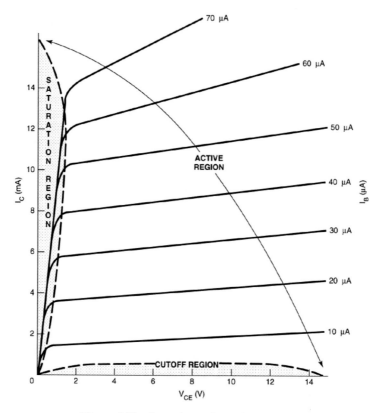

Figure 6.12   Operating regions of a transistor.

value between zero and saturation. **Transistor amplifiers** usually operate in this area.

Figure 6.13(b) shows how a transistor responds when it is in the **cutoff region**. In this condition of operation, the transistor responds as an open switch. With no current flow through the transistor, there is no voltage drop across R_C. V_{CE} of this circuit is, therefore, the same as the source voltage ($V_{CE} = V_{CC}$). A transistor is cut off when no I_C flows.

Figure 6.13(c) shows how a transistor operates in the **saturation region**. In this condition of operation, the transistor responds as a closed switch. V_{CE} is approximately equal to 0 V. Heavy I_C values cause nearly all of V_{CC} to appear across R_C. In a sense, V_{CE} is considered to be at ground. Current passing through the transistor is limited by the value of V_{CC} and R_C. A large value of I_B is needed to cause a transistor to saturate.

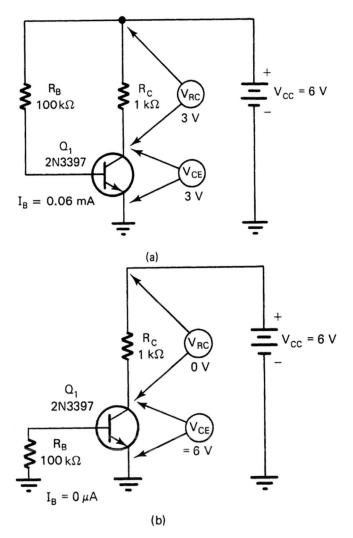

(a)

(b)

Figure 6.13 *Continued.*

Developing Characteristic Curves

Special circuits are used to obtain the data for a **collector family of charac-teristic curves**. **Figure 6.14** shows a test circuit that is used to find the data points of a collector family of characteristic curves. Three meters are used to monitor this information. **Base current** and **collector current** are observed by meters connected in series with the base and collector. **Collector–emitter**

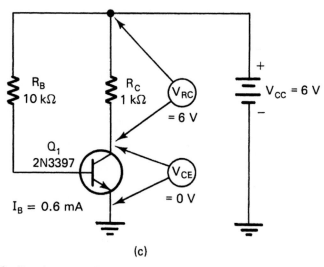

(c)

Figure 6.13   Transistor operation by regions. (a) Active region. (b) Cutoff region. (c) Saturation region.

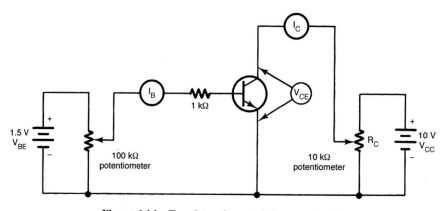

Figure 6.14   Transistor characteristic curve circuit.

voltage is measured with a voltmeter connected across these two terminals. V_{CE} is adjusted to different values by a variable power supply.

The data points for a single curve are developed by first adjusting I_B to a constant value. A representative value for a **small-signal transistor** is 10 μA. Resistor R_B is used to adjust the value of I_B. The V_{CE} is then adjusted through its range, starting at 0 V. As V_{CE} is increased in 0.1-V steps to 1 V, the I_C increases very quickly. Corresponding I_C and V_{CE} values are plotted on the graph. V_{CE} is then increased in 1-V steps while I_C is monitored. The resulting collector current usually levels off to a fairly constant value. The corresponding I_C and V_{CE} values are then added to the first data points. All points are then connected by a continuous line. The completed curve would be labeled 10 μA of I_B.

To develop the second curve, V_{CE} is first returned to zero. I_B is then increased to a new value. A common value for a small-signal transistor would be 20 μA. The second curve is then plotted by recording the I_C values for each V_{CE} value.

To obtain a complete family of curves, the process is repeated for several other I_B values. A representative family of curves may have 8–10 different I_B values. The step values of V_{CE} and the increments of I_B and I_C will change with different transistors. Small-signal transistors may use 10-μA I_B steps with I_C values ranging from 0 to 20 mA. Large-signal transistors may use 1-mA I_B steps with I_C values going to 100 mA or more. Each transistor may require different values of I_B, I_C, and V_{CE} to obtain a suitable curve.

Example 6-5:

Create a family of V_{CE} - I_C characteristic curves for the values of I_B = 10 and 20 μA, for the transistor shown in circuit **Figure 6.14**.

Solution

With I_B = 10 μA		With I_B = 20 μA	
V_{CE} (V)	I_C (mA)	V_{CE} (V)	I_C (mA)
0.1 V		0.1 V	
0.2 V		0.2 V	
0.4 V		0.4 V	
0.6 V		0.6 V	
0.8 V		0.8 V	
1.0 V		1.0 V	
3.0 V		3.0 V	
5.0 V		5.0 V	

Related Problem

Complete the family of V_{CE} - I_C characteristic curves for the values of I_B = 30 and 40 μA, for the transistor shown in circuit **Figure 6.14**. Plot these values on the graph given above.

With $I_B = 30\ \mu A$			With $I_B = 40\ \mu A$	
V_{CE} (V)	I_C (mA)		V_{CE} (V)	I_C (mA)
0.1 V			0.1 V	
0.2 V			0.2 V	
0.4 V			0.4 V	
0.6 V			0.6 V	
0.8 V			0.8 V	
1.0 V			1.0 V	
3.0 V			3.0 V	
5.0 V			5.0 V	

Self-Examination

Answer the following questions.

26. A collector family of characteristic curves for a bipolar transistor has 8–10 individual lines that represent different values of _____ current.
27. The center area of a collector family of characteristic curves is called the _____ region.
28. The area of a collector family of characteristic curves that shows where maximum I_C flows represents the _____ region.
29. The area of a collector family of characteristic curves that show where no I_C occurs is called the _____ region.
30. When a transistor operates in the active region with a moderate value of I_B, the V_{CE} will be (greater, less) _____ than the source voltage V_{CC}.
31. When a transistor operates in the cutoff region, the V_{CE} will be equal to the _____ voltage.
32. When a transistor operates in the saturation region, the V_{CE} will be approximately _____.

6.4 Testing BJTs

Bipolar junction transistor testing is a procedure that is performed periodically when working with semiconductor devices. Lead identification, gain, open leads, shorted conditions, and leakage are some of the tests that can be performed on a transistor. Curve tracers, gain testers, and sound-producing instruments can be used for these tests. These instruments, as a general rule, are not always available. An ohmmeter, however, can be used to perform many of these tests. An ohmmeter is one of the functions of a volt-ohm-milliammeter (VOM) or digital volt-ohm-meter (DVOM). In this section, you will learn how to use an ohmmeter to test a bipolar junction transistor and identify its leads.

6.4 Evaluate the condition of a bipolar junction transistor.

In order to achieve objective 6.4, you should be able to:

- use an ohmmeter to test a bipolar junction transistor and identify its leads;
- use, an ohmmeter to evaluate the forward- and reverse-biased junctions of a transistor.

Junction Testing

A transistor has two P–N junctions in its construction. Both of these junctions can be tested with an **ohmmeter**. **Figure 6.15** shows the junctions of PNP and NPN transistors. Each junction responds as a diode when it is tested. **Forward biasing** occurs when the material polarity of the junction matches the voltage polarity of the ohmmeter. When the ohmmeter polarity is reversed, the same junction becomes **reverse biased**. Recall that forward biasing of a diode indicates a low resistance reading, while reverse biasing indicates an extremely high resistance reading. Both junctions of a transistor will respond to this test in the same way if the transistor is good.

A faulty P–N junction does not show a difference in its **forward** and **reverse resistance** connections. Low resistance in both directions indicates that a junction is shorted. Excessive current generally causes this type of failure in a transistor. Infinite or exceedingly high resistance in both directions indicates an open condition. An open junction is normally caused by a broken internal connection. This may be the result of a current overload or excessive shock. Open and shorted transistor failures, as a general rule, occur suddenly and cannot be corrected.

A rather quick test of both junctions can be made simultaneously with the **ohmmeter**. Connect one ohmmeter probe to the emitter and the other to the collector. Ohmmeter probe polarity in this case is not important. This connection should cause an extremely high resistance reading. Then reverse the two ohmmeter probes. High resistance should also occur in this direction. A good silicon transistor will show an infinite resistance in either direction. Any measurable resistance in either direction indicates leakage. Germanium transistors will show some leakage in this test. Unless the leakage is excessive, it can be tolerated in a germanium transistor. An ideal transistor would have no indication of emitter–collector leakage.

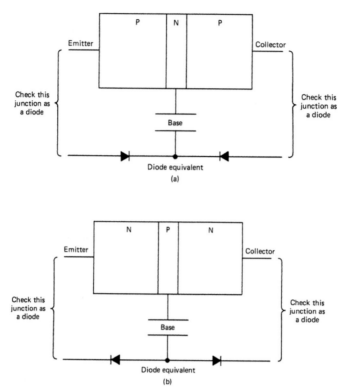

Figure 6.15   Junction polarity of PNP and NPN transistors.

Lead Identification

An **ohmmeter** can also be used to identify the leads of a transistor. The **polarity** of the ohmmeter voltage source should be checked for this test to be meaningful. Straight-polarity ohmmeters have the black or **common probe** negative and the red probe **positive**. Reverse polarity ohmmeters would be connected in the opposite direction for these tests. The polarity of an ohmmeter can be tested with a separate DC voltmeter if it is unknown. The ohmmeter used in this explanation has straight polarity (black-negative and red-positive).

Base Identification

To identify transistor leads, inspect the lead location of the device under test. Pick out the center lead of the transistor. Assume that it is the base lead.

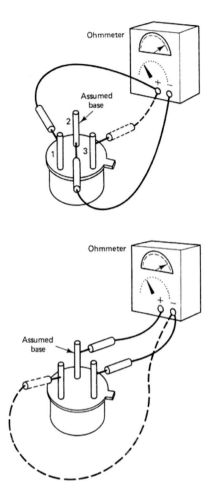

Figure 6.16 Ohmmeter transistor testing.

Connect the negative probe of the ohmmeter to it as shown in **Figure 6.16**. Then, alternately touch the positive ohmmeter probe to the two outside transistor leads. If a **low resistance** indication occurs for each lead, the center lead is actually the base. The test also indicates that the transistor is the **PNP** type. If the resistance is high between the center lead and the two outside leads, reverse the meter polarity. The positive ohmmeter probe should now be connected to the assumed base. Once again, alternately switch the negative ohmmeter probe between the two outside transistor leads. If a low resistance reading is obtained, the transistor is an **NPN** type.

If the center lead does not produce low resistance in either of the two conditions, it is not the base. Then select one of the outside leads as the assumed base. Try the same procedure again with the newly assumed base. If this does not produce results, try the other outside lead. One of the three leads must respond as the base if the transistor is good. If a response cannot be obtained, the transistor must be open or shorted.

Gain Test and Collector and Emitter Identification

Thus far, we have identified the base lead and determined the polarity of the transistor to be NPN or PNP. It is now possible to test the gain and to identify the remaining two leads. A 100,000-Ω resistor and an ohmmeter are needed for this part of the test. The resistor is used to provide base current from the ohmmeter to the transistor. If a power transistor is tested, use the R meter range and a 1000-Ω base resistor.

The process of troubleshooting transistor gain should be more meaningful to you if you know how this test works. **Figure 6.17** shows how PNP and NPN transistors respond to the ohmmeter test. The energy source of the ohmmeter is used to supply the bias voltage to each of the transistor elements. For the **PNP transistor** of **Figure 6.17(a)**, the positive ohmmeter probe is connected to the emitter and the negative lead to the collector. When a resistor is connected between the collector and the base, it causes base current. With the emitter and base forward biased and the collector reverse biased, the transistor becomes low resistant. The ohmmeter responds to this condition by showing a change in resistance. The resistance of the transistor decreases in value. A low resistance reading on the ohmmeter indicates the transistor has gain and correct lead selection.

The ohmmeter test of an **NPN transistor** is shown in **Figure 6.17(b)**. For this type of transistor, the emitter is connected to the negative ohmmeter probe and the collector to the positive ohmmeter probe. When a resistor is connected between the collector–base, it causes the base current to flow. With the emitter–base forward biased and base–collector reverse biased, the transistor becomes low resistant. A low resistance reading on the ohmmeter indicates the transistor has gain and correct lead selection.

The procedure for checking an **NPN transistor** is shown in **Figure 6.18**. Note that the ohmmeter is connected to the two outside leads. The center lead has been previously identified as the base. If another lead were found to be the base, it would be used in place of the center lead. One end of the resistor is connected to the base with the other lead open. In this case, it is assumed that

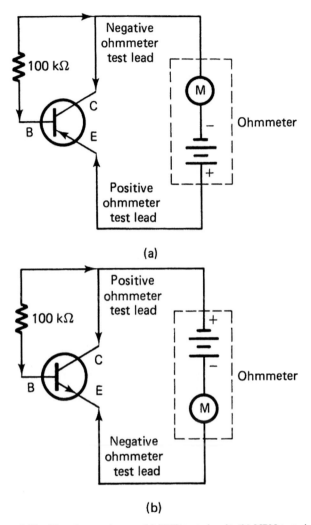

Figure 6.17 Transistor gain test. (a) PNP test circuit. (b) NPN test circuit.

the negative ohmmeter probe is connected to the emitter and the positive lead to the collector. If the assumed leads are correct, the emitter will be forward biased, and the collector is reverse biased. Touching the open end of the base resistor to the positive ohmmeter probe will cause a base current flow. If it does, the ohmmeter will indicate a **low resistance**. In effect, the emitter–base junction is forward biased, and the base–collector junction is reverse biased. If no base current flows, the assumed emitter–collector leads are reversed.

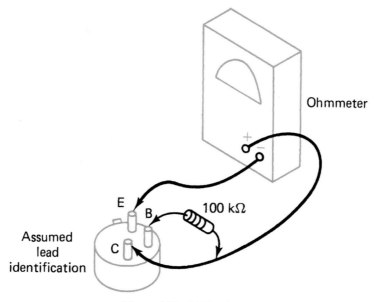

Figure 6.18   NPN gain test.

Simply reverse the ohmmeter's probes and again touch the resistor lead to the positive ohmmeter probe. If the transistor is good, and the assumed leads are correct, the ohmmeter will show a low resistance. If it does not, the transistor has a low gain. In some transistors, low gain is permissible. As a rule, low gain is an indication of some type of transistor problem. The ohmmeter test of transistor gain is only a close approximation of the status of the device.

Figure 6.19 shows the procedure for testing a **PNP transistor**. In this case, the ohmmeter is connected to the two outside leads. We have identified the center lead as the base. The base resistor is now connected to this lead. For a PNP transistor, the positive **ohmmeter** lead goes to the emitter and the negative lead to the collector. If the assumed leads are correct, touching the open end of the base resistor to the negative lead will cause the base current to flow. If it does, the ohmmeter will show a low resistance reading. This indicates that the emitter–base is forward biased, and the base–collector reverse biased as in the circuit. If no current flows, the assumed emitter–collector leads are reversed. Reverse the two ohmmeter probes, and again touch the base resistor to the negative ohmmeter probe. A good transistor with correct lead identification will indicate low resistance. A transistor with low gain will not cause much of a change in resistance.

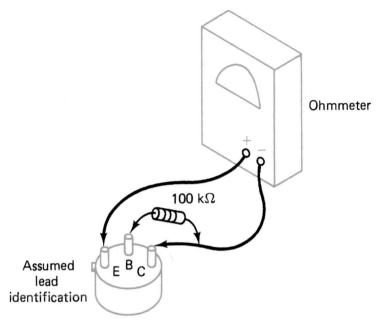

Figure 6.19 PNP gain test.

Self-Examination

Answer the following questions.

33. Each P–N junction of a good transistor will respond as a(n) _____ when tested with an ohmmeter.
34. A(n) _____ P–N junction will not show a difference in its forward and reverse resistance connections.
35. When an ohmmeter is connected between the emitter–collector leads of a good bipolar transistor, it will show (high, low) resistance in either polarity.
36. A good bipolar transistor is connected so that the positive ohmmeter lead goes to the base and the negative lead is switched alternately between the other two leads. If this test shows low resistance between each of the alternate leads, the device is (NPN, PNP).
37. If a bipolar transistor shows low resistance between any two leads in either direction of ohmmeter polarity, it indicates that the device is _____.
38. Infinite or exceedingly high resistance in both directions across either P–N junction of a bipolar transistor indicates that the junction is _____.

6.5 Analysis and Troubleshooting – BJTs

It is important to know the operating **specifications** of a bipolar junction transistor before using it in an electronic circuit. The operating specifications of NPN and PNP BJTs can be obtained by examining the manufacturer **datasheets**.

Data Sheet Analysis

Data sheets for a general purpose **NPN transistor 2N3904** in included at the end of the chapter. Use the data sheets to answer the following:

1. Collector–emitter voltage (V_{CEO}) =
2. Collector–base voltage (V_{CBO}) =
3. Emitter–base voltage (V_{EBO}) =
4. Maximum continuous collector current (I_C) =
5. Total power dissipation (P_D) =
6. DC current gain (h_{FE}) =
7. Collector–emitter saturation voltage ($V_{CE(sat)}$) =
8. Base–emitter saturation voltage ($V_{BE(sat)}$) =

    Data sheets for a general purpose **PNP transistor** 2N4403. Use the data sheets to answer the following:

9. Collector–emitter voltage (V_{CEO}) =
10. Collector–base voltage (V_{CBO}) =
11. Emitter–base voltage (V_{EBO}) =
12. Maximum continuous collector current (I_C) =
13. Total power dissipation (P_D) =
14. DC current gain (h_{FE}) =
15. Collector–emitter saturation voltage ($V_{CE(sat)}$) =
16. Base–emitter saturation voltage ($V_{BE(sat)}$) =

BJT Troubleshooting

The operation of bipolar junction transistors can be tested with an analog or digital ohmmeter. This essentially involves connecting the ohmmeter to two of the three BJT terminals (collector, base, and emitter) and evaluating the resistance. When the ohmmeter is connected across a pair of BJT terminals, it can be either forward or reverse biased. Recall that when a P–N junction is forward biased, it has a low resistance, and a high resistance when it is reverse biased. The polarity of the ohmmeter should be carefully observed. Analog ohmmeters, in general, would be better for evaluating the condition of a BJT. This is because analog meters provide a higher current flow to the circuit being tested.

Typical faults in a BJT involve primarily **short circuits**, and **open circuits** between the terminals. A short circuit between two terminals is indicated by a low resistance (0 Ω) reading in either direction. An open circuit between two terminals is indicated by a high resistance reading in either direction.

Another problem that occurs with BJTs is the development of heat during its operation. It is important to operate the BJT below its **maximum power dissipation rating**.

Summary

- Bipolar junction transistors are three-element devices made of semiconductor materials.
- Bipolar junction transistors have two P–N junctions.
- Current flow is controlled in a bipolar junction transistor by altering the voltage applied to the two P–N junctions.
- A bipolar transistor has a thin layer of doped semiconductor material placed between two layers of doped semiconductor material of the opposite polarity.
- The designations NPN and PNP denote the polarity of the semiconductor material used in its construction.
- A bipolar junction transistor is represented by a schematic symbol that has three leads: emitter, base, and collector.
- When the arrow in a bipolar junction transistor symbol points toward the base (a straight line), it indicates that the transistor is a PNP type.
- When the arrow in a bipolar junction transistor symbol points away from the base, the symbol designates an NPN device.
- The point-contact method of BJT construction has semiconductor materials connected together by pointed wires that are fused to the material.
- Alloy-junction BJTs are formed by attaching two small pieces of metal on opposite sides of a thin piece of semiconductor material.
- The diffusion technique of manufacturing BJTs involves the movement of N-type and P-type impurity atoms into a piece of silicon.
- For a transistor to function, it must have electrical energy applied to its electrodes; the emitter–base junction is forward biased, and the collector–base junction is reverse biased.

- A forward-biased emitter–base junction causes a large amount of current to flow into the base region.
- The term *alpha (α)* is defined as the ratio I_C/I_E.
- The term *beta (β)* is an expression of current gain and is defined as I_C/I_B.
- Transistor testing, lead identification, and polarity of the material from which it is constructed can be determined with an ohmmeter.
- The two junctions of a transistor are tested as diodes.
- Lead and material polarity identification of a transistor is achieved by using the ohmmeter's voltage source to bias the transistor into operation.
- When identifying the lead and material polarity of a transistor, a base lead is assumed; one ohmmeter lead is attached to this lead and the other lead is switched between the two remaining leads.

Formulas

(6-1) $I_E = I_B + I_C$ Emitter current.
(6-2) $\alpha = I_C/I_E$
(6-3) $\beta_{dc} = I_C/I_B$ DC beta.
(6-4) $\beta_{ac} = \Delta I_C/\Delta I_B$ AC beta.

Review Questions

Answer the following questions.

1. The largest current flow of a bipolar transistor occurs:
   a. In the emitter
   b. In the base
   c. In the collector
   d. Through the collector–base
   e. Through the emitter–collector

2. Conventional biasing of a bipolar transistor has the:
   a. Emitter–base forward biased and the collector–base forward biased
   b. Emitter–base reverse biased and the collector–base forward biased
   c. Emitter–base forward biased and the collector–base reverse biased
   d. Emitter–base reverse biased and the collector–base reverse biased

3. The beta of a bipolar transistor is an expression of:

    a. Base current/collector current
    b. Collector current/base current
    c. Collector current/emitter current
    d. Emitter current/base current
    e. Base current/emitter current

4. When a bipolar transistor is saturated:

    a. V_{CE} is very low, and I_C is zero
    b. V_{CE} is low, and I_C is high
    c. V_{CE} equals the source voltage, and I_C is zero
    d. V_{CE} equals the source voltage, and I_C is high

5. When a bipolar transistor is cut off:

    a. V_{CE} equals the source voltage and I_C is high
    b. V_{CE} is low and I_C is high
    c. V_{CE} equals the source voltage and I_C is zero
    d. V_{CE} is high and I_C is low

6. In a properly connected bipolar transistor, an increase in base current will cause an increase in:

    a. Collector current only
    b. Emitter current only
    c. Emitter current and collector current
    d. Leakage current

7. The polarity of the collector voltage of an NPN bipolar transistor is:

    a. Positive
    b. Negative
    c. The same polarity as the emitter
    d. The same polarity as the base

8. When a bipolar transistor responds as an amplifier, it:

    a. Operates in the cutoff region
    b. Operates in the saturation region
    c. Operates in the active region
    d. Switches between the saturation and cutoff regions

9. If a bipolar transistor shows low resistance between any two leads in either direction of ohmmeter polarity, it indicates that the device is:

    a. Shorted

b. Open
c. Leaky
d. Normal

10. If a bipolar transistor shows an infinite or exceedingly high resistance across either P–N junction, it indicates that the junction is:

a. Shorted
b. Open
c. Leaky
d. Normal

Problems

Answer the following questions.

1. A BJT has a collector current, I_C = 25 mA, and a base current, I_B = 125 μA. Determine the value of the emitter current, I_E?
2. A BJT has a collector current, I_C = 50 mA, and β = 400. What is the base current, I_B?
3. A bipolar transistor has a collector current of 9 mA and a base current of 150 μA. In which of the following ranges will the emitter current be found?

a. $1-150\ \mu$A
b. $151-915\ \mu$A
c. $916\ \mu$A to 1.5 mA
d. $1.51-9.15$ mA
e. $9.16-91.5$ mA

4. When a ohmmeter is connected across the emitter–collector junctions of a good bipolar transistor, it will show:

a. Low resistance in both directions of ohmmeter polarity
b. Low resistance in one direction and high resistance in the other
c. Some resistance in both directions of ohmmeter polarity
d. High resistance in both directions of ohmmeter polarity

Glossary

Bipolar

Refers to the conduction by both holes and electrons in bipolar junction transistors.

PNP transistor

A transistor that has a thin layer of N-type material placed between two pieces of P-type material.

Emitter

A section of a bipolar transistor that is responsible for the release of majority current carriers.

Base

A thin layer of semiconductor material between the emitter and collector of a bipolar transistor.

Collector

A section of a bipolar transistor that collects majority current carriers.

NPN transistor

A transistor that has a thin layer of P-type material placed between two pieces of N-type material.

Epitaxial growth

A transistor fabrication technique in which atoms are formed on a surface so that they are an extension of the original crystal structure.

Planar transistor

When the epitaxial growth process is used to create a transistor and the completed structure has a flat top or level plane surface.

Mesa transistor

When the epitaxial process is used to create a transistor and the completed structure rises above the primary surface, forming a plateau.

Gain

A ratio of voltage, current, or power output to corresponding input values.

Beta

A designation of transistor current gain determined by I_C/I_B.

Active region

An area of transistor operation between cutoff and saturation.

Cutoff region

A condition or region of transistor operation in which current carriers cease or diminish.

Saturation region

A condition or region of transistor operation in which a device conducts to its fullest capacity.

FAIRCHILD
SEMICONDUCTOR ™

2N3904   MMBT3904   PZT3904

TO-92

SOT-23
Mark: 1A

SOT-223

NPN General Purpose Amplifier

This device is designed as a general purpose amplifier and switch.
The useful dynamic range extends to 100 mA as a switch and to
100 MHz as an amplifier.

Absolute Maximum Ratings* T_A = 25°C unless otherwise noted

Symbol	Parameter	Value	Units
V_{CEO}	Collector-Emitter Voltage	40	V
V_{CBO}	Collector-Base Voltage	60	V
V_{EBO}	Emitter-Base Voltage	6.0	V
I_C	Collector Current - Continuous	200	mA
T_J, T_{stg}	Operating and Storage Junction Temperature Range	-55 to +150	°C

* These ratings are limiting values above which the serviceability of any semiconductor device may be impaired.

NOTES:
1) These ratings are based on a maximum junction temperature of 150 degrees C.
2) These are steady state limits. The factory should be consulted on applications involving pulsed or low duty cycle operations.

Thermal Characteristics T_A = 25°C unless otherwise noted

Symbol	Characteristic	Max			Units
		2N3904	*MMBT3904	**PZT3904	
P_D	Total Device Dissipation	625	350	1,000	mW
	Derate above 25°C	5.0	2.8	8.0	mW/°C
$R_{\theta JC}$	Thermal Resistance, Junction to Case	83.3			°C/W
$R_{\theta JA}$	Thermal Resistance, Junction to Ambient	200	357	125	°C/W

* Device mounted on FR-4 PCB 1.6" X 1.6" X 0.06".

** Device mounted on FR-4 PCB 36 mm X 18 mm X 1.5 mm, mounting pad for the collector lead min. 6 cm².

PNP Data Sheets

2N3904 / MMBT3904 / PZT3904

NPN General Purpose Amplifier

(continued)

Electrical Characteristics $T_A = 25°C$ unless otherwise noted

Symbol	Parameter	Test Conditions	Min	Max	Units
OFF CHARACTERISTICS					
$V_{(BR)CEO}$	Collector-Emitter Breakdown Voltage	$I_C = 1.0$ mA, $I_B = 0$	40		V
$V_{(BR)CBO}$	Collector-Base Breakdown Voltage	$I_C = 10$ µA, $I_E = 0$	60		V
$V_{(BR)EBO}$	Emitter-Base Breakdown Voltage	$I_E = 10$ µA, $I_C = 0$	6.0		V
I_{BL}	Base Cutoff Current	$V_{CE} = 30$ V, $V_{EB} = 3$V		50	nA
I_{CEX}	Collector Cutoff Current	$V_{CE} = 30$ V, $V_{EB} = 3$V		50	nA
ON CHARACTERISTICS*					
h_{FE}	DC Current Gain	$I_C = 0.1$ mA, $V_{CE} = 1.0$ V	40		
		$I_C = 1.0$ mA, $V_{CE} = 1.0$ V	70		
		$I_C = 10$ mA, $V_{CE} = 1.0$ V	100	300	
		$I_C = 50$ mA, $V_{CE} = 1.0$ V	60		
		$I_C = 100$ mA, $V_{CE} = 1.0$ V	30		
$V_{CE(sat)}$	Collector-Emitter Saturation Voltage	$I_C = 10$ mA, $I_B = 1.0$ mA		0.2	V
		$I_C = 50$ mA, $I_B = 5.0$ mA		0.3	V
$V_{BE(sat)}$	Base-Emitter Saturation Voltage	$I_C = 10$ mA, $I_B = 1.0$ mA	0.65	0.85	V
		$I_C = 50$ mA, $I_B = 5.0$ mA		0.95	V
SMALL SIGNAL CHARACTERISTICS					
f_T	Current Gain - Bandwidth Product	$I_C = 10$ mA, $V_{CE} = 20$ V, $f = 100$ MHz	300		MHz
C_{obo}	Output Capacitance	$V_{CB} = 5.0$ V, $I_E = 0$, $f = 1.0$ MHz		4.0	pF
C_{ibo}	Input Capacitance	$V_{EB} = 0.5$ V, $I_C = 0$, $f = 1.0$ MHz		8.0	pF
NF	Noise Figure	$I_C = 100$ µA, $V_{CE} = 5.0$ V, $R_S = 1.0$kΩ,$f = 10$ Hz to 15.7kHz		5.0	dB
SWITCHING CHARACTERISTICS					
t_d	Delay Time	$V_{CC} = 3.0$ V, $V_{BE} = 0.5$ V,		35	ns
t_r	Rise Time	$I_C = 10$ mA, $I_{B1} = 1.0$ mA .		35	ns
t_s	Storage Time	$V_{CC} = 3.0$ V, $I_C = 10$mA		200	ns
t_f	Fall Time	$I_{B1} = I_{B2} = 1.0$ mA		50	ns

*Pulse Test: Pulse Width ≤ 300 µs, Duty Cycle ≤ 2.0%

Spice Model

NPN (Is=6.734f Xti=3 Eg=1.11 Vaf=74.03 Bf=416.4 Ne=1.259 Ise=6.734 Ikf=66.78m Xtb=1.5 Br=.7371 Nc=2 Isc=0 Ikr=0 Rc=1 Cjc=3.638p Mjc=.3085 Vjc=.75 Fc=.5 Cje=4.493p Mje=.2593 Vje=.75 Tr=239.5n Tf=301.2p Itf=.4 Vtf=4 Xtf=2 Rb=10)

2N4403

C
B
E   TO-92

MMBT4403

C

E

SOT-23   B
Mark: 2T

PNP General Purpose Amplifier

This device is designed for use as a general purpose amplifier
and switch requiring collector currents to 500 mA.

Absolute Maximum Ratings*   TA = 25°C unless otherwise noted

Symbol	Parameter	Value	Units
V_{CEO}	Collector-Emitter Voltage	40	V
V_{CBO}	Collector-Base Voltage	40	V
V_{EBO}	Emitter-Base Voltage	5.0	V
I_C	Collector Current - Continuous	600	mA
T_J, T_{stg}	Operating and Storage Junction Temperature Range	-55 to +150	°C

*These ratings are limiting values above which the serviceability of any semiconductor device may be impaired.

NOTES:
1) These ratings are based on a maximum junction temperature of 150 degrees C.
2) These are steady state limits. The factory should be consulted on applications involving pulsed or low duty cycle operations.

Thermal Characteristics   TA = 25°C unless otherwise noted

Symbol	Characteristic	Max		Units
		2N4403	*MMBT4403	
P_D	Total Device Dissipation	625	350	mW
	Derate above 25°C	5.0	2.8	mW/°C
$R_{\theta JC}$	Thermal Resistance, Junction to Case	83.3		°C/W
$R_{\theta JA}$	Thermal Resistance, Junction to Ambient	200	357	°C/W

*Device mounted on FR-4 PCB 1.6" X 1.6" X 0.06."

7

Bipolar Transistor Amplification

Electronic systems must perform a variety of basic functions to accomplish a particular operation. Understanding how these functions are performed is vital to becoming a successful electronics technician or electronics engineer. In this chapter, you will study the amplification function. Amplification is achieved by devices that produce an increase in signal amplitude. Bipolar junction transistors, with which you are already familiar, are frequently used for this purpose.

The first part of this chapter covers amplification in general. Then, how the bipolar junction transistor achieves amplification is discussed. By the end of this chapter, you will have learned about circuit operation, biasing methods, circuit configurations, classes of amplification, and operational conditions. Keep in mind as you read this chapter that it is directed toward small-signal amplification. Power amplification is achieved in the same manner, with the only difference being the amount of current flow controlled. Power amplification is covered in Chapter 11 − Amplifying Systems.

Objectives

After studying this chapter, you will be able to:

7.1 explain the meaning of amplification with respect to voltage, current, and power;

7.2 analyze the operation of a basic amplifier;

7-3 analyze the operation of a basic amplifier by the load-line method;

7.4 analyze common-emitter, common-base, and common-collector circuit configurations;

7.5 analyze and troubleshoot a BJT amplifier circuit.

Chapter Outline

7.1 Amplification Principles
7.2 Basic BJT Amplifiers
7.3 Load-Line Analysis
7.4 Transistor Circuit Configurations
7.5 Analyzing and Troubleshooting of BJT Amplifier Circuits
7-1 Transistor Load-Line Analysis

In this activity, the DC operating conditions of a bipolar junction transistor will be evaluated. With this information, a transistor amplifier can be easily designed, and its circuit operation can be predicted.

7-2 DC Transistor Amplifier

In this activity, a test circuit is evaluated for the amplification of DC signals applied to a bipolar junction transistor.

7-3 AC Transistor Amplifier

In this activity, a test circuit is constructed for evaluating the amplification of AC signals applied to a bipolar junction transistor.

Key Terms

alpha
bypass capacitor
current amplifier
current gain
emitter biasing
fixed biasing
linear
load line
nonlinear distortion
overdriving
power gain
Q point
self-biasing
self-emitter bias
static state
thermal stability

voltage amplifier
voltage gain

7.1 Amplification Principles

To understand the operation of amplifiers and the purpose of various amplifier configurations, you need to be familiar with some basic amplification principles. Amplification is similar to the step-up capability of a transformer in that the input signal is increased by a certain ratio. In both cases, the amplification of voltage, current, and power can be calculated through a series of ratio formulas. In this section, you will learn about the differences between reproduction and amplification and will use several ratio formulas to determine the voltage, current, and power amplification of an amplifier.

7.1 Explain the meaning of amplification with respect to voltage, current, and power.

In order to achieve objective 7.1, you should be able to:

- explain the difference between signal reproduction and signal amplification;
- identify the formulas for determining voltage gain, current gain, and power gain;
- define *voltage amplifier, voltage gain, current amplifier, current gain,* and *power gain.*

Reproduction and Amplification

Amplification is achieved by an amplifying device and its associated components. As a general rule, the amplifying device is placed in a circuit. The components of the circuit usually have about as much influence on amplification as the amplifying device. The circuit and the amplifying device must be supplied electrical energy to function. Typically, amplifiers are energized by direct current. This may be derived from a battery or an electronic power supply. The amplifier then processes a signal of some type when it is placed in operation. The signal may be either AC or DC, depending on the application.

The signal to be processed by an amplifier is first applied to the input part of the circuit. After being processed, the signal appears in the output circuitry. The output signal may be reproduced in its exact form, amplified, or both amplified and reproduced. The value and type of input signal, operating source energy, device characteristics, and circuit components influence the output signal.

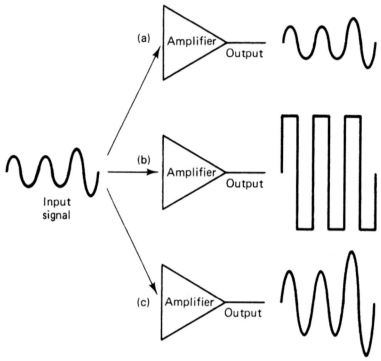

Figure 7.1   Amplification and reproduction. (a) Reproduction. (b) Amplification. (c) Reproduction.

Figure 7.1 illustrates the process of reproduction, amplification, and the combined amplification-reproduction functions of an amplifier. The amplifier at the top of **Figure 7.1** performs only the reproduction function. Note that the input and output signals have the same size and shape. The amplifier in the middle shows only the amplification function. In this case, the input signal is increased in amplitude. The output signal is amplified but does not necessarily resemble the input signal. In many applications, amplification and reproduction must be achieved simultaneously. This function is shown by the amplifier at the bottom of **Figure 7.1**. Depending on the application, an amplifier must be capable of developing any of these three output signals.

Voltage Amplification

A system designed to develop an output voltage that is greater than its input voltage is called a **voltage amplifier**. A voltage amplifier has **voltage gain**, or

voltage amplification, which is defined as a ratio of the output signal voltage to the input signal voltage. In equation form, voltage amplification (A_V) is expressed as

$$A_V = V_{out}/V_{in} \qquad (7.1)$$

or

$$A_V = \Delta V_{out}/\Delta V_{in}. \qquad (7.2)$$

The uppercase letters in V_{out} and V_{in} denote DC voltage values. AC voltages are often expressed by lowercase letters. The Greek capital letter delta (Δ) indicates a changing value. Voltage amplifiers are capable of a wide range of amplification values.

Current Amplification

An amplifying system designed to develop output current that is greater than the input current is called a **current amplifier**. A current amplifier has current gain, which is defined as the ratio of output current to input current. In equation form, **current gain**, or **current amplification**, is expressed as

$$A_i = I_{out}/I_{in} \qquad (7.3)$$

or

$$A_i = \Delta I_{out}/\Delta I_{in}. \qquad (7.4)$$

Current gain takes into account the beta of a transistor and all the associated components that make the device operational.

Power Amplification

Power gain, or **power amplification**, is a ratio of the developed output signal power to the input signal power. Note that this refers only to the power gain of the signal. All amplifiers consume a certain amount of power from the energy source during operation. This is usually not included in an expression of power gain. An equation of power gain shows that power amplification (A_P) equals output signal power (P_{out}) divided by the input signal power (P_{in}):

$$A_P = P_{out}/P_{in}. \qquad (7.5)$$

Power signal gain can also be expressed as the product of voltage gain and current gain. In this regard, the equation is

$$A_P = A_v \times A_i. \qquad (7.6)$$

It should be apparent from the A_P equation that A_v and A_i do not both need to be large to have power gain. Typical power amplifiers may have a large current gain and a voltage gain less than 1. Power gain must take into account both signal voltage and current gain to be meaningful.

Self-Examination

Answer the following questions.

1. The signal to be processed by an amplifier is first applied to the _____ part of the circuit.
2. After an input signal has been processed by an amplifier, it appears in the _____ circuitry.
3. _____ occurs when the input signal retains its size and shape in the output circuitry.
4. _____ occurs when the input signal is increased in amplitude in the output circuitry.
5. An expression of output voltage divided by input voltage is _____ amplification.
6. An expression of output current divided by input current is _____ amplification.
7. An expression of output signal power divided by input signal power is _____ amplification.

7.2 Basic BJT Amplifiers

The primary function of an amplifier is to provide voltage gain, current gain, or power gain. The resulting gain depends a great deal on the device used, its circuit components, and its circuit configuration. Proper component selection and circuit bias are needed to maintain a stable operating point. A stable operating point ensures that amplifier output will not enter cutoff or saturation due to changes in temperature. In this section, you will learn about the basic circuitry and operation of an amplifier circuit. You will also learn about several different methods of achieving bias, while maintaining a stable operating point.

7.2 Analyze the operation of a basic amplifier.

In order to achieve objective 7.2, you should be able to:

- calculate the base current, collector current, load resistor voltage, and collector–emitter voltage of a basic amplifier;

- explain how a bipolar junction transistor achieves amplification of an AC signal;
- explain the relationship between static-state operation and signal amplification;
- identify beta-dependent and beta-independent biasing methods;
- define *static state, thermal stability, fixed biasing, self-biasing, emitter biasing, bypass capacitor*, and *self-emitter bias*.

For a **bipolar junction transistor** to respond as an amplifier, the emitter−base junction must be **forward biased**, and the collector−base junction must be **reverse biased**. The same criteria are used to make a transistor operational. Specific operating voltage values must be selected that will permit amplification.

Normally, an amplifier circuit is energized by DC voltage. An AC signal is then applied to the input of the amplifier. After passing through the transistor, an amplified version of the signal appears in the output. Operating conditions of the transistor and the circuit determine the level of amplification to be achieved.

If both reproduction and amplification are to be achieved, the device must operate in the center of its active region. Remember that this is between the saturation and cutoff regions of the collector family of characteristic curves. Proper circuit component selection permits a transistor to operate in the active region.

Basic Amplifier Circuitry

Look at the circuitry of a basic **amplifier** shown in **Figure 7.2**. Note that a schematic symbol of the NPN transistor is used in this diagram. While you study this circuit, keep in mind the crystal structure of the device represented by the symbol.

The basic amplifier has a number of parts that are needed to make it operational. V_{CC} is the DC energy source. The negative side of V_{CC} is connected to ground. The emitter is also connected to ground. This type of circuit configuration is called a **grounded-emitter amplifier**, or **common-emitter amplifier**. The emitter, one side of the input, and one side of the output are commonly connected together. In an actual circuit, these points are usually connected to ground.

The **base resistor (R_B)** of the basic amplifier is connected to the positive side of V_{CC}. This connection makes the base positive with respect to the emitter. The emitter−base junction is forward biased by R_B. Resistor R_L connects

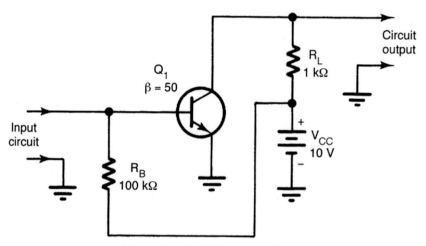

Figure 7.2   Basic bipolar amplifier.

the positive side of V_{CC} to the collector. This connection reverse biases the collector. Through the connection of R_B and R_L, the transistor is properly biased for operation. A transistor connected in this manner is considered to be in its **static state**, or **DC operating state**. It has the necessary DC energy applied to be operational. No signal is applied to the input for amplification.

Static-State Operation

Let us now consider the operation of the basic amplifier in its **static state**. With the given values of **Figure 7.3**, base current (I_B) can be calculated. I_B, in this case, is limited by the value of R_B and the base–emitter **junction resistance**. When the base–emitter junction is forward biased, its resistance becomes very small. The value of R_B is, therefore, the primary limiting factor of I_B. I_B can be determined by an application of Ohm's law:

$$I_B = V_{CC}/R_B. \qquad (7.7)$$

For our basic transistor amplifier, this is determined to be

$$I_B = V_{CC}/R_B$$
$$= 10 \text{ V}/100,000 \text{ }\Omega$$
$$= 0.0001 \text{ A, or } 100 \text{ }\mu\text{ A.}$$

The value of base current, in this case, is extremely small. Remember that only a small amount of I_B is needed to produce I_C.

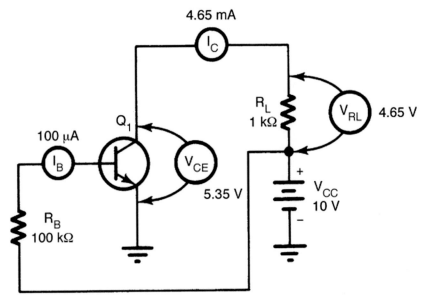

Figure 7.3 Static operating condition of a basic amplifier and calculation of base current (I_B), collector current (I_C), collector–emitter voltage, and load resistor value.

The **beta** of the transistor used in **Figure 7.3** has a given value. With beta and the calculated value of I_B, it is possible to determine the collector current of the circuit. **Beta** is the current gain of a common-emitter amplifier. Recall that beta is described by the following formula:

$$\beta = I_C/I_B.$$

In our transistor amplifier, I_B has been determined and beta has a given value of 50. By transposing the beta formula, I_C can be determined by the following equation:

$$I_C = \beta \times I_B. \tag{7.8}$$

For the amplifier circuit, collector current is determined to be

$$
\begin{aligned}
I_C &= \beta \times I_B \\
&= 50 \ \times 100 \ \mu A \\
&= 50 \ \times 0.0001 \ A \\
&= 0.005 \ A, \text{ or } 5 \ mA.
\end{aligned}
$$

This means that 5 mA of I_C will flow through R_L when an I_B of 100 μ A flows.

In any electric circuit, we know that current flow through a resistor causes a corresponding **voltage drop**. In a basic transistor amplifier, I_C causes a voltage drop across R_L. Take a look at **Figure 7.3**. In the preceding step, I_C was calculated to be 5 mA. By using Ohm's law again, we can determine the voltage drop across R_L from the following equation:

$$V_{RL} = I_C \times R_L. \tag{7.9}$$

For the amplifier circuit, the voltage drop across R_L is

$$
\begin{aligned}
V_{RL} &= I_C \times R_L \\
&= 5\,\text{mA} \times 1\,\text{k}\,\Omega \\
&= 0.005\,\text{A} \times 1000\,\Omega \\
&= 5\,\text{V}.
\end{aligned}
$$

This means that half or 5 V of V_{CC} will appear across R_L. With a V_{CC} value of 10 V, the other 5 V will appear across the collector–emitter of the transistor. This is determined by the following equation:

$$V_{CE} = V_{CC} - V_{RL}. \tag{7.10}$$

Thus,

$$
\begin{aligned}
V_{CE} &= V_{CC} - V_{RL} \\
V_{CE} &= 10 - 5\,\text{V} \\
&= 5\,\text{V}.
\end{aligned}
$$

A V_{CE} voltage of 5 V from a V_{CC} of 10 V means that the transistor is operating near the center of its **active region**. Ideally, the transistor should respond as a linear amplifier. When a suitable signal is applied to the input, it should amplify and reproduce the signal in the output.

The operational voltage and current values of the basic transistor amplifier in its **static state** are summarized in **Figure 7.4**. Note that an I_B of 100 μA causes an I_C of 5 mA. With a beta of 50, the amplifier has the capability of a rather substantial amount of output current. The collector current passing through R_L causes V_{CC} to be divided. This establishes operation in the approximate center of the collector family of characteristic curves. Ideally, this **static** condition should cause the amplifier to respond as a linear device when a signal is applied.

Example 7-1:

Take a look at **Figure 7.3**. From the values provided in **Figure 7.3(a)**, determine base current (I_B), collector current (I_C), load resistor voltage (V_{RL}), and collector–emitter voltage (V_{CE}).

Solution

1. Determine the base current (I_B) by dividing V_{CC} by the base resistor (R_B).

$$I_B = V_{CC}/R_B$$
$$= 12 \text{ V}/200 \text{ K}\Omega$$
$$= 0.00006 \text{ A}$$
$$= 60 \text{ }\mu\text{A}.$$

2. Using the given beta (β) value and the calculated base current value, determine the collector current.

$$I_C = \beta \times I_B$$
$$= 100 \times 60 \mu\text{A}$$
$$= 0.006\text{A}$$
$$= 6 \text{ mA}.$$

3. Determine the voltage across the load resistor (V_{RL}) by multiplying collector current (I_C) by the load resistor (I_B) value.

$$V_{RL} = I_C \times R_L$$
$$= 0.006 \text{ A} \times 1000 \text{ }\Omega$$
$$= 6 \text{ V}.$$

4. Determine the collector–emitter voltage (V_{CE}) by subtracting the load resistor voltage (V_{RL}) from V_{CC}.

$$V_{CE} = V_{CC} - V_{RL}$$
$$= 12 - 6 \text{ V}$$
$$= 6 \text{ V}.$$

Related Problem

From the values provided in **Figure 7.3(b)**, determine base current (I_B), collector current (I_C), load resistor voltage (V_{RL}), and collector–emitter voltage (V_{CE}).

Signal Amplification

For the basic transistor circuit of **Figure 7.3** to respond as an amplifier, it must have a signal applied. This signal is made up of voltage and current components. The applied signal causes the transistor to change from a static condition to a dynamic condition. **Dynamic** conditions involve changing values. All AC amplifiers respond in the dynamic state. The output of this should develop an AC signal.

Figure 7.4 shows a basic transistor amplifier with an AC signal applied. A capacitor is used in this case to couple the AC signal source to the amplifier. Remember, AC passes easily through a capacitor, whereas DC is blocked. As a result, the AC signal is injected into the base–emitter junction. DC does not flow back into the signal source. The AC signal is, therefore, added to the DC operating voltage. The emitter–base voltage is a DC value that changes at an AC rate.

Consider how the applied AC signal alters the emitter–base junction voltage. **Figure 7.5** shows some representative voltage and current values that appear at the emitter–base junction of the transistor.

They can be measured with a voltmeter or observed with an oscilloscope. **Figure 7.5(a)** shows the DC operating voltage and base current. This occurs when the amplifier is in its **steady or static state**. **Figure 7.5(b)** shows the AC signal that is applied to the emitter–base junction. Note that the amplitude change of this signal is a very small value. This also shows how the resulting

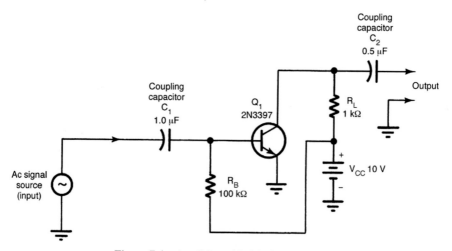

Figure 7.4   Amplifier with AC signal applied.

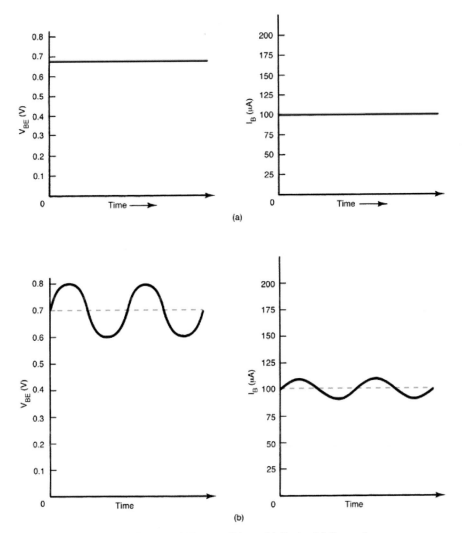

Figure 7.5 I_B and V_{BE} conditions. (a) Static. (b) Dynamic.

voltage and base current change with an AC signal applied. The AC signal essentially rides on the DC voltage and current.

Refer to the schematic diagram of the basic amplifier in **Figure 7.6.** Note, in particular, the waveform inserts that appear in the diagram. These show how the current and voltage values respond when an AC signal is applied. They also show where instruments are connected to measure and observe these values.

The AC signal applied to the input of our basic amplifier rises in value during the positive alternation and falls during the negative alternation. Initially, this causes an increase and a decrease in the value of V_{BE}. This voltage has a DC level with an AC signal riding on it. The indicated I_B is developed as a result of this voltage. The changing value of I_B causes a corresponding change in I_C. Note that I_B and I_C both appear to be the same. There is, however, a very noticeable difference in values. I_B is in microamperes, while I_C is in milliamperes. The resulting I_C passing through R_L causes a corresponding voltage drop across R_L. V_{CE} appearing across the transistor is the reverse of V_{RL}. These signals are both AC values riding on a DC level. The output, V_{out}, is changed to a **pure AC** value. Capacitor C_2 blocks the DC component and passes only the AC signal.

It is interesting to note that in **Figure 7.6**, the input and output signals of the amplifier are reversed. When the positive alternation of the AC input

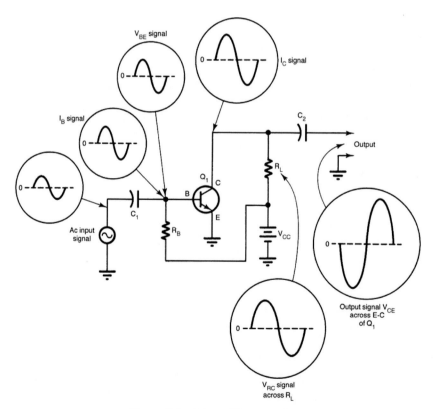

Figure 7.6   AC amplifier operation.

signal rises in value, it causes the output to fall in value. The negative alternation of the input causes the output to rise in value. This condition of operation is called **phase inversion**. It is shown by the output or V_{CE} waveform appearing across the emitter–collector. Phase inversion is a distinguishing characteristic of the **common-emitter amplifier**.

Amplifier Bias

If a transistor is to operate as a **linear amplifier**, it must be properly biased and have a suitable operating point. Its steady state of operation depends a great deal on base current, collector voltage, and collector current. Establishing a desired operating point requires proper selection of bias resistors and a load resistor to provide proper input current and collector voltage.

Stability of operation is a very important consideration. If the operating point of a transistor is permitted to shift with temperature changes, unwanted distortion may be introduced. In addition to distortion, operating point changes may also damage a transistor. Excessive collector current, for example, may cause the device to develop too much heat.

The method of biasing a transistor has a great deal to do with its **thermal stability.** Several different methods of achieving bias are commonly used. One method of biasing is considered to be beta dependent. **Bias voltages** are largely dependent on transistor beta. A problem with this method of biasing is transistor response. Transistor beta is rarely ever the same for a specific device. Biasing set up for one transistor will not necessarily be the same for another transistor.

Biasing that is independent of beta is very important. This type of biasing responds to fixed voltage values. Beta does not alter these voltage values. As a general rule, this form of biasing is more reliable. The input and output of a transistor are very stable, and the results are very predictable.

Beta-Dependent Biasing

Four common methods of **beta-dependent biasing** are shown in **Figure 7.7**. The static states of operation are a function of the transistor's beta. These methods are rather easy to achieve with a minimum of parts. They are not, however, very widely used.

The circuit in **Figure 7.7(a)** is a very simple form of biasing for a **common-emitter amplifier**. This method of biasing was used for our basic transistor amplifier. The base is simply connected to V_{CC} through R_B.

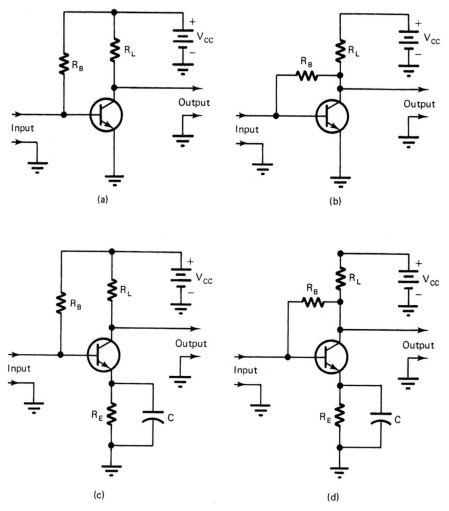

Figure 7.7   Methods of beta-dependent biasing.

This causes the emitter–base junction to be forward biased. The resulting collector current is beta times the value of I_B. The value of I_B is V_{CC} divided by R_B.

As a general rule, biasing of this type is very sensitive to changes in temperature. The resulting output of the circuit is rather difficult to predict. This method of biasing is often called **fixed biasing**. It is primarily used because of its simplicity.

Example 7-2:

How much voltage is at the output (V_{CE}) of the circuit in **Figure 7.4?**
Solution

1. Determine the base current (I_B) by dividing V_{CC} by the base resistor (R_B).

$$I_B = V_{CC}/R_B$$
$$= 10 \text{ V} \times 100 \text{ K}\Omega$$
$$= 0.0001 \text{ A, or} 100 \ \mu\text{A}.$$

2. Using the given beta (β) value and the calculated base current value, determine the collector current (I_C).

$$I_C = \beta \times I_B$$
$$= 50 \times 0.0001 \text{ A}$$
$$= 0.005 \text{ or} 5 \text{ mA}.$$

3. Determine the voltage across the load resistor (V_{RL}) by multiplying collector current (I_C) by the load resistor (I_B) value.

$$V_{RL} = I_C \times R_L$$
$$= 0.005 \text{ A} \times 1000 \ \Omega$$
$$= 5 \text{ V}.$$

4. Determine the collector–emitter voltage (V_{CE}) by subtracting the load resistor voltage (V_{RL}) from V_{CC}.

$$V_{CE} = V_{CC} - V_{RL}$$
$$= 10 \text{ V} - 5 \text{ V}$$
$$= 5 \text{ V}.$$

Related Problem

How much voltage is at the output (V_{CE}) of the circuit in **Figure 7.4** if the beta were 75?

The circuit in **Figure 7.7(b)** was developed to compensate for the temperature sensitivity of the circuit in **Figure ??(a)**. Bias current is used to counteract changes in temperature. In a sense, this method of biasing has **negative feedback**. R_B is connected to the collector rather than to V_{CC}. The voltage available for base biasing is that which is leftover after the voltage drop across the load resistor. This is indicated as V_{CE}. If the temperature rises, it causes an increase in I_B and I_C. With more I_C, there is more voltage drop across R_L. Reduced V_{CE} voltage causes a corresponding drop in I_B. This, in turn, brings I_C back to normal. The opposite reaction occurs when transistor temperature becomes less. This method of biasing is called **self-biasing**.

The circuit in **Figure 7.7(c)** is an example of **emitter biasing**. **Thermal stability** is improved with this type of construction. I_B is again determined by the value of R_B and V_{CC}. An additional resistor is placed in series with the emitter of this circuit. Emitter current passing through R_E produces emitter voltage (V_E). This voltage is of opposite polarity to the base voltage developed by R_B. Proper values of R_B and R_E are selected so that I_B and I_E will flow under ordinary operating conditions. If a change in temperature should occur, V_E will increase in value. This action opposes the base bias. As a result, collector current drops to its normal value.

The capacitor (C) connected across R_E is called a **bypass capacitor**. A bypass capacitor provides an alternate path around a component or to ground. In this case, it provides an AC path for signal voltages around R_E. With C in the circuit, amplifier gain is reduced. The value of C is dependent on the frequencies being amplified. At the lowest possible frequency being amplified, the capacitive reactance (X_C) should be 10 times smaller than the resistance of R_E. Remember that $X_C = 1/(2\pi fC)$; so as frequency increases, X_C decreases.

The circuit in **Figure 7.7(d)** is a combination of circuits in **Figures 7.7(b)** and **7.7(c)**. It is often called **self-emitter bias**. In the same regard, the circuit in **Figure 7.7(c)** could be called **fixed-emitter bias**. As a general rule, emitter biasing is not very effective when used independently. The circuit in **Figure 7.7(d)** has good thermal stability. The output has reduced gain because of the base resistor connection.

Beta-Independent Biasing

Two methods of biasing a transistor that are independent of beta are shown in **Figure 7.8**. These circuits are extremely important because they do not change operation with beta. As a general rule, these circuits have very reliable operating characteristics. The output is very predictable, and the **stability** is excellent.

The circuit in **Figure 7.8(a)** is described as the **divider method** of biasing. It is used widely. The base voltage (V_B) is developed by a voltage-divider

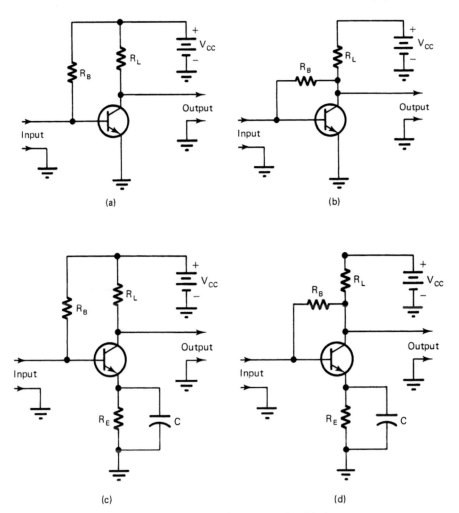

Figure 7.8 Methods of beta-dependent biasing.

network made of R_B and R_1. This network makes the circuit independent of beta changes. Voltages at the base, emitter, and collector depend on external circuit values. By proper selection of components, the emitter–base junction is forward biased and the collector is reverse biased. Normally, these bias voltages are referenced to ground. The base, in this case, is made slightly positive with respect to ground. This voltage is somewhat critical. A voltage that is too positive, for example, will drive the transistor into saturation. With proper selection of bias voltage, however, the transistor can be made to operate in any part of the active region. The temperature stability of the circuit is excellent. With proper R_E bypass capacitor selection, this method of biasing produces very high **gain**. The divider method of biasing is often a universal biasing circuit. It can be used to bias all transistor amplifier circuit configurations.

The circuit in **Figure 7.8(b)** is very similar in construction and operation to that of **Figure 7.8(a)**. One less resistor is used. The power supply requires two voltage values with reference to ground. A **split power supply** is used as an energy source for this circuit. Note the indication of $+V_{CC}$ and $-V_{CC}$. R_B is connected to ground. In this case, the value of R_B determines the value of I_B with only half of the total supply voltage. The values of R_L and R_E are usually made larger to accommodate the increased supply voltage. If R_E is properly bypassed, the gain of this circuit is very high. **Thermal stability** is excellent.

Self-Examination

Answer the following questions.

8. For a bipolar transistor to respond as an amplifier, the emitter–base junction must be (forward, reverse) biased and the collector–base must be (forward, reverse) biased.

9. The _____ of a bipolar transistor is equal to the beta times the base current.

10. If a bipolar transistor has a beta of 100 and a base current of 80 μA, the collector current is _____ mA.

11. When a transistor has DC voltage supplied to its elements and no signal applied, it is considered to be in its _____ state of operation.

12. When a bipolar transistor is energized by DC and an AC signal is applied for amplification, it is considered to be in a _____ state of operation.

13. In a common-emitter transistor circuit configuration, one of the leads must be common to both the _____ and the _____ of the circuit.

14. The method of biasing a bipolar transistor has a great deal to do with its _____ stability during operation.

15. In bipolar transistor operation when base current increases, _____ current increases.

16. In bipolar transistor operation when collector current increases, _____ voltage decreases.

7.3 Load-Line Analysis

Earlier in this chapter, you looked at the operation of an amplifier with respect to its beta. Current and voltage were calculated, and transistor operation was related to these values. Although this method of analysis is very important, amplifier operation can also be determined graphically. This method employs a collector family of characteristic curves graph. A load line is developed for this graph and used for load-line analysis. Through load-line analysis, it is possible to predict how the circuit will respond. A great deal about the operation of an amplifier can be observed through this method. In this section, you will create a load line for a collector family of characteristic curves graph. You will then use this graph and load line to analyze an amplifier circuit.

7-3 Analyze the operation of a basic amplifier by the load-line method.
In order to achieve objective 7.3, you should be able to:

- construct a power dissipation curve;
- construct a load line for an amplifier circuit;
- describe the linear and nonlinear operations of an amplifier circuit;
- define *load line*, *Q point*, *linear*, *nonlinear distortion*, and *overdriving*.

The load-line method of circuit analysis is widely used. Engineers use this method in designing new circuits. The operation of a specific circuit can also be visualized by this method. In circuit design, a specific transistor is selected for an amplifier. Source voltage, load resistance, and input signal levels may be given values in the design of the circuit. The transistor is made to fit the limitation of the circuit.

For our application of the load line, assume that the amplifier circuit in **Figure 7.9** is to be analyzed. A **collector family of characteristic curves** for the transistor is also shown in **Figure 7.9**. Note that the **power dissipation** rating of the transistor is included in the diagram.

Power Dissipation Curve

As discussed in previous chapters, **maximum power dissipation** determines the operational limits of electronic devices. A common practice in load-line analysis for transistors is first to develop a power dissipation curve. This gives some indication of the maximum operating limits of the transistor. Power dissipation (P_D) refers to maximum heat that can be given off by the base−collector junction. Usually, this value is rated at 25°C. P_D is the product of I_C and V_{CE}. In our circuit, the P_D rating for the transistor is 300 mW.

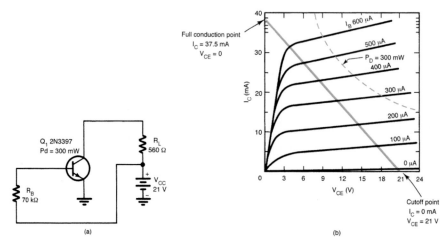

Figure 7.9   Circuit and characteristic curves of a circuit to be analyzed. (a) Circuit. (b) Characteristic curves.

To develop a P_D curve, each value of V_{CE} is used with the P_D rating to determine an I_C value. The formula is

$$I_C = P_D/V_{CE}. \tag{7.11}$$

Using this formula, the I_C value is calculated for each of the V_{CE} values of the collector family of curves. Then, the V_{CE} value and the corresponding calculated I_C value is located and noted. These points are connected together to form the 300-mW **power dissipation curve**. In practice, the load line must be located to the left of the established P_D curve. Satisfactory operation without excessive heat generation can be assured in that area of the curve.

Example 7-3:

Using the transistor characteristic curve of **Figure 7.9**, calculate the values of collector current for a 300-mW power dissipation curve for Y_{CE} values of 9 and 12 V.

Solution

$$I_C = P_D/V_{CE} \qquad\qquad I_C = P_D/V_{CE}$$
$$= 300\,\mathrm{mW}/9\,\mathrm{V} \qquad = 300\,\mathrm{mW}/12\,\mathrm{V}$$
$$= 33.33\,\mathrm{mA} \qquad\qquad = 25\,\mathrm{mA}.$$

Related Problem

Complete the power dissipation curve of **Figure 7.9** by calculating I_C values for V_{CE} points of 15, 18, and 21 V.

Static Load Line

The **load line** of a transistor amplifier represents two extreme conditions of operation. One of these is in the **cutoff region**. When the transistor is cut off, no I_C flows through the device. V_{CE} equals the source voltage of 21 V with zero I_C.

The second load-line point is in the **saturation region**. This point assumes full conduction of I_C. Ideally, when a transistor is fully conductive

$$V_{CE} = 0$$

and

$$I_C = V_{CC}/R_L.$$

The two load-line construction points for the analysis circuit are shown on the collector family of characteristic curves of **Figure 7.13**. The cutoff point is located at the zero I_C, 21 V_{CE} point. The value of V_{CC} determines this point. At cutoff, V_{CE} is equal to V_{CC}. The saturation, or full-conduction, point is located at 37.5 mA of I_C at 0 V_{CE}. The I_C value is calculated using V_{CC} and the value of R_L. The formula is

$$I_C = V_{CC}/R_L. \tag{7.12}$$

Therefore,

$$
\begin{aligned}
I_C &= V_{CC}/R_L \\
&= 21 \text{ V}/560 \ \Omega \\
&= 0.0375 \text{ A, or} 37.5 \text{ mA.}
\end{aligned}
$$

These two points are connected with a straight line. The load line makes it possible to determine the operating conditions of the amplifier. For linear amplification, the operating point should be located near the center of the load line. For the circuit in **Figure 7.13**, an operating point of 30 μA of base current is selected. In the circuit diagram, the value of R_B determines I_B. It is calculated by the following equation:

$$I_B = V_{CC}/R_B. \tag{7.13}$$

The value is

$$I_B = V_{CC}/R_B$$
$$= 21 \text{ V}/700\text{k } \Omega$$
$$= 0.00003\text{A, or} 30 \text{ } \mu\text{A}.$$

The operating point for this value is located at the intersection of the load line and the 30-μA I_B curve. It is indicated as point Q. Knowing this much about an amplifier shows how it will respond in its static state. The **Q point** shows how the amplifier will respond without an applied signal.

The collector family of curves in **Figure 7.9** displays the operation of the amplifier in its **static state**. Projecting a line from the Q point to the I_C scale shows the resulting collector current. In this case, the I_C is 17.5 mA. The **DC beta** for the transistor at this point is determined by the following formula:

$$\beta_{dc} = I_C/I_B.$$

This value is

$$\beta_{dc} = I_C/I_B$$
$$= 17.5 \text{ mA}/30\mu\text{A}$$
$$= 0.0175 \text{ A}/0.000030 \text{ A}$$
$$= 583.3.$$

The resulting V_{CE} that will occur for the amplifier can also be determined graphically. Projecting a line directly down from the Q point shows the value of V_{CE}. In our circuit, V_{CE} is approximately 11 V. This means that 10 V will appear across R_L when the transistor is in its static state.

Example 7-4:

Using the transistor characteristics curve of Figure 7-9, determine the values of R_B and DC β at point A on the load line.

Solution

$$R_B = V_{CC}/I_B \qquad\qquad \beta_{dc} = I_C/I_B$$
$$= 21 \text{ V}/20 \text{ } \mu\text{A} \qquad\qquad = 12 \text{ mA}/20 \text{ } \mu\text{A}$$
$$= 1.05 \text{ M } \Omega \qquad\qquad = 600.$$

Related Problem

Using **Figure 7.9**, calculate R_B and DC β values for points B and C on the load line.

Dynamic Load Line

Dynamic load-line analysis shows how an amplifier responds to an AC signal. In this case, the collector family of curves and the circuit of **Figure 7.10** are used. A load line and Q point for the circuit have been developed on the curves. This establishes the **static** operation of the amplifier. Note the values of V_{CE} and I_C in the static state.

Refer to **Figure 7.1** for the following example. Assume that a 0.1-V peak-to-peak AC signal is applied to the input of the amplifier and the signal causes a 20-μA peak-to-peak change in I_B. During the positive alternation of the input shown on the right of **Figure 7.14**, I_B changes from 30 to 40 μA. This is shown as point P on the load line. During the negative alternation, I_B drops from 30 to 20 μA. This is indicated as point N on the load line. In effect, this means that 0.1 V peak-to-peak causes I_B to change 20 μA peak-to-peak. The I_B signal extends to the right of the load line. Its value is shown as ΔI_B (ΔI_B = 40 − 20 μA = 20 μA).

To show how a change in I_B influences I_C, lines are projected to the left side of the load line of **Figure 7.14**. Note the projection of lines P, Q, and N toward the I_C values. The changing value of I_C is indicated as ΔI_C. An increase and decrease in I_B causes a corresponding increase and decrease in I_C. This shows that I_C and I_B are in phase (ΔI_C = 7.6 − 4.2 mA = 3.4 mA).

The **AC beta** (β_{ac}) of the amplifier can be determined by ΔI_C and ΔI_B. First, determine the peak-to-peak ΔI_C and ΔI_B values.

These values are approximations from the characteristic curve load line:

$$\Delta I_C = 3.4 \text{ mA}$$
$$\Delta I_B = 20 \text{ } \mu\text{A}.$$

Then divide ΔI_C by ΔI_B to determine the AC beta of the amplifier circuit:

$$\beta_{ac} = \Delta I_C / \Delta I_B$$
$$= 3.4 \text{ mA}/20 \text{ } \mu\text{A}$$
$$= 0.0034 \text{ A}/0.00002 \text{ A}$$
$$= 170.$$

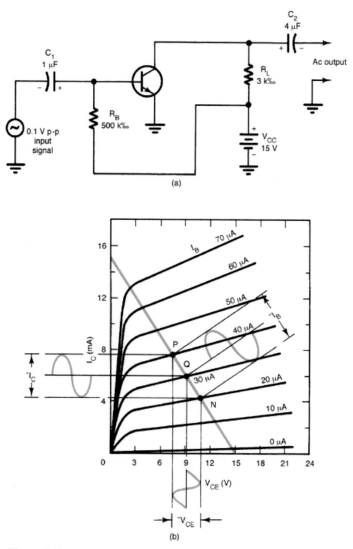

Figure 7.10   Dynamic load line. (a) Circuit. (b) Characteristic curves.

The same procedure can be used to determine the **DC beta** of the transistor at point Q:

$$\beta_{dc} = I_C/I_B$$
$$= 5\,\text{mA}/30\,\mu\text{A}$$

$$= 0.005 \text{ A}/0.00003 \text{ A}$$
$$= 167.$$

Note that the AC and DC beta values of this amplifier are similar values. These values should be similar within the normal operating range of the transistor.

Projecting points P, Q, and N downward from the load line shows how V_{CE} changes with I_B. The value of V_{CE} is indicated as ΔV_{CE}. Note that an increase in I_B causes a decrease in the value of V_{CE}. A decrease in I_B causes V_{CE} to increase. This shows that I_B and V_{CE} are 180° out of phase. The difference in V_{CE} at any point appears across R_L.

The **AC voltage gain** of the amplifier can be determined from the dynamic load line. Remember that 0.1-V peak-to-peak input caused a change of 20-μA peak-to-peak in the I_B signal. The AC voltage gain can be determined by dividing ΔV_{CE} by ΔV_B. The ΔV_B value is 0.1-V peak-to-peak. Using the ΔV_{CE} value from the graph and ΔV_B, determine the AC voltage gain of the amplifier circuit ($\Delta V_{CE} = 11 - 7.5 \text{ V} = 3.5 \text{ V}$):

$$A_{V(ac)} = \Delta V_{CE}/\Delta V_B$$
$$= 3.5 \ V_{pp}/0.1 \ V_{pp}$$
$$= 35.$$

Example 7-5:

Using the characteristics curve of **Figure 7.10**, determine ΔI_C, ΔI_B, β_{ac}, ΔV_{CE}, and $A_{V(ac)}$ with a 0.1-V peak-to-peak AC signal that causes I_B to change from point A to point Q.

Solution

$$\Delta I_C = 9.5 - 6 \text{ mA} = 3.6 \text{ mA} \qquad \beta_{ac} = \Delta I_C/\Delta I_B$$
$$\Delta I_B = 50 - 30 \mu A = 20 \mu A \qquad\qquad = 3.5 \text{ mA}/20 \ \mu A$$
$$\Delta V_{CE} = 9.2 - 7.5 \text{ V} = 1.7 \text{ V} \qquad\qquad = 180$$
$$A_{V(ac)} = \Delta V_{CE}/\Delta V_B = 1.7 \text{ V}/0.1 \text{ V} = 17.$$

Related Problem

Using **Figure 7.10**, calculate ΔI_C, ΔI_B, β_{ac}, ΔV_{CE}, and $A_{V(ac)}$ with a 0.1-V peak-to-peak AC signal that causes I_B to change from point B to point Q.

Linear and Nonlinear Operation

The V_{CE} or output of an amplifier should be a duplicate of the input with some gain. When a sine wave is applied, the output should develop a sine wave. When an amplifier operates in this manner, it is considered to be **linear**. For this to be achieved, the amplifier must operate in or near the center or linear area of the collector curves. As a rule, nonlinearity occurs near the **saturation** and **cutoff regions**. If the operating point is adjusted near these regions, it usually causes the output to be distorted. Normally, this is called **nonlinear distortion**.

Figure 7.11 shows how an amplifier will respond at three different operating points. **Figure 7.11(a)** shows **linear** operation. Note the input signal (ΔI_B) on the right and the output (ΔI_C) on the left. The input and output

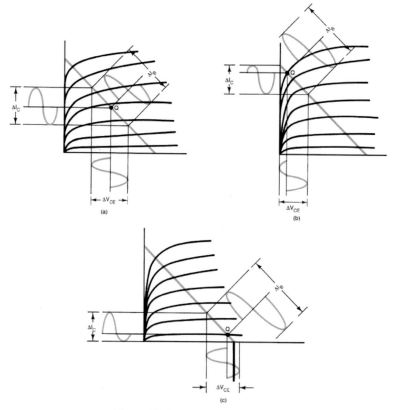

Figure 7.11   Amplifier operation.

signals are duplicate sine waves in this case. **Figure 7.11(b)** shows operation in the **saturation region**. Note the top of the input wave (ΔI_B) and the distortion of the negative alternation of the output (ΔI_C). **Figure 7.11(c)** shows operation near the **cutoff region**. The lower part of the input wave (ΔI_B) on the right extends beyond the load line, causing **distortion** of the positive alternation of the output (ΔI_C). As a general rule, nonlinear distortion can be tolerated in some applications. In other applications, nonlinear distortion is very obvious, such as sound reproduction.

An input signal that is too large can also produce **distortion**. **Figure 7.12** shows how an amplifier responds to a large input signal. Note that the input signal swings into the saturation region during the positive alternation and into the cutoff region during the negative alternation. This condition distorts both alternations of the output. Normally, this condition is described as **overdriving**.

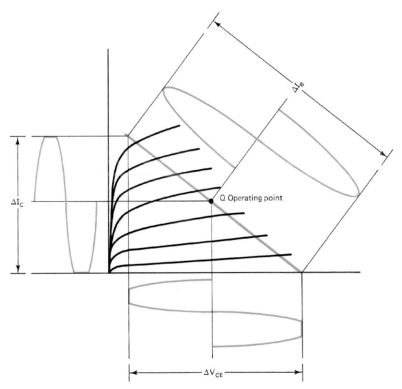

Figure 7.12 Overdriven amplifier.

In a radio receiver or stereo amplifier, overdriving causes speech or music to sound bad. This usually occurs when the volume control is turned too high. The audio or sound signal has its peaks clipped. The volume level may be higher, but the quality of reproduction is usually very poor when this occurs. It is interesting to note that the operating point is near the center of the active region. **Overdriving** can occur even with a properly selected operating point.

Classes of Amplification

Transistor amplifiers may be classified according to their **bias operating point**. This means of classification describes the shape of the output wave. Three general groups of amplifiers are **class A**, **class B**, and **class C**. **Figure 7.13** shows a graphical display of these three amplifier classes.

Class A amplifiers generally have linear operation. The bias operating point is set near the center of the active region. With a sine wave applied to the input, the output is a complete sine wave. **Figure 7.13(a)** shows the input—output waveforms of a class A amplifier. This type of amplifier is used when a true reproduction of the input signal is required.

Figure 7.13(b) shows the input—output waveforms of a **class B amplifier**. The operating point of this amplifier is adjusted near the cutoff point. With a sine wave applied to the input, only one alternation of the signal is reproduced. When using class B amplifiers, it is possible to get a large change in output for one alternation. Each of two class B amplifiers working together can amplify one alternation of a sine wave. When the wave is restored, a complete sine wave is developed. Class B amplifiers are commonly used in push—pull audio output circuits. These amplifiers are usually concerned with power amplification.

A **class C amplifier** is shown in **Figure 7.13(c)**. In this amplifier, the bias operating point is below cutoff. With a sine wave applied to the input, the output is less than half of one alternation. The primary application of class C amplifiers is radio frequency circuits. The operational efficiency of this amplifier is quite high. It consumes energy only for a small portion of the applied sine wave.

Self-Examination

Answer the following questions.

17. With the load-line method of bipolar transistor analysis, it is possible to predict how a device will respond _____.

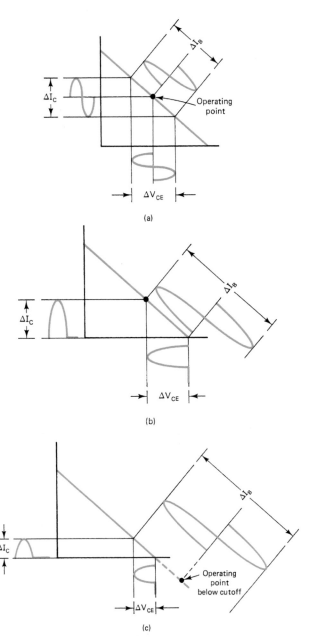

Figure 7.13   Classes of amplification.

18. The _____ rating of a bipolar transistor is the product of collector−emitter voltage and collector current.
19. The _____ rating of a bipolar transistor refers to the maximum heat that can be given off by the base−collector junction.
20. When a load line is set up for a bipolar transistor, the two extreme conditions of operation are _____ and _____.
21. When a bipolar transistor is used as a linear amplifier, operation should be near the _____ of the load line.
22. Nonlinear amplification occurs when operation takes place in or near the _____ and _____ regions.
23. _____ amplification is best achieved when a transistor operates near the center of the load line.
24. If the input signal of a bipolar amplifier is too large, it causes the output signal to be _____.
25. A class (A, B, C) amplifier has sine-wave input and full, undistorted sine-wave output.
26. The bias operating point of a class (A, B, C) amplifier is near the center of the active region.
27. A class (A, B, C) amplifier has sine-wave input and half-sine wave or one alternation of output.
28. A class (A, B, C) amplifier has its bias operating point adjusted below cutoff.

7.4 Transistor Circuit Configurations

In previous sections of this chapter, we have examined the use of transistors as amplifying devices. Transistor circuits are used to amplify voltage, current, and power. The basic operation of amplifiers was discussed and analyzed by the use of characteristic curves and load line calculations. In this section, we will expand the knowledge of transistor circuit operation by looking at additional types of circuits.

7.4 Analyze common-emitter, common-base, and common-collector circuit configurations.

In order to achieve objective 7.4, you should be able to:
- explain the characteristic of common-emitter, common-base, and common-collector circuit configurations;
- identify a common-emitter, common-base, and common-collector circuit configuration;
- define *alpha*.

The elements of a transistor can be connected in one of three different circuit configurations. These are usually described as common-emitter, common-base, and common-collector. Of the three transistor leads, one is connected to the input and one to the output. The third lead is commonly connected to both the input and output. The common lead is generally used as a reference point for the circuit. It is usually connected to the circuit ground or common point. This has brought about the terms **grounded emitter**, **grounded base**, and **grounded collector**. The terms *common* and *ground* mean the same thing in the field of electronics.

In some circuit configurations, the emitter, base, or collector may be connected directly to the ground. When this occurs, the lead is at both DC and AC ground potential. When the lead goes to ground through a battery or resistor that is bypassed, by a capacitor, it is now considered to be at an AC ground only. The capacitor provides a low impedance path for the amplified AC signal but blocks the DC component. With no bypass capacitor, the emitter bias voltage and, thus, the operating point of the transistor fluctuate at the frequency of the amplified signal, causing distortions.

Common-Emitter Amplifier

The **common-emitter amplifier** is a very important amplifier circuit. A very high percentage of all amplifiers in use are of the common-emitter type. The input signal of this amplifier is applied to the base, and the output is taken from the collector. The emitter is the common or grounded element.

Figure 7.14 shows a circuit diagram of the common-emitter amplifier. This circuit is very similar to the basic amplifier introduced in the first part of this chapter. In effect, this circuit was used to acquaint you with amplifiers in general. It is presented here to make a comparison with the other circuit configurations.

The signal amplified by the **common-emitter amplifier** is applied to the emitter–base junction. This signal is superimposed on the DC bias of the emitter–base. The base current then varies at an AC rate. This action causes a corresponding change in collector current. The output voltage developed across the collector–emitter is inverted 180°. The current gain of the circuit is determined by the beta. Typical beta values are about 50. Voltage gain ranges from 250 to 500. Power gain is about 2000. Input impedance is moderately high, typically around 100 Ω. Output impedance is moderate; a typical value is about 2 kΩ. Remember that impedance is the opposition to alternating

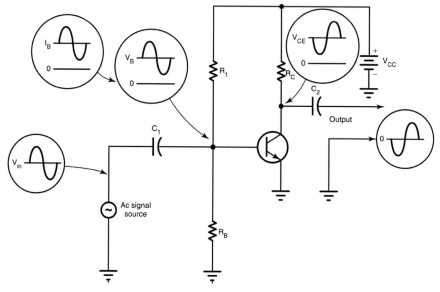

Figure 7.14   Common-emitter amplifier.

current, in this case, at the input and output terminals of the device. In general, common-emitter amplifiers are used in **small-signal** applications or as voltage amplifiers.

Common-Base Amplifier

A **common-base amplifier** is shown in **Figure 7.15.** In this type of amplifier, the emitter–base junction is forward biased, and the collector–base junction is reverse biased. The emitter is the input. An applied input signal changes the circuit value of I_{E}. The output signal is developed across R_{L} by changes in collector current. For each value change in I_{E}, there is a corresponding change in I_{C}. In a common-base amplifier, the current gain is called **alpha**. Alpha is determined by the following formula:

$$\alpha = I_{\mathrm{C}}/I_{\mathrm{E}.} \tag{7.14}$$

The value of the gain in a common-base amplifier is always less than 1. Remember that

$$I_{\mathrm{E}} = I_{\mathrm{B}} + I_{\mathrm{C}.}$$

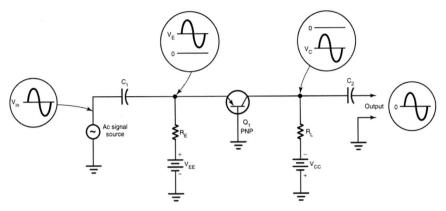

Figure 7.15 Common-base amplifier.

The I_C will, therefore, always be slightly less than I_E by the value of I_B. I_E is the signal input but is the source input for I_B and I_C. Typical values of alpha are 0.98–0.99.

In **Figure 7.15**, V_{EE} forward biases the emitter–base, whereas V_{CC} reverse biases the collector–base. Resistor R_E is an emitter **current-limiting resistor**, and R_L is the **load resistor**. Note that the transistor is a PNP type.

When an AC signal is applied to the input, it is added to the DC operating value of V_E. The positive alternation, therefore, adds to the forward-bias voltage of V_E. This condition causes an increase in I_E. A corresponding increase in I_C also takes place. With more I_C through R_L, there is an increase in its voltage drop. The collector, therefore, becomes less negative, or swings positive. In effect, the positive alternation of the input produces a positive alternation in the output. The input and output of this amplifier are in phase. See the waveform inserts in the diagram. The negative alternation causes the same reaction, the only difference being a reverse in polarity.

A **common-base amplifier** has a number of unique characteristics that should be considered when this amplifier is selected for an application. Its current gain is called alpha. The **current gain** is always less than 1. Typical values are 0.98–0.99. **Voltage gain** is usually very high. Typical values range from 100 to 2500 depending on the value of R_L. **Power gain** is the same as voltage gain. The input impedance is very low. Values of 10–200 Ω are very common. The output impedance of the amplifier is somewhat moderate. Values range from 10 to 40 kΩ. This amplifier does not invert the applied signal.

Common-base amplifiers are used primarily to match a **low-impedance input** device to a circuit. This type of circuit configuration is used in radio-frequency amplifier applications. As a general rule, the common-base amplifier is not used very commonly.

Common-Collector Amplifier

A **common-collector amplifier** is shown in **Figure 7.16**. In this circuit, the base serves as the signal input point. The input of this amplifier is primarily the same as the common-emitter circuit. The collector is connected to ground through V_{CC}. Note that the input, output, and collector are all commonly connected. The unique part of this circuit is the output. It is developed across the load resistor (R_L) in the emitter. There is no resistor in the collector circuit.

When an AC signal is applied to the input, it adds to the base current. The positive alternation increases the value of I_B above its static operating point. An increase in I_B causes a corresponding increase in I_E and I_C. With

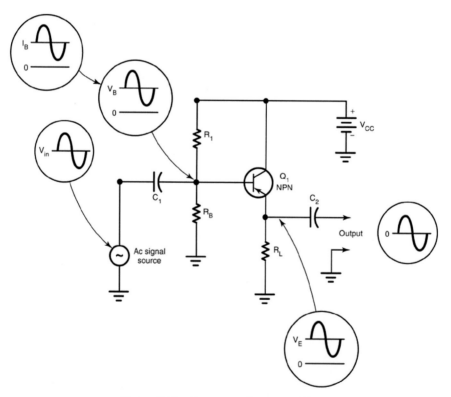

Figure 7.16   Common-collector amplifier.

an increase in I_E, there is more voltage drop across R_L. The topside of R_L and the emitter becomes more positive. A positive input alternation, therefore, causes a positive output alternation across R_L. Essentially, this means that the input and output are in phase. The negative alternation reduces I_B, I_E, and I_C. With less I_E through R_L, the output swings negative. The output is again in phase for this alternation.

The common-collector amplifier is capable of **current gain**. A small change in I_B causes a large change in I_E. This current gain does not have a descriptive term such as **alpha** or **beta**. In general, the current gain is assumed to be greater than 1. Values approximate those of the common-emitter amplifier.

Other factors of importance are voltage gain, power gain, input impedance, and output impedance. **Voltage gain** is less than 1. **Power gain** is moderately high, with typical values in the range of 50. **Input impedance** is in the range of 50 kΩ. The **output impedance** is quite low because of the location of R_L. Typical values are about 50 Ω.

Common-collector amplifiers are used primarily as **impedance-matching** devices. They can match a high-impedance device to a low-impedance load. Practical applications include preamplifiers operated from a high-impedance microphone or phonograph pickup. Common-collector amplifiers are also used as **driver transistors** for the last stage of amplification. In this application, power transistors generally require large amounts of input current to deliver maximum power to the load device. Since the emitter output follows the base, this type of amplifier is often called an **emitter follower**.

Self-Examination

Answer the following questions.

29. The input of a common-emitter amplifier is connected to the _____.
30. The output of a common-emitter amplifier is connected to the _____.
31. The input and output of a common-emitter amplifier are _____ out of phase.
32. The current gain of a common-emitter amplifier is determined by the transistor's _____.
33. The input signal of a common-base amplifier is applied to the _____, and the output is removed from the _____.
34. The input and output of a common-base amplifier are _____ phase.

35. The current gain of a common-base amplifier is called _____.
36. The current gain of a common-base amplifier is always (less, more) than 1.
37. The input of a common-collector amplifier is applied to the _____, and the output is derived from the _____.
38. The input and output of a common-collector are _____ phase.
39. The voltage gain of a common-collector amplifier is (less, more) than 1.

7.5 Analysis and Troubleshooting – Amplifier Circuits

Prior to using a BJT in an electronic circuit, the configuration in which it is placed should be considered. The three primary circuit configurations of BJTs are the common emitter (CE), common base (CB), and common collector (CC). BJTs, when configured in the common-emitter configuration, provide a moderate input and output impedance, high voltage gain, and high current gain. They are used extensively in intermediate stage amplification circuits. The common-base configuration features a low input impedance, high output impedance, a unity current gain, and high voltage gain. It is used as an input impedance matching device in RF circuits. A low impedance device (such as an antenna) is connected to the input of the CB configuration. The output of the CB configuration is then connected to a high impedance device (such as a tuning circuit). The common-collector features a high input impedance, a low output impedance, a high current gain, and a unity voltage gain. It is used as an output impedance matching device in audio applications. A **low impedance**, high current output device (such as a speaker) can be connected to the output of the CC configuration. When connected in this way, maximum power is transferred to the output. **Matching** the application to the transistor configuration is an important circuit consideration.

7.5 Analyze and troubleshoot amplifier circuits.
   In order to achieve objective 7.5, you should be able to:

- match the configuration of the circuit to its intended application;
- distinguish between normal and faulty circuit operation.

Amplifier Circuit Troubleshooting

The operation of a DC BJT amplifier is shown in **Figure 7.2**. It uses only DC sources that are applied to a BJT in a CE configuration. This circuit does not have any capacitors that are needed for passing or blocking AC. When a DC voltage is applied to the transistor, with sufficient voltage to forward bias the base–emitter (V_{BE}), a measure of whether it is working satisfactorily is determined by the value of the DC across the collector–emitter voltage, V_{CE}. If V_{CE} is very low (0 V), it indicates that there might be an internal short circuit between the collector and emitter terminals. On the other hand, if V_{CE} equals the supply voltage, it indicates that there might be an open circuit between the collector and emitter terminals.

The operation of a DC BJT amplifier is shown in **Figure 7.4**. It uses both DC and AC sources which are applied to a BJT in a CE configuration. This circuit has capacitors that are needed for passing or blocking AC. The DC voltage applied sets up the operating conditions. When an AC voltage is applied to the transistor, with sufficient voltage to forward bias the base–emitter (V_{BE}), a measure of whether it is working satisfactorily is determined by the value of the AC across the collector–emitter voltage, V_{CE}. If the AC value of V_{CE} is very low (0 V), it indicates that there might be an internal short or open circuit between the collector and emitter terminals. In either case, the transistor is defective. An oscilloscope can be used to observe the amplification of the AC input signal. If the AC output appearing across V_{CE} is distorted, it indicates that the amplifier is not functioning properly.

Summary

- The reproduction function produces an output signal with the same size and shape as that of the input signal.
- The amplification function produces an output signal that is greater in amplitude than the input signal; however, the output signal does not resemble the input signal in shape.
- Voltage amplification is a process where the output signal voltage is made greater than the input signal voltage.
- Current amplification is a process where the output signal current is made greater than the input signal current.
- Power amplification deals with a combination of voltage and current gain in a transistor amplifier.
- In the operation of a bipolar transistor amplifier, the emitter–base junction must be forward biased and the collector–base reverse biased.
- A bipolar transistor amplifier that is properly biased has the necessary DC energy applied to be operational and is considered to be in a static, or DC operating, state.

- Specific operating voltages are selected for an amplifier circuit that will permit amplification.
- If amplification and reproduction are to be achieved, the amplifying device must operate in the center of its active region.
- When an AC signal is applied to the input of a bipolar transistor amplifier, it causes the DC base current and DC collector current to change at an AC rate.
- Capacitors placed in the signal path block DC and pass AC, permitting only the AC signal to be amplified.
- The operation of an amplifier can be analyzed graphically through the use of a load line drawn on a family of collector curves.
- A load line represents two extreme conditions of operation: saturation and cutoff.
- Amplifier operation can be predicted by establishing a Q point and projecting lines to the collector current and collector—emitter voltage values.
- Selection of the operating, or Q, point shows if amplification will be linear or if distortion will occur.
- Bipolar transistor amplifiers are classified according to their bias operating point.
- Three general groups of amplifiers are class A, class B, and class C.
- Amplifier classification is related to the shape of the resulting output waveform.
- The application of a specific amplifier determines the classification used.
- The common-emitter circuit has the input connected to the emitter—base junction and the output taken from the emitter—collector.
- The output voltage developed across the collector of a common-emitter circuit is inverted 180°.
- The common-base circuit has the input connected to the emitter—base junction and the output connected to the base—collector junction.
- The current gain of a common-base circuit is called alpha.
- The input and output of a common-base circuit are in phase.
- The common-collector circuit has the input connected to the base—emitter junction and the output removed from the emitter—collector junction.
- The input and output of a common-collector circuit are in phase.

Formulas

Gain

(7-1) $A_V = V_{out}/V_{in}$ Voltage gain.
(7-2) $A_V = \Delta V_{out}/\Delta V_{in}$ Change in voltage gain.
(7-3) $A_i = I_{out}/I_{in}$ Current gain.
(7-4) $A_i = \Delta I_{out}/\Delta I_{in}$ Change in current gain.
(7-5) $A_P = P_{out}/P_{in}$ Power gain.
(7-6) $A_P = A_v \times A_i$ Power gain.
(7-7) $I_B = V_{CC}/R_B$ Base current.

Basic amplifier analysis

(7-8) $I_C = \beta \times I_B$ Collector current.
(7-9) $V_{RL} = I_C \times R_L$ Load resistor voltage.
(7-10) $V_{CE} = V_{CC} - V_{RL}$ Collector–emitter voltage.

Power dissipation curve

(7-11) $I_C = P_D/V_{CE}$ Collector current.

Static load line

(7-12) $I_C = V_{CC}/R_L$ Collector current.
(7-13) $I_B = V_{CC}/R_B$ Base current.

Alpha

(7-14) $\alpha = I_C/I_E$ Alpha.

Beta

(7-15) $\beta_{dc} = I_C/I_B$
(7-16) $\beta_{ac} = \Delta I_C/\Delta I_B$

Review Questions

Answer the following questions.

1. When a bipolar transistor is used as an amplifier, it is:

    a. Energized by AC

    b. Energized by DC

    c. Energized from the signal being processed

    d. Not energized from an external source

2. When the input of a bipolar transistor shows an output signal that is the same but has an increase in amplitude, it shows:

    a. Reproduction only

    b. Amplification only

    c. Combined amplification and reproduction

    d. No signal change

3. Which of the following is not considered to be a beta-dependent method of biasing a bipolar transistor?

    a. Fixed biasing

    b. Self-biasing

    c. Emitter biasing

    d. Resistor divider biasing

4. Which of the following biasing methods is considered to be independent of beta?

    a. Fixed biasing

    b. Self-biasing

    c. Resistor divider biasing

    d. Self-emitter biasing

5. The load line of a bipolar transistor represents the extreme operating conditions of:

    a. Full conduction to saturation

    b. Cutoff to the active region

    c. Saturation to the active region

    d. Saturation to cutoff

6. Dynamic operation of a bipolar transistor represents:

    a. DC operation

    b. AC operation

    c. Active region operation

    d. Nonlinear operation

7. Class A bipolar transistor amplifiers have the bias operating point set:

    a. Near the center of the active region

    b. At cutoff

    c. At saturation

    d. In the lower third of the active region

8. Class B bipolar transistor amplifiers have the bias operating point set:

   a. Near the center of the active region
   b. At cutoff
   c. At saturation
   d. In the lower third of the active region

9. In a common-emitter amplifier, the output signal would be:

   a. Reduced because of reverse biasing
   b. The same phase as the input
   c. Unchanged when compared with the input
   d. A 180° inversion of the input signal

10. In a common-emitter circuit configuration, an increase in base current would cause:

    a. An increase in V_{CE}
    b. An increase in I_C
    c. A decrease in I_C
    d. An increase in V_E

11. A 180° phase inversion between the input and output is accomplished in:

    a. All transistor circuit configurations
    b. Only the common-base circuit configuration
    c. Only the common-collector circuit configuration
    d. Only the common-emitter circuit configuration

12. The current gain or alpha of which amplifier is less than 1?

    a. Common-base
    b. Common-emitter
    c. Common-collector
    d. All transistor circuit configurations

13. The current gain of a common-emitter amplifier is:

    a. I_C/I_E
    b. Alpha
    c. Always less than 1
    d. I_C/I_B

14. The output of a common-collector amplifier is developed:

    a. Across the emitter−collector
    b. At the collector
    c. Across the collector resistor
    d. Across the emitter resistor

15. The voltage gain of a common-collector amplifier is:

   a. Less than 1
   b. Equivalent to the beta
   c. Typically 100
   d. Always greater than 1

16. In a common-base amplifier, the input impedance is:

   a. Very low
   b. Very high
   c. The same as the other two circuit configurations
   d. Variable because of the biasing circuit

17. In the three transistor amplifier configurations, the input circuit is:

   a. The same for all
   b. The same for the common-emitter and the common-collector circuits
   c. Different for only the common-emitter circuit
   d. Different for only the common-collector circuit

Problems

Answer the following questions.

1. The input voltage of a bipolar transistor amplifier is 0.1-V peak-to-peak, and the output is 6-V peak-to-peak. The A_V of this amplifier is in the range of:

   a. 0–0.02
   b. 0.021–1.0
   c. 1.01–10
   d. 10.01–100
   e. 100.01 or more

2. The base current of a bipolar transistor is 100 μA, and the collector current is 10 mA. In what range does the value of beta fall?

   a. 0.1–1.0
   b. 1.01–10
   c. 10.01–100
   d. 100.1–1000

3. A bipolar transistor has a power dissipation rating of 300 mW. What is the maximum collector current that can be achieved with a collector–emitter voltage of 15 V?

a. 1.0–15 mA
b. 15.1–30 mA
c. 30.1–45 mA
d. 45.1 or more

4. If the DC base current of a BJT common-emitter DC amplifier is increased, what effect will it have on the DC output voltage measured across the load resistor?
5. If the AC base current of a BJT common-emitter AC amplifier is increased, what effect will it have on the AC output voltage measured across the load resistor?

Answers

Examples

7-1. I_B = 0.1 mA, I_C = 9 mA, V_{RL} = 4.5V, V_{CE} = 4.5 V
7-2. 2.5 V
7-3. 20 mA, 16.66 mA, 14.29 mA
7-4. 2.1 MΩ, 700 (approximately)
7-5. ΔI_C = 3.5 mA, ΔI_B = 20 μA, β_{ac} = 180, ΔV_{CE} = 3 V, $A_{V(ac)}$ =30

Self-Examination

7.1

1. input
2. output
3. Reproduction
4. Amplification
5. voltage
6. current
7. power

7.2

8. forward, reverse
9. collector current, *or* I_C
10. 8 mA
11. static
12. dynamic
13. input, output
14. thermal

15. collector
16. collector
7.3
17. Graphically
18. power dissipation, *or* P_D
19. power dissipation, *or* P_D
20. saturation, cutoff (any order)
21. center
22. saturation, cutoff (any order)
23. Linear
24. distorted
25. A
26. A
27. B
28. C
7.4
29. base
30. collector
31. 180°
32. beta
33. emitter, collector
34. in
35. alpha
36. less
37. base, emitter
38. in
39. less
7.5
40. 40.

Glossary

Voltage amplifier

A system designed to develop an output voltage that is greater than its input voltage.

Voltage gain

A ratio of the output signal voltage to the input signal voltage, which is expressed as $A_v = V_{out}/V_{in}$.

Current amplifier

An amplifying system designed to develop output current that is greater than the input current.

Current gain

The ratio of output current to input current, which is expressed as $A_i = I_{out}/I_{in}$.

Power gain

A ratio of the developed output signal power to the input signal power, which is expressed as $A_P = P_{out}/P_{in}$.

Static state

A DC operating condition of an electronic device with operating energy but no signal applied.

Thermal stability

The condition of an electronic device that indicates its ability to remain at an operating point without variation due to temperature.

Fixed biasing

A type of biasing that is very sensitive to changes in temperature. The resulting output of the circuit is difficult to predict.

Self-biasing

A type of biasing in which bias current is used to counteract changes in temperature.

Emitter biasing

A type of biasing that improves thermal stability.

Bypass capacitor

A capacitor that provides an alternate path around a component or to ground.

Self-emitter bias

A type of biasing that is a combination of self-biasing and emitter biasing. The output has reduced gain and the circuit, good thermal stability. However, this type of emitter biasing is not very effective when used independently.

Load line

A line drawn on a collector family of characteristic curves which shows how a device will respond in a circuit with a specific value of load resistor.

Q point

An operating point for an electronic device that indicates its DC or static operation with no signal applied.

Linear

An amplifying circuit that operates in the active region of a collector family of characteristic curves. It provides signal amplification and duplication.

Nonlinear distortion

A characteristic of an output signal in which the amplifying circuit is operating near the saturation or cutoff regions.

Overdriving

The effect of inputting a signal that swings into the saturation region during the positive alternation and the cutoff region during the negative alternation. This distorts both alternations of the output.

Alpha

Current gain of a common-base amplifier, which is expressed as $\alpha = I_C/I_E$.

8

Field-Effect Transistors — (FETs)

A transistor that has only one P–N junction in its construction is called a **unipolar transistor**. Unlike the bipolar transistor, which has two P–N junctions, a unipolar transistor conducts current through a single piece of semiconductor material. Current carriers passing through this material can be either electrons or holes depending on the polarity of the semiconductor. **Field-effect transistors (FETs)** are unipolar devices. These devices represent a unique part of the semiconductor field.

An FET is primarily a three-terminal device that is capable of amplification, switching, and most of the functions of a bipolar transistor. FETs are used as discrete devices or can be fabricated into integrated circuits. FETs fall into two general classifications: **junction field-effect transistor (JFET)** and **insulated gate field-effect transistor (IGFET)**. The JFET was developed first. The IGFET is more commonly referred to as a **metal-oxide semiconductor field-effect transistor (MOSFET)**. Both classifications of the FET play an important role in solid-state electronics and are studied in this chapter.

Objectives

After studying this chapter, you will be able to:

8.1 analyze the operation of a junction field-effect transistor;
8.2 analyze the operation of E-MOSFET, D-MOSFET, and V-MOSFET transistors;
8.3 analyze and troubleshoot field-effect transistors.

Chapter Outline

8.1 Junction Field-Effect Transistors
8.2 Metal-Oxide Semiconductor FETs
8.3 Analysis and Troubleshooting – Field-Effect Transistors

Key Terms

channel
depletion metal-oxide semiconductor field-effect transistor (D-MOSFET)
drain
dynamic transfer curve
enhancement metal-oxide semiconductor field-effect transistor (E-MOSFET)
gate
junction field-effect transistor (JFET)
ohmic region
pinch-off region
source
substrate
transconductance
vertical metal-oxide semiconductor field-effect transistor (V-MOSFET)

8.1 Junction Field-Effect Transistors

The junction field-effect transistor (JFET) is a three-element electronic device. Its operation is based on the conduction of current carriers passing through a single piece of semiconductor material instead of a junction. The two major types of JFETs are N-channel and P-channel. JFET circuit operation can be predicted with a family of characteristic curves chart and a dynamic transfer curve. In this section, you will learn about the characteristics and operation of N-channel and P-channel JFETs. You will use a family of characteristic curves and a dynamic transfer curve to predict the operation of a JFET and will learn how to read and interpret a JFET data specifications sheet.

8.1 Analyze the operation of a junction field-effect transistor.

In order to achieve objective 8.1, you should be able to:

- interpret JFET characteristic curves;
- interpret a dynamic transfer curve;
- explain the relationship between drain current, gate voltage, and drain−source voltage;
- describe the physical construction of a JFET;
- define junction field-effect transistor, channel, gate, substrate, drain, source, dynamic transfer curve, and transconductance.

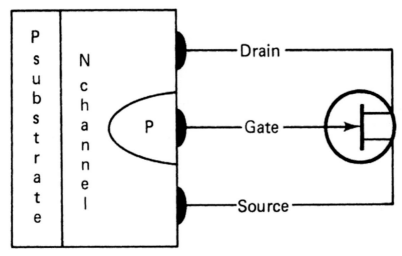

Figure 8.1 Basic JFET crystal structure and symbol.

The operation of a **junction field-effect transistor (JFET)** is based on the conduction of current carriers through a single piece of semiconductor material called a **channel**. **Figure 8.1** shows a cross-sectional view and symbol of a basic JFET crystal that serves as the channel. Note that an additional piece of semiconductor material is diffused into the channel. This element is called the **gate.** The gate (G) surrounds the channel. The assembled device is housed in a package. The physical appearance of a JFET in its housing is very similar to that of a bipolar transistor (see **Figure 8.2**).

N-Channel JFETs

Figure 8.3 shows the crystal structure, element names, and schematic symbol of an **N-channel JFET**. It is constructed of a block of N-type material. The P-type material of the gate is then diffused into the N channel. Lead wires attached to each end of the channel are the **source (S)** and **drain (D)** connections. The **gate** lead is attached to the diffused p-material.

The schematic **symbol** of an N-channel JFET is somewhat representative of its construction. The bar part of the symbol refers to the channel. The **drain** and **source** leads are attached to the channel. The **gate (G)** has an arrow. This shows that it forms a P–N junction. Note that the arrowhead of an N-channel symbol <u>P</u>oints i<u>N</u> toward the channel. This indicates that it is **a**

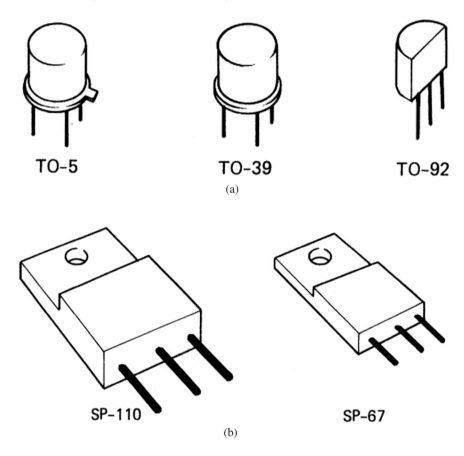

TO-5   TO-39   TO-92

(a)

SP-110   SP-67

(b)

Figure 8.2   JFET packages (Courtesy: Fairchild Semiconductor).

P–N junction. The arrowhead or gate is a P-type material, and the channel is an N-type material. This part of the device responds as a P–N junction diode.

The operation of a JFET is somewhat unusual when compared with that of a bipolar transistor. **Figure 8.3(a)** shows a cross-sectional view of a JFET in a circuit. In **Figure 8.3(b)**, the source and drain leads of the JFET are connected to a DC voltage source. In this case, maximum current flows through the channel. This is called drain current or I_D and is shown by the arrows of the diagram. The value of V_{DD} and the **internal resistance** of the channel determine the amount of channel current flow or I_D. Typical source–drain resistance values of a JFET are several hundred ohms. Essentially, this means

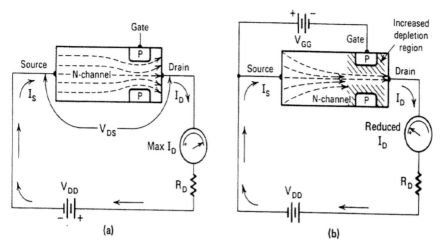

Figure 8.3 N-channel JFET: (a) crystal structure, element names; (b) Biasing.

that full conduction takes place in the channel even when the gate is open. A JFET is, therefore, considered to be normal on a device.

The amount of bias voltage applied to the gate controls the current carriers passing through the channel of a JFET. In normal circuit operation, the gate is reverse biased with respect to the source (see **Figure 8.3(b)**). Reverse biasing of the gate–source of a P–N junction increases the size of its depletion region. In effect, this restricts or depletes the number of majority carriers that can pass through the channel. This means that I_D is controlled by the value of a gate–source voltage (V_{GS}). If V_{GS} becomes great enough, no I_D will be permitted to flow through the channel. The voltage that causes this condition is called the cutoff voltage. I_D can be controlled anywhere between full conduction and cutoff by a small change in gate voltage.

Figure 8.3(b) shows how the JFET responds when V_{DD} and V_{GG} are connected simultaneously. Note that the **gate–source junction** is reverse biased by the V_{GG} battery. This voltage increases the size of the depletion region near the gate. As a result, the number of current carriers in the channel is reduced. An increase in V_{GS} causes a corresponding decrease in I_D. Reducing the value of V_{GS} causes a substantial increase in I_D. Thus, a change in the value of V_{GS} can be used to control the internal resistance of the channel. Variations in V_{GS} are, therefore, used to control channel current, or I_D.

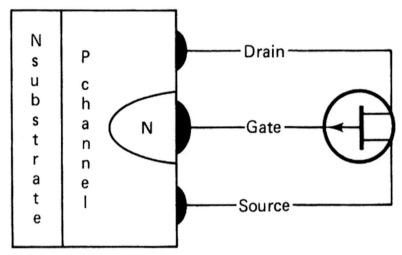

Figure 8.4   P-channel JFET crystal structure, element names, and schematic symbol.

P-Channel JFETs

The crystal structure, element names, and schematic symbol of a **P-channel JFET** are shown in **Figure 8.4**. This device is constructed with a block of P-type material. The N-type material of the gate is diffused into the P-channel. Leads are attached to each end of the channel and to the gate. Other construction details are the same as those of the N-channel device.

The schematic **symbol** of a JFET is different only in the gate element. In a P-channel device, the arrow is Not Pointing toward the channel. This means that the gate is an N-type material, and the channel is a P-type material. In conventional operation of a P-channel JFET, the gate is made positive with respect to the source. Varying values of the reverse-bias gate voltage changes the size of the P–N junction depletion zone. Current flow through the channel can be altered between cutoff and full conduction. P-channel and N-channel JFETs cannot be used in a circuit without changing the polarity of the voltage source.

8.2 JFET Characteristic Curves

The JFET is very different from other solid-state components. A small change in gate voltage, for example, causes a substantial change in drain current. The JFET is, therefore, classified as a voltage-sensitive device. By comparison,

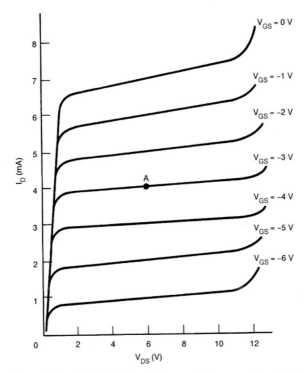

Figure 8.5   A drain family of characteristic curves for a JFET.

bipolar transistors are classified as current-sensitive devices. A JFET has a rather unusual set of characteristics compared with other solid-state devices.

A drain family of **characteristic curves** is shown in **Figure 8.5**. The horizontal part of the graph shows the voltage appearing across the source–drain as V_{DS}. The vertical axis shows the drain current (I_D) in milliamperes. Individual curves of the graph show different values of gate voltage (V_{GS}). The cutoff voltage of this device is approximately -7 V. The control range of the gate is from 0 to -6 V.

A **drain family of characteristic curves** (**Figure 8.5**) tells a great deal about the operation of a JFET. Refer to point A on the -3 V_{GS} curve. If a JFET has -3 V applied to its gate and a V_{DS} of 6 V, approximately 4 mA of I_D flows through the channel. This is determined by projecting a line to the left of the intersection of -3 V_{GS} and 6 V_{DS}. Any combination of I_D, V_{GS}, and V_{DS} can be determined through the drain family of characteristic curves.

Example 8-1:

Refer to **Figure 8.5**. If a JFET has −4 V applied to its gate and a V_{DS} of 6 V, approximately how much drain current (I_D) would flow through the channel?

Solution

1. Mark the intersection of −4 V_{GS} and 6 V_{DS}.
2. Project a line to the left of the intersection of −4 V_{GS} and 6 V_{DS}.

Approximately 3 mA of drain current would flow through the channel.

Related Problem

Refer to **Figure 8.5**. If a JFET has −2 V applied to its gate and a V_{DS} of 4 V, approximately how much drain current (I_D) would flow through the channel?

Developing a JFET Characteristic Curve

A special circuit is used to develop the data for a drain family of characteristic curves. **Figure 8.6** shows a test circuit that is used to identify the different operating regions of an N-channel JFET. Drain current or I_D is an indication of electron flow in the channel. The single characteristic curve of **Figure 8.6(b)** is the result of 0 V_{GS}. This curve shows how the **drain current (I_D)** is affected by drain-to-**source voltage (V_{DS})** when the gate-to-source voltage is 0 V_{GS}. Gate voltage is monitored by a V_{GS} meter connected between the gate−source leads. Drain current (I_D) is measured by a milliampere meter connected in series with the drain. V_{DS} is measured with a voltmeter connected across the source and drain. V_{GS} and V_{DS} are adjusted to different values while I_D is monitored. Two variable DC power supplies are used in the test circuit.

The data points of a single curve are developed by first adjusting V_{GS} to 0 V. Normally, V_{DS} is increased in 0.1-V steps up to 1 V. I_D increases very quickly during this operational time. V_{DS} is then increased in 1-V steps, while the I_D values are recorded. Corresponding I_D and V_{GS} values are then plotted on a graph as the 0-V_{GS} curve.

To develop the second curve, V_{DS} must be returned to zero. V_{GS} is then adjusted to a new value. A value of −1 V_{GS} would be suitable for most JFETS. V_{DS} is again adjusted through its range while I_D is monitored. Data for the second curve is then recorded on the graph.

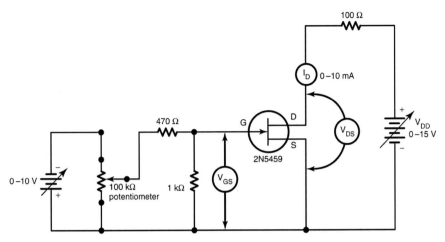

Figure 8.6 A drain family of characteristic curves test circuit for an N-channel JFET.

To obtain a complete drain family of characteristic curves, the process is repeated for several other V_{GS} values. A typical family of curves may have 8–10 different values. The step values of V_{DS}, V_{GS}, and I_D obviously change with different devices. Full conduction is usually determined first. This gives an approximation of representative I_D and V_{DS} values. Normally, I_D levels off to a rather constant value when V_{DS} is increased. V_{DS} can be increased in value to a point where I_D starts a slight increase. Generally, this indicates the beginning of the breakdown region. JFETs are usually destroyed if conduction occurs beyond this area of operation. Maximum V_{DS} values for a specific device are available from the manufacturer. It is a good practice to avoid operation in or near the breakdown region of the device.

Load-Line Analysis

Figure 8.7 shows a **small-signal amplifier** and a drain family of **characteristic curves** for the JFET used in the adjacent circuit. A load line is plotted using the same procedure outlined for the bipolar transistor. The two end points of the load line show how the JFET responds when it is either **fully conductive** or **cut off**. Full conduction occurs when $I_D = V_{DD}/R_L$ and V_{DS} = 0. Cutoff occurs when $V_{DD} = V_{DS}$ and I_D = 0 mA.

The **operating point** of a JFET amplifier is determined by different values of gate–source voltage. For the amplifier of **Figure 8.7**, V_{GS} is supplied by the V_{GG} battery. In this case, V_{GS} is equal to V_{GG} or −2.5 V. As a rule,

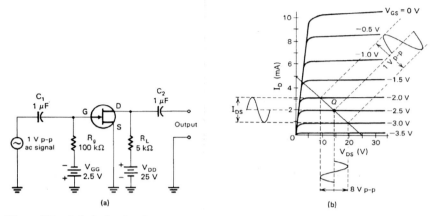

Figure 8.7   A drain family of characteristic curves and JFET amplifier circuit. (a) N-channel JFET amplifier circuit. (b) Drain family of characteristic curves.

the **static operation** of the device is established by the value of V_{GS}. This operating state compares with that of a bipolar transistor. The term **quiescent point**, or **Q point**, is also used to denote this condition of operation.

When an AC signal is applied to a JFET amplifier, it causes the current and voltage values to be in a continuous state of change. An amplifier operating in this manner is considered to be in its dynamic state. An AC signal must be applied to the input of the device for this type of operation to take place. This signal changes the value of V_{GS}. The reverse-bias voltage of the gate then changes in value at an AC rate. This, in turn, changes the value of the gain current passing through the channel. A variation in I_D causes a corresponding change in V_{DS}. The output of the amplifier is based on the changing value of V_{DS}. Note the relationship of V_{GS}, I_D, and V_{DS} in drain family of characteristic curves in **Figure 8.7**.

The **characteristic curve of Figure 8.7(b)** allows us to identify different **operating regions** of a P-channel JFET. Drain current or I_D is an indication of electron flow in the channel. The single I_D/V_{DS} curve of **Figure 8.7(b)** is the result of 0 V_{GS}. This curve shows how the drain current (I_D) is affected by drain-to-source voltage (V_{DS}) when the gate-to-source voltage is 0 V_{GS}. Note that when the voltage source (V_{DD}) of the circuit is increased from 0 V, it causes an increase in the value of V_{DS}. This, in turn, causes a corresponding increase in **drain current**. The curve of **Figure 8.7(b)** shows this as points A and B. In this area of operation, the channel resistance is nearly constant because the depletion region of the channel is not large enough to have any control on I_D. The channel area, therefore, acts as a fixed resistor. It is called the **ohmic region** because Ohm's law says that $I = V/R$ or, in this case, drain current (I_D) = drain−source voltage (V_{DS})/channel resistance (R). The ohmic region is also called the linear area or constant resistance region.

Point B of **Figure 8.7(b)** also represents another parameter called the **pinch-off voltage or V_P**. For the test circuit of **Figure 8.7(a)**, this occurs when $V_{GS} = 0$ V. As shown by the curve, an increase in V_{DS} above the pinch-off voltage produces an almost constant value of drain current. This is shown between points B and C of the curve.

Another parameter of JFET operation is called the constant current area or **saturation region** of operation. This is where channel resistance increases with increasing drain−source voltage and keep I_D fairly constant. **Breakdown** is another parameter of the JFET. This occurs when I_D begins to increase very rapidly with any further increase in V_{DS}. Operating the JFET beyond this point will cause irreversible damage to the channel. This means that JFET operation can only use the areas of the curve between points A and B of **Figure 8.7(b)**.

Example 8-2:

Refer to **Figure 8.7**. If $V_{GS} = -2.0$ V, what are the values of I_{DS} and V_{DS}?

Solution

Locate the -2.0 V_{DS} curve and the intersection of the load line. Project horizontally to find that $I_D = 3.2$ mA. Project vertically from point A to find that $V_{DS} = 8$ V.

Related Problem

Refer to **Figure 8.7**. If $V_{GS} = -3.0$ (Point B), what are the values of I_{DS} and V_{DS}?

Dynamic Transfer Curve

Another way to analyze the operation of a JFET is through the study of its **dynamic transfer curve**. A typical dynamic transfer curve is shown in **Figure 8.8**. This curve plots drain current and gate−source voltage for a given value of drain−source voltage. A constant value of 10 V_{DS} is used for this particular curve.

A dynamic transfer curve is developed from the data of a drain family of characteristic curves. Refer to the 10 V_{DS} line of the curves in **Figure 8.8**. The intersection of each V_{GS} value on the 10 V_{DS} curve represents an I_D value. These I_D values are then plotted on a graph for each value of V_{GS}. The intersecting points are then connected to form the dynamic transfer curve. Essentially, this curve shows how a change in V_{GS} causes a change in I_D.

For the amplifier shown in **Figure 8.7**, the **Q point** is -2.5 V_{GS}. Note the location of this point on the transfer curve of **Figure 8.8**. A 1-Vpp change in V_{GS} causes a change of 2 mA p-p in I_D. This shows a graphic display of the input/output characteristics of a JFET. **Nonlinear distortion** of the output wave can readily be observed by plotting V_{GS} and I_D on the transfer curve.

Example 8-3:

Refer to **Figure 8.8**. If $V_{GS} = -1$ V, what is the value of I_D?
Solution Locate the intersection of the dynamic transfer curve and -1 V_{GS}. Project horizontally to find that $I_D = 6$ mA.
Related Problem If $V_{GS} = -3$V, what is the value of I_D?

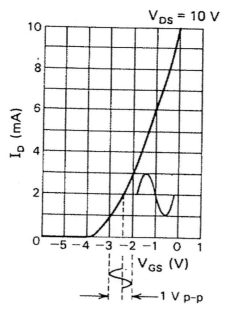

Figure 8.8   Dynamic transfer curve created from the 10-V V_{DS} data from **Figure 8.7**.

Note that the JFET dynamic transfer curve is shaped like a parabola. The value of I_D can be determined for any value of V_{GS} by referring to the data sheet (see **Figure 8.18**) for the JFET by obtaining a value of $V_{GS(off)}$ and I_{DSS}. The equation to calculate I_P is referred to as the "square law" formula. This is due to the squared term in the following formula:

$$I_D = I_{DSS} \left(1 - \frac{V_{GS}}{V_{GS(off)}} \right)^2 \qquad (8.1)$$

where I_{DSS} is the zero gate voltage drain current.

A very useful operating parameter of the JFET can be obtained from a dynamic transfer curve. This property is described as the forward transconductance (gm). **Transconductance** is a measure of the ease with which current passes through the channel of a JFET. It is defined as the ratio of a small change in drain current to a small change in gate–source voltage for a given value of V_{DS} and is expressed by the following formula:

$$\text{gm} = \Delta I_D / \Delta V_{GS}. \qquad (8.2)$$

Transconductance is indicated in **siemens (S)**, which is the fundamental unit of **conductance**. This unit is known as the mho — the reciprocal of an ohm.

Self-Examination

Answer the following questions.

1. A transistor that has only one P–N junction is classified as a(n) _____ device.
2. JFET and MOSFETs are examples of _____ transistors.
3. Current carriers pass through the _____ of a JFET.
4. The bar or straight line of a JFET symbol represents the _____.
5. The _____ and _____ leads of a JFET are attached to the channel.
6. The three elements or leads of a JFET are the _____, _____, and _____.
7. The arrow of a JFET symbol represents the _____.
8. The P–N junction of a JFET represents the _____ and _____.
9. When the gate of a JFET is open or not biased, the channel is _____ conductive.
10. A JFET is considered to be a normally (on, off) device.
11. The gate–source junction of a JFET is _____ biased in normal operation.

12. An increase in gate—source voltage causes a(n) (increase, decrease) in drain current.
13. The two connecting points of a load line for a JFET are _____ conduction and _____.
14. The current flowing through the channel of a JFET is called _____ current.
15. An N-channel JFET is properly biased when the gate is made _____ with respect to the source.
16. When the arrow of a JFET symbol points inward, it indicates a(n) (N, P) channel.
17. A JFET must have the gate _____ biased to achieve control of drain current.
18. The _____ of a JFET can be obtained from a dynamic transfer curve.
19. _____ is a measure of the ease with which current passes through the channel of a JFET.
20. Transconductance of a JFET is measured in _____.

8.3 Metal-Oxide Semiconductor FETs

Metal-oxide semiconductor FETs, or MOSFETs, are a unique variation of the basic transistor family. This type of transistor is classified as an insulated gate field-effect transistor (IGFET) because its gate is completely insulated from the channel by a thin layer of insulating material. As a rule, the term *MOSFET* is more widely used than the term *IGFET*. Like the JFET, operation of a MOSFET is based on the application of a correct voltage value and polarity to the gate. The gate is simply a coating of metal deposited on a layer of insulation attached to the channel. However, voltage applied to the gate causes the gate to develop an electrostatic charge. No current is permitted to flow between the gate and the channel. The control of current carriers through the channel is based on the polarity and value of the gate voltage. E-MOSFETs, D-MOSFETs, and V-MOSFETs are widely used in electronic circuits and are discussed in this section.

8.2 Analyze the operation of E-MOSFET, D-MOSFET, and V-MOSFET transistors.

In order to achieve objective 8.2, you should be able to:

- interpret MOSFET data specifications sheets;
- interpret MOSFET characteristic curves;

- explain the relationship between drain current, gate voltage, and drain−source voltage;
- describe the physical construction of D-MOSFETs, E-MOSFETs, and V-MOSFETs;
- define D-MOSFET, pinch-off region, E-MOSFET, and V-MOSFET.

E-MOSFETs

An **enhancement metal-oxide semiconductor field-effect transistor (E-MOSFET)** is a type of MOSFET in which current carriers are pulled from the substrate to form an induced channel. The channel is produced only when the device is energized. Gate voltage is used to control the size of the induced channel. An **E-MOSFET** is classified as a normally off device.

Figure 8.9 shows a cross-sectional view of the crystal structure, element names, and schematic symbols of the **P-channel** and **N-channel** E-MOSFETs. Note that a defined channel connection does not exist between the source and drain. The entire device is built on a piece of semiconductor material called the substrate. Gate construction is designed to span the entire space between the source and drain. A thin layer of metal is used in the construction of the gate.

The schematic **symbol** shows this by having independent **source, substrate, gate**, and **drain** leads. The middle electrode, which has an arrowhead, shows the polarity of the substrate. When the arrow <u>P</u>oints i<u>N</u> toward the channel, it indicates that the substrate is P-type material and the channel is N-type material, as in (**Figure 8.9(b)**). When the arrow does <u>N</u>ot <u>P</u>oint toward the channel, it shows that the substrate is N-type material and the channel is P-type material, as shown in **Figure 8.9(a)**.

E-MOSFET Operation

The operation of an **E-MOSFET** is primarily based on the voltage value and polarity of its elements. Look at the circuit diagrams in **Figure 8.10**, which illustrate the correct voltage polarity for N-channel and P-channel E-MOSFETs. In circuit, cross-sectional crystal structure views are also included to illustrate the formation of the induced channels.

Note in **Figure 8.10(a)** that when the gate of an **N-channel** device is made positive, electrons are pulled from the P substrate into the source−drain region. This condition causes the channel region to have a high concentration of current carriers. Completion of the channel provides a conduction path for

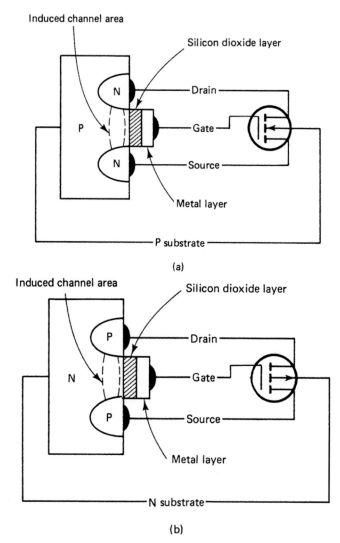

Figure 8.9   E-MOSFET crystal structures, element names, and schematic symbols. (a) P-channel E-MOSFET. (b) N-channel E-MOSFET.

current carriers to flow between the source and drain. The drain current, in effect, is aided or enhanced by the gate voltage (V_G). The absence of gate voltage does not permit the channel to form. The size of the channel and density of the current carriers are directly related to the value of V_G.

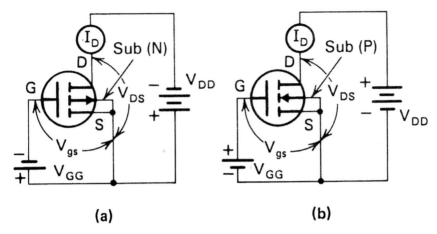

Figure 8.10 E-MOSFET voltage polarities. (a) P-channel. (b) N-channel.

Now, take a look at **Figure 8.10(b)**. Note that the **P-channel** device has the source and drain energized by V_{DD}. The source is positive, and the drain is negative. The gate is made negative with respect to the source by battery V_{GG}. A negative gate pulls holes from the N substrate, forming the induced P-channel. Current flow through the channel is achieved by hole movement. The P–N junction formed by the P-type source and drain and the N-type substrate must be reverse biased. The substrate of both P-channel and N-channel MOSFETs are usually connected to the same **polarity** as the source.

E-MOSFET Characteristic Curves

A drain family of **characteristic curves** for an E-MOSFET is shown in **Figure 8.11**. This set of curves is for an N-channel device. A set of curves for a P-channel device would be very similar. The primary difference is the polarity of the gate voltage. Note that an increase in V_{GS} causes a corresponding increase in drain current. Individual curves represent different values of V_{GS}. Increasing V_{DS} also causes an increase in I_D. The leveling off of I_D indicates the **pinch-off region** of the device. The pinch-off region and the ohmic region of this device are similar to those of the JFET. A drain family of characteristic curves of this type could be used to develop a load line for a representative amplifier. The procedure of creating a **load line** for an **E-MOSFET amplifier** is primarily the same as that for a JFET.

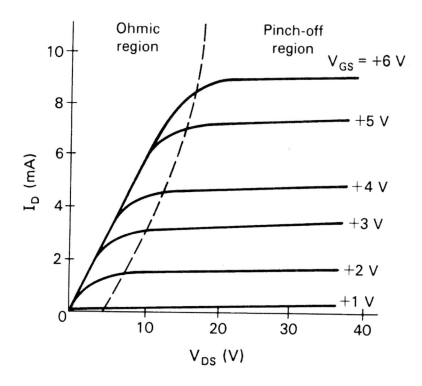

Figure 8.11   A drain family of characteristic curves for an N-channel E-MOSFET. Note that the pinch-off region and the ohmic region of this device are similar to those of the JFET.

D-MOSFETs

A **depletion metal-oxide semiconductor field-effect transistor (D-MOSFET)** has an interconnecting channel built into the substrate. The source and drain are directly connected by the channel. Channel construction of this device is very similar to that of a JFET. The MOSFET responds to the depletion of current carriers that pass through its channel.

Figure 8.12 shows a cross-sectional view of the crystal structure, element names, and symbols of a P-channel and N-channel **D-MOSFET**. In the **N-channel** device shown, a channel of N-type material is formed on a P substrate. The source and drain are connected by this channel. A thin layer of silicon dioxide (SiO2) insulates the gate from the channel. The gate is simply a strip of metal that has been deposited on the SiO2 layer. The entire assembly is built on a P substrate.

The schematic symbol of the device shows the source, substrate, and drain commonly connected together. This indicates that the source and drain are connected by a piece of channel material that is formed on a piece of substrate material. The substrate of the symbol is represented by an arrow. When the arrow <u>P</u>oints i<u>N</u> toward the channel, it indicates that the substrate is P and the channel is N. The current carriers of an N-channel device are electrons.

P-channel device construction is similar to that of the N-channel device. The crystal material of the channel and substrate are of a different polarity. In a P-channel device, the current carriers are holes. The schematic symbol shows that the source, drain, and substrate are commonly connected. The substrate arrow does <u>Not P</u>oint toward the channel. This indicates that the substrate is N-type material, and the channel is P-type material.

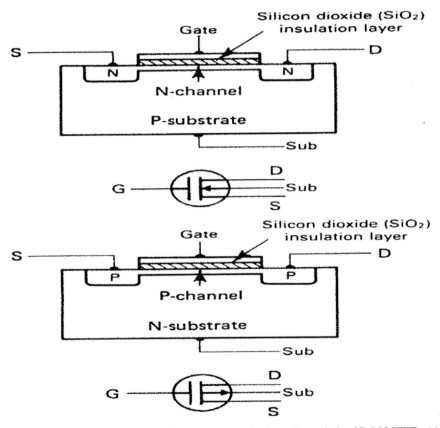

Figure 8.12   Crystal structures, element names, and schematic symbols of D-MOSFETs. (a) N-channel D-MOSFET. (b) P-channel D-MOSFET.

D-MOSFET Operation

The operation of a **D-MOSFET** is quite unusual compared with other FETs. With voltage applied to the source and drain only, drain current flows through the channel. In this condition of operation, the gate does not need voltage to produce conduction. A D-MOSFET, therefore, is classified as a normally on device. Channel current (I_D), however, is controlled by different values of gate voltage. When the channel and gate voltages are of the same polarity, there is a reduction or depletion of current carriers. This condition of operation increases the channel's depletion region. Depleting the number of current carriers in the channel causes a corresponding reduction in I_D. The **depletion mode** of operation represents only one form of control that can be achieved.

If the gate and channel of a D-MOSFET have a different polarity, the number of channel current carriers increases. This means that the device is capable of both depletion and enhancement modes of operation. Making the gate positive in an N-channel device increases the number of current carriers. In effect, this pulls electrons into the channel from the substrate. A negative gate voltage depletes the number of current carriers. This action increases the depletion region of the channel, which reduces I_D. A zero V_{GS} divides the depletion and enhancement modes of operation. In a strict sense, this device could be called an **enhancement-depletion MOSFET**.

D-MOSFET Characteristic Curves

A drain family of characteristic curves for an N-channel **D-MOSFET** is shown in **Figure 8.13**.

A family of curves for a P-channel D-MOSFET would be the same, except that the top V_{GS} lines would be negative values and the gate polarity would be opposite. Note, in this display, that 0 V_{GS} is near the center of the curves. This divides the enhancement and depletion modes of operation. For AC amplification, an ideal operating point is 0 V_{GS}. For the **N-channel** display, the positive alternation of an input signal causes an increase in I_D. The negative alternation produces a decrease in I_D. Can you create an illustration for the above paragraph?

P-channel D-MOSFETs are not as readily available as N-channel devices because P-channel devices are somewhat more difficult to fabricate than N-channel devices and, in general, are more costly to produce. Most applications tend to favor the N-channel D-MOSFET. N-channel devices have better

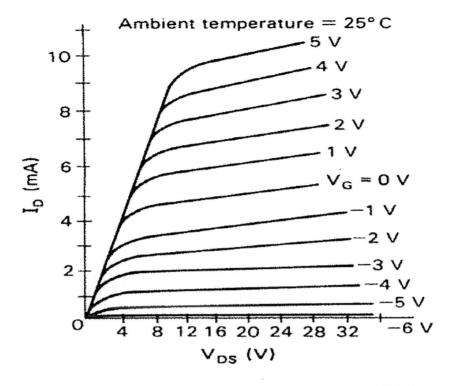

Figure 8.13 A drain family of characteristic curves for an N-channel D-MOSFET.

current mobility. This is a desirable feature in high-frequency AC signal amplification, where these devices are widely used.

V-MOSFETs

Another modification of the field-effect transistor has found its way into solid-state active device technology. This type of device is largely used to replace the bipolar power transistor. It is used in power supplies and solid-state switching applications and is called a **vertical metal-oxide semiconductor field-effect transistor (V-MOSFET), or V-FET.** The V designation refers to vertical-groove MOS technology. This device has a V-shaped groove etched into the substrate. Construction of this type requires less area than a horizontally assembled device. The geometry of a V-MOSFET device also permits greater heat dissipation and high-density channel areas. **V-MOSFETs** have fast switching speeds and lower channel resistance.

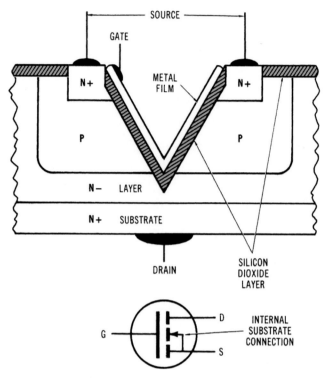

Figure 8.14 A cross-sectional view of an N-channel V-MOSFET with schematic symbol.

Figure 8.14 shows a cross-sectional view of the crystal structure, element names, and schematic **symbol** of an N-channel V-MOSFET. Note that a V-groove is etched in the surface of the structure. From the top, the V-cut penetrates through N^+, P, and N^- layers and stops near the N^+ substrate. The two N^+ layers are heavily doped, and the N^- layer is lightly doped. A thin layer of silicon dioxide covers both the horizontal surface and the V-groove. The source leads on each side of the groove are connected internally. The bottom layer of N^- material serves as a combined substrate and drain. Current carriers move between the source and drain vertically.

V-MOSFET Operation

A **V-MOSFET** responds as an E-MOSFET. No current carriers exist in the source and drain regions until the gate is energized. An **N-channel** device, such as the one in **Figure 8.14**, does not conduct until the gate is made positive with respect to the source. When this occurs, an N channel is induced between the two N^+ areas near the groove. Current carriers can then flow through the vertical channel from source to drain. When the gate of an N-channel device is made negative, no channel exists, and the current carriers cease to flow.

P-channel V-MOSFETs are also available for use in electronic circuit applications. The current carriers of a P-channel device are holes. In general, holes are less mobile than electrons. These devices do not respond as well to high-frequency AC. The primary characteristics and theory of operation are, however, very similar to those of an N-channel V-MOSFET. Differences exist only in voltage polarity and current-carrier flow. As a rule, a wider variety of N-channel devices are manufactured because industrial applications tend to favor the N-channel devices over the P-channel devices. Only N-channel devices are discussed here.

V-MOSFET Characteristic Curves

The **characteristic curves** of a V-MOSFET are very similar to those of an E-MOSFET. **Figure 8.15** shows the drain family of characteristic curves and the dynamic transfer curve of an N-channel V-MOSFET. This particular device is a **power V-MOSFET**. The drain family of characteristic curves in

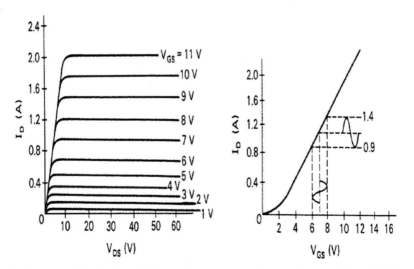

Figure 8.15   A drain family of characteristic curves and a transfer curve for an N-channel V-MOSFET.

Figure 8.15(a) shows that V_{GS} must be positive with respect to the source if there is to be an increase in I_D. The value of V_{GS} also controls I_D for an increase in V_{DS}. Note that V-MOSFETs have ohmic and pinch-off regions very similar to those of an E-MOSFET.

In **Figure 8.15(b)**, the **dynamic transfer curve** of V-MOSFET plots V_{GS} and I_D for a single V_{DS} value. A change in V_{GS} is considered to be a variation in the input voltage. I_D is used to denote an output condition. A transfer curve shows how the output responds to a change in input voltage. This curve is nearly linear over its entire length. The forward transconductance (gm) can be obtained from the data of this curve.

To determine the **forward transconductance** of a V-MOSFET, refer to **Figure 8.15(b)**. In this case, the dynamic transfer curve shows an input voltage change of 2 V_{pp} with an operating point of 7 V_{GS}. A 2-V_{pp} input signal causes the operating point to change from 6 to 8 V. The resulting I_D values are from 0.9 to 1.4 A. This represents a ΔI_D of 0.5 A. The gm is determined to be

$$
\begin{aligned}
\text{gm} &= \Delta I_D / \Delta V_{GS} \\
&= (1.4 - 0.9 \text{ A})/(8 - 6 \text{ V}) \\
&= 0.5 \text{ A}/2 \text{ V} \\
&= 0.25 \text{ S, or } 250 \text{ mS.}
\end{aligned}
$$

Note that the gm of a V-MOSFET is quite large. Compared with other FETs, this is a very significant condition of operation. The voltage gain (A_V) of an FET amplifier is based on gm × R_L. With high values of gm, it is possible to achieve a great deal more voltage gain with a V-MOSFET than with other types of FETs. This feature, its power-handling capability, and high switching speed make the V-MOSFET an extremely attractive amplitude-control device for electronic circuits.

MOSFET Handling Procedures

MOSFETs must receive special care in handling when they are used in an electronic circuit. The silicon dioxide layer that insulates the gate from the source, drain, and substrate is very thin. A static charge or voltage less than 100 V between the gate and channel will cause permanent damage to the insulation material. This generally causes the oxide layer to rupture. As a result, gate leakage current occurs between the gate and channel. This generally causes permanent damage to the device.

The voltage supplied to an operating MOSFET generally comes from a DC power source or supply. The voltage used in most circuit applications must not exceed the breakdown rating of the device. In general, this can be avoided by checking the manufacturer's maximum ratings so that the device is not damaged by excessively high operating voltage values. The correct source voltage values must not be exceeded. The device should not be connected to an energized or operating circuit. As a rule, these precautions apply to nearly all solid-state devices.

Static-charge damage to a MOSFET presents a number of problems that do not occur in other solid-state devices. Friction caused by rubbing the device in a plastic bag or on a piece of fabric during shipping can be damaging. Static buildup on the body of a person handling the MOSFET can also cause it to be damaged. Soldering a device into a circuit with an ungrounded soldering device can cause some form of damage. All of this requires some understanding of the special handling procedure that must be followed. Static damage is typically the most troubling problem in the use of MOSFETs.

Most manufacturers ship MOSFETs with their leads shorted together by means of a shorting ring or wire or pressed into a conducting foam material. The **shorting** method should *not* be removed from the device until the device is installed in its respective circuit. This precaution is extremely important. It is extremely important, however, that the shorting device be completely removed from the device to avoid damage to the circuit. The following precautions should be observed when handling MOSFETs:

- Turn off the power source of the circuit before inserting a MOSFET into it. Voltage transients developed by the circuit will permanently damage the device.
- Use special grounded-tip soldering devices for circuit connections. Do *not* use soldering guns. The tip of a soldering device can be grounded with a clip lead to reduce this problem.
- Neutralize electrostatic body voltage by using a grounding wristband or by connecting a clip lead to your metal watchband.
- Let the shorting ring or conductive foam remain on a MOSFET until all circuit connections are complete and the device is installed in the circuit. Do *not* forget to remove the shorting material from the device.

Most new MOSFETs have **Zener diode** protection built into the device to avoid electrostatic problems. **Figure 8.16** shows how these diodes are connected into the MOSFET internally. Protection is achieved when the

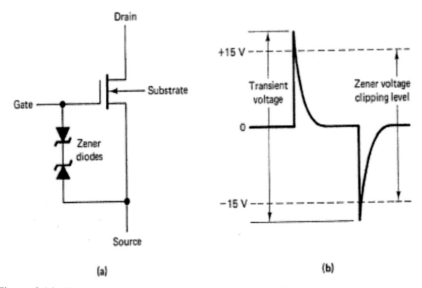

Figure 8.16   Zener diode protection is built into some MOSFETs to avoid electrostatic problems. (a) Zener diode gate protection of a MOSFET. (b) Transient voltage.

diodes become conductive. This occurs when the voltage exceeds 15 V. The Zener diodes are connected in reverse order so that conduction will occur in either polarity of applied voltage. This type of built-in protection reduces most of the in-circuit and out-of-circuit transient problems associated with MOSFET installation. However, it is still recommended that the general precautions for handling MOSFETs be followed to avoid unnecessary damage.

Self-Examination

Answer the following questions.

21. MOSFETs are designed so that the _____ is electrically insulated from the channel.
22. The _____ of a MOSFET is a strip of metal instead of a piece of semiconductor material.
23. Voltage applied to the _____ of a MOSFET causes it to develop an electrostatic charge.
24. In MOSFET operation, no current is permitted to flow between the gate and the _____.

25. A MOSFET is built on a piece of semiconductor material called the _____.

26. The source and drain of an E-MOSFET are not directly connected through a(n) _____.

27. The source and drain of an E-MOSFET are connected by a(n) _____ channel when the device is in operation.

28. A D-MOSFET has an interconnecting channel built in another piece of semiconductor material known as the _____.

29. The schematic symbol of a(n) (enhancement, depletion) MOSFET has independent lines representing the source, substrate, and drain.

30. The majority current carriers of the drain of an N-channel D-MOSFET are _____.

31. The schematic symbol of a(n) (enhancement, depletion) MOSFET has the source, substrate, and gate commonly connected by a line.

32. The majority current carriers passing through the drain of a P-channel E-MOSFET are _____.

33. When (enhancement, depletion) MOSFETs have voltage applied to the source and drain only, drain current will flow through the channel.

34. When the _____ of an E-MOSFET is energized, it pulls current carriers from the substrate to form an induced channel.

35. (Enhancement, Depletion) MOSFETs can operate in either the depletion or enhancement mode of operation.

36. (Enhancement, Depletion) MOSFETs have a 0 V_{GS} value near the centers of their drain family of characteristic curve.

37. In an N-channel E-MOSFET, the polarity of the gate must (match, be opposite) the polarity of the induced channel to produce conduction.

38. V-MOSFETs involve _____ MOS technology.

39. A change of drain current of 1 A caused by a gate−source voltage change of 2 V produces a transconductance of _____ S.

40. V-MOSFETs respond as (enhancement/depletion) MOSFETs.

41. MOSFETs are susceptible to electrostatic damage because of _____ breakdown.

42. New MOSFETs have _____ diodes built in to the structure to protect it from electrostatic damage.

8.4 Analysis and Troubleshooting – Field-Effect Transistors

Field-effect transistors, like bipolar transistors, periodically require testing. The condition of the device can be tested with a curve-tracer oscilloscope, sound-producing instruments, in-circuit/out-of-circuit testers, or the ohmmeter of a multimeter. This equipment is normally used to evaluate a device and to identify leads. Test procedures are rather easy and are a reliable indicator of device condition. This section discusses the use of the ohmmeter to evaluate FETs. This type of instrument is more readily available than specialized testers. Lead identification, shorted devices, open conditions, and gain can be evaluated. If the other equipment listed above is available, become familiar with its operation by reviewing its user manual.

8.3 Analyze and troubleshoot a field-effect transistor.

In order to achieve objective 8.3, you should be able to:

- use an ohmmeter to identify the leads of an FET;
- use an ohmmeter to test the condition of an FET;
- interpret the datasheet of an FET;
- describe how JFETs and MOSFETs respond when tested with an ohmmeter.

In the following procedures, an ohmmeter with straight polarity is described. The black or common lead is negative, and the probe or red lead is positive. If the meter polarity is unknown, test it by connecting a voltmeter across its leads.

Data Sheet Analysis

Data Sheets for field-effect transistors are included at the end of this chapter. The data sheet can be used to derive information for these devices.

Refer to the data sheet for an **N-channel JFET amplifier – BF 245 A**. This particular device is designed for use in very high frequency (VHF) and ultra-high frequency (UHF) amplifiers. Use this data sheet to answer the following:

1) Type of transistor outline (TO) - _____
2) Maximum drain–gate voltage (V_{DG}) = _____ V
3) Maximum gate–source voltage (V_{GS}) = _____ V
4) Maximum gate current (forward bias) (I_{GF}) = _____ mA
5) Gate reverse current (I_{GSS}) = _____ nA = _____ mA

6) Zero gate voltage drain current (I_{DSS}) = _____mA
7) Power dissipation (P_D) at 25°C = _____ mW
8) Gate−source cutoff voltage ($V_{GS(off)}$) = _____V

A data sheet showing the characteristics of a **2N3820 P-channel JFET** is included next. This device is designed to use as a general purpose or audio amplifier and is housed in a **TO-92 package**. Use this data sheet to answer the following:

1) Maximum drain−gate voltage (V_{DG}) = _____V
2) Maximum gate−source voltage (V_{GS}) = _____V
3) Maximum forward gate current (I_{GF}) = _____V
4) Storage temperature range − _____°C to _____ °C
5) Gate reverse current (I_{GSS}) = _____ nA = _____mA
6) Zero gate voltage drain current (I_{DSS}) = _____mA

A data sheet for an **N-channel MOSFET – FQN1N60C** is also included at the end of the chapter. This device may be used for several specialized applications in electronic circuits. Use this data sheet to answer the following:

1. Maximum drain−source voltage (V_{DSS}) = _____V
2. Maximum drain current (I_D) at 25°C = _____mA
3. Maximum gate−source voltage (V_{GSS}) = _____V
4. Maximum power dissipation (P_D) at 25°C = _____mW

FET Testing − JFETs

JFET testing with an ohmmeter is relatively easy. Remember that a JFET is a single bar of silicon with one P–N junction. To identify the leads, first select any two of the three leads. Connect the ohmmeter to these leads. Note the resistance. If it is 100−1000 Ω, the two leads may be the source and drain. Reverse the ohmmeter and connect it to the same leads. If the two leads are actually the source and drain, the same resistance value will be indicated. The third lead is the gate. If the two selected leads show a different resistance in each direction, one must be the gate. Select the third lead and one of the previous leads. Repeat the procedure. A good JFET must show the same resistance in each direction between two leads. This represents the source−drain channel connections.

The gate of a JFET responds as a diode. It will be low resistant in one direction and high resistant in the reverse direction. If the device shows low resistance when the positive probe is connected to the gate and negative to

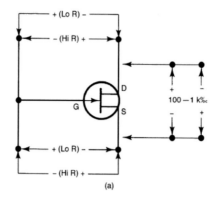

(a)

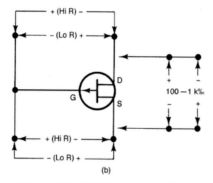

(b)

Figure 8.17   JFET resistance values.

the source or drain, the JFET is an N-channel device. A P-channel device will show low resistance when the gate is made negative and the source or drain positive. The resistance ratio in the forward and reverse direction must be at least 1:1000 for a good device. **Figure 8.17** indicates some representative resistance values for good JFETs.

MOSFET Testing

D-MOSFET testing is very similar to the JFET. The gate of the D-MOSFET is insulated from the channel. It will show infinite resistance between each lead regardless of the polarity of the ohmmeter. This is identified by arrows attached to the gate (G). **Figure 8.18** shows some representative resistance values for good D-MOSFETs. Three-lead devices have the substrate and source internally connected. Four-lead devices have an independent substrate lead.

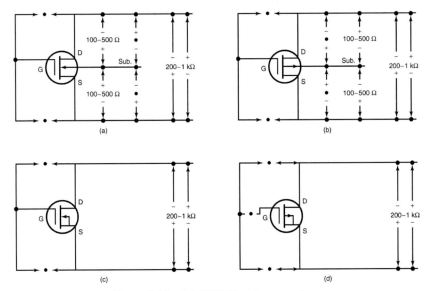

Figure 8.18 D-MOSFET resistance values.

A four-lead device can be identified as an N-channel or a P-channel device. A P-channel device shows low resistance when the positive lead is connected to the substrate and the negative lead is connected to either the source or drain. An N-channel device has low resistance when the negative lead is connected to the substrate and the positive lead to either the source or drain.

E-MOSFET testing with an ohmmeter is not very meaningful. The structure of this device does not interact effectively between different leads. The gate, source, and drain of a good device show infinite resistance between these leads. A four-lead device responds as a diode between the substrate and the source or drain. Polarity of the ohmmeter can be used to identify the substrate and the channel. A three-lead device responds as a diode between the source and drain. **Figure 8.19** shows some representative ohmmeter resistances for good E-MOSFETs.

JFET Troubleshooting

JFET troubleshooting is somewhat easy to diagnose because JFETs have only one junction. Shorts and opens may develop across any two terminals. JFETs can be tested without removing them from their circuits. JFET troubleshooting is similar to bipolar junction transistor troubleshooting.

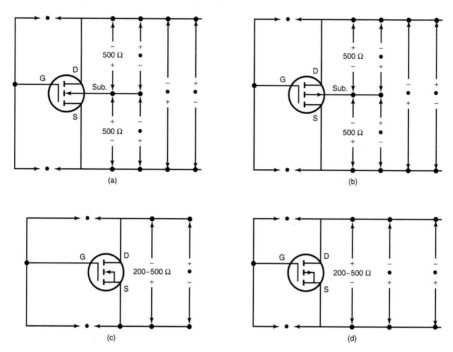

Figure 8.19   E-MOSFET resistance values.

A JFET with a gate−source junction shorted has V_{GS} approximately $= 0$ V. As a result, I_D will approach I_{DSS}. Since there is some gate current, I_S and I_D will not be equal.

A JFET with an open can be tested by checking two simple conditions:

1) V_{GS} does not $= 0$ V;
2) I_D does not $= I_{DSS}$.

The operation of a **JFET amplifier** is shown in **Figure 8.7(a)**. It uses only DC sources that are applied to a JFET for setting up its operating conditions. The DC voltage applied sets up the operating conditions. With DC voltage being applied to the gate of the JFET as shown, a measure of whether it is working satisfactorily is determined by the value of the DC across the **drain−source voltage, V_{DS}**. If the DC value of V_{DS} is very low (0 V), it indicates that there might be an internal **short** between the drain and source terminals. If the DC value of V_{DS} equals the supply voltage, it indicates that there might be an internal **open circuit** between the drain and source terminals. In either case, the JFET is defective.

Summary

- Field-effect transistors (FETs) are unipolar devices.
- The two classifications of field-effect transistors are junction field-effect transistors (JFETs) and metal-oxide semiconductor field-effect transistors (MOSFETs).
- JFETs have three elements: source, drain, and gate.
- The channel of a JFET is a single piece of semiconductor material built on a substrate, and the gate is a piece of semiconductor material diffused into the channel.
- JFET operation is achieved by applying energy to the channel through the source and drain connections.
- A JFET will conduct drain current without gate voltage applied; it is considered a normally on device.
- When the gate on a JFET is reverse biased, it increases the depletion region of the channel and reduces drain current; drain current is, therefore, controlled by the value of reverse-biased gate voltage.
- JFETs are manufactured as N- or P-channel devices.
- A drain family of characteristic curves is used to analyze the operation of field-effect transistors and show drain current and drain−source voltage for different values of gate−source voltage.
- A load line can be developed that shows how an FET operates in a representative circuit; the extreme conditions of operation are full conduction and cutoff.
- A dynamic transfer curve is developed from drain current and drain−source voltage curves to show input characteristics.
- Transconductance can be determined from a dynamic transfer curve.
- Transconductance is a measure of the ease with which current carriers pass through the channel of an FET.
- The unit of transconductance is the siemens (S).
- MOSFETs are a unique variation of the basic field-effect transistor.
- MOSFETs have the gate insulated from the channel by a thin layer of silicon dioxide; the gate is a strip of metal and not a semiconductor material.
- Operation of an E-MOSFET relies on an induced channel; when the gate is energized, it causes current carriers to move into the channel from the substrate.
- An E-MOSFET is considered to be a normally off device; gate voltage must be applied for it to be conductive.

- D-MOSFETs have a channel formed on the substrate; the gate is a piece of metal insulated from the channel by a thin layer of silicon dioxide.
- D-MOSFET operation is based on energizing the channel and applying voltage to the gate; it is considered to have a normally on conduction.
- For an N-channel D-MOSFET, a positive voltage change causes an increase in drain current, and a negative gate−source voltage causes a decrease in drain current.
- A D-MOSFET can be operated as a D-MOSFET or E-MOSFET according to the polarity of the gate voltage.
- V-MOSFETs have a V-groove etched on the surface of the substrate; the channel and gate are then deposited in the groove.
- The construction of a V-MOSFET allows for greater heat dissipation and high-density channel areas.
- The transconductance of V-MOSFETs is quite large compared with that of other FETs; this makes it possible for the V-MOSFET to achieve a great deal more voltage gain.
- MOSFETs are susceptible to damage due to breakdown of the gate insulating material and require special handling.
- MOSFETs are shipped with a shorting strip or placed in conducting foam to prevent electrostatic damage when handling.
- When using MOSFETs, do *not* attach them to energized circuits, solder them into circuits with ungrounded soldering equipment, or handle them without first discharging body static.

Formulas

(8-1) $I_D = I_{DSS} \left(1 - \frac{V_{GS}}{V_{GS(off)}}\right)^2$

(8-2) Transconductance, gm $= \Delta I_D / \Delta V_{GS}$

Answers

Examples

8-1. 5 mA

8-2. $I_D = 1$ mA, $V_{DS} = 9.3$ V

8-3. $I_D = 1$ mA

Self-Examination

8.2

1. unipolar
2. unipolar
3. channel
4. channel
5. source, drain (any order)
6. source, gate, drain (any order)
7. gate
8. gate, channel (any order)
9. fully
10. on
11. reverse
12. decrease
13. full, cutoff
14. drain current (I_D)
15. negative
16. N
17. reverse
18. operation
19. Transconductance
20. mhos

8.3

21. gate
22. gate
23. gate
24. channel
25. substrate
26. channel
27. induced
28. substrate
29. enhancement
30. electrons
31. depletion
32. holes
33. enhancement
34. gate
35. Depletion

36. Depletion
37. be opposite
38. vertical-groove
39. 0.5
40. enhancement
41. insulation
42. Zener

8.4

43. good
44. diode
45. good

Terms

Junction field-effect transistor (JFET)

A three-element electronic device that bases its operation on the conduction of current carriers through a single piece of semiconductor material, called a channel.

Channel

The controlled conduction path of a field-effect transistor.

Gate

The control element of a field-effect transistor.

Substrate

A piece of underlying N or P semiconductor material on which a device or circuit is constructed.

Drain

The output terminal of a field-effect transistor.

Source

The region of a field-effect transistor that is similar to the emitter of a bipolar junction transistor.

Dynamic transfer curve

A graphic display that shows how a change in input voltage (gate−source voltage) causes a change in output current (drain current).

Transconductance

A measure of the ease with which current carriers move through a device.

Enhancement metal-oxide semiconductor field-effect transistor (E-MOSFET)

A type of MOSFET with a conduction characteristic in which current carriers are pulled from the substrate into the channel.

Pinch-off region

The region of an FET where drain-to-source voltage is applied to affect the flow of drain current.

Ohmic region

The linear constant resistance operating region of an FET.

Depletion metal-oxide semiconductor field-effect transistor (D-MOSFET)

A type of MOSFET that has a channel formed on the substrate. The gate is a piece of metal insulated from the channel by a thin layer of silicon dioxide. Operation is based on energizing the channel and applying voltage to the gate.

Vertical metal-oxide semiconductor field-effect transistor (V-MOSFET)

A variation of an E-MOSFET with a higher power handling capability.

Philips Semiconductors

Product specification

N-channel silicon field-effect transistors   BF245A; BF245B; BF245C

FEATURES

- Interchangeability of drain and source connections
- Frequencies up to 700 MHz.

APPLICATIONS

- LF, HF and DC amplifiers.

DESCRIPTION

General purpose N-channel symmetrical junction field-effect transistors in a plastic TO-92 variant package.

CAUTION
The device is supplied in an antistatic package. The gate-source input must be protected against static discharge during transport or handling.

PINNING

PIN	SYMBOL	DESCRIPTION
1	d	drain
2	s	source
3	g	gate

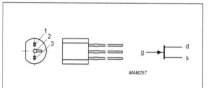

Fig.1   Simplified outline (TO-92 variant) and symbol.

QUICK REFERENCE DATA

SYMBOL	PARAMETER	CONDITIONS	MIN.	TYP.	MAX.	UNIT
V_{DS}	drain-source voltage		–	–	±30	V
V_{GSoff}	gate-source cut-off voltage	I_D = 10 nA; V_{DS} = 15 V	−0.25	–	−8	V
V_{GSO}	gate-source voltage	open drain	–	–	−30	V
I_{DSS}	drain current	V_{DS} = 15 V; V_{GS} = 0				
	BF245A		2	–	6.5	mA
	BF245B		6	–	15	mA
	BF245C		12	–	25	mA
P_{tot}	total power dissipation	T_{amb} = 75 °C	–	–	300	mW
$\|y_{fs}\|$	forward transfer admittance	V_{DS} = 15 V; V_{GS} = 0; f = 1 kHz; T_{amb} = 25 °C	3	–	6.5	mS
C_{rs}	reverse transfer capacitance	V_{DS} = 20 V; V_{GS} = −1 V; f = 1 MHz; T_{amb} = 25 °C	–	1.1	–	pF

N-channel silicon field-effect transistors BF245A; BF245B; BF245C

LIMITING VALUES

In accordance with the Absolute Maximum Rating System (IEC 134).

SYMBOL	PARAMETER	CONDITIONS	MIN.	MAX.	UNIT
V_{DS}	drain-source voltage		–	±30	V
V_{GDO}	gate-drain voltage	open source	–	–30	V
V_{GSO}	gate-source voltage	open drain	–	–30	V
I_D	drain current		–	25	mA
I_G	gate current		–	10	mA
P_{tot}	total power dissipation	up to $T_{amb} = 75\ °C$;	–	300	mW
		up to $T_{amb} = 90\ °C$; note 1	–	300	mW
T_{stg}	storage temperature		–65	+150	°C
T_j	operating junction temperature		–	150	°C

Note

1. Device mounted on a printed-circuit board, minimum lead length 3 mm, mounting pad for drain lead minimum 10 mm · 10 mm.

THERMAL CHARACTERISTICS

SYMBOL	PARAMETER	CONDITIONS	VALUE	UNIT
$R_{th\ j-a}$	thermal resistance from junction to ambient	in free air	250	K/W
	thermal resistance from junction to ambient		200	K/W

STATIC CHARACTERISTICS

$T_j = 25\ °C$; unless otherwise specified.

SYMBOL	PARAMETER	CONDITIONS	MIN.	MAX.	UNIT
$V_{(BR)GSS}$	gate-source breakdown voltage	$I_G = -1\ \mu A$; $V_{DS} = 0$	–30	–	V
V_{GSoff}	gate-source cut-off voltage	$I_D = 10\ nA$; $V_{DS} = 15\ V$	–0.25	–8.0	V
V_{GS}	gate-source voltage	$I_D = 200\ \mu A$; $V_{DS} = 15\ V$			
	BF245A		–0.4	–2.2	V
	BF245B		–1.6	–3.8	V
	BF245C		–3.2	–7.5	V
I_{DSS}	drain current	$V_{DS} = 15\ V$; $V_{GS} = 0$; note 1			
	BF245A		2	6.5	mA
	BF245B		6	15	mA
	BF245C		12	25	mA
I_{GSS}	gate cut-off current	$V_{GS} = -20\ V$; $V_{DS} = 0$	–	–5	nA
		$V_{GS} = -20\ V$; $V_{DS} = 0$; $T_j = 125\ °C$	–	–0.5	μA

Note

1. Measured under pulse conditions: $t_p = 300\ \mu s$; $\delta \leq 0.02$.

N-channel silicon field-effect transistors   BF245A; BF245B; BF245C

DYNAMIC CHARACTERISTICS

Common source; T_{amb} = 25 °C; unless otherwise specified.

SYMBOL	PARAMETER	CONDITIONS	MIN.	TYP.	MAX.	UNIT
C_{is}	input capacitance	V_{DS} = 20 V; V_{GS} = −1 V; f = 1 MHz	–	4	–	pF
C_{rs}	reverse transfer capacitance	V_{DS} = 20 V; V_{GS} = −1 V; f = 1 MHz	–	1.1	–	pF
C_{os}	output capacitance	V_{DS} = 20 V; V_{GS} = −1 V; f = 1 MHz	–	1.6	–	pF
g_{is}	input conductance	V_{DS} = 15 V; V_{GS} = 0; f = 200 MHz	–	250	–	∝S
g_{os}	output conductance	V_{DS} = 15 V; V_{GS} = 0; f = 200 MHz	–	40	–	∝S
$|y_{fs}|$	forward transfer admittance	V_{DS} = 15 V; V_{GS} = 0; f = 1 kHz	3	–	6.5	mS
		V_{DS} = 15 V; V_{GS} = 0; f = 200 MHz	–	6	–	mS
$|y_{rs}|$	reverse transfer admittance	V_{DS} = 15 V; V_{GS} = 0; f = 200 MHz	–	1.4	–	mS
$|y_{os}|$	output admittance	V_{DS} = 15 V; V_{GS} = 0; f = 1 kHz	–	25	–	∝S
f_{gfs}	cut-off frequency	V_{DS} = 15 V; V_{GS} = 0; g_{fs} = 0.7 of its value at 1 kHz	–	700	–	MHz
F	noise figure	V_{DS} = 15 V; V_{GS} = 0; f = 100 MHz; R_G = 1 kΩ (common source); input tuned to minimum noise	–	1.5	–	dB

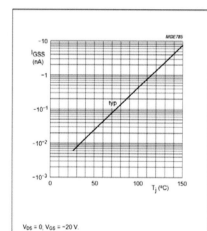

V_{DS} = 0, V_{GS} = −20 V.

Fig.2   Gate leakage current as a function of junction temperature; typical values.

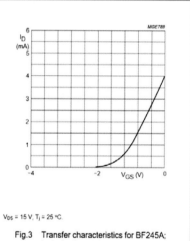

V_{DS} = 15 V, T_j = 25 °C.

Fig.3   Transfer characteristics for BF245A; typical values.

2N3820

P-Channel General Purpose Amplifier

- This device is designed primarily for low level audio and general purpose applications with high impedance signal sources.
- Sourced from process 89.

TO-92

1. Drain 2. Gate 3. Source

Epitaxial Silicon Transistor

Absolute Maximum Ratings* T_C=25°C unless otherwise noted

Symbol	Parameter	Ratings	Units
V_{DG}	Drain-Gate Voltage	-20	V
V_{GS}	Gate-Source Voltage	20	V
I_{GF}	Forward Gate Current	10	mA
T_{STG}	Storage Temperature Range	-55 ~ 150	°C

* This ratings are limiting values above which the serviceability of any semiconductor device may be impaired.

NOTES:
1) These rating are based on a maximum junction temperature of 150 degrees C.
2) These are steady limits. The factory should be consulted on applications involving pulsed or low duty cycle operations.

Electrical Characteristics T_C=25°C unless otherwise noted

Symbol	Parameter	Test Condition	Min.	Typ.	Max.	Units
Off Characteristics						
$V_{(BR)GSS}$	Gate-Source Breakdwon Voltage	$I_G = 10 \times A$, $V_{DS} = 0$	20			V
I_{GSS}	Gate Reverse Current	$V_{GS} = 10V$, $V_{DS} = 0$			20	nA
V_{GS}(off)	Gate-Source Cutoff Voltage	$V_{DS} = -10V$, $I_D = -10 \times A$			8.0	V
On Characteristics						
I_{DSS}	Zero-Gate Voltage Drain Current *	$V_{DS} = -10V$, $V_{GS} = 0$	-0.3		-15	mA
Small Signal Characteristics						
gfs	Forward Transfer Conductance	$V_{DS} = -10V$, $V_{GS} = 0$, f = 1.0KHz	800		5000	∞mhos
C_{iss}	Input Capacitance	$V_{DS} = -10V$, $V_{GS} = 0$, f = 1.0KHz			32	pF
C_{rss}	Reverse Transfer Capacitance	$V_{DS} = -10V$, $V_{GS} = 0$, f = 1.0KHz			16	pF

* Pulse Test: Pulse Width ≤ 300ms, Duty Cycle ≤ 2%

Thermal Characteristics T_A=25°C unless otherwise noted

Symbol	Parameter	Max.	Units
P_D	Total Device Dissipation	350	mW
	Derate above 25°C	2.8	mW/°C
$R_{\theta JC}$	Thermal Resistance, Junction to Case	125	°C/W
$R_{\theta JA}$	Thermal Resistance, Junction to Ambient	357	°C/W

* Device mounted on FR-4 PCB 1.6" : 1.6" : 0.06"

Package Dimensions

TO-92

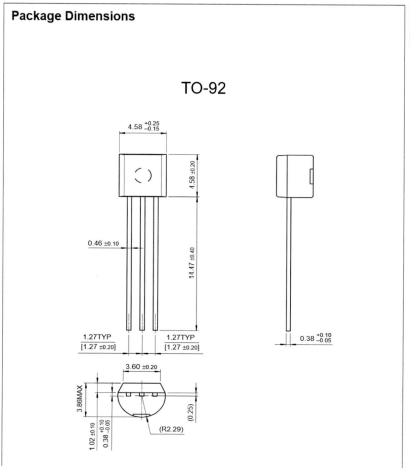

FAIRCHILD
SEMICONDUCTOR®

QFET ®

FQN1N60C
600V N-Channel MOSFET

Features

- 0.3 A, 600 V, $R_{DS(on)}$ = 11.5 Ω @ V_{GS} = 10 V
- Low gate charge (typical 4.8 nC)
- Low Crss (typical 3.5 pF)
- Fast switching
- 100 % avalanche tested
- Improved dv/dt capability

Description

These N-Channel enhancement mode power field effect transistors are produced using Fairchild's proprietary, planar stripe, DMOS technology.
This advanced technology has been especially tailored to minimize on-state resistance, provide superior switching performance, and withstand high energy pulse in the avalanche and commutation mode. These devices are well suited for high efficiency switched mode power supplies, active power factor correction, electronic lamp ballasts based on half bridge topology.

TO-92
SSN Series

G D S

Absolute Maximum Ratings

Symbol	Parameter			FQN1N60C	Units
V_{DSS}	Drain-Source Voltage			600	V
I_D	Drain Current	- Continuous (T_C = 25°C)		0.3	A
		- Continuous (T_C = 100°C)		0.18	A
I_{DM}	Drain Current	- Pulsed	(Note 1)	1.2	A
V_{GSS}	Gate-Source Voltage			± 30	V
E_{AS}	Single Pulsed Avalanche Energy		(Note 2)	33	mJ
I_{AR}	Avalanche Current		(Note 1)	0.3	A
E_{AR}	Repetitive Avalanche Energy		(Note 1)	0.3	mJ
dv/dt	Peak Diode Recovery dv/dt		(Note 3)	4.5	V/ns
P_D	Power Dissipation (T_A = 25°C)			1	W
	Power Dissipation (T_L = 25°C)			3	W
		- Derate above 25°C		0.02	W/°C
T_J, T_{STG}	Operating and Storage Temperature Range			-55 to +150	°C
T_L	Maximum lead temperature for soldering purposes, 1/8" from case for 5 seconds			300	°C

Thermal Characteristics

Symbol	Parameter		Typ	Max	Units
$R_{\theta JL}$	Thermal Resistance, Junction-to-Lead	(Note 6a)	--	50	°C/W
$R_{\theta JA}$	Thermal Resistance, Junction-to-Ambient	(Note 6b)	--	140	°C/W

Package Marking and Ordering Information

Device Marking	Device	Package	Reel Size	Tape Width	Quantity
1N60C	FQN1N60C	TO-92	--	--	2000ea

Electrical Characteristics $T_C = 25°C$ unless otherwise noted

Symbol	Parameter	Test Conditions	Min.	Typ.	Max.	Units
Off Characteristics						
BV_{DSS}	Drain-Source Breakdown Voltage	$V_{GS} = 0$ V, $I_D = 250$ μA	600	--	--	V
$\Delta BV_{DSS}/\Delta T_J$	Breakdown Voltage Temperature Coefficient	$I_D = 250$ μA, Referenced to 25°C	--	0.6	--	V/°C
I_{DSS}	Zero Gate Voltage Drain Current	$V_{DS} = 600$ V, $V_{GS} = 0$ V	--	--	50	μA
		$V_{DS} = 480$ V, $T_C = 125°C$	--	--	250	μA
I_{GSSF}	Gate-Body Leakage Current, Forward	$V_{GS} = 30$ V, $V_{DS} = 0$ V	--	--	100	nA
I_{GSSR}	Gate-Body Leakage Current, Reverse	$V_{GS} = -30$ V, $V_{DS} = 0$ V	--	--	-100	nA
On Characteristics						
$V_{GS(th)}$	Gate Threshold Voltage	$V_{DS} = V_{GS}$, $I_D = 250$ μA	2.0	--	4.0	V
$R_{DS(on)}$	Static Drain-Source On-Resistance	$V_{GS} = 10$ V, $I_D = 0.15$ A	--	9.3	11.5	Ω
g_{FS}	Forward Transconductance	$V_{DS} = 40$ V, $I_D = 0.3$ A (Note 4)	--	0.75	--	S
Dynamic Characteristics						
C_{iss}	Input Capacitance	$V_{DS} = 25$ V, $V_{GS} = 0$ V, $f = 1.0$ MHz	--	130	170	pF
C_{oss}	Output Capacitance		--	19	25	pF
C_{rss}	Reverse Transfer Capacitance		--	3.5	6	pF
Switching Characteristics						
$t_{d(on)}$	Turn-On Delay Time	$V_{DD} = 300$ V, $I_D = 1.1$ A, $R_G = 25$ Ω	--	7	24	ns
t_r	Turn-On Rise Time		--	21	52	ns
$t_{d(off)}$	Turn-Off Delay Time		--	13	36	ns
t_f	Turn-Off Fall Time	(Note 4, 5)	--	27	64	ns
Q_g	Total Gate Charge	$V_{DS} = 480$ V, $I_D = 1.1$ A, $V_{GS} = 10$ V	--	4.8	6.2	nC
Q_{gs}	Gate-Source Charge		--	0.7	--	nC
Q_{gd}	Gate-Drain Charge	(Note 4, 5)	--	2.7	--	nC
Drain-Source Diode Characteristics and Maximum Ratings						
I_S	Maximum Continuous Drain-Source Diode Forward Current		--	--	0.3	A
I_{SM}	Maximum Pulsed Drain-Source Diode Forward Current		--	--	1.2	A
V_{SD}	Drain-Source Diode Forward Voltage	$V_{GS} = 0$ V, $I_S = 0.3$ A	--	--	1.4	V
t_{rr}	Reverse Recovery Time	$V_{GS} = 0$ V, $I_S = 1.1$ A, $dI_F / dt = 100$ A/μs (Note 4)	--	190	--	ns
Q_{rr}	Reverse Recovery Charge		--	0.53	--	μC

Notes:
1. Repetitive Rating : Pulse width limited by maximum junction temperature
2. L = 59mH, I_{AS} = 1.1A, V_{DD} = 50V, R_G = 25 Ω, Starting T_J = 25°C
3. I_{SD} ≤ 0.3A, di/dt ≤ 200A/μs, V_{DD} ≤ BV_{DSS}, Starting T_J = 25°C
4. Pulse Test : Pulse width ≤ 300μs, Duty cycle ≤ 2%
5. Essentially independent of operating temperature
6. a) Reference point of the $R_{θJA}$ is the drain lead
   b) When mounted on 3"x4.5" FR-4 PCB without any pad copper in a still air environment
   ($R_{θJA}$ is the sum of the junction-to-case and case-to-ambient thermal resistance. $R_{θCA}$ is determined by the user's board design)

9

Field-Effect Transistor Amplifiers

In the study of bipolar transistor amplifiers, the primary function of an amplifier is to reproduce an applied signal and provide some level of amplification. **Unipolar transistors** can also be used to achieve this function. JFETs and MOSFETs are examples of unipolar amplifying devices. Amplification using unipolar transistors depends on the FET device used and its circuit components, just like it does with bipolar transistor amplifiers.

The composite circuit of an **FET amplifier** is normally energized by a DC voltage source. An AC signal is then applied to the input of the FET amplifier. After passing through the FET, an amplified version of the signal appears at the output. Operating conditions of the FET and the circuit combine to determine the level of amplification achieved. In this chapter, you will learn about FET amplifier biasing and FET amplifier configurations. It is important to learn the material presented in this chapter because FET amplifiers are used in several types of electronic circuits. FETs are voltage controlled devices that have high gain, low power dissipation in the gate circuit, and a high input impedance. These features enable an FET to function effectively as an amplifier.

Objectives

After studying this chapter, you will be able to:

9.1 describe fixed bias, voltage-divider bias, and self-bias FET circuits;
9.2 analyze common-source, common-gate, and common-drain amplifiers;
9.3 analyze and troubleshoot FET amplifiers.

Chapter Outline

9.1 FET Biasing Methods
9.2 FET Circuit Configurations
9.3 Analyzing and Troubleshooting of FET Amplifier Circuits

9-1 JFET AC Amplifier

In this activity, a test circuit is observed for evaluating the operation of a JFET amplifier.

Key Terms

fixed bias
voltage-divider bias
self-bias
common-drain amplifier
common-gate amplifier
common-source amplifier

9.1 FET Biasing Methods

For an FET operation, biasing refers to the necessary DC voltage needed by the gate with respect to the source lead, that is, the development of a suitable V_{GS} value. By selection of a proper bias voltage, it becomes possible to force a device to operate at a chosen Q point. The V_{DD} voltage applied to the source–drain is generally not considered to be bias voltage. The biasing method selected for a specific circuit depends on the type of device involved. For example, E-MOSFETs and D-MOSFETs require different biasing procedures. This section looks at biasing methods for all FETs.

9.1 Design fixed bias, voltage-divider bias, and self-bias FET circuits.

In order to achieve objective 9.1, you should be able to:

- analyze fixed bias, voltage-divider bias, and self-bias FET circuits;
- describe the characteristics of fixed biasing, voltage-divider biasing, and self-biasing methods;
- recognize fixed biasing, voltage-divider biasing, and self-biasing methods.

Fixed Biasing

The simplest way to bias an FET is with fixed biasing. This method of biasing can be used equally well for JFETs, D-MOSFETs, or E-MOSFETs. In fixed biasing, the correct voltage value and polarity is supplied by a small battery or cell. An electronic power supply could also be used to perform the same operation. **Figure 9.1** shows fixed biasing applied to three different N-channel FET types. If P-channel devices were used, the polarity of V_{GG} and V_{DD} would be reversed. V_{GG} is used as the fixed bias source for the three FET types.

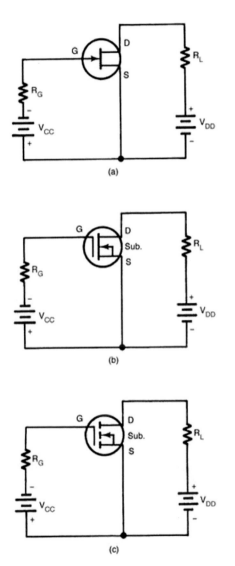

Figure 9.1   Methods of fixed biasing. (a) N-channel JFET. (b) N-channel D-MOSFET. (c) N-channel E-MOSFET.

J-FET Fixed Biasing

Biasing of a JFET circuit is very similar to a bipolar junction transistor (BJT). The circuits of **Figure 9.1(a)** shows fixed biasing of N-channel JFET. Note the polarity of the V_{GG} and V_{DD} power sources.

Example 9-1:

Refer to Figure 9.1(**a**). Assume that $V_{GG} = -5V$, $I_D = 10mA$, $V_{DD} = 12V$, and $R_L = 1k\Omega$. Calculate V_{RL} and V_{DS} for the N-channel FET circuit.

Solution

$$V_{RL} = I_D \times R_L = 10mA \times 1k\Omega = 10V$$
$$V_{DS} = V_{DD} - V_{RL} = 12 - 10V = 2V$$

Related Problem

Refer to **Figure 9.1**. Assume that $V_{GG} = 5V$, $I_D = 5mA$, $V_{DD} = 10V$, and R_L =1.5kΩ. Calculate V_{DS} and V_{RL}.

D-MOSFET Fixed Biasing

D-MOSFET fixed biasing, shown in **Figure 9.2**, is similar to JFETs. In this type of device, the operating point is often $V_{GS} = 0V$ and is referred to as

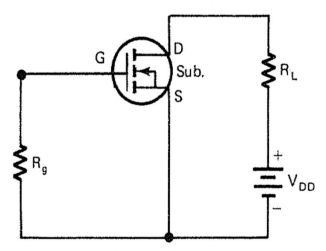

Figure 9.2   Fixed biasing of an N-channel D-MOSFET.

zero bias. No voltage source is needed to establish the operating point. R_g is normally of a very high value. Typically, values of $1-10\text{M}\Omega$ are used. R_g must be included in the circuit. It serves as a return path from the gate to the common connection point.

D-MOSFET circuits are very similar to JFET circuits in terms of biasing. All of the circuit relationships for JFET circuits remain the same for D-MOSFETs.

E-MOSFET Biasing

Enhancement-mode operation requires a positive value of V_{GS}. **Voltage divider bias** is typically used for E-MOSFET circuits. **Figure 9.3(a)** shows an N-channel MOSFET and **Figure 9.3(b)** shows a P-channel MOSFET.

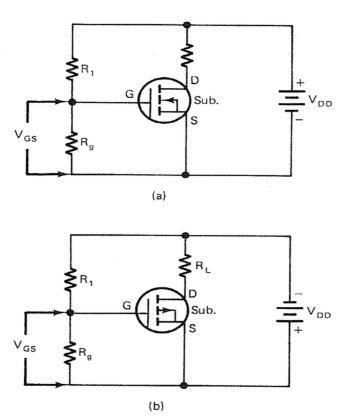

(a)

(b)

Figure 9.3 Voltage-divider method of biasing for E-MOSFETs. (a) N-channel. (b) P-channel.

Voltage-Divider Biasing

Voltage-divider biasing of FETs is rather easy to achieve. It is applicable only to E-MOSFETs. With these devices, V_{GS} and V_{DD} are of the same polarity. As a result of this condition, a single voltage source can be used as a supply. The V_{GS} voltage is only a fraction of the value of V_{DD}. The divider resistors are used to develop the necessary voltage values. **Figure 9.3**shows the divider method of biasing for N-channel and P-channel E-MOSFETs.

The values of R_1 and R_g must be properly chosen to establish a desired operating point. By the voltage-divider equation, we find that the bias voltage is

$$V_{GS} = V_{Rg} = V_G = V_{DD} \times \frac{R_2}{R_1 + R_2}. \tag{9.1}$$

Example 9-2:

Refer to **Figure 9.3(a)**. Assume that $R_1 = 1M\Omega$, $R_2 = 1M\Omega$, $V_{DD} = 24V$, $R_L = 1k\Omega$, and $I_D = 5mA$. Calculate V_G and V_{RL}.

Solution

$$V_G = V_{DD} \times \frac{R_2}{R_1 + R_2}$$

$$V_G = 24\,\text{V} \times \frac{1\,\text{M}\Omega}{1\,\text{M}\Omega + 1\,\text{M}\Omega} = 12\,\text{V}$$

$$V_{RL} = I_D \times R_L$$

$$= 5\text{mA} \times 1\text{k}\Omega$$

$$= 5\text{V}.$$

Related Problem

Refer to **Figure 9.3(b)**. Assume that $R_1 = 5M\Omega$, $R_2 = 2M\Omega$, $V_{DD} = 30V$, $R_L = 2.2k\Omega$, and $I_D = 10mA$. Calculate V_G and V_{RL}.

Self-Biasing

Self-biasing is commonly called **source biasing**. Self-biasing is rather easy to recognize. The source current of an FET is used to develop the bias voltage. The voltage developed is achieved by a resistor (R_S) connected in series with the source. Current flow through the source–drain causes a voltage drop

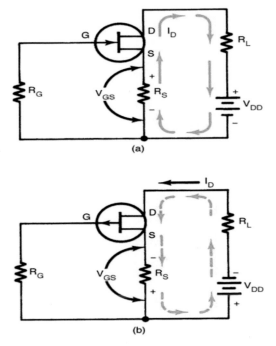

Figure 9.4 Self-biasing of JFETs. (a) N-channel. (b) P-channel.

across R_S. The polarity of the developed voltage is based on the direction of current flow through R_S. **Figure 9.4** shows biasing for N-channel and P-channel **JFETs** using self-biasing.

For the N-channel circuit, current flow through R_S causes the source to be slightly positive with respect to the gate. With R_g connected to the lower side of R_S, R_g is negative and the source positive. This value of R_S and drain current (I_D) determines the bias operating point of the circuit. Note the polarity difference in the P-channel circuit of **Figure 9.4(b)**.

Self-Examination

Answer the following questions.

1. In fixed biasing of a MOSFET, voltage is supplied to the _____ by a battery or DC power supply.
2. The voltage-divider method of biasing applies only to (enhancement, depletion) MOSFETs.

3. Source biasing has a _____ connected in series with the source of a JFET to develop bias voltage.
4. An N-channel JFET with fixed biasing has a _____ polarity applied to the source.
5. A D-MOSFET with $V_{GS} = 0V$ applied is referred to as _____ bias.
6. E-MOSFETs require a _____ polarity applied to the gate.
7. With voltage-divider biasing, V_{GS} and V_{DD} are _____ polarity.
8. Self-biasing is also referred to as _____ biasing.

9.2 FET Circuit Configurations

FET amplifiers, like their bipolar counterparts, can be connected in three different circuit configurations. These are described as common-source, common-gate, and common-drain circuits. One lead of an FET is connected to the input, and a second lead is connected to the output. The third lead is commonly connected to both the input and output. This lead is used as a circuit reference point and is often connected to the circuit ground – hence, the terms *grounded* or *common source*, *grounded* or *common gate*, and *grounded* or *common drain*. The terms *common* and *ground* refer to the same type of connection. In this section, you will study the common-source, common-gate, and common-drain amplifier circuit configurations.

9.2 Analyze common-source, common-gate, and common-drain amplifiers.

In order to achieve objective 9.2, you should be able to:

- describe the characteristics of common-gate, common-source, and common-drain amplifier circuits;
- recognize of common-source, common-gate, and common-drain amplifier circuits;
- define *common-source amplifier*, *common-gate amplifier*, and *common-drain amplifier*.

Common-Source Amplifier

The **common-source amplifier** is the most widely used FET circuit configuration. This circuit is similar in many respects to the common-emitter bipolar amplifier. The input signal is applied to the gate–source, and the output signal is taken from the drain source. The source lead is common to both input and output.

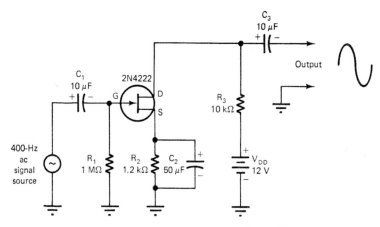

Figure 9.5 Common-source JFET amplifier.

A practical **common-source amplifier** is shown in **Figure 9.5**. This circuit is **self-biased**. The device can be a JFET, a D-MOSFET, or an E-MOSFET. Circuit characteristics are very similar for all three devices. The signal processed by the common-source amplifier is applied to the gate−source. Self-biasing of the circuit is achieved by the source resistor R_2. This voltage establishes the static operating point. The incoming signal voltage is superimposed on the gate voltage. This causes the gate voltage to vary at an AC rate, which causes a corresponding change in drain current. The output voltage developed across the source and drain is inverted 180°. I added the highlighted text about self-biasing. Is this correct and OK?

Voltage gain (A_V) for a common-source JFET amplifier:

$$A_V = V_{DS}/V_{GS}. \tag{9.2}$$

Typical A_V values for a common-source FET amplifier circuit are 5−10. The **input impedance** is extremely high for nearly any signal source. One to several megohms is common. **The output impedance** (Z_{out}) is moderately high. Typical values are in the range of 2−10kΩ. Z_{out} is dependent primarily on the value of R_L.

A common-source amplifier has very high input impedance and relatively high output impedance. Circuits of this type are extremely valuable as **impedance-matching** devices. Common-source amplifiers are used exclusively as voltage amplifiers. They respond well in radio-frequency signal applications.

Common-Gate Amplifier

The **common-gate amplifier** is similar in many respects to the common-base bipolar transistor circuit. The input signal is applied to the source–gate, and the output appears across the drain–gate. Note that the gate is connected to the ground. The common-gate amplifier is capable of a rather significant amount of voltage gain. Current gain is not an important circuit consideration. The gate does not ordinarily have any current flow. JFETs, E-MOSFETs, and D-MOSFETs may all be used as common-gate amplifiers.

A practical **common-gate amplifier** is shown in **Figure 9.6**. This circuit employs an N-channel JFET. The operating voltages are the same as those of the common-source circuit. Self-biasing is achieved by the source resistor R_1. This voltage is used to establish the static operating point. An input signal is applied to R_1 through capacitor C_1. A variation in signal voltage causes a corresponding change in source voltage. Making the source more positive has the same effect on I_D as making the gate more negative. The positive alternation of the input signal makes the source more positive. This, in turn, reduces drain current. With less I_D, there is a smaller voltage drop across the load resistor R_2. The drain or output voltage, therefore, swings positive. The negative alternation of the input reduces the source voltage by an equal amount. This is the same as making the gate less negative. As a result, I_D is increased. The voltage drop across R_2 similarly increases. This, in turn,

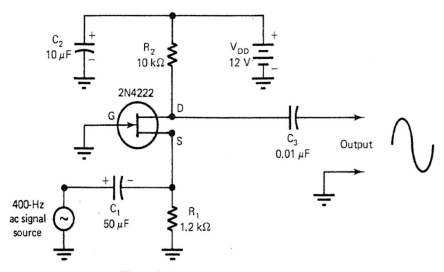

Figure 9.6   Common-gate JFET amplifier.

causes the drain or output to be less positive or negative going. The input signal is, therefore, in phase with the output signal.

The common-gate amplifier has a number of rather unusual character-istics. Its **voltage gain** is somewhat less than that of the common-source amplifier. Representative values are 2–5. The common-gate circuit has very low **input impedance** (Z_{in}). The **output impedance** (Z_{out}) is rather moder-ate. Typical Z_{in} values are 200–1500Ω, with Z_{out} being 5–15kΩ. This type of circuit configuration is often used to amplify **radio-frequency** signals. Amplification levels are very stable for radio frequencies, without feedback between the input and output.

Common-Drain Amplifier

A **common-drain amplifier** has the input signal applied to the gate and the output signal removed from the source. The drain is commonly connected to one side of the input and output. Common-drain amplifiers are also called **source followers**. This circuit has similar characteristics to those of the common-collector bipolar amplifier.

Figure 9.7 shows a practical **common-drain amplifier** using an N-channel JFET. The input of this amplifier is primarily the same as that of

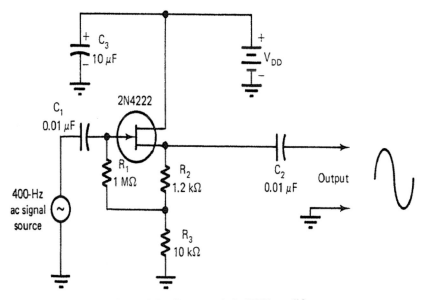

Figure 9.7 Common-drain JFET amplifier.

the common-source amplifier. The **input impedance** is, therefore, very high. Z_{in} is determined largely by the value of R_1. The operating point of our amplifier is determined by the value of R_2. Essentially, this circuit has the same operating point as our other circuit configurations. Resistor R_3 has been switched from the drain to the source in this circuit. Resistors R_2 and R_3 are combined to form the load resistance. The **output impedance** is based on this value.

When an AC signal is applied to the input of the amplifier, it changes the gate voltage. The DC operating point is established by the value of source resistor R_2. The positive alternation of an input signal makes the gate less negative. This causes an N-channel device to be more conductive. With more current through R_3 and R_2, the source swings positive. The negative alternation of the input then makes the gate more negative. This action causes the channel to be less conductive. A smaller current, therefore, causes the source to swing negative. In effect, this means that the input and output voltage values of the amplifier are in step with each other. These signals are in phase in a common-drain amplifier.

Common-drain amplifiers are primarily used as **impedance-matching** devices. It has a high-input impedance and a low-output impedance. They are, thus, frequently used to match a high-impedance device to a low-impedance load. This type of circuit is capable of handling a high-input signal level without causing **distortion**. The input impedance places a minimum load on the signal source. Common-drain amplifiers are used frequently to match high-impedance devices such as microphones and phonograph pickups to the input of an audio amplifier.

Self-Examination

Answer the following questions.

9. In a common-(source, gate, drain) amplifier, the input is applied to the gate−source, and the output is developed across the source−drain.
10. The input and output signals of a common-source amplifier are inverted by _____.
11. The input signal of a common-(source, gate, drain) amplifier is applied to the source−gate, and the output appears across the drain−gate.
12. The input signal of a common-(source, gate, drain) amplifier is applied to the gate, and the output signal is removed from the source.
13. JFET amplifiers are considered to be _____ sensitive devices.

14. When a common source JFET circuit is in cutoff, I_D is at (0 A, full conduction), and V_{DS} is at (ground, supply V_{DD}).

9.3 Analysis and Troubleshooting – JFET Amplifiers

By examining the input and output waveforms of a JFET, its performance as an amplifier is evaluated. Faults in the operation of the amplifier are indicated by the value and shape of the output waveform. Under the proper operating conditions established by the DC sources, when a small AC signal is applied, some level of amplification should occur. If little or no amplification occurs, or the output waveform is distorted, this indicates a faulty condition. In this section, a JFET amplifier will be examined to determine its amplification level and possible faulty conditions.

9.3 Troubleshoot an FET amplifier.

In order to achieve objective 9.3, you should be able to:

- observe the input and output AC waveforms of a JFET amplifier on an oscilloscope;
- distinguish between normal and faulty operations of a JFET amplifier.

JFET Troubleshooting

The operation of a JFET amplifier is shown in **Figure 9.7**. It uses both DC and AC sources that are applied to a JFET. This circuit has capacitors which are needed for passing or blocking AC. The DC voltage applied sets up the operating conditions. When an AC voltage is applied to the gate of the JFET, a measure of whether it is working satisfactorily is determined by the value of the AC across the drain-source voltage, V_{DS}. If the AC value of V_{DS} is very low (0V), it indicates that there might be an internal short or open circuit between the drain and source terminals. In either case, the JFET is defective. An **oscilloscope** can be used to observe the amplification of the AC input signal. If the AC output appearing across V_{DS} is distorted, it indicates that the amplifier is not functioning properly.

Summary

- JFETs and MOSFETs can be used for amplification.
- Biasing of a JFET circuit is very similar to bipolar junction transistor circuits.

- D-MOSFET biasing is similar to JFET circuits.
- E-MOSFET biasing requires a positive value of V_{GS}.
- Voltage divider bias is commonly used for E-MOSFET circuits.
- With voltage-divider biasing, V_{GS} and V_{DD} are of the same polarity.
- Self-biasing is also referred to as source biasing.
- FETs can be connected in three different circuit configurations: common-source, common-gate, and common-drain.
- The common-source amplifier is the most widely used.
- With a common-source amplifier, the signal being processed is applied to the gate−source, and the output signal is developed across the source−drain.
- The output signal of a common-source amplifier is inverted 180°.
- The common-gate amplifier has the input signal applied to the source−gate and the output signal developed across the drain−gate.
- Common-drain amplifiers have the input signal applied to the gate, and the output signal removed from the source.
- Common-drain amplifiers are primarily used as impedance-matching circuits; they can match high-impedance devices to a low-impedance output.
- Common-drain amplifiers are also called *source followers*.

Formulas

(9-1) Voltage-divider biasing, $V_G = V_{DD} \times \frac{R_2}{R_1 + R_2}$

(9-2) Voltage gain for a common-source JFET, $A_V = V_{DS}/V_{GS}$

Review Questions

Answer the following questions.

1. Which circuit configuration of the JFET has a low input impedance:
   - (a) Common-source
   - (b) Common-gate
   - (c) Common-drain

2. Which circuit configuration of the JFET has a low output impedance:
   - (a) Common-source
   - (b) Common-gate
   - (c) Common-drain
   - (d) All configurations

3. Which circuit configuration of the JFET is commonly used as a voltage amplifier:

   (a) Common-source
   (b) Common-gate
   (c) Common-drain
   (d) All configurations

4. Which circuit configuration of the JFET has the gate connected in reverse bias:

   (a) Common-source
   (b) Common-gate
   (c) Common-drain
   (d) All configurations

5. A common-source JFET amplifier configuration has a _____ phase difference between the applied AC signal input and the AC output.

   (a) $0°$
   (b) $90°$
   (c) $180°$
   (d) $270°$

6. In circuit configuration of the JFET, is the input AC signal applied to the gate and the output signal developed at the source:

   (a) Common-source
   (b) Common-gate
   (c) Common-drain
   (d) All configurations

Problems

Answer the following:

Refer to **Figure 9.2(a)** and answer the following Assume V_{GG} =-5V, I_D = 5mA, V_{DD} = 12V, R_L = 1.5kΩ.

1. The value of V_{RL} is:

   (a) 2.5V
   (b) 5V
   (c) 7.5V
   (d) 10V

2. The value of V_{DS} is:
   (a) 1.5V
   (b) 3V
   (c) 4.5V
   (d) 6V

3. The type of biasing used in this circuit is:
   (a) Gate
   (b) Self
   (c) Voltage-divider
   (d) All of the above

4. If the load current I_D decreases to 2.5mA, the voltage across the load resistor V_{RL} will
   (a) Increase
   (b) Decrease
   (c) Be unchanged
   (d) 0V

5. If the voltage applied to the gate changes from -5 to 0V, the output current I_D will:
   (a) Increase
   (b) Decrease
   (c) Be unchanged
   (d) 0A

Answers

Examples

9-1 $V_{RL} = 2.5V$, $V_{DS} = 2.5V$
9-2 $V_G = 8.57V$, $V_{RL} = 22V$

Self-Examination

9.1
   1. gate
   2. enhancement
   3. resistor

4. negativ
5. zero
6. positive
7. same
8. source
9.2
9. source
10. 180°
11. gate
12. source
13. voltage
14. 0A, supply V_{DD}

Terms

Gate biasing of JFETs

Connecting a voltage supply to the gate of a JFET and a resistor for ensuring that the gate−source junction is reverse biased.

Self-biasing of JFETs

Connecting a resistor to the gate of a JFET for developing an $I_D R_S$ voltage across the resistor when the drain current I_D. A resistor is connected in the gate circuit (with or without a voltage), so that the gate−source junction is reverse biased.

Voltage-divider biasing of JFETs

Reverse biasing the gate−source junction of a JFET by using a resistor network in the gate circuit. A resistor is connected between the supply and the gate, and another between the gate and the ground. These resistors reverse bias the gate−source junction.

Common-source amplifier

An amplifier in which the input signal is applied to the gate−source, and the output signal is taken from the drain source.

Common-gate amplifier

An amplifier in which the input signal is applied to the source−gate, and the output appears across the drain−gate.

Common-drain amplifier

An amplifier configuration in which the input signal is applied to the gate, and the output signal is removed from the source. It is also called a *source follower*.

10

Amplifying Systems

Amplifying systems are widely used in the electronic equipment to increase the power, voltage, or current of AC signals. An amplifying system may be only one part or section of a rather large and complex system. A **television** receiver is a good example of this application. Several individual amplifying systems are included in a television receiver, such as audio and video amplifiers. A DVD player, by comparison, has only one amplifying system. Its primary function is to process a sound signal and build it to a level that will drive a speaker. Transistors and integrated circuits are the basic devices used for this type of operation. The function of an amplifier is basically the same regardless of its application: to amplify a signal.

In this chapter, you will learn how amplifying devices are used in an amplifying system. A number of specific circuit functions are discussed, such as voltage and **power gain, coupling, and signal processing**. Input and output **transducers**, such as microphones and speakers, are covered as well since they also play a role in the operation of an amplifying system.

Objectives

After studying this chapter, you will be able to:

10.1 analyze the stages of amplification;
10.2 select the appropriate coupling component for an amplifying system;
10.3 explain the operation of amplifying system transducers;
10.4 design basic amplifier circuits;
10.5 analyze and troubleshoot amplifier systems.

Chapter Outline

10.1 Amplifying System Basics
10.2 Amplifier Coupling

10.3 Input and Output Transducers
10.4 Analysis and Troubleshooting – Amplifying Systems

Activities for Chapter 10

10-1 *RC*-Coupled Amplifier

In this activity a multistage BJT amplifier circuit using *RC* coupling between the stages will be evaluated.

10-2 Darlington Transistor Amplifier

In this activity, a Darlington BJT amplifier circuit will be evaluated.

Key Terms

Bel (B)
bode plot
capacitive coupling
cascade
common logarithm
crystal microphone
Darlington amplifier
decibel (dB)
direct coupling
dynamic microphone
impedance ratio
intermediate range speaker
logarithmic scale
mantissa
microphone
output transformer
piezoelectric effect
stage of amplification
transducer
transformer coupling
tweeter
voice coil
woofer

10.1 Amplifying System Basics

Regardless of its application, an amplifying system has a number of primary functions that must be performed. These functions are energy conversion, amplification, and power distribution. An understanding of these functions is an extremely important part of operational theory. To understand these functions, you should be familiar with concepts such as amplifier gain, decibels, and frequency response. These functions and concepts are covered in this section.

10.1 Analyze the stages of amplification.

In order to achieve objective 10.1, you should be able to:

- determine the gain at various stages of amplification in standard and decibel form;
- describe the stages of amplification;
- identify the functions of an amplifying system;
- define *stage of amplification, transducer, cascade, logarithmic scale, common logarithm, mantissa,* Bel, and *decibel.*

Amplifying System Functions

The major functions of an amplifying system are **energy conversion, amplification**, and **power distribution**. These functions are achieved through the input transducer, amplifier, power supply, and output transducer. A block diagram of an amplifying system is shown in **Figure 10.1**. The triangular-shaped items of the diagram show where the amplification function is performed. A **stage of amplification** is represented by each triangle. Three amplifiers are included in this particular system.

A **stage of amplification** consists of active devices (usually transistor circuits) and all associated components. Small-signal amplifiers are used in the first three stages of this system. The amplifier on the right side of the diagram is an output stage. A rather large signal is needed to control the output amplifier. An output stage is generally a power amplifier, or a large-signal amplifier. In effect, this amplifier is used to control a rather large amount of current and voltage. Remember, power is the product of current and voltage.

In an **amplifying system**, a signal must be developed and applied to the input. The source of this signal varies a great deal with different systems. In a DVD player, an audio signal is produced. Variations in electronic signals

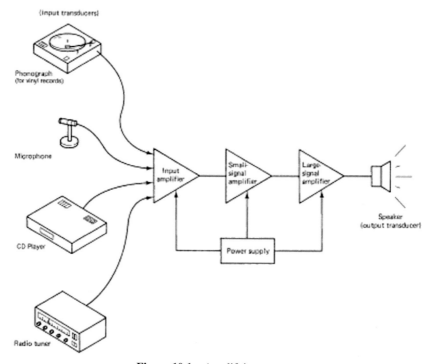

Figure 10.1   Amplifying system.

are changed into sound energy. The **amplitude** level of the electronic signal is increased to a suitable level by amplifying devices. A variety of different input signal sources may be applied to the input of an amplifying system.

In other systems, the signal is also developed by the input. An input transducer is responsible for this function. A **transducer** changes energy of one form into energy of a different form. Several types have been used over the years for home entertainment systems. Microphones, vinyl phonograph pickup cartridges, CD players, and DVD players are input transducers. Input signals may also be received through the air. Antennas may serve as the input transducer for this type of system. An antenna changes electromagnetic waves into radio frequency (RF) voltage signals. The signal is then processed through the remainder of the system.

Signals processed by an amplifying system are ultimately applied to an **output transducer**. This type of transducer changes electrical energy into another form of energy. In a sound system, the speaker is an output transducer. It changes electrical energy into sound energy. The speaker performs work when it achieves this function. Lamps, motors, relays, transformers, and inductors are frequently considered to be output transducers since they perform work. An output transducer is considered to be the load of a system.

For an amplifying system to be operational, it must be supplied with electrical energy. A DC power supply performs this function. A relatively pure form of DC must be supplied to each amplifying device. In most amplifying systems, AC is the primary energy source. AC is changed into DC, filtered, and, in some systems, regulated before being applied to the amplifiers. The reproduction quality of the amplifier depends, to a large extent, on the quality of the DC power supply. Batteries may also be used to energize some portable stereo amplifiers.

Amplifier Gain

You should recall that the gain of an amplifier system can be expressed in a variety of ways. Voltage, current, power, and, in some systems, decibels, are expressed as gain. Nearly, all input amplifier stages are voltage amplifiers. These amplifiers are designed to increase the voltage level of the signal. Several voltage amplifiers may be used in the front end of an amplifier system. The voltage value of the input signal usually determines the level of amplification achieved.

A three-stage voltage amplifier is shown in **Figure 10.2**. These three amplifiers are connected in **cascade**. The term **cascade** refers to a series of amplifiers in which the output of one stage of amplification is connected to the input of the next stage of amplification. The voltage gain of each stage can be observed with an oscilloscope. The waveform shows representative signal levels. Note the voltage-level change in the signal and the amplification factor of each stage.

The first stage has a **voltage gain** of 5. This means that with an input of 0.25 V_{pp} and a total gain of 100, the output is 1.25 V_{pp}. The second stage also has a voltage gain of 5. With 1.25 V_{pp} input, the output is 6.25 V_{pp}. The output stage has a gain factor of 4. With 6.25 V_{pp} input, the output is 25 V_{pp}.

The total gain of the amplifier is 5 × 5 × 4, or 100. Therefore, with 0.25 V_{pp} input, the output is 25 V_{pp}. Note that output is the product of the individual amplifier gains. It is not just the addition of 5 + 5 + 4. Voltage gain (A_V), you should recall, is an expression of output voltage (V_{out}) divided by input voltage (V_{in}). For the amplifier system, A_V is expressed by the

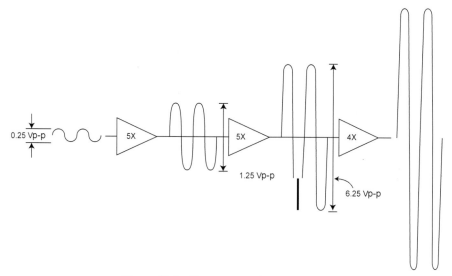

Figure 10.2   Three-stage voltage amplifier.

following formula:

$$A_V = V_{out}/V_{in}$$
$$= 25\ V_{pp}/0.25\ V_{pp}$$
$$= 100$$

Note that the units of voltage cancel each other in the problem. Voltage gain is, therefore, expressed as a unitless value, such as 100.

   Power gain is generally used to describe the operation of the last stage of amplification. Power amplification (A_P), you should recall, is equal to power output (P_{out}) divided by power input (P_{in}). If the last stage of amplification in **Figure 10.2** were a power amplifier, its gain would be expressed in watts rather than volts. In this case, the gain would be

$$A_P = P_{out}/P_{in}$$
$$= 25\ W/6.25\ W$$
$$= 4.$$

Note that the power units of this problem also cancel each other. Power amplification is expressed as a unitless value, such as

Example 10-1:

Solution

$$A_V = V_{out}/V_{in}$$
$$= 12 \ V_{pp}/0.5 \ V_{pp}$$
$$= 24.$$

Related Problem

Calculate A_V with a voltage input of 200 m $V_{(rms)}$ and an output voltage of 24 $V_{(rms)}$.

Decibels

The human ear does not respond to sound levels in the same manner as an amplifying system. An amplifier, for example, has a linear rise in signal level. An input signal level of 1 V can produce, for example, an output of 10 V. The voltage amplification then is 10:1, or 10. The human ear, however, does not respond in a linear manner. It is, essentially, a nonlinear device. As a result of this, sound amplifying systems are usually evaluated on a **logarithmic scale**. This type of scale illustrates how our ears actually respond to specific signal levels. Gain expressed in logarithms, therefore, is much more meaningful than linear gain relationships.

The logarithm of a given number is the power to which another number, called the base, must be raised to equal the given number. Basically, a logarithm has the same meaning as an exponent. A **common logarithm** is expressed in powers of 10. This is illustrated by the following:

$$10^3 = 1000$$
$$10^2 = 100$$
$$10^1 = 10$$
$$10^0 = 1$$

This means that the logarithm of any number between 1000 and 9999 has a **characteristic** value of 3. The characteristic is an expression of the magnitude range of the number. Numbers between 100 and 999 have a characteristic of 2. Numbers between 10 and 99 have a characteristic of 1. Between 1.0 and 9.0, the characteristic is 0. Numbers less than 1.0 have a negative characteristic. It is generally not customary in electronics to use negative characteristic values.

When a number is not an even multiple of 10, its log is a decimal. The decimal part of the logarithm is called the **mantissa**. The log of a number such as 4000 is expressed as 3.6021. The characteristic is 3 because 4000 is between 1000 and 9999. The mantissa of 4000 is 0.6021. To find this value, simply use the [log] key and proper key sequence on a scientific calculator. The log of 4000 is 3.6021, expressed as $\log_{10} 4000 = 3.6021$, or simply as log $4000 = 3.6021$. The characteristic is 3 and the mantissa is 0.6021.

The mantissa is always the same for numbers that are identical, except for the location of the decimal point. For example, the mantissa is the same for 1630, 163.0, 16.3, 1.63. The only difference in these values is the characteristic. The mantissa for 1.63 is 0.2122, and the characteristic is 0. The log is, therefore, 0.2122. The logs of the five values given are 3.2122, 2.2122, 1.2122, and 0.2122, respectively.

Example 10-2:

What is the log of 1590?

Solution

To determine the log of 1590, enter the proper key sequence into a scientific calculator.

The answer is 3.201.

Related Problem

What is the log of 603.8?

Consider now the gain of a sound system with several stages of amplification. **Gain** is best expressed as a ratio of two signal levels. Specifically, gain is expressed as the output level divided by the input level. This is determined by the following expression:

$$A_P = \log_{10}(P_{out}/P_{in}) \text{(in Bels)} \qquad (10.1)$$

where the **Bel (B)** is the fundamental unit of sound-level gain. For an amplifier with 0.1 W of input and 100 W of output,

$$A_P = \log_{10}(100 \text{ W}/0.1 \text{ W})$$
$$= \log_{10} 1000$$
$$= 3B.$$

To find this value with a scientific calculator, key in the following sequence.

As you can see, the Bel represents a rather large ratio in sound level. A **decibel (dB)** is a more practical measure of sound level. A decibel is one-tenth of a Bel. The Bel is named for Alexander Graham Bell, the inventor of the telephone.

The gain of a single stage of amplification within a system can be determined in decibels. A single amplifier stage might have an input of 10 mW and an output of 150 mW. The power gain in decibels is determined by the following formula:

$$A_P = 10 \log_{10}(P_{out}/P_{in})dB. \tag{10.2}$$

Therefore,

$$
\begin{aligned}
A_P &= 10 \log_{10} (P_{out}/P_{in}) \text{ dB} \\
&= 10 \ \log_{10}(150 \text{ mW}/10 \text{ mW}) \\
&= 10 \ \log_{10} 15 \\
&= 10 \ \times 1.1761 \\
&= 11.761 \text{ dB}.
\end{aligned}
$$

To find this value with a scientific calculator, key in the following sequence.

If the input power applied to an amplifier and the power gain is known, the output power can be calculated. Consider, for example, an amplifier with $P_{in} = 10$ mW and $A_P = 11.761$ dB.

$$
\begin{aligned}
A_p &= 10 \log_{10}\left(\frac{P_{out}}{P_{in}} \right) \\
\frac{A_p}{10} &= \log_{10}\left(\frac{P_{out}}{P_{in}} \right) \\
10^{\frac{A_P}{10}} &= \frac{P_{out}}{P_{in}} \\
P_{in} \times 10^{0.1 \times A_P} &= P_{out}
\end{aligned}
$$

$$P_{out} = P_{in} \times 10^{0.1 \times A_P} \tag{10.3}$$

$$
\begin{aligned}
P_{out} = P_{in} \times 10^{0.1 \times A_P} &= 10\text{mW} \times 10^{0.1 \times 11.761} \\
&= 10 \text{ mW} \times 15.0 \\
&= 150 \text{ mW}.
\end{aligned}
$$

The voltage gain of an amplifier can also be expressed in decibels. To do this, the power-level expression must be adapted to accommodate voltage values. The voltage gain formula is

$$A_V = 20 \log_{10}(V_{out}/V_{in}) \text{ dB.} \qquad (10.4)$$

Note that the logarithm of V_{out}/V_{in} is multiplied by 20 in this equation. Power is expressed as V^2/R. Power gain using voltage and resistance values is, therefore, expressed as

$$= 10 \log_{10} \left(\frac{P_{out}}{P_{in}} \right) \text{ dB}$$
$$= 10 \log_{10} \left(\frac{V_{out}^2/R_{out}}{V_{in}^2/R_{in}} \right) \text{ dB.}$$

If the values of R_{in} and R_{out} are equal, the equation that now shows the voltage gain can be expressed as

$$A_V = 10 \log_{10} V_{out}^2 / V_{in}^2 \, dB.$$

The squared voltage values can be expressed as two times the log of the voltage value. Decibel **voltage gain**, therefore, becomes

$$A_V = 2 \times 10 \log_{10}(V_{out}/V_{in}) dB$$

or

$$A_V = 20 \log_{10}(V_{out}/V_{in}) dB.$$

If the input voltage applied to an amplifier and the voltage gain is known, the output power can be calculated. Consider, for example, an amplifier with $V_{in} = 5$ mV and $A_V = 15$dB.

$$A_V = 20 \log_{10} \left(\frac{V_{out}}{V_{in}} \right)$$
$$\frac{A_V}{20} = \log_{10} \left(\frac{V_{out}}{V_{in}} \right)$$
$$10^{\frac{A_V}{20}} = \frac{V_{out}}{V_{in}}$$
$$V_{in} \times 10^{0.05 \times A_V} = V_{out}$$

$$V_{out} = V_{in} \times 10^{0.05 \times A_V} = 5 \, mV \times 10^{0.05 \times 15}$$
$$= 5 \text{ mV} \times 5.623$$
$$= 28.11 \text{ mV.}$$

It is important to remember that decibel voltage gain assumes the values of R_{in} and R_{out} to be equal. **Decibel gain** is primarily an expression of power levels. Voltage gain is, therefore, only an adaptation of the power-level expression.

Example 10-3:

What is the decibel voltage gain of the first amplifier in **Figure 10.2**?

Solution

The input voltage is 0.25 V_{pp} and the output voltage is 1.25 V_{pp}. The voltage gain in dB is, therefore,

$$A_V = 20 \log_{10}(V_{out}/V_{in})dB$$
$$= 20 \log_{10}(1.25V_{pp}/0.25V_{pp})$$
$$= 20 \log_{10}5$$
$$= 20 \times 0.6989$$
$$= 13.979.$$

Related Problem

What is the decibel voltage gain of the second amplifier in **Figure 10.2**?

Example 10-4:

What is the total decibel voltage gain of the three amplifiers in **Figure 10.2**?

Solution

1. Determine the decibel voltage gain of the third amplifier. The input voltage is 6.25 V_{pp} and the output voltage is 25 V_{pp}. The voltage gain in dB is, therefore,

$$A_V = 20 \log_{10}(V_{out}/V_{in})$$
$$= 20 \log_{10}(25V_{pp}/6.25V_{pp})$$
$$= 20 \log_{10}4$$
$$= 20 \times 0.6021$$
$$= 12.041dB.$$

2. Determine the total decibel voltage gain by adding the decibel gain of all three amplifiers.

$$A_{V(total)} = 13.979 + 13.979 + 12.041$$
$$= 40 \text{ dB.}$$

Related Problem

If the voltage gain of the final stage of amplification is changed to 15 dB, what will be the new overall voltage gain in dB?

Frequency Response

The gain of an amplifier does not remain constant over all of the frequencies that are applied to its input. This means that the power available at the output of the amplifier will change with the applied input frequency. This would make the amplifier unacceptable in applications that require that the output remain constant over a wide range of input frequencies, such as those in a high-quality audio amplifier. The range of frequencies over which the gain of the amplifier remains relatively constant is termed as the **bandwidth**. The values of the current (i), voltage (V), and power (P) gain (A) of an amplifier are called its midbands $A_{i(midband)}$, $A_{V(midband)}$, and $A_{P(midband)}$, respectively. The frequency at which the output power becomes half of the input power is designated as a **3dB-down** point, since

$$\text{Power gain/loss (dB)} = 10 \log_{10} \left(\frac{P_{out}}{P_{in}} \right)$$
$$= 10 \log_{10} \left(\frac{P_{in}/2}{P_{in}} \right)$$
$$= 10 \log_{10}(1/2) = 10 \times -0.301 = -3.01 \text{ dB.}$$

Amplifiers generally have two 3dB-down points: one which occurs at a low frequency and another at a high frequency. The range of frequencies between these two points is the bandwidth. The lower 3dB-down point is called the **low cutoff frequency** and the high 3dB-down point is called the **high cutoff frequency**. For proper operation, the frequency of the input applied to the amplifier should stay within the upper and lower 3dB-down points. The bandwidth, low and high cutoff frequencies are shown in**Figure 10.3**.

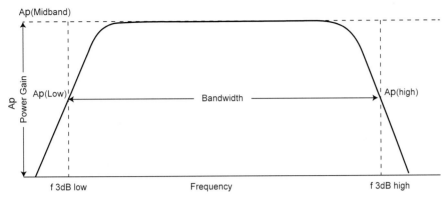

Figure 10.3 Frequency response curve.

The **bandwidth** or BW of an amplifier can be calculated using the following formula:

$$BW = f_{\text{3dB}-\text{down (High)}} - f_{\text{3dB}-\text{down (Low)}} \qquad (10.5)$$

Example 10-5:

What is the bandwidth of an amplifier which has $f_{3dB-down(Low)}$ = 10 kHz and $f_{3dB-down(High)}$ = 50 kHz?

Solution

$$BW = f_{\text{3dB}-\text{down(High)}} - f_{\text{3dB}-\text{down(Low)}}$$
$$= 50 - 10\text{kHz}$$
$$= 40\text{kHz}.$$

Related Problem

If the low cutoff frequency is 20 Hz and the upper cutoff frequency is 20 kHz in an audio amplifier, calculate its bandwidth.

Displaying the range for frequencies ranging from a few Hz to several thousand Hz in the case of an audio amplifier requires use of the **logarithmic scale**, rather than that of the **linear scale**. The reason is that a liner scale which counts of equal values of frequency, starting, say, at 20 Hz, would need to have as its smallest unit 20 Hz. On the same linear frequency scale, if one were to denote 20,000 Hz, that would require 20,000/20 or 1000 equally spaced divisions of 20 Hz. Such a plot would be cumbersome to sketch. Using a logarithmic or log scale makes it possible to display a wide range of frequencies as shown in the following table:

Linear scale	Log scale
1	log (1) = 0
10	log (10) = 1
100	log (100) = 2
1000	log (1000) = 3
10,000	log (10,000) = 4
100,000	log (100,000) = 5

As can be seen by using the log scale, even very high frequencies can be conveniently sketched. On the log scale, each time the frequency changes by a factor of 10, it is called a decade. So if the frequency changes from 100 to 10 Hz, this would be a decade down; or if it changes from 100 to 1000 Hz, that would be a decade up. Amplifiers may have a specification such as "roll-off rate" in terms of dB/decade. For example, an amplifier with a roll-off rate of 20 dB/decade beyond the low (20 Hz) and upper cutoff (20 kHz) indicates that if the frequency becomes either 2 or 200,000 Hz, then the gain will drop by an additional 20 dB.

Bode Plots

Instead of simply plotting the gain vs. frequency, it is often convenient to plot the gain (in dB) as compared to the midband gain vs. frequency, as is done in a **Bode plot**. By doing so, the range of midband frequencies over which the $A_P = A_{P(\text{midband})}$ is plotted on the horizontal axis, since

$$\text{Power loss (dB) with ref. to midband gain} = 10 \log_{10} \left(\frac{A_P}{A_{P(\text{midband})}} \right)$$

$$= 10 \log_{10} \left(\frac{A_{P(\text{midband})}}{A_{P(\text{midband})}} \right)$$

$$= 10 \log_{10}(1) = 10 \times 0 = 0 \text{ dB}.$$

As we saw earlier, the gain of an amplifier remains constant when operated within the upper and lower cutoff frequencies (bandwidth). The gain of an amplifier, thus, stays at 0 dB for the entire range of frequencies specified between the cutoff points. Beyond the upper cutoff point,

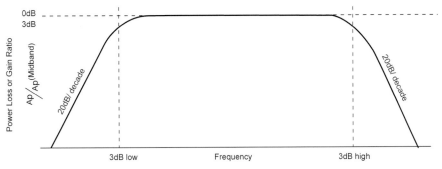

Figure 10.4 Bode frequency plot.

when the frequency increases (or decreases in the case of lower cutoff point) by a factor of 10 times or a decade, the gain changes by the roll-off factor. The power gain of the amplifier, for example, may be specified to drop by 20 dB/decade beyond the midrange frequencies. This is shown in **Figure 10.5**.

Bode plots are used extensively for plotting the **frequency response** of amplifiers, as it conveniently shows the power gained or lost in an amplifier as a function of the change in the input frequency. When amplifiers are cascade connected, the combined bandwidth of the system can be obtained by plotting the 3-dB lower and upper cutoff points of each stage. Recall that beyond the cutoff points, for each decade change in the frequency, in general, the gain drops by a factor of 20 dB. For multistage cascade connected amplifiers that have overlapping **bandwidth** regions, only the portion of frequencies which is completely overlapped will function as the midband region (0-dB portion) for the overall amplifier system in the bode plot. As an example, consider the first stage of an amplifier system with the $f_{3dB-down\ (Low)}$ = 1kHz, and $f_{3dB-down\ (High)}$ = 1MHz, cascaded into the second stage of an amplifier with the $f_{3dB-down\ (Low)}$ = 10kHz and $f_{3dB-down\ (High)}$ = 100kHz. The only portion where the normalized gain essentially stays at 0 dB would be in the regions that overlap completely. In this case, it would be the bandwidth specified by the second stage amplifier.

The **frequency response** of a BJT amplifier depends on its internal and external capacitances of the circuit. Refer to **Figure 10.5** for a multistage *RC*-coupled BJT amplifier system. At a low frequency, the coupling capacitances used in multistage amplifiers offer some impedance to the applied input signal, since $X_C = \frac{1}{2\pi fC}$. So, at low frequencies, in order to keep the value X_C low, a large capacitor value is needed. However, at higher frequencies, X_C

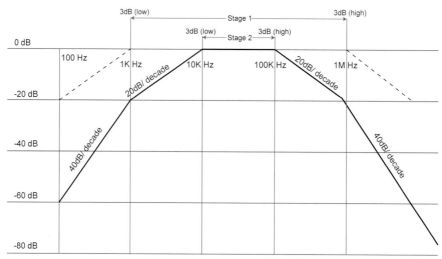

Figure 10.5   Frequency response of a two-stage amplifier.

is low and almost equal to 0 Ω. Also, the capacitors used at high frequencies could be made smaller in value.

The value of the emitter bypass capacitor in a BJT common-emitter configuration circuit and the frequency of the AC input together determine the extent to which the capacitor is bypassed during AC amplification. Ideally, the capacitor should completely bypass the resistor connected between the emitter and the ground, so that the operating point of the BJT does not vary at the frequency of the applied AC input.

All P–N junctions within a transistor have some measurable amount of capacitance and, thus, capacitive reactance, X_C. Since one major concern of an amplifier is its frequency response, the internal capacitance of a capacitor is an important consideration that must be taken into account. BJTs have two P–N junctions and, thus, two capacitances: one between the base and emitter denoted as C_{BE}; and another between the base and collector denoted as C_{BC}. These **internal capacitances** are significant enough to be taken into consideration in high-frequency amplifier applications. In many cases, this capacitance might be identified on the transistor manufacturer datasheet. JFETs and MOSFETs are semiconductor devices that also have P–N junctions in their construction. The internal capacitances across these junctions must be taken into account, along with the AC input frequency for determining its response.

Self-Examination

Answer the following questions.

1. A(n) _____ of amplification consists of active devices and all associated components.
2. The last stage of an amplifying system is a(n) _____ amplifier.
3. A(n) _____ changes energy from one form to another.
4. When the output of one amplifier is connected to the input of the next, they are said to be connected in _____.
5. If the individual voltage gains for each stage of an amplifying system are 5, 4, and 3, the total voltage gain is _____.
6. The ratio of an amplifier's power output to its power input is called

   _____.
7. The power gain in decibels for a power amplifier with an input of 20 mW and an output of 250 mW is _____ dB.
8. Bandwidth is the range of frequencies between the _____ and the _____.
9. A change in frequency by a factor of 10 times is called a(n)

   _____.
10. If the gain is reduced by 20 dB when the frequency changes by a factor of 10, this may be expressed as a 20 dB/decade _____.

10.2 Amplifier Coupling

In many amplifying systems, one stage of amplification does not provide the desired level of signal output. Two or more stages of amplification are, therefore, coupled (connected) together to increase the overall gain. In most amplifying devices, the output voltage of an amplifier is greater than the input voltage. If the output of one stage is connected directly to the input of the next stage, this voltage difference can cause a problem. Signal distortion and component damage might take place. Proper coupling procedures reduce this type of problem. This section covers three methods of coupling: *capacitive coupling*, *direct coupling*, and *transformer coupling*.

10.2 Select the appropriate coupling component for an amplifying system.

In order to achieve objective 10.2, you should be able to:

- explain the advantages and disadvantages of capacitive coupling, direct coupling, and transformer coupling methods.

• define *capacitive coupling*, *direct coupling*, *Darlington amplifier*, *transformer coupling*, *output transformer*, and *impedance ratio*.

Capacitive Coupling

Capacitive coupling is particularly useful when amplifier systems are designed to pass AC signals. You should recall that a capacitor passes AC signals and blocks DC voltages. The capacitor selected must have low capacitive reactance (X_C) at its lowest operating frequency. This is done to ensure amplification over a wide range of frequency. Remember, **capacitive reactance** is an opposition to AC signals. It is also inversely related to frequency as shown by the following formula:

$$X_C = 1/2\pi fC.$$

This means that if frequency is high, capacitive reactance will be low. If frequency is low, capacitive reactance will be high. As you can see, capacitive coupling has some difficulty in passing low-frequency AC signals because low-frequency AC signals create high impedance values, thus opposing the AC signal.

Large capacitance values must be selected when good low-frequency response is desired. This is because the value of capacitor selected is inversely related to capacitive reactance. Therefore, the capacitor selected must be of a value that keeps capacitive reactance low at the lowest operating frequency.

As you can see, the value of a **coupling capacitor** is a very important circuit consideration. In low-frequency amplifying systems, large values of electrolytic capacitors are normally used. These capacitors respond well to low-frequency AC. In high-frequency amplifier applications, small capacitor values are very common. Selection of a specific coupling capacitor for an amplifier is dependent on the frequency being processed. The range of frequencies within which a transistor can operate can usually be determined by examining the datasheets of the device.

A **two-stage amplifier** employing capacitor coupling is shown in **Figure 10.6**. Transistor Q_1 is the input amplifier. Its collector voltage is 8.5 V. At the base of the second amplifier stage, Q_2, the voltage is approximately 1.7 V. The voltage difference across C_1 is 8.5 − 1.7 V, or 6.8 V. The capacitor isolates these two operating voltages. It must have a DC working voltage value that withstands this difference in potential. For example, a capacitor with a DC working voltage of 50 V signifies the maximum voltage that can be applied to it in a specific application. The working for the capacitor being

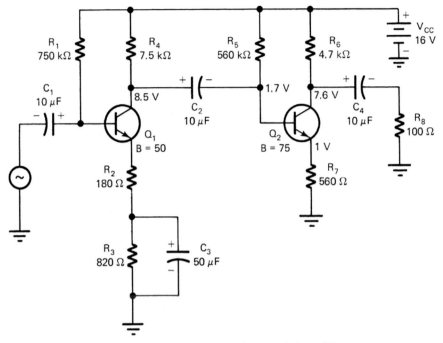

Figure 10.6 Two-stage capacitor-coupled amplifier.

used for coupling the two stages in **Figure 10.6** should have a value in excess of 6.8 V.

Direct Coupling

In **direct coupling**, the output of one amplifier is connected directly to the input of the next stage. In a circuit of this type, a connecting wire or conductor couples the two stages. Circuit design must take into account device voltage values. This type of coupling is an outgrowth of the divider method of biasing. A transistor acts as one resistor in a divider network. The output voltage of one amplifier is the same as the input voltage of the second amplifier. When the circuit is designed to take this into account, the two can be connected together without isolation.

Figure 10.7 shows a **two-stage direct-coupled transistor amplifier**. Note that the signal passes directly from the collector of Q_1 into the base of Q_2. The base current (I_B) of transistor Q_2 is developed without a base resistor. Any I_B needed for Q_2 also passes through the load resistor R_3. The collector voltage V_C of Q_1 remains fairly constant when the two transistors are connected. Note also that the emitter bias voltage of Q_2 is quite large (7.2 V). The base–emitter voltage (V_{BE}) of Q_2 is the voltage difference between V_B and V_E. This is 7.8 V_B – 7.2 V_E or 0.6 V_{BE}. This value of V_{BE} forward biases the base–emitter junction of Q_2. In direct-coupled amplifiers, each stage has a different operating point based on a common voltage source. Several direct-coupled stages supplied by the same source are rather difficult to achieve. As a general rule, only two stages are coupled by this method in an amplifying system.

Direct-coupled amplifiers are very sensitive to changes in ttemperature. The beta of a transistor, for example, changes rather significantly with temperature. An increase in temperature causes an increase in beta and leakage current. This tends to shift the operating point of a transistor. All stages that follow amplify according to the operating point shift. Changes in the operating point can cause nonlinear distortion or a lack of **stability**.

When two transistors are directly coupled, it is often called a **Darlington amplifier**. The transistors are usually called a *Darlington pair*.

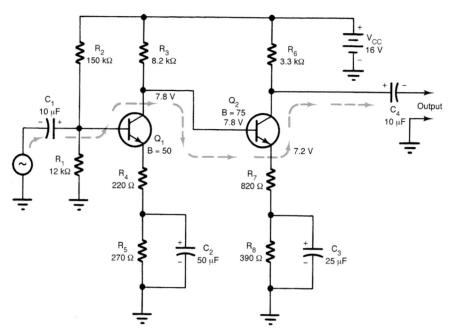

Figure 10.7   Two-stage direct-coupled amplifier.

Two transistors connected in this manner are also manufactured in a single case. This type of unit has three leads. It is generally called a *Darlington transistor*. The gain produced by this device is the product of the two transistor beta values. If each transistor has a current gain (β) of 100, the total current gain is 100×100, or 10,000. **Figure 10.8** shows a Darlington transistor amplifier circuit.

Darlington amplifiers are used in a system where high **current gain** and high **input impedance** are needed. Only a small input signal is needed to control the gain of a Darlington amplifier. In effect, this means that the amplifier does not create a heavy load for the input signal source. The output impedance of this amplifier is quite low. It is developed across the emitter resistor, R_3. A Darlington amplifier has an emitter-follower output (common collector circuit). Several different transistor combinations are used in Darlington amplifiers.

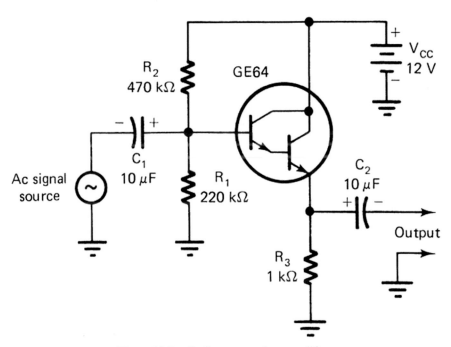

Figure 10.8   Darlington transistor amplifier.

Transformer Coupling

In **transformer coupling**, the output of one amplifier is connected to the input of the next amplifier by mutual inductance. Depending on the frequency being amplified, a coupling transformer may use a metal core or an air core. The output of one stage is connected to the primary winding, and the input of the next stage is connected to the secondary winding. The number of primary and secondary turns determines the impedance ratio of the respective windings. The input and output impedance of an amplifier stage can be easily matched with a transformer. AC signals easily pass through the transformer windings. DC voltages are isolated by the two windings.

Figure 10.9 shows a **two-stage transformer-coupled radio frequency (RF) amplifier**. This particular circuit is used in an **amplitude-modulated** (AM) radio receiver. Circuits of this type are generally built on a compact **printed circuit boards** (PCBs). The operational frequency of this amplifier is in the kHz to MHz range.

The operation of a transformer-coupled circuit is very similar to that of a capacitive-coupled amplifier. Biasing for each transistor element is achieved by resistance and transformer impedance. The primary impedance of T_2 serves as the load resistor of Q_1. The secondary winding of the transformer and R_4 serves as the input impedance for Q_2. The low output impedance of Q_1 is matched to the high input impedance of Q_2 by the transformer. The primary tap connection is used to assure proper load impedance of Q_1.

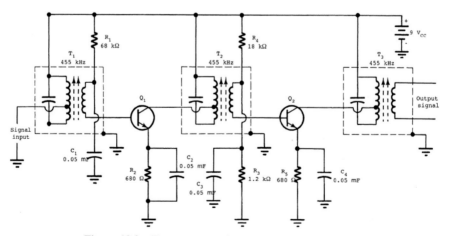

Figure 10.9   Two-stage transformer-coupled RF amplifier.

This particular **transformer-coupled amplifier** is tuned to pass a specific frequency. Selection of this frequency is achieved by coil inductance (*L*) and capacitance (*C*). Tuned transformer-coupled amplifiers are widely used in radio receivers and TV circuits. The dashed line surrounding each transformer indicates that it is housed in a metal can. This is done purposely to isolate the transformers from one another.

Transformers are also used to couple an amplifier to a load device. **Figure 10.10** shows a transistor circuit with a coupling transformer. In this application, the coupling device is called an **output transformer**. The primary transformer winding serves as a collector load for the transistor. The secondary winding couples the transistor output to the load device. The

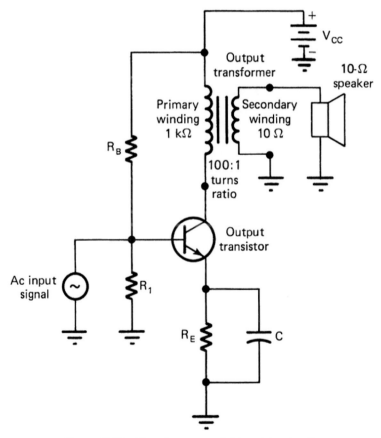

Figure 10.10 Transformer-coupled transistor circuit.

transformer serves as an impedance-matching device. Its primary impedance matches the collector load. In a common-emitter amplifier, typical load resistance values are in the range of 1000 Ω. The output device in this application is a speaker. The speaker impedance is 10 Ω. To get the maximum transfer of power from the transistor to the speaker, the impedances must match. The primary winding matches the transistor output impedance. The secondary winding matches the speaker impedance. Maximum power is transferred from the transistor to the speaker through the transformer impedance.

The number of turns of wire on a particular coil determines its impedance. Assume that the output transformer in **Figure 10.10** has 1000 turns on the primary (N_{pri}) and 100 turns on the secondary (N_{sec}). Its turns ratio (n) is

$$n = \frac{N_{pri}}{N_{sec}} \tag{10.6}$$
$$= 1000/10$$
$$= 100.$$

The **impedance ratio** of a transformer is the square of its turns ratio where Z_{in} is the primary impedance and Z_{out} is the secondary impedance. This is expressed as

$$n^2 = \frac{Z_{pri}}{Z_{sec}}. \tag{10.7}$$

For the output transformer being considered in this example, the impedance ratio is 10^2, or 100 Ω. This means the input or impedance of the transformer primary connected in the collector portion of BJT circuit is 100 times more than the impedance of the load device. If the load (speaker) of this circuit has an impedance of 10 Ω, the input impedance is 100 $\times$ 10 Ω, or 1 kΩ.

Example 10-6:

Calculate the input (primary) impedance of a transformer having a turns ratio of 25, which is connected to a load impedance of 4 Ω on the output (secondary) side.

Solution

$$n^2 = \frac{Z_{pri}}{Z_{sec}}$$
$$25^2 = \frac{Z_{pri}}{4}$$

$$2500 \ \Omega = Z_{\text{pri}}.$$

Related Problem

Calculate the input (primary) impedance of a transformer having a turns ratio of 20, which is connected to a load impedance of 6 Ω on the output (secondary) side.

Transformer coupling has a number of problems. In sound system applications, the impedance increases with frequency. As a result of this, the high-frequency response of a sound signal is rather poor. Better-quality sound systems do not use transformer coupling. The physical size of a transformer must also be quite large when high-power signals are amplified. Because of this, low-power and medium-power amplifying circuits only use transformer coupling. In addition to this, good-quality transformers are rather expensive. These disadvantages tend to limit the number of applications of transformer coupling.

Self-Examination

Answer the following questions.

11. Three common methods of amplifier coupling are _____, _____, and _____.
12. In low-frequency amplifying systems that use capacitive coupling, (large, small) values of capacitors are normally used.
13. In high-frequency amplifying systems that use capacitive coupling, (large, small) values of capacitors are normally used.
14. Two directly coupled transistors are referred to as a(n) _____.
15. _____ amplifiers are used where high current gain and high input impedance are needed.
16. An output transformer is used to couple an amplifier to a _____ device.

10.3 Input and Output Transducers

A *transducer* is a device that changes energy of one form to energy of an entirely different form. When energy conversion occurs, work is performed. In audio systems, an *input transducer* is used to change sound energy into electrical energy. The output of an amplifying system is always applied to a transducer. The load device of a system is an *output transducer*. Functionally, the load may be resistive, inductive, and, in some cases, capacitive. Most of our applications have used resistive loads. Resistors generally change electrical energy into heat energy.

10.3 Explain the operation of amplifying system transducers.

In order to achieve objective 10.3, you should be able to:

- identify common input and output transducers.
- define *microphone, crystal microphone, piezoelectric effect, dynamic microphone, voice coil, tweeter, woofer,* and *intermediate stage speaker.*

Input Transducers

The **input transducer** of an amplifying system is primarily responsible for the development of the signal that is processed by the system. In audio systems, an input transducer is used to change sound energy into electrical energy. Systems that respond to frequencies above the human range of hearing are called radio frequency (RF) systems. Electromagnetic waves must be changed into electrical energy in this type of system. Radio and television receivers respond to RF signals. The type of input transducer used depends on the signal to be processed. RF systems pick up signals that travel through the air or through an interconnecting cable network. Audio frequency (AF) systems develop a signal where a system is being operated. Microphones, compact discs (CDs), and DVDs serve as input transducers.

The primary function of a **microphone** is to change variations in air pressure into electrical energy. These variations in air pressure are called *sound waves* and are a form of energy. The pressure changes of a sound wave are easily converted into voltage, current, or resistance. Microphones generally change sound energy into a changing voltage signal. Crystal microphones and dynamic microphones are widely used.

A **crystal microphone** generates voltage through a mechanical stress placed on a piece of crystal material. This action is called the **piezoelectric effect.** Rochelle salt crystals produce this effect. Synthetic ceramic materials respond in the same way. Both materials respond to changes in pressure by producing a voltage. Sound waves striking the crystal cause it to bend or squeeze together. Metal plates attached to opposite sides of the crystal develop a potential difference in charge (voltage). The output voltage is then routed away from the crystal by connecting wires.

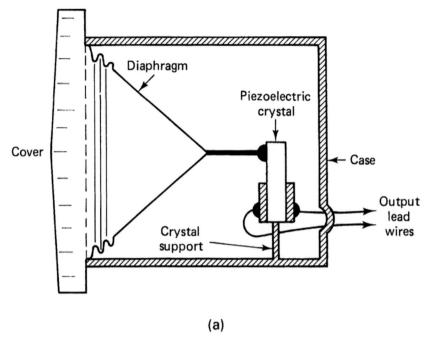

(a)

Figure 10.11 Crystal microphone.

Figure 10.11 shows the construction of a simplified crystal microphone internal structure. A **dynamic microphone** is considered to be mechanical generator of electrical energy. The mechanical energy of a sound wave causes a small wire coil to move through a magnetic field. This action causes electrons to be set into motion in the coil. A difference in potential charge or voltage is induced in the coil. The resulting output of a dynamic microphone is AC voltage.

A simplified dynamic microphone is shown in **Figure 10.12**. Note that a lightweight coil of wire is attached to a cone-shaped piece of material. The cone piece is made of flexible plastic or thin metal. The cone is generally called a diaphragm. The diaphragm is attached to the outside frame of the microphone. Its center is free to move in and out when sound waves are applied. Sound waves striking the diaphragm cause it and the coil to move through a permanent magnetic field. Through this action, sound energy is changed into voltage. Flexible wires attached to the moving coil transport signal voltage out of the microphone. This voltage is applied to the input of a sound system.

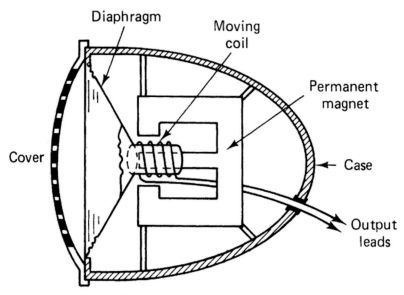

Figure 10.12   Structure of a dynamic microphone.

Magnetic Tape Example

Magnetic tape was, for many years, a common input for amplifying systems. As an example of converting electronic input variations to sound, let us look at how this was accomplished. Cassette tapes, which preceded compact discs (CDs) for recorded music, used this principle. A long strip of plastic tape is coated with iron oxide so that individual molecules of this material can be easily magnetized. A magnetic tape can, therefore, be used to store information in the form of minute charged areas.

Figure 10.13 shows how a sound signal can be placed on a **magnetic tape**. This is considered to be the recording function. Essentially, a sound signal is being applied to the electromagnetic head. Variations in the applied signal will cause a corresponding change in the electromagnetic field. Tape passing under the air gap of the head (such as a cassette tape) will cause oxide molecules to arrange themselves in a specific order. In a sense, a changing magnetic field is being transferred to the tape. The recorded signal will remain stored on the tape for a long period of time. Amplifying systems are used to place information on a tape. This process is achieved by speaking into a microphone. The signal is then processed by the amplifier. The output of the system supplies signal current to the electromagnetic recording head. A

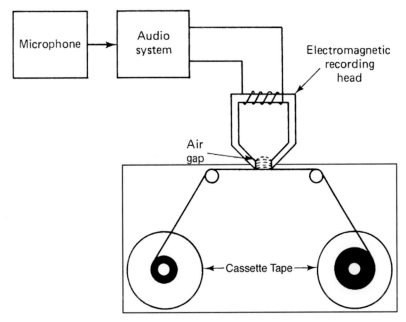

Figure 10.13   Tape recording function.

motor-drive mechanism is needed to move the tape at a selected recording speed. The tape head responds as a load device when the system is recording sound information for reproduction.

The recovery of information is achieved by the **read head**. This part of the system is like a dynamic microphone, except the electromagnetic coil is stationary. Voltage is induced in the coil by a moving magnetic field. As the magnetic tape moves under the read head, voltage is induced into the coil. Changes in the field cause variations in voltage values. **Figure 10.14** shows a simplification of the tape-reading function. The tape head responds as an input transducer when the system responds as *a playback* device.

Compact Disc (CD) Players

The example of the magnetic tape device in the previous section showed how electromagnetic signal variations could be amplified and reproduced as sound. Compact disc (CD) players use an advanced form of this process. Laser diodes and photodiodes (discussed in Chapter 13) are used in the **pick-up system** of CD players.

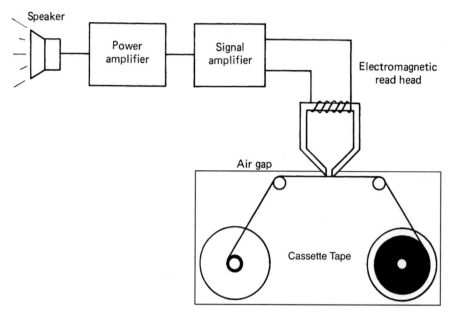

Figure 10.14   Tape reading or playback function.

In a CD player, sound is recorded on the surface of a CD to form "pits" and "flats." A laser beam is focused onto the CD surface. As the CD rotates, the laser light varies according to "pits" and "flats" on the CD surface. The signal variations caused by the "pits" and "flats" of a CD track are amplified and reproduced as sound.

Output Transducers

In an audio amplifying system, the **output transducer**, or load device, is a speaker. It changes electrical energy into sound energy. Variations in current cause the mechanical movement of a stiff paper cone. Movement of the cone causes alternate compression and decompression of air molecules. This causes sound waves to be set into motion. The human ear responds to these waves.

The operation of a **speaker** is based on the interaction of two magnetic fields. One field is usually developed by a permanent magnet. The second field is electromagnetic. Permanent-magnet (PM) speakers are of this type. The electromagnetic part of the speaker is generally called a **voice coil**.

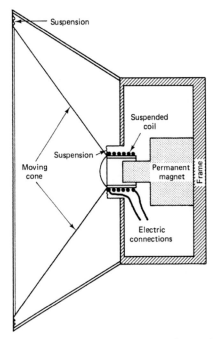

Figure 10.15 Cross-sectional view of a speaker.

The voice coil is attached to the cone and suspended around the permanent magnet. See the cross-sectional view of a speaker in **Figure 10.15**.

When current flows through the voice coil, it produces an electromagnetic field. The polarity of the field (north or south) depends on the direction of current flow. If AC flows in the voice coil, the field varies in both strength and polarity. The **power amplifier** of a sound system supplies AC to the voice coil. The changing field reacts with the permanent magnetic field. This causes the voice coil to move. With the voice coil attached to the cone, the cone also moves. This action causes air molecules to be set into motion. Sound waves are emitted from the **speaker cone.**

The frequency of the applied AC signal determines how slow or fast the cone of a speaker responds. Operational frequency is based on the rate or speed change of the electromagnetic field. The loudness of the developed sound wave is based on the moving distance of the cone. This depends on the amount of current supplied to the voice coil by the power amplifier. The primary function of an amplifying system is to develop electrical power to drive a speaker.

When selecting a **speaker** for a specific application, one must take into account a number of considerations. Generally, it takes a large speaker to properly develop low-frequency sounds. Large volumes of air must be set into motion for low-frequency reproduction. Small speakers cannot effectively move enough to produce low tones. Small speakers respond better to high-frequency tones. High-frequency reproduction requires rapid development of air pressure. Small cones can move very rapidly. Large speakers with a big cone and voice coil cannot react quickly enough to produce high-frequency tones. A speaker obviously cannot be large and small at the same time.

In a **high-fidelity** sound system, at least two speakers are needed to reproduce a typical audio signal: small and large. A small high-frequency speaker is commonly called a **tweeter.** The cone of this speaker is generally made of a rather stiff material. Some units employ thin metal cones. A large low-frequency speaker is called a **woofer**. The cone of this speaker is usually quite flexible. Some systems may also employ an **intermediate range speaker**, or **midrange speaker**. Speakers of this type are designed to respond efficiently to frequencies at the center of the audio range. A great majority of the sound being reproduced falls in this range. These speakers are frequently housed in a wooden enclosure.

Self-Examination

Answer the following questions.

17. Examples of input transducers are _____ and _____.
18. Two types of microphones are _____ and _____.
19. The operation of a(n) _____ microphone is based on the mechanical energy of a sound wave causing small wire coil to move through a magnetic field.
20. The output transducer of a stereo system is a(n) _____.
21. The electromagnetic part of a PM speaker is called a(n) _____.
22. Three types of speakers according to frequency range are _____, _____, and _____.

10.4 Analysis and Troubleshooting – Amplifying Systems

The performance of an amplifying system is evaluated by examining the applied input and observing the resulting output waveform. Some degree of amplification should be displayed by the overall output. For systems with multistage amplifiers, the overall gain depends on the proper functioning of individual stages. In the presence of faults, the output of the amplifier system could be distorted or have reduced output. For isolating the fault in a **multistage amplifier** system, analysis can begin preferably with the first stage of amplification and progress through to the last stage. Alternatively, analysis of a faulty system could start from the output of the last stage and move back to the initial stage. In this section, a multistage amplifier will be examined to determine its amplification level and possible faulty conditions.

10.4 Troubleshoot amplifier systems.

In order to achieve objective 10.4, you should be able to:

- observe the input and output AC waveforms of an amplifying systems on an oscilloscope;
- examine how coupling elements of an amplifying system influence its operation;
- distinguish between normal and faulty operation of an amplification system.

Different stages of an amplifying system may be coupled to each other directly or by capacitors and transformers. If the coupling element is faulty, it will have an adverse effect on the operation of the amplifier. This could result in distortion of the final output, reduction of its amplification, or even damage to circuit components due to excessive current.

Summary

- The major functions of an amplifying system are energy conversion, amplification, and power distribution.
- Amplifying systems typically have many stages of amplification.
- A stage of amplification consists of an amplifying device and associated components.
- The last stage of amplification in an amplifying system is typically a power amplifier.
- A transducer is a device that changes energy of one form to energy of another form.
- The term *cascade* refers to a series of amplifiers where the output of one stage of amplification is connected to the input of the next stage of amplification.
- The total gain of an amplifying system is the product of the individual amplifier gains.

- Power gain is generally used to describe the operation of the last stage of amplification.
- Sound amplifying systems are usually evaluated on a logarithmic scale because the human ear does not respond to sound levels in the same manner as an amplifying system.
- A logarithmic scale is a nonlinear measurement scale where each major division is a whole-number multiple of the previous major division.
- A common logarithm is expressed in powers of 10.
- The Bel (B) is the fundamental unit of sound-level gain.
- The decibel is one-tenth of a Bel.
- Power gain in decibels is determined by the formula $A_P = 10 \log(P_{out}/P_{in})$S.
- The range of frequencies over which the gain remains relatively constant is designated as the bandwidth.
- The gain of an amplifier over the range of frequencies specified by its bandwidth is designated as the midband gain.
- The frequencies at which the power gain becomes half are designated as the 3dB-down frequencies.
- Amplifiers generally have two 3dB-down frequencies – a lower frequency designated as the low cutoff frequency, and a high frequency designated as the high cutoff frequency.
- A Bode plot displays the power loss/gain relative to the midband gain as a function of the frequency. By doing so, the midband gain is plotted on the 0 dB or horizontal axis.
- In a multistage cascade connected amplifier, the bandwidth of the overall system is obtained by identifying the range of frequencies for which the gain of each individual amplifier remains at 0 dB. The BW of the overall system is the portion of the combined frequency response curve where the individual amplifier bandwidths completely overlap.
- Three methods of coupling amplifier stages are capacitive coupling, direct coupling, and transformer coupling.
- To ensure amplification over a wide range of frequency, the capacitor to be used as a coupling capacitor must have low capacitive reactance (X_C) at its lowest operating frequency.
- In a direct-coupled amplifier system, the output voltage of one amplifier is the same as the input voltage of the second amplifier.
- Direct-coupled amplifiers are very sensitive to changes in temperature.
- Typically, no more than two stages of amplification are direct-coupled.

- In a transformer-coupled amplifier system, the output of one amplifier stage is connected to the primary winding, and the input of the next amplifier stage is connected to the secondary winding.
- A transformer that is used to couple an amplifier to a load device is called an output transformer.
- An input transducer is used to change sound energy into electrical energy.
- An output transducer is used to change electrical energy into sound energy.
- Small high-frequency speakers are called tweeters.
- The cone of a tweeter is generally made of a rather stiff material.
- Large low-frequency speakers are called woofers.
- The cone of a woofer is usually flexible.

Formulas

(10-1) $A_p = \log(P_{out}/P_{in})$ (in Bels) Power gain.

(10-2) $A_p \text{ (dB)} = 10\log_{10}(P_{out}/P_{in})$ Power gain in decibels.

(10-3) $P_{out} = P_{in} \times 10^{0.1 \times A_P}$

(10-4) $A_V = 20\log_{10}(V_{out}/V_{in})$ Voltage gain in decibels.

(10-5) $\text{BW} = f_{3dB-down \text{ (High)}} - f_{3dB-down \text{ (Low)}}$

(10-6) $n = \dfrac{N_{pri}}{N_{sec}}$

(10-7) $n^2 = \dfrac{Z_{pri}}{Z_{sec}}$

Review Questions

Answer the following questions.

1. An antenna may serve as an input _____ in an amplification system.

   (a) Transistor
   (b) Transducer
   (c) Capacitor
   (d) All of the above

1. The term cascade with reference to an amplifier system signifies that the output of one stage of amplification is:

   (a) Very high
   (b) Very low

(c) Directly connected to the output of the final stage

(d) Connected to the input of the next stage of amplification

2. The gain of an amplifier _____ over the entire range of frequencies that are applied to its input.

(a) Remains constant

(b) Does not remain constant

(c) Is 0

(d) All of the above

3. The 3dB-down points with reference to the gain of an amplifier signify the frequency at which the power:

(a) Reduces to half

(b) Increase to double

(c) Always reduces by 3 W

(d) Always increases by 3 W

4. A P–N junction of a transistor has _____ capacitive reactance at high frequencies.

(a) Infinite

(b) Significant

(c) 0

(d) All of the above

5. Capacitive coupling between multistage amplifiers is designed to:

(a) Block DC and AC

(b) Pass DC and AC

(c) Pass AC only

(d) Pass DC only

6. Two directly coupled BJTs housed in a single enclosure is called a(n):

(a) Emitter follower

(b) Push–pull amplifier

(c) Complementary-symmetry amplifier

(d) Darlington amplifier

7. The bandwidth of a two-stage cascade connected amplifier system, in which the upper and lower cutoff frequencies of the first amplifier are completely contained within that of the second, in general is:

(a) The same as that of the first amplifier

(b) The same as that of the second amplifier

(c) Is the sum of the bandwidths of the two amplifiers

(d) Is sum of the bandwidths of the two amplifiers

8. The bypass capacitor in a transistor circuit can be replaced by a short circuit at:

(a) Very low frequencies

(b) Very high frequencies

(c) At all frequencies

(d) Can never be replaced by a short circuit

9. A 3-dB power gain signifies that the output power is:

(a) A quarter of the input power

(b) Half the input power

(c) Twice the input power

(d) Quadruple the input power

Problems

Answer the following questions.

1. An amplifier with a P_{in} of 0.25 W, and a P_{out} of 1.5 W, has a power gain within the range:

   a. 0-2

   b. 2.1-4

   c. 4.1-6

   d. 6.1-above

2. The gain of an amplifier is specified to be 3 B. In terms of dB, this will be within the range:

   a. 1-10 dB

   b. 10.1-20 dB

   c. 20.1-40 dB

   d. 40.1-60 dB

3. An amplifier with a $V_{in(pp)}$ of 0.5 V and a $V_{out(pp)}$ of 2 V has a voltage gain in terms of dB within the range:

   a. 0-4 dB

   b. 4.1-8 dB

   c. 8.1-12 dB

   d. 12.1-above dB

4. Four amplifiers with voltage gains of 1, 2, 3, and 4 dB, respectively, are connected in series (cascaded) as part of a multistage amplifier system. The overall gain of the combined amplifier system would be in the range:

   a. 0-3 dB
   b. 3.1-6 dB
   c. 6.1-9 dB
   d. 9.1-above dB

5. The bandwidth of amplifier P is specified to be 3 MHz and that of amplifier Q as 300 kHz. In which amplifier does the midband gain remain constant over a wider range of frequencies:

   a. P
   b. Q
   c. Both amplifiers
   d. Neither amplifier

6. The bandwidth of an amplifier is 25 kHz, and the $f_{3dB-down\ (Low)}$ is 5 kHz. The $f_{3dB-down\ (High)}$ would lie within the range:

   a. 0-10 kHz
   b. 10.1-20 kHz
   c. 20.1-30 kHz
   d. 30.1-above kHz

7. The log of 30,000 is

   a. 0-4 dB
   b. 4.1-8 dB
   c. 8.1-12 dB
   d. 12.1-above dB

8. An amplifier has power gain of 10 dB, and the input power is 1 W. How much times larger is the output power as compared to the input power:

   a. 0-5
   b. 5.1-10
   c. 10.1-15
   d. 15.1-above

9. An amplifier with a rated output of 16 W is connected to a load of 4 Ω. The voltage across the load and the current through it are:

   a. 4 V, 4 A
   b. 8 V, 2 A

    c. 12 V, 1.33 A

    d. 16 V, 1 A

10. An amplifier has a voltage gain of 20, and a voltage output of 8V. Calculate the input voltage:

    a. 0-0.5 V

    b. 0.6-1 V

    c. 1.1-1.6 V

    d. 1.6-above V

Answers

Examples

10-1. 120

10-2. 2.78

10-3. 40

10-4. 42.958 dB

10-5. 19,980 or 19.98 kHz

10-6. 2400 Ω

Self-Examination

10.1

  1. stage

  2. power

  3. transducer

  4. cascade

  5. 60

  6. . power gain, *or* A_P

  7. 10.97

  8. high and the low cutoff, or between the 3dB-down (High) and 3dB-down (Low) frequency

  9. octave

 10. loss

10.2

 11. capacitive, direct, transformer (any order)

 12. large

 13. small

14. Darlington amplifier
15. Darlington
16. load
10.3
17. microphones, antennas (any order, other transducer possible)
18. crystal, dynamic (any order)
19. dynamic
20. speaker
21. voice coil
22. tweeter, intermediate range (*or* midrange), woofer (any order)

Terms

Stage of amplification

A transistor or *IC* amplifier and all the components needed to achieve amplification.

Transducer

A device that changes energy from one form to another. A transducer can be on the input or output.

Cascade

A method of amplifier connection in which the output of one stage of amplification is connected to the input of the next stage of amplification.

Logarithmic scale

A nonlinear scale of measurement where each major division is a whole-number multiple of the previous major division.

Common logarithm

A logarithm that is expressed in powers of 10.

Mantissa

Decimal part of a logarithm.

Bel (B)

A measurement unit of gain that is equivalent to a 10:1 ratio of power levels.

Decibel (dB)

One-tenth of a Bel. Used to express gain or loss.

Midband gain

The gain of an amplifier where it remains relatively constant over a certain range of frequencies.

Cutoff or 3-dB down frequency

The frequency where the power is reduced to half of the midband gain. Usually, amplifiers have two cutoff frequencies – one is the low cutoff frequency, and the other is the high cutoff frequency.

Bandwidth

The range of frequencies between the high and the low cutoff frequencies.

Bode plot

A plot of the overall gain of an amplifier compared to the midband gain for a certain range of frequencies.

Capacitive coupling

A method of coupling amplifier stages with capacitors.

Direct coupling

A method of coupling amplifier stages in which the output of one amplifier stage is connected directly to the input of the next amplifier stage.

Darlington amplifier

Two transistor amplifiers cascaded together for increasing the overall values of the current gain and input impedance.

Transformer coupling

A method of coupling amplifier stages in which the output of one amplifier is connected to the input of the next amplifier by mutual inductance.

Output transformer

A transformer that is used to couple an amplifier to a load device.

Impedance ratio

The AC resistance ratio between the input and output of a transformer.

Microphone

A device that changes sound energy into electrical energy.

Crystal microphone

Input transducer that changes sound into electrical energy by the piezoelectric effect.

Piezoelectric effect

The property of a crystal material that produces voltage by changes in shape or pressure.

Dynamic microphone

An input transducer that changes sound into electrical energy by moving a coil through a magnetic field.

Voice coil

The electromagnetic part of a speaker.

Tweeter

A speaker designed to reproduce only high frequencies.

Woofer

A speaker designed to reproduce low audio frequencies with high quality.

Intermediate range speaker

A speaker that is designed to respond efficiently to frequencies at the center of the audio range.

11

Power Amplifiers

The last stage of an amplifying system is nearly always a **power amplifier** to drive some type of output device such as speakers. This type of amplifier is generally a low impedance device that controls a great deal of current and voltage. You should recall that power is the product of current and voltage. Normally, a power amplifier dissipates a great deal more heat than does a comparable signal amplifier. A power amplifier can typically handle 1 W or more of power and is generally designed to operate over its entire active region. Both the power amplifier and the output device typically have low impedance, providing proper matching.

There are various types of power amplifier circuit configurations. Each type exhibits different operating characteristics. With careful selection of a power amplifier and circuit components, it is possible to achieve **power gain** with minimum **distortion**. This chapter discusses the **single-ended** power amplifier, **push—pull** amplifier, and **complementary-symmetry** amplifier.

Objectives

After studying this chapter, you will be able to:

11.1 analyze the operational efficiency of a single-ended power amplifier;
11.2 analyze the operational efficiency of class B and class AB push—pull power amplifiers;
11.3 analyze the operational efficiency of a complementary-symmetry power amplifier;
11.4 analyze and troubleshoot power amplifiers.

Chapter Outline

11.3 Complementary-Symmetry Power Amplifiers

11.4 Analysis and Troubleshooting – Power Amplifiers

11.1 Single-Ended Transistor Power Amplifier

In this activity, a single low distortion power amplifier will be evaluated.

11.2 Complementary-Symmetry Power Amplifier

In this activity, a high efficiency power amplifier will be evaluated using two complementary (NPN and PNP) transistors for amplifying an AC input signal.

Key Terms

complementary-symmetry amplifier
crossover distortion
efficiency
single-ended amplifier
transformer-coupled push–pull transistor amplifier

11.1 Single-Ended Power Amplifiers

A single-ended power amplifier is very similar in many respects to a small-signal amplifier. It is normally operated as a class A amplifier. You should recall that the operating point of a class A amplifier is near the center of its active region. This causes the AC signal applied to the input to be fully reproduced in the output. As a general rule, this type of amplifier operates with a minimum of distortion. This section discusses the characteristics and design of the single-ended power amplifier.

11.1 Analyze the operational efficiency of a single-ended power amplifier.
   In order to achieve objective 11.1, you should be able to:

   • calculate the operational efficiency of a single-ended power amplifier;
   • describe the characteristics of single-ended amplifier;
   • define single-ended amplifier and efficiency.

An amplifier system that has only one amplifying device that develops output power is called a **single-ended amplifier**. Efficiency levels for a single-ended amplifier are in the range of 30%. **Operational efficiency** refers to a ratio of developed output power to DC supply power. This is often denoted by the

Greek symbol η (Eta). The equation for efficiency is

$$\eta = \frac{P_{\text{out}}}{P_{\text{in}}} \times 100\%. \tag{11.1}$$

For example, a class A amplifier, with 9 W of DC input power develops 3 W of DC output power. The amplifier efficiency would be 33%.

$$
\begin{aligned}
\eta &= (P_{\text{out}}/P_{\text{in}}) \times 100\% \\
&= (3\text{W}/9\text{W}) \times 100\% \\
&= 0.33 \times 100\% \\
&= 33\%.
\end{aligned}
$$

Signal-ended amplifiers, because of their low efficiency rating, are generally used only to develop low-power output signals.

Figure 11.1 shows a single-ended audio output amplifier. The term **audio** refers to the range of human hearing. **Audio frequency** (AF) refers to frequencies that are from 15 to 15 kHz. This particular amplifier responds to signals somewhere within this range.

A collector family of characteristic curves for the transistor used in the single-ended power amplifier is shown in **Figure 11.2**. Note that the power dissipation (P_D) curve of this amplifier is 5 W. Remember that $P_D = I_C \times V_{CE}$ and $I_C = P_D/V_{CE}$. The points of I_C along the curve are plotted at V_{CE} points such as $I_C = P_D/V_{CE} = 5$ W/25 V = 0.2 mA. The developed load lines

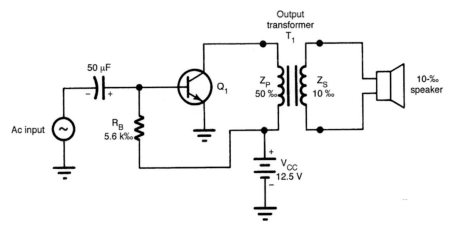

Figure 11.1　Single-ended audio amplifier.

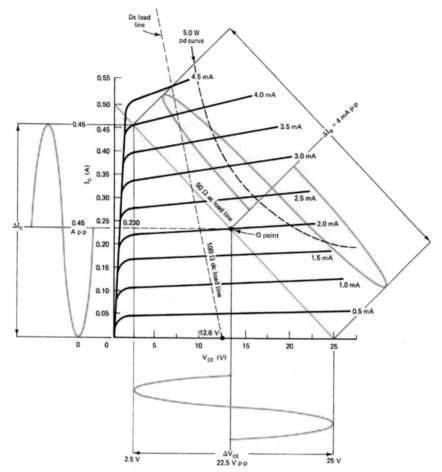

Figure 11.2   Collector family of characteristic curves for the single-ended audio amplifier in **Figure 11.1**.

for the amplifier are well below the indicated P_D curve. This means that the transistor can operate effectively without overheating.

Investigation of the collector family of characteristic curves of this example shows that two load lines are plotted: DC and AC. The static or DC load line is based on the resistance of the transformer primary of 50 Ω. The resistance of the DC load line is 100 Ω, based on 12.5 V_{CC} and 1.25 mA of I_C. This means that the DC load line extends almost vertically from the 12.5-V_{CC} point on the curve. The AC load line is determined with the primary winding impedance (50 Ω) and twice the value of V_{CE} (25 V) compared to the DC load line. The AC output signal is, therefore, twice the value of V_{CC}. The operational points are 25 V_{CC} and 500 mA of I_C. Note the location of these operating points on the AC load line on the example.

When an AC signal is applied to the input, it can cause a swing 12.5 V above and below V_{CC}. The transformer, in this case, accounts for the voltage difference. A transformer is an inductor. When the electromagnetic field of an inductor collapses, a voltage is generated. This voltage is added to the source voltage. In an AC load line, V_{CE}, therefore, swings to twice the value of the source voltage. This usually occurs only in transformer-coupled amplifier circuits. Note that the operating point (Q point) is near the center of the AC load line.

With the load line of the amplifier is established, the base resistance value must be determined. For the circuit in **Figure 11.1**, the emitter voltage is zero. The emitter–base junction of the transistor, being silicon, has 0.7 V across it when it is in conduction. The base resistor, therefore, has 12.5 − 0.7 V, or 11.8 V, across it at the Q point of the 50-Ω AC load line. By Ohm's law, the value of R_B is 11.8 V/2 mA, or 5900 Ω. A standard resistor value of 5.6 kΩ would be selected for R_B. Note the location of this resistor in the circuit.

With this data, it is possible to look at the operational efficiency of the single-ended amplifier. The DC power supplied to the circuit for operation is

$$P_{in} = V_{CC} \times I_C$$
$$= 12.5 \text{ V} \times 0.230 \text{ A}$$
$$= 2.87 \text{ W}.$$

The developed AC output power (P_{out}) is a peak-to-peak value. The V_{CE} peak-to-peak value is 2.5−25 V, or 22.5 V_{PP}. The I_C peak-to-peak value is 0−0.45 A, or 0.45 A_{PP}.

To make a comparison in power efficiency, the peak-to-peak values must be changed to RMS values. Remember that an RMS AC value and a DC value do the same effective work. Conversion of peak-to-peak voltage to RMS is done by the following expression:

$V_{CE \text{ (rms)}} = (V_{PP}/2) \times 0.707$.

Therefore,

$$V_{CE \text{ (rms)}} = (V_{PP}/2) \times 0.707$$
$$= (22.5 \ V_{PP}/2) \times 0.707$$
$$= 11.25 \ V_{PP} \times 0.707$$
$$= 7.954 \text{ V}.$$

For the RMS I_C value, the conversion is

$$I_{C\ (\text{rms})} = (I_{PP}/2) \times 0.707$$
$$= (0.45\ A_{PP}/2) \times 0.707$$
$$= 0.225\ \times 0.707$$
$$= 0.159\ \text{A}.$$

The developed AC output power (P_{out}) is determined by the following formula:

$$P_{\text{out}} = V_{CE\ (\text{rms})}\ \times I_{C(\text{rms})}$$
$$= 7.954\ \text{V} \times 0.159\ \text{A}$$
$$= 1.265\ \text{W}.$$

The operational efficiency of the single-ended amplifier of **Figure 11.1** is, therefore, determined by the ratio of the output power to the input power expressed as a percentage.

$$\eta = \frac{P_{\text{out}}}{P_{\text{in}}} \times 100\%$$
$$= (1.265\ \text{W}/2.898\ \text{W}) \times 100\%$$
$$= 0.4365 \times 100\%$$
$$= 43.65\%,\ or\ 44\%.$$

This shows that an amplifier of this type is only 44% efficient. It takes 2.898 W of DC supply power to develop 1.2652 W of AC output power. This low operational efficiency does not make the single-ended amplifier very suitable for high-power applications. As a general rule, amplifiers of this type are used only where the required output is 10 W or less. The power wasted by a single-ended amplifier appears in the form of heat. Obviously, it would be very desirable to get more sound output and less heat during operation.

Example 11-1:

In the amplifier circuit given in **Figure 11.1**, if 2.5 W of input power is used for developing 1 W of output power, calculate the efficiency of the amplifier.
Solution

$$\eta = \frac{P_{\text{out}}}{P_{\text{in}}} \times 100$$
$$= \frac{1\,\text{W}}{2.5\,\text{W}} \times 100$$
$$= 40.0\%.$$

Related Problem

In the amplifier circuit given in **Figure 11.1**, if the input power is 2.5 W and the efficiency is 80%, calculate the output power developed by this amplifier.

Self-Examination

Answer the following questions.

1.   The efficiency of a class A amplifier is very high: True/False.
2. A class A amplifier is also called a _____.
3. In a class A BJT amplifier, the operating point is located:

   a. In the middle of the load line
   b. In the saturation region
   c. In the cutoff region
   d. Below the cutoff region

4. What is the function of an output transformer in a class A amplifier?

   a. Pass DC signal
   b. Block amplified AC signal
   c. Impedance matching with load
   d. Collector biasing

5. If the load connected to the output of a class A amplifier is shorted, the AC output will

   a. Be 0 V.
   b. Will have reduced current.
   c. Function normally
   d. Be amplified.

11.2 Push – Pull Power Amplifiers

Power amplifiers can also be connected in a push–pull circuit configuration. In this configuration, two amplifying devices are used and are biased at or near cutoff. The circuit is designed so that each amplifying device handles only one alternation of the sine-wave input signal. This produces an operational efficiency that is much higher than that of a single-ended amplifier. This section discusses two types of push–pull amplifiers: class B and class AB. Class B amplifiers are biased at cutoff and class AB amplifiers are biased just above cutoff.

11.2 Analyze the operational efficiency of class B and class AB push—pull power amplifiers.

In order to achieve objective 11.2, you should be able to:

- calculate the operational efficiency of class B and class AB push—pull amplifiers;
- describe the characteristics of class B and class AB push—pull amplifiers;
- define transformer-coupled push—pull transistor amplifier and crossover distortion.

Class B Push—Pull Amplifiers

A simplified **transformer-coupled push—pull transistor amplifier** is shown in **Figure 11.3**. The circuit is somewhat like two single-ended amplifiers placed back to back. The input and output transformers both have a common connection. The V_{CC} supply voltage is connected to this common point. For operation, the base voltage must rise slightly above zero to produce conduction. The turn-on voltage is usually about 0.6 V. Base voltage is developed by the incoming signal between the center tap and outside ends of the transformer. The transistor does not conduct until the base voltage rises above 0.6 V. The secondary winding resistance of the input transformer (T_1) is several thousand ohms. The divided transformer winding means that each

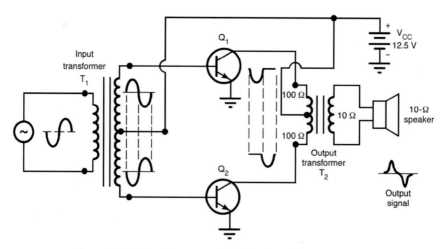

Figure 11.3   Transformer-coupled class B push—pull amplifier.

transistor is alternately fed an input signal. The signals are of the same value but 180° out of phase.

A class B push−pull amplifier produces an output signal only when an input signal is applied. For the first alternation of the input, the base of Q_1 swings positive and the base of Q_2 swings negative. Only Q_1 goes into conduction with this input. Note the resulting waveform at the collector of Q_1. At the same time, Q_2 is at cutoff. The negative alternation drives it further into cutoff. The resulting I_C from Q_1 flows into the top part of the primary winding of T_2. This causes a corresponding change in the electromagnetic field. Voltage is induced in the secondary winding through this action. Note the polarity of the secondary voltage for this alternation and the resulting output signal.

For the next alternation of the input signal, the process is reversed. The base of Q_1 swings negative. It becomes nonconductive. The base of Q_2 swings positive simultaneously. Q_2 goes into conduction. Note the indicated I_C of Q_2 for this alternation. The I_C of Q_2 flows into the lower side of T_2. It, in turn, causes the electromagnetic field to change. Voltage is again induced into the secondary winding. Note the polarity of the secondary voltage for this alternation. It is reversed because the primary current is in an opposite direction to that of the first alternation.

Figure 11.4 shows a combined collector family of characteristic curves for Q_1 and Q_2. These transistors are operated as class B amplifiers. With each transistor operating at cutoff (point Q), only one alternation of output occurs. The entire active region of each transistor can be used in this case to amplify the applied alternation. The combined alternations make a noticeable increase in output power.

Class B Push−Pull Amplifier Analysis

Using the collector family of characteristic curves data of **Figure 11.4**, we will analyze the power output and operational efficiency of the push−pull amplifier. The combined change in I_C for both Q_1 and Q_2 is 0.25 + 0.25 A, or 0.5 A_{PP}. The combined change in collector voltage is 12 + 12 V or 24 V_{PP}. The effective AC current and voltage values are

$$V_{CE \text{ (rms)}} = (V_{PP}/2) \times 0.707$$
$$= (24 \ V_{PP}/2) \times 0.707$$
$$= 12 \ V \times 0.707$$
$$= 8.484 \ V$$
$$I_{C(\text{rms})} = (I A_{PP}/2) \times 0.707$$

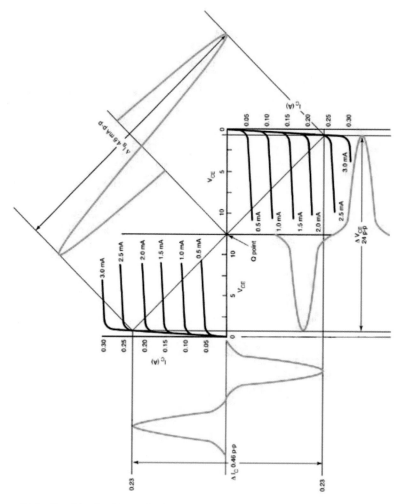

Figure 11.4 Combined collector family of characteristic curves for the push−pull amplifier in **Figure 11.3**.

$$= (0.5\ A_{PP}/2) \times 0.707$$
$$= 0.25\ A \times 0.707$$
$$= 0.177\ A.$$

The effective power output is

$$P_{out} = V_{CE\ (rms)} \times I_C(rms)$$

$$= 8.484\text{V} \times 0.177\text{A}$$
$$= 1.499\text{W, or } \sim 1.5 \text{ W}$$

The operating DC power of this push–pull amplifier must be determined to evaluate its operating efficiency. The DC supply voltage, V_{CC}, is 12.5 V. The value of I_C changes from 0 to 0.25 A in each transistor. The value of $I_{C \text{ (rms)}}$ is, therefore,

$$I_{C \text{ (rms)}} = \Delta I_C \times 0.707$$
$$= 0.25 \text{ A} \times 0.707$$
$$= 0.177 \text{ A.}$$

The average DC operating power (P_{in}) is equal to the DC voltage times the current flow. This value is

$$P_{in} = V_{dc \text{ (in)}} \times I_{dc \text{ (in)}}$$
$$= 12.5 \text{ V} \times 0.177$$
$$= 2.21 \text{ W.}$$

The operational efficiency of the push–pull amplifier in **Figure 11.3** is, therefore,

$$\eta = (P_{out}/P_{in}) \times 100\%$$
$$= (1.5 \text{ W}/2.21 \text{ W}) \times 100\%$$
$$= 0.679 \times 100\%$$
$$= 67.9\%.$$

This means that it takes nearly 2 W of DC input power to develop 1.5 W of AC output power. This type of operation is very good. The operational efficiency of a push–pull amplifier is much better than that of a comparable class A power amplifier. Its efficiency does not effectively change with the value of the signal-level input. The power output of a push–pull amplifier can be nearly four times the output of a single-ended class A power amplifier. Push-pull output circuitry is used primarily to develop high-power signal levels.

Example 11-2:

Determine the efficiency of a push–pull amplifier with V_{CC} = 9 V, power output = 1.75 W, and I_C varying from 0 to 300 mA.

Solution

$$I_{C(rms)} = \Delta I_C \times 0.707 \qquad P_{in} = V_{dc\ (in)} \times I_{dc\ (in)} \qquad \eta = (P_{out}/P_{in}) \times 100\%$$
$$= 0.3\ A \times 0.707 \qquad\quad = 9\ V \times 0.212 \qquad\qquad = 1.75\ W/1.91\ W$$
$$\times 100\%$$
$$= 0.212\ A \qquad\qquad\quad = 1.91\ W \qquad\qquad\qquad = 91.6\%.$$

Related Problem

Calculate the operational efficiency of a push−pull amplifier with V_{CC} = 12 V, power output = 1.6 W, and I_C varying from 0 to 0.2 A.

Crossover Distortion

Class B push−pull amplification has a very good operating efficiency and does a good job in reproducing high-power signals. It does, however, have a distortion problem when conduction changes back and forth between each transistor. This distortion is at the midpoint of the signal. The term **crossover distortion** is used to describe this condition. **Figure 11.4** shows crossover distortion near the center of the I_C and V_{CE} waveforms. In sound-amplifying systems, the human ear is able to detect crossover distortion.

Crossover distortion is due to the nonlinear characteristic of a transistor when it first goes into conduction. The emitter−base junction of a transistor responds as a diode. A silicon diode does not go into conduction immediately when it is forward biased. It takes approximately 0.6 V to produce conduction. This means that no current flowed between 0 and 0.6 V of base−emitter voltage (V_{BE}). **Figure 11.5** shows the input characteristic of a silicon transistor. Note the nonlinear area between 0 and 0.6 V.

When a transistor is cut off, there is no conduction of base current or collector current. When it is forward biased, conduction does not start immediately. There is a slight delay until 0.6 V_{BE} is reached. In a push−pull amplifier circuit, this initial delay causes distortion of the input signal. Distortion occurs when the input signal crosses from the conducting transistor to the off-transistor. One transistor is turning off and the other starts to conduct. The trailing edge of the off-transistor and the leading part of the on-transistor cause some distortion of the waveform. As a general rule, this distortion is very noticeable in small input signals. It is less noticeable in large signals.

Class AB Push−Pull Amplifiers

Crossover distortion can be reduced by changing the operating point of a transistor. If each transistor is forward biased by 0.6 V, there will not be

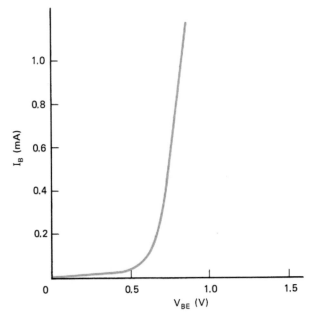

Figure 11.5 Input characteristic of a silicon transistor.

any noticeable crossover distortion. A transistor biased slightly above cutoff is considered to be a class AB amplifier. Operation of this type prevents the base–emitter voltage from reaching the nonlinear part of the curve. **Figure 11.6** shows the operating point of a class AB amplifier on an input curve and its location on a load line. The load line also shows operating points for class A and class B operations.

The operational efficiency of a class AB amplifier is not quite as good as that of a class B amplifier. There is some conduction when no signal is applied. The operating point of a class AB amplifier is usually held to an extremely low conduction level. Its operational efficiency is much better than that of a class A amplifier. In a sense, class AB operation is a design compromise to minimize distortion at a reasonable efficiency level.

A class AB push–pull amplifier is shown in **Figure 11.7**. The circuit is very similar to that of the class B amplifier of **Figure 11.3**. A divider network composed of R_{B1} and R_{B2} is used to bias transistors Q_1 and Q_2. The secondary winding resistance is rather low compared to the values of R_{B1} and R_{B2}. Resistors R_{E1} and R_{E2} are used to establish emitter bias. The circuit actually has divider bias and emitter bias combined. Emitter biasing

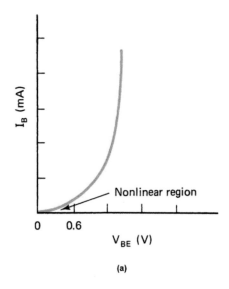

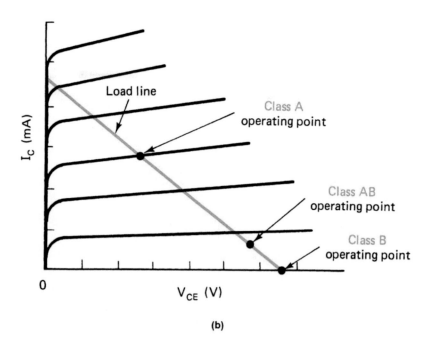

Figure 11.6   Transistor characteristics and operating point. (a) Input characteristic for a silicon transistor. (b) Bias operating points on a load line.

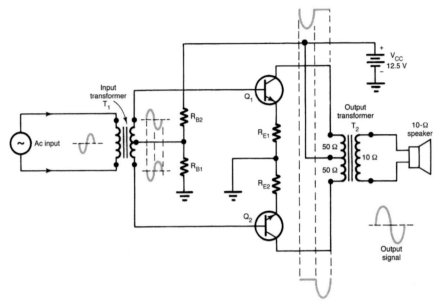

Figure 11.7 Transformer-coupled class AB push–pull amplifier.

is generally used to provide thermal stability for the two transistors. An amplifier connected in this manner is used for high-power audio systems. Low-power versions of this circuit are widely used in small transistor radios.

Self-Examination

Answer the following questions.

6. The main advantage of a class B push–pull amplifier over a class A amplifier is:

   a. Greater power output can be obtained
   b. Harmonic distortion
   c. Output transformer saturation reduction
   d. Reduced distortion of signal

7. The main disadvantage of a class B push–pull amplifier over a class AB push–pull amplifier is:

   a. Frequency response reduction
   b. Harmonic content reduction

    c. Output transformer saturation reduction
    d. Crossover distortion

8. The un-bypassed emitter resistor in an audio frequency amplifier causes

    a. Increased amplification
    b. Reduced gain
    c. Distortion
    d. No change in frequency response

9. Which class of amplifiers has the lowest operating efficiency?

    a. Class A
    b. Class B
    c. Class AB
    d. All amplifier classes

10. Which class of amplifiers has the best operating efficiency?

    a. Class A
    b. Class
    c. Class AB
    d. All amplifier classes

11.3 Complementary-Symmetry Power Amplifiers

The complementary-symmetry power amplifier is designed with complementary transistors. Complementary transistors are PNP and NPN types with similar characteristics. The availability of complementary transistors has made the design of transformerless power amplifiers possible. Power amplifiers designed with these transistors have an operating efficiency and linearity that is equal to that of a conventional push–pull circuit. Since the complementary-symmetry amplifier does not use transformers, component cost and weight are reduced.

11.3 Analyze the operational efficiency of a complementary-symmetry power amplifier.

    In order to achieve objective 11.3, you should be able to:

- describe the characteristics of a complementary-symmetry amplifier;
- define complementary-symmetry amplifier.

    An elementary **complementary-symmetry amplifier** is shown in **Figure 11.8**. This circuit responds as a simple voltage divider. Transistors Q_1 and Q_2 serve as variable resistors with the load resistor (R_L) connected

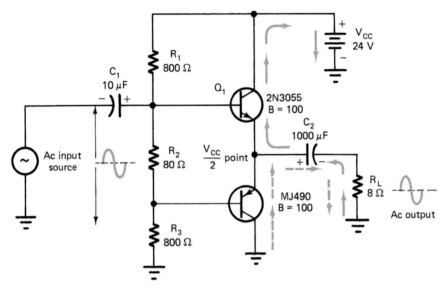

Figure 11.8 Class B complementary-symmetry amplifier.

to their common center point. When the positive alternation of the input signal occurs, it is applied to both Q_1 and Q_2. This alternation causes Q_1 to be conductive and Q_2 to be cut off. Conduction of Q_1 causes it to be low resistant. This connects the load resistor to the $+V_{CC}$ supply. Capacitor Q_2 charges in the direction indicated by the solid-line arrows. Note the direction of the resulting current through the load resistor. The polarity of voltage developed across R_L is the same as the input alternation. This means that the input and output signals are in phase. This is typical of a common-collector, or emitter-follower, amplifier.

When the negative alternation of the input occurs, it goes to both Q_1 and Q_2. The negative alternation reverse biases Q_1. This causes Q_1 to become nonconductive. Q_2, however, becomes conductive. A positive voltage going to the base of an NPN transistor causes it to become low resistant. The positive charge on C_2 from the first alternation forward biases the emitter of Q_2. With Q_2 being low resistant, C_2 discharges through it to ground. The dashed arrows show the discharge current of C_2. The resulting current flow through R_L causes the negative alternation to occur. The negative alternation of the input, therefore, causes the same polarity to appear across R_L. Q_2 is also connected as an emitter-follower. In effect, conduction of Q_1 causes C_2 to charge through R_L, and conduction of Q_2 causes it to discharge. The applied input is reproduced across R_L.

Biasing for Q_1 and Q_2 is achieved by a voltage-divider network consisting of R_1, R_2, and R_3. Each transistor is biased to approximately 0.6 V above cutoff. This biasing must make the emitter junction of Q_1 and Q_2 one-half of the value of V_{CC} when no signal is applied. The voltage drop across R_2 should be 1.2 V if the two transistors are silicon. The voltage drop across R_2 is quite small compared to that across R_1 and R_3. The voltage across R_1 and R_3 must be of an equal value to set the emitter junction at $V_{CC}/2$. To assure proper current output, the values of R_1 and R_3 are based on transistor beta (β) and R_L. This is determined by the following formula:

$$R_1 = R_3 = \beta \times R_L \qquad\qquad (11.2)$$

$$R_1 = R_3 = \beta \times R_L$$
$$= 100 \times 8\ \Omega$$
$$= 800\ \Omega.$$

Therefore, R_1 and R_3 are 800 Ω. The current flow through this network ($I_{n\ (in)}$) is determined by the following formula:

$$I_{n\ (in)} = V_{CC}/(R_1 + R_3). \qquad\qquad (11.3)$$

Therefore,

$$I_{n\ (in)} = V_{CC}/(R_1 + R_3)$$
$$= 24\ \text{V}/(800 + 800\ \Omega)$$
$$= 24\ \text{V}/1600\Omega$$
$$= 15\ \text{mA}.$$

The value of R_2 is then determined by the Ohm's law expression

$$R_2 = V_2/I_{n\ (in)}$$
$$= 1.2\ \text{V}/15\ \text{mA}$$
$$= 80\Omega.$$

V_2 is the voltage value used when Q_1 and Q_2 are silicon transistors. The base−emitter voltage (V_{BE}) is 0.6 V for each transistor. V_2 is, therefore, 0.6 V_{BE} plus 0.6 V_{BE}, or 1.2 V. The value of V_2 is 0.4 V if germanium transistors were used in the circuit.

If the NPN and PNP transistors used in a complementary-symmetry amplifier were interchanged, the polarity of the supply voltage would also need to be reversed. This will ensure that the normal biasing conditions needed for transistor operation are met.

Example 11-3:

In the complementary-symmetry amplifier shown in **Figure 11.8**, if a load resistance is changed from 8 to 10 Ω, determine the values of the biasing resistors R_1, R_2, and R_3 so that the amplifier will respond properly.
Solution

1) $R_1 = R_3 = \beta \times R_L$
   $= 100 \times 10 \ \Omega$
   $= 1000 \ \Omega.$
2) The current flow through this network ($I_{n \ (in)}$) is
   $I_{n \ (in)} = V_{CC}/(R_1 + R_3)$
   $= 24 \ V/(1000 \ \Omega + 1000 \ \Omega)$
   $= 24 \ V/2000 \ \Omega$
   $= 12 \ mA.$
3) The value of R_2 is then determined by the Ohm's law expression
   $R_2 = V_2/I_{n \ (in)}$
   $= 1.2 \ V/12 \ mA$
   $= 100 \ \Omega.$

Related Problem
In the complementary-symmetry amplifier shown in **Figure 11.8**, if the supply voltage is changed from 24 to 12 V, determine the values of the biasing resistors R_1, R_2, and R_3 so that the amplifier will respond properly.

Self-Examination

Answer the following questions.

11. The complementary-symmetry amplifier is also known an a(n) _____.

12. An amplifier with a PNP and an NPN transistor coupled together is called a(n) _____.

13. The term "symmetrical" when used with an amplifier means that _____ _____

14. The complementary-symmetry amplifier circuit configuration uses:
    a. 1 NPN and 1 PNP transistor
    b. 2 NPN transistors
    c. 2 PNP transistors
    d. 1 NPN transistor

15. If the characteristics of the PNP and NPN transistors are different, how would a complementary-symmetry amplifier respond:
    a. No change in amplified signal.
    a. Asymmetrical amplification of AC input waveform
    a. No output
    a. 180° phase shift

11.4 Analysis and Troubleshooting – Power Amplifiers

The performance of a power amplifier is evaluated by examining the applied input and observing the resulting output waveform. Ideally, there should be some level of amplification. Faults in an amplifier may occur when its input is overdriven (excessive amplitude) or when the value of the load impedance connected to it changes excessively. Faults in the operation of the amplifier are indicated by the value and shape of the output waveform. If little or no amplification occurs, or the output waveform is distorted, this indicates a faulty condition. In this section, a power amplifier will be examined to determine its amplification level and possible faulty conditions.

11.4 Troubleshoot power amplifiers.
   In order to achieve objective 11.4, you should be able to:

   • observe the input and output AC waveforms of a power amplifier on an oscilloscope;
   • distinguish between normal and faulty operation of a power amplifier.

Power Amplifier Troubleshooting

An amplifier can be damaged by excessive load current. This occurs when there are changes in the value of the load being driven by the amplifier. Load mismatches in audio amplifier systems for example occur when additional speakers are connected to the existing system. Generally, this causes excessive current to be drawn from the amplifier, which could cause overheating of the components. Extended periods of such overload conditions may cause permanent damage to the amplifier.

Distortion occurs in amplifier systems when the amplitude of AC input waveform applied is in excess of its rated input value. This type of distortion causes overdriving of the transistors used in the amplifier, causing parts of the amplified waveform to be clipped. Examining the output amplified signal on an oscilloscope will indicate when this distortion occurs.

Faulty components in an amplifier will cause distortion, excessive heating, reduced efficiency, and even complete loss of amplification. When replacing the components of an amplifier, the replacements should have similar operating characteristics. Mismatched transistors, for example, in a complementary-symmetry power amplifier will cause asymmetrical amplification.

Summary

- A single-ended amplifier operates as a class A amplifier and has an operating efficiency of approximately 30%.
- Efficiency refers to a ratio of developed output power to DC supply power and is expressed by the formula $\eta = (P_{out}/P_{in}) \times 100\%$.
- A single-ended audio amplifier responds to frequencies from 15 to 15 kHz; this frequency range is known as audio frequency.
- The term *audio* refers to the range of human hearing.
- A push−pull amplifier can operate as a class B or class AB amplifier and has an operating efficiency of about 80%.
- The bias operating point of the two transistors used in a class B push−pull amplifier is at cutoff; therefore, each transistor handles only one alternation of the sine-wave input signal.
- A class AB push−pull amplifier is biased slightly above cutoff; this helps to reduce distortion.
- Complementary-symmetry amplifiers use a PNP and an NPN transistor with similar characteristics.
- Complementary-symmetry amplifiers have an operating efficiency and linearity that is equal to that of a conventional push−pull circuit.
- Component cost and weight are reduced with complementary-symmetry amplifiers because they do not incorporate a transformer.

Formulas

(11-1) $\eta = (P_{out}/P_{in}) \times 100\%$ Efficiency.

(11-2) $R_1 = R_3 = \beta \times R_L$ R_1 and R_3 values (complementary-symmetry amplifier).

(11-3) $I_{n \, (in)} = V_{CC}/(R_1 + R_3)$ Network current (complementary-symmetry amplifier).

Review Questions

Answer the following questions.

1. A transistor bias slightly above cutoff is a class _____ amplifier.

    a. A
    b. B
    c. AB
    d. All of the above

2. If the AC input applied to an amplifier is excessively large, the AC output voltage will be:

    a. Normal
    b. Distorted
    c. 0 V
    d. Supply voltage

3. If the load connected to the amplifier is very high (low impedance), the current drawn from it will:

    a. Remain unchanged
    b. Increase
    c. Decrease
    d. Be the full supply voltage

4. In the complementary-symmetry amplifier shown in **Figure ??**, if the capacitor C_1 to which the in the AC input is applied, becomes open, how will the AC output be affected:

    a. 0 V
    b. The negative half of the waveform will not be present
    c. The positive half of the waveform will not be present
    d. Have reduced amplitude

5. In the complementary-symmetry amplifier shown in **Figure ??**, if the capacitor C_2 through which output is connected becomes open, how will the AC output be affected:

    a. 0 V
    b. The negative half of the waveform will not be present

c. The positive half of the waveform will not be present

d. Have reduced amplitude

6. In the class B push–pull amplifier characteristic curves shown in **Figure 11.4**, the reason that crossover distortion occurs is that:

a. The transistors are biased close to the middle of operating region

b. The transistors are biased close to the cutoff region

c. The transistors are biased in the saturation region

d. There is no biasing

7. In the class B push–pull amplifier characteristic curves shown in **Figure 11.4**, the crossover distortion can be reduced by:

a. Changing the transistor biasing point by 0.6 V

b. Biasing the transistor below cutoff

c. Biasing the transistor at full supply voltage

d. Biasing the transistor in the saturation region

Problems

Answer the following questions.

1. The input voltage of an amplifier is 0.5 V and the output voltage is 50 V. The voltage gain of this amplifier is:

   a. 0–5

   b. 5.1–10

   c. 10.1–50

   d. 50.1–above

2. The input power of an amplifier is 5 W is used to develop 4 W of output power. The efficiency of this amplifier is:

   a. 0%–20%

   b. 20.1%–60%

   c. 60.1%–80%

   d. 80.1%–100%

3. Referring to the complementary-symmetry amplifier circuit in **Figure 11.8**, if the value of the current gain changes from 100 to 80, what will be the values of the biasing resistors R_1 and R_3?

   a. 90 Ω

   b. 180 Ω

    c. 360 Ω

    d. 500 Ω

4. Referring to the push−pull amplifier characteristic curves given in **Figure 11.4**, if the peak-to-peak value of the amplified output voltage changes from 24 to 20 V, determine the RMS of the output:

    a. 0.5−2 V

    b. 2.1−5 V

    c. 5.1−8 V

    d. 8.1−above

5. Referring to the push−pull amplifier characteristic curves given in **Figure 11.4**, if the RMS of the collector current flow changes from 0.17 to 0.25 A, calculate the peak-to-peak current, $I_{C(PP)}$:

    a. 0.1−0.5 A

    b. 0.51 V to 0.8 A

    c. 0.81−1 A

    d. 1.1 A - above

Answers

Examples

11-1 2 W

    11-2 $I_{C(rms)}$ = 0.141 A, P_{in} = 1.697 W, η = 94.29%

    11-3 $R_1 = R_3 = 1000$ Ω, $R_2 = 200$ Ω

Self-Examination

11.1

1. False, the efficiency is low for class A amplifiers
2. Single-ended amplifier
3. a. In the middle of the load line
4. d. Impedance matching with the load
5. a. 0 V when the output is shorted

11.2

6. a. Greater efficiency and, hence, greater output
7. d. Class B has increased crossover distortion
8. c. Distortion owing to change in the biasing

9. a. Class A amplifiers have the lowest efficiency
10. b. Class B amplifiers have the highest operating efficiency

11.3
11. Emitter follower
12. Complementary-symmetry amplifier
13. Amplification of the input AC signal is identical for both halves of the waveform
14. a. 1 NPN and 1 PNP transistor
15. b. Non-symmetrical amplification would occur if the transistor characteristics are not matched exactly.

Terms

Single-ended amplifier

A power amplifier that has only one active device operates as a class A amplifier and has an operating efficiency of about 30%.

Efficiency

A ratio of developed output power to DC supply power.

Transformer-coupled push—pull transistor amplifier

A power amplifier that is somewhat like two single-ended amplifiers placed back to back. The input and output transformers both have a common connection. It operates as a class B amplifier and has an operational efficiency of approximately 80%.

Crossover distortion

The distortion of an output signal at the point where one transistor stops conducting and another conducts.

Complementary-symmetry amplifier

A power amplifier that uses NPN and PNP transistors having similar operating characteristics, for amplifying each half (alternation) of the AC input waveform. It operates as a class B amplifier and has an operational efficiency of approximately 80%.

12

Thyristors

The term **thyristor** refers to a rather general classification of solid-state devices that are used as electronic switches and power control devices. Thyristors can be two-, three-, or four-terminal devices. Thyristors operate through a type of regenerative feedback. When conduction is initiated, the device latches, or holds in, its on-state. Momentarily removing or reducing the energy level of the source stops conduction or switches to the off-state.

In general, thyristors are classified as **voltage controlled switches**. The two major classifications of thyristors are **unidirectional** and **bidirectional**. Unidirectional refers to conduction in only one direction, and bidirectional refers to conductions in two directions. Bidirectional thyristors provide some very efficient power control capabilities. The **diac** and **triac** are examples of bidirectional thyristors. The SCR, UJT, PUT, SCS, and GTO are examples of unidirectional thyristors. This chapter discusses the conductivity capabilities and applications of all of these thyristors.

Thyristors should not be confused with bipolar junction transistors or field-effect transistors. Transistors, in general, are capable of performing switching operations. As a rule, these devices are not as efficient as thyristors and do not have the power-handling capability. Thyristors are used as **power control** devices, whereas transistors are primarily used in amplifying applications.

Objectives

After studying this chapter, you will be able to:

12.1 analyze the operation of an SCR in DC and AC power control circuits;
12.2 analyze the operation of a triac in DC and AC power control circuits;
12.3 analyze the operation of a diac in AC power control circuits;
12.4 analyze the operation of a UJT in pulse-triggered circuits;
12.5 analyze the operation of a PUT in pulse-triggered circuit;

12.6 analyze the operation of SCS and GTO thyristors;

12.7 analyze and troubleshoot thyristors.

Chapter Outline

12.1 Silicon Controlled Rectifiers

12.2 Triacs

12.3 Diacs

12.4 Unijunction Transistors

12.5 Programmable Unijunction Transistors

12.6 Gate Turn-Off Thyristors

12.7 Analysis and Troubleshooting – Thyristors

12-1 Silicon Controlled Rectifier (SCR) Testing

In this activity, the leads of an SCR will be identified and its condition evaluated.

13-2 SCR Power Control

In this activity, an SCR power control circuit will be analyzed.

13-3 Triac Testing

In this activity, the leads of a triac will be identified and its condition evaluated.

13-4 Triac Power Control

In this activity, a triac power control circuit will be analyzed.

13-5 Programmable Unijunction Transistor (PUT) Testing

In this activity, the leads of PUT will be identified, and its condition evaluated.

Key Terms

alternation
bidirectional triggering
conduction time
contact bounce
forward breakover voltage

holding current
interbase resistance
internal resistance
intrinsic standoff ratio
latching
negative resistance region
off-state resistance
triggering
turn-on time

12.1 Silicon Controlled Rectifiers

One of the first thyristors to be developed is the silicon controlled rectifier (SCR). The SCR is a solid-state counterpart of the older thyratron gaseous tube. The term *thyristor* is derived from the words *thyratron* and *transistor*. An SCR is classified as a reverse-blocking triode thyristor. The term *triode* means three electrodes or leads. Reverse blocking refers to conduction in only one direction. An SCR is, therefore, a three-electrode device with unidirectional power control capabilities. It is similar in size to a comparable silicon power diode and is used primarily as a switching device.

12.1 Analyze the operation of an SCR in DC and AC power control circuits.

In order to achieve objective 12.1, you should be able to:

- interpret the *I–V* characteristics of an SCR;
- describe the physical construction of an SCR;
- explain how an SCR is biased to make it operational;
- explain the fundamental operation of an SCR;
- identify an SCR in a specific circuit application;
- define *internal resistance, off-state resistance, triggering, latching, alternation, turn-on time, forward breakover voltage, holding current,* and *conduction time.*

SCR Construction

An SCR is a solid-state device made of four alternating layers of P-type and N-type material. Three P–N junctions are formed by the structure. A schematic symbol and the crystal structure of an SCR are shown in

Figure 12.1. Note that the device has a **PNPN structure** from anode to cathode and that three distinct P–N junctions are formed.

Each SCR has three leads or terminals. The **anode** and **cathode** terminals are similar to those of a regular silicon diode. The third lead is called the gate. The gate determines when the device switches from its off-state to its on-state. An SCR will usually not go into conduction by simply forward biasing

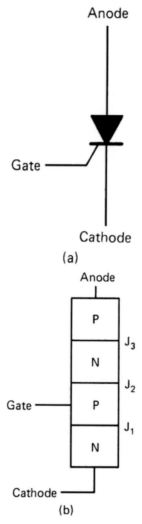

Figure 12.1   SCR crystal. (a) Symbol. (b) Structure.

the anode−cathode. The gate must be forward biased at the same time. When these conditions occur, the SCR becomes conductive. The **internal resistance** of a conductive SCR is less than 19 Ω. Its **off-state resistance** or reverse resistance is generally in excess of 1 MΩ.

When the anode is made positive and the cathode negative, junction 1 (J_1) and junction 3 (J_3) are forward biased, and junction 2 (J_2) is reverse biased. Reversing the polarity of the source alters this condition: J_1 and J_3 become reverse biased, and J_2 becomes forward biased. This condition does not permit conduction. Conduction occurs only when the anode, cathode, and gate are forward biased simultaneously.

SCRs are usually small, are rather inexpensive, waste very little power, and require practically no maintenance. The SCR is available in a full range of types and sizes to meet nearly any power control application and in current ratings from less than 1 A to over 1400 A. Voltage values range from 15 to 2600 V. Some representative SCRs are shown in **Figure 12.2**. Only a few of the more popular packages are shown here. As a general rule, the anode is connected to the largest electrode if there is a difference in their physical size. The gate is usually smaller than the other electrodes. This is because only a small gate current is needed to achieve control. In some packages, the SCR symbol is used for lead identification.

SCR Operation

Operation of an SCR is not easily explained by using the four-layer **PNPN structure** of **Figure 12.1(b)**. A rather simplified method has been developed which describes the crystal structure as equivalent to two interconnected transistors. **Figure 12.3** shows this imaginary division of the crystal. Note that the top three crystals form a PNP device. Two parts of each transistor are interconnected.

Figure 12.4 shows how the two-transistor **equivalent circuit** and the PNPN structure of an SCR respond when voltage is applied. Note that the anode is made positive and the cathode negative. This condition forward biases the emitter of each transistor and reverse biases the collector. Without base voltage applied to the NPN transistor, the equivalent circuit is nonconductive. No current flows through the load resistor. Reversing the polarity of the voltage source will not alter the conduction of the two circuits. The emitter of each transistor is reverse biased by this condition. This means that no conduction takes place when the anode and cathode are reverse biased. With only anode−cathode voltage applied, no conduction occurs in either direction.

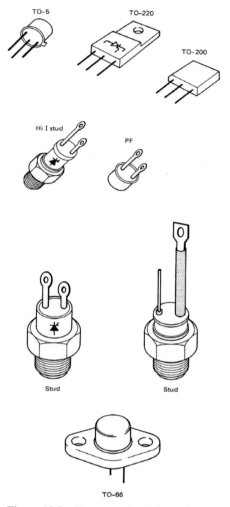

Figure 12.2   Representative SCR packages.

Assume that the anode and cathode of an SCR are forward biased, as indicated in **Figure 12.4**. If the gate is momentarily made positive with respect to the cathode, the emitter−base junction of the NPN transistor becomes forward biased. This action, called **triggering**, immediately causes collector current to flow in the NPN transistor. As a result, base current also flows into the PNP transistor. This, in turn, causes a corresponding collector current to flow. PNP collector current now causes base current to flow into the

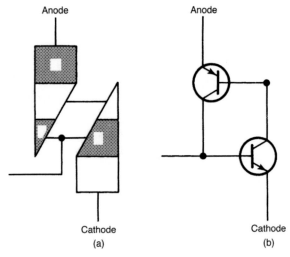

Figure 12.3 Equivalent SCR.

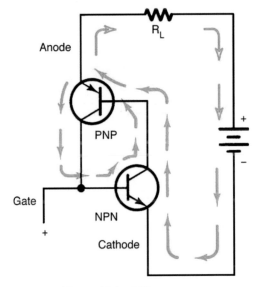

Figure 12.4 SCR response.

NPN transistor. The two transistors, therefore, hold their state when the gate is disconnected. This is called **latching**. In effect, when conduction occurs, the device latches in the on-state. Current continues to flow through the SCR as long as the anode–cathode voltage is of the correct polarity.

To turn off a conductive SCR, it is necessary to momentarily remove or reduce the anode–cathode voltage. This action turns off the two transistors. The device then remains in this state until the anode, cathode, and gate are forward biased again. With AC applied to an SCR, it automatically turns off during one **alternation** of the input. For 60-Hz AC, the SCR is switched on and off 60 times per second. Control is achieved by altering the **turn-on time** during the conductive or on alternation.

SCR I–V Characteristics

The **I–V characteristics** of an SCR tell a great deal about its operation. In **Figure 12.5**, the *I–V* characteristics show that an SCR has two conduction states. Quadrant I shows **conduction** in the forward direction. This shows how conduction occurs when the **forward breakover voltage** is exceeded. Forward breakover voltage (V_{BO}) is the voltage at which a device such as an SCR or triac goes into conduction in quadrant I of its *I–V* characteristics.

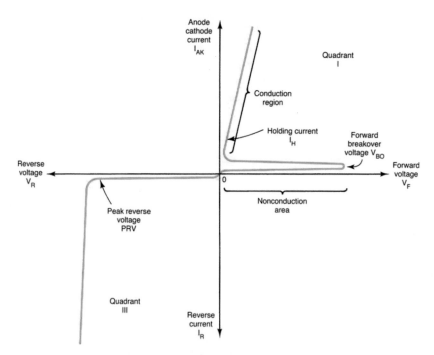

Figure 12.5   *I–V* characteristics of an SCR.

Note that the curve returns to approximately zero after the V_{BO} has been exceeded. When conduction occurs, the internal resistance of the SCR drops to an extremely small value. This value is similar to that of a forward-biased silicon diode. The conduction current, or anode–cathode current (I_{AK}), must be limited by an external resistor. This current, however, must be great enough to maintain conduction when it starts. The **holding current** level must be exceeded for this to take place. The holding current (I_H) is the current level that must be achieved to latch or hold an SCR or triac in the conductive state. Note that the I_H level is just above the knee of the I_{AK} curve after it returns to the center.

Quadrant III of the *I–V* characteristics shows the **reverse breakdown** condition of operation. This characteristic of an SCR is very similar to that of a silicon diode. Conduction occurs when the peak reverse voltage (PRV)

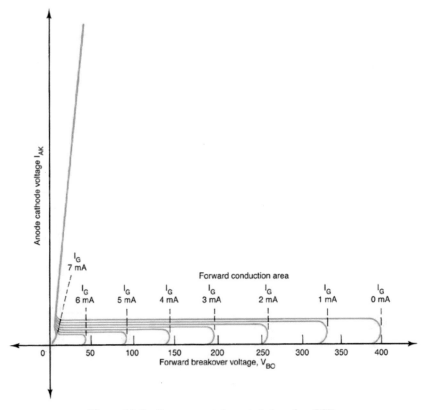

Figure 12.6 Gate-current characteristics of an SCR.

value is reached. Normally, an SCR will be permanently damaged if the PRV is exceeded. SCRs have PRV ratings of 25–2000 V.

For an SCR to be used as a **power control** device, the forward V_{BO} must be altered. Changes in gate current cause a change in the V_{BO}. This occurs when the gate is forward biased. An increase in I_G causes a large reduction in the forward V_{BO}. An enlargement of quadrant I of the *I–V* characteristics is shown in **Figure 12.6**. This shows how different values of I_G change the V_{BO}. With zero I_G, it takes a V_{BO} of 400 V to produce **conduction**. An increase in I_G reduces this quite significantly. With 7 mA of I_G, the SCR conducts as a forward-biased silicon diode. Lesser values of I_G cause an increase in the V_{BO} needed to produce conduction.

The **gate-current characteristics** of an SCR show a very important electrical operating condition. For any value of I_G, there is a specific V_{BO} that must be reached before conduction can occur. This means that an SCR can be turned on when a proper combination of I_G and V_{BO} is achieved. This characteristic is used to control conduction when the SCR is used as a power control device.

Example 12-1:

Refer to **Figure 12.6**. With a gate current of 2 mA, what is the forward breakover voltage of the SCR?
Solution

(1) Find the value of gate current (I_G = 2 mA) along the horizontal axis of the *I–V* characteristic curve.
(2) Project vertically to the horizontal axis to find forward breakover voltage - V_{BO} = 250 V.

Related Problem

Refer to **Figure 12.6**. Find the V_{BO} values for (1) I_G = 6 mA, (2) I_G = 4 mA (3) I_G = 1 mA, and (4) I_G = 0 mA.

SCR Power Control

When an SCR is used as a **power control** device, it responds primarily as a switch. When the applied source voltage is below the forward breakover voltage, control is achieved by increasing the gate current. Gate current is usually made large enough to ensure that the SCR will turn on at the proper time. Gate current is generally applied for only a short period of time. In many

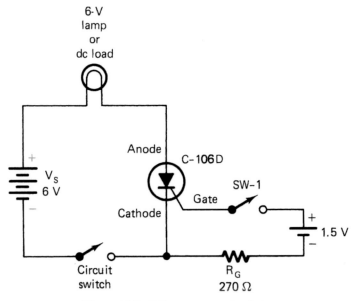

Figure 12.7 DC power control switch.

applications, this may be in the form of a short-duration pulse. Continuous I_G is not needed to trigger an SCR into conduction. After conduction occurs, the SCR will turn off only when I_{AK} drops to zero.

DC Power Control

Figure 12.7 shows an SCR used as a **DC power control switch**. In this circuit, a small gate current controls a rather high load current. Note that the electrical power source, V_S, is controlled by the SCR. This is achieved by making the anode positive and the cathode negative.

When the circuit switch is initially closed, the load is not energized. In this situation, the V_{BO} is greater than V_S. Power control is achieved by closing SW-1. This forward biases the gate. If a suitable value of I_G occurs, V_{BO} decreases and the SCR turns on. When the I_G is removed, the SCR remains in conduction. To turn the circuit off, the circuit switch is momentarily opened. With the circuit switch closed again, the SCR remains in the **off-state**. The SCR will go into conduction again by closing SW-1.

SCR DC power control applications require two switches to achieve control. This application of the SCR is not very practical. The circuit switch

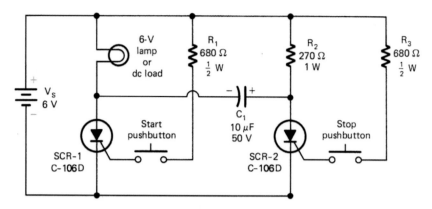

Figure 12.8   DC power control circuit.

would need to be capable of handling the load current. The gate switch could be rated at an extremely small value. If several switches were needed to control the load from different locations, this circuit would be more practical.

More practical DC power circuits can be achieved by adding a number of additional components. **Figure 12.8** shows a **DC power control** circuit with one SCR controlled by a second SCR. SCR-1 controls the DC load current. SCR-2 controls the conduction of SCR-1. In this circuit, switching of a high-current load is achieved with two small low-current switches. SCR-1 can be rated to handle the load current. SCR-2 can have a rather small current-handling capacity. Control of this type can probably be achieved with a smaller current than a circuit employing a large electrical contactor switch.

Operation of the DC control circuit of **Figure 12.8** is based on the conduction of SCR-1 and SCR-2. To turn on the load, the **start pushbutton** is momentarily closed. This forward biases the gate of SCR-1. The V_{BO} is reduced, and SCR-1 goes into its conduction state. This action also causes C_1 to charge to the indicated polarity. The load remains energized as long as power is supplied to the circuit.

Turn-off of SCR-1 is achieved by pushing the stop button. This momentarily applies I_G to SCR-2 and causes it to be conductive. This, in turn, causes C_1 to be connected across SCR-1. The charge on C_1 is momentarily applied to the anode−cathode of SCR-1. This reduces the V_{AK} of SCR-1 and causes it to turn off. The circuit will remain in the off-state until the start button energizes it. An **SCR power circuit** of this type can be controlled with two small pushbuttons. As a rule, control of this type is more reliable and less expensive than a DC electrical contactor circuit.

Example 12-2:

Describe the procedure to turn on the load of the SCR circuit of **Figure 12.8**.

Solution

(1) Press START PB to momentarily close.
(2) Gate of SCR-1 is forward biased.
(3) V_{BO} of SCR-1 is reduced.
(4) SCR-1 goes into conduction.
(5) C_1 charges to cause the 6-V lamp load to remain energized.

Related Problem

Describe the step-by-step procedure to turn off the load of the SCR circuit of **Figure 12.8**.

AC Power Control

AC power control applications of an SCR are very common. As a rule, control is very easy to achieve. The SCR automatically turns off during one alternation of the AC input. This eliminates the turnoff problem with the DC circuit. The load of an AC circuit sees current only for one alternation of the AC input cycle. In effect, an SCR power control circuit has half-wave output. The **conduction time** of an alternation can be varied with an SCR circuit. Variable output is achieved through this method of control.

A simple **SCR power control switch** is shown in **Figure 12.9**. Connected in this manner, conduction of AC only occurs when the anode is positive, the cathode is negative, and SW-1 is closed. When SW-1 is closed, current flows through the gate of the SCR. The value of I_G lowers the V_{BO} value to the point where the SCR becomes conductive. The gate resistor (R_G) limits the peak value of I_G. Diode D_1 prevents reverse voltage from being applied between the gate and cathode of the SCR. With SW-1 closed, the gate is forward biased for only one alternation. This is the same alternation that forward biases the anode–cathode. With a suitable value of I_G, and correct anode–cathode voltage (V_{AK}), the SCR becomes conductive.

The AC power control switch of **Figure 12.9** is designed primarily to take the place of a mechanical switch. With a circuit of this type, it is possible to control a rather large amount of electrical power with a relatively small switch. Control of this type is very reliable. The switch does not have mechanical contacts that spark and arc when changes in load current occur. Control of this type, however, is only an on–off function.

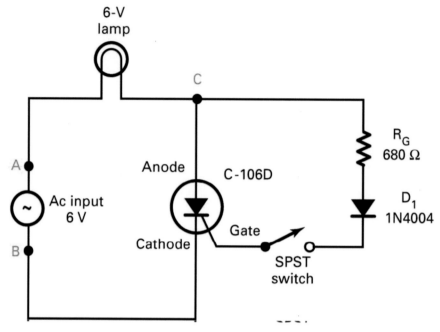

Figure 12.9　SCR AC power control switch.

SCRs are used widely to control the amount of electrical power supplied to a load device. Circuits of this type respond very well to 60-Hz AC. This type of circuit is called a **half-wave SCR phase shifter**. **Figure 12.10** shows a simplified version of the variable power control circuit.

Operation of the variable power control circuit is based on the charge and discharge of capacitor C_1. Assume that the positive alternation occurs. Point A will be positive, and point B will be negative. This forward biases the SCR. The SCR will not turn on immediately; however, because there is no gate current. Capacitor C_1 begins to charge through R_1. In a predetermined amount of time, the gate voltage builds to where the gate current is great enough to turn on the SCR. D_1 is forward biased by the capacitor voltage. I_G then flows through D_1. Once the SCR goes into conduction, it continues for the remainder of the alternation. Current flows through the load. In effect, turn-on of the SCR is delayed by the values of R and C. Remember that time (t) = R × C and is referred to as the **RC time constant**.

When the negative alternation occurs, point A becomes negative and point B goes positive. This reverse biases the SCR. No current flows through the

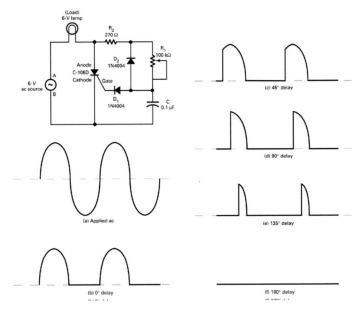

Figure 12.10 Phase-control SCR circuit and waveforms. (a) Applied AC. (b) $0°$ delay. (c) $45°$ delay. (d) $90°$ delay. (e) $135°$ delay. (f) $180°$ delay.

load for this alternation. This voltage, however, forward biases D_2. Capacitor C_1 then discharges very quickly through D_2. This bypasses the resistance of R_1. The capacitor is reset or in a ready state for the next alternation. The process repeats itself for each succeeding cycle.

A change in the resistance of R_1 alters the conduction time of the SCR power control circuit. When R_1 is increased, C_1 cannot charge as quickly during the positive alternation. It, therefore, takes more time for C_1 to build its charge voltage. This means that the turn-on of I_G is delayed for a longer time. As a result, conduction of the SCR is delayed at the beginning of the alternation. See waveform (c) of **Figure 12.10**. The resulting **load current** for this condition is somewhat less than that of the waveform in (b). Variable control of the load is achieved by altering the conduction time of the SCR.

A further increase in the resistance of R_1 causes more delay in the turn-on of the SCR during the positive alternation. Waveforms in (d), (e), and (f) show the result of this change. The waveform in (d) has a $90°$ **delay** of SCR conduction time. The waveform in (e) shows a $135°$ delay in the conduction time. If the value of R_1 is great enough, the SCR will delay its turn-on for the full alternation. The waveform in (f) shows a $180°$ turn-on delay. Note that the load does not receive any current for this condition. Control of this circuit is from **full conduction**, as in waveform (b), to 180° delay, as in waveform (f).

Power control with an SCR is very efficient. No power is lost in the negative portion of the AC alternation when the SCR is not conducting. By controlling or delaying the **switching time** of the gate during the positive portion of the AC alternation, varying amounts of power will be developed by the load. If no delay of the **conduction time** occurs, the load develops power during the entire positive alternation. This is equivalent to 50% of the power developed by the load. Recall that 100% of the power is developed by the load if it receives both halves of the AC waveform. If conduction is delayed 90°, the load develops only a quarter, or 25%, of its potential power. If the gate of the SCR is triggered during the negative half alternation of the AC input, its operation is not affected since the SCR is reverse biased. It is important to note that nearly all the power controlled by the circuit is applied to the load device. Very little of it is consumed by the SCR. In variable power control applications, it is extremely important that the load be supplied to its full value of power. With an SCR control circuit, very little power is consumed by the other components. By delaying the turn-on time of conduction, power can be controlled with the highest level of **efficiency**.

Example 12-3:

In the SCR circuit given in **Figure 12.10** if the conduction is delayed by 270° (which is in the negative half-cycle), how much power is developed by the load?

The load will not develop any power. This is because the SCR is reverse biased in the 180°−270° range and triggering the gate will not cause it to conduct.

Related Problem

In the SCR circuit given in **Figure 12.10**, if the conduction is delayed by 180°, how much power is developed by the load?

Self-Examination

Answer the following questions.

1. An SCR has three leads, which are known as the _____, _____, and _____.

2. An SCR will not go into conduction by simply (forward, reverse) biasing the anode–cathode.
3. An SCR has _____ junctions and _____ semiconductor layers.
4. The internal resistance of an SCR is (high, low) after triggering.
5. The internal resistance of an SCR is (high, low) before it is triggered.
6. To turn off a conductive SCR, the _____ voltage must be momentarily removed or reduced.
7. When AC is applied to the anode–cathode of an SCR, it will automatically turn off during one _____.
8. Conduction of an SCR occurs during quadrant (I, II, III, IV) of the *I–V* characteristics.
9. Conduction of an SCR occurs when the _____ voltage is exceeded.
10. Reverse breakdown of an SCR occurs in quadrant (I, II, III, IV) of the *I–V* characteristics.
11. A(n) _____ in gate current will cause a decrease in the forward breakover voltage of an SCR.
12. The gate–cathode of an SCR must be _____ biased to produce gate current.
13. A suitable value of _____ current lowers the forward breakover voltage of an SCR to produce conduction.
14. The _____ loses control over the anode–cathode after an SCR has been triggered into conduction.
15. When the _____ current of an SCR is reached or exceeded, the SCR latches or continues conduction.

12.2 Triacs

Alternating current power can be efficiently controlled with a special device that switches conduction on and off during each alternation. Control of this type is accomplished with a thyristor known as a triac. A triac is classified as a bidirectional triode thyristor. A triac is better described as a triode AC switch. Conduction is triggered by a gate signal like it is with the SCR. A triac, however, is the equivalent of two reverse-parallel-connected SCRs with one common gate. Conduction can be achieved in either direction with an appropriate gate current. AC can be controlled efficiently and accurately with this device.

12.2 Analyze the operation of a triac in DC and AC power control circuits.

In order to achieve objective 12.2, you should be able to:

- interpret the *I–V* characteristics of a triac;
- describe the physical construction of a triac;
- explain how a triac is biased to make it operational;
- explain the fundamental operation of a triac;
- identify a triac in a specific circuit application;
- define *bidirectional triggering* and *contact bounce*.

Triac Construction

Figure 12.11 shows the crystal structure, junction diagram, and schematic symbol of a **triac**. Note that the crystal structure of this device forms an $N_1P_1N_2P_2$ structure between T_1 and T_2. The gate is commonly connected to the P_1-N_4 junction. From T_2 to T_1, the crystal structure is $N_3P_2N_2P_1$. The gate of this structure uses N_4 of the P_1-N_4 junction. This shows that T_1, T_2, and the gate are in either an NPNP or a PNPN combination, depending on the voltage of T_1 and T_2.

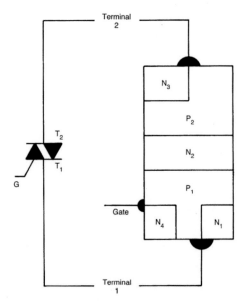

Figure 12.11   Junction diagram and schematic symbol of a triac.

Selection of the crystal combination is based on the polarity of the applied voltage. When T_1 is negative, N_1 is forward biased and P_1 is reverse biased. Current carriers flow easily through the forward-biased material and not into the reverse-biased material. This means that the NPNP or PNPN combination and the appropriate gate voltage for initiating **conduction** are selected by source voltage polarity. For AC, the polarity changes with each alternation. Thus, one structure combination responds for one alternation and the other structure combination responds for the other alternation.

Triac Operation

Functionally, a **triac** is classified as a **gate-controlled AC switch**. Conduction is achieved by selecting an appropriate crystal combination. Selection depends on the polarity of the source. During one alternation, conduction is through a PNPN combination. Conduction for the next alternation is through an NPNP combination. The crystal selection process is achieved automatically. P-type and N-type materials are jointly connected to each terminal. Selection is made according to the **polarity** of a specific alternation.

In **normal operation**, the circuit of a triac must be designed so that it goes into conduction only by action of the gate. The **gate** of a triac responds as a nonlinear low-impedance junction similar to that of a forward-biased diode. Triggering of the gate must, therefore, be achieved by some type of low-impedance source. The gate junction usually requires a rather sizable current to produce **triggering**. This means that the G-T_1 junction is current-sensitive rather than voltage-sensitive. Most triacs, for example, require a hundred or more milliamperes of I_G for a few microseconds to be triggered into conduction. Several things can be done to develop this amount of I_G. A capacitor is usually discharged into the gate circuit to initiate the triggering process. As a rule, this necessitates some type of low-power, on−off switching device to produce a suitable value of I_G. The device must respond equally well to both alternations of the AC input.

The applied source voltage for a triac circuit must, therefore, be of a value that is somewhat less than that normally needed to overcome the operating V_{BO}. This characteristic, therefore, becomes a critical factor when selecting a triac for a specific control function. **Transient pulses** or uncontrolled line voltage spikes must also be limited to avoid uncontrolled triggering of a triac. A special **bidirectional triggering** device known as a diac is generally used in triac circuits to limit voltage spikes to amplitude levels that will not damage or trigger the gate junction. These devices are discussed in the next section of this chapter.

Triac I–V Characteristics

When AC is applied to a triac, T_2 **is positive** and T_1 **is negative** for one alternation. For the next alternation, T_2 is negative and T_1 is positive. The *I–V* characteristics of a triac must, therefore, show how it responds during both alternations of the AC source. **Figure 12-12** shows the typical **I–V characteristics** of a triac. Operation in quadrant I occurs when T_2 is positive and T_1 is negative. This represents the forward conduction mode of operation. Quadrant III denotes reverse conduction. T_2 is made negative and T_1 positive for operation in quadrant III. The **breakover voltage (V_{BO})** in quadrant I or quadrant III is usually quite large when the gate current is zero. Ordinarily, a triac is not designed for conduction when the gate is zero.

When **gate current** is applied to a triac, it lowers the V_{BO}. This action permits conduction to be achieved at a lower voltage. In quadrant I, either a positive or negative gate voltage produces gate current. Positive gate voltage tends to be more sensitive to triggering than negative values. When a suitable value of I_G occurs, the triac is triggered into conduction. I_T flows through T_1,

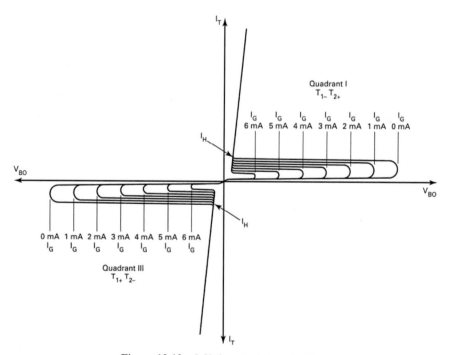

Figure 12.12   *I–V* characteristics of a triac.

the P_1 material, and the N_2-P_2 region, which is forward biased by T_2. This current continues for the remainder of the alternation when it moves into the **holding current** region of the *I–V* characteristics.

During the negative alternation of the source, T_1 **is positive** and T_2 **is negative**. Quadrant III displays this condition of operation. With no gate current, the reverse V_{BO} is quite large. No conduction occurs unless an extremely high value of V_{BO} is reached. This voltage value, like that of quadrant I, is usually avoided in normal circuit operation. When I_G occurs, it lowers the reverse V_{BO} to a value that can cause conduction. In some triacs, either a positive or negative gate voltage may be used to bias the gate. Ordinarily, a positive V_G is avoided in most applications. A negative V_G is, therefore, used to bias the gate and produce I_G. Conduction is triggered and I_G flows from T_2, through N_3, P_2, N_2, P_1, and T_1. The N_2-P_1 junction sees a significant increase in current when I_G flows in the gate. This current added to that of the N_3-P_2 junction lowers the V_{BO} and causes **full conduction** to occur. The resulting *I–V* characteristic is thus a mirror image of conduction in quadrant I.

Due to the dual polarity of the gate, T_1 and T_2 of a triac have four possible trigger polarity combinations. The table shown in **Figure 12.13** lists these combinations as **trigger modes** and representative gate current (I_G) triggering values. **Quadrant I** operation has T_1 negative and T_2 positive. Note that the positive gate voltage produces an I_G that is smaller than that of the negative voltage I_G. This indicates that the triac is more sensitive to the positive voltage value in quadrant I. **Quadrant III** operation occurs best when T_1 is positive and T_2 is negative. In this quadrant, the triac is more sensitive to a negative voltage polarity. Ordinarily, the positive V_G mode of operation is avoided in most circuit applications. Some manufacturers recommend that their triacs not be used in this triggering mode. A common

Quadrant	Element polarities			Required I_G
I +	T_2 +	T_1 −	G +	5–10 mA
I −	T_2 +	T_1 −	G −	10–20 mA
III −	T_2 −	T_1 +	G −	7–15 mA
III +	T_2 −	T_1 +	G +	> 40 mA

Figure 12.13   Trigger modes of a triac.

practice used in selecting the most sensitive mode of operation is to match the polarity of the gate with the polarity of T_2.

Some manufacturers have special-function triacs that are designed for operation in the III+ operational mode. This device generally has a special number designation or code to denote its unusual sensitivity. For example, a manufacturer lists a standard triac as an SC141B. A device designed for III+ operation is numbered SC141BI3. The specialized triac is usually more expensive than a standard triac.

Example 12-4:

Refer to the triac *I–V* characteristics of **Figure 12.12** and triac trigger modes of **Figure 12.13**. What value of gate current is required for triac conduction in quadrant I – with T_2 positive, T_1 negative, and gate negative?

Solution

Refer to **Figure 12.13** and locate required I_G = 10−20 mA.

Related Problem

Refer to **Figures 12.12** and **12.13**. What values of I_G are required for the following: 1) quadrant III−, T_2−, T_1+, and G−; 2) quadrant *I*+, T_2+, T_1−, and G+.

Triac DC Power Control

DC control of a triac is achieved the same way as it is for an SCR. In general, DC gate control of a triac is not very practical. **Conduction** can be delayed for only the first 90° of one alternation. After this point of the alternation, conduction will be either on or off for the last 90°, depending on the gate current. The resulting output of the two alternations is quite different. Quadrants I and III do not have the same sensitivity to gate current. In quadrant III, the gate current must be greater to cause triggering. This, in effect, causes a slightly different triggering level for each alternation. In most devices, a positive gate voltage in quadrant III will not permit any conduction during the negative alternation. **Figure 12.14** shows a circuit of this type.

As a general rule, there is no reason to use DC control to change the operation of a triac. The variable DC source would need to be supplied from

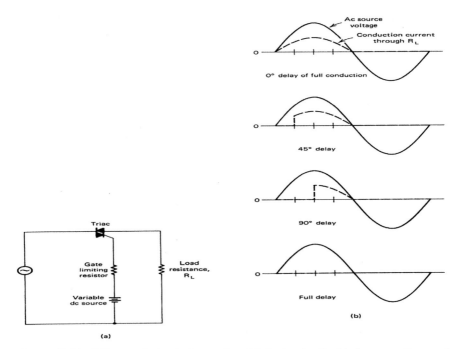

Figure 12.14 DC control circuit for a triac. (a) Basic circuit. (b) Source voltage and conduction current waveforms.

a battery or by a rectified power supply and would still produce half-wave DC output. Control of this type is only about 25% effective. AC control of a triac, by comparison, is 100% effective and can be supplied from the same source.

Triac AC Power Control

Triacs are widely used to control AC power. In this application, the triac is designed to respond as a **switch**. Through this switching action, it is possible to control the AC source for a portion of each alternation. If a triac is in **conduction** for both alternations of the source voltage, 100% of source power is delivered to the load device. Conduction for half of each alternation permits 50% control of the power. When conduction is for one-fourth of each alternation, the load receives only 25% of its normal power. It is possible through this device to control from 0% to 100% of the electrical power applied to a load device. Electrical power control with this device is very efficient. Very little power is actually consumed by a triac when it is used as a **power control** device.

Phase Shifter

With AC control of a triac, it is possible to achieve a 180° delay of the conduction time. For this to be achieved, two distinct AC signals must be applied to the triac. The AC source represents one of these signals. It is applied to T_1 and T_2 and must not be shifted in phase or altered. The second AC signal is used to control the conduction time of a triac by altering $+V_{BO}$ and $-V_{BO}$. This signal is controlled by a **phase shifter** and can be changed from 0° to 90° or 0° to 180°, depending on its design. This type of control produces a change in the phase relationship that exists between the two signals. The gate control signal of a triac must be made to lag behind the source voltage to be used effectively as a control factor.

   Figure 12.15 shows two distinct **phase-control** circuits used to alter the conduction time of a triac. The single-leg circuit of **Figure 12.15(a)** achieves approximately 90° of **conduction time delay**. The bridge phase-shifter circuit of **Figure 12.15(b)** permits 100% control from full conduction to 180° of delay, or nonconduction. A **diac** is used in the gate of both circuits.

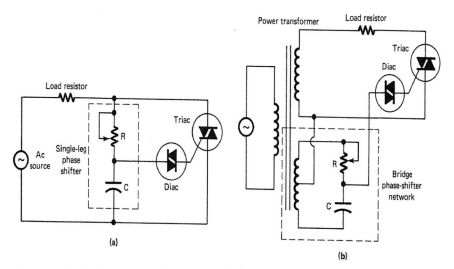

Figure 12.15   AC phase control of a triac. (a) With single-leg phase shifter. (b) With bridge phase shifter.

This device permits the gate to be triggered into conduction in either direction when the applied voltage reaches a specific value.

Figure 12.16 shows six different conditions of operation of an **AC bridge-controlled triac**. In the 0° delay condition of operation, a full sine wave is developed as an output signal. The gate voltage and terminal voltage are in phase, which causes immediate conduction without **delay**. The power output (P_{out}), which is determined by multiplying the terminal voltage (V_T) times the terminal current (I_T), is at its maximum output level during this condition of operation.

If the gate voltage applied to a triac is shifted 45° behind V_T, the output current is delayed. At the 45° point, the gate signal reaches the zero voltage level and causes conduction. Conduction current in this case is reduced somewhat because I_G occurs for less than the full sine wave.

The same process achieves **delay** of conduction time for the remaining four conditions of operation. In the 180° delay condition, however, no output is achieved because I_T is zero. The triac is nonconductive for this condition of operation. By shifting the phase of the gate voltage with respect to T_1 and T_2, it is possible for a triac to achieve full control of the AC power output. Control of this type is easy to achieve, efficient, and can be done effectively with a minimum of costly components.

Example 12-5:

Refer to **Figure 12.16** – AC control conditions of a triac. Describe the conduction of a triac with 0° delay.

Solution

Refer to **Figure 12.16(a)** and note that a full sine wave is developed as the output since gate voltage and terminal voltage are in phase.

Related Problem

Refer to **Figure 12.16**. Describe the conduction of the triac with 90° delay.

Static Switch

The use of a triac as a **static switch** in AC power control is primarily an on−off operation. Control of this type has a number of advantages over mechanical load-switching techniques. A large load current can, for example, be controlled with a small gate-current switch. This type of switching has no

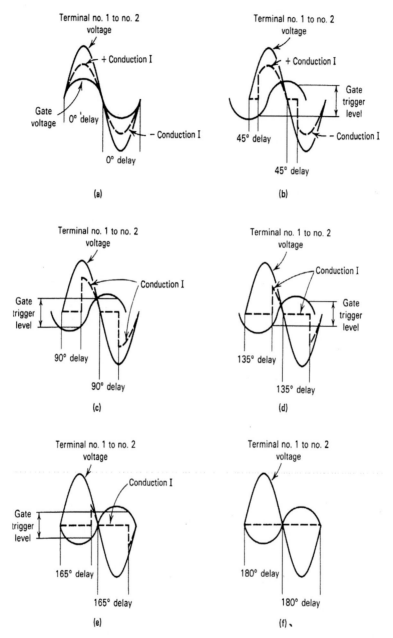

Figure 12.16 AC control conditions of a triac. (a) 0° delay. (b) 45° delay. (c) 90° delay. (d) 135° delay. (e) 165° delay. (f) 180○ delay.

contact bounce problems. Contact bounce occurs when a mechanical switch is turned on and the contacts are forced together. This causes the contacts to open and close several times before making firm contact. With a static switch, contact arcing and switch deterioration are reduced to a minimum. Static switching is usually very easy to achieve. Only a small number of components are needed to achieve control. In general, **static triac switching** has no moving parts, is very efficient, and can be achieved at a nominal cost. It is a very important alternative to power control switching.

Two very elementary **triac static switch** applications are shown in **Figure 12.17**. The circuit in **Figure 12.17(a)** shows a load being controlled by a triac that is energized into conduction by a small SPST toggle switch. This switching operation could be accomplished with a wide variety of small switching devices. When the switch is closed, AC is applied to the gate. Resistor R_2 limits the gate current to a reasonable operating value. With AC applied to the gate, current flows during both alternations. During the positive alternation, T_2 and G are positive, and T_1 is negative. **Gate current** and T_2-T_1 voltage occur at the same time. The triac is triggered into conduction without a delay. **Conduction** continues for the remainder of the alternation. At the zero crossover point, the triac is turned off. The negative alternation then causes T_2 and G to be negative and T_1 to be positive. Conduction is again initiated without a delay. Conduction continues for the remainder of the negative alternation. At the **zero crossover point** or end of the negative alternation, the triac is again turned off. The process is then repeated for each succeeding alternation. Control is continuous as long as the gate switch is on and T_1 and T_2 are energized. Turn-off is achieved by opening the SPST switch.

Figure 12.17(b) is a very simple **three-position static switch**. In position 1, the gate is open and the power to the load is off. In position 2, gate current flows only during the positive alternation. The triac is triggered into conduction only during the positive alternation. The load is energized for only one alternation or has a half-wave output. Only 50% of the source energy is supplied to the load in this switch position. In position 3, gate current flows for both alternations of the AC source. The load is energized for both alternations and receives full power from the source. The load of this circuit could be one or more incandescent lamps, a heating element, or a universal motor.

Example 12-6:

Refer to **Figure 12.17(b)** – Triac static switch. Describe the operation of the circuit with the three-position switch in position 3.

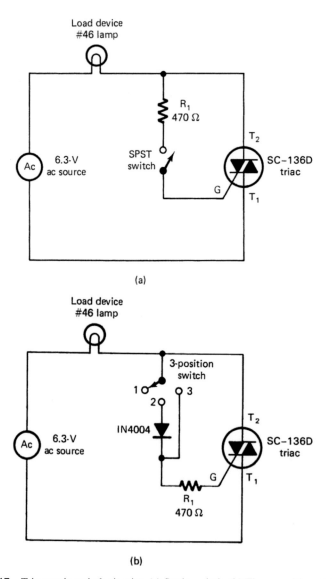

(a)

(b)

Figure 12.17   Triac static switch circuits. (a) Static switch. (b) Three-position static switch.

Solution

The gate current (I_G) flows during both alternations of the AC source. The load is energized during both alternations and receives full power from the source.

Related Problem

Refer to **Figure 12.17(b)** and describe the operation of the triac static switch with the three-position switch in position 2.

Self-Examination

Answer the following questions.

16. A triac is the equivalent of two reverse-parallel-connected _____ with a common gate.
17. In a triac, crystal selection is achieved automatically according to the _____ of the applied source voltage.
18. According to **Figure 12.11**, the crystal structure from T_1 to T_2 is _____.
19. According to **Figure 12.11**, the crystal structure from T_2 to T_1 is _____.
20. The breakover voltage of a triac in quadrants I and III is at its highest value when the gate current is _____.
21. Gate current (lowers, raises) the breakover voltage of a triac in either direction of operation.
22. According to the trigger mode chart of **Figure 12.13**, triac operation is more sensitive when T_2 is +, T_1 is −, and G is (+, −).
23. According to the trigger mode chart of **Figure 12.13**, triac conduction in quadrant III is best when T_2 is −, T_1 is +, and G is (+, −).
24. By shifting the phase of the _____ voltage with respect to T_1 and T_2, it is possible to achieve full control of AC power output.
25. When a triac is not conducting, the voltage across T_1 and T_2 is equal to the _____ voltage.

12.3 Diacs

Special AC diodes, known as diacs, have been developed for triac triggering. A diac is a bidirectional diode that can be triggered into conduction by reaching a specific voltage value. A diac goes into conduction at precisely the same positive and negative voltage level. For voltage values less than the trigger level, the device is nonconductive. This section discusses diac construction and operation and how the diac is used to trigger a triac into conduction.

12.3 Analyze the operation of a diac in AC power control circuits.

In order to achieve objective 12.3, you should be able to:

- interpret the *I–V* characteristics of a diac;
- describe the physical construction of a diac;
- explain how a diac is biased to make it operational;
- explain the fundamental operation of a diac;
- identify a diac in a specific circuit application.

Diac Construction and Operation

The crystal structure of this device is basically the same as an NPN transistor with no base connection. N_1 and N_2 are identical in nearly all respects so that the two junctions match operating characteristics in either direction.

Figure 12.18 shows the **crystal structure** and schematic **symbol** of a diac. Note that the two leads are labeled terminal 1 and terminal 2 instead of the conventional anode–cathode designations. When T_1 is negative and T_2 is positive, the N_1-P_2 junction is forward biased and N_2-P_2 junction is reverse biased. When the breakover voltage of N_2-P_2 is reached, the entire structure is triggered into conduction. This is the equivalent of reaching the Zener voltage of a Zener diode.

Reversing the polarity of the source voltage causes T_1 to be positive and T_2 negative. The N_2-P_2 junction is now forward biased and the N_1-P_2 junction is reverse biased. When the breakover voltage in this direction is reached, the structure again becomes conductive. Ordinarily, the **breakover voltage** is symmetrical in either direction.

Diac I–V Characteristics

The *I–V* characteristics of a diac are shown in **Figure 12.19**. Note that the breakover voltage and resulting current are the same in quadrant I and quadrant III. When T_1 is made positive and T_2 is made negative, this device does not conduct until the breakover voltage is reached. For example, for a type ST-2 diac, **triggering** occurs at 28 V ±10%. Reversing the polarity of T_1 and T_2 causes an identical triggering action as shown in quadrant III.

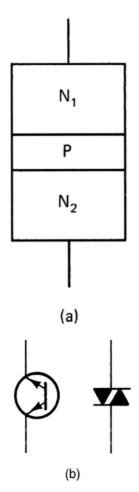

Figure 12.18   Crystal structure and schematic symbol of a diac.

After the **breakover voltage** of a diac has been reached, current conduction occurs very quickly. A resistor placed in series with the device limits the conduction current to a safe operating level in either direction as determined by the manufacturer's specifications. The voltage across a diac also decreases quite readily as it increases in conduction. This conduction is known as the **negative-resistance characteristic** of the diac. The pronounced curve that takes place after V_{BO} is reached shows this characteristic in both quadrants.

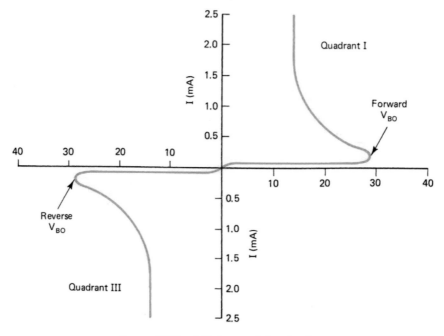

Figure 12.19   *I–V* characteristics of a diac.

Example 12-7:

Refer to **Figure 12.19** – *I–V* characteristics of a diac. Describe the conduction of the diac in quadrant I.

Solution

When T_1 is positive and T_2 is negative, the diac does not conduct until V_{BO1} is reached. After the V_{BO1} value is reached, the diac begins to conduct.

Related Problem

Refer to **Figure 12.19** and describe the conduction of the diac in quadrant III.

Diac—Triac Power Control

An elementary full-wave AC power control circuit employing a diac and a triac is shown in **Figure 12.20**. The diac is used in an RC circuit to trigger the triac into conduction. Operation is based on the polarity of the applied AC source. The circuit is connected so that T_2 and the gate of the triac are of the same polarity. The gate is connected to capacitor C_1 through the diac. Gate

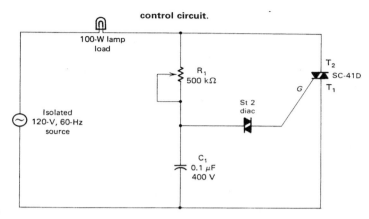

Figure 12.20 Full-wave triac power control circuit.

current is developed when the capacitor voltage builds to a value that triggers the breakover voltage of the diac. Discharging C_1 causes gate current. The polarity of the charge on C_1 is determined by the alternation of the source.

Assume that the **positive alternation** of the source occurs first. This makes point A positive and B negative. Capacitor C_1 begins to charge to the voltage source through R_1. At the same time, T_2 is made positive and T_1 negative. When the charge voltage of the capacitor exceeds the breakover voltage on the diac, C_1 discharges. The values of C_1, R_1, and the AC source determine the amount of gate current and the duration of the discharge pulse. Typical values of I_G are 50 mA or more for approximately 30 ms. The polarity of the charge on the top plate of C_1 is positive for this alternation. The discharge of C_1 causes current to flow from the bottom plate of C_1 to T_1 through the gate, the diac, and to the top plate of C_1. This discharge current or I_G triggers the triac into conduction. Current then flows into the load during this alternation. **Delay** of the conduction time is controlled by the resistance value of R_1. A low-resistance setting causes zero or no delay, whereas a high-resistance setting may cause delay for some value up to the end of the alternation.

For the negative alternation of the source, A is negative and B is positive. Capacitor C_1 again charges to its source voltage through R_1. In this case, the charge polarity is reversed because of the alternation change in the source. At the same moment, T_2 is made negative and T_1 positive. When the charge voltage of the capacitor exceeds the breakover voltage of the diac, C_1 discharges. Discharge current flows from the top plate of C_1, through the diac, the gate, T_1, and returns to the lower plate of C_1. This I_G triggers the triac into conduction during the negative alternation. Current flows through the lamp and into the triac from the source. **Delay** of conduction time is again based on the value setting of R_1. Control of this alternation is nearly the same as that of the positive alternation.

Example 12-8:

In the triac circuit given in **Figure 12.20** if the conduction is delayed by 90° (in the positive half-cycle), and at 270° (in the negative half-cycle), how much power is developed by the load?

The load will develop a half or 50% of the power possible. This is because the triac will begin conducting at 90° which is in the positive half-cycle and will switch off when the positive half of the alteration ends, 90° later (at 180°). This will cause the load to develop 1/4th or 25% of the power. The triac will begin conducting again at 270° which is in the negative half-cycle and will switch off when the negative half of the alternation ends, 90° later (at 360°). This also will cause the load to develop 1/4th or 25% of the power. Taken together, the power developed by the load over an entire cycle will be 25% + 25% = 50%.

Related Problem

In the triac circuit given in **Figure 12.20**, if the conduction is delayed by 0° and 180°, how much power is developed by the load?

Self-Examination

Answer the following questions.

26. A(n) _____ is a bidirectional diode that can be triggered into conduction with voltage.
27. The two leads of a diac are labeled _____ and _____.
28. The breakover voltage of a diac is the same in quadrants _____ and _____ of the *I–V* characteristics.
29. In **Figure 12.20**, the diac is connected so that the _____ and T_2 are of the same polarity.
30. In **Figure 12.20**, the gate voltage triggering time is determined by the values of _____ and _____.
31. In **Figure 12.20**, when the capacitor voltage (V_C) exceeds the V_{BO} of the diac, it triggers the gate of the triac to produce _____.

12.4 Unijunction Transistors

A unijunction transistor is a three-terminal, single-junction solid-state device that has unidirectional conductivity. UJTs are commonly used to trigger an SCR into conduction and in oscillator applications. A unique characteristic of the UJT is its negative resistance characteristic. The term UJT refers to the single P–N junction in the UJT's basic construction.

12.4 Analyze the operation of a UJT in pulse-triggered circuits.
   In order to achieve objective 12.4, you should be able to:

- interpret the *I–V* characteristics of a UJT;
- describe the physical construction of a UJT;
- explain how a UJT is biased to make it operational;
- explain the fundamental operation of a UJT;
- identify a UJT in a specific circuit application;
- define *interbase resistance*, *negative resistance region*, and *intrinsic standoff ratio*.

UJT Construction

Figure 12.21 shows the crystal structure and schematic symbol of a **unijunction transistor**. A small bar of N-type silicon is mounted on a ceramic base. Leads attached to the silicon base are called **base 1 (B_1)** and **base 2 (B_2)**. The **emitter (E)** is formed by fusing an aluminum wire to the opposite side of the silicon bar. The emitter is oriented so that it is closer to B_2 than to B_1. The emitter and the silicon bar form a P–N junction. The arrow of the UJT symbol Points <u>iN</u>, which indicates that the emitter is a P-type material and the silicon bar is an N-type material. The arrow of the symbol is slanted to distinguish the UJT from the N-channel JFET symbol.

   The **triggering** of a UJT depends on changes in the conductivity of the E-B_1 junction. When the E-B_1 junction is forward biased, electrons leave the silicon bar and move into the P-type material of the emitter. This action causes holes to appear in the silicon bar at the B_1 region. Electrons pulled from the negative side of the source attached to B_1 fill the injected holes. In effect, this causes an increase in the conduction of the E-B_1 junction. This region, referred to as R_{B1}, then becomes low resistant. A sudden change in R_{B1} causes a corresponding change in B_1-B_2 current.

   To understand the **internal resistance** of a UJT, look at **Figure 12.22**, which shows an equivalent circuit of the UJT. When the E-B_1 junction is reverse biased, R_{B1} and R_{B2} are of practically the same resistance value.

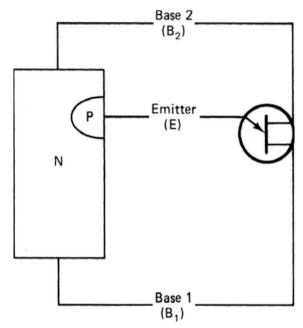

Figure 12.21 Unijunction transistor crystal structure and schematic symbol.

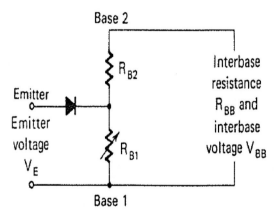

Figure 12.22 Equivalent of a UJT.

Forward biasing of the E-B_1 junction, however, causes the resistance of the junction to change to a rather small value. The E-B_2 junction resistance is not changed by the action of the E-B_1 junction. As a result, E-B_1 responds as a voltage-controlled variable resistor, whereas E-B_2 remains at a fixed value.

The **interbase resistance**, or R_{BB}, is a combination of R_{B1} and R_{B2}. It ranges from 4 Ω to 10 kΩ. The **interbase voltage**, or V_{BB}, appears across B_1 and B_2. Emitter voltage (V_E) causes an emitter current (I_E) when E-B_1 is forward biased. No I_E flows when E-B_1 is reverse biased.

UJT Operation

The operation of a **UJT** is quite different from other solid-state devices. When the emitter is reverse biased, B_1-B_2 responds as a simple voltage divider. The resistance of R_{BB} is at its maximum value. Any I_E that flows at this time is due to the leakage current of E-B_1. When the emitter is forward biased, R_{BB} drops to a very small value. A change in E-B_1 voltage, therefore, causes a significant change in R_{BB}. The current flow between B_1 and B_2 increases in value, and V_{BB} decreases in value.

Normal operation of the UJT has B_1 grounded or connected to the negative side of the source voltage and B_2 connected to the positive side. This voltage causes an interbase current flow between B_1 and B_2. Internally, V_{BB} divides across the **interbase resistance** of R_{B1} and R_{B2}. For example, if R_{B1} is 4 kΩ and R_{B2} is 6 kΩ, the voltage across R_{B1} is 0.4 of V_{BB} and R_{B2} is 0.6 of V_{BB}. It is not possible to measure these two voltage values because the junction point of R_{B1} and R_{B2} does not exist. However, the fractional value of voltage that appears across R_{B1} is called the **intrinsic standoff ratio**. It is represented by the η symbol (Greek letter eta) and is determined by the following formula:

$$\eta = R_{B1}/(R_{B1} + R_{B2}) \tag{12.1}$$

$$\begin{aligned}
\eta &= R_{B1}/(R_{B1} + R_{B2}) \\
&= 4\text{ k}\Omega/(4\text{ k}\Omega + 6\text{ k}\Omega) \\
&= 4\text{ k}\Omega/10\text{ k}\Omega \\
&= 0.4.
\end{aligned}$$

Typical intrinsic standoff ratio values are in the range from 0.45 to 0.85 of V_{BB}. This value determines the emitter **triggering voltage** for a constant V_{BB}. If V_E is less than ηV_{BB}, the emitter junction is reverse biased and only leakage current (I_{EO}) flows. If V_E exceeds ηV_{BB} by a value slightly greater than the voltage drop across the E-B_1 diode, triggering occurs. R_{B1} is then reduced to a few ohms, and the interbase current I_{BO} increases.

UJT I–V Characteristics

A **characteristics curve** for a typical UJT is shown in **Figure 12.23**. The vertical part of this characteristic curve indicates different values of **emitter voltage**. An increase in the value of V_E causes a sharp vertical rise in the curve. **Emitter current** is represented by the horizontal part of the graph. Note that the first application of V_E is to the left of a vertical line. This indicates the reverse bias condition of operation. Any current flow that occurs here is due to the leakage of the E-B_1 diode junction. After an extremely small leakage current (I_{EO}), there is a significant increase in I_E shown by the curve extending to the right. Forward biasing of E-B_1 is shown where the vertical line is crossed. The peak emitter voltage point (V_P) shows where the UJT is triggered into conduction. The peak current point (I_P) is the minimum I_E needed for triggering. A further increase in I_E causes V_E to drop to the valley voltage (V_V) point. The area between V_P and V_V is called the **negative resistance region**. An increase in I_E causes a decrease in V_E in this part of the graph. A device that has a negative resistance region is capable of regeneration or oscillation. This part of the characteristic curve is an important operating area for the UJT in pulse-trigger generation

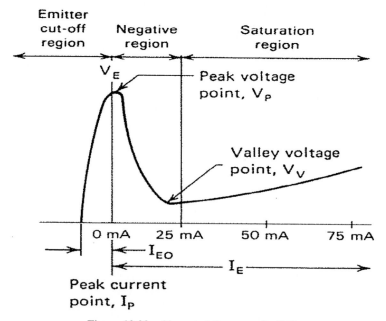

Figure 12.23   Characteristic curve of a UJT.

applications. Note that voltage beyond the valley voltage point beings to rise with a corresponding increase in I_E. This corresponds to a positive resistance region of the characteristic curve.

Example 12-9:

Refer to **Figure 12.23** – *I–V* characteristics of a UJT. Describe the operation of the UJT up to the minimum emitter current needed for triggering.

Solution

As V_E increases up to the peak voltage (V_P) point, the value of I_E is very small, consisting primarily of leakage current. This represents reverse biasing of the emitter−base1 junction, until V_P is reached.

Related Problem

Refer to **Figure 12.23** and describe the negative resistance region of operation of the UJT.

SCR Pulse Triggering with a UJT

A positive pulse of energy can be used to trigger an SCR into conduction instead of a continuous form of AC or DC. **Triggering** by this method takes advantage of the latching characteristic of the SCR. When conduction is initiated, it is made continuous for the remainder of the alternation. The duration of the triggering pulse may be extremely short. This method of triggering in general consumes very little energy from the source. The operational efficiency of a pulse-triggered power control circuit is usually much higher than that of other gate-triggering methods.

    Pulse triggering of an SCR can be accomplished with a UJT. This device is commonly described as a **voltage-controlled diode**. When used to trigger an SCR into conduction, UJTs can be designed to respond to low-level signals from other inputs. The output of photocells, thermocouples, and other transducers can be used to energize the input of a UJT, which, in turn, triggers the conduction of an SCR. Through this type of circuitry, it is possible to control a large SCR current with an extremely small input signal.

    To **trigger** an SCR into conduction, the magnitude of the pulse must be several times greater than the normal I_G triggering level. A high-amplitude voltage pulse discharges into the gate and causes an I_G. For pulses above 20 μs in duration, the triggering level is approximately the same as that of a DC

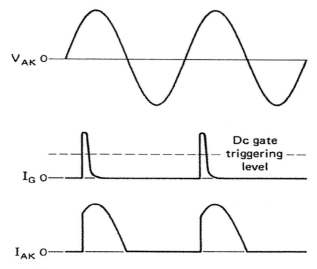

Figure 12.24   Waveforms of a pulse-triggered SCR.

source. Below 20 μs, the magnitude of the pulse should be at least equal to the DC gate-triggering level. Pulse-widths of less than 1 μs duration will not trigger a device into conduction.

The waveforms of a **pulse-triggered SCR** are shown in **Figure 12.24**. The anode–cathode voltage (V_{AK}) is AC. The gate current triggering pulse is labeled I_G. The triggering level of I_G is for DC control values. Note that the amplitude of the pulse exceeds the triggering level by a substantial amount. The delay of conduction is based on the position of the pulse with respect to the positive alternation.

Load current (I_{AK}) flows during the remainder of the alternation after the gate pulse initiates conduction. By varying the time or position of the pulse, conduction can be altered from 0° to 180°.

UJT Relaxation Oscillator

Figure 12.25 shows a **UJT relaxation oscillator** that is frequently used to trigger an SCR. In this circuit, two parallel paths are formed by the components. Resistors R_1, R_2, and C_1 form the charge path, whereas R_3 and the UJT form the discharge path. When the switch is turned on, approximately 4 V appears across R_3 and 6 V appears across the UJT and R_2. This voltage reverse biases the E-B_1 junction.

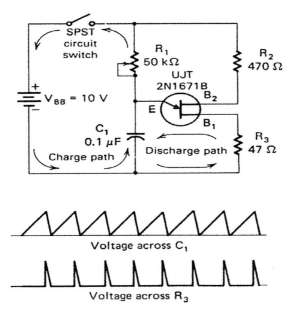

Figure 12.25   UJT sawtooth oscillator.

Resistor R_1 and capacitor C_1 receive energy from the source at the same time that the base path does. Capacitor C_1 begins to develop voltage across it at a rate based on the RC values of R_1 and C_1. When the charge voltage at C_1 reaches 6 V, it overcomes the reverse biasing of the E-B_1 junction. When this occurs, the junction becomes very low resistant and C_1 discharges very quickly. The E-B_1 junction immediately becomes reverse biased because of the reduced emitter voltage. The capacitor then recharges and the process is repeated. A **sawtooth waveform** appears across the capacitor as indicated in **Figure 12.25**. The discharge voltage across R_3 produces a spiked pulse that shows the discharge time of C_1. The rise time of the sawtooth wave is adjusted to some extent by different values of R_1. When the charging action of the capacitor is reached in the first time constant and maintained, this circuit can achieve very accurate **pulse generation**.

UJT Pulse-Trigger Control Circuit

A **UJT pulse-trigger control** circuit for an SCR is shown in **Figure 12.26**. Adjustment of the variable resistor, R_1, determines the pulse-triggering rate of the SCR. A value change in R_1 alters the RC value of the UJT trigger circuit.

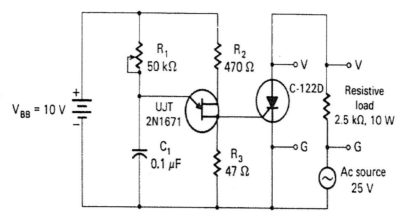

Figure 12.26   UJT pulse-trigger control circuit.

When the charge voltage of C_1 reaches the V_P value, it discharges through the E-B_1 junction of the UJT and R_3. The top of the capacitor is positive, and the bottom is negative. This forward biases the gate–cathode of the SCR and triggers it into conduction. The load receives current only during the positive alternation of the applied AC. A pulse-triggered circuit of this type can be used to effectively **delay** the conduction time of the positive alternation from 0° to 180°.

Self-Examination

Answer the following questions.

32. A unijunction transistor is described as a voltage-controlled _____.
33. A UJT is a(n) _____-terminal device.
34. UJTs are commonly used to _____ the conduction of an SCR.
35. Two leads connected to the common N-type silicon bar of a UJT are called _____ and _____.
36. The lead attached to the P-type element of a UJT is called the _____.
37. When the _____ junction is open or reverse biased, the resistances of B_1 and B_2 are practically the same.
38. When the E-B_1 junction of a UJT is forward biased by an appropriate voltage value, the internal base resistance (decreases, increases) in value.
39. The area between V_P and V_V of a UJT characteristic curve is called the _____ region.

40. The fractional value of source voltage that appears across R_{B1} of a UJT is called the _____ ratio or eta.
41. In the UJT circuit of **Figure 12.25**, when the voltage across capacitor C_1 builds up to a value that exceeds the intrinsic standoff ratio, it causes _____ of the E_{B1} junction.

12.5 Programmable Unijunction Transistors

A programmable unijunction transistor (PUT) is frequently used to generate trigger pulses for SCR control applications. A PUT is actually a thyristor that responds as a UJT that has a variable trigger voltage. This voltage can be adjusted to a desired value by changing two external voltage-divider resistors. The trigger voltage level can, therefore, be set, or programmed, to respond to a specific value.

12.5 Analyze the operation of a PUT in pulse-triggered circuits.

In order to achieve objective 12.5, you should be able to:

- interpret the *I–V* characteristics of a PUT;
- describe the physical construction of a PUT;
- explain how a PUT is biased to make it operational;
- explain the fundamental operation of a PUT;
- identify a PUT in a specific circuit application.

PUT Construction

Figure 12.27 shows the crystal structure, schematic symbol, and element names of a **programmable unijunction transistor (PUT)**. The crystal structure and schematic symbol of the PUT are very similar to the crystal structure and schematic symbol of the SCR.

The gate junction is the primary difference. In a PUT, the **gate (G)** is connected to the N-type material nearest to the anode. A P–N junction is formed by the anode–gate. **Conduction** of the device is controlled by the bias voltage of *A-G.*

The polarity of the PUT bias voltage is referenced with respect to the cathode. The cathode is usually connected to the ground or the negative side of the power source. The gate is then made positive relative to the cathode. This is the gate voltage (V_G). The anode voltage (V_A) is also made positive with respect to the cathode.

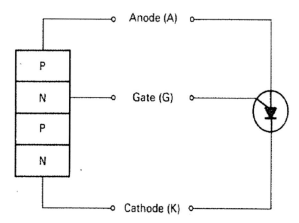

Figure 12.27   Programmable unijunction transistor.

PUT Operation

Conduction of the PUT is based on the difference in positive voltage between the gate and the anode. When the gate is more positive than the anode, the *A-G* junction is reverse biased. This condition causes the device to be nonconductive, or in its **off-state**, and is shown in **Figure 12.28**. The anode–cathode has infinite resistance, and the device responds as an **open switch**. When the anode becomes more positive than the gate by 0.5 V, it forward biases the *A-G* junction. This condition causes gate current to flow, and the device is triggered into conduction, or in its on-state, and is shown in **Figure 12.28**. In the on-state, the anode–cathode resistance drops to a very low value. The device then responds as a switch in the **on-state**.

PUT I–V Characteristics

Refer to the **PUT** circuit and characteristics curve in **Figure 12.28**. **Gate voltage (V_G)** is developed by resistors R_1 and R_2 connected across V_{GG}. The anode–cathode voltage (V_{AK}) is supplied by the source voltage (V_S). When V_{AK} is made more positive than V_G, the PUT is triggered into conduction. The characteristics curve shows this by the changing value of V_{AK}. Triggering occurs when V_{AK} reaches the peak voltage point (V_P). The voltage then drops to the valley voltage point (V_V). Conduction current, I_{AK}, is indicated by expansion of the curve to the right. The PUT has a **negative resistance region** between V_P and V_V. The *I–V* characteristics of a PUT are very similar to that of a UJT.

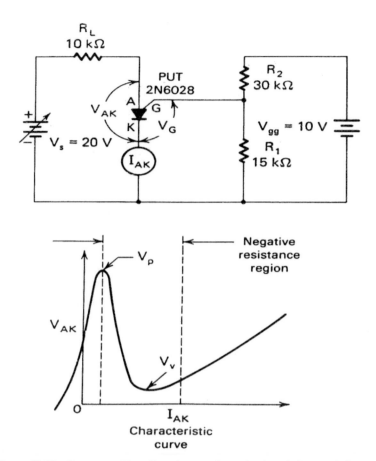

Figure 12.28 Programmable unijunction transistor circuit and characteristic curve.

The anode of a PUT responds as the emitter of a UJT. Resistors R_1 and R_2 are similar to the **interbase resistance** of the UJT. The **intrinsic standoff ratio (η)** of a PUT is determined by the value of the gate voltage with respect to the anode voltage. V_G can be altered externally by changing the values of R_1 and R_2. This permits the PUT to be programmed or adjusted to different trigger voltage values.

Example 12-10:

Refer to **Figure 12.28** – PUT characteristic curve. Describe the operation of the PUT up to the peak voltage (V_P) value.

Solution

As V_{AK} increases up to the peak voltage (V_P) point, the value of I_{AK} is very small, consisting primarily of leakage current. This represents reverse biasing of the anode–cathode junction, until V_P is reached.

Related Problem

Refer to **Figure 12.27** and describe the negative resistance region of the PUT characteristic curve.

PUT Pulse Triggering

A **programmable unijunction transistor** is frequently used to control the conduction of an SCR. The PUT is primarily responsible for generating trigger pulses. These pulses are developed by a **relaxation oscillator**. The pulses are then applied to the gate of an SCR to initiate the conduction process. Only one pulse is needed during a selected alternation to initiate conduction. A circuit of this type synchronizes the trigger pulse with the alternation that forward biases the SCR. This is accomplished by energizing the SCR and PUT from the same AC source. Conduction of the SCR occurs only when the anode–cathode is forward biased. Control by synchronized **pulse triggering** is very precise and has a wide range of adjustment capabilities.

Figure 12.29 shows a **PUT relaxation oscillator** that can be used to trigger an SCR. In this circuit, two parallel paths are formed by the components and the device. Resistor R_3 and capacitor C_1 form the charge path for the capacitor and the anode voltage. The PUT and resistor R_4 form a discharge path for C_1. A **trigger pulse** is developed across R_4 when C_1 discharges through the anode–cathode of the PUT.

When the circuit of **Figure 12.29** is turned on, it energizes the PUT from the power source V_S. Resistors R_1 and R_2 form a voltage-divider network across the V_{GG} source. In this case, R_1 is 15 kΩ, and R_2 is 30 kΩ. The **intrinsic standoff ratio** is determined by

$$\eta = R_1/(R_1 + R_2)$$
$$= 15 \text{ k}\Omega/(15 \text{ k}\Omega + 30 \text{ k}\Omega)$$
$$= 15 \text{ k}\Omega/45 \text{ k}\Omega$$
$$= 0.333.$$

The gate voltage (V_G) of the PUT is then determined by

$$V_G = \eta \times V_S. \tag{12.2}$$

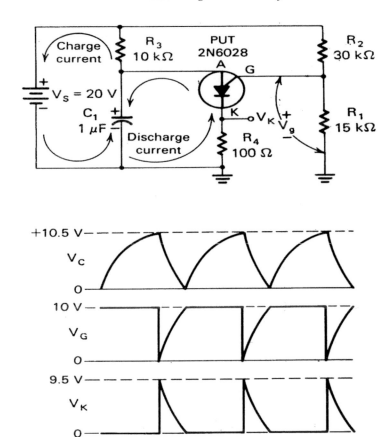

Figure 12.29 PUT relaxation oscillator.

Therefore, V_G for this circuit is

$$V_G = \eta \times V_S$$
$$= 0.333 \times 20 \text{ V}$$
$$= 6.66 \text{ V}.$$

Initially, V_G makes the gate more positive than the anode by 6.66 V. The PUT is in its **nonconductive state** or off-state. After a short period of time, capacitor C_1 begins to charge to the source voltage through R_1. When the charge potential exceeds 6.66 V, it makes the anode more positive than the gate. This condition triggers the PUT into conduction. C_1 then discharges through the low resistance of the PUT and R_4. Discharge takes place very quickly due to the low resistance. The resulting current produces a positive-going voltage pulse across R_4. **Triggering** the PUT into conduction also causes a drop in the value of V_G. This is due to the low resistance of the forward-biased *A-G* junction. The waveforms of **Figure 12.29** show how the **relaxation oscillator** responds for three operational cycles.

Example 12-11:

Refer to **Figure 12.29** – PUT relaxation oscillator. Describe the action that occurs up to the time that the PUT conducts.

Solution

When voltage is applied, C_1 begins to charge through R_1. When the anode of the PUT becomes more positive than the gate, the PUT conducts.

Related Problem

Refer to **Figure 12.29** and describe the action that occurs **after** the PUT conducts.

PUT Pulse-Trigger Control Circuit

A **PUT pulse-triggered SCR** power control circuit is shown in **Figure 12.30**. The outer part of the circuit is controlled by the SCR. Note the location of the load device, SCR, and AC power source. The center part of the circuit is used to develop **trigger pulses** that control the conduction of the SCR. The PUT, in this case, is used as a relaxation oscillator. Operation of the oscillator is essentially the same as **Figure 12.29**.

The PUT is energized by a DC source. A 15-V **Zener diode** is used, in this case, as a rectifier-regulator. It is energized by the same AC source that supplies the SCR load. A series-dropping resistor (R_S) connects the AC to the Zener diode. During the **positive alternation**, the Zener diode clips the source voltage to a peak value of 15 V. The **negative alternation** causes forward conduction of the diode, and the voltage drops to approximately zero. The DC supply is essentially a pulsating DC of 15 V. See the V_Z DC supply and the AC source waveforms. Note the phase relationship of the AC source and the PUT supply voltage. When the positive alternation and the DC supply pulse occur at the same time, this relationship permits the trigger pulse to have synchronized control of the SCR. The time of the trigger pulse can be changed, however, by altering the resistance of R_1.

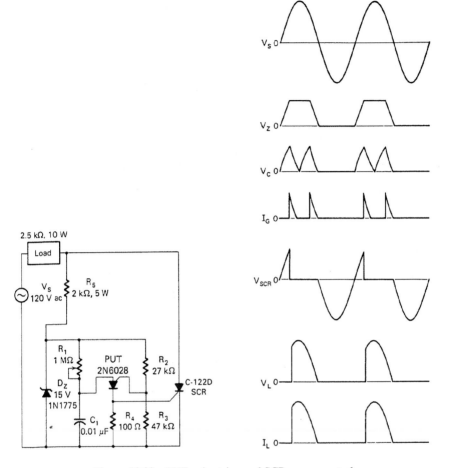

Figure 12.30   PUT pulse-triggered SCR power control.

The waveforms of **Figure 12.30** show how the circuit responds for two complete operational cycles. Note that more than one triggering pulse may be generated by the PUT for a given DC source pulse. The number of pulses produced is based on the time constant of R_1 and C_1. A low-resistance setting causes several pulses to occur. Higher resistance values may permit only one pulse to be generated. The SCR is **triggered** into conduction by only the first pulse. With a suitable value of **holding current**, conduction continues for the remainder of the alternation. The SCR is nonconductive during the negative alternation. Refer to the SCR voltage (V_{SCR}), load current (I_{L}), and load voltage (V_L) waveforms. V_{AK} of the SCR and V_L are primarily the reverse of each other. Conduction causes the amplitude of V_{AK} to drop to a straight line. Conduction of I_L and V_L is shown as a rise in amplitude. **Delay** of conduction time is altered by changing the value of R_1.

Self-Examination

Answer the following questions.

42. A(n) _____ is a thyristor that responds as a UJT.
43. The crystal structure of a PUT is similar to that of a(n) _____.
44. In a PUT, the _____ is connected to a piece of N-type material next to the anode.
45. Conduction of a PUT is based on the difference in positive voltage between the _____ and _____.
46. When the anode becomes more (positive, negative) than the gate by 0.5 V, it forward biases the anode–gate junction.
47. When a PUT is triggered into conduction, its internal resistance is very (low, high).
48. Before a PUT is triggered into conduction, its internal resistance is very (low, high).
49. The _____ of a PUT is determined by the value of the gate voltage with respect to the anode voltage.
50. The intrinsic standoff ratio of a PUT can be _____ by altering the resistance of a divider network.
51. A PUT is commonly used as a(n) _____ source for an SCR.

12.6 Gate Turn-Off Thyristors

Gate turn-off thyristors are unique power control devices that can be triggered into conduction or turned off by a gate signal. Characteristically, this type of thyristor responds as an SCR. It has unidirectional current conductivity and responds only in quadrant I of the *I–V* characteristics. The unique feature of this type of device is its gate turn-off capability. The gate can be used to extinguish conduction and turn off the device when it is reverse biased. The gate turn-off principle of operation is found in two distinct types of thyristors: silicon-controlled switch (SCS) and gate turn-off (GTO) switch. The SCS is a low-power, low-current control device. The GTO is used to control larger amounts of power than the SCS. These devices are primarily used in computer logic circuits, oscillators, and as trigger devices for other thyristors.

12.6 Analyze the operation of SCS and GTO thyristors.

In order to achieve objective 12.6, you should be able to:

- interpret the *I–V* characteristics of SCS and GTO thyristors;
- describe the physical construction of SCS and GTO thyristors;
- explain how an SCS and a GTO are biased to make them operational;
- explain the fundamental operation of SCS and GTO thyristors;
- identify an SCS or a GTO in a specific circuit application.

Silicon-Controlled Switches

The **silicon-controlled switch (SCS)**, like the SCR, is a four-layer PNPN thyristor. External connection is made to all four layers of the SCS. It has an anode, cathode, anode−gate, and a cathode−gate. The crystal structure, schematic symbol, and two-transistor equivalent of an SCS are shown in **Figure 12.31**.

The **I–V characteristics** of an SCS are primarily the same as that of an SCR. Conductivity occurs in quadrant I. This is achieved by making the anode positive and the cathode negative. Quadrant III represents the reverse-biased condition of operation. Ordinarily, the SCS is not designed to operate in this condition. The peak reverse voltage (PRV) rating of the device is also represented by quadrant III.

Operation of the SCS is quite different from the SCR. An SCS can be triggered into conduction or turned off by an appropriate signal applied to either gate. A positive voltage or **pulse** applied to the cathode−gate with respect to the cathode causes gate current to flow. This cathode−gate current lowers the **forward breakover voltage** and permits conduction. A negative voltage with respect to the anode forward biases the anode−gate. This reaction causes a corresponding anode−cathode current to flow. I_{GA} then lowers the V_{BO} and causes triggering.

The polarity of the gate-triggering voltage is more meaningful if we study the two-transistor equivalent of the SCS in **Figure 12.30(c)**. Note that the cathode−gate is connected to the base of the NPN transistor Q_2. It should be obvious that the base of Q_2 must be positive to cause forward biasing. When this occurs, some of the output of Q_2 is supplied to the base of Q_1, thus causing regeneration. The entire assembly then becomes conductive. A negative voltage or trigger pulse applied to anode−gate also causes forward biasing of the PNP transistor Q_1. Some of this output is then returned to the base of Q_2, which causes regeneration and the assembly to be conductive. In effect, a negative voltage forward biases the anode−gate,

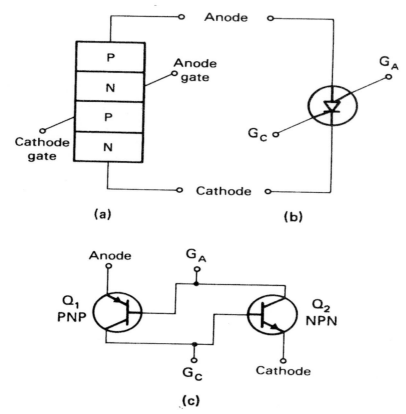

Figure 12.31   Silicon-controlled switch. (a) Crystal structure. (b) Schematic symbol. (c) Two-transistor equivalent.

and a positive voltage forward biases the cathode–gate. In most SCS devices, the cathode–gate responds to a lower **triggering current** than the anode–gate.

An SCS can be turned off or brought out of conduction in three different ways. The most obvious method is to reduce the anode current below the **holding current (I_H)** level. This response is the same as that of an SCR. The second method is to apply a **negative** pulse to the cathode–gate. This reverse biases the NPN transistor (Q_2) of the equivalent circuit. The third method is to apply a **positive-going pulse** to the anode–gate. This condition reverse biases the base of the PNP transistor (Q_1). In most SCS thyristors, it takes more gate current to bring the device into conduction. The off-triggering current must overcome a larger anode–cathode current to stop conduction.

Ordinarily, this is not a real problem in most SCS applications. Maximum I_{AK} values are usually rather small. Typical I_{AK} values are $100-300$ mA, with power dissipation ratings of $100-500$ mW.

Gate Turn-Off Switches

The **gate turn-off (GTO) switch** is a thyristor that is similar in construction to the SCR. Its operation is different in that conduction can be triggered on or turned off by a single gate. A positive gate pulse causes the device to be conductive. Conduction continues if the holding current is maintained. A negative gate signal turns off a conductive GTO.

The crystal structure and schematic symbol of a **GTO** are shown in **Figure 12.32**. The crystal structure is primarily the same as an SCR. The gate of a GTO, however, normally is made thicker than that of an SCR. Reverse biasing of the gate increases the size of the depletion region to a point that it will extinguish conduction. The schematic symbol of a GTO differs from that of an SCR by employing a small line across the gate lead. This denotes the **on-off switching** function of the gate. Any operation other than this is primarily the same as that for an SCR. It has unidirectional conductivity and will latch when made conductive.

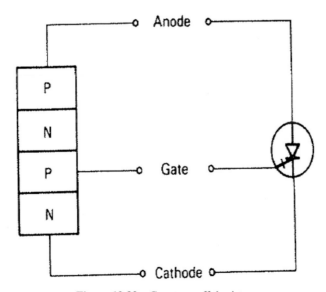

Figure 12.32 Gate turn-off thyristor.

The most obvious advantage of a GTO over an SCR is its **gate turn-off** capability. A consequence of this function is an increase in the amount of I_G needed to produce triggering. For an SCR and GTO with similar I_{AK} ratings, an SCR can be triggered with 1 mA, while the I_G of the GTO must be 25 mA. The turn-off current of a GTO is significantly greater than the triggering current. GTOs are available that can control up to 3 A with a power dissipation of 20 W.

The switching time of a GTO is another important characteristic. The **trigger time** of a GTO is primarily the same as that for an SCR. Typical trigger-time values are 1 μs. The turn-off time of a GTO is, however, approximately the same as its trigger time. An SCR must be turned off by reducing I_{AK} below the holding current level. Typical turn-off values may take from 5 to 30 μs. The rapid on$-$off characteristic of a GTO permits it to be used in high-speed switching applications.

GTO Sawtooth Generator

Figure 12.33 shows a GTO used as a **sawtooth generator**. When the source voltage is applied, the GTO turns on immediately. The gate is made positive with respect to the cathode by the value of the Zener voltage (V_Z). Conduction of the GTO permits capacitor C_1 to charge to the source voltage. When the voltage across C_1 reaches V_Z, it reverse biases the gate. This action increases the depletion region of the cathode$-$gate junction. The holding current (I_H) drops in value to turn off the GTO. C_1 then discharges through the combination of R_2 and R_3. The **discharge time** is determined by

$$t = (R_2 + R_3) \times C_1. \tag{12.3}$$

It takes five **time constants** for a capacitor to fully charge (to within 99%) and discharge (to within 1%) of its rated value. Proper selection of the R and C values results in a sawtooth waveform. When the value of V_{C1} drops below V_Z, the GTO turns on and the process repeats.

Example 12-12:

Refer to **Figure 12.32** – GTO sawtooth waveform generator. Calculate the time taken to discharge the capacitor C_1 (0.1 μF), through resistor R_2 (set at 50 kΩ) which is in series with R_3 (1 kΩ).

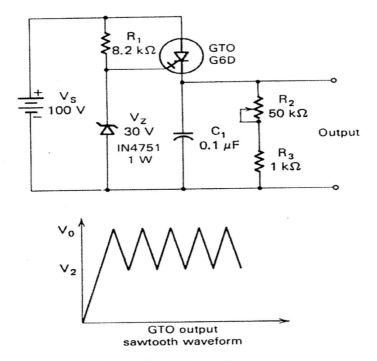

Figure 12.33   GTO sawtooth generator.

Solution

The time constant "t" = $(R_2 + R_3) \times C_1$ = (50, 000 Ω + 1000 Ω) $\times$ 0.1 $\times$ 10^{-6} = 0.0051 s = 5.1 ms.

 Total time to discharge the capacitor (T) = 5 $\times$ t = 5 $\times$ 5.1 ms = 25.5 ms.

Related Problem

Referring to **Figure 12.33**, calculate the time taken to discharge the capacitor C_1 (0.1 μF), through resistor R_2 (set at 25 kΩ) which is in series with R_3 (1 kΩ).

Self-Examination

Answer the following questions.

 52. A gate turn-off thyristor is similar in construction to a _____.
 53. A silicon-controlled switch has _____ gates.

54. A $(+, -)$ voltage pulse applied to the cathode–gate will trigger an SCS into conduction.
55. A $(+, -)$ voltage pulse applied to the anode–gate will trigger an SCS into conduction.
56. An SCS can be turned off in _____ ways.
57. In a GTO, a $(+, -)$ voltage pulse will cause triggering and a $(+, -)$ voltage pulse will cause turn-off.
58. A GTO generally takes more _____ current to produce triggering than an equivalent SCR.

Analysis and Troubleshooting – Thyristors

Thyristors are a family of devices that are used primarily for **power control**. The **datasheets** of various types of thyristors will be examined so that an appropriate device can be selected for a specific application. These include AC and DC power control circuits. Some of these devices directly control the electrical load, whereas others are used indirectly in power control circuits for generating **trigger pulses**. Direct power control devices normally have high power ratings since the entire load current flows through the device, and indirect power control devices have a lower power rating. By varying the conduction time, the performance of a thyristor as a power control device can be evaluated. Loss of power control occurs when the conduction time of the device cannot be varied. The performance of an individual device can be evaluated when it is connected in the circuit by examining the voltage waveforms and values. Its operational status can also be evaluated when it is taken out of the circuit. The thyristor gate is susceptible to damage by **voltage spikes** and surges. This could cause the thyristor to malfunction.

12.7 Analyze and troubleshoot thyristors.

In order to achieve objective 12.7, you should be able to:

1. use an ohmmeter to test the condition and identify the leads of a thyristor;
2. observe the input and output waveforms of a power control device on an oscilloscope;
3. examine a thyristor to determine when it is triggered;
4. distinguish between normal and faulty operation of a thyristor.

Data Sheet Analysis

The **specifications** of a thyristor can be obtained from a manufacturer's datasheet. The circuit symbol, pin assignment, sample test circuits, and rated operational values are often readily available from the datasheet.

A datasheet of a sample **SCR, 2N6394**, manufactured by ON Semiconductor is included at the end of the chapter. Use this datasheet to answer the following questions:

1. The number of terminals = _____.
2. Maximum on-state RMS current = _____A.
3. Peak forward gate current = _____A.
4. Operating junction temperature range = _____ °C to _____°C.
5. Blocking voltage = _____V.
6. Peak forward on-state voltage = _____V.
7. The tab or terminal 4 of the device functions as the _____.

A datasheet of a sample **Triac, FKN0PN60**, manufactured by Fairchild is included at the end of the chapter. Use this datasheet to answer the following questions:

1. The number of terminals = _____.
2. What pin on the device acts as the gate = _____.
3. Maximum on-state RMS current = _____A.
4. Peak gate voltage = _____V.
5. Peak gate current = _____A.
6. Operating junction temperature range = _____ °C to _____°C.
7. Peak forward on-state voltage = _____V.
8. The gate trigger voltage = _____V.

A datasheet of a sample **Diac, ST2** is included at the end of the chapter. Use this datasheet to answer the following questions:

1. The number of terminals = _____
2. Minimum switching voltage for an ST2 = _____V.
3. Maximum switching voltage for an ST2 = _____V.
4. Maximum switching current for an ST2 = _____A.
5. Peak forward gate current = _____A.
6. Package outline "B" is for (ST2, ST4).

A datasheet of a sample **UJT, 2N2646**, manufactured by Philips Semiconductors is included at the end of the chapter. Use this datasheet to answer the following questions:

1. The number of terminals = _____.
2. Which terminal of the UJT is the emitter = _____.
3. With respect to the tab how can the emitter of the UJT be identified =

   _____.
4. Peak emitter current = _____A.
5. Static interbase resistance = _____kΩ.
6. Maximum emitter−base 2 voltage = _____V.
7. Maximum interbase voltage = _____V.
8. Total power dissipation = _____V.

A datasheet of a sample **PUT, BRY56A**, manufactured by Philips Semiconductors is included at the end of the chapter. Use this datasheet to answer the following questions:

1. The number of terminals = _____.
2. Which terminal of the PUT is the gate = _____.
3. With respect to the flat side of the PUT, how can the gate be identified =

   _____.
4. Maximum total power dissipation = _____mW.
5. Maximum gate anode voltage = _____V.

A datasheet of a sample **SCS, BR101**, manufactured by Philips Semiconductors is included at the end of the chapter. Use this datasheet to answer the following questions:

1. The number of terminals = _____.
2. What do the terms "a_g" and "k_g" denote = _____.
3. List a few applications where SCSs can be used = _____
4. Maximum total power dissipation = _____mW.
5. When used as an PNP transistor, with an open collector, what is the maximum emitter−base voltage = _____V.
6. When used as an NPN transistor, with an open emitter, what is the maximum collector−base voltage = _____V.

A datasheet of a sample **GTO, MGTO1000**, manufactured by Motorola is included at the end of the chapter. Use this datasheet to answer the following questions:

1. The number of terminals = _____.
2. Maximum on-state current flow at an operating temperature of 65°C = _____A.
3. Repetitive peak off-state voltage = _____V.
4. Operating junction temperature range = _____ °C to _____°C.

5. Maximum non-repetitive surge current = _____A.
6. Primary applications of a GTO thyristor are = _____.

Troubleshooting

The **faults** that can occur in an **SCR** include internal **open** or **short** circuits between the terminals. If the gate is open circuit, then the device will not go into conduction, and no output voltage will appear across the load. If there is a direct electrical connection or short between the anode and the cathode, then the device will have continuous conduction. This will cause it to supply power to the load for either alteration of the AC input, whereas it would normally conduct during one alternation. In this case, the power through the load cannot be controlled by triggering the SCR gate.

The faults that can occur in a **triac** are similar to those which occur in an SCR, except that they will influence the behavior of the system in both alternations of the AC cycle. If the gate of a triac is **open** circuited, then the device will not go into conduction in either direction, and no output voltage will appear across the load. If there is a direct electrical connection or short between the anode and the cathode, then the device will have continuous conduction. In this case, the power through the load cannot be controlled by triggering the triac gate.

Internal **open circuit** and **short circuit** faults can occur between the terminals of other types of thyristors. In a PUT, an open circuit fault condition between the anode and cathode will cause it to be inoperative. Referring to the PUT pulse generation circuit shown in **Figure 13-29**, an open circuit between the anode and cathode terminals will prevent the capacitor C_1 from discharging through the load resistor R_4. On the other hand, if there is a short between the anode and cathode terminals, the voltage across the load resistor R_4 will be very small. Most of the supply voltage will appear across the 10-kΩ resistor. If the gate were to have an open circuit fault, the capacitor C_1 would not be able to discharge through the load R_4. Hence, there will be no output pulse generated by the PUT circuit.

Summary

- A thyristor is a general classification for solid-state devices that are used for power control applications.
- The term *thyristor* is a contraction of the words (thyr)atron transistor.

- Thyristors can be two-, three-, or four-element devices.
- Two major classifications of thyristors are reverse blocking and bidirectional.
- Thyristors classified as reverse blocking are silicon controlled rectifiers, unijunction transistors, programmable unijunction transistors, gate turn-off switches, and silicon-controlled switches.
- Thyristors classified as bidirectional are triacs and diacs.
- A silicon controlled rectifier (SCR) is a solid-state device made of four alternate layers of P-type and N-type silicon.
- The leads of an SCR are the anode, cathode, and gate.
- The gate of an SCR is used to trigger conduction current between the anode and cathode.
- When conduction in the SCR reaches the holding current level, the SCR latches and holds its conduction.
- To stop conduction in the SCR, the anode−cathode current must be lowered below the holding current level or the voltage must be momentarily removed.
- When an SCR is used as a power control device, it primarily responds as a switch.
- DC power control using an SCR requires two switches to achieve control: one turns on the source voltage, and the second switch controls the gate.
- AC power control using an SCR has automatic commutation or turn-off because of the reversal of each alternation.
- The conduction time of an alternation can be changed to make a circuit have variable AC power control; this is achieved with a phase shifter, an RC circuit, or a pulse-control device.
- A triac is functionally classified as a gate-controlled AC switch; it responds as two reverse-parallel-connected SCRs with one common gate.
- Each terminal connection of the triac is jointly connected to an N-P material combination; the leads of a triac are called terminal 1, terminal 2, and gate.
- Conduction in a triac is achieved by selecting an appropriate crystal combination; selection depends on the polarity of the source.
- The dual polarity of each connection causes the triac to have four possible trigger combinations; the most sensitive modes of operation occur when the gate polarity matches the polarity of terminal 2.

- A triac has bidirectional conductivity; therefore, conduction in quadrants I and III of its *I–V* characteristics chart is identical.
- When a triac has AC control of the gate, it is possible to achieve 180° delay of the conduction time for each alternation; conduction time of the AC waveform can be varied to achieve 100% control of a power source.
- A diac is a bidirectional diode used to trigger a triac.
- The diac is the equivalent of an NPN transistor with no base connection.
- Conduction for the diac is the same in each direction.
- The *I–V* characteristic of a diac shows that conduction occurs when the breakover voltage is exceeded; this is the same in quadrants I and III.
- Pulse triggering of an SCR can be accomplished with a unijunction transistor.
- A unijunction transistor (UJT) is commonly described as a voltage-controlled diode; it responds as a three-terminal, single-junction solid-state device.
- A UJT is constructed with a small bar of N-type silicon, which has base 1 and base 2 attached to the ends of the bar; an emitter is formed by fusing an aluminum wire to the approximate center of the bar.
- Operation of a UJT is based on the bias voltage applied to the emitter–base 1 junction; when E-B_1 is forward biased, it lowers the internal B_1B_2 resistance.
- The bias voltage needed to achieve conduction in a UJT is determined by the intrinsic standoff ratio.
- UJTs are normally used in pulse-generator circuits that are attached to an SCR; the pulse rate of the circuit is an RC function.
- A programmable unijunction transistor (PUT) is commonly used to generate trigger pulses; it is actually a thyristor that responds as a UJT.
- The intrinsic standoff ratio voltage of a PUT can be altered or programmed externally by changing the resistance ratio of a voltage-divider network.
- The crystal structure of a PUT has its gate connected to the N-type material nearest to the anode; a P–N junction is formed by the anode–gate.
- Conduction in a PUT is controlled by the bias voltage of anode–gate.
- Gate turn-off thyristors are unique power control devices that can be triggered into conduction or turned off by a gate signal.
- The silicon-controlled switch (SCS) is a four-layer PNPN thyristor; an external lead connection is made to all four layers of the device.
- An SCS has an anode, cathode, anode–gate, and a cathode–gate.

- The *I–V* characteristics for a silicon-controlled switch are the same as those of an SCR; conductivity occurs in quadrant I and occurs when the anode is positive and the cathode is negative.
- Triggering with an SCS is achieved by making the anode–gate negative or the cathode–gate positive.
- An SCS can be turned off in three ways: bring the anode current below the holding current level; apply a negative pulse to the cathode–gate; apply a positive-going pulse to the anode–gate.
- A gate turn-off thyristor (GTO) is very similar in construction to the SCR – a positive gate pulse causes the device to be conductive, and a negative gate pulse signal turns off the conduction.

Formulas

(12-1) $\eta = R_{B1}/(R_{B1} + R_{B2})$ Intrinsic standoff ratio.
(12-2) $V_G = \eta \times V_S$ Gate voltage (V_G) of a PUT.
(12-3) $t = (R_2 + R_3) \times C_1$ Discharge time of a sawtooth generator.

Review Questions

Answer the following questions.

1. Thyristors are best described as:

    a. Power supply rectifiers
    b. Amplifying devices
    c. Pulse-generating devices
    d. Voltage-controlled switches

2. How many P–N junctions are there between the anode and cathode of an SCR?

    a. One
    b. Two
    c. Three
    d. Four

3. An SCR energized by DC is triggered into conduction. The conduction can be turned off by:

    a. Momentarily disconnecting the gate
    b. Reducing the positive gate voltage
    c. Applying negative gate voltage
    d. Lowering the holding current below a minimum level

4. An SCR energized by AC is triggered into conduction. This conduction will occur for:

    a. The entire AC cycle of operation
    b. Only the positive alternation
    c. Only the negative alternation
    d. Part of each alternation

5. An ohmmeter connected across two leads of an SCR shows low resistance in one direction and high resistance in the opposite direction. This represents:

    a. The anode–cathode leads
    b. The anode–gate leads
    c. The gate–cathode leads
    d. A faulty response for an SCR

6. The best method of triggering an SCR into conduction is accomplished by:

    a. Increasing the forward breakover voltage
    b. Increasing the peak reverse voltage
    c. Increasing the gate current
    d. Decreasing the gate current

7. Increasing the gate current of an SCR causes:

    a. A reduction in the forward breakover voltage
    b. An increase in the peak reverse voltage
    c. A lowering of the holding current level
    d. The forward breakover voltage to increase in value

8. A device that is triggered into conduction with either a positive or negative gate current is:

    a. A triac
    b. A diac
    c. An SCR
    d. A UJT
    e. A PUT

9. Before an SCR is triggered into conduction, the V_{AK}:

    a. Equals the source voltage
    b. Is less than the source voltage
    c. Is zero

d. Is the source voltage less the voltage drop across the series load resistor

10. The gate of an SCR loses control after being triggered into conduction because it:

   a. Loses the applied voltage
   b. Is reverse biased
   c. Has a large amount of current flow
   d. Has an I_{AK} that is great enough to cause continuous conduction

11. When an SCR goes into conduction, the current needed to cause continuous conduction is called the:

   a. Trigger current
   b. Breakover current
   c. Reverse breakdown current
   d. Forward current
   e. Holding current

12. An ohmmeter connected across T_1 and T_2 of a good triac will:

   a. Show high resistance in either direction of polarity
   b. Show low resistance in either direction of polarity
   c. Show low resistance in one direction and high resistance in the opposite direction
   d. Not tell anything about the condition of an SCR

13. A triode bidirectional conduction device is:

   a. An SCR
   b. A triac
   c. A UJT
   d. A PUT
   e. A silicon-controlled switch

14. An ohmmeter connected across two leads of a triac shows the same low resistance in both directions of polarity. This represents:

   a. T_1 and T_2
   b. T_2 and G
   c. G and T_1
   d. G and T_2
   e. A faulty response for the triac

15. A device that is triggered into conduction by either a positive or negative source voltage change is:

    a. An SCR
    b. A triac
    c. A diac
    d. A UJT
    e. A PUT

16. The operation of a conductive triac is best described by saying that the T_1-T_2 voltage is:

    a. High, and the load current is low
    b. High, and the load current is high
    c. Low, and the load current is low
    d. Low, and the load current is high
    e. Zero with maximum load current

17. If a silicon diode is used in the gate circuit of a triac power control circuit instead of a diac, the:

    a. Circuit will not work
    b. Diode will be damaged
    c. Output will not be effectively changed
    d. Gate will be controlled by DC gate current
    e. Output will be AC with reduced delay capabilities

18. An ohmmeter connected across T_1 and G of a good triac will:

    a. Read high resistance in either direction of polarity
    b. Read low resistance in either direction of polarity
    c. Read low resistance in one direction and high resistant in the opposite direction
    d. Not tell anything about the condition of the junctions

19. The most sensitive triggering polarities of a standard triac occur when:

    a. G and T_2 are of the same polarity
    b. G and T_2 are of opposite polarity
    c. G and T_1 are of the same polarity
    d. G and T_1 are of the opposite polarity

20. Triggering or conduction of a UJT occurs when the emitter is:

    a. Positive with respect to B_1
    b. Negative with respect to B_1

  c. Zero with respect to V_{BB}
  d. Positive with respect to B_2
  e. Negative with respect to B_2

21. The negative resistance region of a UJT refers to a characteristic that occurs:

  a. Before it is triggered
  b. Between the peak and valley points after triggering has occurred
  c. Beyond the valley point after triggering
  d. In the saturation region

22. A programmable unijunction transistor is a three-junction thyristor that is triggered into conduction when the gate is:

  a. Less positive than the anode
  b. More positive than the anode
  c. More positive than the cathode
  d. More negative than the cathode

Problems

Answer the following questions.

1. If the intrinsic standoff ration of a UJT is 0.45, and the applied V_{BB} is 15 V, the amount of V_E needed to produce triggering is in the range of:

  a. 1–3 V
  b. 3.1–6 V
  c. 6.1–8 V
  d. 8.1–10 V
  e. 10.1 or above

2. An ohmmeter connected across two terminal of a UJT shows a resistance of 5 kΩ in either direction of polarity. This represents:

  a. B_1 and B_2
  b. B_1 and E
  c. E and B_2
  d. A faulty condition for a UJT

3. In operation, a nontriggered UJT responds as:

  a. An open circuit
  b. An infinite resistance

    c. A low resistant diode

    d. A 4- to 10-kΩ resistance

Answers

Examples

12-1. (1) V_{BO} = 48 V for I_G = 6 mA; (2) V_{BO} = 148 V for I_G = 4 mA; (3) V_{BO} = 325 V for I_G = 1 mA; (4) V_{BO} = 400 V for I_G = 0 mA

12-2. Assuming that the lamp load has already been switched on. To switch it off:

    (1) Momentarily press the STOP PB

    (2) Gate of SCR-2 will be triggered

    (3) V_{BO} of SCR-2 is reduced

    (4) SCR-2 goes into conduction

    (5) Capacitor C_1 discharges through SCR-2

    (6) The anode of SCR-1 is momentarily grounded

    (7) SCR-1 is switched off

    (8) The lamp load is switched off.

12-3. No power is developed by the load as the SCR never goes into conduction.

12-4. 1) 7−15 mA 2) 5−10 mA

12-5. Refer to **Figure 12.26(c)**, to see that the triac begins conducting at 90° in the positive alteration of the AC input. This will cause the output to be available from 90° through 180° of this alternation. After this, the triac will be switched off, until 270° of the negative alteration of the AC input. The triac begins conducting at 270° and will produce to 360°. This process will then be repeated.

12-6. In switch position 2, the diode D_1 will be connected to the gate during the positive alteration of the AC input, and this triggers the triac into conduction, energizing the load. In the negative alteration of the AC input, the diode is reverse biased and does not trigger the triac into conduction. The load is thus de-energized during the negative alteration of the AC input.

12-7. In quadrant III, when T_1 is negative and T_2 is positive, the diac does not conduct until V_{BO2} is reached. After the V_{BO2} value is reached, the diac begins to conduct.

12-8. Full power will be developed by the load for this condition of operation. The triac is triggered at the earliest possible instant (at $0°$) in the positive half alteration causing it to conduct for the entire alteration. It is triggered again at the earliest possible instant (at $180°$) of the negative alteration causing conduction.

12-9. When the voltage applied across the emitter–base 1 junction reaches V_P, it is forward biased. At this point, there is an increase in the current I_E. V_P drops to V_V (the emitter–base 1 voltage at the "valley voltage point") even as the current increases. This portion of the UJT characteristic curve represents the negative resistance region.

12-10. When the voltage applied across the anode–cathode junction reaches V_P, it is forward biased. At this point, there is an increase in the current I_{AK}. V_P drops to V_V (the anode–cathode voltage at the "valley voltage point") even as the current increases. This portion of the PUT characteristic curve represents the negative resistance region.

12-11. When the PUT goes into conduction, it causes the capacitor C_1 to discharge through R_4 causing a momentary pulse to be generated across R_4. The voltage across the capacitor reduces and this causes the anode voltage of the PUT to reduce as well. The PUT turns off which, in turn, causes C_1 to charge through the resistor R_3.

12-12. $t = 2.5$ ms; $T = 12.5$ ms

Self-Examination

12.1

1. anode, cathode, gate (in any order)
2. forward
3. three, four
4. low
5. high
6. anode–cathode voltage
7. alternation
8. I
9. forward breakover voltage, *or* V_{BO}
10. III
11. increase
12. forward
13. gate

14. gate
15. holding

12.2

16. SCRs
17. polarity
18. $N_1P_1N_2P_2$
19. $N_3P_2N_2P_1$
20. zero
21. lowers
22. +
23. −
24. gate
25. . source, *or* supply

12.3

26. diac
27. T_1, T_2 (any order)
28. I, III (any order)
29. gate
30. R_1, C_1 (any order)
31. conduction, *or* current

12.4

32. resistor
33. three
34. trigger
35. base 1, base 2 (any order)
36. emitter
37. E-B_1
38. decreases
39. negative resistance
40. intrinsic standoff
41. triggering

12.5

42. programmable unijunction transistor, *or* PUT
43. silicon controlled rectifier, *or* SCR
44. gate
45. gate, anode (any order)
46. positive
47. low
48. high

49. intrinsic standoff ratio
50. programmed, *or* changed
51. trigger

12.6

52. silicon controlled rectifier, *or* SCR
53. two
54. +
55. −
56. three
57. +, −
58. gate

Terms

Internal resistance

The resistance between the anode and cathode of an SCR or between terminals 1 and 2 of a triac. It is also called the dynamic resistance.

Off-state resistance

The resistance of an SCR or triac when it is not conducting.

Triggering

The process of causing an NPNP device to switch states from off to on.

Latching

The process of placing an SCR or triac in a holding state in which the device turns on and stays in conduction when the gate current is removed.

Alternation

Half of an AC sine wave. There is a positive alternation and a negative alternation for each AC cycle.

Turn-on time

The time of an AC waveform when an alternation occurs.

Forward breakover voltage

The voltage at which an SCR or triac goes into conduction in quadrant I of its *I–V* characteristics.

Holding current

A current level that must be achieved when an SCR or triac latches or holds in the conductive state.

Conduction time

The time when a solid-state device is turned on or is in its conductive state.

Bidirectional triggering

The triggering or conduction that can be achieved in either direction of an applied AC wave. Diacs and triacs are of this classification.

Contact bounce

When a mechanical switch is turned on, the contacts are forced together. This causes the contacts to open and close several times before making firm contact.

Interbase resistance

The resistance between base 1 and base 2 of a unijunction transistor.

Negative resistance region

A UJT characteristic where an increase in emitter current causes a decrease in emitter voltage.

Intrinsic standoff ratio (η)

The ratio of the resistance of base 1 (R_{B1}) to the total resistance of the bases ($R_{B1} + R_{B2}$) of a UJT.

MOTOROLA
SEMICONDUCTOR TECHNICAL DATA

Order this document
by 2N6394/D

Silicon Controlled Rectifiers
Reverse Blocking Triode Thyristors

... designed primarily for half-wave ac control applications, such as motor controls, heating controls and power supplies.

- Glass Passivated Junctions with Center Gate Geometry for Greater Parameter Uniformity and Stability
- Small, Rugged, Thermowatt Construction for Low Thermal Resistance, High Heat Dissipation and Durability
- Blocking Voltage to 800 Volts

**2N6394
thru
2N6399**
Motorola preferred devices

**SCRs
12 AMPERES RMS
50 thru 800 VOLTS**

CASE 221A-07
(TO-220AB)
STYLE 3

*MAXIMUM RATINGS (T_J = 25°C unless otherwise noted.)

Rating		Symbol	Value	Unit
Peak Repetitive Forward and Reverse Blocking Voltage[1] (Gate Open, T_J = −40 to 125°C) 2N6394 2N6395 2N6397 2N6398 2N6399		V_{DRM}, V_{RRM}	50 100 400 600 800	Volts
RMS On–State Current (T_C = 90°C) (All Conduction Angles)		$I_{T(RMS)}$	12	Amps
Peak Non-Repetitive Surge Current (1/2 Cycle, Sine Wave, 60 Hz, T_J = 125°C)		I_{TSM}	100	Amps
Circuit Fusing (t = 8.3 ms)		I^2t	40	A^2s
Forward Peak Power		P_{GM}	20	Watts
Forward Average Gate Power		$P_{G(AV)}$	0.5	Watt
Forward Peak Gate Current		I_{GM}	2	Amps
Operating Junction Temperature Range		T_J	−40 to +125	°C
Storage Temperature Range		T_{stg}	−40 to +150	°C

THERMAL CHARACTERISTICS

Characteristic	Symbol	Max	Unit
Thermal Resistance, Junction to Case	$R_{\theta JC}$	2	°C/W

*Indicates JEDEC Registered Data.

1. V_{DRM} and V_{RRM} for all types can be applied on a continuous basis. Ratings apply for zero or negative gate voltage; however, positive gate voltage shall not be applied concurrent with negative potential on the anode. Blocking voltages shall not be tested with a constant current source such that the voltage ratings of the devices are exceeded.

Preferred devices are Motorola recommended choices for future use and best overall value.

2N6394 thru 2N6399

ELECTRICAL CHARACTERISTICS ($T_C = 25°C$ unless otherwise noted.)

Characteristic	Symbol	Min	Typ	Max	Unit
*Peak Repetitive Forward or Reverse Blocking Current (V_{AK} = Rated V_{DRM} or V_{RRM}, Gate Open) T_J = 25°C T_J = 125°C	I_{DRM}, I_{RRM}	— —	— —	10 2	μA mA
*Forward "On" Voltage (I_{TM} = 24 A Peak)	V_{TM}	—	1.7	2.2	Volts
*Gate Trigger Current (Continuous dc) (V_D = 12 Vdc, R_L = 100 Ohms)	I_{GT}	—	5	30	mA
*Gate Trigger Voltage (Continuous dc) (V_D = 12 Vdc, R_L = 100 Ohms) (V_D = Rated V_{DRM}, R_L = 100 Ohms, T_J = 125°C)	V_{GT} V_{GD}	— 0.2	0.7 —	1.5 —	Volts
*Holding Current (V_D = 12 Vdc, Gate Open)	I_H	—	6	40	mA
Turn-On Time (I_{TM} = 12 A, I_{GT} = 40 mAdc, V_D = Rated V_{DRM})	t_{gt}	—	1	2	μs
Turn-Off Time (V_D = Rated V_{DRM}) (I_{TM} = 12 A, I_R = 12 A) (I_{TM} = 12 A, I_R = 12 A, T_J = 125°C)	t_q	— —	15 35	— —	μs
Critical Rate-of-Rise of Off-State Voltage Exponential (V_D = Rated V_{DRM}, T_J = 125°C)	dv/dt	—	50	—	V/μs

*Indicates JEDEC Registered Data.

FIGURE 1 — CURRENT DERATING

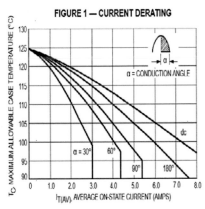

FIGURE 2 — MAXIMUM ON-STATE POWER DISSIPATION

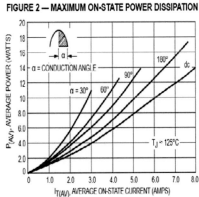

2N6394 thru 2N6399

FIGURE 3 — ON–STATE CHARACTERISTICS

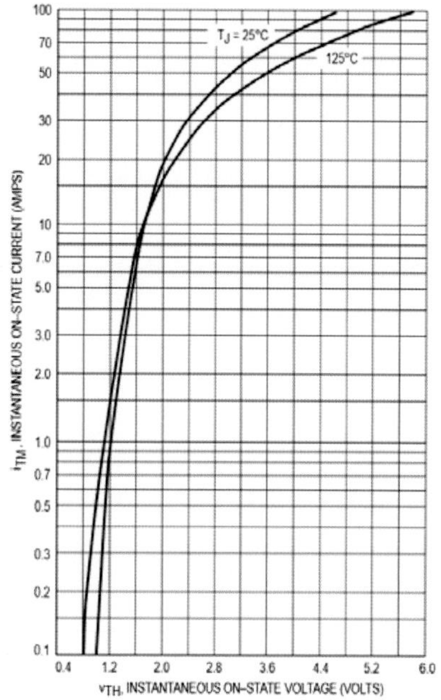

FIGURE 4 — MAXIMUM NON-REPETITIVE SURGE CURRENT

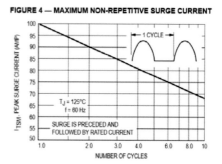

FIGURE 5 — THERMAL RESPONSE

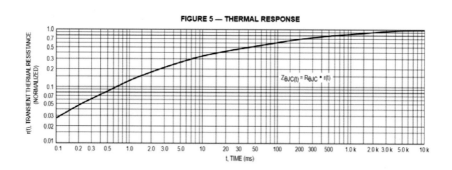

2N6394 thru 2N6399

TYPICAL CHARACTERISTICS

FIGURE 6 — PULSE TRIGGER CURRENT

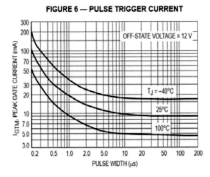

FIGURE 7 — GATE TRIGGER CURRENT

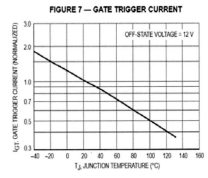

FIGURE 8 — GATE TRIGGER VOLTAGE

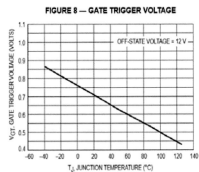

FIGURE 9 — HOLDING CURRENT

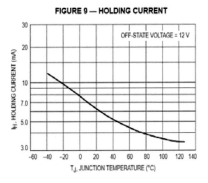

2N6394 thru 2N6399

PACKAGE DIMENSIONS

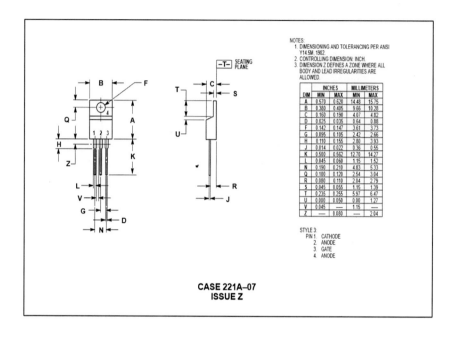

NOTES:
1. DIMENSIONING AND TOLERANCING PER ANSI Y14.5M, 1982.
2. CONTROLLING DIMENSION: INCH.
3. DIMENSION Z DEFINES A ZONE WHERE ALL BODY AND LEAD IRREGULARITIES ARE ALLOWED.

DIM	INCHES MIN	INCHES MAX	MILLIMETERS MIN	MILLIMETERS MAX
A	0.570	0.620	14.48	15.75
B	0.380	0.405	9.66	10.28
C	0.160	0.190	4.07	4.82
D	0.025	0.035	0.64	0.88
F	0.142	0.147	3.61	3.73
G	0.095	0.105	2.42	2.66
H	0.110	0.155	2.80	3.93
J	0.014	0.022	0.36	0.55
K	0.500	0.562	12.70	14.27
L	0.045	0.060	1.15	1.52
N	0.190	0.210	4.83	5.33
Q	0.100	0.120	2.54	3.04
R	0.080	0.110	2.04	2.79
S	0.045	0.055	1.15	1.39
T	0.235	0.255	5.97	6.47
U	0.000	0.050	0.00	1.27
V	0.045	—	1.15	—
Z	—	0.080	—	2.04

STYLE 3:
PIN 1. CATHODE
2. ANODE
3. GATE
4. ANODE

CASE 221A–07
ISSUE Z

FAIRCHILD
SEMICONDUCTOR®

FKN08PN60
TRIAC (Silicon Bidirectional Thyristor)

Application Explanation
- Switching mode power supply, light dimmer, electric flasher unit, hair drier
- TV sets, stereo, refrigerator, washing machine
- Electric blanket, solenoid driver, small motor control
- Photo copier, electric tool

TO-92

1 2 3

1: T₁
2: Gate
3: T₂

Absolute Maximum Ratings $T_a = 25°C$ unless otherwise noted

Symbol	Parameter	Value		Rating	Units
V_{DRM} V_{RRM}	Peak Repetitive Off-State Voltage	Sine Wave 50 to 60Hz, Gate Open		600	V
$I_{T (RMS)}$	RMS On-State Current	Commercial frequency, sine full wave 360° conduction, Tc= 70°C		0.8	A
I_{TSM}	Surge On-State Current	Sinewave 1 full cycle, peak value, non-repetitive	50Hz	8	A
			60Hz	9	A
I^2t	I^2t for Fusing	Value corresponding to 1 cycle of halfwave, surge on-state current, tp=8.4ms		0.33	A^2s
P_{GM}	Peak Gate Power Dissipation			5	W
$P_{G (AV)}$	Average Gate Power Dissipation			0.1	W
V_{GM}	Peak Gate Voltage			5	V
I_{GM}	Peak Gate Current			1	A
T_J	Junction Temperature			– 40 ~ 125	°C
T_{STG}	Storage Temperature			– 40 ~ 125	°C

Thermal Characteristics

FKN08PN60 TRIAC (Silicon Bidirectional Thyristor)

Electrical Characteristics $T_C = 25°C$ unless otherwise noted

Symbol	Parameter		Test Condition		Min.	Typ.	Max.	Units
I_{DRM} I_{RRM}	Repetitive Peak Off-State Current		V_{DRM}/V_{RRM} applied		-	-	100	µA
V_{TM}	On-State Voltage		$T_C = 25°C$, $I_{TM} = 1.12A$ Instantaneous measurement		-	-	1.8	V
V_{GT}	Gate Trigger Voltage	I	$V_D = 12V$, $R_L = 100\Omega$	T2(+), Gate (+)	-	-	2.0	V
		II		T2(+), Gate (-)	-	-	2.0	V
		III		T2(-), Gate (-)			2.0	V
I_{GT}	Gate Trigger Current	I	$V_D = 12V$, $R_L = 100\Omega$	T2(+), Gate (+)	-	-	5	mA
		II		T2(+), Gate (-)	-	-	5	mA
		III		T2(-), Gate (-)	-	-	5	mA
V_{GD}	Gate Non-Trigger Voltage		$T_J = 125°C$, $V_D = 1/2V_{DRM}$		0.2	-	-	V
I_H	Holding Current	(I, II, III)	$V_D = 12V$, $I_{TM} = 200mA$		-	-	15	mA
I_L	Latching Current	I, III	$V_D = 12V$, $I_G = 10mA$		-	-	15	mA
		II			-	-	20	mA
dv/dt(s)	Critical Rate of Rise of Off-State Voltag		$V_{DRM} = 63\%$ Rated, $T_J = 125°C$, Exponential Rise		20	-	-	V/µs
dv/dt(c)	Critical-Rate of Rise of Off State Commutating Voltage (di/dt=-0.7A/uS)				3.0	-	-	V/µs

Commutation dv/dt test

V_{DRM} (V)	Test Condition	Commutating voltage and current waveforms (inductive load)
FKN08PN60	1. Junction Temperature $T_J = 125°C$ 2. Rate of decay of on-state commutating current $(di/dt)_C$ 3. Peak off-state voltage $V_D = 300V$	Supply Voltage — Time / Main Current — Time $(di/dt)_C$ / Main Voltage — Time $(dv/dt)_c$ V_D

TRIAC TRIGGERS

The ST2 (diac) is a silicon bi-directional diode which may be used for triggering triacs or SCRs. It has a three layer structure with negative resistance switching characteristics in both directions.

The ST4 is an asymmetrical AC trigger integrated circuit for use in triac phase control applications. This device reduces the snap-on effects that are present in conventional trigger circuits by eliminating control circuit hysteresis. This performance is possible with a single RC time constant where as a symmetrical circuit of comparable performance would require at least three more passive components.

GE Type	V_{S2} Switching Voltage		V_{S1} Switching Voltage		I_{S2}, I_{S1} Switching Current Max. (µA)	Pulse Output Min. (V)	Package Outline No.
	Min. (V)	Max. (V)	Min. (V)	Max. (V)			
ST2	28	36	28	36	200	3.0	B
ST4	7	9	14	18	80	3.5	A

For ST2, $V_{S2} = V_S \pm 10\%$

Data sheet	
status	Preliminary specification
date of issue	December 1990

2N2646
Silicon unijunction transistor

QUICK REFERENCE DATA

SYMBOL	PARAMETER	CONDITIONS	MIN.	TYP.	MAX.	UNIT
$-V_{EB2}$	emitter-base 2 voltage		–	–	30	V
I_{EM}	emitter current	peak value	–	–	2	A
P_{tot}	total power dissipation		–	–	300	mW
T_j	junction temperature		–	–	125	°C
R_{BB}	static inter-base resistance	$V_{B2B1} = 3$ V $I_E = 0$	–	7	–	kΩ
V_{EB1sat}	emitter-base 1 saturation voltage	$V_{B2B1} = 10$ V $I_E = 50$ mA	–	3.5	–	V
$I_{E(V)}$	emitter valley point current		4	6	–	mA
$I_{E(P)}$	emitter peak point current		–	1	5	µA

PINNING - TO-18
Base 2 connected to case.

PIN	DESCRIPTION
1	emitter
2	base 1
3	base 2

PIN CONFIGURATION

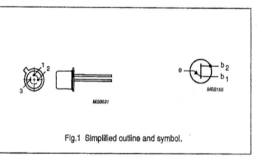

MSB031

MBB155

Fig.1 Simplified outline and symbol.

Philips Semiconductors

Preliminary specification

Silicon unijunction transistor

2N2646

LIMITING VALUES

In accordance with the Absolute Maximum System (IEC 134).

SYMBOL	PARAMETER	CONDITIONS	MIN.	MAX.	UNIT
$-V_{EB2}$	emitter-base 2 voltage		–	30	V
V_{B2B1}	inter-base voltage		–	35	V
I_E	emitter current	average value	–	50	mA
I_{EM}	emitter current (note 1)	peak value	–	2	A
P_{tot}	total power dissipation (note 2)	$T_{amb} \leq 25\ °C$	–	300	mW
T_{stg}	storage temperature range		–65	150	°C
T_j	junction temperature		–	125	°C

Notes

1. Capacitor discharge $\leq 10\ \mu F$ at $\leq 30\ V$.

2. Must be limited by external circuit.

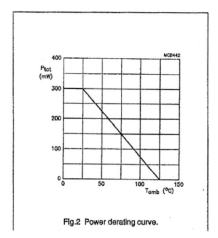

Fig.2 Power derating curve.

Philips Semiconductors

Preliminary specification

Silicon unijunction transistor

2N2646

CHARACTERISTICS

T_{amb} = 25 °C unless otherwise specified.

SYMBOL	PARAMETER	CONDITIONS	MIN.	TYP.	MAX.	UNIT
R_{BB}	static inter-base resistance	V_{B2B1} = 3 V I_E = 0	4.7	7	9.1	kΩ
TC_{RBB}	inter-base resistance temperature coefficient	V_{B2B1} = 3 V I_E = 0 T_{amb} = −55 to 125 °C	0.1	–	0.9	%/K
$-I_{EB2O}$	emitter cut-off current	$-V_{EB2}$ = 30 V I_{B1} = 0	–	–	12	V
V_{EB1sat}	emitter-base 1 saturation voltage	V_{B2B1} = 10 V I_E = 50 mA	–	3.5	–	V
I_{B2mod}	inter-base current modulation	V_{B2B1} = 10 V I_E = 50 mA	–	15	–	mA
η	input/output ratio (note 1)	V_{B2B1} = 10 V	0.56	–	0.75	
$I_{E(V)}$	emitter valley point current	V_{B2B1} = 20 V R_{B2} = 100 Ω	4	6	–	mA
$I_{E(P)}$	emitter peak point current	V_{B2B1} = 25 V	–	1	5	µA
V_{OB1M}	base 1 impulse/output voltage		3	5	–	V

Note

1. $\eta = \dfrac{(V_{E(P)} - V_{EB1})}{V_{B2B1}}$, when $V_{E(P)}$ = emitter peak point voltage, V_{EB1} = emitter-base 1 breakdown voltage, (approximately 0.5 V at 10 µA), and V_{B2B1} = inter-base voltage.

DATA SHEET

BRY56A
Programmable unijunction
transistor

Philips Semiconductors

Product specification

Programmable unijunction transistor

BRY56A

DESCRIPTION

Planar PNPN trigger device in a TO-92; SOT54 plastic package.

APPLICATIONS

- Switching applications such as:
  - Motor control
  - Oscillators
  - Relay replacement
  - Timers
  - Pulse shapers, etc.

PINNING

PIN	DESCRIPTION
1	gate
2	anode
3	cathode

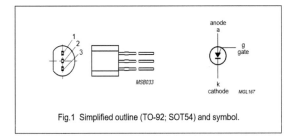

Fig.1 Simplified outline (TO-92; SOT54) and symbol.

LIMITING VALUES

In accordance with the Absolute Maximum Rating System (IEC 134).

SYMBOL	PARAMETER	CONDITIONS	MIN.	MAX.	UNIT
V_{GA}	gate-anode voltage		–	70	V
$I_{A(AV)}$	average anode current		–	175	mA
I_{ARM}	repetitive peak anode current	$t_p = 10 \propto s$; $\delta = 0.01$	–	2.5	A
I_{ASM}	non-repetitive peak anode current	$t_p = 10 \propto s$	–	3	A
dI_A/dt	rate of rise of anode current	$I_A \leq 2.5$ A	–	20	A/$\propto$s
P_{tot}	total power dissipation	$T_{amb} \leq 75\ °C$	–	300	mW
T_{stg}	storage temperature		–65	+150	°C
T_j	junction temperature		–	150	°C
T_{amb}	operating ambient temperature		–65	+150	°C

Programmable unijunction transistor   **BRY56A**

THERMAL CHARACTERISTICS

SYMBOL	PARAMETER	CONDITIONS	VALUE	UNIT
$R_{th\,j-a}$	thermal resistance from junction to ambient	in free air	250	K/W

CHARACTERISTICS

T_{amb} = 25 °C unless otherwise specified.

SYMBOL	PARAMETER	CONDITIONS	MIN.	TYP.	MAX.	UNIT
I_P	peak point current	V_S = 10 V; R_G = 10 kΩ; see Fig.7	–	–	200	nA
		V_S = 10 V; R_G = 100 kΩ; see Fig.7	–	–	60	nA
I_V	valley point current	V_S = 10 V; R_G = 10 kΩ; see Fig.7	2	–	–	∝A
		V_S = 10 V; R_G = 100 kΩ; see Fig.7	1	–	–	∝A
V_{offset}	offset voltage	typical curve; I_A = 0; see Fig.7	–	$V_P - V_S$	–	V
I_{GAO}	gate-anode leakage current	I_K = 0; V_{GA} = 70 V; see Fig.5	–	–	10	nA
I_{GKS}	gate-cathode leakage current	V_{AK} = 0; V_{KG} = 70 V; see Fig.6	–	–	100	nA
V_{AK}	anode-cathode voltage	I_A = 100 mA	–	–	1.4	V
V_{OM}	peak output voltage	V_{AA} = 20 V; C = 10 nF; see Figs 8 and 9	6	–	–	V
t_r	rise time	V_{AA} = 20 V; C = 10 nF; see Fig.9	–	–	80	ns

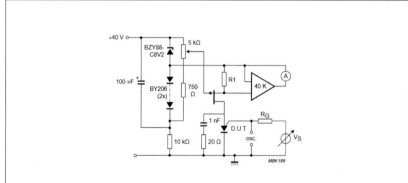

I_P and I_V determined by value of R1.

$R1 = \dfrac{1}{I_A}$; i.e. maximum voltage drop over R1 = 1 V.

Internal resistance of oscilloscope = 10 MΩ.

Fig.2 Measuring circuit for peak and valley point currents.

Programmable unijunction transistor   BRY56A

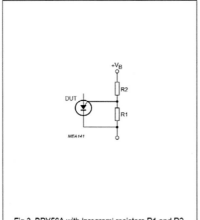

Fig.3 BRY56A with 'program' resistors R1 and R2.

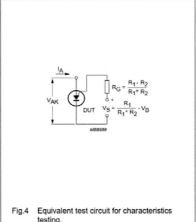

Fig.4 Equivalent test circuit for characteristics testing.

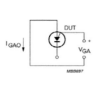

Fig.5 Equivalent test circuit for gate-anode leakage current.

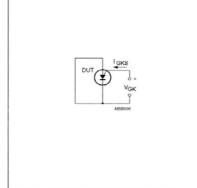

Fig.6 Equivalent test circuit for gate-cathode leakage current.

Programmable unijunction transistor   BRY56A

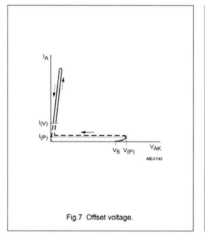

Fig.7 Offset voltage.

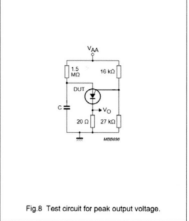

Fig.8 Test circuit for peak output voltage.

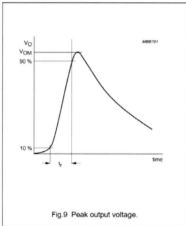

Fig.9 Peak output voltage.

Programmable unijunction transistor   **BRY56A**

PACKAGE OUTLINE

Plastic single-ended leaded (through hole) package; 3 leads   **SOT54**

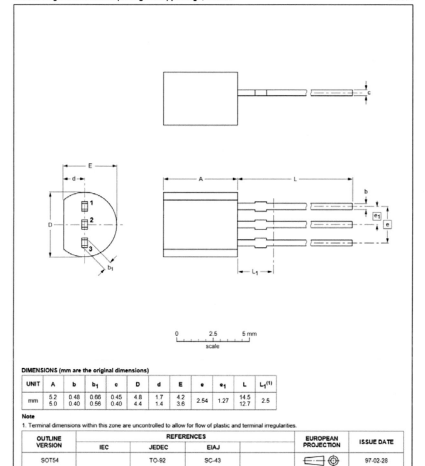

```
0        2.5        5 mm
|...|...|...|...|...|
         scale
```

DIMENSIONS (mm are the original dimensions)

UNIT	A	b	b₁	c	D	d	E	e	e₁	L	L₁(1)
mm	5.2 5.0	0.48 0.40	0.66 0.56	0.45 0.40	4.8 4.4	1.7 1.4	4.2 3.6	2.54	1.27	14.5 12.7	2.5

Note

1. Terminal dimensions within this zone are uncontrolled to allow for flow of plastic and terminal irregularities.

OUTLINE	REFERENCES				EUROPEAN	ISSUE DATE
VERSION	IEC	JEDEC	EIAJ		PROJECTION	
SOT54		TO-92	SC-43		⊏⊐⊙	97-02-28

Philips Semiconductors   Product specification

Silicon controlled switch   **BR101**

DESCRIPTION

Silicon planar PNPN switch in a
TO-72 metal package. It is an
integrated PNP/NPN transistor pair,
with all electrodes accessible.

APPLICATIONS

- Time base circuits
- Switching in television circuits
- Trigger device for thyristors.

PINNING

PIN	DESCRIPTION
1	cathode
2	cathode gate
3	anode gate (connected to case)
4	anode

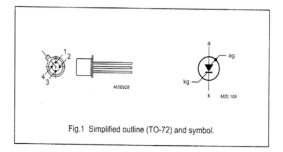

Fig.1 Simplified outline (TO-72) and symbol.

QUICK REFERENCE DATA

SYMBOL	PARAMETER	CONDITIONS	MAX.	UNIT
PNP transistor				
V_{EBO}	emitter-base voltage	open collector	−50	V
NPN transistor				
V_{CBO}	collector-base voltage	open emitter	50	V
I_{ERM}	repetitive peak emitter current		−2.5	A
P_{tot}	total power dissipation	$T_{amb} \leq 25\ °C$	275	mW
T_j	junction temperature		150	°C
V_{AK}	forward on-state voltage	$I_A = 50$ mA; $I_{AG} = 0$; $R_{KG-K} = 10$ kΩ	1.4	V
I_H	holding current	$I_{AG} = 10$ mA; $V_{BB} = -2$ V; $R_{KG-K} = 10$ kΩ	1	mA

Philips Semiconductors

Product specification

Silicon controlled switch

BR101

LIMITING VALUES

In accordance with the Absolute Maximum Rating System (IEC 134).

SYMBOL	PARAMETER	CONDITIONS	MIN.	MAX.	UNIT
NPN transistor					
V_{CBO}	collector-base voltage	open emitter	–	50	V
V_{CER}	collector-emitter voltage	$R_{BE} = 10\ k\Omega$	–	50	V
V_{EBO}	emitter-base voltage	open collector; note 1	–	5	V
I_C	collector current (DC)	note 2	–	175	mA
I_{CM}	peak collector current		–	175	mA
I_E	emitter current (DC)		–	–175	mA
I_{ERM}	repetitive peak emitter current	$t_p = 10\ \mu s;\ \delta = 0.01$	–	–2.5	A
PNP transistor					
V_{CBO}	collector-base voltage	open emitter	–	–50	V
V_{CEO}	collector-emitter voltage	open base	–	–50	V
V_{EBO}	emitter-base voltage	open collector	–	–50	V
I_E	emitter current (DC)		–	175	mA
I_{ERM}	repetitive peak emitter current	$t_p = 10\ \mu s;\ \delta = 0.01$	–	2.5	A
Combined device					
P_{tot}	total power dissipation	$T_{amb} \leq 25\ ^\circ C$	–	275	mW
T_{stg}	storage temperature		–65	+150	°C
T_j	junction temperature		–	150	°C
T_{amb}	operating ambient temperature		–65	+150	°C

Notes

1. It is permitted to exceed this voltage during the discharge of a capacitor of max. 390 pF, provided the charge does not exceed 50 nC.

2. Provided the I_E rating is not exceeded.

THERMAL CHARACTERISTICS

SYMBOL	PARAMETER	CONDITIONS	VALUE	UNIT
$R_{th\,j\text{-}a}$	thermal resistance from junction to ambient	in free air	0.45	K/mW

Philips Semiconductors

Product specification

Silicon controlled switch

BR101

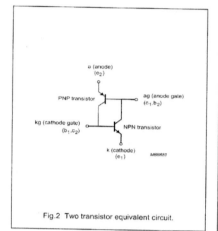

Fig.2 Two transistor equivalent circuit.

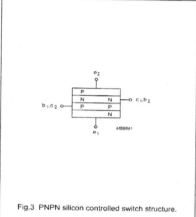

Fig.3 PNPN silicon controlled switch structure.

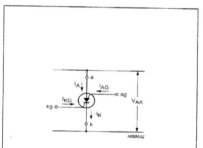

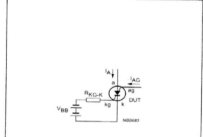

Gate Turn-Off Thyristors

**GTOs
18 AMPERES RMS
1000 and 1200 VOLTS**

The GTO is a family of asymmetric gate turn-off thyristors designed primarily for dc power switching applications such as motor drives, switching power supplies, inverters, or wherever a need exists for high surge current capabilities and fast switching speeds.

- Fast Turn-Off With Reverse Gate Pulse
- High Voltage — V_{DRXM} = 1000 and 1200 Volts
- Momentary Forward Pulse For Turn-On
- Minimizes Drive Losses
- Interdigitated Emitter Geometry Aids Turn-On Current Spreading and Improves Turn-On di/dt
- Clip and Current Spreading Ring for Reliable High Surge Capability — I_{TSM} = 200 A

**CASE 221A-04
(TO-220AB)
STYLE 3**

MAXIMUM RATINGS

Rating	Symbol	Value	Unit
Repetitive Peak Off-State Voltage (T_J = −40 to +125°C, 1/2 Sine Wave 50 to 60 Hz) Note 1	V_{DRXM}	1000 1200	Volts
Repetitive Peak Reverse Voltage, Gate Open (T_J = −40 to +125°C), Note 2	V_{RRM}	15	Volts
Repetitive Peak Reverse Gate Voltage, Note 3	V_{GRM}	15	Volts
On-State Current at T_C = 65°C (1/2 Cycle Sine Wave, 50 to 60 Hz)	$I_{T(RMS)}$	18	Amps
Peak Nonrepetitive Surge Current (8.3 ms Conduction, Half Sine Wave T_C = 65°C)	I_{TSM}	200	Amps
Circuit Fusing (t = 8.3 ms)	I^2t	167	A^2s
Repetitive Controllable On-State Current, Note 4	I_{TCM}	50	Amps
Nonrepetitive Maximum Interruptable On-State Current, Note 5	I_{TCSM}	70	Amps
Peak Forward Gate Power	P_{GFM}	10	Watts
Average Forward Gate Power	$P_{GF(AV)}$	3	Watts
Peak Reverse Gate Power	P_{GRM}	400	Watts
Average Reverse Gate Power	$P_{GR(AV)}$	5	Watts
Operating Junction Temperature Range	T_J	−40 to +125	°C
Storage Temperature Range	T_{stg}	−40 to +150	°C

Notes: 1. V_{DRXM} for all types can be applied on a continuous basis without damage. Ratings apply for R = 39 Ω or shorted gate conditions or negative voltage on the gate. Devices should not be tested for blocking voltage such that the supply voltage exceeds the rating of the device.
2. This is an asymmetric anode shorted part with a blocking gate-cathode junction. The ability to support a reverse voltage depends on the gate-cathode terminal conditions. Gate-cathode reverse bias increases V_{BRM}.
3. Instantaneous voltage at turn-off may exceed rated V_{GRM} provided P_{GRM} is not exceeded.
4. V_D Maximum Peak = V_{DRXM} − 300 V, T_J < 125°C, L_G = 2 μH, V_{GR} = 12 V (See Figure 2)
C_S = 0.1 μF for MGTO1000
C_S = 0.05 μF for MGTO1200
5. V_D Maximum Peak = V_{DRXM} − 300 V, T_J < 125°C, L_G = 2 μH, V_{GR} = 12 V (See Figure 2)
C_S = 0.2 μF for MGTO1000
C_S = 0.1 μF for MGTO1200

MOTOROLA THYRISTOR DEVICE DATA

13

Optoelectronic Devices

Optoelectronics is a branch of physical science that is concerned with the study of light. Optoelectronic devices have electronic properties that are affected by light energy. The term **light** is used in a general way to include visible, infrared, and ultraviolet regions of the frequency spectrum.

The role of optoelectronics has expanded through the years. Optoelectronics was once called **photoelectricity**. Photocells and phototubes were used to achieve electronic control. The field of optoelectronics expanded to include a number of unique solid-state devices. These devices are capable of converting light energy into electrical energy or vice versa. They can be used as a source of light energy or as a detector of light energy. As a rule, these devices are combined in a system that is used to control light energy. Light-control systems are largely responsible for the rapid expansion of this field.

Light-emitting diodes (LEDs), **photodiodes**, **phototransistors**, **lasers**, and **optocouplers** are some of the solid-state devices included in the area of optoelectronics. These devices and others are covered in this chapter. This chapter begins with an explanation of the nature of light and a description of related terms.

Objectives

After studying this chapter, you will be able to:

13.1 describe some of the basic characteristics of light energy;
13.2 evaluate the performance of a device that radiates light energy;
13.3 evaluate the performance of a device that detects light energy;
13.4 analyze and troubleshoot optoelectronic systems.

Chapter Outline

13.1 The Nature of Light
13.2 Radiation Sources

13.3 Optoelectronic Detectors
13.4 Troubleshooting Optoelectronic Systems
13.5 Analysis and Troubleshooting – Optoelectronic Systems

13-1 Light-Emitting Diode (LED) Testing

In this activity, the leads of an LED will be identified using an ohmmeter, and its status will be evaluated.

13-2 LED Operation

In this activity, circuits will evaluate the forward- and reverse-bias operation of an LED.

13-3 Photoresistive Cells

In this activity, a photoresistive cell circuit will be constructed and evaluated.

13-4 Phototransistors

In this activity, a phototransistor circuit will be evaluated.

13-3 Photovoltaic Cells

In this activity, a photovoltaic cell circuit will be evaluated.

Key Terms

angstrom
candela (cd)
dark current
dynodes
electromagnetic spectrum
heterochromatic source
illuminance
infrared emitting diode (IRED)
intensity
light-emitting diode (LED)
lumen
luminous exitance
luminous flux
luminous intensity
monochromatic source
panchromatic source
photoconductive device
photoelectric emission

photoemissive device
photometric system
photoresistive cell
photovoltaic cell
photovoltaic device
quantum theory
radiance
radiant exitance
radiant flux
radiant incidence
radiant intensity
radiometric system
steradian (sr)
transverse wave
wavelength

13.1 The Nature of Light

Light is a type of radiant energy that travels through space in the form of electromagnetic waves. Infrared, ultraviolet, and visible light are types of radiant energy classified as light. Radio waves, microwaves, and X-rays are forms of radiant energy that are not classified as light. This section discusses the basic characteristics of radiant energy classified as light. The content of this discussion serves as a foundation to understanding the characteristics and operation of the optoelectronic devices to be discussed in the following sections.

13.1 Describe some of the basic characteristics of light energy.

In order to achieve objective 13.1, you should be able to:

- define *electromagnetic spectrum, transverse wave, wavelength, quantum theory, photometric system, radiometric system, radiant flux, radiant intensity, steradian, radiant incidence, radiant exitance, radiance, intensity, luminous intensity, candela, luminous flux, lumen, illuminance,* and *luminous exitance.*

Electromagnetic Spectrum

Radiant energy is usually described by a graph that shows the location of different frequencies. This graph is called an **electromagnetic spectrum**.

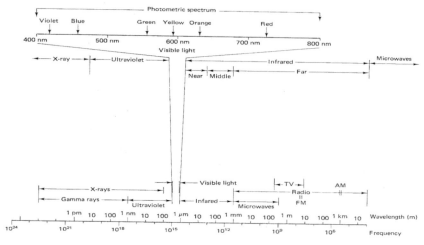

Figure 13.1   Electromagnetic spectrum.

The electromagnetic spectrum of **Figure 13.1** shows frequencies for radio, television, infrared, visible light, ultraviolet, X-rays, and gamma rays. These differ only with respect to their frequencies or wavelengths. **Wavelength** (see **Figure 13.2**) is a measurement of the distance a wave travels in space and the value is determined by the frequency of the vibrating source that produces it. Waves are motions that carry energy from one place to another, such as sound.

Wave Theory

The **wave theory** of light states that light consists of waves that spread out from the primary source of light. This process is called **radiation**. These waves are similar to those that occur in water when a pebble is dropped in a pool. Light waves are considered to be transverse. A **transverse wave** is one that causes the particles of a medium to vibrate at right angles to the direction in which the wave is moving. Each wave consists of a crest and a trough. As the wave moves forward, there is an alternate rise and fall in its **amplitude**.

   Light is emitted from a source in the form of **electromagnetic waves** that are of a specific wavelength. All electromagnetic waves, regardless of wavelength, travel at 186,000 mi/s, or 300,000,000 m/s through vacuum. Since light is an electromagnetic wave, its velocity of **propagation** is the same.

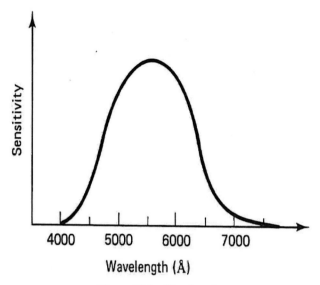

Figure 13.2   Wavelength.

To understand the wave theory of light transmission, it is necessary to relate the speed of travel, angular velocity, and shape of the radiated signal. Since the speed of light is a constant, there is an inverse relationship between the number of cycles per second (**frequency**) of the wave and its **wavelength**, as shown in **Figure 13.2**. If the frequency is doubled, one cycle, or one wavelength, will occupy one-half as much space. In other words, the higher the frequency, the shorter the wavelength, and the lower the frequency, the longer the wavelength. The symbol most commonly used to denote wavelength is the Greek letter lambda (λ). The relationship between wavelength and frequency is expressed mathematically as follows:

$$\lambda = 300,000,000/f \qquad (13.1)$$

where λ is the wavelength in meters, f is the frequency in hertz, and 300,000,000 m/s is the speed of light (c) in meters per second or 3×10^{17} nm.

An **Angstrom (Å)** is an extremely small unit that is also used to measure the wavelength of light. One Angstrom unit equals one hundred-millionth of a centimeter. The **visible light** range is often expressed in nanometers (nm). A nanometer is equal to 1×10^{-9} m. The visible region is approximately 400−770 nm.

Example 13-1:

The AM radio band occupies frequencies in the range of 535–1620 kHz. The center of this frequency range is near 1000 kHz. What is the wavelength of this radio frequency?

Solution

$$\lambda = 300,000,000/f$$
$$= 300,000,000 m/s/1,000,000 Hz$$
$$= 300m.$$

Related Problem

The range of the FM frequency band is 88–108 MHz. The approximate center of this band is 100 MHz. What is the wavelength of this frequency?

Example 13-2:

A wavelength for red light is 7000 Å. What is the wavelength of this light in nanometers?

Solution

Since 1 Å = 10^{-10} m and 1 nm = 10^{-9} m, the Angstrom unit is divided by 10
$$\text{Wavelength (in nm)} = Å\frac{}{10}$$

$$= \frac{7000}{10}$$
$$= 700 \text{ nm.}$$

Related Problem

A wavelength of violet light is 420 nm. What is the wavelength of this violet light in Angstroms?

   As you can see from the previous examples, the **wavelength** of light is extremely short compared to radio frequencies. Wavelengths are more commonly used than frequencies to describe light. The wavelengths of visible light range from approximately 4000 Å (400 nm) for violet to 7700 Å (770 nm) for red.

   The human eye responds to **electromagnetic waves** in the visible light band of frequencies between 400 and 770 nm. Each color of light is associated with a different frequency or wavelength. In the order of increasing frequency or decreasing wavelength, colors range from red, orange, yellow, green, blue, to violet. The response of the human eye to visible light is frequency sensitive, as shown in **Figure 13.2**. The greatest **sensitivity** on this response curve is nearly 5500 Å or 550 nm. The poorest sensitivity is around 4000 Å (400 nm) at the lower wavelengths and 7700 Å (770 nm) at the higher wavelengths. Our eyes perceive various degrees of brightness due to their response to different intensities of energy at different wavelengths of light.

The **wave theory** of light transmission is used to analyze the response of the human eye to light energy. It is also used to explain how light bends when it passes through glass or water and many other phenomena. We know that light also behaves as though it consists of bundles of energy when it interacts with matter. This model, which is known as the **quantum theory**, is used to explain how light energy is transferred to an optoelectronic device.

Quantum Theory

The **quantum theory** states that light is emitted by the atoms of a luminous body in separate packets of energy called **quanta** or **photons**. The energy of each bundle depends on the wavelength of the light. Each atom has a distinctive **spectrum** of wavelengths that it can radiate. These spectrums depend on the atom's distribution of electrons. **Photons** travel at the same speed, namely the speed of light, but they may have different amounts of energy. The energy content of a photon determines the color that is perceived by the brain when light strikes the eye.

Earlier, we learned that discrete amounts of energy applied to an atom can cause it to release electrons. A loss of electron energy can cause the electron to fall into a lower energy level. It is believed that photons are released from atoms when electrons fall to a lower energy level. The photon has energy that is directly related to the frequency of the light wave it produces. This energy is expressed by the following equation:

$$E = hf \qquad (13.2)$$

where E is the **quantum** of energy (joules), h is **Planck's universal constant** (6.626×10^{-34} joules/s), and f is the **frequency** (hertz).

The quantum theory shows that **light transmission** has some distinct characteristics that are important in optoelectronics. It is made of discrete bundles of energy that travel from the source in a wave pattern. The wave pattern has a distinct frequency that determines its color.

Terms and Units of Measurement

The **optoelectronics** field is unusual in that it uses two distinct systems of measurement to evaluate the operation of an optoelectronic device or system: photometric and radiometric. The **photometric system** deals with visible light energy and applies specifically to the response of the human eye. The **radiometric system** deals with the entire optical spectrum, including ultraviolet, visible, and infrared light. Wavelengths in the radiometric spectrum range from 0.005 to 4000 μm. Visible light is only a small part of the total radiometric spectrum. The two systems are similar in many respects but cannot be used interchangeably. It is often difficult to convert units of measure from one system to the other.

Radiometric Systems

A beam of **radiant energy** has several measurable characteristics that are important in the optoelectronic field. Optical energy can be radiated through space in a wide range of patterns, can be emitted from a material, or can be reflected from a surface. The radiation process must be specified in order for measurements to be meaningful. Terms such as **flux, illumination, intensity**, and **luminance** are used. These terms apply to both radiometric and photometric measurements. Radiometric terms are distinguished from photometric by being preceded by the word *radiant.*

In the radiometric system, energy travels from an energy source in the form of electromagnetic waves. **Radiant flux**, or **radiant power**, is the flow rate of radiant energy per unit of time and is represented by the Greek letter phi (Φ). It is measured in joules per second, or watts.

The term **radiant intensity** is used to indicate the amount of electromagnetic energy produced by a specific source. Radiant intensity is a measure of the radiant flux per solid angle unit or watts per steradian. The solid angle of a sphere is shown in **Figure 13.3**. A **sphere** contains 47π steradians. A **steradian (sr)** is the unit of measure of a solid angle originating at the center of a sphere with a radius of 1 m subtended by 1 m^2 on the surface of the sphere. A steradian is generally considered to be a dimensionless unit like the **radian**. It is best expressed as a solid angular measurement. Keep in mind that this measurement has a light source at the center of the sphere, and it indicates the amount of intensity produced on 1 sr of outer surface.

Radiant incidence, or **irradiance**, is a measure of the radiant energy that strikes the surface of a specific area. It is measured in watts per unit area.

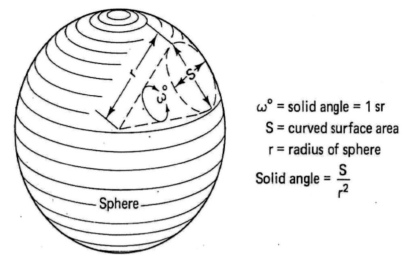

ω° = solid angle = 1 sr
S = curved surface area
r = radius of sphere

Solid angle = $\dfrac{S}{r^2}$

Figure 13.3 Solid angle of a sphere.

The unit area most commonly used is the square meter or centimeter. Radiant incidence is primarily used to describe the amount of radiation applied to the active surface of a detector. If the sensitivity of the detector is known, radiant incidence indicates if a particular signal will be detected.

Radiant exitance is a measure of the radiant power emitted or released by a specific surface. Radiant exitance is also called **radiant emittance** or, simply, emittance. It is measured in watts per square meter or square centimeter. Emittance is an indication of the optical power reflected from a surface. The reflectance of a surface varies with wavelength. For example, a radiant exitance of 3.2 mW/cm^2 results when a surface with 50% reflectance is irradiated with 6.4 mW/cm^2.

Radiance is a measure of the radiant intensity that is leaving, passing through, or arriving at a specific surface area. It is measured in watts per steradian meter squared and is determined by dividing the radiant energy intensity from a source by the projected area.

Photometric Systems

As mentioned previously, **photometric systems** deal with the measurement of electromagnetic energy that falls in the visible part of the frequency spectrum. Photometric terms are distinguished from radiometric terms by the prefix *luminous*. Luminous intensity, luminous flux, luminous exitance, and luminous power are some of the common photometric units of measurement.

One characteristic of light that is important in many applications is intensity. **Intensity** is a measure of the amount of energy contained in an electromagnetic wave. In the photometric system, the amount of light produced by a source is called its **luminous** *intensity*. The unit of luminous intensity is a standard light source called a **candela (cd)**. One candela is equal to one lumen per steradian (see **Figure 13.5**). The candela is an alternative measurement for the foot-candle measurement of light intensity.

Luminous flux is another important unit of the photometric system. Luminous flux is sometimes called *luminous power*. It specifically refers to the rate at which light energy flows per unit of time. The unit of luminous flux is the lumen (lm) and is denoted by the Greek letter phi (Φ). The **lumen** is the basic unit of measurement for the photometric system. The lumen is the SI unit that has replaced the candela to describe visible light energy.

Illuminance, also called **luminous incidence**, is the density of luminous power that it takes to illuminate a surface. The unit of illuminance is the lux. One lux is equal to 1 lm/m^2. Illuminance is frequently used to express the amount of light received by a solid-state device. When one lumen falls on a surface with an area of one square meter, the illuminance is one lux. Other units of illuminance are the lumen per square foot or foot-candle and the lumen per square centimeter or phot.

Luminous exitance is a measure of the amount of luminous flux given off or reflected by a surface. It is measured in lumens per square meter and is used to denote the reflecting capability of a specific surface. The reflectance of the surface varies with the wavelength of the illuminance. The unit of luminous exitance was once called **luminous emittance** or, simply, **emittance**.

Self-Examination

Answer the following questions.

1. Optoelectronics is concerned with the study of _____.
2. _____ is the process of emitting or releasing energy in the form of electromagnetic waves.
3. A graph or chart that shows the location of different electromagnetic wave frequencies and wavelengths is called a(n) _____.
4. A(n) _____ wave causes particles to move at right angles to the direction of motion.

5. Electromagnetic waves travel at _____ mi/s or _____ m/s.
6. One _____ is a measure of the distance that an electromagnetic wave travels in space during one cycle of operation.
7. Higher frequencies cause the wavelength to be _____.
8. Visible light ranges from a wavelength of _____ (Å) for violet to _____ (Å) for red.
9. The color _____ has the lowest frequency, whereas the color _____ has the highest frequency.
10. The color _____ has the longest wavelength, whereas the color _____ has the shortest wavelength.
11. The quantum theory is based on the emission of energy packets called _____.
12. The _____ system deals with the entire electromagnetic spectrum.
13. The _____ system deals with visible light.
14. Radiant _____ is measured in joules per second, or watts.
15. Radiant _____ is used to measure the amount of electromagnetic energy produced by a source.
16. Radiant _____ is a measure of the radiant energy applied to a surface.
17. Radiant _____ is a measure of the radiant power released from a surface.
18. Luminous _____ is a measure of the amount of visible light produced by a source.
19. The unit of luminous intensity is the _____.
20. The unit of luminous flux is the _____.
21. The unit of illuminance is the _____.

13.2 Radiation Sources

13.2 Evaluate the performance of a device that radiates light energy.
In order to achieve objective 13.2, you should be able to:

- interpret a data sheet for a device that detects light energy;
- describe the characteristics of incandescent lamps;
- describe the characteristics of light-emitting diodes;
- define *panchromatic source, heterochromatic source, monochromatic source, light-emitting diode*, and *infrared emitting diode*.

Photometric Source Classifications

Photometric sources are frequently classified by the amount or range of visible light emitted. In this regard, sources are described as being *panchromatic*, *heterochromatic*, or *monochromatic*. The term *chromatic* in these expressions refers to a specific color or various colors of the spectrum.

An incandescent lamp is an example of a **panchromatic source**. Its radiation extends over a very large portion of the optical spectrum. A wide range of visible colors can be produced by a panchromatic source.

A **heterochromatic source** produces a very limited number of different colors. Mercury arc lamps, for example, produce colors that are predominantly red or orange. Specifically, this particular source generates light in the region of 6500 Å.

A **monochromatic source** radiates energy of only one specific wavelength or in a very narrow part of the spectrum. A sodium vapor lamp, for example, radiates energy at 5000 Å.

Figure 13.4 shows the spectral response of several different light sources. Note the chromatic differences in these sources.

Incandescent Lamps

An **incandescent lamp** is frequently used as the radiation source of an optoelectronic system. This type of source is readily available, is reliable, and has a consistent level of operation. Tungsten filament lamps are used widely for

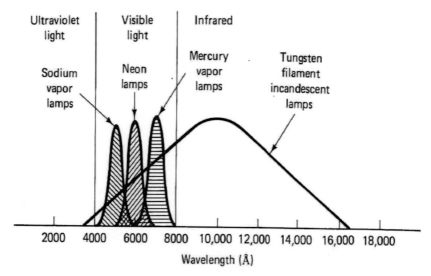

Figure 13.4   Spectral response of radiation sources.

this application. The radiation output of this source peaks at approximately 10,000 Å, as shown in **Figure 13.4**. It also radiates electromagnetic energy in the visible light region.

When incandescent lamps are used as a light source, they are generally operated at the lowest voltage value that is consistent with good performance. Low-voltage operation extends the life span of the lamp.

Lamp life in an optoelectronic system ranges from 5000 hours to something in excess of 50,000 hours, depending on the lamp and its operating voltage.

Operation of an incandescent lamp is based on the passage of current through an electrical conductor. When electrons pass through a conductor, they encounter a certain amount of resistance. This opposition to electron flow produces heat in the conductor. When enough heat is produced, the conductor changes its appearance. At low values of heat, there may be no change in the outward appearance of the material. As the temperature rises, the color of the material changes. It is possible to see a dull orange glow, bright orange, yellow, or a very intense white glow called **incandescence**. Tungsten lamps operating at 6332°F (3500°C) produce a very high-level light intensity. These lamps have a very short life expectancy. Operating the same lamp at a lower temperature increases its life expectancy. The wavelengths of light produced by a tungsten filament lamp are shown in the spectral response of **Figure 13.4**.

Light-Emitting Diodes

A semiconductor optoelectronic device that is used rather extensively as a light source is the **light-emitting diode (LED)** or **solid-state lamp (SST)**. The LED contains a P-N junction that emits light when forward biased. A schematic symbol, crystal structure, and typical package of an LED are shown in **Figure 13.5**. The housing or lens of an LED is transparent and focuses on the P-N junction.

When an LED is **forward biased**, electrons cross the junction and combine with holes, which cause them to fall out of conduction and return to the valence band. The energy possessed by each free electron is then released. Some of this appears as heat energy, and the rest of it is given off as light energy. Special materials such as gallium arsenide (GaAs), gallium phosphide (GaP), and gallium arsenide phosphide (GaAsP) cause this reaction when used in a P-N junction. These materials are purposely used because they permit various forms of radiant energy to be produced.

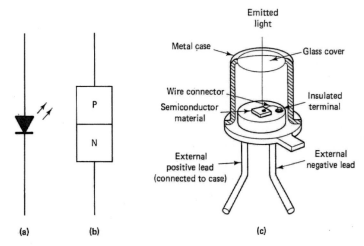

Figure 13.5   Light-emitting diode. (a) LED symbol. (b) LED crystal structure. (c) LED package.

When an LED is forward biased, the applied electrical energy causes the radiation of electromagnetic energy. The **spectral response** of an LED varies according to the type of semiconductor material used in its construction. Combinations of certain materials and varying doping levels are used to produce LEDs that emit many different wavelengths of energy and, consequently, different colors. Some of these combinations and resulting wavelengths are listed in the following table.

Material	Wavelength	Color
Gallium arsenide (GaAs)	8800 Å (880 nm)	Infrared
Gallium phosphide (GaP)	5500 Å (550 nm)	Green to red
Gallium arsenide phosphide (GaAsP)	5800–6600 Å (580–660 nm)	Amber to red

When GaAs is used in the construction of an LED, it emits infrared energy, which is not visible to the human eye. Devices of this type are called **infrared emitting diode (IRED)**. IREDs are used in systems that must not be influenced by ambient light or normal room light. If the environment where the device is being used contains traces of infrared energy, it will not function as intended.

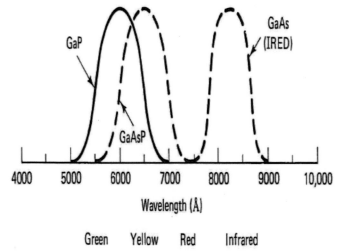

Figure 13.6 LED spectral response curves.

Spectral response curves for different LEDs are shown in **Figure 13.6**. Compare these curves and their peak response with the color response curve shown in **Figure 13.4**. This comparison shows that the **wavelengths** of radiated energy range from the visible to the infrared region of the frequency spectrum. Specific combinations of different phosphors are added to produce the desired color effects. LEDs are very sensitive to temperature changes, which cause the spectral response of the LED to shift to longer wavelengths. An increase in temperature also reduces the amount of radiant output power.

A unique feature of the LED as a light source is its rapid **switching time**. Compared with an incandescent lamp, an LED can be switched on and off rapidly. Switching rates of several thousand hertz can be achieved with no significant delay time. High-speed switching of an LED can be achieved with a square-wave generator or a trigger pulse. No heat is required by an LED to produce light. Radiation is produced only when the P-N junction is forward biased, and holes and electrons recombine at the surface of the junction.

LED Characteristics

Figure 13.7 shows the **I–V characteristics** for a GaAs LED. Note that the operational characteristics of an LED are very similar to those of a typical diode. The current increases very rapidly, with the voltage remaining nearly constant after forward conduction occurs.

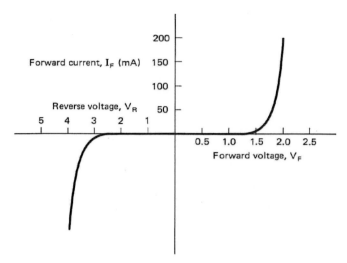

Figure 13.7   *I-V* graph of a GaAs LED.

There is a limit to the forward current and reverse voltage that an LED can withstand without being damaged. Most LEDs can withstand a **forward current** of 50–100 mA. Higher current values require a heat sink to prevent thermal damage to the junction. The **PRV rating** of an LED is usually quite low compared with a conventional silicon diode. Only a few volts can be tolerated without causing permanent damage to the junction. Typical PRV ratings are less than 50 V for an LED.

The light-output characteristic of an LED is an important operational consideration. **Figure 13.8** shows a graph of the output of a red GaAsP LED. This graph was made by pointing an LED into a calibrated photometer and applying current to the LED. The current was then monitored and recorded for each level of light intensity. Note that the light output is linear until approximately 65 mA is reached. Beyond 80 mA, the light output falls well below the peak value due to the development of excess heat across the junction. This particular LED is intended for low current applications. Its forward current rating is 50 mA. Operation is permitted beyond the upper current rating with reduced efficiency.

LED Applications

LEDs are used primarily as indicating devices for electronic circuits. When the LED is **forward biased**, it produces light. **Reverse biasing** does not

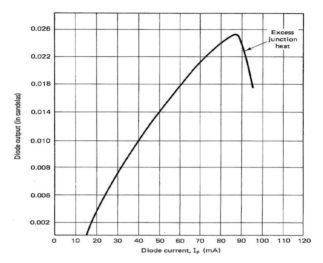

Figure 13.8 Light output versus forward current characteristic of a GaAsP LED.

produce light. In an actual circuit, very little current flows or light is produced until the forward-bias voltage is greater than the forward voltage (V_F) of the LED. Typical V_F values are in the range of $1-3$ V. In the reverse-bias direction, very little current flows and no light is produced.

Most LED applications require that the LED be protected from excessive forward current flow. This is achieved by connecting the LED in series with a **current-limiting resistor**. **Figure 13.9** shows an LED connected to a voltage source and a series resistor. The value of the series resistor can be calculated from Ohm's law. The forward voltage (V_F) of the LED is fairly constant when it goes into conduction. The voltage across the series resistor (V_{RS}) is the difference between the source voltage (V_{CC}) and the forward voltage (V_F) of the diode:

$$V_{RS} = V_{CC} - V_F. \tag{13.3}$$

For example, find the voltage across a series resistor if a circuit has a V_{CC} of 5 V and an LED with a forward voltage of 1.8 V.

$$V_{RS} = V_{CC} - V_F$$
$$= 5 - 1.8V$$
$$= 3.2V.$$

After V_{RS} is determined, a safe forward current (I_F) can be chosen that will produce a reasonable level of light. This current, which also flows

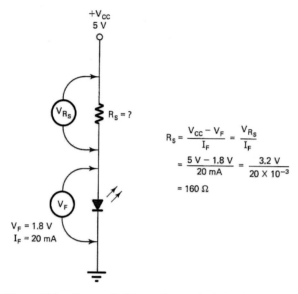

Figure 13.9   Current-limiting resistor calculation for an LED.

through R_S, is divided into V_{RS} to find the value of R_S, as expressed by the following formula:

$$R_S = V_{RS}/I_F. \tag{13.4}$$

Therefore, if the diode in our previous example has a safe forward current of 20 mA, the series resistor value would be calculated as follows:

$$R_S = V_{RS}/I_F$$
$$= 3.2 \text{ V}/20 \text{ mA}$$
$$= 160 \text{ } \Omega.$$

Example 13-3:

Find the value of series resistor needed for an LED circuit with $V_{CC} = 5$ V and forward voltage = 1.75 V and forward current = 12.5 mA.
Solution

$$V_{RS} = V_{CC} - V_F \qquad\qquad R_S = V_{RS}/I_F$$
$$= 5 - 1.75 \text{ V} \qquad\qquad = 3.25 \text{ V}/12.5 \text{ mA}$$
$$= 3.25 \text{ V} \qquad\qquad = 260\Omega.$$

Related Problem

Forward voltage of an LED is 2.2 V and I_F = 5 mA. If V_{CC} = 5 V, what value of series resistor is needed?

LEDs are commonly used in **seven-segment** and **5 × 7 dot-matrix displays**. The LEDs of these devices produce visible light when forward biased and no light when reverse biased. As a result of this two-state condition, discrete segments or dots can be illuminated when diodes are energized. Typically, the positive side of the energy source is applied to the anode of each diode through a current-limiting resistor. The cathode of the respective diode is then grounded by switching. When the circuit is complete, the diode is energized, thus producing light.

Seven-segment LED displays often contain four or more discrete diodes connected in parallel to form a segment. This type of construction usually requires only one current-limiting resistor for each segment. The amount of voltage needed to produce illumination is typically 3.5−5 V DC.

Figure 13.10 shows the circuitry of **seven-segment** and 5 × 7 **dot-matrix LED display** devices. The LEDs in both circuits are similar in all respects. The switching method needed to energize specific diodes is somewhat different. In the seven-segment device of **Figure 13.10(b)**, a single switch controls each segment. By comparison, two or more switches control the dot-matrix circuit of **Figure 13.10(d)**. A discrete diode can be energized by two or more switches, such as row 4, column 5. A complete vertical row would require one column switch and all seven row switches. A complete horizontal row would be energized by one row switch and all five column switches. As a general rule, dot-matrix display devices are used to produce letter displays more so than numbers. LED display devices are used more frequently in circuit applications than all other devices combined.

Another application combines an LED with other light-sensitive devices such as **photodiodes** and **phototransistors** to form an optocoupler, as shown in **Figure 13.11**. **Optocouplers**, which are sometimes referred to as **optoisolators**, provide a one-way transfer path for electrical signals.

Optocouplers are commonly manufactured in **dual-in-line packages** as shown in **Figure 13.11(a)**. Note the equivalent circuit of the optocoupler in **Figure 13.11(b)**. An LED or an IRED is combined with a transistor amplifier in the same package. The input of this device is the LED; the output is the transistor circuit. The collector current of the transistor varies in direct proportion to the current through the LED. Therefore, when an input signal is applied to the LED, it emits light in proportion to the applied

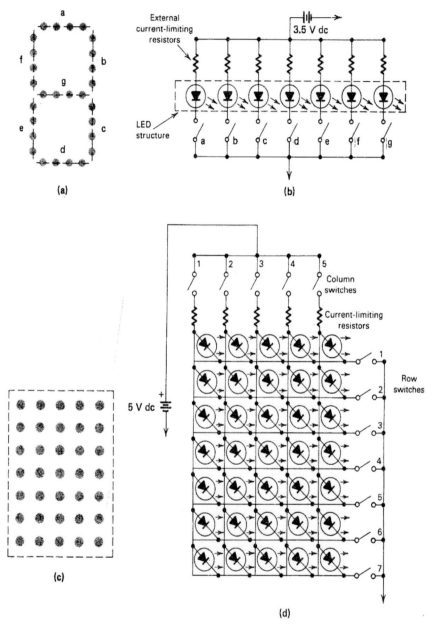

Figure 13.10   LED display devices. (a) Seven-segment LED display with four diodes per segment. (b) Diagram showing connections of segments. (c) LED 5 × 7 dot-matrix display. (d) LED dot-matrix diagram.

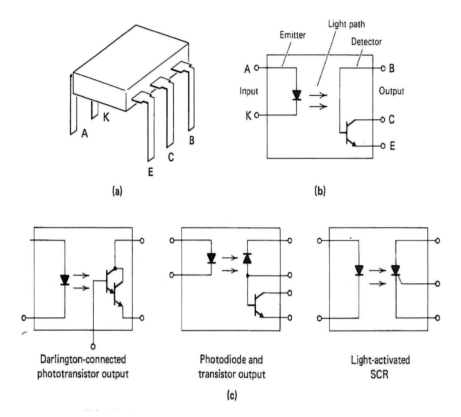

(a)   (b)

Darlington-connected   Photodiode and   Light-activated
phototransistor output   transistor output   SCR

(c)

Figure 13.11   Optocouplers. (a) Package. (b) Equivalent circuit. (c) Other configurations.

signal voltage. The light causes the transistor to conduct, thus increasing the collector current. This is achieved with no electrical connection between the devices that comprise the coupler. Isolation of the two circuits is, therefore, achieved.

Self-Examination

Answer the following questions.

22. The source of an optoelectronic system is responsible for generating some form of _____ energy.
23. An incandescent lamp is an example of a(n) _____ source.
24. A(n) _____ lamp produces light when current passes through the conductive material of the filament.

25. An LED produces light when _____ biased.
26. When GaAs, GaP, or GaAsP is used in the construction of an LED, the LED produces different _____ of radiant energy.
27. IREDs produce _____ energy.
28. A unique operational feature of a(n) _____ source of radiant energy is that it can be switched on and off very rapidly.
29. Most LEDs can withstand a forward current of _____ mA to _____ mA.
30. When an excessive amount of forward current passes through an LED, its output intensity is reduced because of _____.
31. When an LED is connected to an energy source, it must be protected from excessive current with a(n) _____ resistor.
32. A(n) _____ display has several LEDs formed into bars or segments.
33. A(n) _____ display has individual LEDs formed in a 5 × 7 matrix.
34. In an optocoupler, an LED serves as a(n) _____ of light energy.

13.3 Optoelectronic Detectors

The detector of an optoelectronic system is classified as a transducer because it is designed to change radiant energy into electrical energy. This is the opposite function than that of radiation sources. A number of detection devices are available. They are divided into three general categories: photoemissive, photoconductive, and photovoltaic. This section explains the characteristics of each of the three general categories and introduces their related devices.

13.3 Evaluate the performance of a device that detects light energy.
   In order to achieve objective 13.3, you should be able to:

- interpret a data sheet for a device that detects light energy;
- describe the characteristics of photoemissive, photoconductive, and photovoltaic radiation detectors;
- define *photoemissive devices, photoelectronic emission, photoconductive devices, photoresistive cells, dark current, photovoltaic device,* and *photovoltaic cell.*

Photoemissive Devices

A **photoemissive device** emits electrons in the presence of light. Vacuum and gaseous **phototubes** are typical photoemissive devices. The phototube shown in **Figure 13.12** is similar in appearance to a **vacuum tube** or **gaseous tube**. When light strikes the cathode of a phototube, photons of light are absorbed

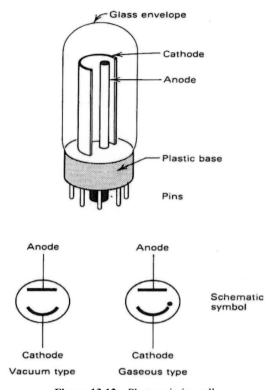

Figure 13.12   Photoemissive cell.

by the surface of the cathode. The absorption of this energy causes electrons on the surface of the cathode to gain enough energy to leave the cathode. This phenomenon is called **photoelectric emission**. The energy of the electrons depends on the wavelength of the light striking it. A spectral response curve is plotted by the manufacturer for each type of device produced. A representative spectral response curve is shown in **Figure 13.13**. Note that the **phototube** responds better to some ultraviolet and visible violet and blue wavelengths. The vacuum and gaseous phototube technology has largely been superseded by solid-state devices.

Example 13-4:

What is the approximate % response of the phototube of **Figure 13.16** for a wavelength of 600 nm?

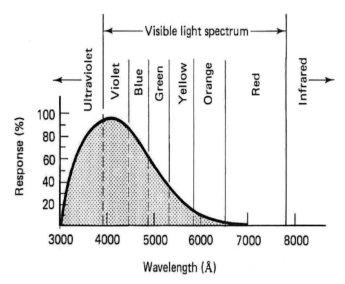

Figure 13.13   Spectral response curve for a phototube.

Solution

Locate 6000 Å on the horizontal axis of **Figure 13.13** (600 nm = 6000 Å). Project vertically to the response curve (point A) and across the vertical axis on the left. Response is approximately 10%.

Related Problem

What is the approximate response of the phototube of **Figure 13.13** for a wavelength of 400 nm?

A **phototube** may be connected in a circuit as shown in **Figure 13.14**. When light of the proper wavelength is focused on the cathode, electrons are emitted and travel to the positively charged anode or plate where they cause a plate current (I_P) to flow. This produces a voltage drop across the load (R_L). The plate current, caused by various combinations of light and plate voltage, may be determined by using a phototube characteristic curve supplied by the manufacturer. Such a curve is shown in **Figure 13.15**.

Example 13-5:

Refer to **Figure 13.15**. What is the approximate plate current (I_P) with a plate voltage of 100 V and light intensity of 0.04 lumens?

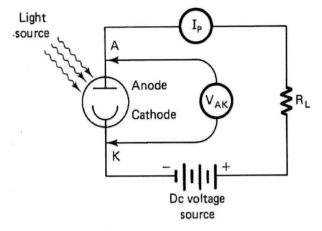

Figure 13.14 Phototube circuit action.

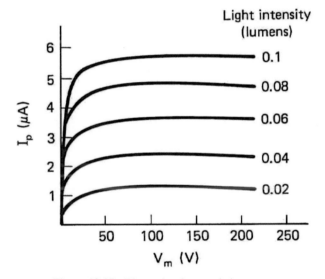

Figure 13.15 Phototube characteristic curves.

Solution

Locate 100 V on the horizontal axis. Project upward to the 0.04 lumens line (point A) and across the vertical plate current line. Plate current ~ 2.5 μA.

Related Problem

What is the approximate I_P at $V_P = 200$ V and light intensity $= 0.08$ lumens?

Photoemissive tubes may be of the vacuum type or filled with a gas. A **gas phototube** is usually more sensitive and, thus, requires less light to produce a given amount of anode current. A phototube circuit may be calibrated to measure specific values of light. This permits the tube to be used in instruments such as light-exposure meters.

At one time, nearly all **photoelectric control** was achieved by photoemissive tubes. The unusually high voltage needed to energize a phototube is rather uncharacteristic of most electronic devices. In general, this type of device has been replaced by solid-state photodiodes. As a rule, circuit control can be achieved more efficiently with lower voltage values. The **photoemissive principle** is important, however, because it is the basis of **photomultiplier tube** operation.

Most phototubes have a very small output when the light intensity is low. The **photomultiplier tube** shown in **Figure 13.16** overcomes this

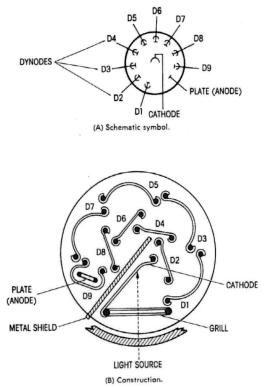

Figure 13.16   Multiplier phototube. (a) Schematic symbol. (b) Construction.

disadvantage. Note the internal construction of this tube. Light that strikes the cathode causes electrons to be emitted and focused toward positive plates called dynodes. Each **dynode** possesses a successively higher positive charge to attract these electrons. The anode has the highest positive potential. Through the principle of secondary emission, a large number of electrons are produced by each successive dynode.

Most multiplier phototubes use 9−14 **dynodes** for producing a high output at low light levels. Thus, a small amount of light can cause enough electrons to be emitted to produce a significant output. Photomultiplier tubes have a fairly high level of **signal multiplication**, which is extremely important in detecting low levels of light.

Photoconductive Devices

Photoconductive devices are designed so that when light intensity increases, their resistance decreases, and when light intensity decreases, their resistance increases. In effect, a photoconductive device changes light intensity to electrical conductivity. You should recall that conductivity is a measure of the ease with which current carriers pass through a material. **Photoresistive cells**, **photodiodes**, **PIN diodes**, **phototransistors**, and **light-activated SCRs** are common examples of photoconductive devices.

Photoresistive Cells

A **photoresistive cell**, or **light-dependent resistor (LDR)**, is essentially a semiconductor device. The words **resistive** and **resistors** are used in their terminology because of the relationship between conductance and resistance (conductance is the reciprocal of resistance). Photoresistive cells change their resistance level and thus their conductivity based on the response of its material to light energy.

Light-sensitive materials, such as cadmium sulfide (CdS), cadmium selenide (CdSe), and cadmium telluride (CdTe), are used in the construction of **photoresistive cells**. The material used in device construction determines how the device responds to different levels of light and to different wavelengths of radiant energy. CdS cells have a response very similar to that of the human eye. They respond best to yellow-green light, which has a wavelength of 5500 Å. CdSe cells are more sensitive to red, or 7000 Å, whereas CdTe is best suited for infrared light, or 8000 Å.

Cadmium sulfide cells are currently used in more applications than the other materials. This popularity is primarily related to its high sensitivity to

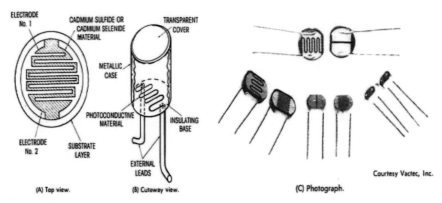

(A) Top view.   (B) Cutaway view.   (C) Photograph.

Courtesy Vactec, Inc.

Figure 13.17   Cadmium sulfide photoconductive cell. (a) Top view. (b) Cutaway view. (c) Photograph.

light. The resistance of a CdS crystal in the dark may be from 10,000 to 100,000 times greater than its resistance when exposed to an intense light. The sensitivity of the material is improved by increasing its surface area. Most CdS cells are constructed with a geometric pattern of the material on a glass substrate. **Figure 13.17** shows a pattern designed to have maximum surface area coverage. This wafer is then hermetically sealed in a glass or metal housing. A glass or plastic window covers the CdS pattern area. Light energy passing through the window causes the material to change resistance.

The operation of a **photoconductive cell** is based on the response of its material to light energy. When photons of light strike the light-sensitive material, they cause valence electrons to break their atomic bonding. These electrons are then free to take part in the conduction process. For each free electron produced, a corresponding hole is established in the covalent-bonded structure.

When conduction takes place, electrons move in the reverse direction. Conduction is based on the number of current carriers moving in the material. High levels of light intensity cause the material to be low resistant. A typical **photoconductive cell** will change resistance to 100 Ω when the light intensity is 100 lux, which is the equivalent of 9 foot-candles. Reduced light intensity causes an increase in material resistance. In total darkness, the resistance of the cell may be 100 kΩ. When the light intensity is reduced, electrons fall out of conduction and recombine with holes. The intensity of the light and its wavelength control the conductivity of the material. The total change in resistance due to a small change in light intensity is the unique advantage of this device.

The symbol for a CdS cell is shown in **Figure 13.18**. This **transducer** is represented by two different symbols. The symbol with arrows shows light energy being directed to the resistive material. The alternate symbol shows a resistor with the Greek letter lambda (λ), which is a common designation for **wavelength**. This symbol shows that the resistor responds to light wavelengths. When these devices are connected in a circuit as indicated, an increase in light intensity causes a reduction in resistance. This causes a corresponding increase in device current and a reduction in voltage across the device. In a sense, this type of device responds as a variable resistor that is controlled by light intensity.

An application of the photoconductive cell is shown in **Figure 13.19**. In this circuit, the SCR conducts when light is focused on the **photoresistive cell**. When light strikes the photoresistive cell, its resistance decreases. The potential at point A becomes more positive and causes gate current (I_G) to

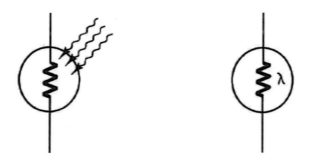

Figure 13.18 Schematic symbols of a CdS cell.

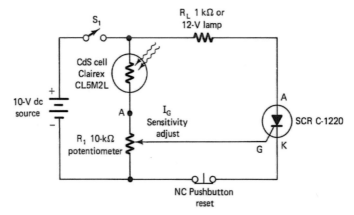

Figure 13.19 Photoconductive cell application.

flow. A sufficient amount of gate current **triggers** the SCR into conduction. When the SCR conducts, the load device is activated. The SCR then conducts until its anode circuit is opened. The variable resistor (R_1) is used as a sensitivity adjustment to control the level of light required to cause the SCR to conduct.

Photodiodes

Photodiodes are also classified as **photoconductive** devices. This type of device has light-sensitive P-type and N-type materials in its construction as shown in **Figure 13.20(a)**. Note that the two materials form a P-N junction. Normally, a photodiode is connected in the reverse-biased mode of operation. Without light, there is an extremely high reverse resistance. Any conduction current that flows is called **dark current**. As a rule, dark current is due to the thermal generation of current carriers. Only a few nanoamperes (nA) of dark current occur in a reverse-biased photodiode at low light levels.

 Photodiodes are energized in a circuit with reverse-biased DC. When a photodiode is illuminated, radiant energy causes the valence electrons of the P-type and N-type materials to go into conduction. In effect, this causes electrons and holes to be returned to the current carrier depletion region near the reverse-biased P-N junction. The return of current carriers increases conductivity and lowers junction resistance. A photodiode changes

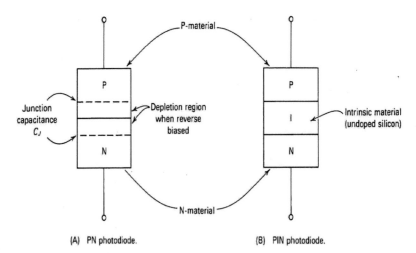

(A) PN photodiode.   (B) PIN photodiode.

Figure 13.20   Comparison of a photodiode and PIN photodiode. (a) Photodiode. (b) PIN photodiode.

its conductivity with light intensity only when the junction is reverse biased. Forward biasing the same P-N junction causes conductivity that does not change with light intensity. Photodiodes find use in **smoke detectors** where the amount of light reaching the photodiode is continuously monitored. If the light is blocked, which may happen due to smoke, the conductivity of the photodiode changes and this can be used to trigger an alarm. Photodiodes are also used in the receiver circuitry of a VCR or television that is remotely controlled. Other applications include **light meters** in digital cameras for indicating whether additional lighting is needed for taking a photograph. In the case of DVDs and CDs, photodiodes are used to sense the presence or absence of light which is reflected by the surface of the disk during its operation.

An application of the **photodiode** is shown in **Figure 13.19**. A photodiode is connected in series with a DC power source and a load resistor (R_L). The DC power source applies a reverse bias to the photodiode. When light intensity is increased, the photodiode becomes conductive. Increased current flow causes the voltage across R_L to increase.

PIN Diodes

Another type of the photodiode is the **PIN diode** shown in **Figure 13.20**. **Figure 13.20(a)** shows that a regular photodiode has a P-N junction in its construction. Light applied to the junction causes current carriers to return to the depletion region. This type of diode has a rather significant junction capacitance (C_j), which tends to slow down the response of the diode to changes in light intensity.

The **PIN diode**, as shown in **Figure 13.20(b)**, has a layer of undoped semiconductor material between the P-N junction materials. The letter *I* denotes this intrinsic, or undoped, semiconductor material. The added *I* layer physically increases the width of the depletion region of the P-N junction, which lowers the junction capacitance of the device. A PIN diode, therefore, has lower **junction capacitance** and a faster response to changes in light intensity. As a rule, this type of construction also reduces the dark current of the device to a much lower value.

PIN diodes may be used in communication systems as DC-controlled **microwave switches**, as a **modulation** device or as photodetectors in **fiber optic systems**. During high speed optical data transmission, it is important that detector circuitry responds only to the signal and not to ambient optical energy. PIN photodiodes with the intrinsic region limit the flow of thermally generated charge carriers which could affect the reception of the data.

Phototransistors

The **phototransistor** is a photoconductive device that has two P-N junctions in its construction, similar to that of a conventional bipolar transistor. The collector–base junction, therefore, responds as a photodiode. Light energy applied to this part of the device generates current carriers in the base region. This photocurrent controls the emitter–collector current. Essentially, a small change in base current is used to control a larger collector current. A phototransistor is, therefore, capable of **amplification**. This makes the phototransistor more sensitive to changes in light intensity than the photodiode.

A cross-sectional view of a **phototransistor** is shown in **Figure 13.21**. In this structure, light energy must be directed toward the base area. The entire assembly is housed in an enclosure that has a lens or window centered over the base region. Many devices of this type have the base lead omitted. A floating-base device has only emitter and collector lead connections. Base current is generated by light energy and controls current flow between the emitter and collector. With an external base lead, the phototransistor has additional control over the collector current. A three-lead phototransistor can also be used as a **photodiode** by disconnecting the emitter and using only the base–collector leads. **Figure 13.22** shows the schematic symbols of a **floating-base** and a three-lead NPN phototransistor.

A **phototransistor** equivalent circuit and characteristics curve for a three-lead device are shown in **Figure 13.23**. The collector–base junction is reverse

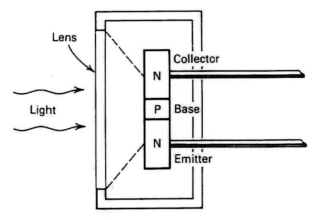

Figure 13.21   Cross-sectional view of a phototransistor.

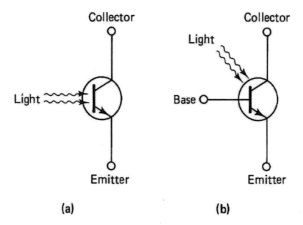

Figure 13.22 Schematic symbols of a phototransistor. (a) Floating base. (b) Three-lead.

biased. When light is focused on the collector−base junction, base current flows. This base current is amplified by the transistor. It is possible to bias the base to control the sensitivity of the device.

A **photo-Darlington transistor** is composed of two directly coupled phototransistors on a single structure. It is characterized as having high input impedance, low output impedance, and high sensitivity. The response time of a photo-Darlington transistor is slower than that of the phototransistor and photodiode.

Phototransistors, in general, are not very widely used as discrete-component control devices. As a rule, other optoelectronic devices can achieve equal and, in many cases, much better control than a single phototransistor. Phototransistors are, however, commonly used in **optocoupling devices**. This application permits the interfacing of an electrical output from a low power device to a high-current or high-voltage device along with offering with excellent isolation. In this case, the low power device may be used to switch on an **LED**. Light energy from the LED is directed to the base of a phototransistor which is on an independent circuit. The light falling on the base of the phototransistor causes it to go into conduction. This permits electrical signal transfer between devices without a direct electrical connection.

Light-Activated SCRs

Light-*activated silicon controlled rectifiers (LASCRs)* are SCRs that have been adapted for use with optical applications. Refer to **Figure 13.24**. These devices permit the flow of current in one direction when light of a certain

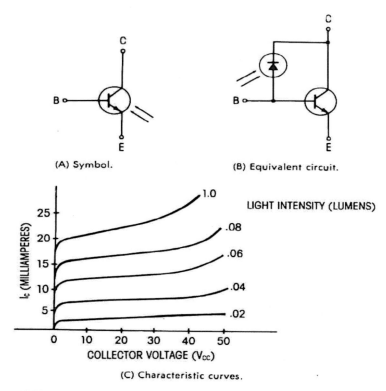

(A) Symbol.

(B) Equivalent circuit.

(C) Characteristic curves.

Figure 13.23   Phototransistor. (a) Symbol. (b) Equivalent circuit. (c) Characteristic curves.

intensity falls on the gate of the device, triggering the SCR. Once triggered into operation, the device is "**latched**" and will continue conduction, even if the light source is removed. The device can be taken out of conduction by decreasing the current flowing through the device below its holding value or by reverse biasing the LASCR. Many LASCRs have a gate terminal as well so that the device can be controlled not only with a light source but also with an electrical gate trigger circuit.

Figure 13.25 shows an **LASCR** circuit being used as a controlled recti-fier. In the positive alteration of the AC input, when light of sufficient intensity falls on the gate of an LASCR, it goes into conduction. It will continue to conduct for the rest of this alteration. During the negative AC input alteration, it will go out of conduction as the LASCR is reverse biased. As long as light falls on the gate of the device, the LASCR will continue to act as a controlled half-wave rectifier.

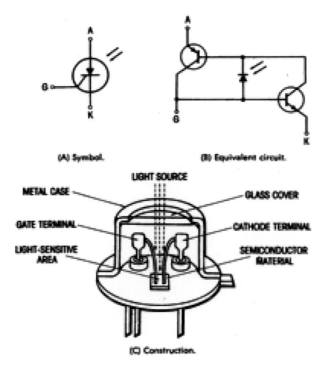

(A) Symbol.

(B) Equivalent circuit.

LIGHT SOURCE

METAL CASE

GLASS COVER

GATE TERMINAL

CATHODE TERMINAL

LIGHT-SENSITIVE
AREA

SEMICONDUCTOR
MATERIAL

(C) Construction.

Figure 13.24   Light-activated SCRs. (a) Symbol. (b) Equivalent circuit. (c) Construction.

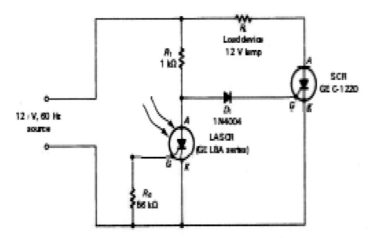

Figure 13.25   Light-activated SCR (LASCR) circuit.

Example 13-6:

In **Figure 13.25**, if the AC input source is replaced by a DC input source, how will the device respond when light is applied to the gate?

Solution

With DC of the proper polarity (anode connected to positive, cathode to negative) is applied to the LASCR, and the gate is triggered by light, the device will go into conduction. It will continue to conduct even if the light source is now removed. This indicates that the device is in a latched state of conduction.

Related Problem

If the polarity of the DC input source were to be reversed, how will the device respond when light is applied to the gate?

Photovoltaic Devices

Photovoltaic devices convert light energy into electrical energy. When a photovoltaic device is illuminated, the device creates an electrical potential. The **photovoltaic cell**, commonly called a **solar cell**, is used to convert light energy into electrical energy. Since this process represents a direct energy conversion, a great deal of research has been conducted in an attempt to convert large amounts of light energy into electrical energy. A common application of the solar cell is in photographic **light-exposure meters**. The electrical output of the solar cell is proportional to the amount of light falling onto its surface. The output of the solar cell is used to energize a light intensity meter.

The construction of a photovoltaic cell is shown in **Figure 13.26**. This **selenium cell** has a layer of selenium deposited on a metal base and then a layer of cadmium. In the fabrication, one layer of cadmium selenide (CdSe) and another layer of cadmium oxide (CdO) are produced. A transparent conductive film is placed over the CdO, and a section of conductive alloy is then placed on the film. External leads are connected to the conductive material around the CdO layer and the metal base. When light strikes the CdO layer, electrons are emitted and move toward the external load device. A deficiency of electrons is then created in this region, which is filled by electrons from the CdSe. Electrons from the metal contact then flow into the CdSe, causing the metal contact plate to become positive. Thus, light energy causes a difference in potential between the two external leads.

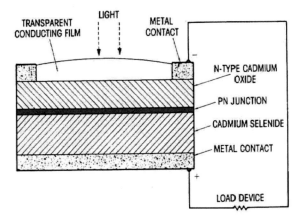

Figure 13.26 Selenium photovoltaic cell.

Selenium cells have a very low efficiency for converting light energy into electrical voltage. Typical efficiency ratings are less than 1%. A great deal of light and an extremely large active area are needed to produce electrical energy of any significant value. Selenium photovoltaic cells, however, have a spectral response that is very similar to that of the human eye. In many applications, the **spectral response** is much more significant than a high level of output efficiency.

Silicon photovoltaic cells are much more commonly used than the selenium units. Silicon cells have low efficiencies, typically in the 10%−20% range. The more common silicon photovoltaic cell is shown in **Figure 13.27**.

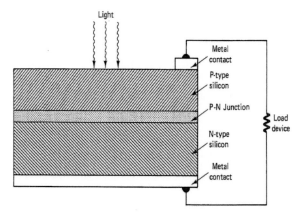

Figure 13.27 Silicon photovoltaic cell.

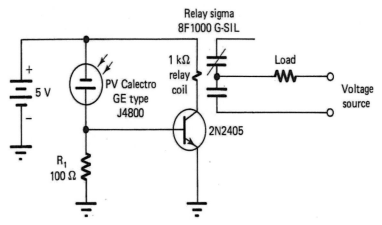

Figure 13.28   Photovoltaic cell relay control circuit.

When no light is focused onto the silicon cell, it operates similarly to a conventional P-N junction diode. When light strikes the cell, a voltage develops across the external leads. The more intense the light, the greater the potential difference across the cell.

Photovoltaic cells are used in a variety of applications. Although their electrical output is low, they may be used with amplifying devices to develop an output that drives a relay or some other load device. One such application is shown in **Figure 13.28**. In this circuit, the output of the photovoltaic cell is amplified by transistor Q_1 so that when light strikes the photovoltaic cell, the base current of the transistor increases. The increase in base current is amplified by Q_1, causing a collector current that is sufficient to activate the relay. The load connected across the relay terminals could be turned either off or on by the presence of light on the photovoltaic cell.

Since **photovoltaic cells** are a DC voltage source, they may be connected in either series or parallel. Remember from your study of DC circuits that cells connected in series increase voltage. Cells connected in parallel increase current-output capability.

Self-Examination

Answer the following questions.

35. An optoelectronic _____ is designed to change light energy into electrical energy.

36. Optoelectronic detectors are divided into three general categories: _____, _____, and _____.
37. When photons of light energy strike the _____ of a photoemissive tube, it emits electrons.
38. A(n) _____ device changes light intensity into electrical conductivity.
39. _____ is the ease with which current carriers pass through a material.
40. Cadmium sulfide (CdS) cells are _____ devices.
41. A high level of light intensity applied to a CdS cell will cause it to be _____ resistant.
42. A low level of light intensity applied to a CdS cell will cause it to be _____ resistant.
43. A photoresistive cell is classified as a(n) _____ device.
44. Photodiodes are classified as _____ devices.
45. For a photodiode to be responsive to changes in light intensity, it must be _____-biased.
46. Light energy applied to a phototransistor generates current carriers in the _____ region.
47. A solar cell is an example of a(n) _____ device.
48. The efficiency of a silicon photovoltaic cell is _____ than that of a selenium photovoltaic cell.
49. The output voltage of a photovoltaic cell circuit configuration increases when they are connected in _____.
50. The current capability of a photovoltaic cell circuit configuration increases when they are connected in _____.

13.4 Analysis and Troubleshooting – Optoelectronic Systems

Optoelectronic devices are commonly used to control electrical power and to isolate different components of a system. The **datasheets** of various optoelectronic devices will be examined in order to match its characteristics to a specific control application. The semiconductor components used in the construction of these devices must be sensitive to optical energy. Electrical properties of the device such as resistance and voltage change significantly with the application of light of a certain **wavelength** and intensity. By varying the optical energy applied to the device, its performance can be evaluated. These devices are generally rather fragile and can be physically damaged easily. These devices should not be used in an environment that is subjected to some form of particle contamination.

13.4 Troubleshoot an optoelectronic system.
   In order to achieve objective 13.4, you should be able to:

- examine datasheets for determining important operating conditions of various optoelectronic devices;
- distinguish between normal and faulty operation of various optoelectronic devices.

Data Sheet Analysis

The **specifications** of optoelectronics devices can be obtained from manufacturer datasheets. The circuit symbol, pin assignment, sample test circuits, and rated operational values are often readily available from the datasheet.
   The **datasheet** of a sample **infrared light-emitting diode, QEC121**, manufactured by Fairchild Semiconductor is included at the end of the chapter. Use this datasheet to answer the following questions:

1. The number of terminals = _____.
2. Sketch the schematic symbol for an LED = _____.
3. Wavelength of light emitted by LED = __.
4. What material is used for manufacturing this LED = __.
5. Maximum rated reverse voltage at 25 °C = _____.
6. Maximum rated power dissipation (P_D) = _____ mW.
7. Maximum rated continuous forward current = _____mA.
8. Emission angle = _____ř.

The *datasheet* **of** *light-dependent resistor*, **CL5M, manufactured by Clairex is included at** the end of the chapter. Use this datasheet to answer the following questions:

1. The number of terminals = _____.
2. Maximum power in air (no heat sink) = _____W.
3. Maximum power (with heat sink) = _____W.
4. Operating temperature range = _____°C to _____°C.
5. The material used for manufacturing the CL5M2L LDR = _____.
6. Minimum dark resistance = _____ Ω.
7. Resistance with 2 ft-c (foot-candle) light falling on the CL5M2L LDR = _____Ω.

The *datasheet* of a sample *photodiode, QSD2030*, manufactured by Fairchild Semiconductor is included at the end of the chapter. Use this datasheet to answer the following questions:

1. The number of terminals = _____.
2. Sketch the schematic symbol for a photodiode = _____.
3. Maximum rated reverse breakdown voltage (V_{BR})= _____V.
4. Maximum rated power dissipation (P_D) = _____ mW.
5. Wavelength sensitivity range = _____nm to _____nm.
6. Peak sensitivity wavelength = _____nm.

The *datasheet* of a sample *PIN photodiode*, QSB34GR, manufactured by Fairchild Semiconductor is included at the end of the chapter. Use this datasheet to answer the following questions:

1. The number of terminals = _____.
2. Sketch the schematic symbol for a PIN photodiode = _____.
3. Maximum rated reverse breakdown voltage (V_R) = _____ V.
4. Maximum rated power dissipation (P_C) at 25°C = _____ mW.
5. Spectral (wavelength) sensitivity range for the QSB34GR PIN photodiode = _____nm to _____nm.
6. Peak sensitivity wavelength = _____nm.

The **datasheet** of a sample **phototransistor, QSC112**, manufactured by Fairchild Semiconductor is included at the end of the chapter. Use this datasheet to answer the following questions:

1. The number of terminals = _____.
2. Sketch the schematic symbol for a phototransistor = _____.
3. Maximum rated collector−emitter voltage = _____ V.
4. Maximum rated power dissipation = _____ mW.
5. Peak sensitivity wavelength = _____ nm.
6. Collector−emitter dark current with a V_{CE} of 10 V = _____ nA.
7. On-state collector current flow for the QSC112 phototransistor, at V_{CE} of 5 V, minimum = _____mA; maximum = _____mA.

The *datasheet* of a sample *light activated silicon controlled rectifier, IS6051*, manufactured by Fairchild Semiconductor is included at the end of the chapter. Use this datasheet to answer the following questions:

1. The total number of pins on the device = _____.
   Pin numbers used for the LED (input diode) portion of the LASCR = _____.

   Pin numbers used for the SCR (detector) portion of the LASCR = _____.

2. Maximum forward current flow for the input diode = _____ mA.
3. Power dissipation of the input diode = _____ mW.

4. RMS on-state current of the detector = _____ mA.
5. Power dissipation of the detector = _____ mW.

The *datasheet* of a sample *photovoltaic cell, BPX79*, manufactured by Siemens is included at the end of the chapter. Use this datasheet to answer the following questions:

1. The number of terminals = _____.
2. Operating and storage temperature range = _____°C to _____°C.
3. Suitable optical wavelength range for applications = _____ nm to _____ nm.
4. Wavelength of maximum sensitivity = _____ nm.
5. Open-circuit voltage = _____ mV.
6. Applications of the BPX79 photovoltaic cell = _____.

The *datasheet* of a *reflective objective sensor, QRD1113*, manufactured by Fairchild Semiconductor is included at the end of the chapter. Use this datasheet to answer the following questions:

1. The total number of pins on the device = _____.
   Pin numbers used for the LED (emitter) = _____.
   Pin numbers used for the phototransistor (sensor) = _____.
2. Maximum continuous current flow of the LED (emitter) = _____ mA.
3. Maximum rated power dissipation (P_D) of the LED = _____ mW.
4. Peak emission wavelength of the LED = _____ nm.
5. Maximum rated collector–emitter (sensor) current = _____ V.
6. Maximum rated power dissipation (P_D) of the phototransistor = _____ mW.
7. Dark current with a V_{CE} of 10 V = _____ nA.
8. Referring to Figure 5 of the **QRD1113 datasheet**, determine the approximate distance in mils (1/1000th of an inch), when the normalized collector current is maximum = _____.

Summary

- Optoelectronic devices have electrical properties that are affected by light.
- The term *light* includes visible, infrared, and ultraviolet regions of the frequency spectrum.

- Light is a form of radiant energy.
- Radiant energy is made up of the entire optical spectrum, which includes ultraviolet, visible, and infrared light.
- One theory of radiant energy transmission considers light to travel as transverse waves that cause particles of a medium to vibrate at right angles to the direction of motion of the waves.
- Light travels at 186,000 mi/s, or 300,000,000 m/s.
- There is an inverse relationship between the length of a wave and its frequency.
- Wavelength is a measure of how far the wave travels during one cycle of operation.
- The quantum theory of light considers light to be emitted in discrete packets of energy called photons.
- The energy of a photon depends on the wavelength of the light.
- Two systems of measurement used by the optoelectronic field are radiometric and photometric.
- The radiometric system deals with the entire optical spectrum.
- The photometric system deals with electromagnetic energy that falls in the visible part of the electromagnetic spectrum.
- Photometric terms are preceded by the word luminous, and radiometric terms are preceded by the word radiant.
- The unit of luminous intensity is the candela.
- An optoelectronic system is distinguished from other systems by its source, which must generate some form of radiant energy, and its detector, which responds to or receives radiant energy.
- Most optoelectronic systems respond to radiant energy that falls in the visible part of the spectrum.
- The source of an optoelectronic system is usually classified according to the amount of visible light emitted; sources can be panchromatic, heterochromatic, or monochromatic.
- An incandescent lamp is an example of a panchromatic source, which generates light over a large part of the visible spectrum.
- A mercury arc lamp is an example of a heterochromatic source; this lamp radiates energy in a narrow part of the spectrum that is red to orange and peaks at 6500 Å.
- A sodium vapor lamp is an example of a monochromatic source; it generates light that is predominantly of one wavelength.
- Light-emitting diodes (LEDs) are solid-state sources that have a P-N junction that emits light when forward biased.

- The material used to make the P-N junction of an LED determines the wavelength of energy released.
- An LED must be connected in series with a current-limiting resistor to prevent excessive current from damaging the junction.
- The detector of an optoelectronic system is designed to change radiant energy into electrical energy.
- The three general categories of detectors are photoemissive, photoconductive, or photovoltaic.
- Photoemissive detectors give off or emit electrons when the light-sensitive material of the cathode absorbs photons of light energy.
- Photomultiplier tubes are photoemissive devices that respond to the secondary emission of several dynode plates, which increase the output of the device.
- Photoconductive detectors change light intensity into electrical conductivity.
- Cadmium sulfide (CdS) cells change resistance when exposed to light.
- Photodiodes are photoconductive devices that have a light-sensitive P-N junction.
- A photodiode is normally connected in the reverse-bias direction; when light is applied to the junction, current carriers return to the depletion region and cause conduction.
- A PIN photodiode has a layer of intrinsic semiconductor material between the P-type and N-type materials; this construction lowers the junction capacitance, which causes it to have a faster response time.
- Phototransistors are photoconductive devices that have two P-N junctions; the collector−base junction responds as a photodiode.
- A phototransistor has an amplification capability, which makes it more sensitive to changes in light intensity.
- Light-activated silicon controlled rectifiers (LASCRs) can be triggered into conduction when light energy is applied to the gate, and a voltage of proper polarity is applied to the anode and cathode terminals. Once triggered, it will stay latched until the current drops below the holding level.
- Photovoltaic devices are designed to change light energy directly into electrical energy.
- Selenium photovoltaic cells have an efficiency of 1%, whereas silicon devices have an efficiency of 15%.

Formulas

(13-1) $\lambda = 300{,}000{,}000/f$ Wavelength in meters.
(13-2) $e = hf$ Energy.
(13-3) $V_{RS} = V_{CC} - V_F$ Voltage value of the current-limiting resistor.
(13-4) $R_S = V_{RS}/I_F$ Resistance value of a current-limiting resistor.

Review Questions

Answer the following questions.

1. Discrete bundles of energy produced by a source of light are called:

   a. Ions
   b. Electrons
   c. Neutrons
   d. Photons

2. Light is best described as a form of:

   a. Work
   b. Energy
   c. Waves
   d. Force
   e. Heat

3. Radiation is described as:

   a. Energy transfer
   b. Energy conversion
   c. A form of work
   d. A frequency
   e. A series of waves

4. A photometric system deals with:

   a. The response of the human eye
   b. All electromagnetic radiation
   c. Only infrared energy
   d. Only ultraviolet energy
   e. A combination of infrared and ultraviolet energy

5. Radiometric systems deal with:

   a. Only visible light
   b. All electromagnetic energy

    c. Only AM, FM, and TV electromagnetic energy

    d. Only gamma rays, X-rays, and microwaves

    e. Only the invisible part of the electromagnetic spectrum

6. The human eye responds to wavelengths of electromagnetic energy that are in the range of:

    a. 300–3000 Å

    b. 3000–4000 Å

    c. 4000–7700 Å

    d. 7700–10,000 Å

    e. 10,000 Å and above

7. The sensitivity of the human eye peaks at the:

    a. Red part of the visual region of the spectrum

    b. Violet region

    c. Green area of the spectrum

    d. White part of the spectrum

    e. Black or absence of light

8. An LED source produces light when:

    a. Holes and electrons combine in the depletion region

    b. The P-N junction is reverse biased

    c. The depletion region becomes wider

    d. Electrons are emitted from the junction surface

    e. The P-N junction becomes hot

9. A photomultiplier tube responds to:

    a. The conductivity of current through a specific material

    b. The emission of electrons from a cathode

    c. The secondary emission of electrons from dynodes

    d. Thermionic emission

    e. Electrostatic excitation

10. Which of the following is not classified as a photoconductive device?

    a. A P-N photodiode

    b. A PIN photodiode

    c. A phototransistor

    d. A light-dependent resistor

    e. A photovoltaic cell

11. Phototransistors respond much like a regular transistor except that light energy:

    a. Acts as a switch to turn on the transistor
    b. Produces base current
    c. Alters the leakage current
    d. Changes the base voltage
    e. Alters the emitter current

12. Light striking a properly connected photodiode will cause an increase in:

    a. Forward current
    b. Resistance
    c. Reverse current
    d. Reverse voltage
    e. Forward voltage across the device

Problems

Answer the following questions.

1. An LED with a V_F of 2 V is to be operated from a 10-V source at 10 mA. In which range of resistance values will the current-limiting resistor be found?

    a. $1-100 \ \Omega$
    b. $101-300 \ \Omega$
    c. $301-500 \ \Omega$
    d. $501-700 \ \Omega$
    e. $701-900 \ \Omega$
    f. $901 \ \Omega$ or higher

2. Under normal operating conditions, a GaAs LED will have a forward voltage between:

    a. 0 and 1 V
    b. 1.1 and 3 V
    c. 3.1 and 5 V
    d. 5.1 and 7 V
    e. 7.1 and 9 V

3. An ohmmeter connected across an LED shows 100 kΩ of resistance in the forward direction and infinite resistance in the reverse direction. This indicates:

    a. That the device is shorted
    b. An open condition
    c. A faulty device
    d. A good device
    e. That the condition of the device is questionable

Answers

Examples

13-1. 3 m
13-2. 4000 Å
13-3. 560 Ω
13-4. 95% approx.
13-5. 4.75 μA approx.
13-6. When the LASCR is reverse biased, triggering it by a light signal applied to the gate will not cause it to go into conduction.

Self-Examination

13.1

    1. light
    2. Radiation
    3. electromagnetic spectrum
    4. transverse
    5. 186,000, 300,000,000
    6. wavelength
    7. shorter
    8. 4000, 7700
    9. red, violet
   10. red, violet
   11. photons
   12. radiometric
   13. photometric
   14. flux

15. intensity
16. incidence
17. exitance
18. intensity
19. candela, *or* cd
20. lumen
21. lux

13.2

22. radiant
23. panchromatic
24. incandescent
25. forward
26. wavelengths
27. infrared
28. LED
29. 50, 100
30. heat
31. current-limiting
32. seven-segment
33. dot-matrix
34. source

13.3

35. detector
36. photoemissive, photoconductive, photovoltaic (any order)
37. cathode
38. photoconductive
39. Conductivity
40. photoconductive
41. low
42. high
43. photoconductive
44. photoconductive
45. reverse
46. base
47. photovoltaic
48. higher
49. series
50. parallel

Terms

Electromagnetic spectrum

A graph that describes radiant energy by showing the location of different frequencies.

Wavelength

The distance that an electromagnetic or light wave travels in one cycle.

Transverse wave

A wave that causes the particles of a medium to vibrate at right angles to the direction in which the wave is moving.

Angstrom

A radiometric measuring unit used to describe the wavelength of an electromagnetic wave. An Angstrom is 10^{-10} m.

Quantum theory

A theory based on the absorption and emission of discrete amounts of energy.

Photometric system

A system of describing or quantifying light associated with the visible part of the frequency spectrum.

Radiometric system

A method of describing or quantifying electromagnetic energy that includes both visible and invisible radiations.

Radiant flux

The flow rate of radiant energy per unit of time. It is represented by the Greek letter phi (Φ) and is measured in joules per second, or watts.

Radiant intensity

The measure of the radiant power per solid angle unit, or watts per steradian.

Radiant incidence

A measure of the radiant energy that strikes the surface of a specific area. It is measured in watts per unit area.

Steradian (sr)

The unit of measure of a solid angle originating at the center of a sphere with a radius of 1 m subtended by 1 m^2 on the surface of the sphere.

Radiant exitance

A measure of the radiant power emitted or released by a specific surface. It is measured in watts per square meter or square centimeter.

Radiance

A measure of the radiant intensity that is leaving, passing through, or arriving at a specific surface area. It is measured in watts per steradian meter squared and is determined by dividing the radiant energy intensity from a source by the projected area.

Intensity

A measure of the amount of energy contained in an electromagnetic wave.

Luminous intensity

The amount of light produced by a photometric source.

Candela (cd)

The unit of luminous intensity equal to one lumen per steradian.

Luminous flux

The rate at which light energy flows per unit time. The unit of luminous flux is the lumen (lm) and is denoted by the Greek letter phi (Φ).

Lumen

The unit used to express the amount of light falling on a surface. One lumen falling on a surface area of one square meter producing an illuminance of one lux.

Lux

A unit of illuminance equal to one lumen per meter squared.

Illuminance

The density of luminous power that it takes to illuminate a surface. The unit of illuminance is the lux.

Luminous exitance

A measure of the amount of luminous flux given off or reflected by a surface. It is measured in lumens per square meter and is used to denote the reflecting capability of a specific surface.

Panchromatic source

A radiation source whose radiation extends over a very large portion of the optical spectrum and, thus, produces a wide range of visible colors.

Heterochromatic source

A radiation source that produces a very limited number of different colors.

Monochromatic source

A radiation source that radiates energy of only one specific wavelength or in a very narrow part of the spectrum.

Light-emitting diode (LED)

A semiconductor optoelectronic device that contains a P-N junction that emits light when forward biased.

Infrared emitting diode (IRED)

A type of LED that uses GaAs in its construction and emits infrared energy, which is not visible to the human eye.

Photoemissive device

A device that emits electrons in the presence of light.

Photoelectric emission

The phenomenon of electrons on the surface of the cathode gaining enough energy to leave the cathode due to the cathode absorbing photons of light.

Dynode

The positive plate of a phototube.

Photoconductive devices

A device designed so that its resistance decreases when light becomes more intense and increases when light intensity decreases.

Light-activated silicon controlled rectifiers (LASCRs)

A light-sensitive SCR that is triggered into conduction when light energy is applied to the gate. It must also have proper voltage values and polarity is applied to the anode and cathode terminals.

Photoresistive cell

A device that changes its resistance level and, thus, its conductivity based on its response to light energy.

Dark current

Any conduction current that flows in a photodiode when no light is applied.

Photovoltaic devices

A device that converts light energy into electrical energy. When light is applied to a photovoltaic device, the device creates an electrical potential.

Photovoltaic cell

It converts light energy into electrical energy.

SEMICONDUCTOR®

PLASTIC INFRARED
LIGHT EMITTING DIODE

QEC121   **QEC122**   **QEC123**

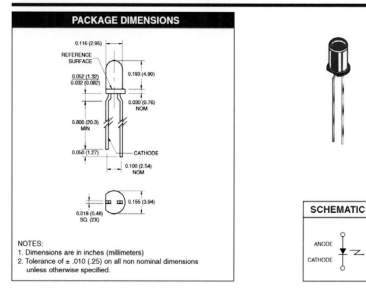

PACKAGE DIMENSIONS

NOTES:
1. Dimensions are in inches (millimeters)
2. Tolerance of ± .010 (.25) on all non nominal dimensions
   unless otherwise specified.

SCHEMATIC

ANODE

CATHODE

DESCRIPTION

The QEC12X is an 880 nm AlGaAs LED encapsulated in a clear purple tinted, plastic T-1 package.

FEATURES

- λ = 880 nm
- Chip material = AlGaAs
- Package type: T-1 (3mm lens diameter)
- Matched Photosensor: QSC112/113/114
- Narrow Emission Angle, 16°
- High Output Power
- Package material and color: Clear, purple tinted, plastic

PLASTIC INFRARED
LIGHT EMITTING DIODE

QEC121   QEC122   QEC123

ABSOLUTE MAXIMUM RATINGS (T_A = 25°C unless otherwise specified)

Parameter	Symbol	Rating	Unit
Operating Temperature	T_{OPR}	-40 to +100	°C
Storage Temperature	T_{STG}	-40 to +100	°C
Soldering Temperature (Iron)[2,3,4]	T_{SOL-I}	240 for 5 sec	°C
Soldering Temperature (Flow)[2,3]	T_{SOL-F}	260 for 10 sec	°C
Continuous Forward Current	I_F	50	mA
Reverse Voltage	V_R	5	V
Power Dissipation[1]	P_D	100	mW

NOTES
1. Derate power dissipation linearly 1.33 mW/°C above 25°C.
2. RMA flux is recommended.
3. Methanol or isopropyl alcohols are recommended as cleaning agents.
4. Soldering iron 1/16" (1.6mm) minimum from housing.

ELECTRICAL / OPTICAL CHARACTERISTICS (T_A = 25°C)

PARAMETER	TEST CONDITIONS	SYMBOL	MIN	TYP	MAX	UNITS
Peak Emission Wavelength	I_F = 100 mA	λ_{PE}	—	880	—	nm
Emission Angle	I_F = 100 mA	$2\theta_{1/2}$	—	16	—	Deg.
Forward Voltage	I_F = 100 mA, tp = 20 ms	V_F	—	—	1.7	V
Reverse Current	V_R = 5 V	I_R	—	—	10	µA
Radiant Intensity QEC121	I_F = 100 mA, tp = 20 ms	I_E	14	—	—	mW/sr
Radiant Intensity QEC122	I_F = 100 mA, tp = 20 ms	I_E	27	—	94	mW/sr
Radiant Intensity QEC123	I_F = 100 mA, tp = 20 ms	I_E	39	—	—	mW/sr
Rise Time	I_F = 100 mA	t_r	—	800	—	ns
Fall Time	I_F = 100 mA	t_f	—	800	—	ns

PLASTIC INFRARED
LIGHT EMITTING DIODE

QEC121   **QEC122**   **QEC123**

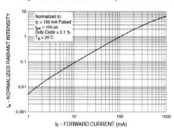

Fig.1 Normalized Radiant Intensity vs. Forward Current

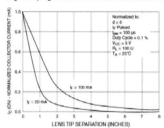

Fig.2 Coupling Characteristics of QEC12X And QSC11X

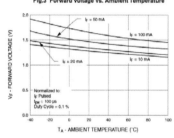

Fig.3 Forward Voltage vs. Ambient Temperature

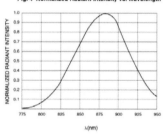

Fig. 4 Normalized Radiant Intensity vs. Wavelength

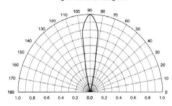

Fig. 5 Radiation Diagram

CL5M SERIES
HERMETICALLY SEALED TO-8 CASE

- 9 Types
- Maximum Power — Air: ½ Watt @ 25° C
- Maximum Power — Heat Sink: 2 Watts @ 25° C Case
- Resistance Tolerance at 2 ft-c: ± 33-1/3%
- 6 Photoconductive Materials
- Temperature Range: − 50° C to + 75° C

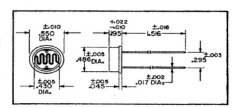

TYPE	Sensitive Material	Peak Spectral Response (Angstroms)	Resistance @ 2ft-c (Ohms)	Min. Dark Resistance 5 sec. After 2 ft-c	Maximum Voltage Rating (Peak A.C.)	Measurement Voltage
CL5M2	Type 2 CdS	5150	55K	3.6 Meg	250V	10V
CL5M2L			9K	600K	170V	12V
CL5M3	Type 3 CdSe	7350	7.2K	48 Meg	250V	12V
CL5M4	Type 4 CdSe	6900	1.5K	400K	250V	10V
CL5M4L			.25K	67K	170V	1.35V
CL5M5	Type 5 CdS	5500	9K	600K	250V	12V
CL5M5L			1.5K	100K	170V	10V
CL5M7	Type 7 CdS	6150	7.2K	1.4 Meg	250V	12V

QSD2030
PLASTIC SILICON PHOTODIODE

PACKAGE DIMENSIONS

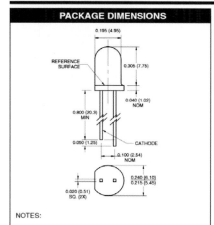

NOTES:

1. Dimensions for all drawings are in inches (mm).
2. Tolerance of ± .010 (.25) on all non-nominal dimensions unless otherwise specified.

FEATURES

- PIN Photodiode
- Package type: T-1 3/4 (5mm lens diameter)
- Wide Reception Angle, 40°
- Package material and color: Clear epoxy
- High Sensitivity
- Peak Sensitivity λ = 880 nm

1. Derate power dissipation linearly 1.33 mW/°C above 25°C.
2. RMA flux is recommended.
3. Methanol or isopropyl alcohols are recommended as cleaning agents.
4. Soldering iron 1/16" (1.6mm) minimum from housing.

SCHEMATIC

ABSOLUTE MAXIMUM RATINGS (T$_A$ = 25°C unless otherwise specified)

Parameter	Symbol	Rating	Unit
Operating Temperature	T$_{OPR}$	-40 to +100	°C
Storage Temperature	T$_{STG}$	-40 to +100	°C
Soldering Temperature (Iron)[2,3,4]	T$_{SOL-I}$	240 for 5 sec	°C
Soldering Temperature (Flow)[2,3]	T$_{SOL-F}$	260 for 10 sec	°C
Reverse Breakdown Voltage	V$_{BR}$	50	V
Power Dissipation[1]	P$_D$	100	mW

ELECTRICAL / OPTICAL CHARACTERISTICS (T$_A$ =25°C)

PARAMETER	TEST CONDITIONS	SYMBOL	MIN	TYP	MAX	UNITS
Peak Sensitivity Wavelength		λ$_{PS}$	—	880	—	nm
Wavelength Sensitivity Range		λ$_{SR}$	400	—	1100	nm
Reception Angle		Θ	—	±20	—	Deg.
Forward Voltage	I$_F$ = 80 mA	V$_F$	—	1.3	—	V
Reverse Dark Current	V$_R$ = 20 V, E$_e$ = 0	I$_D$	—	—	5	nA
Reverse Light Current	E$_e$ = 0.5 mW/cm², V$_R$ = 5 V, λ = 950 nm	I$_L$	15	25		μA
Capacitance	V$_R$ = 0, f = 1 MHz, E$_e$ = 0	C	—	60	—	pF
Rise Time	V$_R$ = 5 V, R$_L$ = 50 Ω	t$_r$	—	5	—	ns
Fall Time	λ = 950 nm	t$_f$	—	5	—	ns

SURFACE MOUNT SILICON PIN PHOTODIODE
QSB34

QSB34GR, QSB34ZR

PACKAGE DIMENSIONS, QSB34GR

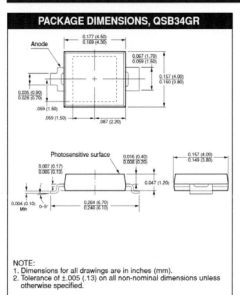

NOTE:
1. Dimensions for all drawings are in inches (mm).
2. Tolerance of ±.005 (.13) on all non-nominal dimensions unless otherwise specified.

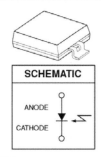

SCHEMATIC

ANODE

CATHODE

FEATURES

- Daylight Filter
- Surface Mount Packages:
  - QSB34GR for overmount board
  - QSB34ZR for undermount board
- Fast PIN Photodiode
- Wide Reception Angle, 120°
- Large Chip Size = .014 in^2 (9 mm^2)
- High Sensitivity
- Low Capacitance
- Available in 0.470" (12mm) width tape on 7" (178mm) diameter reel; 1,000 units per reel

ABSOLUTE MAXIMUM RATINGS (T_A = 25°C unless otherwise specified)

Parameter	Symbol	Rating	Unit
Operating Temperature	T_{OPR}	-40 to +85	°C
Storage Temperature	T_{STG}	-40 to +85	°C
Soldering Temperature (Reflow)[2,3]	T_{SOL-F}	240 for 5 sec	°C
Reverse Voltage	V_R	32	V
Power Dissipation[1]	P_D	150	mW

Notes:
1. Derate power dissipation linearly 2.50 mW/°C above 25°C.
2. Solder iron (15W max temp 260°C for 5 max sec.)
3. Methanol or isopropyl alcohols are recommended as cleaning agents.
4. Light source is an GaAs LED which has a peak emission wavelength of 940 nm.

SURFACE MOUNT SILICON PIN PHOTODIODE
QSB34

QSB34GR, QSB34ZR

ELECTRICAL / OPTICAL CHARACTERISTICS (T_A =25°C unless otherwise specified)

Parameter	Test Conditions	Symbol	Min	Typ	Max	Units
Reverse Voltage	I_R = 0.1 mA	V_R	32		—	V
Dark Reverse Current	V_R = 10 V	$I_{R(D)}$	—		30	nA
Peak Sensitivity	V_R = 5 V	λ_{PK}		940		nm
Reception Angle @ 1/2 Power		Θ		±60		Degrees
Photo Current	Ee = 1.0 mW/cm^2, V_{CE} = 5 V[4]	I_{PH}	25	37	—	µA
Capacitance	V_R = 3 V	C		25		pF
Rise Time	V_R = 10 V, R_L = 50 Ω	t_r		20		ns
Fall Time		t_f		20		ns
Spectral Sensitivity		S_λ		.40		A/W

TYPICAL PERFORMANCE CURVES

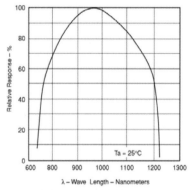

Fig. 1 Relative Spectral Sensitivity vs. Wavelength

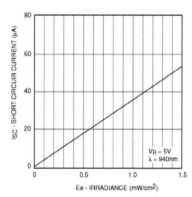

Fig. 2 Short Circuit Current vs. Irradiance

PLASTIC SILICON INFRARED PHOTOTRANSISTOR

QSC112   QSC113   QSC114

PACKAGE DIMENSIONS

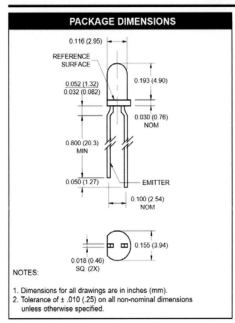

0.116 (2.95)

REFERENCE SURFACE

0.052 (1.32)
0.032 (0.082)

0.193 (4.90)

0.030 (0.76) NOM

0.800 (20.3) MIN

0.050 (1.27)   EMITTER

0.100 (2.54) NOM

0.018 (0.46) SQ. (2X)

0.155 (3.94)

NOTES:

1. Dimensions for all drawings are in inches (mm).
2. Tolerance of ± .010 (.25) on all non-nominal dimensions unless otherwise specified.

SCHEMATIC

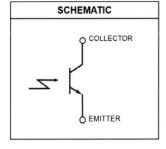

COLLECTOR

EMITTER

DESCRIPTION

The QSC112/113/114 is a silicon phototransistor encapsulated in an infrared transparent, black T-1 package.

FEATURES

• Tight production distribution.
• Steel lead frames for improved reliability in solder mounting.
• Good optical-to-mechanical alignment.
• Plastic package is infrared transparent black to attenuate visible light.
• Mechanically and spectrally matched to the QECXXX LED.
• Black plastic body allows easy recognition from LED.

PLASTIC SILICON INFRARED
PHOTOTRANSISTOR

QSC112   QSC113   QSC114

ABSOLUTE MAXIMUM RATINGS (T_A = 25°C unless otherwise specified)

Parameter	Symbol	Rating	Unit
Operating Temperature	T_{OPR}	-40 to +100	°C
Storage Temperature	T_{STG}	-40 to +100	°C
Soldering Temperature (Iron)[2,3,4]	T_{SOL-I}	240 for 5 sec	°C
Soldering Temperature (Flow)[2,3]	T_{SOL-F}	260 for 10 sec	°C
Collector-Emitter Voltage	V_{CE}	30	V
Emitter-Collector Voltage	V_{EC}	5	V
Power Dissipation[1]	P_D	100	mW

1. Derate power dissipation linearly 1.33 mW/°C above 25°C.
2. RMA flux is recommended.
3. Methanol or isopropyl alcohols are recommended as cleaning agents.
4. Soldering iron 1/16" (1.6mm) minimum from housing.
5. λ = 880 nm, AlGaAs.

ELECTRICAL / OPTICAL CHARACTERISTICS (T_A = 25°C)

PARAMETER	TEST CONDITIONS	SYMBOL	MIN	TYP	MAX	UNITS
Peak Sensitivity Wavelength		λ_{PS}	—	880	—	nm
Reception Angle		Θ	—	±8	—	Deg.
Collector-Emitter Dark Current	V_{CE} = 10 V, Ee = 0	I_{CEO}	—	—	100	nA
Collector-Emitter Breakdown	I_C = 1 mA	BV_{CEO}	30	—	—	V
Emitter-Collector Breakdown	I_E = 100 µA	BV_{ECO}	5	—	—	V
On-State On-State Collector QSC112	Ee = 0.5 mW/cm^2, V_{CE} = 5 V[5]	$I_{C(ON)}$	1	—	4	mA
On-State On-State Collector QSC113			2.40	—	9.60	
On-State On-State Collector QSC114			4.00	—	—	
Saturation Voltage	Ee = 0.5 mW/cm^2, I_C = 0.5 mA[5]	$V_{CE(sat)}$	—	—	0.4	V
Rise Time	V_{CC} = 5 V, R_L = 100 Ω	t_r	—	5.0	—	µs
Fall Time	I_C = 2 mA	t_f	—	5.0	—	

IS6051

LOW INPUT CURRENT INFRA-RED EMITTING DIODE & LIGHT ACTIVATED SCR

APPROVALS
- UL recognised, File No. E91231

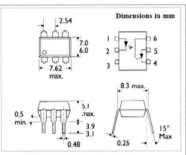

Dimensions in mm

DESCRIPTION
The IS6051 is an optically coupled isolator consisting of infrared light emitting diode and a light activated silicon controlled rectifier in a standard 6pin dual in line plastic package.

FEATURES
- Options :-
  10mm lead spread - add G after part no.
  Surface mount - add SM after part no.
  Tape&reel - add SMT&R after part no.
- High Isolation Voltage (5.3kV$_{RMS}$, 7.5kV$_{PK}$)
- High Surge Anode Current (5.0 A)
- High Blocking Voltage (400V*1)
- Low Turn on Current (5mA typical)
- All electrical parameters 100% tested
- Custom electrical selections available

APPLICATIONS
- 10A, T²L compatible, Solid State Relay
- 25W Logic Indicator Lamp Driver
- 400V Symmetrical transistor coupler

ABSOLUTE MAXIMUM RATINGS
(25°C unless otherwise specified)

Storage Temperature	-55°C to + 150°C
Operating Temperature	-55°C to + 100°C
Lead Soldering Temperature	
(1/16 inch (1.6mm) from case for 10 secs)	260°C

INPUT DIODE

Forward Current	60mA
Forward Current (Peak)	
(1<s pulse, 300pps)	3A
Reverse Voltage	6V
Power Dissipation	100mW

DETECTOR

Peak Forward Voltage	
IS605	400V*1
Peak Reverse Gate Voltage	6V
RMS On-state Current	300mA
Peak On-state Current	
(100<s, 1% duty cycle)	10A
Surge Current (10ms)	5A
Power Dissipation	300mW

*1 IMPORTANT : A resistor must be connected between gate and cathode (pins 4 & 6) to prevent false firing (R$_{GK}$ < 56kΩ)

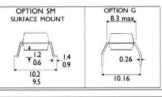

OPTION SM SURFACE MOUNT	OPTION G

ELECTRICAL CHARACTERISTICS (T_A= 25°C Unless otherwise noted)

	PARAMETER	MIN	TYP	MAX	UNITS	TEST CONDITION
Input	Forward Voltage (V_F)		1.2	1.5	V	$I_F = 10mA$
	Reverse Voltage (V_R)	3			V	$I_R = 10\propto A$
Output (note 2)	Peak Off-state Voltage (V_{DM})	400			V	$R_{GK}=10k\Omega, I_D=2\propto A$
	Peak Reverse Voltage (V_{RM})	400			V	$R_{GK}=10k\Omega, I_D=2\propto A$
	On-state Voltage (V_{TM})		1.1	1.3	V	$I_{TM}=300mA$
	Off-state Current (I_{DM})			2	$\propto A$	$R_{GK}=10k\Omega, I_F=0,$ $V_{DM}=400V$
	Reverse Current (I_R)			2	$\propto A$	$R_{GK}=10k\Omega, I_F=0,$ $V_{DM}=400V$
Coupled	Input Current to Trigger (I_{FT}) (note 2)			3	mA	$V_{AK}=100V, R_{GK}=27k\Omega$
	Turn on Time (t_{on})			50	$\propto S$	$R_{GK}=27k\Omega, I_F=30mA,$ $V_{AK}=20V, R_L=200\Omega$
	Coupled dv/dt, Input to Output (dv/dt)	500			V/$\propto$S	
	Input to Output Isolation Voltage V_{ISO}	5300			V_{RMS}	See note 1
		7500			V_{PK}	See note 1
	Input-output Isolation Resistance R_{ISO}	10^{11}			Ω	$V_{IO} = 500V$ (note 1)
	Input-output Capacitance Cf			2	pF	$V = 0, f = 1MHz$

SIEMENS

Silizium-Fotoelement mit erhöhter Blauempfindlichkeit
Silicon Photovoltaic Cell with Enhanced Blue Sensitivity

BPX 79

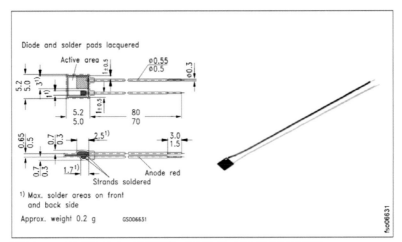

Maße in mm, wenn nicht anders angegeben/Dimensions in mm, unless otherwise specified.

Wesentliche Merkmale

- Speziell geeignet für Anwendungen im Bereich von 350 nm bis 1100 nm
- Kathode = Chipunterseite
- Mit feuchtigkeitsabweisender Schutzschicht überzogen
- Weiter Temperaturbereich

Anwendungen

- für Meβ-, Steuer- und Regelzwecke
- zur Abtastung von Lichtimpulsen
- quantitative Lichtmessung im sichtbaren Licht- und nahen Infrarotbereich

Features

- Especially suitable for applications from 350 nm to 1100 nm
- Cathode = back contact
- Coated with a humidity-proof protective layer
- Wide temperature range

Applications

- For control and drive circuits
- Light pulse scanning
- Quantitative light measurements in the visible light and near infrared range

SIEMENS

BPX 79

Grenzwerte
Maximum Ratings

Bezeichnung Description	Symbol Symbol	Wert Value	Einheit Unit
Betriebs- und Lagertemperatur Operating and storage temperature range	T_{op}; T_{stg}	– 55 ... + 100	°C
Sperrspannung Reverse voltage	V_R	1	V

Kennwerte (T_A = 25°C, Normlicht A, T = 2856 K)
Characteristics (T_A = 25 °C, standard light A, T = 2856 K)

Bezeichnung Description	Symbol Symbol	Wert Value	Einheit Unit
Fotoempfindlichkeit, V_R = 0 V Spectral sensitivity	S	170	nA/lx
Wellenlänge der max. Fotoempfindlichkeit Wavelength of max. sensitivity	$\lambda_{S\,max}$	800	nm
Spektraler Bereich der Fotoempfindlichkeit S = 10 % von S_{max} Spectral range of sensitivity S = 10 % of S_{max}	λ	350 ... 1100	nm
Bestrahlungsempfindliche Fläche Radiant sensitive area	A	20	mm²
Abmessungen der bestrahlungsempfindlichen Fläche Dimensions of radiant sensitive area	$L \cdot B$ $L \cdot W$	4.47 · 4.47	mm
Halbwinkel Half angle	ϕ	± 60	Grad deg.
Dunkelstrom, V_R = 1 V; E = 0 Dark current	I_R	0.3 (≤ 50)	∝A
Spektrale Fotoempfindlichkeit, λ = 400 nm Spectral sensitivity	S_λ	0.19	A/W
Quantenausbeute, λ = 400 nm Quantum yield	η	0.60	Electrons Photon
Leerlaufspannung, E_v = 1000 lx Open-circuit voltage	V_O	450	mV
Kurzschlußstrom Short-circuit current E_e = 0.5 mW/cm², λ = 400 nm	I_{SC}	19 (≥ 14)	∝A

SIEMENS

Kennwerte (T_A = 25°C, Normlicht A, T = 2856 K)
Characteristics (T_A = 25 °C, standard light A, T = 2856 K)

Bezeichnung Description	Symbol Symbol	Wert Value	Einheit Unit
Anstiegs und Abfallzeit des Fotostromes Rise and fall time of the photocurrent R_L = 1 kΩ; V_R = 1 V; λ = 850 nm; I_p = 150 ∝A	t_r, t_f	6	∝s
Temperaturkoeffizient von V_O Temperature coefficient of V_O	TC_V	– 2.6	mV/K
Temperaturkoeffizient von I_{SC} Temperature coefficient of I_{SC}	TC_I	0.2	%/K
Kapazität, V_R = 10 V, f = 1 MHz, E_v = 0 lx Capacitance	C_0	2500	pF

January 2008

QRD1113, QRD1114
Reflective Object Sensor

Features

- Phototransistor Output
- No contact surface sensing
- Unfocused for sensing diffused surfaces
- Compact Package
- Daylight filter on sensor

Description

The QRD1113/14 reflective sensor consists of an infra-red emitting diode and an NPN silicon phototransistor mounted side by side in a black plastic housing. The on-axis radiation of the emitter and the on-axis response of the detector are both perpendicular to the face of the QRD1113/14. The phototransistor responds to radiation emitted from the diode only when a reflective object or surface is in the field of view of the detector.

Package Dimensions

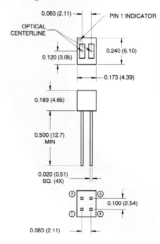

Schematic

PIN 1 COLLECTOR   PIN 3 ANODE
PIN 2 EMITTER   PIN 4 CATHODE

Notes:
1. Dimensions for all drawings are in inches (millimeters).
2. Tolerance of ± .010 (.25) on all non-nominal dimensions unless otherwise specified.
3. Pins 2 and 4 typically .050" shorter than pins 1 and 3.
4. Dimensions controlled at housing surface.

Absolute Maximum Ratings (T_A = 25°C unless otherwise specified)

Stresses exceeding the absolute maximum ratings may damage the device. The device may not function or be operable above the recommended operating conditions and stressing the parts to these levels is not recommended. In addition, extended exposure to stresses above the recommended operating conditions may affect device reliability. The absolute maximum ratings are stress ratings only.

Symbol	Parameter	Rating	Units
T_{OPR}	Operating Temperature	-40 to +85	°C
T_{STG}	Storage Temperature	-40 to +100	°C
T_{SOL-I}	Lead Temperature (Solder Iron)[2,3]	240 for 5 sec	°C
T_{SOL-F}	Lead Temperature (Solder Flow)[2,3]	260 for 10 sec	°C
EMITTER			
I_F	Continuous Forward Current	50	mA
V_R	Reverse Voltage	5	V
P_D	Power Dissipation[1]	100	mW
SENSOR			
V_{CEO}	Collector-Emitter Voltage	30	V
V_{ECO}	Emitter-Collector Voltage		V
P_D	Power Dissipation[1]	100	mW

Electrical/Optical Characteristics (T_A = 25°C)

Symbol	Parameter	Test Conditions	Min.	Typ.	Max.	Units
INPUT (Emitter)						
V_F	Forward Voltage	I_F = 20mA			1.7	V
I_R	Reverse Leakage Current	V_R = 5V			100	µA
λ_{PE}	Peak Emission Wavelength	I_F = 20mA		940		nm
OUTPUT (Sensor)						
BV_{CEO}	Collector-Emitter Breakdown	I_C = 1mA	30			V
BV_{ECO}	Emitter-Collector Breakdown	I_E = 0.1mA	5			V
I_D	Dark Current	V_{CE} = 10 V, I_F = 0mA			100	nA
COUPLED						
$I_{C(ON)}$	QRD1113 Collector Current	I_F = 20mA, V_{CE} = 5V, D = .050"[6,8]	0.300			mA
$I_{C(ON)}$	QRD1114 Collector Current	I_F = 20mA, V_{CE} = 5V, D = .050"[6,8]	1			mA
$V_{CE(SAT)}$	Collector Emitter Saturation Voltage	I_F = 40mA, I_C = 100µA, D = .050"[6,8]			0.4	V
I_{CX}	Cross Talk	I_F = 20mA, V_{CE} = 5V, E_E = 0[7]		.200	10	µA
t_r	Rise Time	V_{CE} = 5V, R_L = 100Ω, $I_{C(ON)}$ = 5mA		10		µs
t_f	Fall Time			50		µs

Notes:
1. Derate power dissipation linearly 1.33mW/°C above 25°C.
2. RMA flux is recommended.
3. Methanol or isopropyl alcohols are recommended as cleaning agents.
4. Soldering iron tip 1/16" (1.6 mm) minimum from housing.
5. As long as leads are not under any stress or spring tension.
6. D is the distance from the sensor face to the reflective surface.
7. Crosstalk (I_{CK}) is the collector current measured with the indicated current on the input diode and with no reflective surface.
8. Measured using Eastman Kodak neutral white test card with 90% diffused reflecting as a reflecting surface.

Typical Performance Curves

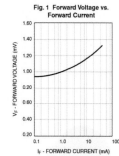

Fig. 1 Forward Voltage vs. Forward Current

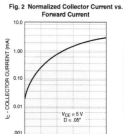

Fig. 2 Normalized Collector Current vs. Forward Current

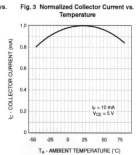

Fig. 3 Normalized Collector Current vs. Temperature

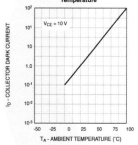

Fig. 4 Normalized Collector Dark Current vs. Temperature

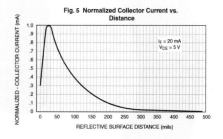

Fig. 5 Normalized Collector Current vs. Distance

14

Integrated Circuits

The expression **integrated circuit**, or **IC**, is a generic term used to describe a group of small electronic components that are constructed and permanently interconnected on or in a piece of semiconductor material called a *substrate*. Integrated circuits (ICs) are actually micro-miniature circuits. Each may contain resistors, conductors, semiconductors, capacitors, and, in some cases, inductors in a single package. The components are classified as either active or passive. **Active** components are semiconductor devices such as transistors and diodes. **Passive** components are resistors, capacitors, and inductors. These components perform the same job that they do in a conventional circuit; however, they do not have the same general appearance when housed in an IC.

Two general classifications of ICs are linear and digital. **Linear ICs** are used to achieve amplification, regulation, and voltage comparisons. **Digital ICs** contain switching circuitry of which input and output voltages are limited to two possible states: high and low. The output of a digital IC is related to the input in some logical way. Digital ICs include logic gates, flip-flops, counters, clock chips, calculators, memory, and microprocessors. Since this textbook provides coverage of analog electronics, linear ICs will be introduced in this chapter, as well as **IC construction** and the advantages and disadvantages of using ICs. A discussion of digital ICs is reserved for a course on digital electronics. The information presented in this chapter serves as a foundation for understanding for applications such as op-amps, communication circuits, and voltage regulators.

Objectives

After studying this chapter, you will be able to:

14.1 explain the construction differences between various families of integrated circuits;

14.2 describe the characteristics of linear ICs;
14.3 based on the advantages and disadvantages of ICs, select to use discrete components or an IC for a given application;
14.4 follow basic procedures to troubleshoot an IC.

Chapter Outline

14.1 IC Construction
14.2 Linear ICs
14.3 Advantages and Disadvantages of ICs

Key Terms

bipolar IC
bonding pads
diffusion
evaporation process
film IC
metallization
monolithic
MOS IC
thick-film IC
thin-film IC
wafer
yield

14.1 IC Construction

The technology used in the construction of ICs is not particularly new to the solid-state field. You have already learned about things such as substrates, epitaxial growth, and diffusion in chapters dealing with diodes and transistors. IC technology is an extension of the manufacturing techniques used in bipolar transistors and FETs. In fact, these two devices are the construction basis of many ICs. This technology is continually improving, which permits extremely large numbers of components to be fabricated on a single chip. The technology is very basic, but the manufacturing techniques used to achieve a specific chip may be held in strict secrecy. As a result, we will only look at some of the basics of IC fabrication.

14.1 Explain the construction differences between various families of integrated circuits.

In order to achieve objective 14.1, you should be able to:

- describe the construction techniques used to develop an integrated circuit;
- recognize basic types of IC packages;
- define *wafer, yield, monolithic, evaporation process, diffusion, bipolar IC, metallization, bonding pads, MOS IC, film IC, thin-film IC,* and *thick-film IC.*

General Construction

The making of an IC usually starts with a piece of circular silicon called a **wafer**. The physical size of the wafer is 1, 2, 4, or 6 in. in diameter with a thickness of 0.006 in. The wafer serves as the **substrate** of the IC. Only a small portion of the wafer is needed for one chip. Many ICs are formed simultaneously on one wafer, as shown in **Figure 14.1**. The ICs are independent chips but of the same type. After individual ICs are formed, the wafer is cut into small squares or **chips**. Each chip in this case is a complete IC. It may contain 50 or more interconnected components. Individual chips are then placed in a **package** or housing.

The number of useable IC chips developed from a production run is called the **yield**. IC yield is usually expressed as a percentage of the total possible number of chips that can be developed in a production run. The

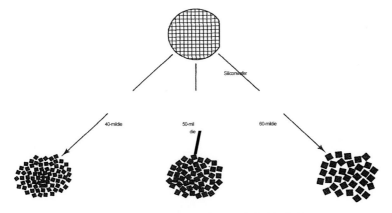

Figure 14.1   Cutting a wafer into individual ICs.

yield of a complex IC may be as low as 25%, which means that 75% of the chips in a particular run may be faulty or may not meet specifications. These chips are discarded, and the materials reclaimed. Detailed differences in the manufacturing process depend on whether an IC is monolithic or thin film.

Monolithic ICs

The term **monolithic** describes the structure of an IC in which elements of the circuit are built in a single crystal of semiconductor material. "Mono" refers to something of a single structure, and "lithic" refers to a printing process called **lithography**. Monolithic, therefore, refers to the silicon-diffused construction procedure in which all components are grown or formed on the surface of a silicon wafer. All contacts or connections to various components are made on the same surface. Interconnections between different components of the chip are made by depositing a metal wiring pattern on the silicon dioxide covered surface of the wafer.

The shapes of individual components of a **monolithic IC** are designed so that they can all be formed simultaneously. This includes transistors, diodes, resistors, and capacitors. Specific elements of a complete circuit can, therefore, be diffused on a single wafer by using the same processes.

A monolithic IC is made by evaporating and depositing different materials on a piece of silicon. Have you ever taken a hot shower and noticed that a layer of water has evaporated and then condensed on a mirror or glass door? This process is similar to the evaporation process used by IC manufacturers to deposit the vapors of such materials as chromium, nickel, or silicon dioxide on a substrate. The **evaporation process** develops a thin layer of N-type silicon on the surface of a P-type silicon substrate. **Figure 14.2** shows an example of the evaporation process. This process was described as epitaxial growth in a previous chapter – *Bipolar Junction Transistors*. **Epitaxial growth** and **evaporation** achieve the same thing in IC construction.

After an N-type layer of material is formed on a P-type **substrate**, the wafer is placed in an oven that has a controlled atmosphere. This causes the N-type material to diffuse or be absorbed by the P-type silicon substrate. **Diffusion** can be compared with the absorbing of water by a dry sponge or paper towel. The towel is obviously many times larger than the silicon chip. The completed chip can easily fit into the lowercase *o* of this printed page.

The amount of heat used and the length of time that the chip is exposed to the heat determine the depth of diffusion. The area of **diffusion** is determined by the area of the N-type or P-type material deposited on the substrate.

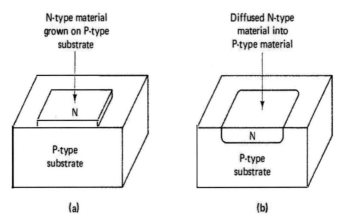

Figure 14.2 Evaporation, or epitaxial growth, process. (a) N-type material grown on P substrate. (b) Sectional view of diffused area.

The addition of the N-type material to the P-type substrate material when completed is shown in **Figure 14.2(b)**.

Component construction of a **monolithic IC** is used to determine the classification or grouping of a particular chip. A **bipolar IC** uses the same technology that is used in the construction of a bipolar transistor. The bipolar transistor is the fundamental component of this family of ICs.

N-type and P-type materials can be evaporated or grown on certain areas of a diffused N-type material to make a bipolar transistor. **Figure 14.3(a)** shows an NPN bipolar transistor with the emitter, base, and collector leads exposed. Note the different levels of diffusion needed to form the transistor. Each **deposition layer** and diffusion operation requires special processing.

Diodes are made by diffusing N-type impurities into a portion of the diffused P-type material. By adding enough N-type impurities into the P-type material, the N-type impurities can dominate and produce an N-type semiconductor in a specific area. **Figure 14.3** shows the crystal formation of a diode on a substrate. Metallized leads are attached to the respective material areas to achieve the diode structure. These may be connected to outside terminals or interconnected with other components on the chip.

Resistors may be made, for example, from doped P-type material having a certain internal resistance as shown in **Figure 14.4**. The area of the P-type material and the doping level of the material determines its resistance. If a small resistance value is required, a relatively large number of impurity atoms are diffused into the substrate. Conversely, larger resistances require fewer

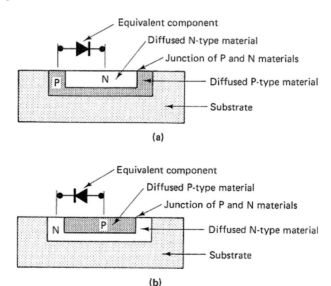

Figure 14.3   P–N junction (diode) formed in substrate. (a) Diffusing N impurities into P-type material. (b) Diffusing P impurities into N-type material.

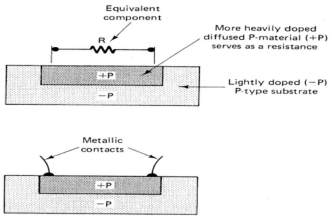

Figure 14.4   Resistor formation by diffusion of semiconductor material.

impurities. **Figure 14.4** also shows a P-type material resistor constructed on a **substrate**. The metallic contacts attached to the P-type material are deposited by a special process called **metallization**. In general, the metal must not be permitted to diffuse into the N-type or P-type material. It is used only to make an electrical connection or to interconnect components.

Capacitors are made by using a layer of silicon dioxide as the dielectric material. A diffused N-layer serves as one plate, and a metallized layer covering the layer serves as the other plate. **Figure 14.5** shows a representation of an IC capacitor. The area of the material and the thickness of the silicon dioxide layer determine the value of the capacitance.

Leads are connected to the IC by a very high temperature solder. A sectional view of a typical IC is shown in **Figure 14.6**. The leads are attached at points called **bonding pads**. In the illustration, these connection points are labeled "evaporated **jumper connections**." Note the location of the capacitor, the transistor, and resistors in this sectional view. The oxide layer serves as an insulating material and a dielectric for the capacitor. This **chip** drawing shows only a few necessary components needed to make a complete IC.

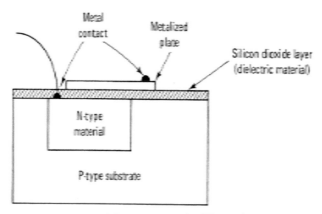

Figure 14.5   Representative IC capacitor.

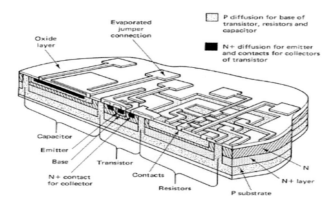

Figure 14.6   Sectional view of an integrated circuit.

MOS ICs

The **monolithic IC** just discussed uses bipolar transistors in its construction. Bipolar ICs represent an important division of the IC market. A second major division of monolithic ICs is the **metal-oxide semiconductor field-effect transistor (MOSFET) IC**, or MOS IC. The **MOS IC** is used in many applications. It is less expensive than the bipolar devices because it requires only one diffusion step to form the source and drain. A bipolar chip requires three and sometimes four diffusion steps to form a transistor. The MOSFET chip does not require isolation between the source−drain and other P–N junctions formed on the substrate. **Figure 14.7** shows where two P-channel enhancement-MOSFETs are formed on the same substrate.

The **package density** of a MOS IC is very high compared with that of a bipolar IC. This means that more devices can be fabricated in a given area of the chip. As a rule, component density of a MOS IC is 10 or more times that of the bipolar IC. **Isolation** between the elements is one reason for the increased component density. MOS resistors occupy less than 1% of the space needed for a conventional diffused resistor. This high-density construction capability makes the MOS IC an extremely important device in complex ICs.

MOS ICs use three basic components in their construction: MOSFETs, MOS capacitors, and MOS resistors. A **resistor** is formed by simply using an E-MOSFET with its gate connected to the drain so that the device is biased in the on-state. The **channel** of the device then serves as a resistor whose value is controlled by its size and doping level.

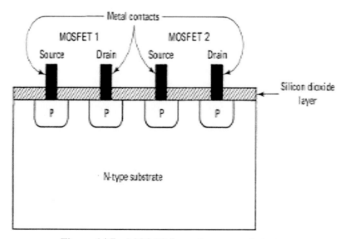

Figure 14.7   MOS IC formation on a substrate.

The principal disadvantage of a MOSFET IC is its **operating speed**. This is because the MOS transistor is a high-impedance device, and it cannot charge the stray circuit capacitance very quickly. The operating speed problem applies primarily to only ultra-high-speed applications. The advantages of low-cost, low-power consumption, and high-density packaging make the MOS IC a very attractive device for **very large scale integration (VLSI)** chips. VLSI chips contain over one thousand components on a chip. These chips are used in calculators, memory, and microprocessors.

Film ICs

A **film IC** is an extension of the earlier discrete-component electronic circuits. These ICs are an assembly of components formed on the surface of an insulating substrate. Conductors and resistors are formed by evaporation and by screen-printed deposition areas. Transistors, diodes, and MOS capacitors are added to the circuit in discrete form. These components are attached to the circuit with special cement. Film ICs are classified as hybrid devices. Hybrid refers to something that is the outgrowth of two different types or styles. In this case, a **hybrid device** is used to denote an IC that employs monolithic and film construction in its fabrication. This grouping of ICs is either of the thin-film or thick-film type.

A **thin-film IC** is so named because conductors, resistors, capacitors, transistors, and diodes are formed as very thin films on a substrate. The film is in the order of a thousand Angstroms thick. You should recall from the previous chapter that an Angstrom is 10^{-10}, or timeHour13Minute10one ten-billionth, of a meter. The thin films are deposited on substrates of glass or aluminum through a vapor-deposition process. **Resistors** are formed by depositing nichrome, tantalum, or tin oxide in thin strips on a glass substrate. Resistance values can be effectively controlled by varying the length, width, or thickness of the deposited material. Values can be made from 100 Ω to 1 MΩ. **Conductors** are made from a thin layer of deposited gold-nichrome. Transistors, diodes, and capacitors are generally added to the circuit in chip form. These components are secured to the substrate by a conductive epoxy. A thin-film IC structure is shown in **Figure 14.8**.

A **thick-film IC** is a hybrid IC that is formed on a film in the order of a mil or 0.001 in. by printing or silk screening conductor patterns onto the substrate. The printed material is a mixture of pulverized glass and aluminum. This type of construction can make a wide range of resistor values and conductor path designs. Active semiconductor components are attached to the structure

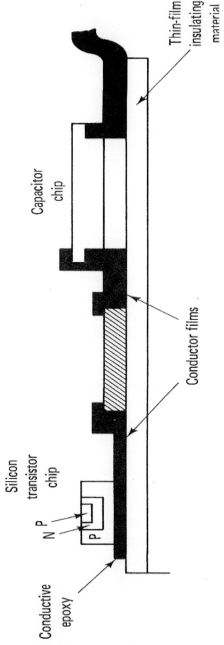

Figure 14.8   Thin-film IC construction.

in discrete form. These components are attached to the chip by conductive **epoxy**. Thick-film ICs are similar in many respects to thin-film ICs.

IC Packages

The outside appearance of an IC is a major concern to the user. In this regard, there are a number of different **package types** and styles used to house ICs. Each package has unique advantages and, in some cases, disadvantages over the other types. **Figure 14.9** shows some of the packages used for ICs. The first form of housing is the **TO-5 package** (see **Figure 14.9(a)**). This package is similar to that used to house transistors. Wherever a small size is important, a package known as the **flat pack** is used (see **Figure 14.9(b)**). This package is 0.26-in. long, 0.15-in. wide, and 0.05-in. thick. The assembled unit is hermetically sealed with metal to glass seals for each lead. The leads of this device are soldered into the circuit.

A popular form of IC housing is the **dual in-line package (DIP)**, which has a row of leads on each side of the package (see **Figure 14.9(c)**). Some common types of ICs are housed in DIPs that have 8, 14, 16, 24, 48, and 64 pins. The leads are on 0.1-in. centers to match mounting holes in sockets or circuit boards. A 14-pin DIP is 0.8-in. long, 0.25-in. wide, and 0.2-in. thick. The molded plastic housing is suitable for most circuit applications. Ceramic units are available for applications where temperature is a problem.

Other types of packaging have come about with the need for more pin connections and reduced size ICs. Names such as **small-outline integrated circuits (SOIC)**, **plastic leaded chip carrier (PLCC)**, **ceramic leaded chip**

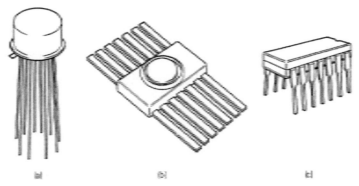

(a) (b) (c)

Figure 14.9 Common IC packages. (a) TO-5 housing. (b) Flat pack. (c) Dual in-line package.

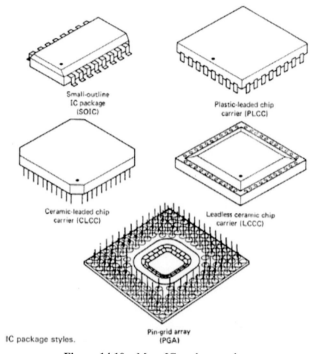

Figure 14.10   More IC package styles.

carrier **(CLCC)**, **leadless ceramic chip carrier (LCCC)**, and **pin-grid array (PGA)** are used to describe these packages. These housing techniques permit ICs to be packaged in smaller units and have more connection leads. **Figure 14.10** shows some of these IC package styles. As a rule, these packages can house more pin connections because they are placed on all sides of the device. This tends to reduce the size of the package so that it more closely conforms to the actual size of the chip. DIP packages, for example, are approximately 50 times larger than the chip they house. The new packages are smaller and can accommodate more leads.

Self-Examination

Answer the following questions.

1. ICs are built on a(n) _____.
2. ICs are actually _____ circuits built on a tiny chip.
3. Construction of an IC starts with a circular piece of silicon called a(n) _____.

4. The small square of material that houses an IC is called a(n) _____.
5. The percentage of good chips developed from a production run is called _____.
6. The construction technique that has components formed on a single piece of silicon is called _____.
7. Through a process of _____, a thin layer of material can be formed on a substrate.
8. _____ is the process in which an evaporated or deposited layer of material is absorbed into another piece of material.
9. Metal contacts formed on deposited materials are achieved by a process called _____.
10. Monolithic ICs that use bipolar transistors in their construction are called _____ ICs.
11. Monolithic ICs that use MOSFETs as an element in their construction are called _____ ICs.
12. Film ICs have components assembled on the surface of a(n) _____ substrate.
13. The most widely used package for housing ICs is the _____.
14. _____ ICs use monolithic and film construction in their fabrication.
15. The component density of a(n) _____ IC is greater than that of a bipolar IC.
16. The value of a(n) _____ in a MOS IC is determined by length, area, and material of a channel.

14.2 Linear ICs

Linear integrated circuits (ICs) are unique electronic components. Linear ICs develop an output signal that is proportional to its input signal. They are commonly used to achieve amplification, regulation, and voltage comparisons. Devices used for these purposes include the op-amp, IC regulator, and 555 timer. These devices are introduced in this section and covered in detail in later chapters.

14.2 Describe the characteristics of linear ICs.

    In order to achieve objective 14.2, you should be able to:

- list the basic functions of linear ICs;
- name some common linear ICs.

Linear amplification refers to the capability of an amplifier to operate in the linear or straight-line portion of the device's operating range. In discrete-component amplifiers, linear amplification is achieved by selecting an **operating point** near the center of the device's dynamic transfer characteristic curve. A bias operating point is selected near the center of the input characteristic line. A signal is then applied to the input and increased in amplitude at the output. The output of an amplifier must resemble the input with the only change being the magnitude of the signal. Linear amplification is widely used in electronic communication circuits and home entertainment equipment.

Linear ICs are essentially self-contained amplifiers placed in an appropriate housing. These chips are usually housed in standard IC packages. Some modification of the DIP is used when a device is used for **power amplification**. The internal structure of the chip is generally quite small and rather complex. The physical size of the chip varies a great deal for different types of ICs. A triangle symbol is used to denote the amplification function. The point of the triangle is the output. The flat side with two leads is the input. A plus sign on the input lead denotes the noninverting input. The negative sign indicates the inverting input. All the components of the amplifier are located inside the triangle. Some chips have several complete amplifiers housed in the same enclosure.

The **LM741C** integrated circuit, with specifications, schematic, and connection diagrams, as shown in **Figure 14.11**, is one type of **operational amplifier**. The LM741C has 20 transistors, 11 resistors, and a capacitor in its package. All of this is housed on a chip that is approximately 1/16 sq. in. Note the **inverting input**, **noninverting input**, and **output** points in the circuit diagram. Also note the pin-outs in the drawings of the different IC packages for this circuit and that a triangle represents the complete operational amplifier **(op-amp)** circuit.

Operational Amplifiers

A large number of the linear ICs are called **operational amplifiers**, or **op-amps**. These devices have high AC and DC gain capabilities with stable operating characteristics. Op-amps are primarily used for **small-signal amplification**, **waveform generation**, and **impedance matching**. Op-amps were originally designed to perform a number of **mathematical operations**, such as summing, subtracting, averaging, differentiation, and integration. The amplification function of this device is an outgrowth of the original math operations.

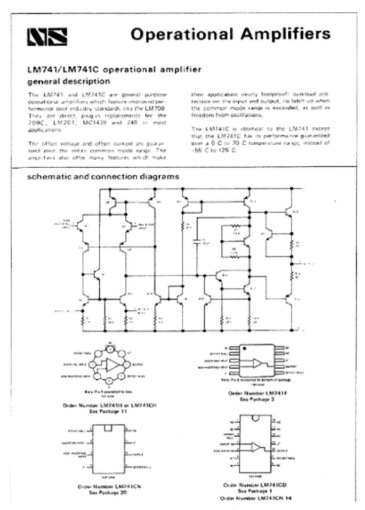

Figure 14.11 Data sheet of an LM741C (Courtesy: National Semiconductor).

The op-amp is essentially a **high-gain amplifier** that is designed to operate over a wide range of voltages. Its internal structure is small and quite complex, as shown in **Figure 14.11**. As a general rule, components connected outside the op-amp determine its operating capabilities. **Gain** can be 100,000 or more when the device is operated without a feedback resistor. This component controls the level of amplification. As a **voltage follower**, gain may be only 1. In either case, the output voltage cannot exceed the value of the supply voltage.

Voltage Regulators

The linear integrated-circuit regulator, or **IC regulator**, accepts an unregulated DC input and develops a regulated output. Some IC regulators are designed to deliver one fixed **output voltage**. A typical series of regulators is available in values of 5, 8, 12, 15, 18, and 24 V. Adjustable regulated voltage outputs can also be obtained with the addition of a few extra components. Current output of 1 A or more can be delivered to a load by a representative IC of this type.

The internal structure of an **IC regulator** is somewhat complex. In general, it contains a series-pass transistor, reference voltage source, feedback amplifier, and short-circuit protection. **Figure 14.12** shows a **block diagram** of the internal structure of a representative three-terminal **linear voltage regulator**.

Operation of the **three-terminal regulator** is based on the applied input voltage and developed output voltage. Output voltage is compared with a **reference voltage** by a high-gain error amplifier. The error amplifier senses

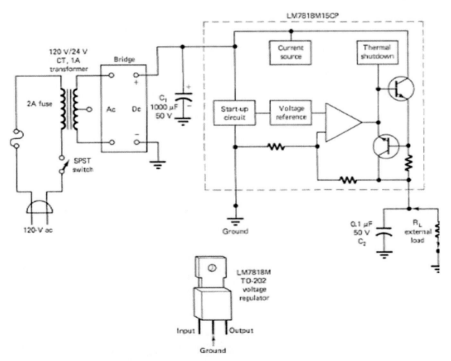

Figure 14.12   Three-terminal IC regulator.

any difference in voltage. The output of the error amplifier then controls conduction of the series-pass transistor. This transistor responds as a variable resistor. It changes inversely with the load current to maintain the output voltage at a constant value. An increase in output voltage causes a corresponding increase in the conduction of the series-pass transistor. This, in turn, causes more voltage drop across the series resistor, which lowers the output voltage. A decrease in output voltage causes less conduction of the series-pass transistor. Reduced current through the series resistor causes less voltage drop. The **output voltage**, therefore, increases in value to compensate for the decrease in voltage.

Linear integrated-circuit regulators are usually equipped with some type of current **overload protection**. This function prevents damage to the series-pass transistor. The three-terminal regulator of **Figure 14.12** has a thermal shutdown circuit. The temperature of the **series-pass transistor** is sensed. If its operation temperature exceeds 347°F (175°C), the regulator turns off. Removing an excessive load lowers the output current and permits the transistor to cool down. Operation is restored when the temperature returns to normal. Protection of this type makes the chip virtually indestructible.

555 Precision Timers

Linear integrated circuits are available, which have the capability of producing a variety of **timing** functions. The 555 precision timer IC was the first in this series. This IC is usually classified as a linear device since linear amplifiers are utilized in its construction. It also has a digital component called a flip-flop in its construction. **Flip-flops** permit the timer to count and have memory. Because of this, the **555 timer** is generally classified as a **hybrid** device that has both linear and digital capabilities. **Figure 14.13** shows a functional block diagram of the precision timer IC.

The operation of a **555 timer** is directly dependent on its internal functions. The three resistors of its block diagram, for example, serve as an internal voltage divider for the source voltage. One-third of the source appears across each resistor. Connections made at the 1/3 V_{CC} and 2/3 V_{CC} points serve as reference voltages for the two comparators. A **comparator** is an op-amp that changes states when one of its inputs exceeds the reference voltage. Comparator 2 is referenced at +1/3 V_{CC}. If a **trigger voltage** applied to the negative input of this comparator drops below 1/3 V_{CC}, it causes a state change. Comparator 1 is referenced at +2/3 V_{CC}. If voltage at the threshold

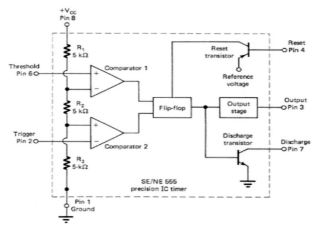

Figure 14.13   Functional block diagram of an SE/NE555 IC.

exceeds this reference voltage, the comparator goes through a state change. Note that the output of each comparator is connected to an input of the flip-flop. **Flip-flops** are used for timing and counting functions in electronic circuits. The flip-flop of the 555 timer controls the operation of the output stage, the reset transistor, and the discharge transistor.

A 555 has two general modes of operation: **astable** and **monostable**. Astable operation is used in the generation of square or rectangular **waveforms**. Monostable operation is used to produce a wave that can be used to achieve **delay** or interval timing operations. The 555 timer is also used as an **oscillator** and will be discussed in Chapter 19.

Self-Examination

Answer the following questions.

17. _____ ICs develop an output that is proportional to the input signal.
18. _____ were originally designed to achieve mathematical operations.
19. An op-amp is essentially a(n) _____ gain amplifier.
20. The two inputs of an op-amp are labeled _____ and _____.
21. In an IC regulator, the _____ protection function prevents damage to the series-pass transistor.
22. A three-terminal voltage regulator is a (linear, digital) IC.
23. The operation of a 555 timer is directly dependent on its (external, internal) functions.
24. The two general modes of 555 timer operation are _____ and _____.

14.3 Advantages and Disadvantages of ICs

This section discusses the advantages and disadvantages of ICs. As you will see, when all factors are considered, the advantages of an IC outweigh the disadvantages. In general, circuits that have a great deal of redundancy or duplication of specific operations are best suited for IC applications. Computers are a classic example of electronic systems that rely on integrated circuits for internal circuitry. Complex circuits that contain a large number of discrete components are also well suited for IC design considerations. The dollar value of ICs far exceeds the value of discrete circuit components in terms of the function they can perform.

14.3 Based on the advantages and disadvantages of ICs, select to use discrete components or an IC for a given application.
    In order to achieve objective 14.3, you should be able to:

  • explain the advantages and disadvantages of ICs.

There are a number of **advantages** that integrated circuits have over standard electronic parts wired in a conventional circuit. These include small size, cost, reliability, low-power operation, and ease of maintenance.

  • **Size.** ICs are becoming so small that hundreds can be manufactured on a piece of material with the size of a shirt button. The industry is predicting much smaller and more complex chips in the future. Chips with a component density of 1 million are available.

  • **Cost.** The cost of an integrated circuit is based largely on its size and component density. By packing more components into less chip area, the cost per component is reduced. Additional savings can be realized when using ICs because fewer parts are needed to order, inventory, and assemble into an operating system.

  • **Reliability**. Perhaps the most important advantage of an IC over discrete-component circuits is reliability. Increased reliability is due to a number of factors. The most significant of these is the reduction of interconnections between components. In earlier discrete-component circuits, up to 50% of circuit failures were related to interconnections between components. IC interconnections are achieved on the chip and formed at the same time during the manufacturing process. If the chip has interconnection problems, they will be detected before it is placed in a completed package.

  • **Low-power operation.** The small size of an IC makes it well suited for low-power operation. The closeness of components in the structure reduces stray electrical pickup, allowing for very small signal operation. Low-power operation means lower internal temperature rise, which improves reliability.

- **Ease of maintenance.** Since the IC is actually a circuit containing many components, servicing can be achieved by plugging in new chips to replace faulty ones. The faulty chip is then discarded since its repair is not practical.

Integrated Circuit Limitations

In spite of the numerous advantages of ICs, they have some very definite **limitations**, such as low power, low voltage, and limited component selection. These limitations can be altered to some extent by more costly manufacturing techniques and circuit design changes. IC limitations are often considered to be some of the tradeoffs used to accomplish unique advantages. The disadvantages of ICs are low power, low voltage, limited component selection, and limited repair.

- **Low power.** The power limitation of an IC is primarily due to its physical size. The current-handling capability of a device determines its heat production. Heat concentrated in a small component may produce temperatures great enough to destroy the device. The size advantage of an IC is actually a tradeoff for its current-handling capacity. A large number of the ICs being manufactured are used only to process information rather than control power. The data output of an IC is often used to control a signal applied to a discrete transistor that is used to control high values of power.
- **Low voltage.** The voltage used by ICs must be kept very low because of the weak insulation between circuit elements. Typical voltage ratings are in the range of 5–30 V. If this voltage is exceeded, it generally causes insulation breakdown in some part of the chip. IC voltage values limit applications to signal processing or data manipulation. Although new developments in IC manufacturing have reduced this limitation to some extent, ICs are still considered to be low-voltage devices.
- **Limited component selection.** The component limitation of an IC is primarily due to the use of silicon in its construction. Silicon is ideal for the fabrication of diodes and transistors, but it does not work very satisfactorily for other components. Resistors, for example, cannot be fabricated to very precise values with silicon. Resistance value tolerances are very high when silicon is used in their production. As a rule,

higher resistance values occupy more space than small values. A 40-kΩ resistor, for example, takes the same amount of space as that needed for five transistors. Capacitance poses an even more significant space problem in IC design than a resistor. Capacitance of any size occupies a rather significant amount of space. A 20-pF capacitor takes the same space needed to fabricate 10 or more transistors. In general, most ICs do not employ capacitors except for those of an extremely small value. Similarly, inductors and transformers are almost impossible to construct on an IC. Component limitations are one of the serious drawbacks of IC use. Unique circuit designs have solved these limitations to some extent.

- **Limited repair.** Another limitation of the IC is based on the repair of its components. When one specific component becomes defective, it cannot be replaced. This means that the entire IC must be replaced when only one component becomes faulty. The large number of good components that remain on the faulty chip cannot be recovered and used for other applications. The faulty chip must be discarded and replaced with a good one to make the circuit functional. This disadvantage is not as bad as it may appear. The process of locating a faulty IC is much easier than locating a faulty component in a complex circuit. As a rule, the additional cost of a new IC is offset by reduced labor costs in the servicing procedure. This simplifies the servicing procedure and reduces the down time of equipment. It also reduces the equipment and spare-part inventory.

Self-Examination

Answer the following questions.

25. Size, cost, reliability, low-power operation, and ease of maintenance are some of the unique (advantages, disadvantages) of an IC.
26. The reduced number of interconnections between components in an IC is a contributing factor to the IC's _____.
27. When a single component of an IC becomes faulty, the entire chip must be (repaired, discarded).
28. The electrical _____ consumed by an IC is generally quite low.
29. ICs generally require low _____ to operate.
30. Most ICs are designed to process _____ instead of controlling power.
31. The operational voltages on an IC are in the range of _____ V to _____ V.

32. The internal circuitry of an IC relies heavily on the use of _____ and _____.

Summary

- Integrated circuits are micro-miniature circuits built on a substrate.
- ICs are made of active and passive components.
- The two general classifications of ICs according to their functions are linear and digital.
- Linear ICs develop an output that is proportional in some way to the input signal; these ICs operate in the straight-line portion of the operating range.
- Monolithic ICs have all components built on a single crystal structure; the components of this IC are all formed on a common substrate at the same time.
- Monolithic ICs that use the same technology as the bipolar transistor in its construction are called bipolar ICs.
- Monolithic ICs that use MOSFETs in their construction are called MOS ICs.
- Film ICs are an assembly of components formed on an insulating material substrate and are classified as hybrid devices.
- Conductors and resistors of a thin-film IC are formed by the evaporation process and are located in a specific area by a screen-printed deposition process; transistors, diodes, capacitors, and MOSFETs are added to the circuit in discrete form.
- Thick-film ICs are similar in many respects to the thin-film ICs; the primary difference is the thickness of the film and the method of material deposition on the film.
- ICs are housed in a number of unique packages; where small size is important for an IC, the flat pack is used.
- A number of packages have been developed to accommodate the reduced size of ICs and to increase the number of pin connections.
- Linear ICs are primarily used to achieve amplification.
- An op-amp is a linear IC that has high gain capabilities.
- A positive sign (+) on the op-amp denotes the noninverting input; a negative sign (−) denotes the inverting input.
- A three-terminal voltage regulator is a linear IC.
- Operation of a voltage regulator IC is based on the applied input voltage and the developed output.

- The output voltage of a voltage regulator IC is compared with a reference voltage by a high gain error amplifier; the output of the error amplifier then controls conduction of the series-pass transistor.
- This series-pass transistor of a voltage regulator IC responds as a variable resistor; it changes inversely with the load current to maintain the output voltage at a constant value.
- A 555 timer contains a three-resistor voltage divider, two comparators, a flip flop, an output, a reset transistor, and a discharge transistor.
- Operation of the 555 timer is based on voltage values applied to the trigger and threshold inputs.
- A 555 has two general modes of operation: astable and monostable.
- Astable operation is used in the generation of square or rectangular waveforms.
- Monostable operation is used to produce a wave that can be used to achieve delay or interval timing operations.
- The advantages of ICs are small size, cost, reliability, low-power operation, and ease of maintenance.
- The disadvantages of ICs are low power, low voltage, and limited component selection.

Review Questions

Answer the following questions.

1. An IC chip is a:

    a. Wafer on which construction is achieved
    b. Fractional part of a wafer
    c. Transistor or diode in the construction of an IC
    d. Logic gate
    e. Substrate

2. The most popular form of digital IC packaging is the:

    a. Flat pack
    b. TO-3 package
    c. Dual in-line package (DIP)
    d. Small outline integrated circuit (SOIC)

3. The foundation on which an IC is built is called a:

    a. Thick film
    b. Thin film

    c. Wafer

    d. Substrate

4. The overall production cost of an IC:

    a. Depends on the number of ICs being manufactured

    b. Is design-dependent

    c. Is lower than a discrete-component circuit

    d. Depends on the size of the wafer

5. Monolithic IC construction uses:

    a. Extensive numbers of resistors and capacitors

    b. High values of resistors and capacitors

    c. Extensive numbers of transistors

    d. Discrete components on a thin film

6. Monolithic refers to:

    a. All circuit elements built on a common substrate

    b. Thin-film construction technology

    c. Active components being built on discrete chips

    d. Circuit elements formed on an insulated substrate

7. MOS and CMOS ICs have an advantage over most other IC families because they:

    a. Are best suited for digital circuit construction

    b. Consume less power

    c. Operate with low voltage and current

    d. Have low fan-in and fan-out capabilities

8. The yield of an IC is:

    a. Greater with a decrease in circuit area

    b. Dependent on the size of the circuit area

    c. Normally very high

    d. Approximately 60% of the potential number of devices on a wafer

9. Normal operation of a linear IC is set:

    a. Near cutoff

    b. Midway between cutoff and saturation

    c. Near saturation

    d. At zero input level

10. In a linear op-amp symbol, the negative input terminal is:

    a. Connected to the power supply

b. The inverting input lead
c. The noninverting input lead
d. Grounded

11. An op-amp essentially responds as a:

    a. Transistor amplifier
    b. Mathematical calculator
    c. High-gain amplifier
    d. Logic gate

12. Fan-out of a logic gate is the:

    a. Total number of package connections
    b. Number of output terminals
    c. Number of circuits the output can drive
    d. Output voltage developed

13. Popular and widely used digital IC families include:

    a. TTL and RTL
    b. MOS and RTL
    c. CMOS and MOS
    d. TTL and MOS
    e. CMOS and TTL

14. A complimentary-symmetry circuit configuration is part of:

    a. The TTL IC family
    b. A bipolar IC
    c. MOS ICs
    d. CMOS ICs

15. Digital ICs primarily respond as:

    a. High-gain amplifiers
    b. Mathematical devices
    c. Switching devices
    d. Voltage controllers

16. A 555 IC used in its astable mode of operation:

    a. Produces repeated square waves
    b. Responds as a gate
    c. Regulates voltage
    d. Is a linear amplifier

Answers

Self-Examination

14.1

1. substrate
2. micro-miniature
3. wafer
4. chip
5. yield
6. monolithic
7. evaporation
8. Diffusion
9. metallization
10. bipolar
11. MOS
12. insulating
13. dual in-line package, *or* DIP
14. Hybrid
15. MOS
16. resistor

14.2

17. Linear
18. Op-amps
19. high
20. inverting, noninverting (any order)
21. overload
22. linear
23. internal
24. astable, monostable (any order)

14.3

25. advantages
26. reliability
27. discarded
28. power
29. voltage
30. information, data, *or* signals
31. 5, 30
32. diodes, transistors (any order)

Terms

Wafer

A piece of circular silicon that serves as the substrate of an IC.

Yield

A percentage ratio of the number of good devices produced to the maximum number of possible devices that can be produced in a production run.

Monolithic

An IC construction procedure in which all circuit elements are formed and interconnected on or within a single piece of silicon.

Evaporation process

A fabrication process in which materials are vaporized and deposited on the surface of another material. Epitaxial growth is an evaporation process.

Diffusion

A process by which the atoms of one material are absorbed or moved into another material when subjected to a controlled atmosphere.

Bipolar IC

An IC that uses the same technology that is used in the construction of a bipolar transistor.

Metallization

The process of attaching metal contacts to the P-type material without permitting the metal to diffuse into the N-type or P-type material.

Bonding pads

The points at which leads are attached to the IC.

MOS IC

An IC made of MOSFET technology. MOS ICs use three basic components in their construction: MOSFETs, MOS capacitors, and MOS resistors. They do not require isolation between the source–drain and other P–N junctions formed on the substrate.

Film IC

An IC with an assembly of components formed on the surface of an insulating substrate. Conductors and resistors are formed by evaporation and by screen-printed deposition areas. Transistors, diodes, and MOS capacitors are added to the circuit in discrete form and are attached to the circuit with special cement.

Thin-film IC

A hybrid assembly in which passive circuit elements and interconnections are formed by evaporation onto a substrate. Active devices are added as discrete components.

Thick-film IC

A hybrid IC in which the passive circuit parts and interconnections are formed on an insulating material substrate by print screening. The active components are added in chip form.

15

Operational Amplifiers (Op-amps)

An **operational amplifier** or **op-amp** is a modular, multistage amplifying device capable of high-gain signal amplification from DC to several million hertz. The amplifier circuits are directly coupled and contain several transistor devices. The entire assembly is built on a small silicon substrate and packaged as an IC. Op-amps are used in many electronic systems, such as communications and home entertainment systems, for amplification and various types of mathematical operations such as integrators, differentiators, and summing.

In this chapter, you will learn the basic characteristics of the op-amp. Op-amps are faster, cheaper, and easier to work with than discrete component circuits. However, power-handling capability is still a consideration. We will take a look "inside" the op-amp to see how it functions.

Objectives

After studying this chapter, you will be able to:
15.1 describe the basic operation of an op-amp;
15.2 describe the function of each op-amp stage;
15.3 evaluate the performance of an op-amp;
15.4 analyze and troubleshoot op-amps.

Chapter Outline

15.1 Introduction to the Op-amp
15.2 Inside the Op-amp
15.3 Op-amp Characteristics
15.4 Analysis and Troubleshooting – Op-amps

15-1 Open-Loop Gain of Operational Amplifiers
In this activity, an open-loop operational amplifier circuit will be evaluated.

Key Terms

closed-loop gain
common mode
common-mode rejection ratio (CMRR)
differential amplifier
differential mode
floating state
input bias current
input offset voltage
intermediate amplifier stage
inverting input
noninverting input
open-loop gain
output stage
single-ended mode
slew rate

15.1 Introduction to the Op-amp

Op-amps are used for a wide variety of applications. In order to understand these applications, the basis functions of the op-amp will be discussed.

15.1 Describe the basic operation of an op-amp.
   In order to achieve objective 15.1, you should be able to:

- describe the basic op-amp modes of operation;
- identify the op-amp symbol and terminals;
- define inverting input, noninverting input, and differential amplifier, common mode, differential mode, and single-ended mode.

Op-amp Schematic Symbol

The **schematic symbol** of an op-amp is generally displayed as a triangle symbol. **Figure 15.1** shows a typical op-amp symbol with its terminals labeled. The triangle denotes the amplification function. An op-amp has at least five terminals or connections in its construction. Two of these are for the power supply voltage, two for differential input, and one for the output.

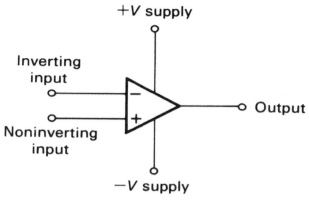

Figure 15.1 Op-amp schematic symbol.

The point or **apex** of the op-amp symbol identifies the output. The two leads labeled − and + identify the differential input terminals. The − sign indicates inverting input and the + sign denotes noninverting input. A signal applied to the **inverting input** is inverted 180° at the output. Standard op-amp symbols usually have the inverting input located in the upper-left corner. A signal applied to the **noninverting input** is not inverted at the output and remains in phase with the input. The + input is located in the lower-left corner of the symbol. In all cases, the two inputs are clearly identified as + and − inside the triangle symbol.

Connections or **terminals** on the sides of the triangle symbol are used to identify a variety of functions. The most significant of these are the two power supply terminals. Normally, the positive voltage terminal (+V) is positioned on the top side and the negative voltage terminal (−V) is positioned on the bottom side. In practice, most op-amps are supplied by a split, or divided, power supply. This supply has +V, ground, and −V terminals. It is important that the correct voltage polarity be supplied to the appropriate terminals or the device may be permanently damaged. A good rule to follow for most op-amps is not to connect the **ground lead** of the power supply to −V. An exception to this rule is the current-differencing amplifier (CDA) op-amp. These op-amps are made to be compatible with digital logic ICs and are supplied by a straight 5-V voltage source.

There may be other terminals in the makeup of this device, depending on its internal construction or intended function. Each terminal is generally attached to a schematic symbol at some convenient location. Numbers located near each terminal of the symbol indicate pin designations. The schematic

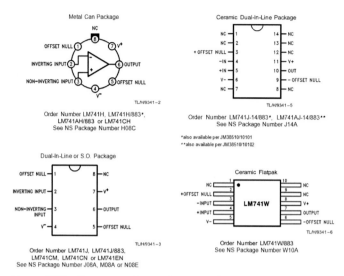

Figure 15.2   Op-amp packages (Courtesy: Fairchild Semiconductor).

symbol of an op-amp generally has its terminals numbered and the element names omitted. A pin-out key identifies the name of each terminal. **Figure 15.2** shows various op-amp packages. Note the pin-out key for each package.

Basic Op-amp Operation

The fundamental operation of an **op-amp** is based on a unique circuit known as a differential amplifier. This circuit is designed to amplify the difference between two voltage values that are applied to its input. A simplification of the **differential amplifier** circuit and equivalent op-amp symbol is shown in **Figure 15.3**. Note that bipolar transistors are used in this construction and that Q_1 and Q_2 respond as a differential amplifier.

The **differential amplifier** of **Figure 15.3** has some rather unusual considerations that must be taken into account. It, for example, has two inputs and two outputs that can be connected in a number of different ways. These are called its **modes of operation.** Single-ended, differential, and common modes of operation will be discussed in this section.

Normally, with no input signal applied to either Q_1 or Q_2, no output appears. If identical values of input are applied to both Q_1 and Q_2, there is still no output. In a sense, each transistor conducts the same amount. The

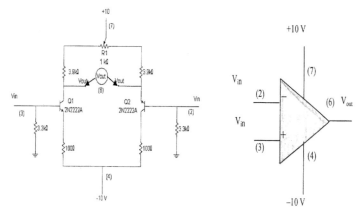

Figure 15.3 Simplified differential amplifier and equivalent op-amp symbol.

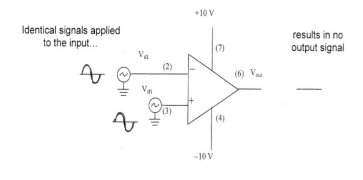

Common Mode

Figure 15.4 Common-mode operation.

output signal is, therefore, of the same value or balanced. There is no potential difference in the two outputs. This means that there is no output signal. This is called **common-mode** operation. It is illustrated in **Figure 15.4**.

When different signals are applied to the input of Q_1 and Q_2, the output becomes unbalanced. The transistor receiving the greater signal conducts more current. With current regulated by Q_3, the alternate transistor conducts less current. A pronounced difference in the two outputs occurs. The resulting output is indicative of the conduction difference in Q_1 and Q_2. This is called **differential-mode** operation. It is illustrated in **Figure 15.5**.

If a signal is applied only to the inverting input of a differential amplifier, the output will be amplified and inverted. A signal applied only to the noninverting input will be amplified without inversion. For the circuit to

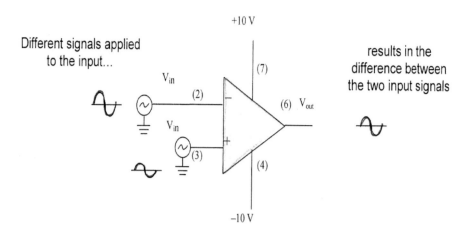

Differential Mode

Figure 15.5   Differential-mode operation.

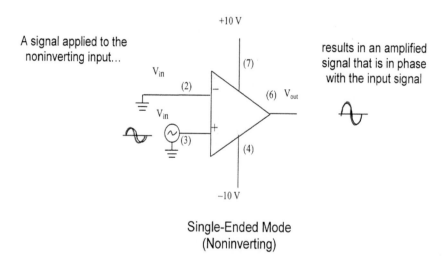

Single-Ended Mode
(Noninverting)

Figure 15.6   Single-ended mode operation, noninverting.

respond in this manner, the input not being used must be grounded. The output will be the difference between input and ground. This is called **single-ended mode** operation. The **noninverting** single-ended mode operation is shown in **Figure 15.6**, and the **inverting** single-ended mode of operation is shown in **Figure 15.7**.

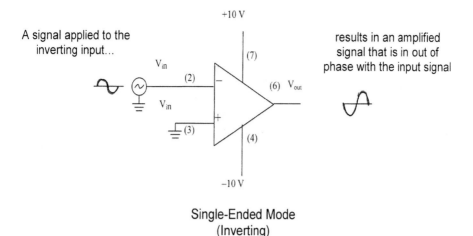

Single-Ended Mode
(Inverting)

Figure 15.7 Single-ended mode operation, inverting.

Self-Examination

Answer the following questions.

1. The triangle symbol in electronics denotes the _____ function.
2. Five terminals of an op-amp are _____, _____, _____, _____, and _____.
3. When identical values of input are applied to both op-amp input terminals, the op-amp is considered to be in _____ mode.
4. When different values of input are applied to both op-amp input terminals, the op-amp is considered to be in _____ mode.
5. When an input signal is applied to one terminal of an op-amp and the other terminal is grounded, the op-amp is considered to be in _____ mode.

15.2 Inside the Op-amp

Op-amps have a number of other circuits in their construction. A high-gain amplifier, for example, follows the differential amplifier. This amplifier raises the signal level so that very small input signals can be amplified. An emitter-follower amplifier comes after the high-gain amplifier. The emitter-follower provides the op-amp with a low-impedance output. This permits it to be connected to a variety of different output devices. The combination of all this makes the circuitry of an op-amp rather complex. As a general rule, the internal construction is not very important to a person using the device. It is, however, helpful for you to have some general understanding of what the internal circuitry accomplishes. This will permit you to see how the device performs and some of its limitations as a functioning unit.

15.2 Describe the function of each op-amp stage.

In order to achieve objective 15.2, you should be able to:

- name the op-amp stages of amplification;
- define intermediate amplifier stage, output amplifier stage, common-mode rejection ratio, and floating state.

Internal Circuitry

The internal circuitry of an **op-amp** can be divided into three functional units, as shown in **Figure 15.8**. Note that each function is enclosed in a triangle. This diagram shows that the op-amp has three basic amplification functions. These functions are generally called **stages of amplification**. A stage of amplification contains one or more active devices and all the associated components needed to achieve amplification.

Figure 15.9 shows the internal circuitry of a **general-purpose op-amp**. The first stage, or input, of an op-amp is usually a **differential amplifier**.

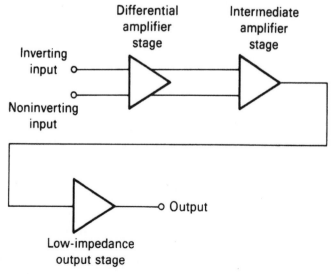

Figure 15.8   Op-amp diagram depicting the stages of amplification.

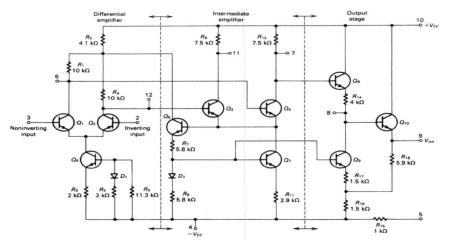

Figure 15.9 Internal circuitry of a general-purpose op-amp.

Note that this amplifier has two inputs, which are labeled inverting input and noninverting input. It provides high gain of the signal difference supplied to the two inputs and low gain for common-mode signals applied to both inputs simultaneously. The input impedance is high to any applied signal. The output of the differential amplifier is generally two signals of equal amplitude and 180° out of phase. This could be described as a push–pull input and output.

One or more intermediate stages of amplification follow the differential amplifier. **Figure 15.9** shows an op-amp with only one **intermediate amplifier stage**. This amplifier is designed to shift the operating point to a zero level at the output and has high current and voltage gain capabilities. Increased gain is needed to drive the output amplifier stage without loading down the input. The intermediate amplifier stage generally has two inputs and a single-ended output.

The **output stage** of an op-amp has rather low output impedance and is responsible for developing the current needed to drive an external load. Its input impedance must be great enough that it does not load down the output of the intermediate amplifier. The output stage can be an emitter-follower amplifier (as shown) or two transistors connected in a complementary-symmetry configuration. Voltage gain is rather low in this stage with a sizable amount of current gain. The following sections take a detailed look at each stage of amplification.

Differential Amplifier Stage

A **differential amplifier** is the operational basis of most op-amps. This amplifier is best described as having two identical or balanced transistors sharing a single emitter resistor. Each transistor has an input and an output. A schematic diagram of a simplified differential amplifier is shown in **Figure 15.10**. Note that the circuit is energized by a dual-polarity, or split-power, supply. The source leads are labeled $+V_{CC}$ and $-V_{CC}$ and are measured with respect to a common ground lead.

Operation of a **differential amplifier** is based on its response to input signals applied to the base. Grounding one base and applying an input signal to the other base produces two output signals. This occurs because the two transistors are controlled by the same emitter current. The output signals have the same amplitude but are inverted 180°. This type of input causes the amplifier to respond in its differential mode of operation.

When two input signals of equal amplitude and polarity are applied to each base at the same time, the resulting output is zero. This type of input causes a difference or canceling voltage to appear across the commonly connected emitter resistor. In a sense, the differential amplifier responds as

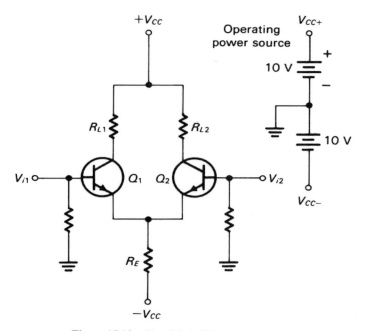

Figure 15.10   Simplified differential amplifier.

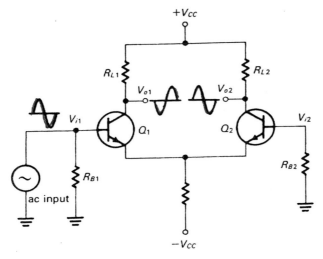

Figure 15.11   AC differential amplifier.

a balanced bridge circuit to identical input signals. There is no output when the circuit is balanced. You should recall that this is referred to as common-mode operation. A differential amplifier is designed to reject signals common to both inputs. The term **common-mode rejection ratio (CMRR)** is used to describe this action of the amplifier. CMRR is a unique characteristic of the differential amplifier. Undesirable noise, interference, or AC hum can be rejected by this operating condition.

In the circuit of **Figure 15.11**, an input signal is applied to the base of Q_1, and the base of Q_2 is left open. This condition causes signals to be developed at both outputs and across the common emitter resistor. The emitter signal, as indicated, is in phase with the input. The two output signals are out of phase with each other and have a substantial degree of amplification. Output V_{out1} is out of phase with the input and V_{out2} is in phase.

A **differential amplifier** produces two output signals when only one input signal is applied. Coupling of the input signal from Q_1 to Q_2 is accomplished through the emitter resistor. The **positive alternation** of the input signal, for example, causes increased forward bias of Q_1. This causes an increase in the conduction of Q_1. With more I_E, there is a greater voltage developed across the emitter resistor. This in turn causes both emitters to be less negative. The conduction of Q_1 is not appreciably influenced by this voltage because it has an external signal applied to its input. Q_2 is, however, directly influenced by the reduced negative voltage to its emitter. This causes the conduction of Q_2 to be reduced. Reduction in current through Q_2 causes less voltage drop across R_{L2}, and the collector voltage to swing in the positive direction. In effect, an input signal applied to the base of Q_1 reduces the V_E of Q_2, which, in turn, increases the value of the output voltage V_{out2}. An input signal is, therefore, coupled to Q_2 by the commonly connected emitter resistor.

The **negative alternation** of the input signal causes a reversal of the action just described. Q_1, for example, will be less conductive, and Q_2 will have increased conduction. This action causes a reduction of I_C through Q_1 and an increase swing in the value of V_{01}. Increased conduction of Q_2 causes a corresponding reduction in V_{out2}. The two output signals continue to be $180°$ out of phase. In effect, both alternations of the input appear in the output. A differential amplifier connected in the differential mode will develop two output signals that are reflective of the entire input signal.

Differential amplifiers respond in primarily the same way when the inputs are reversed. In this case, an input signal is applied to the base of Q_2 with the base of Q_1 open or floating. V_{out1} is out of phase with the input, and V_{out2} is in phase. The amplitude, or output signal level, of the amplifier is still based on the signal difference between the two inputs. With only one signal applied to the input at a time, the amplifier sees a very large differential input and develops a sizable output voltage.

The **differential amplifier** of **Figure 15.11** is rarely used in op-amp construction. Ordinarily, the resistance of R_E needs to be quite large to have good coupling and common-mode rejection capabilities. Large resistance values are rather difficult to fabricate in IC construction. R_E can, however, be replaced with a transistor. This transistor and its associated components are called a **constant-current source**.

Figure 15.12 shows a differential amplifier with a **constant-current source** in the emitter circuit. Transistor Q_3, in this case, has a fixed or constant bias voltage. This voltage maintains the internal resistance of Q_3 at a rather high value. In some op-amps, the constant-current source can be altered by an external base-bias voltage. This is achieved by having an external lead connection to the base of Q_3. The current source of the differential amplifier can then be altered to some extent. The resistance, or impedance, of Q_3 can be adjusted to meet the design parameters of a specific circuit application. When the impedance of Q_3 is high, the common-mode gain of the amplifier is very low and signal coupling is good. The constant-current source of an input differential amplifier is an important part of op-amp construction.

The **differential amplifier** of an op-amp can be made with other transistor devices and connected in a variety of different configurations. Two rather common configurations are shown in **Figure 15.13**.

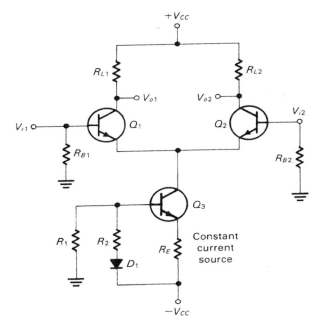

Figure 15.12 Differential amplifier with constant-current source in the emitter.

The amplifier of **Figure 15.13(a)** is achieved with bipolar junction transistors (BJTs) connected in a **Darlington-pair** configuration. The input impedance of this circuit is increased by a factor of 1.5 times the beta. The beta of each transistor branch is also increased by the same value. Darlington-pair devices respond at high speed to a wide range of frequencies.

The differential amplifier of **Figure 15.13(b)** employs JFETs instead of BJTs in its input. The JFET input is extremely high resistant when compared to the BJT. Typical input impedance values are 10 MΩ for this type of differential input. Op-amps employing the **JFET input** are widely used in process control, medical instrumentation, and other applications requiring very low input current values.

Intermediate Amplifier Stage

The **intermediate amplifier stage** of an op-amp follows the input differential amplifier stage and feeds the output stage. This part of the op-amp is primarily responsible for additional gain and DC voltage stabilization. As a rule, the intermediate differential amplifier has rather high gain capabilities compared with the input differential amplifier.

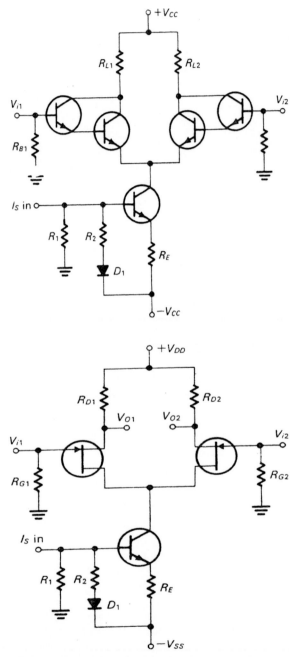

Figure 15.13   Differential amplifiers. (a) Darlington transistor. (b) JFETs.

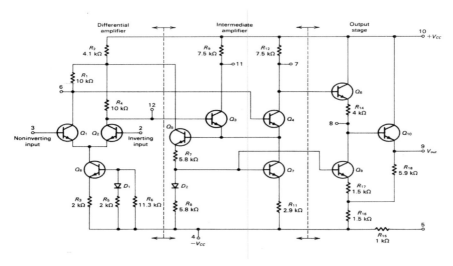

Figure 15.14 Intermediate and output amplifier stages.

The intermediate amplifier stage of **Figure 15.14** is equipped with a **voltage-stabilizing** transistor, Q_5. This transistor evaluates the signal at the emitters of Q_3 and Q_4. Since the intermediate differential amplifier is driven by a push−pull signal, there should be a zero reference level at its input when the first differential amplifier is operating properly. If an operational error occurs, a correcting voltage is developed by Q_5 across R_2 of the input amplifier. Q_5 also feeds the error bias voltage to the constant-current transistor Q_7. This voltage further reduces the error and improves the common-mode rejection capabilities of the amplifier. The output of the intermediate amplifier stage is zero-voltage-level corrected and appears at the collector of Q_4.

Output Stage

The **output stage** of an op-amp is designed primarily to develop the power needed to drive an external load device. In accomplishing this function, there must be a maximum output voltage signal developed across a low-impedance load. The output of an op-amp may be a single transistor or two transistors connected in a complementary-symmetry power amplifier.

The output stage of **Figure 15.14** is a **single-ended emitter-follower amplifier**. This part of the op-amp is on the right side of the schematic diagram. Transistors Q_8, Q_9, and Q_{10} make up the output stage. Q_8 is an emitter-follower driver transistor. Q_9 serves as a constant-current source for the output transistor Q_{10}. The single-ended output of the intermediate differential amplifier stage drives the base of Q_8. This transistor matches the output impedance of Q_4 to the low output impedance of Q_{10}. Maximum signal transfer is accomplished through Q_8.

The **signal gain** of a single-ended output stage must be controlled or limited to some extent to provide good stability. Transistor Q_9 performs this function. This transistor has a dual role. It responds as a constant-current source for the driver transistor Q_8 and as a feedback regulator for the output transistor. **Feedback** is regulated by maintaining the current through R_{16} at a rather constant level by conduction of Q_9. The constant-current function of Q_9 maintains the emitter of Q_8 at a consistent level to reduce level shifting and improve common-mode rejection.

Many op-amps employ two transistors in a **complementary-symmetry** output circuit instead of the single-ended emitter-follower circuit. The emitter resistor of a single-ended emitter-follower output generally consumes a great deal of power at high current levels. The complementary-symmetry output circuit overcomes this problem by using two transistors. The transistors are NPN and PNP complements that have the same or symmetrical characteristics. **Figure 15.15** shows a simplification of the complementary-symmetry power amplifier used in many op-amps.

The **output** of the complementary-symmetry amplifier is developed across an external load connected to pin 6. Note that this terminal is connected

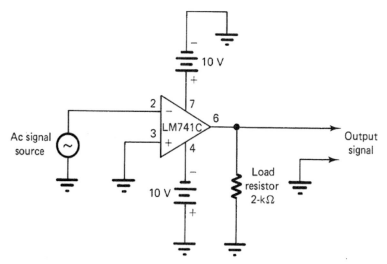

Figure 15.15   Complementary-symmetry output of an op-amp.

directly to the emitters of Q_{10} and Q_{11}. These transistors, connected in a common-collector circuit configuration, have a low-impedance output.

A resistive **divider network** is connected across $+V_{CC}$ and $-V_{CC}$. The base of each transistor is biased at cutoff or slightly above cutoff by this network. A signal applied to the input is fed directly to the base of each transistor. The positive alternation of an AC signal causes the NPN transistor Q_{10} to be conductive. Conduction current flows from ground through R_L, Q_{10}, and to $+V_{CC}$ for this alternation.

Q_{11} is driven further into **cutoff** by this alternation and has no output. For the negative alternation, Q_{11} is forward biased and goes into conduction. Q_{10} is reverse biased by this alternation and goes into a nonconductive state. Conduction of Q_{11} is from $+V_{CC}$, through Q_{11} and R_L, to ground. The negative alternation develops output across R_L. This means that each transistor goes into **conduction** for one alternation. The output current through R_L is a combination of the current flow produced by each transistor. This type of amplifier can develop large amounts of current with good power gain and low output impedance

Self-Examination

Answer the following questions.

6. The input of an op-amp is usually a(n) _____ amplifier.
7. Three stages of an op-amp circuit are _____, _____, and _____.
8. The property of a differential amplifier to reject signals common to both of its inputs is called _____.
9. DC voltage stabilization of an op-amp is caused by the _____ stage.
10. The output stage of an op-amp may be _____ or _____.

15.3 Op-amp Characteristics

Operational amplifiers have several important characteristics. The manufacturer of the device supplies data sheets that identify the pin-out and operating data. The data sheet also shows absolute maximum ratings, electrical characteristics, and typical performance curves.

15.3 Evaluate the performance of an op-amp.

In order to achieve objective 15.3, you should be able to:

• determine the open-loop gain, closed-loop gain, and slew rate of an op-amp;

- describe the electrical characteristics of an op-amp;
- define open-loop gain, closed-loop gain, input offset voltage, input bias current, and slew rate.

Open-Loop Gain

The **open-loop gain** characteristic of an op-amp refers to an output that is developed when only difference voltage is applied to the input. The μA751C op-amp of **Figure 15.16** is connected in an open-loop circuit configuration. The open-loop voltage gain (A_{Vol}) is a ratio of the output voltage (V_{out}) divided by the differential input voltage ($V_{diff(in)}$). This is expressed by the following formula:

$$A_{Vol} = V_{out}/V_{diff(in)}. \tag{15.1}$$

The **open-loop voltage gain** of an op-amp is usually quite large. Typical values are in the range of 10,000–200,000. When the differential input voltage is zero, the output voltage is also zero. If a slight difference in input voltage occurs, the output voltage increases accordingly. The output voltage, however, cannot exceed 90% of the source voltage. When the output voltage reaches this level, the amplifier is saturated. Due to the high A_{Vol} of an op-amp, only a few millivolts of V_{in} are needed to cause it to go into saturation. As a rule, op-amps are rarely used in an open-loop circuit configuration.

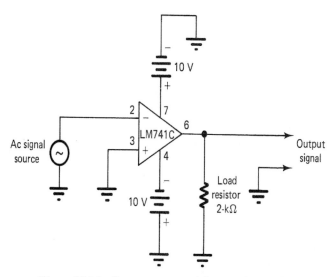

Figure 15.16   Op-amp connected for open-loop gain.

To see how an op-amp responds in the open-loop mode of operation, assume that the op-amp of **Figure 15.16** has an open-loop gain (A_{Vol}) of 100,000. Typical values of A_{Vol} could exceed a million. The supply voltage is ± 10, as indicated. For a representative op-amp, saturation occurs when the output voltage (V_{out}) reaches approximately 90% of the source voltage:

$$V_{sat} = 0.90 \times V_{CC}. \tag{15.2}$$

For this circuit, the saturation voltage is, therefore,

$$V_{sat} = 0.90 \times V_{CC}$$
$$= 0.90 \times 10V$$
$$= +9V.$$

Transposing the previous A_{Vol} formula for input voltage (V_{in}) causes it to be

$$V_{in} = V_{sat} \, A_{Vol}. \tag{15.3}$$

The value of V_{in} needed to cause saturation of the op-amp is approximately

$$V_{in} = V_{sat}/A_{Vol}$$
$$= +9V/100,000$$
$$= 0.00009 \text{ or } 90 \ \mu V.$$

Therefore, **saturation** occurs when the input voltage reaches 90 μV. An AC signal of 90 μV_{pp} would also cause distortion of the output. This shows that the op-amp in the open-loop configuration is easy to distort. **Closed-loop** operation is generally used to alter this problem.

Example 15-1:

An op-amp has a voltage input of 0.2 mV and an output of 9 V. What is the open-loop voltage gain?

Solution

$$A_{Vol} = V_{out} = \underline{9 \text{ V}} = 45,000$$
$$V_{in}0.2mV.$$

Related Problem

An op-amp has the following values:
$V_{in} = 0.6$ mV and $V_{out} = 10$ V. What is the open-loop voltage gain?

Closed-Loop Gain

The **closed-loop gain** characteristic of an op-amp refers to the gain or amplification achieved by an amplifier that has negative feedback between the input and output. When an op-amp is used in a closed-loop configuration, the voltage gain can be adjusted to nearly any desired value. This is achieved by a feedback resistor network connected between the output and the input.

Figure 15.17 shows a simplified op-amp connected to a **feedback** network in a **noninverting** circuit. The output will be the same polarity as the input. Ideally, the two inputs are balanced when resistors R_1 and R_2 are equal. In an actual circuit, R_2 is a calculated value based on the parallel arrangement of R_1 and the feedback resistor R_F:

$$R_2 = (R_1 \times R_F)/(R_1 + R_F). \tag{15.4}$$

For the circuits in **Figure 15.17**, the value of R_2 would be

$$\begin{aligned} R_2 &= (R_1 \times R_F)/(R_1 + R_F) \\ &= (1 \text{ k}\Omega \times 99 \text{ k}\Omega)/(1 \text{ k}\Omega \times 99 \text{ k}\Omega) \\ &= 99 \text{ M}\Omega/100 \text{ k}\Omega \\ &= 990 \text{ }\Omega. \end{aligned}$$

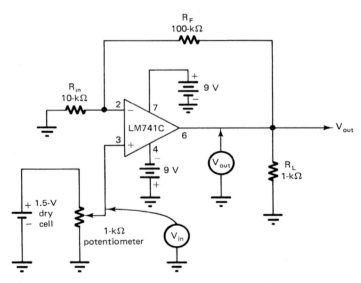

Figure 15.17   Closed-loop op-amp simplification – noninverting amplifier.

Amplifier gain is based on the following formula:

$$A_V = (R_1 + R_F)/R_1. \tag{15.5}$$

For the circuit in **Figure 15.17**, the gain is

$$
\begin{aligned}
A_V &= (R_1 + R_F)/R_1 \\
&= (1\text{ k}\Omega + 99\text{ k}\Omega)/1\text{ k}\Omega \\
&= 100.
\end{aligned}
$$

Example 15-2:

What would the gain of the circuit in **Figure 15.17** be if R_F were changed to 9.9 kΩ and R_1 to 100 Ω?

Solution

$$
\begin{aligned}
A_V &= (R_1 + R_F)/R_1 \\
&= (100\ \Omega + 9.9\text{ k}\Omega)/100\ \Omega \\
&= 10\text{ k}\Omega/100\ \Omega \\
&= 100.
\end{aligned}
$$

Related Problem

What would the gain of the circuit in **Figure 15.17** be if R_F were changed to 2.9 kΩ and R_1 remains at 100 Ω?

Figure 15.18 shows an **inverting** circuit configuration. The input signal is applied to the negative terminal. The feedback network is composed of R_1 and R_F. R_2 returns the noninverting input to ground. The value of R_2 is determined by the parallel value of R_1 and R_F. R_2 is generally smaller than R_1. Changing these values can modify the op-amp gain. The **voltage gain** of this amplifier is determined by the following formula:

$$A_V = R_F/R_1. \tag{15.6}$$

For the inverting op-amp in **Figure 15.18**, the gain is

$$
\begin{aligned}
A_V &= R_F/R_1 \\
&= 100\text{ k}\Omega/1\text{ k}\Omega \\
&= 100.
\end{aligned}
$$

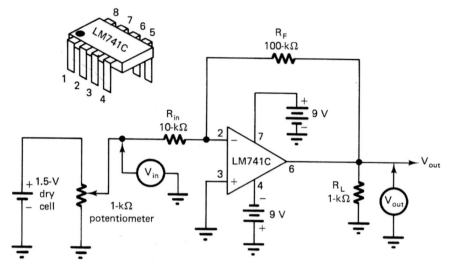

Figure 15.18   Closed-loop op-amp simplification − inverting amplifier.

Common-Mode Rejection Ratio

Signals applied to an op-amp are applied to one input or with opposite polarities on both input lines. The desired input signals are then amplified by the op-amp. Undesired signals (called <u>noise</u>) of the same polarity on both input terminals are canceled and do not appear at the output of the op-amp. The amplifier's ability to reject these **noise signals** is called **common-mode rejection ratio (CMMR)** and is equal to the ratio of the open-loop voltage gain and the common-mode gain.

Example 15-3:

An op-amp has an open-loop voltage gain of 50,000 and a common-mode gain of 0.1. What is the CMMR?
Solution

$$CMMR = A_{\mathrm{OL}}/A_{\mathrm{CM}} = 50,000/0.1 = 500,000.$$

Related Problem

An op-amp has the following values: A_{OL} = 100,000 and A_{CM} = 0.25. What is the common-mode rejection ratio?

Input Offset Voltage

Ideally, the output voltage (V_{out}) of an op-amp should be 0 V when the input voltage is 0 V. In the internal construction of an op-amp, it is extremely difficult to develop perfectly balanced differential amplifiers that eliminate this problem. Thus, there is actually some output voltage even when the input is 0 V. Values range from microvolts to millivolts in a typical circuit.

An **input offset voltage** is used to overcome the unwanted output voltage of an op-amp. The net effect is 0 V at the output when no voltage is applied to the input. This voltage is called the input offset voltage (V_{os}). Typical V_{in} values range from a few microvolts to several millivolts. As a general rule, the lower the input offset voltage, the better the quality of the op-amp.

Input Bias Current

The internal construction of an op-amp generally causes input current values of the differential amplifier to be unequal to some extent. **Input bias current** is the average current flowing into or out of the two inputs. This parameter describes the relationship of the two input current values. In general, smaller values of input bias current (I_{bias}) are used with certain op-amps.

The type of transistor device used in the construction of the input differential amplifier has a great deal to do with its input bias current. Op-amps with JFET inputs have lower I_{bias} values than BJT inputs. Typical I_{bias} values for the μA741 are in the range of 80 nA. For a JFET input op-amp such as the LH0042, I_{bias} values drop to the 5-pA to 10-pA range. Input bias current values are often used when comparing the quality of different devices.

Input Impedance

The **input impedance** of an op-amp is the equivalent resistance that an input source sees when connected to the differential input terminals. Normally, the input impedance (Z_{in}) of a conventional op-amp is quite high. Typical values are in the range from approximately 10 kΩ to 2 MΩ. If the Z_{in} of a specific device is unknown, an estimate of the value would be 250 kΩ. Impedance, in this case, refers to the opposition encountered by either an AC or DC signal voltage. Two ways of specifying input impedance are differentials and common mode. Differential input impedance is the total resistance between the inverting and noninverting inputs of an op-amp. Common-mode input impedance is the resistance between each input and ground.

Input Current

The **input current (I_{in})** of a conventional op-amp is usually quite small. A voltage source connected to the differential input sees extremely high input impedance. Typical I_{in} values are in the range of a few nanoamperes. To illustrate this, the **LM741C op-amp** is used to determine the input current for circuit operation at or near the point of saturation. Assume that the Z_{in} is 250 kΩ. The input voltage that produces saturation was previously determined to be 90 μV. Therefore,

$$I_{in} = V_{in}/Z_{in}$$
$$= 90 \ \mu V/250 \ k\Omega$$
$$= 0.36 \ nA.$$

With input current values being this small, they are often considered to be negligible in most op-amp circuit calculations.

Slew Rate

The **slew rate** of an op-amp refers to the rate at which its output changes from one voltage to another in a given time. This parameter is extremely important at high frequencies because it indicates how the output responds to a rapidly changing input signal. Slew rate depends on such things as amplifier gain, compensating capacitance, and the polarity of the output voltage. The worst case, or slowest slew rate, occurs when the gain is at 1, or **unity**. As a rule, slew rate is generally indicated for unity amplification. Mathematically, slew rate (SR) is expressed as maximum change in output voltage (ΔV_{out}) divided by a change in time (Δ_t):

$$SR = \Delta V_{out}/\Delta_t. \tag{15.7}$$

Slew rate is primarily an indication of how an op-amp responds to different frequencies. A slew rate of 0.5 V/1 μs means that the output has time to rise only 0.5 V in 1 μs. Since frequency is time-dependent, this indicates where certain input frequencies do not have enough time to produce a corresponding output signal. The resulting output is a distorted version of the input. The maximum frequency (f_{max}) at which an op-amp can obtain an undistorted output is determined by the following expression:

$$f_{max} = SR/(6.28 \times V_{out}). \tag{15.8}$$

For the μA741 op-amp, the slew rate is 0.5 V/μs. The undistorted output frequency of the op-amp with an undistorted output voltage of 1 V is, therefore,

$$f_{max} = \frac{SR}{6.28 \times V_{out}}$$
$$= \frac{0.5V/\mu s}{6.28 \times 1}$$
$$= \frac{0.0796178}{10^{-6}}$$
$$= 796,178 \, Hz.$$

The **slew rate** of an op-amp is primarily determined by a compensating capacitor that is either internal or externally connected. Essentially, it takes a certain period of time for the capacitor to develop a charge voltage. The values of the capacitor and the charging current determine the response of the capacitor. Since the op-amp has internal constant-current sources to limit the current, the capacitor can only charge at a specified rate. The input signal must occur at a slower rate than the capacitor charge time, or some other type of distortion will then occur. When the input frequency is higher than the slew rate limit, the square wave appears as a slope and the sine wave becomes a triangle.

Example 15-4:

A signal applied to an op-amp causes the output voltage to vary from +5 V to −5 V in 0.5 μs. What is the slew rate?
Solution

$$\text{Slew rate} = \Delta V_{out} = +5 \, V - (-5V) = 20 \, V/\mu s \quad \Delta_t \quad 0.5\mu s.$$

Related Problem

A signal applied to an op-amp causes the output voltage to vary from +9 V to −9 V in 0.67 μs. What is the slew rate?

Frequency Response

The **slew rate** of an op-amp is one of the major factors in determining frequency response of a circuit. The slew rate is a measure of how fast an op-amp's output voltage can change in volts per microsecond (V/μs). The formula shown previously (f_{max} = SR/6.28 × V_{out}) shows that the peak output voltage limits maximum frequency. Above the f_{max} value, a distorted output will occur. In addition, open-loop voltage gain will decrease.

Bandwidth

The **bandwidth (BW)** of an op-amp is an important consideration. Basically, the higher the gain of an op-amp, the narrower the bandwidth. Likewise, as gain becomes lower, bandwidth is wider. If a wide bandwidth is desired, a lower gain results.

A factor called **gain-bandwidth product** of an op-amp is used to determine the maximum value of closed-loop gain (A_{CL}) at a specified upper cutoff frequency (f_c). The product of A_{CL} and f_c is equal to the unity gain frequency of an op-amp.

Output Short Circuit Current

The **output short circuit current** of an op-amp is the maximum output current measured when load resistance is shorted (R_L = 0 Ω). Op-amps are protected internally from the excess current demand of a shorted load. Once this value is established, the output current of an op-amp will not exceed this value.

Output Impedance

An **ideal amplifier** has no (0 Ω) output impedance, so that there is no voltage drop when its output is connected to a load. The output impedance of a typical op-amp is generally quite low, in the range of 25−100 Ω. While this will cause some voltage drop and power dissipation in the output section of the op-amp, for most applications, this would be acceptable. The output voltage of the op-amp is not significantly affected by the voltage drop caused by the current flowing through its output impedance. Thus, the voltage gain would not be affected by loading.

Self-Examination

Answer the following questions.

11. Op-amps are rarely used in _____ configurations.

12. The ratio of output voltage to differential input voltage of an op-amp is called _____.
13. An op-amp has an open-loop voltage gain of 200,000 with a supply voltage of 12 V. The value of input voltage needed to cause saturation is _____ V.
14. The type of feedback of an inverting op-amp circuit is _____.
15. If the feedback resistor for a noninverting circuit is 78 kΩ and the input resistor is 2 kΩ, the voltage gain is _____.
16. If the feedback resistor for an inverting circuit is 10 kΩ and the input resistor is 2 kΩ, the voltage gain is _____.
17. A(n) _____ is used to overcome unwanted output voltage of an op-amp.
18. The rate at which an op-amp's output voltage changes is called _____.

15.4 Analysis and Troubleshooting – Operational Amplifiers

In order to perform a test on **operational amplifiers**, you must understand their characteristics and be proficient in the use of the test equipment. Operational amplifiers can be tested using analog meters or digital multimeters in some situations. A **defect** such as an **open** circuit is easy to detect when no power is applied to the circuit being tested with an ohmmeter reading of an extremely high resistance. A defective **shorted** device will measure zero or a very low resistance.

Op-amps are complex integrated circuits with many internal failures possible. A technician may evaluate op-amp defects or problems with its associated circuitry. Since you cannot troubleshoot an op-amp, you must replace it as a device rather than internal repairs. If an op-amp is defective, it is replaced as any other type of component.

15.4 Analyze and troubleshoot op-amps.
 In order to achieve objective 15.4, you should be able to:

1. examine the datasheet of op-amps to determine its operational characteristics;
2. distinguish between normal and faulty operation of an op-amp.

Data Sheet Analysis

The type of information found on **data sheets** for operational amplifiers varies among manufacturers and the specific type of device. A typical data sheet for an operational amplifier is included at the end of the chapter. We will use this data sheet to analyze typical information available for operational amplifiers.

Refer to the **data sheet** for an **LM 324**/324A/2902/224/224A operational amplifier (Courtesy: Fairchild Semiconductor). This is a device designed to produce a high DC voltage gain and has four separate op-amps in one integrated circuit package. Use the data sheet at the end of the chapter to answer the following:

1. What two packages are available? _____
2. How many op-amps are in the IC package? _____
3. DC voltage gain = _____dB.
4. Operating voltage supply range for the LM224A = _____ V to _____V.
5. Power supply voltage of an LM 324 op-amp = _____.
6. Input voltage range of an LM 2902 = _____ to _____.
7. Power dissipation of a 14-pin dual-inline package LM 224 op-amp = _____ mW.
8. Operating temperature range of an LM 324 op-amp = _____ °C to _____ °C.
9. Maximum input offset voltage of an LM 224 op-amp = _____ mV.
10. Maximum input offset current of an LM 2902 op-amp = _____ nA = _____ mA.
11. Output source current (typical) of an LM 224 op-amp = _____ mA.

Troubleshooting Op-amps

Op-amps are very versatile devices with stable operating characteristics over a wide range of temperature, frequency, input supply voltage, and output load demands. Most op-amps use a split voltage supply, wherein both positive and negative voltages must be supplied to the op-amp for it to function satisfactorily.

Summary

- The minimum number of terminals an op-amp has is five; these consist of two for the power supply voltage, two for the differential input, and one for the output.
- The input terminal labeled – is the inverting input, and the input terminal labeled + is the noninverting input.
- A signal applied to the inverting input is inverted 180° at the output.

www.fairchildsemi.com

LM2902,LM324/LM324A,LM224/ LM224A

Quad Operational Amplifier

Features

- Internally Frequency Compensated for Unity Gain
- Large DC Voltage Gain: 100dB
- Wide Power Supply Range:
  LM224/LM224A, LM324/LM324A : 3V~32V (or ±1.5 ~ 16V)
  LM2902: 3V~26V (or ±1.5V ~ 13V)
- Input Common Mode Voltage Range Includes Ground
- Large Output Voltage Swing: 0V to V_{CC} -1.5V
- Power Drain Suitable for Battery Operation

Description

The LM324/LM324A,LM2902,LM224/LM224A consist of four independent, high gain, internally frequency compensated operational amplifiers which were designed specifically to operate from a single power supply over a wide voltage range. operation from split power supplies is also possible so long as the difference between the two supplies is 3 volts to 32 volts. Application areas include transducer amplifier, DC gain blocks and all the conventional OP Amp circuits which now can be easily implemented in single power supply systems.

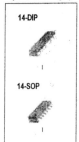

14-DIP

14-SOP

Internal Block Diagram

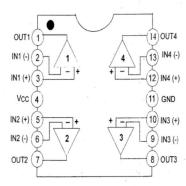

LM2902,LM324/LM324A,LM224/LM224A

Schematic Diagram

(One Section Only)

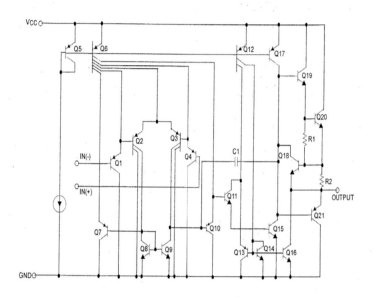

Absolute Maximum Ratings

Parameter	Symbol	LM224/LM224A	LM324/LM324A	LM2902	Unit
Power Supply Voltage	V_{CC}	±16 or 32	±16 or 32	±13 or 26	V
Differential Input Voltage	$V_{I(DIFF)}$	32	32	26	V
Input Voltage	V_I	-0.3 to +32	-0.3 to +32	-0.3 to +26	V
Output Short Circuit to GND Vcc≤15V, TA=25°C(one Amp)	-	Continuous	Continuous	Continuous	-
Power Dissipation, TA=25°C 14-DIP 14-SOP	P_D	1310 640	1310 640	1310 640	mW
Operating Temperature Range	T_{OPR}	-25 ~ +85	0 ~ +70	-40 ~ +85	°C
Storage Temperature Range	T_{STG}	-65 ~ +150	-65 ~ +150	-65 ~ +150	°C

Thermal Data

Parameter	Symbol	Value	Unit
Thermal Resistance Junction-Ambient Max. 14-DIP 14-SOP	$R_{\theta ja}$	95 195	°C/W

Electrical Characteristics

(VCC = 5.0V, VEE = GND, TA = 25°C, unless otherwise specified)

Parameter	Symbol	Conditions	LM224			LM324			LM2902			Unit
			Min.	Typ.	Max.	Min.	Typ.	Max.	Min.	Typ.	Max.	
Input Offset Voltage	V_{IO}	V_{CM} = 0V to VCC -1.5V $V_{O(P)}$ = 1.4V, RS = 0Ω (Note1)	-	1.5	5.0	-	1.5	7.0	-	1.5	7.0	mV
Input Offset Current	I_{IO}	V_{CM} = 0V	-	2.0	30	-	3.0	50	-	3.0	50	nA
Input Bias Current	I_{BIAS}	V_{CM} = 0V	-	40	150	-	40	250	-	40	250	nA
Input Common-Mode Voltage Range	$V_{I(R)}$	Note1	0	-	Vcc -1.5	0	Vcc -1.5	-	0	-	Vcc -1.5	V
Supply Current	I_{CC}	R_L = ∞,VCC = 30V (LM2902,VCC=26V)	-	1.0	3	-	1.0	3	-	1.0	3	mA
		R_L = ∞,VCC = 5V	-	0.7	1.2	-	0.7	1.2	-	0.7	1.2	mA
Large Signal Voltage Gain	G_V	VCC = 15V,RL=2kΩ $V_{O(P)}$ = 1V to 11V	50	100	-	25	100	-	25	100	-	V/ mV
Output Voltage Swing	$V_{O(H)}$	Note1, R_L = 2kΩ	26	-	-	26	-	-	22	-	-	V
		R_L=10kΩ	27	28	-	27	28	-	23	24	-	V
	$V_{O(L)}$	VCC = 5V,RL=10kΩ	-	5	20	-	5	20	-	5	100	mV
Common-Mode Rejection Ratio	CMRR	-	70	85	-	65	75	-	50	75	-	dB
Power Supply Rejection Ratio	PSRR	-	65	100	-	65	100	-	50	100	-	dB
Channel Separation	CS	f = 1kHz to 20kHz (Note2)	-	120	-	-	120	-	-	120	-	dB
Short Circuit to GND	I_{SC}	VCC = 15V	-	40	60	-	40	60	-	40	60	mA
Output Current	I_{SOURCE}	$V_{I(+)}$ = 1V, $V_{I(-)}$ = 0V VCC = 15V $V_{O(P)}$ = 2V	20	40	-	20	40	-	20	40	-	mA
	I_{SINK}	$V_{I(+)}$ = 0V, $V_{I(-)}$ = 1V VCC = 15V $V_{O(P)}$ = 2V	10	13	-	10	13	-	10	13	-	m
		$V_{I(+)}$ = 0V, $V_{I(-)}$ = 1V VCC = 5V,$V_{O(R)}$ = 200mV	12	45	-	12	45	-	-	-	-	μ
Differential Input Voltage	$V_{I(DIFF)}$	-	-	-	Vcc	-	-	Vcc	-	-	Vcc	V

- Op-amps are typically supplied voltage by a split power supply.
- The fundamental operation of an op-amp is based on the differential amplifier.
- When two signals of identical values are applied simultaneously to the op-amp inputs, no output signal is produced; this is referred to as the common mode of operation.
- When different signals are applied simultaneously to the op-amp inputs, the output is the conduction difference in Q_1 and Q_2; this is referred to as the differential mode of operation.
- In the single-ended mode of operation, a signal is applied only to one of the inputs while the other input is grounded.
- A signal applied only to the inverting input of a differential amplifier will be amplified and inverted.
- A signal applied only to the noninverting input of a differential amplifier will be amplified without inversion.
- The internal circuitry of an op-amp consists of three functional units or stages of amplification: differential amplifier, intermediate amplifier, and output.
- The differential amplifier stage provides high gain of the signal difference supplied to the inputs and low gain of common signals supplied to the inputs.
- The constant-current source of a differential amplifier provides for low common-mode gain and good signal coupling.
- A differential amplifier can be made with other transistor configurations or devices, such as with a Darlington pair or JFETs.
- The intermediate amplifier stage provides for additional gain and DC voltage stabilization.
- The low-impedance output stage develops the power needed to drive an external load.
- The output stage of an op-amp can be single-ended or complementary-symmetry.
- Since a single-ended configuration consumes a great deal of power at high current levels, a complementary-symmetry configuration is typically used.
- Open-loop gain refers to the gain of an amplifier connected to a load that does not have a feedback path between the output and input.
- Closed-loop gain refers to the gain of an amplifier that has a feedback path between the output and input.

- When the output voltage of an open-loop op-amp reaches 90% of the source voltage, the op-amp is saturated.
- An input offset voltage is used to overcome the unwanted output of an op-amp.
- Input bias current is the average current flowing into or out of the two op-amp inputs.
- Input impedance is the equivalent resistance that an input source sees when connected to the inputs of an op-amp.
- Slew rate is the rate at which the output of an op-amp changes from one voltage to another in a given time and is an indication of how an op-amp responds to different frequencies.

Formulas

(15-1) $A_{Vol} = V_{out}/V_{in}$ Open-loop voltage gain.

(15-2) $V_{sat} = 0.90 \times V_{CC}$ Op-amp saturation voltage.

(15-3) $V_{in} = V_{sat}/A_{Vol}$ Input voltage to cause op-amp saturation.

(15-4) $R_2 = (R_1 \times R_F)/(R_1 + R_F)$ Value of R_2 in an op-amp feedback network.

(15-5) $A_V = (R_1 + R_F)/R_1$ Gain in a noninverting op-amp feedback network.

(15-6) $A_V = R_F/R_1$ Gain in an inverting op-amp feedback network.

(15-7) $SR = \Delta V_{out}/\Delta_t$ Slew rate.

(15-8) $f_{max} = SR/(6.28 \times V_{out})$ Maximum frequency before output is distorted.

Review Questions

Answer the following questions.

1. The voltage gain figure listed below which is typical of a voltage follower is:

   a. 1
   b. 10
   c. 100
   d. 1000

2. The characteristic that would describe an operational amplifier used as a voltage follower is:

   a. Inverting, high gain
   b. Inverting, low gain

c. Noninverting, high gain
d. Noninverting, low gain

3. Impedance characteristics that describe a voltage follower are:

a. Low input, low output
a. Low input, high output
a. High input, low output
a. High input, high output

4. The voltage gain of an operational amplifier may be calculated by the formula:

a. $A = V_{in} \times V_{out}$
b. $A = V_{in}/V_{out}$
c. $A = V_{out}/V_{in}$
d. None of the above

5. The minimum number of transistors needed to make a discrete component differential amplifier is:

a. One
b. Two
c. Three
d. Four

6. The input stage of almost all operational amplifiers is a(n):

a. NPN stage
b. PNP stage
c. Ground-base stage
d. Differential amplifier

7. Common-mode rejection ratio is described as:

a. Differential gain $\times$ Common-mode gain
b. Differential gain/Common-mode gain
c. Common-mode gain/Differential gain
d. Differential gain/Common-mode gain

8. The circuit shown in **Figure 16-22** is a:

a. Ramp generator
b. Constant-current source
c. Nonsinusoidal oscillator
d. Basic differential amplifier

9. The correct method of finding the gain of an operational amplifier circuit is:

    a. $A = R_1/R_2$
    b. $A = R_2/R_1$
    c. $A = R_1 + 1/R_2$
    d. $A = R_2 + 1/R_1$

10. In the circuit shown in **Figure ??** if R_1 and R_2 were both 1000-Ω resistors, the gain of the operational amplifier circuit would be:

    a. 1.0
    b. 1.5
    c. 2.0
    d. 2.5

11. The gain of the operational amplifier in **Figure ??** is:

    a. +20
    b. −20
    c. +21
    d. −21

12. An inverting op-amp has a 20-kΩ input resistor and a 50-kΩ feedback resistor. With a +2-V input signal applied, the output voltage would be:

    a. +2 V
    b. −2 V
    c. +5 V
    d. −5 V

13. An operational amplifier that has an output of −5 V with an input of +10 V is considered:

    a. Inverting, with a gain of 2
    b. Noninverting ,with a gain of 2
    c. Inverting, with a gain of 0.5
    d. Noninverting, with a gain of 0.5

14. The gain of a *noninverting* operational amplifier may be calculated by the formula:

    a. $A = R_F + 1/R_{in}$
    b. $A = R_F/R_{in}$
    c. $A = R_{in} + 1$
    d. $A = -R_{in}/R_F$

15. The gain of an *inverting* op-amp may be calculated by the formula:

    a. $A = R_F + 1/R_{in}$
    b. $A = -R_F/R_{in}$
    c. $A = R_{in} + 1/R_F$
    d. $A = -R_{in}/R_F$

16. The gain of an operational amplifier with a bandwidth of 10 kHz and a gain bandwidth product of 100,000 is:

    a. 1
    b. 10
    c. 100
    d. 1000

17. A noninverting operational amplifier has a +20 V output when the input is +5 V. If the input resistor is 5 kΩ, the value of the feedback resistor is:

    a. 10 kΩ
    b. 15 kΩ
    c. 20 kΩ
    d. 30 kΩ

18. A noninverting operational amplifier has a 10-kΩ feedback resistor and a 5-kΩ input resistor. The input voltage required to obtain a +10 V output is:

    a. 3.3 V
    b. 5.0 V
    c. 20 V
    d. 30 V

19. The output of the circuit of **Figure ??** when the inputs are +2, −2, and +1 V is:

    a. +1 V
    b. −1 V
    c. +10 V
    d. −10 V

20. The output of the circuit of **Figure ??** when the inputs are +1, +2, and +3 V is:

    a. +3 V
    b. +6 V

  c. −6 V
  d. −60 V

Answers

Examples

15-1. 16,667
15-2. 30
15-3. 400,000
15-4. 26.87 V/μs

Self-Examination

15.1
  1. amplification
  2. +V, −V, inverting input, noninverting input, output (any order)
  3. common
  4. differential
  5. single-ended
15.2
  6. differential amplifier, intermediate amplifier, output (any order)
  7. differential amplifier
  8. common-mode rejection ratio
  9. intermediate amplifier
  10. single-ended, complementary-symmetry (any order)
15.3
  11. open-loop
  12. open-loop voltage gain
  13. 54 μA
  14. negative
  15. 40
  16. −5
  17. input offset voltage
  18. slew rate

Terms

Inverting input

One of two inputs of an operational amplifier or voltage comparator. The inverting input changes the signal 180° in the output and is identified by a − sign on an amplifier symbol.

Noninverting input

One of two inputs of an operational amplifier or voltage comparator. This input accepts a signal and causes it to appear in the output without a change in phase. It is identified by a + sign on an amplifier symbol.

Differential amplifier

An amplifier having a high common-mode rejection capability and output that is proportional to the difference of its two input signals. This type of amplifier is also called a *difference amplifier*.

Common mode

When signals identical in phase and amplitude appear at the inputs of an op-amp simultaneously.

Differential mode

When different signals are applied to the inputs of an op-amp simultaneously.

Single-ended mode

When a signal is applied to one input of an op-amp and the other end is grounded.

Intermediate amplifier stage

The stage of an op-amp that is primarily responsible for additional gain and DC voltage stabilization.

Output stage

The stage of an op-amp that has rather low output impedance and is responsible for developing the current needed to drive an external load.

Common-mode rejection ratio (CMRR)

The ability of a differential amplifier to cancel a common-mode signal.

Open-loop gain

The gain of an amplifier connected to a load that does not have a feedback path between the output and input.

Closed-loop gain

Gain or amplification achieved by an amplifier that has negative feedback between the input and output.

Input offset voltage

Op-amp output voltage that occurs when no differential input voltage is applied.

Input bias current

The average current flowing into or out of the two inputs.

Slew rate

The rate at which an op-amp's output changes from one voltage to another in a given time. It is expressed in volts per microsecond.

16

Linear Op-amp Circuits

A **linear operational amplifier** is a high performance amplifying device made of many discrete components. The assembled amplifier circuit is formed on a small silicon substrate and packaged as an integrated circuit. This device has many applications in electronics. Chapter 16 focuses on linear amplifiers that achieve inverting, noninverting, and summing.

Objectives

After studying this chapter, you will be able to:

16.1 describe the operation of an inverting op-amp circuit;
16.2 describe the operation of a noninverting op-amp circuit;
16.3 evaluate the operation of a summing op-amp circuit;
16.4 analyze and troubleshoot linear op-amp circuits.

Chapter Outline

16.1 Inverting Amplifiers
16.2 Noninverting Amplifiers
16.3 Summing
16.4 Analysis and Troubleshooting – Linear Op-amp circuits

Activities for Chapter 16

16-1 Inverting Operational Amplifiers

In this activity, an inverting op-amp circuit will be evaluated and its gain evaluated.

16-2 Noninverting Operational Amplifiers

In this activity, a noninverting op-amp circuit will be evaluated and its gain evaluated.

Key Terms

averaging amplifier
feedback resistor
input resistor
inverting amplifier
noninverting amplifier
scaling adder
subtractor
summing amplifier
virtual ground
voltage gain

16.1 Inverting Amplifier

A widely used op-amp configuration is the inverting amplifier. The inverting amplifier is a general purpose voltage amplifier that has an output signal 180° out of phase with the input signal. The circuit's voltage gain is simply determined by the ratio of the value of the feedback resistor to the value of the input resistor. The input impedance (Z_{in}) is equal to the value of the input resistor.

16.1 Describe the operation of an inverting op-amp circuit.
   In order to achieve objective 16.1, you should be able to:

   • analyze an inverting op-amp circuit;
   • describe the characteristics of inverting op-amp circuits;
   • define inverting amplifier and virtual ground.

An **inverting amplifier** is a circuit that receives a signal voltage at its input and delivers a large, undistorted version of the signal at its output. The phase or polarity of the output signal is an inversion of the input. Operation does not ordinarily permit the output to reach saturation. The level of amplification is controlled by a feedback resistor connected between the output and inverting input. This causes the amplifier to have negative feedback. The addition of a feedback resistor permits the amplifier to have a controlled level of amplification. Performance is no longer dependent on the open-loop gain (A_{Vol}) of the device. Closed-loop voltage gain (A_{Vcl}), or simply A_V, can be controlled by altering the value of feedback resistor (R_F).
   A typical **inverting op-amp** circuit is shown in **Figure 16.1**. This basic circuit consists of an op-amp and three resistors. The noninverting input

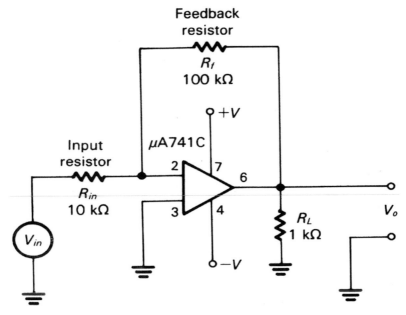

Figure 16.1 Inverting op-amp.

is connected to ground. Input to the amplifier is applied to the inverting input through resistor R_{in}. The output signal is developed across the load resistor (R_L) and ground. A portion of the output signal is also returned to the inverting input through the feedback resistor (R_F). The value of the inverting input signal is, therefore, determined by a combination of V_{in} and the output signal fed back through R_F. Thus, voltage gain (A_r) = $-R_f/R_{in}$ and $Z_{in} = R_{in}$.

To assess the operation of an **inverting amplifier**, we will describe a number of events that occur rather quickly when it is placed into operation. Take a look at **Figure 16.2**. The op-amp is energized by ±15 V. Assume that no signal is initially applied to the input. In this operational state, with no differential input signal applied, the output is zero. This represents the **quiescent** condition, or steady state of operation.

If an input signal of +1 V is applied to the inverting input of the op-amp, the inverting input immediately goes positive. This action causes V_{out} to swing immediately in the negative direction. At the same instant, a negative-going voltage is fed back to the inverting input through R_F. This immediately reduces the original +1 V applied to the inverting input. The **feedback** signal does not completely cancel the V_{in} signal. It simply reduces the value. The +1 V signal is immediately changed to a value of only a few

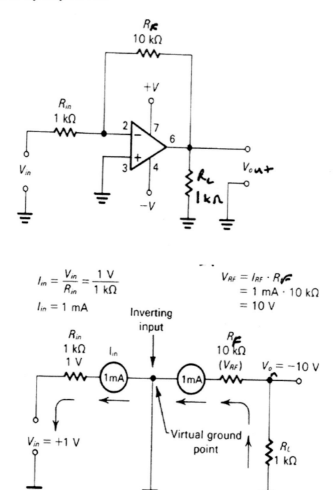

Figure 16.2 Virtual ground point of an inverting op-amp. (a) Inverting op-amp circuit. (b) Equivalent circuit.

microvolts. This means that the inverting input is now controlled or limited to a rather low voltage value. As a general rule, the input is considered to be at approximately zero. Through the feedback loop, the inverting input voltage is held to approximately zero regardless of the value of V_{in}.

To see how an inverting op-amp responds to an input signal, we must consider the virtual ground concept. A **virtual ground** is the point of a circuit that is at zero potential (0 V) but is not actually connected to ground. In an inverting op-amp circuit, a virtual ground appears at the inverting input terminal. With the noninverting input grounded, the voltage at the inverting input is never greater than a fraction of a millivolt. V_{in}, V_{out}, R_{in}, and R_F all tend to hold the voltage of the inverting input to practically zero. With this condition existing, the inverting input responds as if it were grounded. It is a common practice to refer to this point of an op-amp as a virtual ground.

Look at the inverting op-amp and its equivalent circuit in **Figure 16.2**. Keep in mind that the voltage at the inverting input terminal is nearly zero and its input impedance is approximately 1 MΩ. Assume that an input of +1 V is applied to V_{in}. This condition causes 1 mA of current to flow through R_{in}. The **inverting input** is calculated through the following formula:

$$I_{in} = V_{in}/R_{in}. \tag{16.1}$$

Therefore

$$\begin{aligned} I_{in} &= V_{in}/R_{in} \\ &= 1\ V/1\ k\Omega \\ &= 1\ mA. \end{aligned}$$

This shows that a 1-V drop will appear across R_{in}. The inverting input continues to remain at zero or virtual ground. The 1 mA of current entering the equivalent circuit at the V_G point must, therefore, flow through the feedback resistor (R_F). Very little input control flows into the inverting input. This means that, practically, all of the 1 mA flows through the 10-kΩ feedback resistor. The voltage drop across R_F will be 1 mA times 10 kΩ, or 10 V. Note the calculation near R_F of the equivalent circuit.

$$\begin{aligned} V_{RF} &= I_{RF} \times R_F \tag{16.2} \\ V_{RF} &= I_{RF} \times R_F \\ &= 1\ mA \times 10\ k\Omega \\ &= -10\ V. \end{aligned}$$

As indicated, the **polarity** of the output voltage is negative with respect to the V_G point. This shows the inverting characteristic of the op-amp.

Example 16-1:

Calculate the voltage drop across the feedback resistor of an inverting op-amp circuit when $R_f = 100\ k\Omega$ and $I_{RF} = 0.05\ mA$.

Solution

$$V_{RF} = I_{RF} \times R_F$$
$$= 0.05 \text{ mA} \times 100 \text{ k}\Omega$$
$$= 5 \text{ V}.$$

Related Problem

Calculate V_{RF} of an inverting op-amp circuit when $R_f = 10 \text{ k}\Omega$ and $I_{RF} = 0.25 \text{ mA}$.

At this point, you may be wondering how an op-amp controls a -10 V output signal if the inverting input voltage and current are both zero. It should be pointed out that the inverting input terminal is considered to be at a virtual zero. In a circuit, a signal applied to V_{in} always causes some voltage and current to appear at the inverting input terminal. The actual value of it is so small that it is considered to be zero. In practice, these values are usually not measurable. In effect, this means that the right side of R_{in} is considered to be zero and the left side is V_{in}. Across the feedback resistor, the right side is V_{out} and the left side is zero.

The **closed-loop voltage amplification** of an inverting amplifier is determined primarily by the value of R_{in} and R_F. For the op-amp of **Figure 16.2**, this is based on the following formula:

$$A_{Vcl} = -R_F/R_{in}$$
$$= -10 \text{ k}\Omega/1 \text{ k}\Omega$$
$$= -10.$$

The negative sign of the formula of A_v denotes the inversion function of the op-amp. Typical applications of the inverting amplifier have values that range from -1 to $-10,000$. By selecting values of R_f and R_{in}, an inverting op-amp circuit can be easily designed for a specific gain.

Example 16-2:

Calculate the voltage gain of an inverting op-amp with an input resistor (R_{in}) $= 5 \text{ k}\Omega$ and a feedback resistor $R_f = 100 \text{ k}\Omega$. What is the input impedance (Z_{in})?

Solution

$$A_v = -R_f/R_{in} \quad Z_{in} = R_{in}$$

$$= 100 \text{ k}\Omega/5 \text{ k}\Omega = 5 \text{ k}\Omega$$
$$= -20.$$

Related Problem

Calculate the voltage gain (A_v) of an inverting op-amp circuit and the input impedance (Z_{in}) with the following resistance values: $-R_f = 200$ k Ω, $R_{in} =$ 10 k Ω.

Example 16-3:

Select values of R_f for an inverting op-amp circuit to produce a voltage gain of -50, when $R_{in} = 2$ kΩ.

Solution

Since

$$A_v = -R_f/R_{in},$$
$$-50 = -R_f/2 \text{ k}\Omega$$
$$100 \text{ k}\Omega = R_f.$$

Related Problem

Select values of R_f for a noninverting op-amp circuit, with $R_{in} = 2$ kΩ to produce a voltage gain of -1000.

Self-Examination

Answer the following questions.

1. A circuit that receives a signal voltage at its input and delivers a large, undistorted version of the signal $180°$ out of phase at its output is a(n) _____ amplifier.
2. The point of a circuit that is at zero potential (0 V) but is not actually connected to ground is called a _____.
3. An inverting op-amp has a 10-kΩ R_F and a 2-kΩ R_{in}. The potential A_V is _____.
4. An inverting op-amp has a 22-kΩ R_{in} and a 68-kΩ R_F. With a +0.5 V_{pp} input signal, V_{out} is _____.
5. The voltage gain of an inverting op-amp is the ratio of R_f to _____.

16.2 Noninverting Amplifiers

A noninverting on-amp is a general purpose voltage amplifier circuit that has an output signal in phase with the input signal. The voltage gain is determined by the ratio of the feedback resistor value to the input resistor that is connected to the inverting input. The noninverting op-amp circuit has a very high input impedance.

16.2 Describe the operation of a noninverting op-amp circuit.
   In order to achieve objective 16.2, you should be able to:

   • analyze an inverting op-amp circuit;
   • describe the characteristics of noninverting op-amp circuits;
   • define noninverting amplifier.

A **noninverting amplifier** can provide controlled voltage gain with high input impedance and no inversion of the input-output signals. Voltage gain is dependent on the input voltage and feedback resistors. An unusual feature of the noninverting op-amp circuit configuration is the placement of the feedback resistor network. It is placed in the inverting input with a resistor connected to ground. The input voltage (V_{in}) is applied to the noninverting input.
   A representative **noninverting amplifier** is shown in **Figure 16.3**. In this circuit, the signal voltage (V_{in}) to be amplified is applied to the noninverting

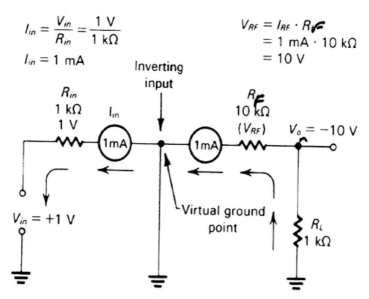

$$I_{in} = \frac{V_{in}}{R_{in}} = \frac{1\ V}{1\ k\Omega}$$

$$I_{in} = 1\ mA$$

$$V_{RF} = I_{RF} \cdot R_F$$
$$= 1\ mA \cdot 10\ k\Omega$$
$$= 10\ V$$

Figure 16.3   Noninverting op-amp circuit.

input (+). A fraction of the output voltage (V_{out}) is returned to the inverting input (−) through a voltage divider network composed of R_F and R_1. In theory, we again assume that very little current flows into the + and − inputs due to the virtual ground concept. This means that the differential input voltage (V_{di}) is essentially zero. The voltage (V_1) developed across resistor R_1 is, therefore, equal to the input voltage (V_{in}). This means that the current (I_{R1}) passing through R_1 is equal to either

$$I_{R_{in}} = V_{in}/R_{in} \tag{16.3}$$

or

$$I_{Rin} = V_{in}/R_{in}. \tag{16.4}$$

Note also that I_{R1} is equal to I_F. The voltage out, therefore, is equal to

$$V_{out} = I_{Rin}(R_{in} + R_f). \tag{16.5}$$

In terms of voltage gain, we can then say

$$A_V = V_{out}/V_{in}.$$

In an actual circuit, the **closed-loop voltage gain** of a noninverting amplifier depends almost entirely on the external circuit components. In this regard, A_V is usually determined by the values of R_1 and R_F. The standard voltage gain formula for a noninverting circuit configuration is expressed as follows:

$$A_V = (R_{in} + R_f)/R_{in} \tag{16.6}$$

or

$$A_V = (R_f/R_{in}) + 1. \tag{16.7}$$

Example 16-4:

Calculate the voltage gain of the noninverting op-amp circuit shown in **Figure 16.3**.

Solution

$$\begin{aligned} A_v &= (R_f/R_{in}) + 1 \\ &= (99 \text{ k}\Omega/1 \text{ k}\Omega) + 1 \\ &= 100. \end{aligned}$$

Related Problem

Calculate the voltage gain of a noninverting op-amp circuit with $R_f = 200$ kΩ and $R_{in} = 1.5$ kΩ.

Next, consider the modified **noninverting amplifier** circuit shown in **Figure 16.3(b)**, in which the resistor R_{in} is removed (making it equivalent to an infinite resistance), and the feedback resistor R_F is replaced by a direct or short connection between the inverting terminal and the output terminal. With this modification, the gain of the noninverting amplifier is calculated by setting $R_{in} = 8$ Ω, $R_f = 0$ Ω

$$A_V = (R_f/R_{in}) + 1 = (0/8) + 1 = 1.$$

Hence, this modified noninverting op-amp circuit configuration provides a unity (**??**) gain and is also called the **unity follower**. The output signal is in phase with the input signal and remains essentially unchanged.

Self-Examination

Answer the following questions.

6. A(n) _____ provides controlled voltage gain with high input impedance and no inversion of the input-output signals.
7. A noninverting op-amp circuit configuration places the feedback resistor in the _____ input with a resistor connected to ground.
8. If the feedback resistor for a noninverting circuit is 78 kΩ and the input resistor is 2 kΩ, $A_V = $ _____.
9. A noninverting op-amp has an R_{in} of 10 kΩ and an R_F of 120 kΩ. With a +0.6 V_{pp} input voltage, V_{out} is _____.
10. Input impedance of a noninverting op-amp circuit is (high, low).

16.3 Summing Amplifiers

A summing operational amplifier circuit can be used to add (calculate the sum) of input signal voltages. The inputs have individual voltage gains and can be either negative or positive. The amplifiers shown in **Figure 16.4** are inverting summing amplifiers.

16.3 Evaluate the operation of a summing op-amp circuit.
    In order to achieve objective 16.3, you should be able to:
    • analyze a summing op-amp circuit;

- describe the characteristics of summing op-amp circuits;
- define summing point and scaling adder.

A slight modification of the inverting op-amp permits it to achieve the mathematical operation of addition. Used in this manner, it can add DC voltages or AC waveforms. Adding or summing operations are very useful in computers and in analog-to-digital conversion functions.

Summing can be achieved when two or more voltage values are applied to the input of an op-amp. Representative schematic diagram of an adder is shown in **Figure 16.4**. Note that the circuit is an inverting op-amp with two inputs. The resistor values are 1 kΩ. The resulting output voltage that appears across R_1 is the sum of the input voltages applied to V_1 and V_2. For simplicity, the circuit shows the addition of DC voltage values. Keep in mind that it can add AC voltage values equally well.

The **voltage gain** of this summing op-amp circuit is 1. This is based on the resistance values of R_{in} and R_F. The output, however, is -1 due to the inversion characteristic of the op-amp. If a positive value is desired, the output of the adder can be followed by an op-amp with a gain of 1.

The voltage applied to each input of a summing amplifier responds as an independent source. Input V_1 does not alter or change the input of V_2. This is primarily due to the virtual ground appearing at the inverting input terminal. In **Figure 16.4**, note that input V_1 causes 3 mA of current to flow through

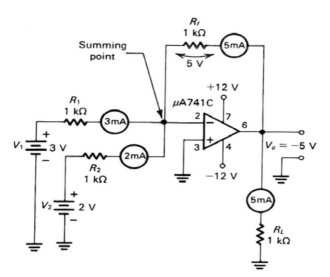

Figure 16.4   Op-amp summing operation.

input resistor R_1. With only this voltage applied, the feedback resistor R_F would have a resulting 3 mA through it. The load resistor R_L would likewise have 3 mA and develop an output voltage of 3 V. An applied input voltage of $+3$ V produces a corresponding -3 V output. Therefore, $3 + 0$ V equals 3 V.

Applying $+3$ V to V_1 and $+2$ V to V_2 causes 3 and 2 mA to flow through the respective resistors. Current flow at the inverting input will be the sum of these two values or 5 mA. This point of the circuit is usually called the **summing point** instead of the virtual ground. In effect, this common point isolates V_1 from V_2. The combined current from this point must pass through the feedback resistor R_F. Thus, R_F has 5 mA of current. The voltage drop across R_F will be 5 V, with V_{out} being -5 V due to the inverting characteristic. A similar current value flows through the load resistor R_L. In effect, an input of $+3$ and $+2$ V causes the output to increase to -5 V. The summing function is expressed by the following formula:

$$V_{out} = -(V_1 + V_2). \tag{16.8}$$

This formula is used only when the input resistors are of the same value. Any number of input voltages may be added by this circuit. The input resistors must be of the same value for each input voltage.

Figure 16.5 shows three variations of the **summing circuit**. **Figure 16.5(a)** shows the standard adding function. This particular circuit simply adds the input voltages and produces the sum at the output. All resistors must be of an equal value for the circuit to respond correctly.

Example 16-5:

A two-input summing op-amp circuit has the following values (see **Figure 16.4**): $R_1 = 5$ kΩ, $R_2 = 5$ kΩ, and $R_f = 10$ kΩ. If $V_1 = 5$ V and $V_2 = 3$ V, what is the voltage gain (A_v)?
Solution

$$V_0 = -R_f[(V_1/R_1) + (V_2/R_2)]$$
$$= -10k\Omega[(5V/5k\Omega) + (3V/5k\Omega)]$$
$$= -16.$$

Related Problem

Calculate the voltage gain of a two-input summing op-amp with the following circuit values: $R_1 = 10$ kΩ, $R_2 = 10$ kΩ, $R_f = 20$ kΩ, $V_1 = 3$ V, and $V_2 = -5$ V.

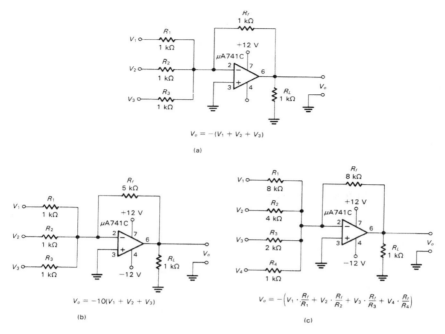

Figure 16.5 Summing amplifiers. (a) Standard adder. (b) Adder with gain. (c) Scaling adder.

The op-amp of **Figure 16.5(b)** achieves both **summing** and **gain**. The input resistors are of the same value. The **feedback resistor** is a multiple of the input resistors. The sum of V_1, V_2, and V_3, in this case, is multiplied by a factor of 5. The formula for this mathematical operation is

$$V_{\text{out}} = R_{\text{f}}(V_1 + V_2 + V_3). \tag{16.9}$$

Example 16-6:

Refer to **Figure 16.5(b)**. Calculate V_0.

Solution

$$\begin{aligned}
V_0 &= -R_{\text{f}}[(V_1/R_1) + (V_2/R_2) + (V_3/R_3)] \\
&= -5\ k\Omega[(5\ V/1\ k\Omega) + (3\ V/1\ k\Omega) + (1\ V/1\ k\Omega)] \\
&= -50.
\end{aligned}$$

Related Problem

Refer to **Figure 16.5(b)**. Calculate V_O with the following circuit values: $R_f = 10$ kΩ, $V_1 = 6$ V, $V_2 = -4$ V, $V_3 = 2$ V, $R_1 = 1$ kΩ, $R_2 = 1$ kΩ, and $R_3 = 1$ kΩ.

Figure 16.5(c) is a **scaling adder**. The input to this circuit sees different weighting factors in the input resistors. An input of $+1$ V applied to V_1 produces a 1-V output. In the same manner, V_2 is 2 V, V_3 is 4 V, and V_4 is 8 V. This means that the inputs are scaled down by the powers of 2. The resulting output of V_4 carries eight times more weight than V_1. Scaling adders are frequently used in digital-to-analog converter functions. The formula for determining the output of a scaled adder notes the difference in the weighting factor.

$$V_{\text{out}} = -\left[V_1 \times \frac{R_f}{R_1} + V_2 \times \frac{R_f}{R_2} + V_3 \times \frac{R_f}{R_2} + V_4 \times \frac{R_f}{R_4} \right] \qquad (16.10)$$

$$V_{\text{out}} = -R_f \left[\left(\frac{V_1}{R_1} \right) + \left(\frac{V_2}{R_2} \right) + \left(\frac{V_3}{R_3} \right) + \left(\frac{V_4}{R_4} \right) \right].$$

It should be apparent that voltages fed into the smaller valued input resistors are more heavily scaled because they produce a larger voltage output.

A modification of the summing op-amp is known as an **averaging amplifier**. **Figure ??** shows an averaging amplifier that is basically a summing op-amp circuit with input resistors of equal value and the value of any input resistor to the feedback resistor equal to the number of inputs.

Another modification of a summing amplifier is the **subtractor** shown in **Figure ??**.

The voltage output of the subtractor circuit is equal to the difference between V_1 and V_2. The V_1 signal is applied to an inverting op-amp that is designed to have a gain of 1.0 (unity gain). The output of this inverting amplifier is equal to $-V_1$. This output is then applied to a summing op-amp with unity gain through R_2. The input to the summing amplifier combines $-V_1$ and V_2 signals. The output voltage (V_O) is calculated as follows: $-V_O = V_2 + (-V_1) = V_2 - V_1$.

Example 16-7:

Refer to **Figure 16.5(c)**. Calculate V_O for this four input summing op-amp circuit.

Solution

$$V_{out} = - \left[V_1 \times \frac{R_f}{R_1} + V_2 \times \frac{R_f}{R_2} + V_3 \times \frac{R_f}{R_2} + V_4 \times \frac{R_f}{R_4} \right]$$

$$V_{out} = - \left[3 \times \frac{1000}{8000} + 2 \times \frac{1000}{4000} + 1 \times \frac{1000}{2000} + 4 \times \frac{1000}{1000} \right]$$

$$= -5.375\ V$$

Related Problem

Refer to **Figure 16.5(c)**. Change the value of R_f to 2000 kΩ and calculate the value of V_{out}.

Self-Examination

Answer the following questions.

11. An inverting op-amp circuit may be used to accomplish the mathematical function of _____.
12. Refer to the summing amplifier of **Figure 16.5(a)**. If V_1 equals +2 V, V_2 equals +3 V, and V_3 equals −1 V, V_{out} is _____.
13. The ratio of the feedback resistor to the input resistor value of an op-amp is _____.
14. Each of the op-amps of a subtractor has _____ gain.
15. The output voltage of a subtractor is the (sum, difference) of the applied input voltages.

16.4 Analysis and Troubleshooting – Linear Op-amps

In order to perform a **test** on operational amplifiers, you must understand their characteristics and be proficient in the use of the test equipment. Operational amplifiers can be tested using analog meters or digital multimeters in some situations. A defect such as an **open** circuit is easy to detect when no power is applied to the circuit being tested with an ohmmeter reading of an extremely high resistance. A defective **shorted** device will measure zero or a very low resistance.

Op-amps are complex integrated circuits with many internal failures possible. A technician may evaluate op-amp defects or problems with its associated circuitry. Since you cannot troubleshoot an op-amp, you must

replace it as a device rather than internal repairs. If an op-amp is defective, it is replaced as any other type of component.

16.4 Analyze and troubleshoot linear op-amp circuits.

In order to achieve objective 16.4, you should be able to:

1. determine the configuration of a given op-amp circuit;
2. distinguish between normal and faulty operation of a linear op-amp circuit.

Troubleshooting Op-amps

The internal construction of the op-amp cannot be altered. However, by connecting external electrical components, its overall operation can be modified. When connected in **open-loop**, the voltage gain of the op-amp is extremely high and is limited by the value of the power supply. In **closed-loop** operation, the **voltage gain** can be controlled by using external feedback components such as resistors, capacitors, and even diodes. During operation, if the feedback element is accidentally disconnected, the op-amp will not function satisfactorily.

   Let us assume that an op-amp circuit is configured to operate as an oscillator or an AC generator. When the circuit is energized, it produces an AC output that drives a speaker load. If the circuit is functioning properly, the speaker will show that sound is being produced. The output AC waveform can also be observed on an oscilloscope.

Summary

1. The inverting input of an op-amp produces an output voltage with a 180° phase difference from input to output.
2. The noninverting input of an op-amp produces an output voltage with no phase difference between input and output voltages.
3. Inverting amplifier voltage gain $(A_v) = R_f/R_{in}$.
4. Noninverting amplifier gain $(A_v) = (R_f/R_{in}) + 1$.
5. A unity follower is a modified noninverting amplifier with a gain of 1.
6. A summing op-amp circuit can add input signal voltages algebraically.
7. A subtractor op-amp circuit can be used to determine the algebraic difference between two input signals.

Formulas

Inverting Op-amp Circuits

(16-1) $I_{in} = V_{in}/R_{in}$ Inverting input current.
(16-2) $V_{RF} = I_{RF} \times R_F$ Feedback resistor voltage.

Noninverting Op-amp Circuits

(16-3) $I_{R1} = V_{in}/R_1$ Current through resistor 1.
(16-4) $I_{R1} = V_1/R_1$ Current through resistor 1.
(16-5) $V_{out} = I_{R1} (R_1 + R_F)$ Output voltage.
(16-6) $A_V = (R_1 + R_F)/R_1$ Gain.
(16-7) $A_V = (R_F/R_1) + 1$ Gain.

Summing Op-amp Circuits

(16-8) $V_{out} = -(V_1 + V_2)$ Output voltage of a standard adder.
(16-9) $V_{out} = -10(V_1 + V_2 + V_3)$ Output voltage of an adder with gain.
(16-10) $V_{out} = -(V_1 \times (R_F/R_1) + V_2 \times (R_F/R_1) + V_3 \times (R_F/R_3) + V_4 \times (R_F/R_4))$ Output voltage of a scaling adder.

Review Questions

Answer the following questions.

1. An op-amp subtraction circuit can be constructed using:

    a. Two noninverting op-amps connected in series
    b. Two inverting op-amps connected in series
    c. Two open-loop op-amps connected in series
    d. Not possible

2. The output of a typical inverting op-amp when no voltage is applied to the input terminals is:

    a. Positive supply voltage
    b. Negative supply voltage
    c. 0 V
    d. Cannot be determined

3. What is the phase difference between the input and output of an inverting amplifier?

    a. $0°$

    b. $45°$

    c. $90°$

    d. $180°$

4. The phase difference between the input and output of a noninverting op-amp is:

    a. $0°$

    b. $45°$

    c. $90°$

    d. $180°$

5. A virtual ground point or terminal of an op-amp circuit is:

    a. Positive supply voltage

    b. When the negative supply is grounded

    c. Point at 0 potential even though it is not grounded

    d. Point at negative supply potential

6. The voltage gain of an op-amp depends:

    a. Only on feedback resistor

    b. Only on input resistor

    c. On both input and feedback resistor

    d. all of the above

7. A general purpose op-amp circuit that has the output and input signals in phase is:

    a. Open-loop amplifier

    b. Inverting amplifier

    c. Noninverting amplifier

    d. Summing amplifier

8. A general purpose op-amp circuit that has the output and input signals out of phase is:

    a. Open-loop amplifier

    b. Inverting amplifier

    c. Noninverting amplifier

    d. Non-feedback amplifier

9. The gain of a unity follower op-amp is:

    a. 1

    b. 10

    c. 100

    d. Infinite

10. The unity follower op-amp configuration is derived from the:

    a. Open-loop op-amp configuration

    b. Inverting op-amp configuration

    c. Noninverting op-amp configuration

    d. Summing op-amp configuration

Problems

Answer the following questions.

1. In an inverting op-amp amplifier circuit, such as that shown in **Figure 16.2**, if R_{in}= 10 kΩ and R_F= 10 kΩ, with an input of V_{in}= 1 V applied, the output voltage will be:

    a. +1 V

    b. 0 V

    c. −1 V

    d. −2 V

2. If the gain of the inverting amplifier shown in **Figure 16.2** is to be adjusted to −5, with an input of V_{in}= 1 V applied, and with R_{in}= 10 kΩ, the new value of R_F should be:

    a. 10 kΩ

    b. 25 kΩ

    c. 50 kΩ

    d. 100 kΩ

3. In a noninverting op-amp amplifier circuit, such as that shown in **Figure 16.3**, if R_{in}= 10 kΩ and R_F= 10 kΩ, with an input of 1 V applied, the output voltage will be:

    a. +1 V

    b. 0 V

    c. −1 V

    d. 2 V

4. In the standard summer circuit shown in **Figure 16.5**, if the resistor input resistor R_3 is removed from the circuit, the output voltage will be:

    a. $-(V_1 + V_2 + V_3)$ V
    b. $-(V_1 + V_2)$ V
    c. $-(V_1)$ V
    d. 0 V

5. In the unity follower circuit shown in **Figure 16.3(b)**, if the input voltage V_{in} is 2 V, the output will be:

    a. $+2$ V
    b. -2 V
    c. 0 V
    d. $+1$ V

Answers

1. Two or more
2. 0 V
3. 180°
4. 0°
5. The voltage difference between inputs

Examples

16-1. 2.5 V
16-2. $A_v = -20$, $Z_{in} = 10$ kΩ
16-3. 2000 kΩ or 2 MΩ
16-4. 134.33
16-5. $+4$ V
16-6. -10.75

Self-Examination

16.1

1. inverting
2. virtual ground
3. -5
4. -1.55 V_{pp}
5. R_{in}

16.2
6. noninverting amplifier
7. inverting
8. 40
9. 7.8 V
10. high

16.3
11. addition
12. −4
13. voltage gain
14. unity
15. difference

Terms

Inverting amplifier

An op-amp circuit that receives a signal voltage at its input and delivers a large undistorted version of the signal at its output.

Virtual ground

The point of a circuit that is at zero potential (0 V) but is not actually connected to ground.

Noninverting amplifier

An op-amp circuit that provides controlled voltage gain with high input impedance and no inversion of the input-output signals.

Unity follower

A modified noninverting amplifier circuit that has a gain of 1, by shorting out the feedback resistor and removing the input resistor in this circuit so that the output is identical to the input.

Summing point

A common point that isolates input voltage values applied to the inverting input.

Scaling adder

A summing amplifier that has different weighted values applied to its input resistors.

17

Specialized Op-amps
and Integrated Circuits

General purpose op-amps are extremely versatile electronic devices that can be used to achieve a variety of different functions. To this point, we have looked at the fundamental operation of the device. It can, however, be used to achieve a variety of very specialized operations that are extremely valuable in electronic circuitry. These operations are primarily derived from the basic op-amp. Specialized op-amp circuitry is used to identify this group of device applications. Included in this classification are things such as **controlled voltage and current sources**, **converters**, **comparators**, **integrators**, **differentiators**, **precision diodes**, **logarithmic devices**, and **instrumentation amplifiers**.

Objectives

After studying this chapter, you will be able to:

17.1 explain the operation of a controlled voltage and current source;
17.2 analyze the operation of a converter op-amp;
17.3 explain the operation of a comparator;
17.4 describe the operation of an integrator;
17.5 explain how a differentiator works;
17.6 analyze the operation of a precision rectifier;
17.7 analyze the operation of an instrumentation op-amp;
17.8 analyze and troubleshoot specialized op-amp circuits.

Chapter Outline

17.1 Controlled Voltage and Current Sources
17.2 Converters

671

17.3 Comparators
17.4 Integrators
17.5 Differentiators
17.6 Precision Rectifiers
17.7 Instrumentation Op-amps
17.8 Analysis and Troubleshooting – Specialized Op-amp Circuits

Key Terms

constant-current sources
constant-voltage source
voltage-controlled voltage source
voltage-controlled constant-current source
converters
voltage-to-current converters
current-to-voltage converters
comparators
adjustable-referenced comparator
active circuit
waveshaper
integrators
differentiator
precision rectifiers
external gain resistor
instrumentation amplifier

17.1 Controlled Voltage and Current Sources

A specialized op-amp circuit application is described as a controlled or constant energy source. This type of circuit has its output voltage or current determined by the magnitude of another independent voltage or current source. The output of this circuit will then remain at a constant value that is independent of load changes. All of this depends on the circuit configuration being utilized by an op-amp.

17.1 Explain the operation of a controlled voltage and current source.
   In order to achieve objective 17.1, you should be able to:

   1. define the terms "controlled voltage source" and "controlled current source";

2. examine the operation of an op-amp controlled voltage source circuit;
3. examine the operation of an op-amp controlled current source circuit.

When an op-amp is connected as a **constant energy source**, it generally utilizes one or more of the characteristics of an **ideal amplifier**. This includes things such as infinite gain or amplification, infinite input impedance, and zero or very low output impedance. When these attributes are properly utilized, the selected output of an op-amp can be used as a constant source of voltage or current for a variety of different circuit applications.

A **voltage-controlled** voltage source is a common op-amp circuit application. This circuit produces an output voltage that is equal to the constant value of another controlling voltage and is independent of the current drawn from it. Both inverting and noninverting configurations of the op-amp can be used to achieve this condition of operation.

Figure 17.1 shows a representative circuit of an op-amp connected as a **constant voltage source**. In this circuit, the op-amp is connected as an inverting amplifier with feedback to control its gain or amplification. This causes the output voltage (V_{out}) to become a fixed or a constant value due to the high level of amplification achieved. Ideally, the **closed-loop amplification (A_{cl})** of this circuit is determined by the resistance of R_{in} and R_f. This is expressed as follows:

$$A_{cl} = R_f/R_{in}. \qquad (17.1)$$

The resulting V_{out} of the circuit is then the product of the input voltage (V_{in}) and A_{cl}. This is expressed as follows:

$$V_{out} = A_{cl} \times V_{in}. \qquad (17.2)$$

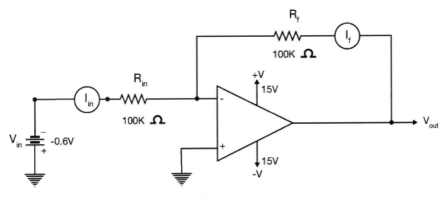

Figure 17.1   Constant voltage source.

Note also that when the op-amp has V_{in} applied to its inverting input through R_{in}, it causes current to flow in both R_{in} and R_f. Since the inverting input of an ideal op-amp has infinite impedance, it causes all of I_{in} to flow into I_f. The values of I_{in} and I_f can now be determined by the expressions $I_{in} = V_{in}/R_{in}$ or $I_f = V_{in}/R_f$. This also means that the value of V_{out} can now be determined as follows:

$$V_{out} = I_f \times R_f.$$

It should also be noted that the value of V_{out} is independent of any current drawn from it. Note also that the **output resistance** of an ideal op-amp is 0 or very low. This means that there will be no voltage division at the output and that (V_{out}) is independent of all load resistance changes. As a result of this, V_{out} becomes a constant voltage source whose value is determined by an independent voltage source that is applied to its input.

Another op-amp energy control circuit is the voltage-controlled **constant-current source**. This type of circuit is primarily designed to supply a constant value of current to a load resistance. The resulting load current remains at a constant value even when the resistance of the load changes. This circuit delivers a load current whose value is equal to a fixed resistance times the voltage supplied by an independent source. The resulting current is also independent of any resistance changes that may occur in the load. **Figure 17.2** shows the circuitry of an inverting op-amp constant-current source. In this circuit, a stable voltage source (V_{in}) is connected to the input resistor (R_{in}) which causes (I_{in}) to flow to the inverting input through R_{in}. This is expressed as follows:

$$I_{in} = V_{in}/R_{in}. \tag{17.3}$$

We know from previous discussions that the **input impedance** of an ideal op-amp is infinite. This means that (I_{in}) must now find an alternate path in which to flow. In previous circuits, a **feedback** resistor (R_f) has provided an alternate path for I_{in}. Note in **Figure 17.2** that the feedback resistor (R_f) is identified as the load resistor (R_L). Now since the input impedance of an ideal op-amp is nearly infinite, R_L provides an alternate path for I_{in}. This means that, practically, all of I_{in} will now flow through R_L which serves as the feedback path. This also means that I_{in} and I_L are now the same. Since I_L and I_{in} are equal, I_L is determined by the same expression as I_{in} or in this case: $I_L = V_{in}/R_{in}$. This means that a constant-current source derives its operating energy from an independent voltage source and causes a **constant-current** flow to the load resistor regardless of its value.

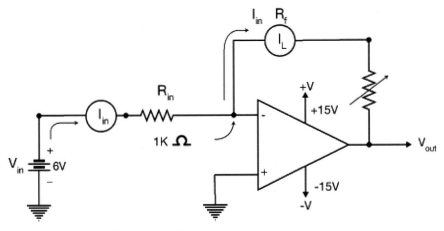

Figure 17.2   Inverting constant-current source.

Example 17-1:

Determine the A_{cl}, I_{in}, I_f, and V_{out} of **Figure 17.1** when -0.6 V is supplied by V_{in}.

Solution

The closed-loop amplification or A_{cl} of the circuit is

$$A_{cl} = R_f/R_{in}$$
$$= 100,000 \; \Omega/100,000 \; \Omega$$
$$= 1.$$

When 0.6 V is applied to R_{in}, it causes I_{in} to be

$$I_{in} = V_{in}/R_{in}$$
$$= 0.6 \text{ V}/100,000 \; \Omega$$
$$= 0.000006 \text{ A or } 6 \; \mu A.$$

Since the inverting input of an ideal op-amp has infinite impedance, it causes all of I_{in} to flow into I_f. This means that I_{in} and I_f have the same value.

As a result of this, V_{out} can be determined as

$$V_{out} = I_f \times R_f$$
$$= 0.000006 \text{ A} \times 100,000 \; \Omega$$
$$= 0.6 \text{ V}.$$

This shows that the output of a voltage-controlled voltage source is the same as the applied input voltage.

When the values of R_{in} and R_f are known, V_{out} can also be determined by the following expression:

$$V_{out} = A_{cl} \times V_{in}$$
$$= 1 \times -0.6 \text{ V}$$
$$= 0.6 \text{ V}.$$

Related Problem

Assume now that R_{in} is 1000 Ω, R_f is 5000 Ω, and V_{in} is 0.8 V. What are the A_{cl}, I_{in}, I_f, and V_{out} of a circuit when this occurs?

Example 17-2:

Determine the amount of load current (I_L) supplied to a 1-kΩ resistor R_L when V_{in} is 6 V. Then determine the value of I_L that occurs when R_L changes to 5 kΩ.

Solution

When a V_{in} of 6 V is applied to R_{in}, it causes I_{in} to be

$$I_{in} = V_{in}/R_{in}$$
$$= 6 \text{ V}/1000 \ \Omega$$
$$= 0.006 \text{ A or 6 mA}$$

Since $I_{in} = I_L$, then

$$I_L = V_{in}/R_{in}$$
$$= 6 \text{ V}/1000\Omega$$
$$= 0.006 \text{ A or 6 mA}.$$

Now if R_L is changed to 5 kΩ, I_L will continue to be 0.006 A as long as V_{in} and R_{in} remain the same value.

Related Problem

Determine the I_L of **Figure 17.2** when R_L and R_{in} are both 5 kΩ and V_{in} is 10 V. What happens to I_L when R_L changes to 1 kΩ?

Self-Examination

Answer the following questions.

1. A constant-current source is a circuit that produces a:

    a. Variable value of current
    b. Constant value of current
    c. AC only
    d. Constant value of current and voltage

2. A voltage-controlled constant-current source generates a constant-value current which:

    a. Does not depend on the value of a controlling voltage
    b. Depends on the value of a controlling voltage
    c. Depends on the open-loop gain
    d. Varies with the load

3. The load resistor of a constant voltage source op-amp circuit of **Figure 17.1** should be connected to:

    a. V_{out}
    b. V_{in}
    c. The inverting op-amp input
    d. The noninverting op-amp input

4. In the constant-current source of **Figure 17.2**, the feedback component is:

    a. The load resistor
    b. The input resistor
    c. Connected to the noninverting op-amp input
    d. The op-amp

5. In the constant-current source of **Figure 17.2**, if the value of the load resistor, R_L, is decreased, the load current will:

    a. Decrease
    b. Increase
    c. Remain unchanged
    d. Be zero

17.2 Converters

A **converter** is a device that transforms the electrical behavior of a system. Op-amps may be used as converters for this operation. In this regard, they can be used for changing voltage to current or from current to voltage. Transforming a voltage response of photovoltaic cells into a current value can be accomplished using this type of circuit. It is also used in metering devices for transforming voltage and current.

17.2 Analyze the operation of a converter op-amp.

In order to achieve objective 17.2, you should be able to:

- describe the characteristics of converter op-amp circuits;
- explain the operation of converter op-amp circuits.

Figure 17.3 shows an op-amp circuit connected as a **current-to-voltage converter**. This circuit is essentially an inverting op-amp without an input resistor. As a result of this, all of the input current (I_{in}) is applied directly to the inverting input. Since the inverting input of an ideal op-amp has infinite impedance, it is considered to be at **virtual ground** or 0 V as indicated. This then means that all of I_{in} must now flow into the feedback resistor (R_f). The path of this current flow is shown by arrows. As a result of this, an output

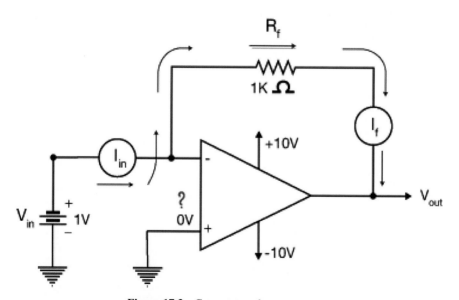

Figure 17.3 Current-to-voltage converter.

voltage (V_{out}) is developed across R_{f}. V_{out} is equal to the product of I_{in} and R_{f} or the following formula:

$$V_{\text{out}} = I_{\text{in}} \times R_{\text{f}}. \tag{17.4}$$

Example 17-3:

Refer now to **Figure 17.3**. Note that when 1 V is supplied by V_{in}, it causes 0.001 A of I_{in} to flow through R_{f}. When this occurs, what is the V_{out} of the circuit?

Solution

When an I_{in} of 0.001 A flows through R_{f}, it causes V_{out} to be

$$V_{\text{out}} = I_{\text{in}} \times R_{\text{f}}$$
$$= 0.001 \text{ A} \times 1000\Omega$$
$$= 1 \text{ V}.$$

Related Problem

What is the V_{out} of the current-to-voltage converter shown in **Figure 17.3**, which has a V_{in} of 2 V, developing 2 mA of I_{in}, that flows through a 2000-Ω load resistor?

Some electronic circuits may use an op-amp to change an applied input voltage into a corresponding current flow. The **voltage-to-current converter** of **Figure 17.4** performs this operation very effectively. In this circuit, the load resistor (R_{L}) is connected in series with the feedback resistor (R_{f}). As a result of this, all of the output current (I_{out}) flows through R_{L} and R_{f}. The resulting I_{out} is then a constant value as long as the input voltage (V_{in}) remains constant. This action is the same for either AC or DC voltage values. Any value change in I_{out} causes a corresponding change in the voltage developed across R_{f}. This means that the **feedback resistor** voltage (V_{Rf}) and feedback voltage (V_{f}) applied to the inverting input have the same value because they are supplied by the same source. Any change in V_{f} causes a corresponding change in the difference voltage (V_{diff}). This means that V_{diff} is equal to $V_{\text{in}} - V_{\text{f}}$ or $V_{\text{f}} - V_{\text{in}}$.

$$V_{\text{diff}} = V_{\text{in}} - V_{\text{f}}. \tag{17.5}$$

The resulting V_{diff} is then amplified by the **open-loop gain** or (A_{ol}) of the circuit. This immediately changes V_{out} and causes it to restore I_{out} to its

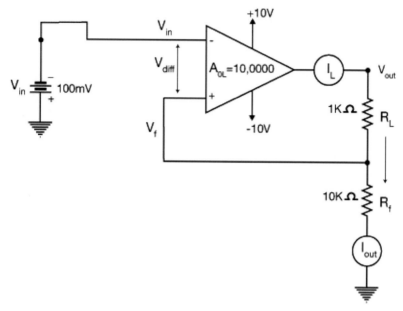

Figure 17.4   Voltage-to-current converter.

original value. Essentially, this means that V_{in} controls V_{out} which, in turn, causes I_{out} to be at the same value.

In **Figure 17.4**, the **converter** has 100 mV applied to V_{in}. This immediately drives V_{out} into saturation or $+10$ V. This is the result of the high open-loop amplification (A_{ol}) of the circuit. Typical A_{ol} values are in the range of 100,000 or more. Now, with $+10$ V applied to R_L, it causes 10 mA of I_{out} to flow through the 1 kΩ and 10 Ω resistors. This, in turn, causes a voltage to be developed across R_L and R_f. The resulting V_{out} or V_{RL}, and V_{Rf} values are equal to I_{out} times the resistance of R_L or R_f. That is,

$$V_{RL} = I_{out} \times R_L \tag{17.6}$$

$$V_{Rf} = I_{out} \times R_f \tag{17.7}$$

$$
\begin{aligned}
V_{RL} &= I_{out} \times R_L \\
&= 10 \text{ mA} \times 1 \text{ k}\Omega \\
&= 0.01 \text{ A} \times 1000 \text{ }\Omega \\
&= 10 \text{ V}
\end{aligned}
\qquad
\begin{aligned}
V_{Rf} &= I_{out} \times R_f \\
&= 10 \text{ mA} \times 10 \text{ }\Omega \\
&= 0.01 \text{ A} \times 10 \text{ }\Omega \\
&= 0.1 \text{ V or} 100 \text{ mV}.
\end{aligned}
$$

As a result of this, V_{Rf} and V_{in} are now 100 mV and the V_{diff} is 0 V. This causes I_{out} to continue to be 10 mA as long as V_{in} remains at the same value.

Example 17-4:

In **Figure 17-4**, when 125 mV is applied to V_{in}, it drives the op-amp into saturation. When this occurs, what are the values of I_{out}, V_{Rf}, and V_{diff} when R_L is 400 Ω and R_f is 5 Ω?

Solution

When the op-amp goes into saturation, V_{out} is +10 V. This causes I_{out} to be

$$I_{out} = V_{sat}/R_L$$
$$= 10 \text{ V}/400 \text{ } \Omega$$
$$= 0.025 \text{ A or} 25 \text{ mA.}$$

V_{Rf} can then be determined by

$$V_{Rf} = I_{out} \times R_f$$
$$= 0.025 \times 5\Omega$$
$$= 0.125\text{V or} 125 \text{ mV.}$$

V_{diff} is then equal:

$$V_{diff} = V_{in} - V_{Rf}$$
$$= 125 \text{ mV} - 125 \text{ mV}$$
$$= 0 \text{ V.}$$

Related Problem

If 50 mV of V_{in} is applied to **Figure 17.4** when R_L is 1 kΩ and R_f is 50 Ω, what are the values of I_{out}, V_{sat}, V_{Rf}, and V_{diff}?

Self-Examination

Answer the following questions.

   6. Functionally, an op-amp converter:

      a. Changes one form of energy into another form of energy
      b. Compares voltage and current values

    c. Amplifies an applied input signal

    d. Changes the shape of an applied input signal

7. The voltage-to-current converter in **Figure 17.4** has:

    a. The load and feedback resistor in series

    b. No feedback resistor

    c. The input signal applied to the inverting op-amp input

    d. No feedback current flow

8. The voltage-to-current converter in **Figure 17.3** has:

    a. The load and feedback resistor in series

    b. No feedback resistor

    c. A feedback resistor and no input resistor

    d. No feedback current flow

9. An op-amp circuit that changes an applied input voltage into a corresponding current flow is called a _____ to_____ converter.

10. In a _____to_____ converter, the applied input voltage controls the output current in such a way as to maintain a constant output voltage.

17.3 Comparators

Operational amplifiers are often used as **comparators** that identify or compare the value of one input voltage to that of another input voltage. One input is considered to be a reference value to which the second input is compared. The resulting output is then an indication of whether the value of the reference input is above or below the value of the alternate input.

17.3 Explain the operation of a comparator.

In order to achieve objective 17.3, you should be able to:

- define comparator and waveshaper;
- describe the characteristics of comparator op-amp circuits;
- analyze comparator op-amp circuits.

Op-amps are rarely used in an **open-loop** circuit configuration. The A_{Vol} is generally so high that it is rather difficult to prevent the output (V_{out}) from being driven into saturation. However, one application of an op-amp in an open-loop circuit configuration is the **differential voltage comparator**.

When an op-amp is used as a comparator, it is generally connected in an open-loop circuit configuration. This means that the circuit does not employ a feedback resistor to control its level of amplification. As a result, the op-amp has extremely high voltage gain capabilities. The open-loop gain A_{ol} in terms of the applied input and generated output voltage is expressed as follows:

$$V_{out} = V_{in} \times A_{ol}. \tag{17.8}$$

A very minute difference between the two inputs will drive the output voltage into saturation or to its maximum voltage value. This value cannot, however, exceed the maximum value of the applied source voltage. This means that a comparator is capable of developing a high level of output voltage when a very minute difference in input voltage occurs. Comparators are commonly used in household appliances and in electronic measuring instruments.

Comparators are frequently used to determine when an applied input voltage exceeds the value of a referenced voltage input. A typical circuit that does this is shown in **Figure 17.5(a)**. Note that the **noninverting input** is connected to ground. This causes it to be 0 V or be referenced to ground. Note also that V_{in} is connected to the inverting input through resistor (R_{in}). If V_{in} swings slightly positive, it causes the output voltage (V_{out}) to go into negative saturation ($-V_{sat}$). A small change in the value of V_{in} in the negative direction also causes the output to go into positive saturation ($+V_{sat}$). Used in this manner, the op-amp compares the voltage difference between its inverting input and a zero-reference value.

Figure 17.5(b) shows how the zero-referenced comparator responds when AC voltage is applied to V_{in}. The waveforms show that an AC input produces a square wave at the output. The output is, however, inverted, or 180° out of phase with the AC input. A comparator circuit of this type is often called a **waveshaper**.

Example 17-5:

Refer to the zero-referenced inverting comparator of **Figure 17.5(a)**. Note the values of A_{ol}, V_{in}, and the polarity of V_{out}. When this occurs, what happens to V_{out}?

Solution

When V_{in} is +0.15 mV, it drives the op-amp into saturation because it is applied to the inverting input through R_{in}. This causes V_{out} to be

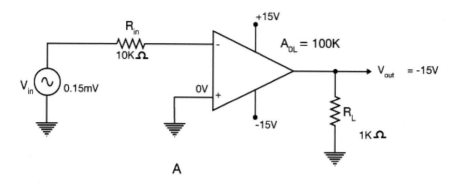

A

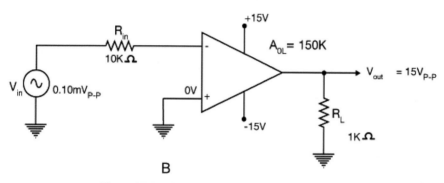

B

Figure 17.5   Zero-referenced inverting comparator.

$$V_{out} = V_{in} \times A_{ol}$$
$$= +0.15 \text{ mV} \times 100,000\Omega$$
$$= +0.00015 \text{ V} \times 100,000\Omega$$
$$= -15 \text{ V}.$$

Related Problem

Refer to the comparator of **Figure 17.5(b)**. Note that this circuit has AC applied to it. How does it respond to the AC input?

A zero-referenced **noninverting comparator** is shown in **Figure 17.6(a)**. In this circuit, the output voltage detects or indicates the polarity of the input

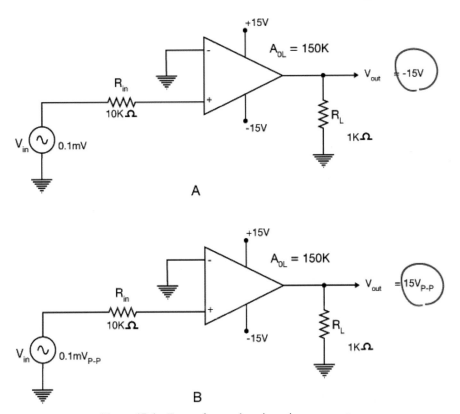

Figure 17.6 Zero-referenced noninverting comparator.

voltage. When the input voltage (V_{in}) is positive, it causes the output voltage to swing into $+V_{sat}$. A negative value of V_{in} likewise causes the output to reach $-V_{sat}$.

When AC voltage is applied to the **noninverting input** of the comparator circuit of **Figure 17.6(b)**, the resulting output is nearly the same as the inverting comparator. In this case, however, the output remains in phase with the input. This type of comparator is used in applications that require a positive relationship or reflection between the input and output values.

Example 17-6:

Refer to the noninverting comparator of **Figure 17.6(a)**. Note that the values of A_{ol} and V_{in} of the circuit have changed. How does V_{out} respond to these changes?

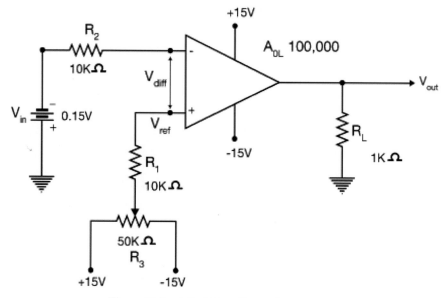

Figure 17.7   Adjustable-referenced comparator.

Solution

When V_{in} is +0.1 mV, it drives the op-amp into positive saturation because it is applied to the noninverting input. This, in turn, causes V_{out} to be

$$V_{out} = V_{in} \times A_{ol}$$
$$= +0.1 \text{ mV} \times 150,000$$
$$= +0.0001 \text{ V} \times 150,000$$
$$= +15 \text{ V}.$$

This shows that the polarities of V_{out} and V_{in} are the same.

Related Problem

Refer to the noninverting comparator of **Figure 17.6(b)**. How does the V_{out} of this comparator respond when AC is applied to V_{in}?

    Figure 17.7 shows the circuit of a variable or **adjustable comparator**. This type of circuit permits an operator to change the value or polarity of the reference voltage (V_{ref}). Note that the reference input is connected to the noninverting input. Its value and polarity are determined by the position setting of potentiometer R3. When the value of R3 is changed, it alters the polarity and value of the **difference voltage** (V_{diff}) because: $V_{diff} = V_{in} - V_{ref}$

or $V_{ref} - V_{in}$. As a result of this, the output voltage (V_{out}) will go positive only if V_{in} is made more positive than the voltage setting of R3. It also swings negative if V_{in} is more negative than the voltage value adjusted by R3. At the center or middle-position of R3, the reference voltage can be set to any value from +15 to 0 to −15 V. Adjustable comparators are commonly used in instruments that measure a wide range of voltage values that change polarity often.

Example 17-7:

Refer to the adjustable-referenced comparator of **Figure 17.7**. As indicated, V_{in} is 0.15 mV and R3 is adjusted to the center of its adjusting range. This causes V_{ref} to be 0 V. When this occurs, what happens to V_{out} and V_{diff}?

Solution

When V_{in} is 0.15 mV and V_{ref} is 0 V, it causes V_{diff} to be

$$V_{diff} = V_{in} - V_{ref}$$
$$= +0.15 \, mV - 0 \, V$$
$$= +0.15 \, mV.$$

This then causes V_{out} to be

$$V_{out} = V_{diff} \times A_{ol}$$
$$= +0.15 \text{ mV} \times 100,000$$
$$= +0.00015 \text{ mV} \times 100,000$$
$$= +15.$$

Related Problem

Refer to **Figure 17.7**. Assume that the potentiometer R3 is adjusted to a position that will cause V_{ref} to be −0.10 mV. When this occurs, what happens to V_{out} and V_{diff}?

Self-Examination

Answer the following questions.

11. The feedback component of an op-amp comparator is a(n):

    a. Resistor
    b. Diode

  c. Capacitor
  d. Open circuit

12. The primary function of an op-amp comparator is:

  a. Waveshaping
  b. Amplification
  c. Making voltage comparisons
  d. Rectification

13. If the input voltage applied to the inverting terminal of an op-amp comparator is greater than that applied to the noninverting terminal, its output will be:

  a. Positive supply voltage
  b. Negative supply voltage
  c. 0 V
  d. None of the above

14. When an op-amp is used as a comparator, it is generally connected in a(n):

  a. Open-loop configuration
  b. Closed-loop inverting configuration
  c. Closed-loop noninverting configuration
  d. Summing configuration

15. The gain of an op-amp when used as a comparator is:

  a. 0
  b. 1
  c. Low
  d. Extremely high

17.4 Integrators

A very important function of the operational amplifier is **signal processing**. This operation deals with circuits that will change or modify some type of signal that is applied to its input. The applied signal may be alternating current, direct current, or some combination of these which occurs at different frequencies. In order to perform this operation, the circuit must employ different passive components such as resistors, capacitors, or inductors. When a signal is forced to pass through an R, C, or L device, it will cause some signal loss to occur. This loss can be minimized when passive components

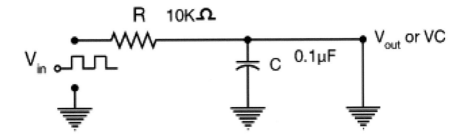

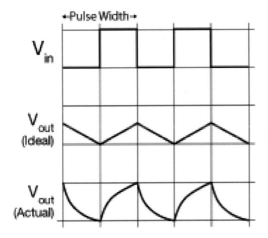

Figure 17.8 A basic integrator circuit.

are connected to an op-amp. This type of configuration is then considered to be an **active** circuit. **Active circuits** permit some level of amplification or at least maintain the amplitude of the input signal at some desired level. Integrators are used to achieve this operation.

17.4 Describe the operation of an integrator.

In order to achieve objective 17.4, you should be able to:

- describe the characteristics of integrator op-amp circuits;
- analyze integrator op-amp circuits.

A simplification of the basic **integrator** is shown in **Figure 17.8**. This circuit has a resistor and a capacitor in its construction. These two passive components are designed to add specific quantities of the input signal being measured over a period of time. The resulting output that is developed across the capacitor shows the time interval of the signal being applied to the input. Note that a square wave is applied to the input of the *RC* integrator and the resulting voltage output (V_{out}) is developed across the capacitor. Ideally, V_{out} should be a triangular-shaped wave as indicated. In an actual circuit, the resulting triangular wave as shown has some curvature instead of being straight lined or linear. This **distortion** of the wave is due to the charge and discharge rates of the capacitor. Ideally, we would expect this action to occur at a **linear** rate. In an actual circuit, however, it charges and discharges at an exponential rate. This causes the resulting wave to be curved as indicated by the actual wave.

When a **square wave** is applied to the input of an integrator, it causes a charge voltage (V_C) to be developed across the capacitor (C). The time that it takes V_C to occur is the time constant of the circuit. In this case, time (t) in seconds equals the resistance (R) in ohms (Ω) times the capacitance in farads (F) or: $t = RC$.

In an actual circuit, it takes five **time constants** for a capacitor to reach the full potential of its charge. The time for this to occur is expressed as follows:

$$t = 5RC. \tag{17.9}$$

Example 17-8:

How long would it take the capacitor of an integrator to become fully charged when R is 100 kΩ and C is 2 μF?
Solution

$$
\begin{aligned}
t &= 5(\text{R} \times \text{C}) \\
&= 5(100 \text{ k}\Omega \times 2 \ \mu \text{ F}) \\
&= 5(100,000 \ \Omega \times 0.000002 \text{ F}) \\
&= 5(0.2 \text{ s}) \\
&= 1 \text{ s}.
\end{aligned}
$$

Related Problem

How long does it take for the capacitor of **Figure 17.8** to reach its full charge potential?

When V_{in} is applied to a basic **integrator** circuit, it takes a certain period of time for the square wave to be developed. This time is an indication of the frequency of the wave. Terms such as cycles per second, hertz (Hz), pulse

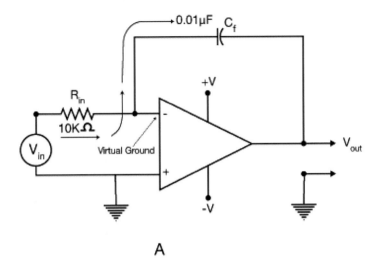

A

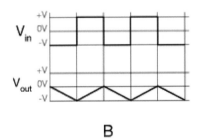

B

Figure 17.9 An op-amp integrator.

repetition rate, and pulse-width time are expressions that show the relation-ship of frequency and time. From this, it can be seen that the developing time of a wave and the charging time of a capacitor are frequency related. How an integrator responds to time then becomes an important consideration. In this regard, if the time constant of an integrator equals the pulse-width time of the applied square wave, the resulting V_C will be nonlinear as shown by the actual V_{out} of **Figure 17.8**. In most integrator applications, **nonlinearity** is an undesirable feature. It can, however, be reduced or eliminated by adding an op-amp to the circuit. This modification causes the actual and ideal outputs of the integrator to both be linear.

An ideal op-amp **integrator** is shown in **Figure 17.9**. Note that the feedback element of this circuit is capacitor (C_f) and it forms an *RC* network with the input resistor (R_{in}). Both of these components are then connected to the inverting input of the op-amp. As a result of this construction, the circuit now responds as a constant-current source for the capacitor when V_{in} is applied. This means the C_f will now be charged at a constant rate which causes V_{out} to be linear. This occurs because the inverting input is considered to be at **virtual ground** and it has a very high input impedance. As a result of this, the input current (I_{in}) is

$$I_{in} = V_{in}/R_{in}.$$

Assume now that V_{in} is a constant value over a period of time and that R_{in} has a fixed value that causes I_{in} to be constant for the same period of time. This means that nearly all of I_{in} flows into and comes from C_f when it is being charged by the constant-current source. When V_{in} remains constant, it causes C_f to charge and discharge at a linear rate. As a result of this, V_{out} will now be linear and have the shape of a triangular ramp as indicated by the ideal V_{out} waveform of **Figure 17.8.**

Self-Examination

Answer the following questions.

16. The feedback component of an op-amp integrator is a:

    a. Resistor
    b. Diode
    c. Capacitor
    d. Transistor

17. The primary function of an integrator is:

    a. Waveshaping
    b. Amplification
    c. Making voltage comparisons
    d. Rectification

18. If a square wave is applied to an integrator, its output will be:

    a. Amplified square wave
    b. Sine wave
    c. Rectified sine wave
    d. Triangular wave

19. In an op-amp integrator circuit in which the applied input voltage remains constant, the charging current of the capacitor will be:

    a. Constant
    b. Zero
    c. Gradually increasing
    d. Gradually decreasing

20. An op-amp integrator circuit in which the applied input voltage remains constant, the discharging current of the capacitor will be:

    a. Constant
    b. Zero
    c. Gradually increasing
    d. Gradually decreasing

17.5 Differentiators

An electronic **differentiator** is designed to produce an output waveform whose value at any instant of time is equal to the rate of change of the input at that point in time. In many respects, a differentiator is just the opposite or the inverse operation of an **integrator**. In this regard, integrators are sometimes called "anti-differentiators." This means that a signal applied to the input of an ideal integrator whose output is connected to the input of an ideal differentiator has exactly the same shape as the original input signal.

17.5 Explain how a differentiator works.

In order to achieve objective 17.5, you should be able to:

- describe the characteristics of differentiator op-amp circuits;
- analyze differentiator op-amp circuits.

An ideal **op-amp differentiator** is shown in **Figure 17.10**. Note the placement of the resistor and capacitor of this circuit and compare it with the op-amp integrator of **Figure 17.9**. If the placement of the resistor and capacitor of an integrator circuit are transposed, we have an op-amp differentiator. This change means that the capacitor is an input device, and the resistor is the feedback element. Because of this circuit change, the differentiator produces an output that is proportional to the rate change of the input voltage. This is determined by the time constant of R_f and C_{in} which is shown to be 5 μs.

In order to see how a **differentiator** works, assume that it has a triangular wave applied to its input. Note that this wave starts at time zero (t_0) and has a linear rise in positive voltage until it reaches time one (t_1). Between t_0

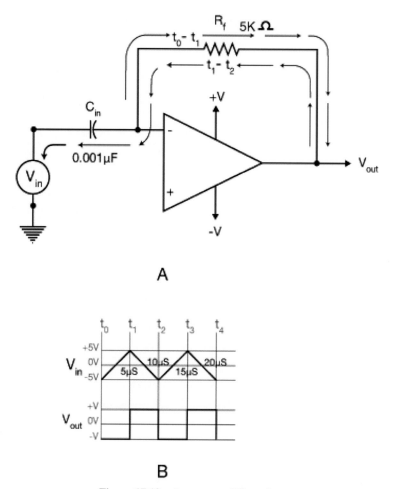

Figure 17.10   An op-amp differentiator.

and t_1, the rate change of the wave is shown to be a constant positive value that occurs in 5 μs. Being connected to the **inverting input** of the op-amp, this constant positive input causes the output to be immediately driven to a constant negative voltage. Between t_1 and t_2, the rate change of the input changes to a constant negative value. This, in turn, causes the output to switch to a positive voltage value. As a result of this, a constant valued triangular input wave will produce an inverted constant valued square wave in 10 μs. This input/output relationship of a differentiator is shown by the V_{in} and V_{out} waves.

Now when the positive going ramp voltage of V_{in} is applied to the input of the differentiator, it charges C_{in} and causes V_C and V_{in} to be equal at the same time. The resulting current flow of this action also causes I_{in}, I_C, and I_{Rf} to be equal because of the virtual ground of the inverting input. The direction of the charging current through R_f is indicated by the t_0-t_1 arrows and the discharging current is shown by the t_1-t_2 arrows.

Self-Examination

Answer the following questions.

21. The feedback component of an op-amp differentiator is a:

    a. Resistor
    b. Diode
    c. Capacitor
    d. Transistor

22. The primary function of a differentiator is:

    a. Waveshaping
    b. Amplification
    c. Making comparisons
    d. Rectification

23. If a triangular wave is applied to an differentiator, its output will be a(n):

    a. Constant DC
    b. Sine wave
    c. Rectified sine wave
    d. Square wave

24. In an op-amp differentiator, the output is :

    a. Constant
    b. Zero
    c. Proportional to the rate of change of the input signal
    d. Proportional to the area of its input waveform

25. If the input signal applied to a differentiator has a constant voltage value, the output will be:

    a. Zero
    b. Square wave
    c. Gradually increasing
    d. Gradually decreasing

17.6 Precision Rectifiers

A **precision rectifier** is an op-amp circuit that is used to modify or alter the shape of a wave applied to its input. Diodes are primarily used to achieve this type of operation. Recall that a diode is a semiconductor device that will only conduct current in one direction.

17.6 Analyze the operation of a precision rectifier.

   In order to achieve objective 17.6, you should be able to:

   • describe the characteristics of rectifier op-amp circuits;
   • explain the operation of a precision rectifier op-amp circuit.

In the operation of a diode rectifier, **forward biasing** causes the diode to be conductive and reverse biasing causes it to be nonconductive. Forward biasing of a silicon diode occurs at approximately 0.7 and at 0.3 V for germanium diodes. Assume now that an AC voltage of 2 V_{p-p} is applied to a silicon diode rectifier as shown in **Figure 17.11**. Note that the positive alternation is +1 V_p and the negative alternation is -1V_p. The diode is forward biased by the positive alternation and reverse biased by the negative alternation. The resulting output of this circuit occurs for only one alternation, and it is considered to be a **half-wave rectifier**. In this case, the rectified output voltage will be +1 V_p - 0.7 V = 0.3 V. Note that most of the +1 V_p applied to the diode is needed to cause it to be conductive. Suppose now that a 200 mV V_{p-p} AC signal is applied to the same silicon diode rectifier. In this case, the output will be +0.100 V_p - 0.7 V = -0.6 V or 0 V because the diode does not go into conduction. This means that any input voltage less than +0.7 V_p will not forward bias a silicon diode. To rectify voltages less that +0.7 V, the simplified diode rectifier can be combined with an op-amp to form a precision rectifier.

   A **precision half-wave op-amp rectifier** is shown in **Figure 17.12**. Note that this circuit has only an op-amp and a diode in its construction. The

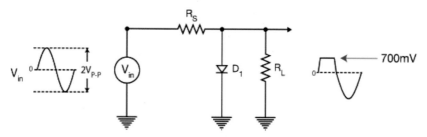

Figure 17.11   Silicon diode rectifier.

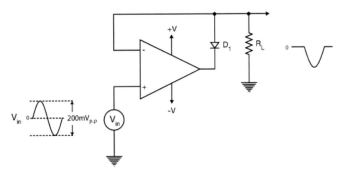

Figure 17.12   Precision rectifier.

rectified output of this circuit comes from the diode. Normally, a silicon diode would not be conductive until 0.7 V is applied. The output of the circuit also has a feedback line connected from the output of the diode back to the inverting input. This means that with no input signal applied, the diode is nonconductive, and the **feedback loop** is open. When AC is applied to the input, the diode goes into conduction when the positive alternation reaches 0.7 V.

A very basic characteristic of an op-amp shows that the voltage difference between the inverting and noninverting inputs is very small or near 0 V. This means that when a positive alternation is applied to the noninverting input, it causes a nearly identical **positive alternation** to appear at the inverting input because of the feedback loop. This is the result of the high open-loop gain of the op-amp. When the AC input first begins to rise in the positive direction, its voltage value is amplified 100,000 times or more by the open-loop gain of the op-amp. This causes the output of the op-amp to instantly be large enough to forward bias the diode. Thus, only a very minute change in positive input voltage will now cause conduction of the diode. For the **negative alternation** of the input, the diode is reverse biased and is nonconductive. Essentially, this means that a precision op-amp rectifier will respond to any positive voltage change applied to its input.

Self-Examination

Answer the following questions.

26. A precision rectifier op-amp and diode circuit conducts at:
    a. Very high diode forward-bias voltages
    b. Very low diode forward-bias voltages

    c. Constant 1 V

    d. Under reverse-bias conditions

27. A precision rectifier op-amp and diode circuit uses:

    a. An open-loop configuration of the op-amp

    b. A diode connected directly to the output of the op-amp

    c. A load resistor connected directly to the output of the op-amp

    d. No feedback component

28. In the op-amp/silicon diode precision rectifier shown in **Figure 17.12**, when an AC input is applied, rectification occurs:

    a. -1 to -0.9 V

    b. 0-0.7 V

    c. 0.7-0.9 V

    d. 1 V or above

29. When an AC input voltage is applied to the precision rectifier circuit of **Figure 17.12**, the output is a(n):

    a. Constant DC signal

    b. Half-wave rectified signal

    c. Full-wave rectified signal

    d. AC signal

1. For measuring small AC signals (few hundred mV in amplitude), instruments that rectify the AC to DC in order to determine the value of the AC signal should use:

    a. Only silicon diodes

    b. Precision rectifier circuits containing op-amps and diodes

    c. Integrator op-amps circuits

    d. Differentiator op-amp circuits

17.7 Instrumentation Amplifiers

When several operational amplifiers are connected together, the resulting circuit configuration is called an **instrumentation amplifier**. Specifically, this type of circuit is designed to measure and process very minute signal voltages that may exist in an abnormal environment. In order to respond under these conditions, the circuit must have high differential **voltage gain**, good **common mode rejection**, high **input impedance**, and low **output impedance**.

17.7 Analyze the operation of an instrumentation op-amp.

In order to achieve objective 17.7, you should be able to:

- describe the characteristics of instrumentation op-amp circuits;
- explain the operation of an instrumentation amplifier.

Previously, we identified these conditions as the characteristics of an **ideal amplifier**. In general, it is rather difficult to achieve these things with a single op-amp. However, when three or more op-amps are connected together, the circuit can readily achieve these conditions in its operation. **Instrumentation amplifiers** are now available that will accomplish all of these conditions in its operation.

The internal construction of a basic **instrumentation amplifier** is shown in **Figure 17.13**. Note that this circuit has three op-amps and six resistors in its construction. Op-amps A1 and A2 are identical noninverting amplifiers that provide high input impedance and voltage gain. Op-amp A3 is a **differential amplifier** that responds to the difference between outputs VA1out and VA2out and has low output impedance. The gain setting resistor R_G is externally connected to the circuit. Assume now that V_{in1} and V_{in2} are applied to the noninverting input (+) of A1 and A2. When this occurs, the amplified voltage output of A1 is

$$VA1out = 1 + (R_1/R_G)Vin1.$$

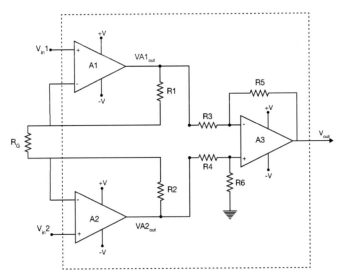

Figure 17.13 Internal construction of an instrumentation amplifier.

At the same time, Vin2 is also applied to the inverting input (-) through A2 and a path formed by R_2 and R_G. This means that Vin2 is also amplified by op-amp A1. Its amplification is equal to

$$VA1out = (R_2/R_G)Vin2.$$

Since the output of A1 is equal to the difference between its two inputs, then VA1out is

$$VA1out = (1 + R_1/R_G)Vin1 - (R_2/R_G)Vin2.$$

As previously noted, op-amps A1 and A2 are identical in construction and operation. As a result of this, the output of A2 can also be expressed as

$$VA2out = (1 + R_2/R_G)Vin2 - (R_1/R_G)Vin1.$$

Since R_1 and R_2 of these two expressions have the same value, they can be combined and identified as 2R. These two expressions then become

$$VA1out = (1 + 2R/R_G)Vin1$$
$$VA2out = (1 + 2R/R_G)Vin2.$$

As a result of this action, op-amp A3 now has VA1out applied to its inverting input and VA2out applied to its noninverting input. The differential input of A3 then becomes VA2out - VA1out. By design, A3 has unity gain because of the equal values of R_3, R_4, R_5, and R_6. This means that the combined output of A3 is then VA3out. Since A3 has unity gain, VA3out is then equal to 1 × (VA2out - VA1out). When Vin1 and Vin2 are applied to the input of the circuit, the final **output** is

$$VA3out = 1 \times (Vin2 - Vin1)or[1 + (2R/RG)]Vin2 - Vin1.$$

Thus, the overall closed-loop gain (A_{cl}) of the instrumentation amplifier is

$$A_{cl} = \frac{VA3out}{Vin2 - Vin1} = \frac{(1 + \frac{2R}{R_G})(Vin2 - Vin1)}{Vin2 - Vin1} = 1 + \frac{2R}{R_G}$$

$$A_{cl} = 1 + \frac{2R}{R_G}. \tag{17.10}$$

This means that the gain of the instrumentation amplifier can now be set by the value of only the external resistor R_G, when the values of R are known.

The value of the external gain setting resistor (R_G) can now be determined for an instrumentation amplifier, by using the value of the closed-loop gain (A_{cl}) and the value of R. This is expressed as

$$A_{cl} = 1 + \frac{2R}{R_G}$$

$$A_{cl} - 1 = \frac{2R}{R_G}$$

$$R_G = \frac{2R}{A_{cl} - 1}.$$

Example 17-9:

What is the closed-loop gain of an instrumentation amplifier that has an external gain resistor of 150 Ω, when $R_1 = R_2 = R = 20$ kΩ?

Solution

$$A_{cl} = 1 + \frac{2R}{R_G}$$
$$= 1 + \frac{2 \times 20,000 \, \Omega}{150 \, \Omega}$$
$$= 1 + 266.66$$
$$\cong 268.$$

Related Problem

What is the closed-loop gain of an instrumentation amplifier that has an external gain resistor of 250 Ω, when $R_1 = R_2 = R = 25$ kΩ?

Example 17-10:

Determine the value of the external gain resistor (R_G) for an instrumentation amplifier with a closed-loop gain of 200, when $R_1 = R_2 = R = 20$ kΩ.
Solution

$$R_G = \frac{2R}{A_{cl} - 1}$$
$$= \frac{2 \times 20,000}{200 - 1}$$
$$\cong 200 \, \Omega.$$

Related Problem

Determine the value of the external gain resistor (R_G) for an instrumentation amplifier with a closed-loop gain of 300, when $R_1 = R_2 = R = 20$ kΩ.

Self-Examination

Answer the following questions.

31. If the same input voltage and polarity are applied to both the inputs, the output of the instrumentation amplifier will be:

    a. 0 V
    b. $+$ Supply voltage
    c. $-$ Supply voltage
    d. Addition of the two applied voltage signals

32. Referring to **Figure 17.13**, the function of the op-amp A3 is:

    a. Open-loop gain
    b. Common mode rejection
    c. Differential amplifier
    d. Differentiator

33. The basic instrumentation amplifier shown in **Figure 17.13** has:

    a. One op-amp and a feedback resistor
    b. Two op-amps and seven resistors
    c. Three op-amps and seven capacitors
    d. Three op-amps and seven resistors

34. An instrumentation amplifier has an external resistor that is used for:

    a. Establishing input impedance
    b. Setting the voltage gain
    c. Setting the current gain
    d. Interfacing with external components

35. Instrumentation amplifiers are primarily used in a(n):

    a. High noise environment
    b. Filter circuit
    c. Test instruments
    d. Audio amplifiers

17.8 Analysis and Troubleshooting – Specialized Op-amp Circuits

Specialized op-amps are used for accomplishing a variety of different applications such as waveshaping, pulse generation, control, and instrumentation amplifiers. The fundamental component of such circuits is the basic **operational amplifier**. Specialized amplifier circuits make use of both open- and closed-loop configurations. Troubleshooting specialized amplifier circuits generally makes a comparison between the expected and actual output when a known input signal is applied. Since the internal structure of an op-amp is not repairable, troubleshooting is usually limited to op-amp replacement if defective, and examination of externally connected circuit components.

17.8 Analyze and troubleshoot specialized op-amp circuits.
   In order to achieve objective 17.8, you should be able to:

- identify the key operating features of a specialized op-amp circuit;
- distinguish between normal and faulty responses of a specialized op-amp circuit.

Data Sheet Analysis

Figure 17.14 shows the data sheet of the **AD620**, a low-cost, low-power instrumentation amplifier. Use this data sheet to answer the following:

- How many pins (terminals) does the IC have? _____
- How is the gain of the instrumentation amplifier determined? _____
- What is the range of the voltage gain? _____ to _____
- What is the maximum supply current? _____
- What is the common-mode rejection ratio for a gain of 10? _____dB
- What is the bandwidth of the amplifier for a gain of 100? _____kHz
- Applications of the instrumentation amplifier include: _____
- Internal power dissipation: _____mW
- Operating temperature ranges from _____ °C to _____ °C.
- Maximum supply voltage: _____V

Low Cost, Low Power Instrumentation Amplifier

AD620

FEATURES
EASY TO USE
Gain Set with One External Resistor
(Gain Range 1 to 1000)
Wide Power Supply Range (±2.3 V to ±18 V)
Higher Performance than Three Op Amp IA Designs
Available in 8-Lead DIP and SOIC Packaging
Low Power, 1.3 mA max Supply Current

EXCELLENT DC PERFORMANCE ("B GRADE")
50 μV max, Input Offset Voltage
0.6 μV/°C max, Input Offset Drift
1.0 nA max, Input Bias Current
100 dB min Common-Mode Rejection Ratio (G = 10)

LOW NOISE
9 nV/$\sqrt{Hz}$, @ 1 kHz, Input Voltage Noise
0.28 μV p-p Noise (0.1 Hz to 10 Hz)

EXCELLENT AC SPECIFICATIONS
120 kHz Bandwidth (G = 100)
15 μs Settling Time to 0.01%

APPLICATIONS
Weigh Scales
ECG and Medical Instrumentation
Transducer Interface
Data Acquisition Systems
Industrial Process Controls
Battery Powered and Portable Equipment

PRODUCT DESCRIPTION
The AD620 is a low cost, high accuracy instrumentation amplifier that requires only one external resistor to set gains of 1 to

CONNECTION DIAGRAM
8-Lead Plastic Mini-DIP (N), Cerdip (Q)
and SOIC (R) Packages

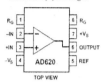

TOP VIEW

1000. Furthermore, the AD620 features 8-lead SOIC and DIP packaging that is smaller than discrete designs, and offers lower power (only 1.3 mA max supply current), making it a good fit for battery powered, portable (or remote) applications.

The AD620, with its high accuracy of 40 ppm maximum nonlinearity, low offset voltage of 50 μV max and offset drift of 0.6 μV/°C max, is ideal for use in precision data acquisition systems, such as weigh scales and transducer interfaces. Furthermore, the low noise, low input bias current, and low power of the AD620 make it well suited for medical applications such as ECG and noninvasive blood pressure monitors.

The low input bias current of 1.0 nA max is made possible with the use of Superβeta processing in the input stage. The AD620 works well as a preamplifier due to its low input voltage noise of 9 nV/$\sqrt{Hz}$ at 1 kHz, 0.28 μV p-p in the 0.1 Hz to 10 Hz band, 0.1 pA/$\sqrt{Hz}$ input current noise. Also, the AD620 is well suited for multiplexed applications with its settling time of 15 μs to 0.01% and its cost is low enough to enable designs with one in-amp per channel.

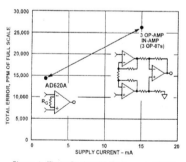

Figure 1. Three Op Amp IA Designs vs. AD620

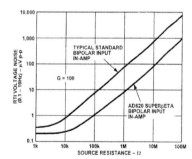

Figure 2. Total Voltage Noise vs. Source Resistance

REV. E

One Technology Way, P.O. Box 9106, Norwood, MA 02062-9106, U.S.A.
Tel: 781/329-4700   World Wide Web Site: http://www.analog.com
Fax: 781/326-8703   © Analog Devices, Inc., 1999

AD620–SPECIFICATIONS (Typical @ +25°C, $V_S = \pm 15$ V, and $R_L = 2$ kΩ, unless otherwise noted)

Model	Conditions	AD620A Min	Typ	Max	AD620B Min	Typ	Max	AD620S[1] Min	Typ	Max	Units
GAIN	$G = 1 + (49.4\,k/R_G)$										
Gain Range		1		10,000	1		10,000	1		10,000	
Gain Error[2]	$V_{OUT} = \pm 10$ V										
G = 1			0.03	0.10		0.01	0.02		0.03	0.10	%
G = 10			0.15	0.30		0.10	0.15		0.15	0.30	%
G = 100			0.15	0.30		0.10	0.15		0.15	0.30	%
G = 1000			0.40	0.70		0.35	0.50		0.40	0.70	%
Nonlinearity,	$V_{OUT} = -10$ V to +10 V,										
G = 1–1000	$R_L = 10$ kΩ		10	40		10	40		10	40	ppm
G = 1–100	$R_L = 2$ kΩ		10	95		10	95		10	95	ppm
Gain vs. Temperature											
G = 1				10			10			10	ppm/°C
Gain >1[2]				-50			-50			-50	ppm/°C
VOLTAGE OFFSET	(Total RTI Error = $V_{OSI} + V_{OSO}/G$)										
Input Offset, V_{OSI}	$V_S = \pm 5$ V to ± 15 V		30	125		15	50		30	125	μV
Over Temperature	$V_S = \pm 5$ V to ± 15 V			185			85			225	μV
Average TC	$V_S = \pm 5$ V to ± 15 V		0.3	1.0		0.1	0.6		0.3	1.0	μV/°C
Output Offset, V_{OSO}	$V_S = \pm 15$ V		400	1000		200	500		400	1000	μV
	$V_S = \pm 5$ V			1500			750			1500	μV
Over Temperature	$V_S = \pm 5$ V to ± 15 V			2000			1000			2000	μV
Average TC	$V_S = \pm 5$ V to ± 15 V		5.0	15		2.5	7.0		5.0	15	μV/°C
Offset Referred to the Input vs. Supply (PSR)	$V_S = \pm 2.3$ V to ± 18 V										
G = 1		80	100		80	100		80	100		dB
G = 10		95	120		100	120		95	120		dB
G = 100		110	140		120	140		110	140		dB
G = 1000		110	140		120	140		110	140		dB
INPUT CURRENT											
Input Bias Current			0.5	2.0		0.5	1.0		0.5	2	nA
Over Temperature				2.5			1.5			4	nA
Average TC			3.0			3.0			8.0		pA/°C
Input Offset Current			0.3	1.0		0.3	0.5		0.3	1.0	nA
Over Temperature				1.5			0.75			2.0	nA
Average TC			1.5			1.5			8.0		pA/°C
INPUT											
Input Impedance											
Differential			10‖2			10‖2			10‖2		GΩ‖pF
Common-Mode			10‖2			10‖2			10‖2		GΩ‖pF
Input Voltage Range[3]	$V_S = \pm 2.3$ V to ± 5 V	$-V_S + 1.9$		$+V_S - 1.2$	$-V_S + 1.9$		$+V_S - 1.2$	$-V_S + 1.9$		$+V_S - 1.2$	V
Over Temperature		$-V_S + 2.1$		$+V_S - 1.3$	$-V_S + 2.1$		$+V_S - 1.3$	$-V_S + 2.1$		$+V_S - 1.3$	V
	$V_S = \pm 5$ V to ± 18 V	$-V_S + 1.9$		$+V_S - 1.4$	$-V_S + 1.9$		$+V_S - 1.4$	$-V_S + 1.9$		$+V_S - 1.4$	V
Over Temperature		$-V_S + 2.1$		$+V_S - 1.4$	$-V_S + 2.1$		$+V_S - 1.4$	$-V_S + 2.3$		$+V_S - 1.4$	V
Common-Mode Rejection Ratio DC to 60 Hz with 1 kΩ Source Imbalance	$V_{CM} = 0$ V to ± 10 V										
G = 1		73	90		80	90		73	90		dB
G = 10		93	110		100	110		93	110		dB
G = 100		110	130		120	130		110	130		dB
G = 1000		110	130		120	130		110	130		dB
OUTPUT											
Output Swing	$R_L = 10$ kΩ,										
	$V_S = \pm 2.3$ V to ± 5 V	$-V_S + 1.1$		$+V_S - 1.2$	$-V_S + 1.1$		$+V_S - 1.2$	$-V_S + 1.1$		$+V_S - 1.2$	V
Over Temperature		$-V_S + 1.4$		$+V_S - 1.3$	$-V_S + 1.4$		$+V_S - 1.3$	$-V_S + 1.6$		$+V_S - 1.3$	V
	$V_S = \pm 5$ V to ± 18 V	$-V_S + 1.2$		$+V_S - 1.4$	$-V_S + 1.2$		$+V_S - 1.4$	$-V_S + 1.2$		$+V_S - 1.4$	V
Over Temperature		$-V_S + 1.6$		$+V_S - 1.5$	$-V_S + 1.6$		$+V_S - 1.5$	$-V_S + 2.3$		$+V_S - 1.5$	V
Short Current Circuit			± 18			± 18			± 18		mA

AD620

Model	Conditions	AD620A			AD620B			AD620S[1]			Units
		Min	Typ	Max	Min	Typ	Max	Min	Typ	Max	
DYNAMIC RESPONSE											
Small Signal –3 dB Bandwidth											
G = 1			1000			1000			1000		kHz
G = 10			800			800			800		kHz
G = 100			120			120			120		kHz
G = 1000			12			12			12		kHz
Slew Rate		0.75	1.2		0.75	1.2		0.75	1.2		V/μs
Settling Time to 0.01%	10 V Step										
G = 1–100			15			15			15		μs
G = 1000			150			150			150		μs
NOISE											
Voltage Noise, 1 kHz	Total RTI Noise = $\sqrt{(e^2_{ni})+(e_{no}/G)^2}$										
Input, Voltage Noise, e_{ni}			9	13		9	13		9	13	nV/√Hz
Output, Voltage Noise, e_{no}			72	100		72	100		72	100	nV/√Hz
RTI, 0.1 Hz to 10 Hz											
G = 1			3.0			3.0	6.0		3.0	6.0	μV p-p
G = 10			0.55			0.55	0.8		0.55	0.8	μV p-p
G = 100–1000			0.28			0.28	0.4		0.28	0.4	μV p-p
Current Noise	f = 1 kHz		100			100			100		fA/√Hz
0.1 Hz to 10 Hz			10			10			10		pA p-p
REFERENCE INPUT											
R_{IN}			20			20			20		kΩ
I_{IN}	$V_{IN+}, V_{REF} = 0$		+50	+60		+50	+60		+50	+60	μA
Voltage Range		–V_S + 1.6		+V_S – 1.6	–V_S + 1.6		+V_S – 1.6	–V_S + 1.6		+V_S – 1.6	V
Gain to Output			1 ± 0.0001			1 ± 0.0001			1 ± 0.0001		
POWER SUPPLY											
Operating Range[3]		±2.3		±18	±2.3		±18	±2.3		±18	V
Quiescent Current	$V_S = ±2.3$ V to ±18 V		0.9	1.3		0.9	1.3		0.9	1.3	mA
Over Temperature			1.1	1.6		1.1	1.6		1.1	1.6	mA
TEMPERATURE RANGE											
For Specified Performance			–40 to +85			–40 to +85			–55 to +125		°C

NOTES

[1] See Analog Devices military data sheet for 883B tested specifications.

[2] Does not include effects of external resistor R_G.

[3] One input grounded. G = 1.

[4] This is defined as the same supply range which is used to specify PSR.

Specifications subject to change without notice.

Figure 17.14   Low-cost, low-power instrumentation amplifier, AD620 (Courtesy: Analog Devices).

Troubleshooting Op-amps

The **closed-loop** operation of specialized op-amp circuits are generally determined by the external components, including feedback resistors and capacitors that are connected to it. If the feedback element gets open-circuited or shorted during operation, the closed-loop behavior of the op-amp circuit will be affected. Being familiar with the expected output waveforms of a given op-amp circuit will be helpful in diagnosing which circuit element is not functioning properly. For example, in a **differentiator** op-amp circuit, if the capacitor connected to the input voltage source is open, no voltage will appear at the inverting input of the op-amp. As a result of this, the differentiator circuit will not generate the expected output. If the circuit is functioning properly, the **output** of the integrator should be a triangular waveform, which can be observed on an **oscilloscope**.

Summary

- A voltage-controlled voltage source generates a constant voltage regardless of the electrical load. The voltage value depends on a controlling voltage source circuit.
- A voltage-controlled constant source generates a constant current regardless of the electrical load. The current value depends on a controlling voltage source circuit.
- A voltage-to-current converter transforms the electrical behavior of the system. The input voltage controls the output voltage in such a way that the output current remains constant.
- A current-to-voltage converter transforms the electrical behavior of the system. The input voltage controls the output current in such a way that the output voltage remains constant.
- In an op-amp comparator circuit, the values of the voltage inputs applied to its inverting and noninverting input terminals are compared. It can generate a very high output voltage even when there is a minute voltage difference between the applied inputs.
- An op-amp integrator is a waveshaping circuit that generates an output voltage corresponding to the addition of specific quantities of the input signal over a period of time.
- A differentiator is a waveshaping circuit that develops an output voltage corresponding to the rate of change of the applied input. This operation is the inverse of an integrator.

- A precision op-amp/diode rectifier circuit is used for measuring and detecting minute AC input voltages.
- An instrumentation amplifier generally consists of at least three op-amps that are connected in such a way as to minimize noise and have high gain capabilities.

Formulas

(17-1) $A_{cl} = R_f/R_{in}$
(17-2) $V_{out} = A_{cl} \times V_{in}$
(17-3) $I_{in} = V_{in}/R_{in}$
(17-4) $V_{out} = I_{in} \times R_f$
(17-5) $V_{diff} = V_{in} - V_f$
(17-6) $V_{RL} = I_{out} \times R_L$
(17-7) $V_{Rf} = I_{out} \times R_f$
(17-8) $V_{out} = V_{in} \times A_{ol}$
(17-9) $t = 5\,RC$
(17-10) $A_{cl} = 1 + \frac{2R}{R_G}$
(17-11) $R_G = \frac{2R}{A_{CL}-1}$

Review Questions

Answer the following questions.

1. The characteristics of an ideal amplifier are (8 or 1) gain, (8 or low) input impedance, and (high or low) output impedance.
2. A circuit that produces an output voltage that is equal to the constant value of another controlling voltage and is independent of the current drawn from it is a (voltage or current)-controlled source.
3. The closed-loop amplification (A_{cl}) of a voltage-controlled voltage source is determined by:

   a. $R_{in} \times R_f$
   b. R_f/R_{in}
   c. R_{in}/R_f
   d. V_{in}/V_{out}

4. A (voltage or current)-controlled constant (voltage or current) source is designed to supply constant current to a load resistance.

5. The output of a voltage-controlled current source remains constant even when the load resistance is:

   a. 8 Ω
   b. Open
   c. 0 Ω
   d. Variable

6. In an integrator, the feedback element is a(n)_____, whereas in a differentiator, the feedback element is a(n) _____.

7. When a square wave is applied to the input of an integrator, the output will be a(n) _____ wave.

8. When a triangular wave is applied to the input of a differentiator, the output will be a(n) _____ wave.

9. If the output of an integrator is connected to the input of a differentiator, and a square wave signal is applied to the integrator, the final output will be a(n) _____ wave.

   a. Triangular
   b. Square
   c. Constant DC
   d. None of the above

10. The function of the external resistor in an instrumentation amplifier is for setting the _____.

11. A(n) _____ amplifier is a specialized high-gain op-amp circuit that is best suited for use in an environment having electrical noise.

12. A precision rectifier is used to _____ a signal that is applied to its input.

13. A precision rectifier op-amp and diode circuit responds to (low or high) value forward-bias voltages.

14. A zero-level detector is an application of a(n):

    a. Differentiator
    b. Integrator
    c. Comparator
    d. Instrumentation amplifier

15. How long does it take to fully (99%) charge the capacitor of an *RC* network:

    a. $0 \times RC$
    b. $1 \times RC$

     c. $2 \times RC$
     d. $5 \times RC$

Problems

Answer the following questions.

1. Refer to the circuit diagram of the instrumentation amplifier shown in **Figure 17.13**. The value of the external gain resistor (R_G), which will set the closed-loop gain (A_{cl}) to 100, when the values of $R_1 = R_2 = R = 10$ kΩ, is:

    a. 0-0.9 Ω
    b. 1-9.9 Ω
    c. 10-99.99 Ω
    d. 100 Ω or above

2. Refer to the circuit diagram of the instrumentation amplifier shown in **Figure 17.13**. What is the closed-loop gain (A_{cl}), which has an external gain resistor (R_G) of 200 Ω, when $R_1 = R_2 = R = 30$ kΩ:

    a. 0-99
    b. 100-299
    c. 300-399.99
    d. 400 or above

3. Refer to the circuit diagram of the instrumentation amplifier shown in **Figure 17.13**. What is the closed-loop gain (A_{cl}), which has an external gain resistor (R_G) of 10 MΩ, when $R_1 = R_2 = R = 30$ kΩ:

    a. 0-0.9
    b. 11.9
    c. 2-2.9
    d. 3 or above

4. Refer to the circuit diagram of the noninverting comparator shown in **Figure 17.6(b)**, which has an A_{ol} of 150,000, and a V_{in} of +0.1 mV$_{p-p}$. The value of V_{out} is:

    a. 0-4.9 V
    b. 5-9.9 V
    c. 10-14.9 V
    d. 15 V or above

5. Refer to the circuit diagram of the integrator shown in **Figure 17.9**. The *RC* time constant is:

    a. 0-0.9 μs
    b. 1-9.9 μs
    c. 10-99.9 μs
    d. 100 μs or above

Answers

Examples

17-1. $A_{cl} = 5$, $I_{in} = 0.8$ mA, $I_f = 0.8$ mA, $V_{out} = 4$ V
17-2. $I_L = 2$ mA; I_L remains unchanged
17-3. $V_{out} = 4$ V
17-4. $I_{out} = 10$ mA, $V_{sat} = 10$ V, $V_{Rf} = 0.5$ V, $V_{diff} = 0.45$ V
17-5. On the application of an AC sine wave input, with $V_{p-p} = 0.15$ mV, the output will be a square waveform, with $V_{p-p} = 30$ V. The output waveform is $180°$ out of phase with the applied AC input signal.
17-6. On the application of an AC sine wave input, with $V_{p-p} = 0.15$ mV, the output will be a square waveform, with $V_{p-p} = 30$ V. The output waveform is in phase with the applied AC input signal.
17-7. $V_{out} = 5$ V, $V_{diff} = 0.05$ mV
17-8. 5 ms
17-9. $A_{cl} = 201$
17-10. $R_G \ddot{Y} 134 \ \Omega$

Self-Examination

17.1

1. b. constant-value of current
2. b. depends on the value of the controlling voltage
3. a. V_{out}
4. a. the load resistor
5. c. remain unchanged

17.2

6. a. changes one form of energy into another form of energy
7. a. the load and feedback resistor in series
8. c. a feedback resistor and no input resistor.
9. voltage, current

10. current, voltage

17.3

11. d. open circuit
12. c. making voltage comparisons
13. b. negative supply voltage
14. a. open-loop configuration
15. d. extremely high

17.4

16. c. capacitor
17. a. waveshaping
18. d. triangular wave
19. a. constant
20. a. constant

17.5

21. a. resistor
22. a. waveshaping
23. d. square wave
24. proportional to the rate of change of the input signal
25. a. zero

17.6

26. b. very low diode forward-bias voltages
27. b. a diode connected directly to the output of the op-amp
28. b. 0 to 0.7 V
29. b. half-wave rectified signal
30. b. precision rectifier circuits containing op-amps and diodes

17.7

31. a. 0 V
32. c. differential amplifier
33. d. three op-amps and seven resistors
34. b. setting the voltage gain
35. a. high noise environment

Terms

Voltage-controlled voltage source

An electronic circuit that generates a constant voltage regardless of the electrical load. The voltage value depends on a controlling voltage source circuit.

Voltage-controlled current source

An electronic circuit that generates a constant current regardless of the electrical load. The current value depends on a controlling voltage source circuit.

Converters

An electronic circuit that transforms the electrical behavior of the system.

Voltage-to-current converter

An electronic circuit in which the input voltage controls the output voltage in such a way that the output current remains constant.

Current-to-voltage converter

An electronic circuit in which the input voltage controls the output current in such a way that the output voltage remains constant.

Comparator

An op-amp function that compares the voltage applied to one input, to that applied to another input, that is of a predetermined value or reference.

Waveshaper

An electronic circuit that alters the shape of the waveform over a given time duration.

Active circuit

An electronic circuit that has passive components such as resistors and capacitors, connected to amplifying or control devices such as op-amps and transistors.

Integrator

An op-amp circuit used for generating output voltage corresponding to the addition of specific quantities of the input signal over a period of time.

Differentiator

An op-amp circuit used for developing an output voltage corresponding to the rate of change of the applied input signal.

Precision rectifier

An op-amp/diode circuit that has the ability to conduct at very low diode forward-bias voltages.

External gain resistor

The resistor used to set the gain of an instrumentation amplifier.

Instrumentation amplifier

An instrumentation amplifier has high gain, high common mode rejection and is used for amplifying small signals in an environment that has electrical noise.

18

Voltage Regulator Circuits

A basic **power supply** is made up of an input transformer, rectifier, filter circuit, and an output load device. A block diagram of a basic power supply was discussed in Chapter 4 and is shown in **Figure 18.1**. The output of a basic power supply is normally applied to the load device. In normal usage, a power supply may experience changes in the applied input which will have some effect on the resulting output. Similarly, changes in the output load current can also affect the output voltage of the power supply.

Voltage regulators are generally described as being linear or switching based on whether the current flow to the load device is continuous or interrupted during operation. If the flow of current is uninterrupted, it is considered to be a **linear power supply**. On the other hand, if the flow of current is momentarily interrupted during its normal operation, the circuit is then considered to be a **switching regulator**. Linear regulators are further classified as being either series or a shunt (parallel), based on how certain electronic components are connected to the load device. Switching regulators are classified as being either **step-down** or **step-up**. When the switching electronic component is in series with the load, it will reduce or step down the voltage. When the switching component is in parallel with the load, it will increase or step up the voltage.

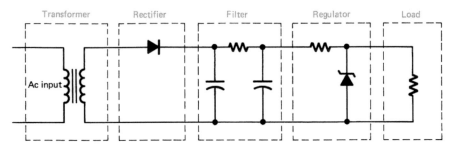

Figure 18.1   Basic power supply.

A number of voltage regulators today are packaged as integrated circuits (ICs). This permits regulation to be achieved by using a smaller number of components, reducing the overall size of the power supply.

Objectives

After studying this chapter, you will be able to:

18.1 explain the basic concept of voltage regulation;

18.2 describe how a linear voltage regulator responds to input/output changes;

18.3 describe how a switching voltage regulator responds to input/output changes;

18.4 interpret the datasheet of an IC voltage regulator;

18.5 analyze and troubleshoot voltage regulator circuits.

Chapter Outline

18.1 Voltage Regulation

18.2 Linear Voltage Regulators – Series and Shunt

18.3 Switching Voltage Regulator – Step-up and Step-down

18.4 IC Voltage Regulators

18.5 Troubleshooting Voltage Regulators

Key Terms

regulator
line regulator
load regulator
thermal overload
linear regulator
switching regulator
series regulator
shunt regulator
step-up regulator
step-down regulator
foldback regulation
pulse-width modulation

18.1 Voltage Regulation

The variation of the **output voltage** as the load current increases from the minimum to the maximum rated value. These problems arise due to changes in the applied input voltage or the output **load current** and, thus, affect the output voltage of the power supply. A regulator circuit is typically inserted between the **filter** section and the **load** of a power supply. **Figure 18.2** shows where a voltage regulator is inserted in a power supply. It then maintains the output voltage at a constant value, regardless of changes in the input and output conditions, within the prescribed limits. Such a supply is commonly called a voltage regulated power supply or a **voltage regulator**.

An **ideal** voltage regulator is designed to maintain a constant output voltage regardless of changes in either the input voltage or its load current demand. Typically, voltage regulators will exhibit some change in the resulting voltage output owing to changes in the input voltage or load current demand. Specifically, if the input voltage V_{in} changes, the regulator voltages should maintain output V_{out} within a certain range. Similarly, if the load demand changes, the regulated output voltage should be maintained within a certain range.

18.1 Explain the basic concept of voltage regulation.

In order to achieve objective 18.1, you should be able to:

- define line regulation;
- determine line regulation;
- define load regulation;
- determine load regulation.

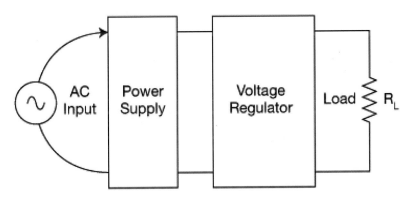

Figure 18.2 Regulated power supply.

Line Regulation

The ability of a regulator to maintain an output voltage at a constant value regardless of variations in the input voltage is called **line regulation**. **Figure 18.3** shows a representative **voltage regulator circuit** in which the output voltage is being maintained at a constant value when the input voltage either increases or decreases in value. **Figure 18.3(a)** shows a 5-V regulator with a 10-V DC applied to its input. **Figure 18.3(b)** shows the same regulator with 7.5-V DC applied to its input. In both cases, the resulting output V_{out} should ideally be 5 V across the 1-kΩ resistor.

In an actual **voltage regulator** circuit, the input voltage V_{in} may see a change in the value of its input. This change is designated as ΔV_{in} with the Δ indicating a change in the value. This typically causes the output V_{out} to change its value. The change in output is designated as ΔV_{out}. The line regulation of a voltage regulator is defined as the ratio of the change in the value of the output voltage due to corresponding change in the value of the input voltage.

$$\text{Line regulation} = \frac{\Delta V_{out}}{\Delta V_{in}}. \tag{18.1}$$

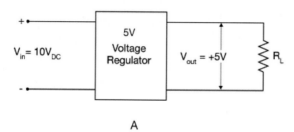

A

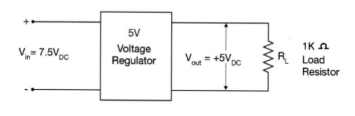

B

Figure 18.3   Line regulation.

Example 18-1:

Refer to **Figure 18.3**. The output of a 5-V voltage regulator changes by 7.5 μV when the input voltage changes by 2.5 V. Determine the value of the line regulation.

Solution

$$\text{Line regulation} = \frac{\Delta V_{\text{out}}}{\Delta V_{\text{in}}} = \frac{7.5\,\mu V}{2.5\,V} = 3\,\mu V/V.$$

A line regulation of 3 μV/V indicates that when the input voltage changes by 1 V, the output voltage will change by 3 μV.

Related Problem

The output of a 10-V voltage regulator changes by 20 μV when the input voltage changes by 5.5 V. Determine the value of the line regulation.

When comparing **voltage regulators**, it is important to determine the amount by which the output voltage will change with respect to changes in the applied input voltage. Ideally, there should be no change in output voltage when the regulator circuit functions properly. The value of the line regulation for an ideal regulator will be very small, preferably 0.

Line regulation can be expressed in a variety of different ways such as:

- μV/V – The change in output voltage measured in micro-volts when the input voltage changes by an incremental value such as 1 V.
- %/V – The percentage of change in the output voltage which occurs when the input voltage changes by an incremental value such as 1 V.
- % – The total percentage of change in the output voltage when the input voltage changes over its permitted operating range.
- mV or μV – The actual change in the output voltage measured in mV or μV when the input voltage changes over its permitted operating range.

Load Regulation

The ability of a regulator to maintain an **output voltage** at a constant value regardless of variations in the load current demand is called load regulation. **Figure 18.4** shows a representative voltage regulator circuit in which the output voltage is being maintained at a constant value when the load current demand either increases or decreases in value. As the value of the load resistor changes, the amount of load current will vary accordingly. **Figure 18.4(a)** shows a 5-V regulator with no-load resistor connected. This causes the load current $I_{\text{L(NoLoad)}}$ to be 0 A and the output voltage $V_{\text{out}} = V_{\text{NoLoad}} = 5$ V.

Figure 18.4(b) shows the same regulator with a 1-kΩ load resistor connected. In both cases, the resulting output V_{out} should ideally be 5 V.

In an actual **voltage regulator** circuit, the output load current I_{out} may see a change in the value of its output demand. This change is designated as ΔI_L with the Δ indicating a change in the value. This typically causes the output V_{out} to change its value. The change in output is designated as ΔV_{out}. As the load current demand increases from a minimum value to the full rated load current demand, the output voltage will be reduced from V_{NoLoad} to $V_{FullLoad}$.

$$\text{Load regulation} = \frac{\Delta V_{out}}{\Delta I_L} = \frac{V_{\text{No Load}} - V_{\text{Full Load}}}{I_{\text{Full Load}} - I_{\text{No Load}}}. \tag{18.2}$$

Example 18-2:

Refer to **Figure 18.4**. A voltage regulator has an output of 5 V under no-load conditions (I_{NoLoad}= 0 mA). However, under rated full-load conditions of 25 mA, the voltage may drop to a value of 4.9996 V. When this occurs, determine the value of the load regulation.

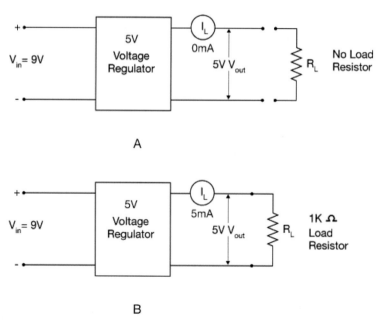

Figure 18.4   Load regulation.

Solution

$$\text{Load regulation} = \frac{V_{\text{No Load}} - V_{\text{Full Load}}}{I_{\text{Full Load}} - I_{\text{No Load}}}$$

$$= \frac{5\,\text{V} - 4.9996\,\text{V}}{25\,\text{mA} - 0\,\text{mA}} = \frac{0.0004}{25} = 0.000016 = 16\,\mu\text{V}/\text{mA}.$$

The load regulation of 16 μV/mA indicates that when the output load current demand changes by 1 mA, the output voltage will change by 16 μV.

Related Problem

A voltage regulator has an output of 10 V under no-load conditions. However, under rated full-load conditions of 75 mA, the voltage drops to 9.9993 V. Determine the value of the load regulation.

When comparing voltage regulators, it is important to determine the amount by which the output voltage will change with respect to changes in the load current demand. Ideally, there should be no change in output voltage when the regulator circuit functions properly. The value of the load regulation for an ideal regulator will be very small, preferably 0.

Load regulation can be expressed in a variety of different ways such as the following:

- μV/mA – The change in output voltage measured in micro-volts when the load current changes by 1 mA.
- %/mA – The percentage change in the output voltage when the load current changes by 1 mA.
- % – The total percentage change in the output voltage when the load current changes from no-load to rated full-load.
- mV or μV – The total change in the output voltage measured in mV or μV which can occur when the load current demand changes from no-load to rated full-load.

Example 18-3:

Refer to the voltage regulator block diagram of **Figure 18.4**. The load regulation of two voltage regulators, I and II, is to be compared. The specification of the voltage regulators is given below:

Regulator	I	II
Rated voltage	5 V	5 V
Maximum load current	25 mA	25 mA
Load regulation	16 μV/mA	0.5%

Determine which voltage regulator has a better load regulation.

Solution

For voltage regulator "I," the regulation is specified in terms of $\mu V/mA$.

$$\text{Load regulation} = \frac{V_{\text{No Load}} - V_{\text{Full Load}}}{I_{\text{Full Load}} - I_{\text{No Load}}}$$

$$\frac{5 - V_{\text{Full Load}}}{25\,\text{mA}} = 16\,\mu V/mA$$

$$5 - V_{\text{Full Load}} = 16\frac{\mu V}{mA} \times 25\,\text{mA} = 400\,\mu V$$

$$V_{\text{Full Load}} = 5 - 0.0004 = 4.9996.$$

For voltage regulator "II," the regulation is specified in terms of %.

$$\text{Load regulation} \% = \frac{V_{\text{No Load}} - V_{\text{Full Load}}}{V_{\text{No Load}}} \times 100$$

$$0.5 = \frac{5\,\text{V} - V_{\text{Full Load}}}{5\,\text{V}} \times 100$$

$$0.5 \times 5 = 5\,\text{V} - V_{\text{Full Load}} \times 100$$

$$V_{\text{Full Load}} = 5 - 0.025 = 4.975\,\text{V}.$$

This shows that over the full-load current range, the value of the output voltage of regulator II reduces by a larger amount (4.975 V) as compared to the output voltage of regulator I (4.9996 V). Thus, regulator I has a better load regulation than regulator II.

Related Problem

Refer to **Figure 18.4**. A 10-V voltage regulator "I," with a rated maximum current of 75 mA, has a load regulation of 20 $\mu V/mA$. A second 10-V voltage regulator "II," also with a rated maximum load current of 75 mA, has a load regulation of 0.5%. Determine which voltage regulator has a better load regulation.

It should be noted that the entire load current flows through the regulator. As the load current demand increases, there is a corresponding amount of heat developed by the regulator. This could cause a thermal overload condition to occur. As a result of this, most regulators should have some form of heat dissipation. This will include things such as heat sinks, special packaging, and even liquid cooling.

Self-Examination

Answer the following questions.

1. In an ideal 5-V voltage regulator, if the applied input voltage increases in value, the output voltage will:

   a. Increase
   b. Decrease
   c. Remain constant
   d. Be 0 V

2. In an actual 5-V voltage regulator, with a line regulation of 0.05%, if the applied input voltage increases in value by 10%, the output voltage will:

   a. Increase
   b. Decrease
   c. Remain constant
   d. Be 0 V

3. In an ideal 5-V voltage regulator, if the load resistor is removed, the output voltage will:

   a. Increase
   b. Decrease
   c. Remain constant
   d. Be 0 V

4. In an actual 5-V voltage regulator, if the load resistor is shorted, the output current flow will:

   a. Increase
   b. Decrease
   c. Remain constant
   d. Be 0 A

5. Under which conditions would a voltage regulator be subjected to excessive thermal overload?

   a. Increase voltage, no-load demand
   b. Decrease voltage, constant current demand
   c. Increase voltage, increase current demand
   d. Decrease voltage, no-load demand

18.2 Linear Voltage Regulators – Series and Shunt

Typically, a **voltage regulator** circuit is designed to control variations in the output of a power supply. We have discussed, functionally, the operation of a regulator with respect to its ability to control input voltage and output current demand variations. At this time, we are ready to examine the internal construction of a regulator. Generally, it consists of a control element, an output voltage sampling circuit, and a reference voltage circuit. The control element responds to the difference between the reference voltage and the sampled output. The term linear associated with a voltage regulator circuit refers to the control element being maintained in the linear region of operation. In this regard, it is not completely switched off and on during its operation. The primary component of a regulator is the control element that changes its amount of conduction in response to changes in the applied **input voltage** or the **load current** demand.

The positioning of the control element with respect to the load device determines the specific type of **linear voltage regulator**. If the control element is connected in series with the load device, it is called a series linear voltage regulator. On the other hand, if the control element is connected in parallel (shunt) with the load, the regulator is called a shunt linear voltage regulator. Both the series and shunt voltage regulators have voltage reference and sampling circuits that alter the conduction of the control element.

18.2 Describe how a linear voltage regulator responds to input/output changes.

In order to achieve objective 18.2, you should be able to:

- describe the operating principle of a linear voltage regulator;
- discuss the operation of a series linear voltage regulator;
- describe the overload conditions of a series regulator;
- discuss the operation of a shunt linear voltage regulator;
- describe the overload conditions of a shunt regulator.

Series Linear Voltage Regulator

Figure 18.5(a) shows the location of a **series linear voltage regulator** in a power supply. A **block diagram** of the regulator is shown in **Figure 18.5(b)**. Note that the main component of the regulator, the control element, is connected in series with the load. The input side of the regulator is used to generate a reference voltage. The output voltage of the regulator is sampled

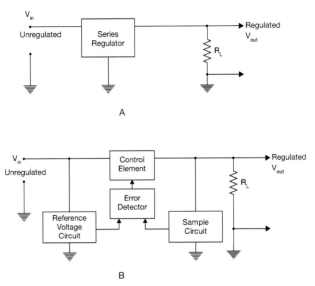

Figure 18.5 Block diagram of a series regulator.

and compared with the reference voltage using an error detector circuit. The output of the error detector circuit is then used to control the conduction of the control element. The conduction of the control element is adjusted so that the output voltage is maintained at some desired level.

Figure 18.6(a) shows the circuit diagram of a **series linear voltage regulator**. The unregulated input voltage, V_{in}, is applied to a Zener diode, D_1, through series resistor R_S. As a result of this, the Zener diode develops a voltage, V_Z, which is the reference voltage for the regulator. The **output voltage** of the regulator is developed across the R_1-R_2 which is a voltage divider network. The values of the R_1 and R_2 determine the amount of sample voltage, V_S. The reference voltage V_Z and the feedback voltage V_S are compared by the operational amplifier. The output of the op-amp, V_B, is applied to the base of the control transistor Q_1. An increase in the value of V_B will cause Q_1 to be more conductive, which reduces its collector-emitter voltage V_{CE}. A corresponding decrease of V_B will cause Q_1 to be less conductive, which increases the value of V_{CE}.

When voltage is first applied to the regulator circuit shown in **Figure 18.6**, the **series resistor** R_S and the **Zener diode** D_1 are used to develop the fixed reference voltage V_Z. Note that the Zener diode should operate in the reverse breakdown region to develop its Zener voltage. For this to occur, a minimum

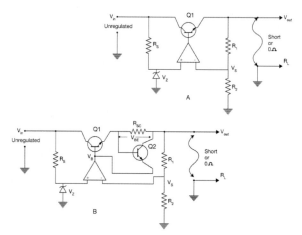

Figure 18.6   Series voltage regulator circuit.

amount of current must pass through the Zener diode. This also represents the current passing through the series resistor R_S. I_Z can be determined by the following expression:

$$I_Z = \frac{V_{in} - V_Z}{R_S}. \tag{18.3}$$

V_Z is applied to the noninverting (+) input of the op-amp which serves as the error detector. When the value of the unregulated input voltage V_{in} remains constant, V_Z will be constant (assuming it is already in conduction). If there is an increase or decrease in V_{in}, the Zener diode will continue maintaining its Zener voltage V_Z. Since $V_Z = V_{REF}$, this causes a constant voltage to be applied to the op-amp.

The output voltage of the regulator is sampled by the voltage divider network of R_1 and R_2. The voltage developed across resistor R_2 is the sample voltage V_S. This voltage is applied to the noninverting (V−) terminal of the op-amp. The value of V_S is determined by first calculating the value of the current I which flows through the R_1 and R_2.

$$I \times (R_1 + R_2) = V_{out}$$

$$I = \frac{V_{out}}{R_1 + R_2}.$$

The sample voltage developed across resistor R_2 can be determined by the following expression:

$$V_S = I \times R_2 = \frac{V_{out}}{R_1 + R_2} \times R_2.$$

If the value of the **load resistor** stays constant, then the value of the sampled voltage V_S will remain constant as well. Changes in the resistance of the load will cause a change in the current flow through the resistor network. This will cause the V_S to change accordingly. If the value of the load resistance decreases, the current drawn by the load will increase. A smaller amount of current will then flow through the resistor network. The value of the voltage developed across R_2, which is also the sampled voltage V_S, will decrease. On the other hand, if the value of the load resistance increases, the current drawn by the load will decrease. A larger amount of current will then flow through the resistor network. The value of the voltage developed across R_2, which is V_S, will also increase. If the input voltage V_{in} increases or decreases, V_{out} will be immediately influenced by it. For regulation to be achieved, V_{out} should be maintained at a constant value.

The **reference voltage (V_Z)** is applied to the noninverting input (+) of the op-amp and the sample voltage (V_S) is applied to the inverting input (-). If the value of V_S is reduced, which occurs either when the load resistance is reduced or the input voltage is decreased, the error voltage appearing at the output of the op-amp will increase instantaneously. This voltage represents V_B, which is applied to the base of the transistor Q_1. This, in turn, will cause the conduction of the transistor to increase, which will reduce V_{CE}. The output voltage of the series regulator circuit, V_{out}, is expressed as: $V_{out} = V_{in} - V_{CE}$. With instantaneous changes in the values of V_{CE}, the output voltage will continue to be maintained at a constant value. On the other hand, if the value of V_S is increased, which occurs either when the load resistance is increased or the input voltage is increased, the error voltage appearing at the output of the op-amp will decrease. This, in turn, will cause the conduction of the transistor to decrease, which will increase V_{CE}. The instantaneous change in V_{CE} will maintain the value of V_{out} at a constant value.

The **output voltage** of the regulator can be determined by the values of the circuit components. These include resistors R_1 and R_2 of the sampling circuit, and the value of the Zener diode V_Z of the voltage reference circuit. Ideally, the sampled voltage and the Zener voltage applied to the inputs of the op-amp should be equal.

$$V_Z = V_S$$

$$V_Z = \frac{V_{out}}{R_1 + R_2} \times R_2$$

$$V_Z \times \frac{R_1 + R_2}{R_2} = V_{out}$$

$$V_Z \times \left(1 + \frac{R_1}{R_2}\right) = V_{out}. \tag{18.4}$$

Example 18-4:

Refer to **Figure 18.6**. Design a series voltage regulator using a 5-V Zener which is needed to maintain a regulated output of 10 V across the load resistor. The value of resistor R_1 in the feedback network is 10 kΩ. Assume that the unregulated input voltage varies between 12 and 20 V, and the current through the Zener should be at least 20 mA for Zener breakdown operation. Determine the value of the resistor R_2 in the feedback network, and the series resistor R_S in the reference voltage circuit.

Solution

$$V_Z \times \left(1 + \frac{R_1}{R_2}\right) = V_{out}$$

$$5 \times \left(1 + \frac{10,000}{R_2}\right) = 10; 1 + \frac{10,000}{R_2} = 2$$

$$R_2 = 10,000\,\Omega.$$

The current (I_Z) through the series resistor R_S should be at least 20 mA. This will occur when the input voltage is at its lowest permissible value. When the input voltage increases, I_Z will increase.

$$I_Z = \frac{V_{in}(min) - V_Z}{R_S}; 0.020 = \frac{12 - 5}{R_S}$$

$$R_S = \frac{7}{0.020} = 350\,\Omega.$$

Related Problem

Refer to **Figure 18.6**. Design a series voltage regulator using a 6-V Zener needed to maintain a regulated output of 10 V. The value for the resistance R_1 in the feedback resistor network is 20 kΩ. Assume that the unregulated input voltage varies between 14 and 20 V, and the current through the Zener should be at least 25 mA for Zener breakdown operation.

Series Linear Voltage Regulators with Current Limiting Protection

Most **series linear voltage regulator** circuits employ some form of overload protection to prevent the series control element from being damaged. **Figure 18.7(a)** shows a linear series regulator without **overload protection**. Note that the load resistor of this circuit is indicated as a short, having zero or very low resistance. Should this condition occur, an excessive amount of current, $I_L(\text{max})$, will pass through the load device and the control transistor Q_1. As a result of this, the transistor may become overheated. If the transistor is operated for any extended time under these conditions, it could be permanently damaged. To prevent this from happening, a regulator circuit must employ some form of sensing circuit that will detect when this condition occurs.

 Figure 18.7(a) shows a transistor Q_2 and **short-circuit sensing resistor** added. It is important to note that when an overload occurs, the output voltage will no longer be regulated. The voltage developed across the load will be lower than the normal regulated voltage and is given by the expression: $V_{\text{out}} = I_{L(\text{max})} \times R_L$. This circuit provides **overload protection** by reducing the current conduction of Q_1 under overload conditions. When this occurs, an excessive amount of current flows through the load, denoted by $I_L(\text{max})$, and also through R_{SC}, the short-circuit resistor. This causes the voltage across R_{SC} and the base-emitter junction voltage, V_{BE}, of transistor Q_2 to increase. When the voltage V_{BE} reaches 0.7 V, Q_2 will go into conduction. Under these conditions:

$$V_{BE} = 0.7 = R_{SC} \times I_{L(\text{max})}$$

$$I_{L(\text{max})} = \frac{0.7}{R_{SC}}. \tag{18.5}$$

 When Q_2 goes into conduction, it diverts some of the current which was originally flowing into the base of Q_1. This reduces the current conduction of Q_1. As a result of this, less current will pass through Q_1, which reduces the heat developed by the transistor, thus avoiding a thermal overload. As long as the current does not exceed $I_{L(\text{max})}$, the output voltage will be regulated and remain at a nearly constant value regardless of changes in the load. If the load current exceeds $I_{L(\text{max})}$, the output voltage will be reduced, and the transistor Q_1 will be protected from a thermal overload. **Figure 18.7(c)** shows the variation of the output voltage with changes in the load current I_L. In the **current limiting region**, the value of the load current may continue to increase beyond $I_{L(\text{max})}$. It should be noted that a large current continues to

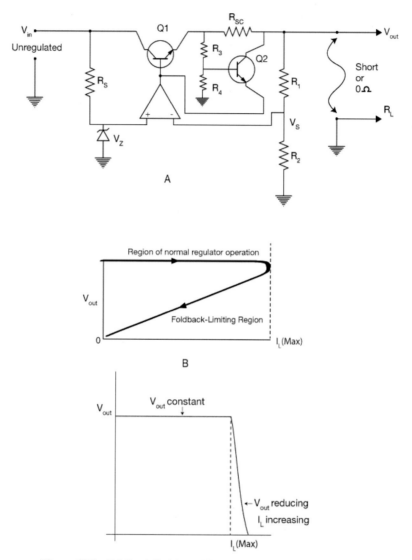

Figure 18.7   Foldback limiting with an overloaded voltage regulator.

flow through the load and, thus, through the control element under current limiting conditions. This may cause excessive heat to be developed by the control element. Hence, a method of reducing the current flow through the control element under these conditions would improve the performance.

Example 18-5:

Refer to **Figure 18.7**. Determine the value of the short-circuit resistance (R_{SC}) which will be the maximum current drawn by the load to 0.25 A.

Solution

$$I_{L(\text{max})} = \frac{V_{BE(Q2)}}{R_{SC}}$$

$$0.25 = \frac{0.7}{R_{SC}}$$

$$R_{SC} = 3\Omega.$$

Related Problem

Refer to **Figure 18.7**. Determine the value of the short-circuit resistance (R_{SC}) which will be the maximum current drawn by the load to 0.125 A?

Foldback Current Limiting in Series Voltage Regulators

A **current limiting circuit** can be modified to reduce the amount of current flow if the load exceeds the maximum permitted current flow $I_{L(\text{max})}$. The term "foldback" current limiting is used to describe this type of circuit behavior. **Figure 18.7(a)** shows a **series regulator** circuit with **foldback current limiting**. It is similar to the general current limiting circuit of **Figure 18.6(b)**. The difference in the two circuits is the manner in which voltage is applied to the base of transistor Q_2. In **Figure 18.7(a)**, resistors R_3 and R_4 form a voltage divider; so only a part of the voltage is applied to the base of Q_2. Applying **Kirchoff's voltage law** around the loop formed by the short-circuit resistor R_{SC}, the base-emitter of transistor Q_2, and resistor R_3, we have the following expression:

$$V_{BE(Q2)} + V_{R3} = V_{RSC}$$
$$V_{BE(Q2)} = V_{RSC} - V_{R3}.$$

When the **load current** increases to $I_{L(\text{max})}$, its maximum permissible value, the voltage across the short-circuit resistor, will be: $V_{RSC} = R_{SC} \times I_{L(\text{max})}$. The value of V_{RSC} will increase as the current through it increases. An increase in the load current causes a decrease in the output voltage V_{out}. When V_{out} is reduced, V_{R3} will likewise be reduced. The transistor Q_2 begins to conduct when the base-emitter voltage $V_{BE(Q2)}$ reaches 0.7 V. This voltage will remain almost constant at 0.7 V once the current limiting condition has been reached.

If the **load resistance** is made any smaller, V_{out} will decrease further. However, if this happens, V_{R3} will decrease. Thus, a smaller value of V_{RSC} will now be used for maintaining $V_{BE(Q2)}$ at 0.7 V since $V_{BE(Q2)} = V_{RSC} - V_{R3}$. As a result of this, $I_{RSC} = \frac{V_{RSC}}{R_{SC}}$ is reduced, which causes the value of the load current I_L to be reduced as well. This is shown in the **foldback region** of **Figure 18.7(b)**. Thus, the amount of current flowing through the series control element never reaches values higher than $I_{L(max)}$. As shown in **Figure 18.7(c)**, the value of I_L under short-circuit conditions is actually much smaller than the maximum permitted current. If the load resistance increases in value so that the load current is no longer in the foldback current region, the circuit resumes normal voltage regulation.

Shunt Linear Voltage Regulators

Figure 18.8(a) shows the location of a **shunt linear voltage regulator** in a power supply. A **block diagram** of this regulator is shown in **Figure 18.8(b)**. Note that the main component of the regulator, the **control element**, is connected in parallel or shunt with the load. The input side of the regulator is used to generate a **reference voltage**. The output voltage of the regulator is sampled and compared with the reference voltage using an error detector circuit. The output of the **error detector** circuit is then used to control the conduction of the control element. The conduction of the control element is adjusted so that the output voltage is maintained at some desired level. Resistor R_{in} on the input side of the regulator is connected in series with the parallel combination of the control element and the load R_L. The load voltage V_{out} is maintained at a constant value by varying the amount of current which is being shunted away from the load by varying the conduction of the control element.

Figure 18.9 shows the circuit diagram of a **shunt linear voltage regulator**. The functioning of the reference voltage generation circuit and the output sampling voltage circuit is similar to that of the series linear voltage regulator. As before, the reference voltage V_Z and the feedback voltage V_S are compared by the operational amplifier. The output of the op-amp, V_B, is applied to the base of the control transistor Q_1. An increase in the value of V_B will cause Q_1 to be more conductive, thereby shunting more current away from R_L since the effective resistance of the control element has been reduced. A corresponding decrease of V_B will cause Q_1 to be less conductive,

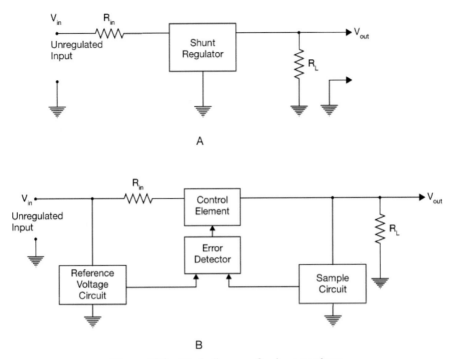

Figure 18.8 Block diagram of a shunt regulator.

thereby shunting less current away from R_L since the effective resistance of the control element has been increased.

The **reference voltage (V_Z)** is applied to the inverting input (-) of the op-amp and the sample voltage (V_S) is applied to the noninverting input (+). When the load resistance is reduced or the input voltage is decreased, the value of V_S will change accordingly. If the value of V_S is reduced, the error voltage appearing at the output of the op-amp will decrease instantaneously. This voltage represents V_B, which is applied to the base of the transistor Q_1. This, in turn, will cause the conduction of the transistor to decrease, in effect increasing its resistance; so less current is shunted away from the load. The parallel combination of the resistances of the control element and R_L will remain unchanged. With instantaneous changes in the values of the resistance of the control element, the output voltage will continue to be maintained at a constant value. On the other hand, if the value of V_S is increased, which occurs either when the load resistance is increased or the input voltage is increased, the error voltage appearing at the output of the op-amp will

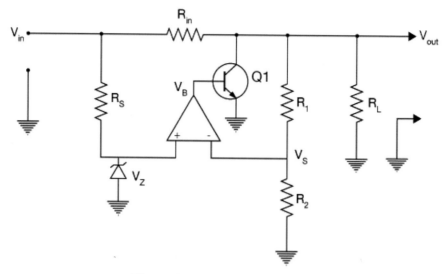

Figure 18.9   Shunt voltage regulator.

increase. This, in turn, will cause the conduction of the transistor to increase, in effect decreasing its resistance; so more current is shunted away from the load. The parallel combination of the resistances of the control element and R_L will remain unchanged. As before, with instantaneous changes in the values the resistance of the control element, the **output voltage** will continue to be maintained at a constant value.

Over-Voltage Protection in Shunt Linear Voltage Regulators

Voltage regulator circuits generally require some form of overload and over-voltage protection. In the shunt regulator circuit, the input resistor R_{in} provides short-circuit protection. If the load resistor R_L becomes shorted, the load current will flow through R_{in}, thereby providing protection from an over-load. The circuit also requires protection from an over-voltage condition. This is because when the unregulated voltage V_{in} increases, the conduction of the control element Q_1 increases as it tries to maintain a constant output voltage. The increased current flow through the transistor may cause it to dissipate more power. By selecting a transistor that has a higher **maximum power dissipation** rating $P_{D(max)}$ than that needed by the circuit, the possibility of thermal overload conditions is reduced.

Self-Examination

Answer the following questions.

1. In a series voltage regulator, the control element is placed in _____ with the load.

    a. Parallel
    b. Shunt
    c. Series
    d. Across

2. If the load resistance of a series voltage regulator decreases, the conduction of the control element _____.

    a. Increases
    b. Decreases
    c. Remains unchanged
    d. Will not be affected

3. If the input voltage applied to a series voltage regulator decreases, the conduction of the control element _____.

    a. Increases
    b. Decreases
    c. Remains unchanged
    d. Will not be affected

4. In a shunt linear voltage regulator, the control element is placed in _____ with the load.

    a. Line
    b. Parallel
    c. Series
    d. The same path

5. If the input voltage applied to a shunt voltage regulator increases, conduction of the control element will _____.

    a. Increase
    b. Decrease
    c. Remains unchanged
    d. Will not be affected

18.3 Switching Voltage Regulators – Step-Down and Step-Up

In the previous section, we discussed linear voltage regulators of the series and shunt type. In this type of regulator, the control element is continuously operated in the linear region. The power dissipated by a control element of this type, typically, a transistor, is high. The power depends on the voltage that appears across the collector-emitter V_{CE} and the current flowing through the transistor, I_C. For linear region operation, both V_{CE} and I_C are relatively high, with some of the power being provided to the circuit being wasted in the form of heat. This causes the **linear voltage regulator** to be inefficient. One way to improve the efficiency of a voltage regulator is by operating the control element as a switching device. This occurs when the control element transistor is rapidly switched on and off. When the transistor is switched on, it goes into saturation. When this occurs, V_{CE} drops to a very low value, and the collector current flow is large. The power dissipation then is nominal owing to the low value of V_{CE}. When the transistor is switched off, it goes into cutoff. When this occurs, V_{CE} rises to its maximum value, and the collector current flow is small. The power dissipation continues to be nominal owing to the low value of I_C. Thus, whenever the transistor is operated exclusively in saturation or cutoff, the power dissipation of the control element is at a nominal value. Since the transition between these states is not instantaneous, some power continues to be dissipated by the device.

In order for a **control element** to achieve its switching operation a **pulse-width modulator (PWM) circuit** is employed. This type of circuit typically generates a pulsating DC waveform in which the width of the pulses can be adjusted. The time period (T) of this waveform consists of an on-portion (T_{on}) and an off-portion (T_{off}). When the output of a PWM is applied to the control element, it causes the conduction time to vary accordingly. The duration of the on- and off-times (pulse-width) or the time period of the waveform (pulse repetition rate) in a PWM can be either fixed or variable.

When the control element is rapidly switched on and off, it causes the output to have a similar response. This means that the value of the output will fluctuate according to the switching rate. Some form of filtering is now needed to correct this condition. Switching regulators circuits also contain voltage reference and output voltage sampling circuits. The output of these circuits is applied to an error detection amplifier that is used to control the PWM circuit for altering the conduction of the control element.

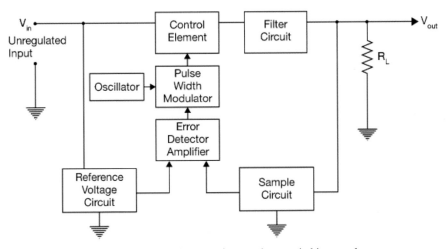

Figure 18.10   Block diagram of a step-down switching regulator.

Switching voltage regulators are classified according to the value of the output voltage (V_{out}) generated with respect to the applied input voltage (V_{in}). If V_{out} is smaller than or equal to V_{in}, the configuration is termed as a step-down switching regulator. The block diagram of a **step-down switching regulator** is shown in **Figure 18.10**. If V_{out} is larger than V_{in}, the configuration is termed as a step-up switching regulator. It should be noted that these configurations of switching regulators are somewhat similar to that of a transformer which can step down or step up the applied input voltage.

18.3 Describe how a switching voltage regulator responds to input/output changes.

In order to achieve objective 18.3, you should be able to:

- describe the operating principle of a switching voltage regulator;
- discuss the operation of a step-up voltage regulator;
- discuss the operation of a step-down voltage regulator;
- compare the output response of step-up and step-down voltage regulators.

Step-Down Switching Regulators

A **step-down switching regulator** circuit is shown in **Figure 18.11**. This regulator is similar to the series linear voltage regulator circuit of **Figure 18.6**.

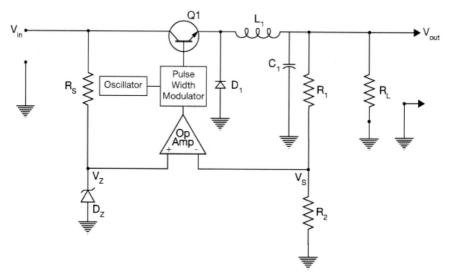

Figure 18.11   Step-down switching voltage regulator.

The control element, transistor Q_1, is connected in series with the load resistor R_L. The regulator also develops a reference voltage V_Z across Zener diode D_Z. It has an output sampling circuit consisting of resistor R_1 and R_2. The sampled voltage V_S is compared with the reference voltage V_Z by the op-amp.

The construction of the **step-down switching regulator** in **Figure 18.11** has some additional components that are unique to its design. These include a **pulse-width modulator (PWM)**, an **oscillator**, and a **filter** circuit. The output of the error detector operational amplifier and the pulses generated by the oscillator are applied to the PWM circuit. This portion of the circuit generates a pulsating DC that is used to drive the transistor. The filter circuit consists of a capacitor, an inductor, and a diode. The capacitor C_1 is connected across the load resistor R_L to prevent rapid changes in voltage. An inductor L_1 is connected in series with R_L to prevent rapid changes in current through the load. When the control transistor is switched on and off, the resulting current is interrupted, causing a high voltage to be generated across the inductor. Diode D_1 is used to suppress the voltage spikes that occur during switching.

Figure 18.12 shows the **switching intervals** of transistor Q_1. The interval for which the Q_1 is switched on is designated as t_{on}. The interval for which Q_1 is switched off is designated as t_{off}. The capacitor C_1 shown in **Figure 18.11** charges during the on-time of Q_1, and it discharges through the load during

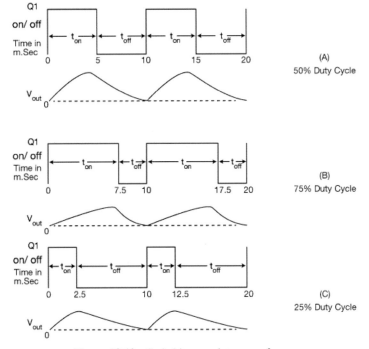

Figure 18.12 Switching regulator waveforms.

the off-time. The total time period (T) of the waveform is the sum of t_{on} and t_{off}.

$$T = t_{on} + t_{off}.$$

The frequency of the waveform (f) is defined as the number of waveforms completed within a given unit of time:

$$f = \frac{1}{T} Hz.$$

The ratio of the on-time (t_{on}) to the time period (T) is designated as the duty cycle. It is usually expressed as a percentage.

$$\% \text{ Duty Cycle} = \frac{t_{on}}{T} \times 100 = \frac{t_{on}}{t_{on} + t_{off}} \times 100. \qquad (18.6)$$

As t_{on} increases in proportion to t_{off}, the **duty cycle** will increase. With an increased on-time, the capacitor will develop a larger charge. The output voltage V_{out} of the regulator that appears across the capacitor will increase as well. Thus, a higher duty cycle results in a higher output voltage. Alternatively, if t_{on} decreases in proportion to t_{off}, the duty cycle will decrease. With a decreased on-time, the capacitor will develop a smaller charge. The output voltage V_{out} of the regulator which appears across the capacitor will decrease as well. Thus, a lower duty cycle results in a lower output voltage. This is expressed as

$$V_{out} = \frac{t_{on}}{T} \times V_{in}. \tag{18.7}$$

When the **load resistance** is reduced or the input voltage is decreased, the value of the sample voltage V_S shown in **Figure 18.12** will change accordingly. If the value of V_S is reduced, the error voltage appearing at the output of the op-amp will increase instantaneously. A larger voltage is, thus, applied to the pulse-width modulator. It will produce a waveform that has a higher duty cycle. This waveform is applied to the base of transistor Q_1, turning it on for a longer time in proportion to the off-time. This charges the capacitor C_1 to a higher value and, thus, increases the output voltage, maintaining it at a constant value. On the other hand, if the value of V_S is increased, it will affect the operation of the op-amp. An increase in V_S will occur when either the load resistance is increased or the input voltage is increased. The error voltage appearing at the output of the op-amp will decrease. A smaller voltage is, thus, applied to the pulse-width modulator. It will produce a waveform that has a lower duty cycle. This waveform is applied to the base of transistor Q_1, turning it on for a smaller time in proportion to the off-time. This charges the capacitor C_1 to a lower value and, thus, decreases the output voltage, maintaining it at a constant value.

Example 18-6:

Refer to **Figure 18.12(a)**. Calculate the frequency and the duty cycle of the waveform. If V_{in} is 12 V, calculate V_{out}.

Solution

$$t_{on} = 5ms$$
$$t_{off} = 5ms$$
$$T = t_{on} + t_{off} = 5\text{ms} + 5\text{ms} = 10\text{ms}$$
$$f = \frac{1}{10\,\text{ms}} = 100\,\text{Hz}$$

$$\% \text{ Duty Cycle} = \frac{5\,\text{ms}}{10\,\text{ms}} \times 100 = 50\%$$

$$V_{\text{out}} = \frac{t_{\text{on}}}{T} \times V_{\text{in}} = \frac{5\,\text{ms}}{10\,\text{ms}} \times 12\,\text{V} = 6\,\text{V}.$$

Related Problem

Refer to **Figure 18.12(c)**. Calculate the frequency and the duty cycle of the waveform. If V_{in} is 14, calculate V_{out}.

Step-Up Switching Regulators

A **step-up switching regulator** is used for stepping up or increasing the input voltage V_{in}. In order to achieve this, an inductor is first connected in series with the input voltage. When the current flowing through this inductor is suddenly interrupted by switching action, the inductor generates a high voltage across its terminals. This voltage is added to the input voltage and appears as the combined output voltage V_{out}.

Figure 18.13 shows the circuit diagram of a **step-up switching voltage regulator**. The main component of this regulator is the control transistor Q_1. Q_1 is connected in parallel or shunt with the applied input voltage. The input voltage source V_{in} for the regulator is connected to the series inductor L_1. Diode D_1 is also connected in series with L_1 and the load R_L. Capacitor C_1 is a filter component for V_{in} that is used to stabilize the output V_{out}. In addition

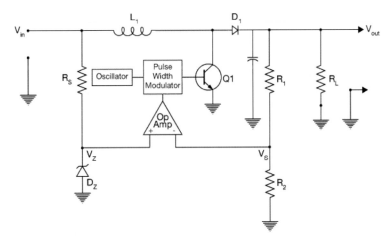

Figure 18.13 Step-up switching voltage regulator.

to this, the regulator consists of other parts. The input side of the regulator is used to generate a reference voltage. The output voltage of the regulator is sampled and compared with the reference voltage using an error detector circuit. The output of the error detector circuit is then applied to a PWM circuit used to control the conduction of the control element.

The switching of transistor Q_1 controls the current flowing through inductor L_1. When Q_1 is on, the inductor is connected directly to the applied input voltage V_{in}. The voltage across the inductor (V_L) increases instantaneously with the applied input voltage but drops in value when Q_1 is on. The current flowing through L_1 increases linearly, which causes energy to be stored in its magnetic field. With Q_1 switched on, the anode of diode D_1 is grounded through the transistor, thereby reverse biasing it. During this time, the load is effectively disconnected from the input side of the circuit and capacitor C_1 discharges through the load.

When Q_1 is switched from on to off, the energy stored in the magnetic field of L_1 causes a current flow when the magnetic field collapses. This causes the stored energy of L_1 to be transferred to capacitor C_1 through diode D_1 which is now forward biased. If Q_1 is kept on for a longer duration, the value of the inductor voltage will reduce accordingly. When Q_1 is switched off, this will cause a higher voltage to be developed by the inductor. This voltage is then added to the input voltage and charges C_1, developing a higher output voltage. This means that as the duty cycle of Q_1 controls the **output voltage**. In an actual step-up regulator, the output voltage can be increased as much as five times the applied input voltage, by varying the duty cycle.

The response of the regulator to changes in either the applied input voltage or the load resistance is similar to that of the step-down voltage regulator. The value of the output voltage is maintained at a constant value. Varying the duty cycle of Q_1 will cause the voltage developed by the inductor L_1 to change accordingly. For example, an increase in the duty cycle will cause the inductor to develop a higher voltage. This voltage will be added to the input voltage, which causes the **output voltage** to be maintained at a constant value.

Self-Examination

Answer the following questions.

11. The control transistor used in a switching voltage regulator operates predominantly in the _____ region(s).
    a. Linear
    b. Cutoff

c. Saturation

d. Alternating between cutoff and saturation

12. A duty cycle of a pulse-width modulator (PWM) is 50%. This indicates that:

    a. $t_{on} > t_{off}$
    b. $t_{on} < t_{off}$
    c. $t_{on} = t_{off}$
    d. $T = t_{on}$
    e. $T = t_{off}$

13. In a step-down switching voltage regulator, the control transistor is placed in _____ with the load.

    a. Series
    b. Shunt
    c. Parallel
    d. Across

14. In a step-up switching voltage regulator, the control transistor is placed in _____ with the input voltage.

    a. Series
    b. Shunt
    c. Parallel
    d. Across

15. In a step-down switching regulator, if the load increases, the duty cycle of the pulse-width modulator applied to control transistor _____.

    a. Remains constant
    b. Decreases
    c. Increases
    d. Becomes 0

18.4 IC Voltage Regulators

A number of manufactures are now producing integrated circuits (ICs) that perform voltage regulation. IC regulators are now available for both linear and switching applications. These ICs have the essential regulator components built on a single chip instead of using discrete components. Based on the specific type of **IC regulator** used, it may include the voltage reference, sampling, error detection, overload protection, control transistor, oscillator, and pulse-width modulation circuitry. As a result of this, when using a regulator IC in a circuit, manufacturer datasheets should be examined to determine how the device is to be connected. The datasheets are also used to identify operating characteristics and various configurations in which the device can be used. **Block diagrams** are often used to show the functional operation of an IC voltage regulator. This permits the user to see how the device will respond in an actual circuit application.

18.4 Interpret the datasheet of an IC voltage regulator.

In order to achieve objective 18.4, you should be able to:

- identify the function of an IC voltage regulator;
- describe the input/output response of a fixed linear IC voltage regulator;
- describe the input/output response of an adjustable linear IC voltage regulators;
- identify the primary function of an IC switching voltage regulator.

Fixed Linear IC Voltage regulator

A representative **three-terminal fixed IC voltage regulator** is shown in **Figure 18.14**. This circuit has input, ground, and output terminals, identified by the numbers 1, 2, and 3, respectively. Unregulated input voltage is applied at terminal 1, and a fixed regulated output appears at terminal 3.

Voltage regulators are identified by the numbers corresponding to the regulated voltage value and polarity. The **7800 series** is a representative type of positive voltage regulators, and the **7900 series** for negative voltage regulators. Both the 7800 and 7900 series can supply output currents exceeding 1 A while maintaining the regulated voltage output. This can result in considerable heat being generated by the regulator. An adequate cooling system should be provided for the regulator operating under these conditions. Typically, heat sinks are used for this purpose.

Another consideration in the operation of these regulators is based on the value of the voltage applied to its input. The input voltage must exceed the output regulated voltage by 2 V or more in order to function properly. This is generally referred to as the dropout voltage. These regulators generally require some form of **filtering**. Filter capacitor C_1 is connected to the input terminal of the regulator. This capacitor is designed to reduce unwanted oscillations or noise that may appear at the input when it is some distance

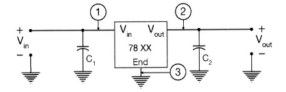

IC#	Output Voltage	Minimum Input V
7805	+5.0 V	7.3
7809	+9.0 V	11.3

(A) 7800 Series Positive Voltage Regulators

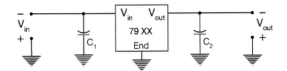

IC#	Output Voltage	Minimum Input V
7905	-5.0 V	-7.3
7909	-9.0 V	-11.3

(B) 7900 Series Positive Voltage Regulators

Figure 18.14   7800/7900 voltage regulators.

away from the power supply. Another filter capacitor C_2 is connected to the output terminal of the regulator to reduce high-frequency noise.

Note the polarity of the input/output voltages, and the capacitor connections for the 7800 series regulators, as compared to those of the 7900 series. **Figure 18.15** shows the **datasheet** of **KA78XX** manufactured by Fairchild Semiconductor. The XX denotes the specific voltage that will be regulated by the device. The KA7805 for example is a +5 V regulator. It has a dropout voltage of approximately 2 V and can supply up to 1 A of output current without a heat sink.

Adjustable Linear IC Voltage Regulator

A number of adjustable **three-terminal IC voltage regulators** are available in IC form today. Regulators of this type are needed in applications where a non-standard regulator voltage values are needed. **Figure 18.16** shows the circuit diagram of LM317 and LM337 which are positive and negative adjustable voltage regulators, respectively. These regulators have three terminals which are identified as the input, output, and adjust. Capacitors are used for decoupling purposes and do not affect the operation of the regulator circuit.

www.fairchildsemi.com

KA78XX/KA78XXA
3-Terminal 1A Positive Voltage Regulator

Features

- Output Current up to 1A
- Output Voltages of 5, 6, 8, 9, 10, 12, 15, 18, 24V
- Thermal Overload Protection
- Short Circuit Protection
- Output Transistor Safe Operating Area Protection

Description

The KA78XX/KA78XXA series of three-terminal positive regulator are available in the TO-220/D-PAK package and with several fixed output voltages, making them useful in a wide range of applications. Each type employs internal current limiting, thermal shut down and safe operating area protection, making it essentially indestructible. If adequate heat sinking is provided, they can deliver over 1A output current. Although designed primarily as fixed voltage regulators, these devices can be used with external components to obtain adjustable voltages and currents.

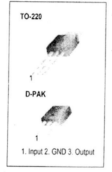

TO-220

D-PAK

1. Input 2. GND 3. Output

Internal Block Digram

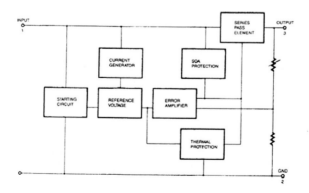

KA78XX/KA78XXA

___

Absolute Maximum Ratings

Parameter	Symbol	Value	Unit
Input Voltage (for V_O = 5V to 18V)	V_I	35	V
(for V_O = 24V)	V_I	40	V
Thermal Resistance Junction-Cases (TO-220)	$R_{\theta JC}$	5	$^\circ$C/W
Thermal Resistance Junction-Air (TO-220)	$R_{\theta JA}$	65	$^\circ$C/W
Operating Temperature Range (KA78XX/A/R)	T_{OPR}	0 ~ +125	$^\circ$C
Storage Temperature Range	T_{STG}	-65 ~ +150	$^\circ$C

Electrical Characteristics (KA7805/KA7805R)

(Refer to test circuit ,0°C < T$_J$< 125°C, I$_O$ = 500mA, V$_I$ =10V, C$_I$= 0.33µF, C$_O$=0.1µF, unless otherwise specified)

Parameter	Symbol	Conditions		KA7805			Unit
				Min.	Typ.	Max.	
Output Voltage	V_O	T_J =+25 $^\circ$C		4.8	5.0	5.2	V
		5.0mA ≤ Io ≤ 1.0A, P$_O$ ≤ 15W V_I = 7V to 20V		4.75	5.0	5.25	
Line Regulation (Note1)	Regline	T_J=+25 $^\circ$C	V_O = 7V to 25V	-	4.0	100	mV
			V_I = 8V to 12V	-	1.6	50	
Load Regulation (Note1)	Regload	T_J=+25 $^\circ$C	I$_O$ = 5.0mA to1.5A	-	9	100	mV
			I$_O$ =250mA to 750mA	-	4	50	
Quiescent Current	I$_Q$	T_J =+25 $^\circ$C		-	5.0	8.0	mA
Quiescent Current Change	ΔI$_Q$	I$_O$ = 5mA to 1.0A		-	0.03	0.5	mA
		V$_I$= 7V to 25V		-	0.3	1.3	
Output Voltage Drift	ΔV$_O$/ΔT	I$_O$= 5mA		-	-0.8	-	mV/$^\circ$C
Output Noise Voltage	V$_N$	f = 10Hz to 100KHz, T$_A$=+25 $^\circ$C		-	42	-	µV/V$_O$
Ripple Rejection	RR	f = 120Hz V_O = 8V to 18V		62	73	-	dB
Dropout Voltage	V$_{Drop}$	I$_O$ = 1A, T$_J$ =+25 $^\circ$C		-	2	-	V
Output Resistance	r$_O$	f = 1KHz		-	15	-	mΩ
Short Circuit Current	I$_{SC}$	V$_I$ = 35V, T$_A$ =+25 $^\circ$C		-	230	-	mA
Peak Current	I$_{PK}$	T_J =+25 $^\circ$C		-	2.2	-	A

Note:

1. Load and line regulation are specified at constant junction temperature. Changes in V$_O$ due to heating effects must be taken into account separately. Pulse testing with low duty is used.

Figure 18.15   Datasheet for 78XX voltage regulator.

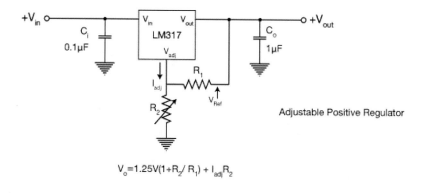

$$V_o = 1.25V(1 + R_2/R_1) + I_{adj}R_2$$

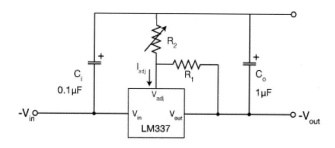

Adjustable Negative Regulator

Figure 18.16   Circuit diagram of adjustable voltage regulators. (a) Adjustable positive regulator. (b) Adjustable negative regulator.

Figure 18.16(a) shows the circuit diagram of an **LM317 positive adjustable voltage regulator**. It uses externally connected resistors in its operation. R_1 is a fixed resistor that is connected between the output and the adjust terminals. A fixed voltage reference V_{REF}= 1.25 V is developed by the regulator between these terminals. Thus, a constant current I_{REF} always flows through R_1.

$$I_{REF} = \frac{V_{REF}}{R_1} = \frac{1.25}{R_1}.$$

A variable resistor R_2 is connected between the adjust terminal and the ground. A small constant current I_{ADJ} flows through the adjust terminal. The combined current of I_{REF} and I_{ADJ} then flows through R_2. The value of the

SEMICONDUCTOR®   www.fairchildsemi.com

LM317
3-Terminal Positive Adjustable Regulator

Features

- Output Current In Excess of 1. 5A
- Output Adjustable Between 1. 2V and 37V
- Internal Thermal Overload Protection
- Internal Short Circuit Current Limiting
- Output Transistor Safe Operating Area Compensation
- TO-220 Package

Description

This monolithic integrated circuit is an adjustable 3-terminal positive voltage regulator designed to supply more than 1.5A of load current with an output voltage adjustable over a 1.2 to 37V. It employs internal current limiting, thermal shut-down and safe area compensation.

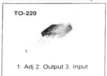

TO-220

1. Adj 2. Output 3. Input

Internal Block Diagram

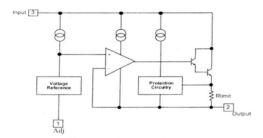

output voltage is the sum of the voltages developed across resistors R_1 and R_2.

$$V_{\text{out}} = V_{\text{REF}} \times \left(1 + \frac{R_2}{R_1}\right) + I_{\text{ADJ}} \times R_2. \qquad (18.8)$$

Once the value of the resistor R_1 is selected, the range of output voltages can be determined by adjusting R_2. When R_2 is at a minimum value (0 Ω), the regulator develops its lowest output voltage, which will be approximately 1.25 V. When R_2 is at a maximum value, the regulator develops its highest output voltage. This value cannot exceed 37 V as specified in the **datasheet** of the **LM317** given in **Figure 18.17**.

Figure 18.16(b) shows the circuit diagram of an **LM337** negative adjustable voltage regulator, and its **datasheet** is shown in **Figure 18.18**.

LM317

Absolute Maximum Ratings

Parameter	Symbol	Value	Unit
Input-Output Voltage Differential	$V_I - V_O$	40	V
Lead Temperature	T_{LEAD}	230	°C
Power Dissipation	P_D	Internally limited	W
Operating Junction Temperature Range	T_j	0 ~ +125	°C
Storage Temperature Range	T_{STG}	-65 ~+125	°C
Temperature Coefficient of Output Voltage	$\Delta V_O/\Delta T$	±0.02	%/°C

Electrical Characteristics

(V_I-V_O=5V, I_O= 0.5A, 0°C ≤ T_J ≤ + 125°C, I_{MAX} = 1.5A, P_{DMAX} = 20W, unless otherwise specified)

Parameter	Symbol	Conditions	Min	Typ.	Max.	Unit
Line Regulation (Note1)	Rline	T_A = +25°C 3V ≤ V_I - V_O ≤ 40V	-	0.01	0.04	% / V
		3V ≤ V_I - V_O ≤ 40V	-	0.02	0.07	% / V
Load Regulation (Note1)	Rload	T_A = +25°C, 10mA ≤ I_O ≤ I_{MAX} V_O< 5V V_O ≥ 5V	-	18 0.4	25 0.5	mV% / V_O
		10mA ≤ I_O ≤ I_{MAX} V_O < 5V V_O ≥ 5V	-	40 0.8	70 1.5	mV% / V_O
Adjustable Pin Current	IADJ	-	-	46	100	μA
Adjustable Pin Current Change	ΔIADJ	3V ≤ V_I - V_O ≤ 40V 10mA ≤ I_O ≤ I_{MAX} Pd ≤ P_{MAX}	-	2.0	5	μA
Reference Voltage	VREF	3V ≤ V_{IN} - V_O ≤ 40V 10mA ≤ I_O ≤ I_{MAX} P_D ≤ P_{MAX}	1.20	1.25	1.30	V
Temperature Stability	ST$_T$	-	-	0.7	-	% / V_O
Minimum Load Current to Maintain Regulation	$I_{L(MIN)}$	V_I - V_O = 40V	-	3.5	12	mA
Maximum Output Current	IO(MAX)	V_I - V_O ≤ 15V, P_D ≤ P_{MAX} V_I - V_O ≤ 40V, P_D ≤ P_{MAX} T_A=25°C	1.0	2.2 0.3	-	A
RMS Noise, % of V_{OUT}	eN	T_A= +25°C, 10Hz ≤ f ≤ 10KHz	-	0.003	0.01	% / V_O
Ripple Rejection	RR	V_O = 10V, f = 120Hz without C_{ADJ} C_{ADJ} = 10μF (Note2)	66	60 75	-	dB
Long-Term Stability, T_J = T_{HIGH}	ST	T_A = +25°C for end point measurements, 1000HR	-	0.3	1	%
Thermal Resistance Junction to Case	RθJC	-	-	5	-	°C / W

Note:

1. Load and line regulation are specified at constant junction temperature. Change in V_D due to heating effects must be taken into account separately. Pulse testing with low duty is used. (P_{MAX} = 20W)
2. C_{ADJ}, when used, is connected between the adjustment pin and ground.

Figure 18.17 Datasheet for LM317 positive adjustable voltage regulator.

www.fairchildsemi.com

LM337
3-Terminal 1.5A Negative Adjustable Regulator

Features

- Output current in excess of 1.5A
- Output voltage adjustable between -1.2V and - 37V
- Internal thermal overload protection
- Internal short circuit current limiting
- Output transistor safe area compensation
- Floating operation for high voltage applications
- Standard 3-pin TO-220 package

Description

The LM337 is a 3-terminal negative adjustable regulator. It supplies in excess of 1.5A over an output voltage range of -1.2V to - 37V. This regulator requires only two external resistor to set the output voltage. Included on the chip are current limiting, thermal overload protection and safe area compensation.

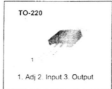

TO-220

1. Adj 2. Input 3. Output

Internal Block Diagram

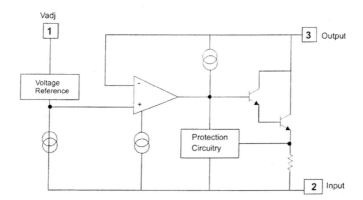

Example 18-7:

Refer to **Figure 18.16**. Determine the range of output voltage for the LM317 voltage regulator, assuming I_{ADJ}= 46 μA, V_{REF}= 1.25 V, and fixed resistor R_1= 470 Ω, and the variable resistor R_2= 10 kΩ.

LM337

Absolute Maximum Ratings

Parameter	Symbol	Value	Unit
Input-Output Voltage Differential	$|V_I - V_O|$	40	V
Power Dissipation	P_D	Internally limited	W
Operating Temperature Range	T_{OPR}	0 ~ +125	°C
Storage Temperature Range	T_{STG}	-65 ~ +125	°C

Electrical Characteristics

($V_I - V_O = 5V$, $I_O = 40mA$, $0°C \le T_J \le +125°C$, $P_{DMAX} = 20W$, unless otherwise specified)

Parameter	Symbol	Conditions	Min	Typ.	Max.	Unit
Line Regulation (Note1)	R_{line}	$T_A = +25°C$ $3V \le |V_I - V_O| \le 40V$	-	0.01	0.04	%/ V
		$3V \le |V_I - V_O| \le 40V$	-	0.02	0.07	
Load Regulation (Note1)	R_{load}	$T_A = +25°C$ $10mA \le I_O \le 0.5A$	-	15	50	mV
		$10mA \le I_O \le 1.5A$	-	15	150	
Adjustable Pin Current	I_{ADJ}	-	-	50	100	μA
Adjustable Pin Current Change	ΔI_{ADJ}	$T_A = +25°C$ $10mA \le I_O \le 1.5A$ $3V \le |V_I - V_O| \le 40V$	-	2	5	μA
Reference Voltage	V_{REF}	$T_A = +25°C$	-1.213	-1.250	-1.287	V
		$3V \le |V_I - V_O| \le 40V$ $10mA \le I_O \le 1.5A$	-1.200	-1.250	-1.300	
Temperature Stability	ST_T	$0°C \le T_J \le +125°C$	-	0.6	-	%
Minimum Load Current to Maintain Regulation	$I_{L(MIN)}$	$3V \le |V_I - V_O| \le 40V$	-	2.5	10	mA
		$3V \le |V_I - V_O| \le 10V$	-	1.5	6	
Output Noise	e_N	$T_A = +25°C$ $10Hz \le f \le 10KHz$	-	0.003	-	$V/10^6$
Ripple Rejection Ratio	RR	$V_O = -10V$, $f = 120Hz$	-	60	-	dB
		$C_{ADJ} = 10μF$ (Note2)	66	77	-	
Long Term Stability	ST	$T_J = 125°C$,1000Hours	-	0.3	1	%
Thermal Resistance Junction to Case	$R_{θJC}$	-	-	4	-	°C/ W

Note:

1. Load and line regulation are specified at constant junction temperature. Change in V_O due to heating effects must be taken into account separately. Pulse testing with low duty is used.
2. C_{ADJ}, when used, is connected between the adjustment pin and ground.

Figure 18.18   Datasheet for LM337 negative adjustable voltage regulator.

Solution

The output voltage depends on the value of the variable resistor as given by the following expression:

$$V_{out} = V_{REF} \times \left(1 + \frac{R_2}{R_1}\right) + I_{ADJ} \times R_2.$$

The output voltage will be smallest when the resistor R_2 is set to its minimum value, i.e., 0 Ω.

$$V_{out} = 1.25 \times \left(1 + \frac{0}{470}\right) + 46\mu \times 0 = 1.25 + 0 = 1.25 \text{ V}.$$

The output voltage will be largest when the resistor R_2 is set to its maximum value, i.e., 10 kΩ.

$$V_{out} = 1.25 \times \left(1 + \frac{10,000}{470}\right) + 46 \times 10^{-6} \times 10,000$$

$$= 27.85 + 0.46 = 28.30 \text{ V}.$$

The range of output voltage is 1.25-28.30 V.

Related Problem

Refer to **Figure 18.16**. Determine the range of output voltage for the LM317 voltage regulator, assuming I_{ADJ}= 46 μA, V_{REF}= 1.25 V, and fixed resistor R_1= 330 Ω, and the variable resistor R_2= 5 kΩ.

Switching IC Voltage Regulator

A representative **IC switching regulator** is shown in **Figure 18.19.** It is a DC-to-DC converter and can be used in various configurations such as step-up and step-down switching regulators. It contains a built-in voltage reference regulator, comparator, oscillator, and switching control element circuitry. External components are needed to achieve various regulator configurations.

Instead of using a PWM controller for adjusting the duty cycle as discussed earlier, this regulator uses a variable off-time modulation of the oscillator. Variable off-time regulation is accomplished by altering the off-time (T_{off}) of the oscillator's output, while keeping the **on-time (T_{on}) duration constant**. Thus, the time period (T) of the oscillator can be altered. A longer off-time will result in the control transistor being switched off for a longer time.

SEMICONDUCTOR®

www.fairchildsemi.com

MC34063A/MC33063A
SMPS Controller

Features

- Operation from 3.0 to 40V input
- Short circuit current limiting
- Low standby current
- Output switch current of 1.5A without external transistors
- Output voltage adjustable
- Frequency of operation from 100Hz to 100KHz
- Step up, Step down or inverting switching regulators

Description

The MC34063A/MC33063A is a monolithic regulator sub system intended for use as DC to DC converter. This device contains a temperature compensated bandgap reference, a duty cycle control oscillator, driver and high current output switch. It can be used for step down, step up or inverting switching regulators as well as for series pass regulators.

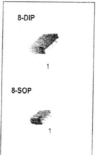

8-DIP

8-SOP

Internal Block Diagram

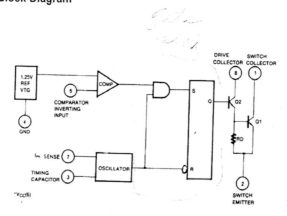

MC34063A/MC33063A

Absolute Maximum Ratings

Parameter	Symbol	Value	Unit
Supply Voltage	V_{CC}	40	V
Comparator Input Voltage Range	$V_{I(COMP)}$	$-0.3 \sim +40$	V
Switch Collector Voltage	$V_{C(SW)}$	40	V
Switch Emitter Voltage	$V_{E(SW)}$	40	V
Switch Collector To Emitter Voltage	$V_{CE(SW)}$	40	V
Driver Collector Voltage	$V_{C(DR)}$	40	V
Switch Current	I_{SW}	1.5	A
Storage Temperature Range	T_{STG}	$-65 \sim +150$	°C

Electrical Characteristics

(V_{CC} = 5.0V, T_A = 0°C to +70°C for the MC34063, T_A= -40°C to the +85°C for the MC33063, unless otherwise specified)

Parameter	Symbol	Conditions	Min.	Typ.	Max.	Unit
OSCILLATOR						
Charging Current	I_{CHG}	V_{CC} = 5 to 40V T_A = 25°C	22	31	42	µA
Discharging Current	I_{DISCHG}	V_{CC} = 5 to 40V T_A = 25°C	140	190	260	µA
Oscillator Amplitude	$V_{(OSC)}$	T_A = 25°C	-	0.5	-	V
Discharge To Charge Current Ratio	K	$V_7 = V_{CC}$, T_A = 25°C	5.2	6.1	7.5	-
Current Limit Sense Voltage	$V_{SENSE(C.L)}$	$I_{CHG} = I_{DISCHG}$ T_A = 25°C	250	300	350	mV
OUTPUT SWITCH						
Saturation Voltage 1 (Note)	$V_{CE(SAT)}1$	I_{SW} = 1.0A V_C(driver) = $V_{C(SW)}$	-	0.95	1.3	V
Saturation Voltage 2 (Note)	$V_{CE(SAT)}2$	I_{SW} = 1.0A, V_C(driver) = 50mA	-	0.45	0.7	V
DC Current Gain (Note)	$G_{I(DC)}$	I_{SW} = 1.0A, V_{CE} = 5.0V, T_A = 25°C	50	180	-	-
Collector off State Current (Note)	$I_{C(OFF)}$	V_{CE} = 40V, T_A = 25°C	-	0.01	100	µA
COMPARATOR						
Threshold Voltage	V_{TH}	-	1.21	1.24	1.29	V
Threshold Voltage Line Regulation	ΔV_{TH}	V_{CC} = 3 to 40V	-	2.0	5.0	mV
Input Bias Current	I_{BIAS}	V_I = 0V	-	50	400	nA
TOTAL DEVICE						
Supply Current MC34063	I_{CC}	V_{CC} = 5 to 40V C_T = 0.001uF $V_7 = V_{CC}$, $V_5 > V_{TH}$ pin2 = GND	-	-	4.0	mA
MC33063			-	-	5.0	

Note :
Output switch tests are performed under pulsed conditions to minimize power dissipation

Figure 18.19   Switching IC voltage regulator.

The **comparator** of **Figure 18.19** determines whether the internally generated 1.25 V input reference voltage is greater than the output sample voltage. If the reference voltage is greater than the sampled output voltage, then the output of the comparator goes high; otherwise, the output is low. This is designated as the **error voltage**. An oscillator that generates a **square waveform** is used for providing a constant train of pulses. The frequency of the oscillator waveform is determined by an externally connected resistor and capacitor network. Variable off-time modulation of the oscillator output is achieved using a gated latch as a switching circuit. The circuit consists of a logic gate for combining the inputs, and a **set-reset (S-R) latch** for maintaining the duration of switching levels. Inputs from the error detector and an oscillator circuit are applied to the gated latch. As long as the error voltage is high, the square wave output from the oscillator is applied directly to the base of the control transistor. If the **error voltage** is low, the output of the oscillator is no longer applied to the base of the control transistor, and the control transistor is switched off. When the error voltage is high, the switching state of the control transistor is the same as the switching state of the oscillator.

Refer to **Figure 18.20(a)**, which shows a **step-down switching voltage regulator** constructed using the **MC34063A**, which is a **switching mode power supply (SMPS)** controller IC. **Figure 18.20(b)** shows the voltage waveform of the **oscillator** (V_{Osc}), the error voltage (V_{Error}) developed by the op-amp after comparing the sampled output voltage with the reference voltage, and, finally, the voltage applied to the base of the control element (V_B). **Variable off-time modulation** of the oscillator is achieved by varying the duration of the off-time. Doing so changes the overall time period of the pulses applied to the base of the control element. As a result of this, the duration of current flow through the inductor and the charging time of the capacitor are controlled.

Self-Examination

Answer the following questions.

16. The 7800 series of voltage regulators ICs are _____terminal devices that can provide a fixed _____output voltage.

    a. 2, positive
    b. 3, negative
    c. 3, positive

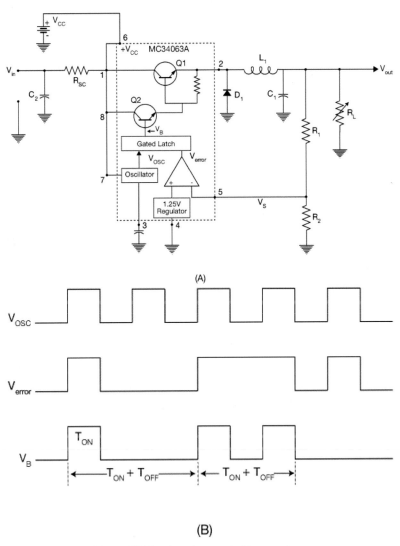

(A)

(B)

Figure 18.20   Step-down switching regulator.

d. 2, negative

17. In order to maintain regulated voltages under heavy load conditions _____ are used with regulators for providing adequate cooling.

a. High currents
b. Heat sinks

c. High voltages
d. Low output resistance

18. Adjustable voltage regulators use _____resistor(s) for determining the value of the output voltage.

    a. Zero
    b. One
    c. Internal
    d. External

19. IC switching regulators that use PWM usually change the _____ of the waveform.

    a. Duty cycle
    b. Time period
    c. Frequency
    d. Peak amplitude

20. IC switching regulators that use variable off-time modulation change the _____ of the waveform.

    a. On-time
    b. Time period
    c. Peak amplitude
    d. Reference level

18.5 Analysis and Troubleshooting - Voltage Regulators

The performance of a **voltage regulator** is evaluated by observing its output voltage under varying load conditions and input voltage variations. Ideally, the regulated output voltage should be maintained at a desired level when either the current drawn by the load or the applied input voltage changes. When troubleshooting a voltage regulator, it is common practice to compare its performance with its rated specifications. These specifications are shown in the **datasheet** of the device.

18.5 Analyze and troubleshoot voltage regulator circuits.

In order to achieve objective 18.5, you should be able to:

- examine the datasheet of an IC voltage regulator;
- distinguish between normal and faulty operations of a voltage regulator circuit.

Datasheet Analysis - Linear IC Voltage Regulator

Figure 18.17 shows the data sheet of the LM317 which is a positive adjustable regulator. Use this data sheet to answer the following:

1. How many pins (terminals) does the IC have? _____
2. Identify the pin used for adjusting the output voltage of the IC:

   _____
3. What is the range of the adjustable output voltage? _____ to

   _____
4. What is the typical value of line regulation at 25°C? _____%/V
5. What is the typical value of load regulation at 25°C with the output voltage $V_O = 5$ V? _____mV%/V_O
6. What is the typical value of the internal reference voltage? _____V
7. What is the typical value of the minimum load current, $I_{L(MIN)}$, required to maintain regulation? _____A
8. What is the typical value of the maximum output current, $I_{O(MAX)}$, when the difference between the input and output voltages (V_I- V_O) is =15 V, provided the power dissipation P_D is = maximum power dissipation $P_{D(MAX)} = 20$ W, at 25°C? _____A

Linear IC Troubleshooting

Linear IC voltage regulator troubleshooting is rather easy to accomplish. Essentially, the IC can be open, shorted, or good. An ohmmeter test of the IC generally does not tell a great deal about its condition. For example, if a linear IC regulator is shorted, there will be a low resistance between the input and output. An open condition cannot be tested with an ohmmeter. This means that the IC must be tested in circuit with meaningful power applied. In this regard, the IC must have voltage applied to the input and be capable of developing a regulated output voltage. This can be observed with a voltmeter. As a rule, this test must be performed with varying load values. An extremely heavy load may cause the regulator to shut down when its power dissipation rating is exceeded. Reducing the load should cause the IC to turn on and restore the regulation function.

Summary

- Voltage regulators operating within specified conditions maintain a constant output voltage regardless of changes in either the input voltage or its load current demand.

- Voltage regulators are generally described as being linear or switching based on whether the current flow to the load device is continuous or interrupted during operation respectively.
- The ability of a regulator to maintain an output voltage at a constant value regardless of variation in the input voltage is called line regulation.
- The ability of a regulator to maintain an output voltage at a constant value regardless of variations in the load current demand is called load regulation.
- Linear regulators are further classified as being either series or a shunt (parallel), based on how certain electronic components are connected to the load device.
- Most series linear voltage regulator circuits employ some form of over-load protection to prevent the series control element from being damaged, whereas most shunt linear voltage regulators require protection from input over-voltage conditions.
- Switching regulators are classified as being either step-down or step-up. When the switching electronic component is in series with the load, it will reduce or step down the voltage. It steps up the voltage when the switching electronic component is in parallel with the load.
- The efficiency of a voltage regulator can be improved by operating the control element as a switching device by rapidly switching it on and off.
- A pulse-width modulator (PWM) circuit is frequently employed in switching the control element of a regulator. It typically generates a pulsating DC waveform in which the width of the pulses can be adjusted.
- ICs voltage regulators have the essential regulator components built on a single chip instead of using discrete components.
- The 7800 series is a representative type of fixed positive voltage regulators, and the 7900 series for fixed negative voltage regulators.
- Adjustable IC voltage regulators are used in applications where non-standard regulator voltage values are needed.

Formulas

(18-1) Line regulation $= \frac{\Delta V_{\text{out}}}{\Delta V_{\text{in}}}$

(18-2) Load regulation $= \frac{\Delta V_{\text{out}}}{\Delta I_L} = \frac{V_{\text{No Load}} - V_{\text{Full Load}}}{I_{\text{Full Load}} - I_{\text{No Load}}}$

(18-3) Zener current: $I_Z = \frac{V_{\text{in}} - V_Z}{R_S}$

(18-4) Voltage output of developed by an op-amp when comparing sample voltage using resistor networks R_1 and R_2 with a Zener voltage:
$$V_{\text{out}} = V_Z \times \left(1 + \frac{R_1}{R_2}\right)$$

(18-5) Maximum load current in series voltage regulator with current limiting: $I_{L(max)} = \frac{0.7}{R_{SC}}$

(18-6) % Duty Cycle $= \frac{t_{on}}{T} \times 100 = \frac{t_{on}}{t_{on}+t_{off}} \times 100$

(18-7) Average value of output voltage for a square wave pulsating DC waveform: $V_{out} = \frac{t_{on}}{T} \times V_{in}$

(18-8) Output voltage developed by the adjustable voltage regulator LM317:

$$V_{out} = V_{REF} \times \left(1 + \frac{R_2}{R_1}\right) + I_{ADJ} \times R_2$$

Review Questions

Answer the following questions.

1. A _____ regulator has a continuous flow of current supplied to the load device.

    a. Switching
    b. Step-down
    c. Linear
    d. Non-linear

2. Which regulator specification identifies the variation of the output voltage due to changes in the load current.

    a. Switching time
    b. Line regulation
    c. Duty cycle
    d. Load regulation

3. If the input voltage applied to a series voltage regulator decreases, the conduction of the control element _____.

    a. Increases
    b. Decreases
    c. Remains unchanged
    d. Will not be affected

4. If the input voltage applied to a shunt voltage regulator decreases, the conduction of the control element _____.

    a. Increases
    b. Decreases
    c. Remains unchanged
    d. Will not be affected

5. The control element of a shunt voltage regulator is connected in _____ with respect to the load.

    a. Series

    b. An open circuit

    c. Parallel

    d. PWM

6. A(n) _____circuit is used for protecting the control element of a series voltage regulator from excessive load currents.

    a. Output sampling

    b. Reference input voltage

    c. Thermal overload protection

    d. Comparator

7. The _____circuit of a shunt regulator compares the sampled output voltage with the reference input voltage.

    a. Control element

    b. Load

    c. Thermal overload protection

    d. Error detector

8. The operation of a step-up or step-down switching voltage regulator is similar to that of a _____.

    a. Resistor

    b. Transformer

    c. Diode

    d. Oscillator

9. The waveform generated by the oscillator used in a step-down switching regulator is a _____.

    a. Positive DC

    b. Negative DC

    c. Square wave

    d. AC

10. The adjust terminal of a LM337 negative voltage regulator controls the _____.

    a. Voltage reference

    b. Protection circuitry

c. Input current
d. Feedback voltage

Problems

Answer the following questions.

1. Refer to **Figure 18.3**. The output of a 5-V voltage regulator changes by 2 mV when the input voltage changes over its permitted operating range. Determine the value of the line regulation as a %.

   a. 0%-0.09%
   b. 0.1%-0.9%
   c. 1%-9.99%
   d. 10% or above

2. Refer to **Figure 18.4**. The output of a 5-V voltage regulator changes by 100 mV when the load current changes from no-load to rated full-load. Determine the value of the load regulation as a %.

   a. 0%-0.09%
   b. 0.1%-0.9%
   c. 1%-9.99%
   d. 10% or above

3. Refer to **Figure 18.6**. Determine the output voltage of the series voltage regulator using a 5-V Zener which uses resistor R_1 in the feedback network of 10 kΩ and R_2 of 25 kΩ.

   a. 0-4.9 V
   b. 5-9.9 V
   c. 10-14.9 V
   d. 15 V or above

4. Refer to **Figure 18.8**. If the short-circuit resistance (R_{SC}) is 1.4 Ω, determine the maximum permissible current that can be drawn by the load:

   a. 0-0.9 A
   b. 1-2.9 A
   c. 3-4.9 A
   d. 5 A or above

5. The PWM output within a switching voltage regulator has an on-time (t_{on}) of 3 ms and an off-time (t_{off}) of 5 ms. Determine the percentage duty cycle of the waveform.

    a. 0%-24.9 %
    b. 25%-49.9 %
    c. 50%-74.9 %
    d. 75% or above

6. The PWM output within a switching voltage regulator has an on-time (t_{on}) of 3 ms and an off-time (t_{off}) of 5 ms. If V_{in} is 10 V, determine the average value of the output of the waveform.

    a. 0-3.9 V
    b. 3-5.9 V
    c. 6-8.9 V
    d. 9 V or above

7. Refer to **Figure 18.17**. Determine the maximum voltage of the LM317 voltage regulator, assuming I_{ADJ}= 46 μA, V_{REF}= 1.25 V, and fixed resistor R_1= 770 Ω, and the variable resistor R_2= 17 kΩ.

    a. 0-14.9 V
    b. 15-24.9 V
    c. 25-39.9 V
    d. 40 V or above

8. Refer to **Figure 18.17(a)** which shows the LM317 adjustable positive voltage regulator. If the resistor R_2 were to become shorted momentarily, what effect will it have on the operation of the regulator?

    a. No effect on regulator operation
    b. Output voltage will become maximum
    c. Output voltage can no longer be adjusted
    d. Output voltage will become 0 V

9. Refer to **Figure 18.12** which shows a step-down switching voltage regulator. If the control element is open-circuited, how will the output respond?

    a. $V_{out} = V_{in}$
    b. $V_{out} = 0$
    c. $V_{out} = V_{out} - V_{in}$
    d. $V_{out} = -V_{in}$

Answers

Examples

18-1. 3.63 μV/V

18-2. 0.0093333 V/mA = 9.33 mV/mA

18-3. Regulator I would be better since the output voltage under full load is 9.9985 V, whereas that for regulator II is 9.95 V.

18-4. R_2= 30 kΩ, R_S= 320 Ω

18-5. R_{SC}= 5.6 Ω

18-6. f = 100 Hz, % duty cycle = 25%, V_{out}= 3.5 V

18-7. V_{out} range: 1.25-20.42 V

Self-Examination

18.1

1. remain constant
2. increase
3. remain constant
4. increase
5. Increase voltage, increase current demand

18.2

6. series
7. increases
8. decreases
9. parallel
10. increase

18.3

11. alternating between cutoff and saturation
12. t_{on} = t_{off}
13. series
14. shunt
15. increases

18.4

16. 3, positive
17. heat sinks
18. external
19. duty cycle
20. Time period

Terms

Line Regulation

The ability of a regulator to maintain an output voltage at a constant value regardless of variations in the input voltage is called line regulation.

Load Regulation

The ability of a regulator to maintain an output voltage at a constant value regardless of variations in the load current demand is called load regulation.

Linear voltage regulator

A voltage regulator in which the control element, usually a transistor, is operated in the linear region and is never completely switched off and on during its operation.

Switching voltage regulator

A voltage regulator in which the control element, usually a transistor, is completely switched off and on as part of its regulation normal operations.

Series linear voltage regulator

A voltage regulator in which the control element is connected in series with the load device and operates in its linear region.

Shunt linear voltage regulator

A voltage regulator in which a control element is connected in shunt or parallel with the load device and operates in its linear region.

Foldback current limiting

A current limiting circuit in a linear voltage regulator which is used for reducing the amount of current flow if the load exceeds the maximum permitted current flow through the load.

Step-up switching voltage regulators

A type of switching regulator in which the value of the output voltage (V_{out}) generated is smaller than or equal to the applied input voltage (V_{in}).

Step-down switching voltage regulators

A type of switching regulator in which the value of the output voltage (V_{out}) can be increased beyond the applied input voltage (V_{in}).

Pulse-width modulation (PWM)

The process of generating a pulsating DC waveform in which the width of the pulses can be adjusted.

Duty cycle

The ratio of the on-time (t_{on}) to the time period (T) of a pulsating DC waveform usually expressed as a percentage.

Fixed IC voltage regulator

An integrated circuit voltage regulator that generates a fixed positive or negative voltage output when operated within rated specifications.

Adjustable IC voltage regulator

An integrated circuit voltage regulator that generates a variable positive or negative voltage output when operated within rated specifications. The voltage is usually varied by adjusting the values of externally connected components.

19

Filter Circuits

In communication electronics, there is a need to pass certain frequencies, while limiting or rejecting the passage of other frequencies. **Filter circuits** are primarily responsible for achieving this function. Specifically, this type of circuit allows only those signals which have frequencies in a certain range to pass through it to the output. All other frequencies are attenuated. This condition of operation is also referred to as **resonance**. An application of this would be in a transistor radio when selecting a specific radio frequency station and rejecting all other stations.

Filter circuits are primarily constructed using passive components such as resistors, capacitors, and inductors. For more precise control of a selected range of frequencies, passive components can be connected to an active component such as an op-amp. This type of configuration is termed as an **active filter**.

Filter circuits have a specific **frequency response curve** associated with their operation. In a typical frequency response curve, the frequency is graphed on the horizontal axis and the voltage output on the vertical axis. Frequency response curves for each type of filter, including active filters, and resonant circuits are discussed in the sections that follow.

Objectives

After studying this chapter, you will be able to:

19.1 explain the operation of low-pass, high-pass, band-pass, and band-stop filter circuits;

19.2 plot frequency response curves for low-pass, high-pass, band-pass filter, and band-stop filter circuits;

19.3 explain the use of decibels to measure voltage and power gains in filter circuits;

19.4 analyze the operation of resonant circuits;

19.5 explain the operation of an active filter circuit;
19.6 analyze and troubleshoot filter circuits.

Chapter Outline

19.1 Filter Circuits
19.2 Filter Circuit Power and Voltage Gain
19.3 Resonant Circuits
19.4 Active Filters
19.5 Analysis and Troubleshooting – Filter Circuits

Key Terms

attenuation
bandwidth
low-pass filter
high-pass filter
band-pass filter
frequency response
quality factor (Q)
power gain/loss
power-loss ratio
voltage-loss ratio
multiple-order filters
center frequency
3 dB-down frequency
half-power frequency
resonant frequency
resonant circuit
active filter
dB/decade
dB/octave
selectivity

19.1 Filter Circuits

Filter circuits are designed to respond in some way to specific or variable AC frequencies. These circuits are designed to pass certain frequencies from input to output and block other frequencies. You should be able to:

19.1. Explain the operation of low-pass, high-pass, band-pass, and band-stop filter circuits.

19.2. Plot frequency response curves for low-pass, high-pass, band-pass, and band-stop filter circuits.

In order to achieve objectives 19.1 and 19.2, you should be able to:

- define the terms low-pass, high-pass, band-pass, and band-reject filter circuits;
- calculate the critical (or cutoff) frequencies of various types of filter circuits;
- develop the frequency response curve for various types of filter circuits;
- identify the output response of different types of filter circuits.

The basic types of filter circuits are shown in **Figure 19.1**. Filter circuits are used to separate one range of frequencies from another. **Low-pass filters** pass low AC frequencies and block higher frequencies. **High-pass filters** pass high frequencies and block lower frequencies. **Band-pass filters** pass a mid-range of frequencies and block lower as well as higher frequencies. **Band-stop filters** block a mid-range of frequencies and pass lower as well as higher frequencies. The ranges of frequencies that are permitted to pass through a certain type of filter are considered to be the pass-band frequencies. Filter circuits have resistance and capacitance or inductance. The reactance of capacitors or inductors makes possible the frequency selection characteristic of filter circuits.

Low-Pass Filters

An ideal **low-pass filter** has a constant gain for all low frequencies in its pass-band and zero gain for all frequencies that lie outside the pass-band. In an ideal filter, there is a specific frequency value below which all signals with frequencies below it are passed and above which all frequencies are blocked. This is generally called the cutoff or critical frequency (f_c). In an actual low-pass filter, however, higher frequency signals are gradually attenuated instead of having an abrupt cutoff transition for frequencies above the pass-band.

Figure 19.2 shows the circuits used for **low-pass filters** and a typical frequency response curve. Many low-pass filters are series RC circuits (see **Figure 19.2(a)**). Output voltage (V_{out}) is taken across a capacitor. As frequency increases, capacitive reactance (X_C) decreases, since $X_C = \frac{1}{2\pi fC}$. The voltage drop across the output is equal to current (I) times the capacitive reactance (X_C), with X_C measured in ohms. So, as the frequency increases, the X_C decreases and the voltage output decreases.

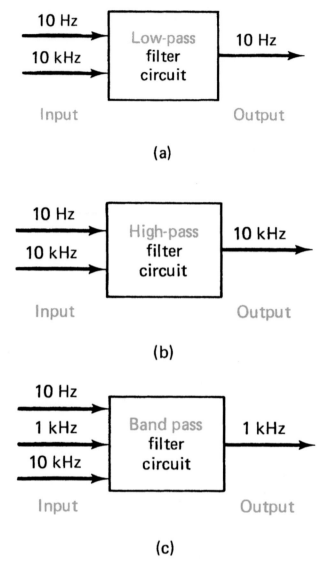

Figure 19.1   Types of filter circuits. (a) Low-pass filter – passes low frequencies and block high frequencies. (b) High-pass filter – passes high frequencies and block low frequencies. (c) Band-pass filter – passes a mid-range of frequencies and blocks high and low frequencies.

In a series RC circuit, the cutoff frequency (f_c) occurs when $X_C = R$. Thus,

$$X_C = \frac{1}{2\pi f_c C} = R. \tag{19.1}$$

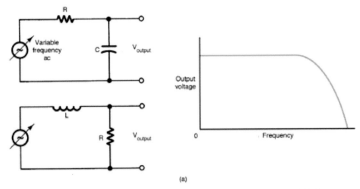

Figure 19.2 Low-pass filters. (a) Series RC low-pass filter. (b) Series RL low-pass filter. (c) Frequency response curve of low-pass filter.

Transposing the frequency and resistance terms, the resonant frequency is

$$f_c = \frac{1}{2\pi RC}. \tag{19.2}$$

Series RL circuits such as that shown in **Figure 19.2(b)** may also be used as low-pass filters. As frequency increases, the inductive reactance (X_L) also measured in ohms increases since $X_L = 2\pi fL$. Therefore, any increase in X_L reduces the circuit's current. The voltage output taken across the resistor is equal to current (I) times the resistance (R), or I × R. So, when I decreases, V_{out} also decreases. As frequency increases, X_L increases, I decreases, and V_{out} decreases. Note the shape of the frequency response curve of **Figure 19.2(c)**, showing that as frequency increases, voltage output decreases. In a series RL circuit, the cutoff frequency (f_c) occurs when $X_L = R$. Thus,

$$X_L = 2\pi f_c L = R. \tag{19.3}$$

Transposing the frequency and resistance terms, the resonant frequency is

$$f_c = \frac{R}{2\pi L}. \tag{19.4}$$

Example 19-1:

Calculate the capacitive reactance (X_C) and capacitance of a low-pass filter circuit shown in **Figure 19.2**. Assume that the value of R = 1 kΩ and f_c= 3 kHz.

Solution

In a series RC circuit, the cutoff frequency (f_c) occurs when $X_C = R$

$$X_C = \frac{1}{2\pi f_C C} = R$$
$$= 1000\ \Omega = 1\,\text{k}\Omega.$$

Under these conditions, the value of the capacitance (C) is

$$X_C = \frac{1}{2\pi f_c C}.$$

Transposing for "C," the expression becomes

$$C = \frac{1}{2\pi f_c X_C}$$
$$= \frac{1}{6.28 \times 3000 \times 1000}$$
$$= 0.05 \times 10^{-6} F = 0.05\,\mu\text{F}.$$

Related Problem

Calculate the capacitive reactance (X_C) and capacitance of a low-pass filter circuit shown in **Figure 19.2**. Assume that the value of R = 1 kΩ and f_c= 100 Hz.

Example 19-2:

Calculate the cutoff frequency of a low-pass filter circuit shown in **Figure 19.2**. Assume that the value of R = 10 kΩ and C = 1 μF.

Solution

$$f_C = \frac{1}{2\pi RC}$$
$$= \frac{1}{6.28 \times 10,000 \times 0.000001}$$
$$= 15.92\,\text{Hz}.$$

Related Problem

Calculate the cutoff frequency of a low-pass filter circuit shown in **Figure 19.2**. Assume that the value of R = 50 kΩ and C = 0.05 μF.

High-Pass Filters

Figure 19.3 shows two types of high-pass filters and a typical frequency response curve. The series RC circuit of **Figure 19.3(a)** is a common type of RC filters. The voltage output (V_{out}) is taken across the resistor (R). As frequency increases, X_C decreases. A decrease in X_C causes current flow to increase. The voltage output across the resistor (V_{out}) is equal to current (I) times the resistance (R), or I × R. So, as I increases, V_{out} increases. As frequency increases, X_C decreases, I increases, and V_{out} increases.

As with the low-pass series RC filter, the cutoff frequency (f_c) for a high-pass filter occurs when $X_C = R$. Thus,

$$X_C = \frac{1}{2\pi f_c C} = R. \tag{19.5}$$

Transposing the frequency and resistance terms, the resonant frequency is

$$f_c = \frac{1}{2\pi RC}. \tag{19.6}$$

Note that the positions of the resistor and capacitor in the low- and high-pass filter circuits have been transposed.

A series RL circuit such as that shown in **Figure 19.3(b)** may also be used as a high-pass filter. In this circuit, the V_{out} is taken across the inductor. As the applied frequency increases, X_L increases, and V_{out} also increases. In

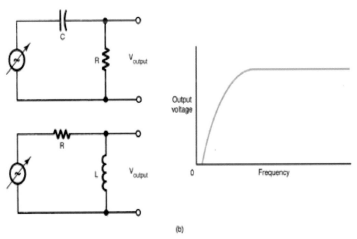

(b)

Figure 19.3 High-pass filters. (a) Series RC high-pass filter. (b) Series RL high-pass filter. (c) Frequency response curve of high-pass filter.

this circuit, as frequency increases, X_L increases and V_{out} increases. Note the shape of the frequency response curve of **Figure 19.3(c)**, showing that as the frequency increases, the voltage output increases. Again, as with the series RL low-pass filter, the cutoff frequency (f_c) occurs when $X_L = R$. Thus,

$$X_L = 2\pi f_c L = R. \tag{19.7}$$

Transposing the frequency and resistance terms, the resonant frequency is

$$f_c = \frac{R}{2\pi L}. \tag{19.8}$$

Note that the positions of the resistor and inductor in the low- and high-pass filter circuits have been transposed.

Example 19-3:

Calculate the cutoff frequency of a high-pass filter circuit shown in **Figure 19.3**. Assume that the value of R = 650 Ω and L = 10 mH.

Solution

$$2\pi f L = R$$

$$\frac{R}{2\pi L} = f_c$$

$$f_c = \frac{650}{6.28 \times 10 \times 10^{-3}} = 10,350.3\,\text{Hz}$$

$$= 10.35\,\text{kHz}.$$

Related Problem

Calculate the cutoff frequency of a low-pass filter circuit shown in **Figure 19.3**. Assume that the value of R = 100 Ω and L = 5 mH.

Band-Pass Filters

The **band-pass filter** of **Figure 19.4(a)** is a combination of low-pass and high-pass filter sections. It is designed to pass only an intermediate band of frequencies that lie between a designated low- and high-frequency limit. It also blocks the remaining low and high frequencies.

In **Figure 19.4(a)**, R_1 and C_1 form a **low-pass filter** and R_2 and C_2 form a **high-pass filter**. The range of frequencies to be passed is determined by

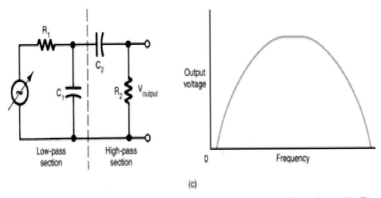

Figure 19.4 Band-pass filters. (a) Band-pass filter circuit configuration. (b) Frequency response curve of band-pass filter.

calculating the values of resistance and capacitance. Note the shape of the **frequency response curve** of **Figure 19.4(b)**, showing that on the low-frequency end, as frequency goes down, voltage output goes down, and on the high-frequency end, as frequency increases, voltage output decreases. The band-pass filter, thus, has a lower critical frequency (f_{cl}) that corresponds to the initial low-pass filter section, and an upper critical frequency (f_{cu}) that corresponds to the subsequent high-pass filter section. The bandwidth of the filter is defined as the range of frequencies between the lower and upper critical frequencies.

$$BW = f_{cu} - f_{cl}. \tag{19.9}$$

The **center frequency** (f_{cf}) of a band-pass filter is defined as the frequency about which the pass-band of the filter is centered. It is defined as the geometric mean of the upper and lower critical frequencies and is expressed as follows:

$$f_{cf} = \sqrt{f_{cu} \times f_{cl}}. \tag{19.10}$$

The **quality factor (Q)** of a band-pass filter is the ratio of the center frequency (f_{cf}) to the bandwidth (BW).

$$Q = \frac{f_{cf}}{f_{cu} - f_{cl}}. \tag{19.11}$$

The narrower the **bandwidth**, the higher is the quality factor. A high quality factor is desirable as it is an indication of the selectivity of the band-pass filter. Selectivity is important in radio receiver tuning circuits where multiple stations use the same or similar operating frequencies. This helps in the selection or distinguishing a particular radio station from others.

Example 19-4:

Calculate the center frequency (f_{cf}) of a band-pass filter circuit shown in **Figure 19.4**. Assume that the lower critical frequency (f_{cl}) = 10 Hz and the upper critical frequency (f_{cu}) = 20 kHz.

Solution

$$f_{cf} = \sqrt{f_{cu} \times f_{cl}}$$
$$= \sqrt{20,000 \times 10}$$
$$= \sqrt{200,000}$$
$$= 447.21 Hz.$$

Related Problem

Calculate the center frequency (f_{cf}) of a band-pass filter circuit shown in **Figure 19.4**. Assume that the lower critical frequency (f_{cl}) = 500 Hz and the upper critical frequency (f_{cu}) = 10 kHz.

Band-Stop/Notch Filters

A **band-reject filter** is designed to block only an intermediate band of frequencies that lie between designated low- and high-frequency limits. It permits the remaining low and high frequencies to pass through the output. The **band-stop filter**, thus, functions as a dual of the band-pass filter. These filters are also called band-reject filters.

The circuit diagram of a **band-stop filter** is shown in **Figure 19.5(a)**. It is a combination of low-pass and high-pass filter sections that are connected in parallel with each other. The low-pass filter permits only frequencies below the lower critical frequency (f_{cl}) to pass through it, whereas the high-pass section permits only frequencies above the high critical frequency (f_{cu}) to pass through it. With the parallel circuit connection of the low-pass and high-pass filters and for f_{cu} greater than f_{cl}, the range of frequencies between f_{cl} and f_{cu} are attenuated. The cumulative effect is that there is a range of frequencies that is blocked while others are permitted to go through the band-stop filter. The frequency response of a **band-reject filter** is shown in **Figure 19.5(b)**. The bandwidth of the band-stop filter is the range of frequencies that are blocked, i.e., between f_{cu} and f_{cl}.

$$BW = f_{cu} - f_{cl}. \tag{19.12}$$

The **center frequency (f_cf)** of a band-stop filter is defined as the frequency about which the blocking band of the filter is centered. It is defined as the geometric mean of the upper and lower critical frequencies and is expressed as

$$f_{cf} = \sqrt{f_{cu} \times f_{cl}}. \tag{19.13}$$

Example 19-5:

Calculate the bandwidth (BW) of a band-stop filter circuit shown in **Figure 19.5**. Assume that the lower critical frequency (f_{cl}) = 1 kHz and the upper critical frequency (f_{cu}) = 2.5 kHz.

Solution

$$BW = f_{cu} - f_{cl}$$
$$= 2.5 - 1 kHz$$
$$= 1.5 kHz.$$

Related Problem

Calculate the bandwidth (BW) of a band-stop filter circuit shown in **Figure 19.5**. Assume that the lower critical frequency (f_{cl}) = 15 MHz and the upper critical frequency (f_{cu}) = 15.1 MHz.

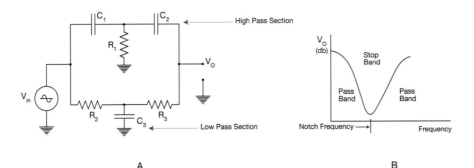

Figure 19.5 Band-stop filters. (a) Band-stop filter circuit configuration. (b) Frequency response curve of band-stop filter.

Self-Examination

Answer the following questions.

1. A _____ filter will pass low-frequency signals and will block high-frequency signals.

    a. Low-pass
    b. High-pass
    c. Band-pass
    d. Band-stop

2. A _____ filter will block low-frequency signals and will pass high AC signals.

    a. Low-pass
    b. High-pass
    c. Band-pass
    d. Band-stop

3. A _____ filter will pass a mid-range of frequency signal and will block all other frequency signals.

    a. Low-pass
    b. High-pass
    c. Band-pass
    d. Band-stop

4. A _____ filter will block a mid-range frequency signal and will pass all other frequency signals.

    a. Low-pass
    b. High-pass
    c. Band-pass
    d. Band-stop

5. As the frequency increases, _____ reactance decreases, whereas _____ reactance increases.

    a. Inductive, capacitive
    b. Capacitive, inductive
    c. Resistive, inductive
    d. Capacitive, resistive

19.2 Filter Circuit Power and Voltage Gain

The term **decibel or dB** is commonly used to show the relationship of power or voltage level changes. In filter circuits, decibels are used to show the power or voltage gain/loss at various frequencies. This information is often displayed graphically as a **frequency response curve** for various filter circuits.

19.3 Explain the use of decibels to measure voltage and power gains in filter circuits.

In order to achieve objective 19.3, you should be able to:

- define what is meant by the term 3 dB-down frequency;
- calculate the power and voltage gains/losses;
- develop the frequency response curve of a filter circuit that shows the relationship between the voltage loss (dB) and the frequency changes.

The fundamental unit is the **bel**, which is derived from the surname of Alexander Graham Bell. The bel (B), however, is a rather large unit defined by the following equation:

$$\text{Power gain loss (B)} = \log_{10} \frac{P_{out}}{P_{in}}$$

where P_1 and P_2 are comparisons of output power over input power and P_1 is greater than P_2.

A more practical expression of changes is represented by the **decibel (dB)**. The term deci is the metric equivalent of 0.1 or 1/10 of the fundamental unit. A dB is, therefore, 1/10 of a bel, or 10 dB = 1 bel. The mathematical expression for the power gain or loss in terms of decibels (dB) is

$$\text{Power gain (dB)} = 10 \log_{10} \frac{P_{out}}{P_{in}}; \text{when } P_{out} > P_{in} \qquad (19.14)$$

$$\text{Power loss (dB)} = -10 \log_{10} \frac{P_{out}}{P_{in}}; \text{when } P_{out} < P_{in}. \qquad (19.15)$$

For calculating the **voltage gain**, an expression that compares the output voltage to the input voltage is needed. The power and voltage are related by the following expression:

$$P = V \times I = V \times \frac{V}{R} = \frac{V^2}{R}$$

$$\text{Gain (dB)} = 10 \log_{10} \frac{P_{out}}{P_{in}}$$

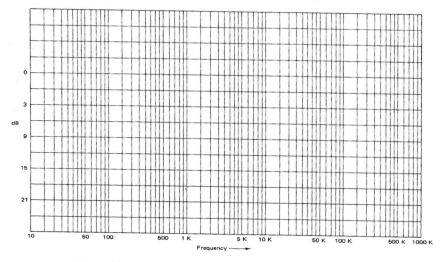

Figure 19.6   Decibel values used to plot frequency response.

$$= 10 \log_{10} \frac{V_{out}^2}{V_{in}^2}$$

$$= 10 \log_{10} \left(\frac{V_{out}}{V_{in}} \right)^2$$

$$\text{Voltage gain (dB)} = 20 \log_{10} \frac{V_{out}}{V_{in}}. \qquad (19.16)$$

Decibel values are also used to plot frequency response curves, as shown in **Figure 19.6**. This unit of measure is determined by a **logarithmic scale**. Our ears, for example, respond to logarithmic changes in sound. A sound level that physically increases by 10 times (10 dB) appears only as an increase of one time to the human ear, or a doubling in loudness. Likewise, a sound level that increases 100 times (20 dB) appears as an increase of twice as loud as the previous value, or as an increase by a factor of four times the initial loudness level. In this system, a sound level or power level of 0 dB is considered to be at the threshold of hearing or reference level.

A sample problem using decibels follows. The gain of an amplifier that has an input of 25 mW and an output of 300 mW is determined as follows. The **power gain** is found by using the following formula:

$$\text{Power gain (dB)} = 10 \log_{10} \frac{P_{out}}{P_{in}}$$

$$= 10 \log_{10} \frac{300 \, \text{mW}}{25 \, \text{mW}}$$

$$= 10 \log_{10} 12$$

$$= 10 \times 1.0791$$

$$= 10.791.$$

The **voltage gain** of an amplifier may also be expressed in decibel values. The decibel voltage gain formula is

$$\text{dB} = 20 \log V_0/V_{\text{in}}.$$

Note that the **logarithm** of V_0/V_{in} is multiplied by 20 in this equation.

As an example of the decibel voltage gain equation, assume that an amplifier has an input voltage of 0.45 V_{p-p} and the output voltage is 0.86 V_{p-p}. The voltage gain in decibels is

$$\text{Voltage gain (dB)} = 20 \log_{10} \frac{V_{\text{out}}}{V_{\text{in}}}$$

$$= 20 \log_{10} \frac{0.86 \, V_{p-p}}{0.45 \, V_{p-p}}$$

$$= 20 \log 1.91$$

$$= 20 \times 0.2812$$

$$= 2.812.$$

When the decibel value of an amplifier is known, the power gain or voltage gain may be determined by using inverse logarithms or anti-logarithms. An inverse logarithm is the number from which a logarithm is derived. The process of finding an inverse logarithm is the reverse of finding a logarithm. These values may also be easily determined by using the **inverse logarithm** function on a scientific calculator.

The power gain ratio or $P_{\text{out}}/P_{\text{in}}$ of a device, which has a power gain in decibels of +3 dB, can be found as follows:

$$dB = 10 \log_{10} \frac{P_{\text{out}}}{P_{\text{in}}}$$

$$3 = 10 \log_{10} \frac{P_{\text{out}}}{P_{\text{in}}}$$

$$0.3 = \log_{10} \frac{P_{\text{out}}}{P_{\text{in}}} (\text{dividebothsidesoftheequationby10})$$

$$\text{inv}(\log_{10} 0.3) = P_{out}/P_{in} \text{(find inverse logarithm)}$$
$$10^{0.3} = P_{out}/P_{in}$$
$$1.995 = P_{out}/P_{in}.$$

The value of $1.995 \approx 2$ obtained in the example is the power ratio. An amplifier with a power gain of $+3$ dB, thus, has a power gain ratio of approximately 2 to 1.

Decibels are also used to express reduction in power or voltage levels. Reduction of input signal level in a circuit is called **attenuation**. A circuit that attenuates a signal is compared to an amplifier circuit in **Figure 19.7**. Note that the decibel value is marked with a minus sign when the circuit attenuates the input signal. A common example of attenuation occurs in coaxial cable or other signal transmission cables in which a reduction of signal occurs from input to output.

Decibels are commonly used to plot **frequency response curves** for filter circuits. One example of a low-pass filter circuit is shown in **Figure 19.8(a)**. The procedure for plotting a frequency response curve for the low-pass circuit is shown in **Figure 19.8(b)**.

The selection of decibel values of 3, 9, 15, and 21 dB is standard for plotting frequency response in terms of voltage output of a circuit. First, locate the 3-dB line. Note that with a **3-dB reduction** of a signal (denoted as -3 dB), the power output is approximately 0.5, or 50%, of the 0-dB reference

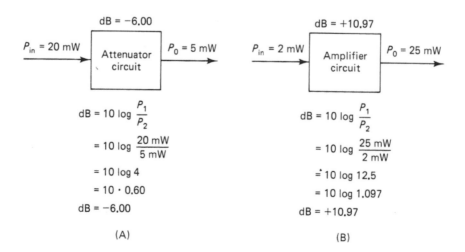

(A)   (B)

Figure 19.7   Comparison of (a) attenuator and (b) amplifier circuits.

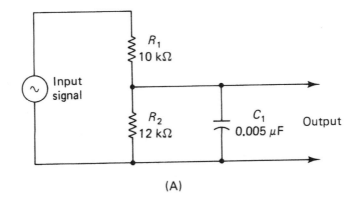

(A)

1. Find the Thevenin equivalent resistance (R_{Th}) of R_1 and R_2:

$$R_{Th} = R_1 \| R_2 = \frac{10\ K \times 12\ K}{10\ K + 12\ K} = 5.45\ k\Omega$$

2. Determine the 3-dB frequency using this formula:

$$f_{3\ dB} = \frac{1}{2\pi \cdot C \cdot R_{Th}} = \frac{1}{6.28 \cdot (0.005 \times 10^{-6}) \cdot (5.45 \times 10^3)}$$

$$= \frac{1}{1.7113 \times 10^{-4}} = 5843\ Hz$$

3. Find the 9-, 15-, and 21-dB frequencies:

$$f_{9\ dB} = 2 \cdot f_{3\ dB} = 2 \cdot 5843\ Hz = 11,686\ Hz$$
$$f_{15\ dB} = 4 \cdot f_{3\ dB} = 4 \cdot 5843\ Hz = 23,372\ Hz$$
$$f_{21\ dB} = 8 \cdot f_{3\ dB} = 8 \cdot 5843\ Hz = 46,744\ Hz$$

4. Label the points on a sheet of frequency response paper (see Figure 9-14).

5. Connect each of the points to form a low-pass frequency response curve.

(B)

Figure 19.8   Frequency response of a low-pass filter circuit. (a) Circuit. (b) Procedure for plotting frequency response curve.

level and the voltage is approximately 0.707, or 70.7%, of the 0-dB level. Since the power output of a circuit reduces to about 50% of its original value (0 dB), the 3-dB frequency is called the half-power point.

The **power loss ratio** or P_{out}/P_{in} of a device, which has a power loss in decibels of -3 dB, can be found as follows:

$$\text{Power gainloss (dB)} = 10\log_{10}\frac{P_{out}}{P_{in}}$$

$$-3 = 10\log_{10}\frac{P_{out}}{P_{in}}$$

$$-0.3 = \log_{10}\frac{P_{out}}{P_{in}}(\text{dividebothsidesoftheequationby10})$$

$$inv(\log_{10} -0.3) = P_{out}/P_{in}(\text{findinverselogarithm})$$

$$10^{-0.3} = P_{out}/P_{in}$$

$$0.5012 = P_{out}/P_{in}.$$

The value of $0.5012 \approx 0.5$ obtained in the example is the power loss ratio. A device with a loss of -3 dB, thus, has a power loss ratio of approximately 0.5 to 1.

In general, the power gain/loss ratio of a device can be derived from the expression for the power gain/loss in decibels as shown by the following:

$$\text{dB} = 10\log_{10}\frac{P_{out}}{P_{in}}$$

$$\frac{\text{dB}}{10} = \log_{10}\frac{P_{out}}{P_{in}}$$

$$10^{\frac{\text{dB}}{10}} = \frac{P_{out}}{P_{in}}.$$

Thus, the **power loss ratio** of a device can be expressed as follows:

$$\frac{P_{out}}{P_{in}} = 10^{\frac{\text{dB}}{10}}. \tag{19.17}$$

The voltage loss ratio or V_{out}/V_{in} of a device, which has a voltage loss in decibels of -3 dB, can be found as follows:

$$\text{Voltage gainloss (dB)} = 20\log_{10}\frac{V_{out}}{V_{in}}$$

$$-3 = 20\log_{10}\frac{V_{out}}{V_{in}}$$

$$-0.15 = \log_{10} \frac{V_{\text{out}}}{V_{\text{in}}} \text{(dividebothsidesoftheequationby10)}$$
$$inv(\log_{10} -0.15) = V_{\text{out}}/V_{\text{in}} \text{(findinverselogarithm)}$$
$$10^{-0.15} = V_{\text{out}}/V_{\text{in}}$$
$$0.7071 = V_{\text{out}}/V_{\text{in}}.$$

The value of 0.707 obtained in the example is the voltage loss ratio. A device that has a loss of -3 dB, thus, has a voltage loss ratio of approximately 0.707 to 1.

The power and voltage loss ratio values of the other decibel points (-9, -15, and -21 dB) can be found by the same procedure. These are listed in the table below:

Decibels(dB)	Power loss ratio$P_{\text{out}}/P_{\text{in}}$	Voltage loss ratio$V_{\text{out}}/V_{\text{in}}$
-3	0.5012	0.7071
-9	0.1259	0.3548
-15	0.03162	0.1778
-21	0.00794	0.0891

Using a same procedure as that used for deriving the power gain/loss express, the voltage gain/loss ratio of a device is expressed as follows:

$$\frac{V_{\text{out}}}{V_{\text{in}}} = 10^{\frac{dB}{20}}. \tag{19.18}$$

While developing the **frequency response curve** of a filter (see **Figure 19.6**), the first step is to determine the critical (3 dB-down) frequency or frequencies. For example, in the case of the low-pass filter circuit shown in **Figure 19.8(a)**, the critical frequency is determined by the formula: $f_c = \frac{1}{2\pi RC}$ to be 5843 or 5.8 kHz. The power gain/loss of the filter remains at approximately 0 dB starting from 0 Hz (DC) through its critical frequency of 5.8 kHz. The -9 dB frequency is obtained next by doubling the -3 dB frequency, i.e., $f_{-9dB} = 2 \times f_{-3dB} = 2 \times 5.8\,\text{kHz} = 11.6\,\text{kHz}$. The -15 dB frequency is approximately four times the critical frequency (23.2 kHz), and the -21 dB frequency is eight times the critical frequency (46.4 kHz). For plotting the **frequency response curve**, the gain/loss values in decibels are plotted according to specific multiples of the critical frequency, for example (-3 dB, 5.8 kHz) or (-9 dB, 11.6 kHz). These points are connected forming the frequency response curve for the low-pass filter of **Figure 19.8(a)**.

A **high-pass filter** circuit and the procedure for plotting its frequency response curve are shown in **Figure 19.9(a)**. The procedure for developing a frequency response curve is shown in **Figure 19.9(b)**. This includes identification of the 3 dB-down frequency and those frequencies corresponding to the -9, -15, and -21 dB voltage loss values. These points are plotted to form the frequency response curve of **Figure 19.9(c)**.

Band-pass filter circuits are a combination of low-pass and high-pass filter circuits. An example of a band-pass filter circuit is shown in **Figure 19.10(a)**. The procedure for plotting a frequency response curve for

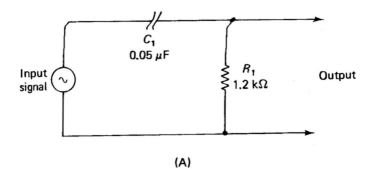

(A)

1. Determine the 3-dB frequency:

$$f_{3\ dB} = \frac{1}{2\pi \cdot R \cdot C} = \frac{1}{6.28 \times (1.2 \times 10^3) \times (0.05 \times 10^{-6})}$$

$$= \frac{1}{3.768 \times 10^{-4}} = 2654 \text{ Hz}$$

2. Find the 9-, 15-, and 21-dB frequencies:

$$f_{9\ dB} = f_{3\ dB} \div 2 = 2654 \div 2 = 1327 \text{ Hz}$$

$$f_{15\ dB} = f_{3\ dB} \div 4 = 2654 \div 4 = 663.5 \text{ Hz}$$

$$f_{21\ dB} = f_{3\ dB} \div 8 = 2654 \div 8 = 332 \text{ Hz}$$

3. Label each of the points on a sheet of frequency response paper.

4. Connect each of the points to form a high-pass frequency response curve.

(B)

Figure 19.9   Frequency response of a high-pass filter circuit. (a) Circuit. (b) Procedure for plotting frequency response curve.

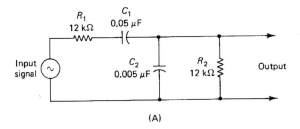

(A)

1. Determine the 3-, 9-, 15-, and 21-dB frequencies for the high-pass section:

$$f_{3\,dB} = \frac{1}{2\pi \cdot R \cdot C} = \frac{1}{6.28 \times (24 \times 10^3) \times (0.05 \times 10^{-6})}$$

$$= \frac{1}{7.536 \times 10^{-3}} = 133 \text{ Hz } (f_{LC})$$

$f_{9\,dB} = f_{3\,dB} \div 2 = 66.5$ Hz

$f_{15\,dB} = f_{3\,dB} \div 4 = 33.25$ Hz

$f_{21\,dB} = f_{3\,dB} \div 8 = 16.6$ Hz

2. Determine the 3-, 9-, 15-, and 21-dB frequencies for the low-pass section:

$$f_{3\,dB} = \frac{1}{2\pi \cdot R_{Th} \cdot C} = \frac{1}{6.28 \times (6 \times 10^3) \times (0.005 \times 10^{-6})}$$

$$= \frac{1}{1.884 \times 10^{-4}} = 5308 \text{ Hz } (f_{HC})$$

$f_{9\,dB} = f_{3\,dB} \times 2 = 10,616$ Hz

$f_{15\,dB} = f_{3\,dB} \times 4 = 21,232$ Hz

$f_{21\,dB} = f_{3\,dB} \times 8 = 42,464$ Hz

3. Label each of the points on a sheet of frequency response paper.

4. Connect each of the points to form a bandpass frequency response curve.

(B)

Figure 19.10 Frequency response of a band-pass filter circuit. (a) Circuit. (b) Procedure for plotting frequency response curve.

this filter is given in **Figure 19.10(b)**. The 3-dB frequency on the low-frequency end of the response curve is called the **low cutoff frequency (f_{LC})**. The high-frequency 3-dB point is called the **high cutoff frequency (f_{HC})**.

Example 19-6:

Calculate the voltage loss ratio V_{out}/V_{in} of a device, which has a voltage loss of -27 dB.

Solution

$$\text{Voltage gainloss (dB)} = 20 \log_{10} \frac{V_{out}}{V_{in}}$$

$$-27 = 20 \log_{10} \frac{V_{out}}{V_{in}}$$

$$-1.35 = \log_{10} \frac{V_{out}}{V_{in}} (\text{dividebothsidesoftheequationby10})$$

$$inv(\log_{10} -1.35) = V_{out}/V_{in} (\text{findinverselogarithm})$$

$$10^{-1.35} = V_{out}/V_{in}$$

$$0.04467 = V_{out}/V_{in}.$$

Related Problem

Calculate the voltage loss ratio V_{out}/V_{in} of a device, which has a voltage loss of -33 dB.

Self-Examination

Answer the following questions.

6. The 3-dB point on a frequency response curve indicates the_____-power point(s).

    a. Quarter
    b. Half
    c. Full
    d. Zero

7. The cutoff frequency or frequencies of a filter circuit signifies the _____-power point(s).

    a. Quarter
    b. Half
    c. Full
    d. Zero

8. When the power output (P_{out}) is less than the power input (P_{in}), the power ratio (P_{out}/P_{in}) will be:

    a. Greater than 1
    b. Less than 1

c. Equal to 1
d. Zero

9. The 0 dB portion on a frequency response curve indicates where the frequencies for which the output is:

a. Greater than the input
b. Smaller than the input
c. Equal to the input
d. Zero

10. When the output of a high-pass filter is connected to the input of a low-pass filter, the resultant circuit will function as a _____ filter.

a. Low-pass
b. High-pass
c. Band-pass
d. Band-stop

19.3 Resonant Circuits

Resonant circuits are designed to pass a certain range of frequencies and block other frequencies. Such circuits have resistance, inductance, and capacitance. **Figure 19.11** shows the two types of resonant circuits – series resonant circuits and parallel resonant circuits – and their frequency response curves, which show how circuit current and impedance vary with frequency.

19.4 Analyze the operation of resonant circuits.

In order to achieve objective 19.4, you should be able to:

• compare the operation of series and parallel resonant circuits;
• develop frequency response curves for resonant circuits.

Series Resonant Circuits

Series resonant circuits are series circuits that have inductance, capacitance, and resistance. Series resonant circuits offer a low impedance to some AC frequencies and a high impedance to other frequencies. They are used to select or reject frequencies of a frequency range applied to the input of a circuit.

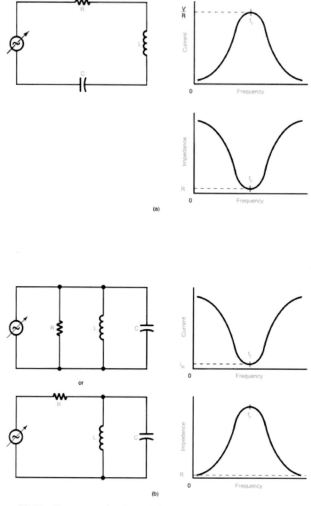

Figure 19.11   Resonant circuits. (a) Series resonant. (b) Parallel resonant.

In a series circuit, the current flow through all components is the same. However, the voltage across each component may be different. The voltage across an inductor and a capacitor in an AC series circuit are in direct opposition to each other (180° out of phase) and cancel each other out. The frequency applied to a series resonant circuit affects inductive reactance (X_L) and capacitive reactance (X_C). At a specific input frequency, X_L of a series *RLC* circuit equals X_C. The voltages across the inductor (V_L) and capacitance

(V_C) in the circuit are then equal. The total reactive voltage (V_X) is 0 V at this frequency. The reactance of the inductor and the capacitor cancel each other at this frequency. The total reactance (X_T) of the circuit (X_L- X_C) is zero. The impedance (Z) of the circuit is then equal to the resistance (R).

The frequency at which X_L= X_C in a circuit is called **resonant frequency** (**f$_r$**). Thus,

$$X_L = X_C$$

$$2\pi f_r L = \frac{1}{2\pi f_r C}$$

$$2\pi f_r L \times 2\pi f_r C = 1$$

$$4\pi^2 f_r^2 LC = 1$$

$$f_r^2 = \frac{1}{4\pi^2 LC}$$

$$f_r = \sqrt{\frac{1}{4\pi^2 LC}}$$

$$f_r = \frac{1}{2\pi\sqrt{LC}}. \tag{19.19}$$

In this formula, L is in henries, C is in farads, and f_r is in hertz. Note that as either inductance or capacitance increases, resonant frequency decreases. When the resonant frequency is applied to a circuit, a condition called resonance exists. Resonant frequency is calculated in the same way for series and parallel circuits.

Resonance for a series circuit causes the following conditions to exist:

(1) $X_L = X_C$
(2) X_T is equal to zero
(3) $V_L = V_C$
(4) Total reactive voltage, $V_X = 0$
(5) $Z = R$ and is minimum
(6) Total current (I_T) is maximum

These are the basic characteristics of all series resonant circuits. Remember that these characteristics apply to series *RLC* circuits.

The ratio of reactance (X_L or X_C) to resistance (R) at resonant frequency of a circuit is called **quality factor (Q)**. This ratio is used to determine the range of frequencies or bandwidth (BW) a resonant circuit will pass.

A sample series resonant circuit problem is shown in **Figure 19.12(a)**. The resonant frequency, bandwidth, lower- and upper-cutoff frequencies, quality

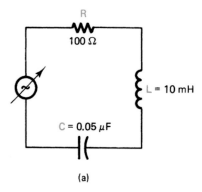

(a)

Finding circuit values:

1. Find resonant frequency (f_r):

$$f_r = \frac{1}{2\pi \times \sqrt{L \times C}} = \frac{1}{6.28 \times \sqrt{(10 \times 10^{-3}) \times (0.05 \times 10^{-6})}} =$$

$$= \frac{1}{6.28 \times \sqrt{0.5 \times 10^{-9}}} = \frac{1}{6.28 \times (2.23 \times 10^{-5})} = \frac{1}{1.4 \times 10^{-4}} = 7121 \text{ Hz}$$

2. Find X_L^* and X_C at resonant frequency:

$X_L = 2\pi \cdot f \cdot L = 6.28 \times 7121 \times (10 \times 10^{-3}) = 447 \ \Omega$

*Easier to calculate.

3. Find quality factor (Q):

$$Q = \frac{X_L}{R} = \frac{447 \ \Omega}{100 \ \Omega} = 4.47$$

4. Find bandwidth (BW):

$$BW = \frac{f_r}{Q} = \frac{7121 \text{ Hz}}{4.47} = 1593 \text{ Hz}$$

5. Find low-frequency cutoff (f_{lc}):

$f_{lc} = f_r - \frac{1}{2} BW = 7121 \text{ Hz} - 797 \text{ Hz} = 6324 \text{ Hz}$

6. Find high-frequency cutoff (f_{hc})

$f_{hc} = f_r + \frac{1}{2} BW = 7121 + 797 = 7918 \text{ Hz}$

(b)

Figure 19.12   Series-resonant circuit problem. (a) Circuit with component values. (b) Procedure to determine circuit values.

factor, and component voltage values can be determined by the procedure shown in **Figure 19.12(b)**.

The **frequency range** that a resonant circuit will pass is the bandwidth (BW). The frequency cutoff points are at approximately 70% of the maximum output voltage. These are called the low-frequency cutoff (f_{LC}) and high-frequency cutoff (f_{HC}). The bandwidth of a resonant circuit is determined by the Q of the circuit. The circuit Q is determined by the ratio of X_L and X_C to R (X_L/R). Resistance is the primary factor that determines bandwidth, as summarized by the following:

(1) When R is increased, Q decreases, since

$$Q = X_L/R. \tag{19.20}$$

(2) When Q decreases, bandwidth increases, since

$$BW = f_r/Q. \tag{19.21}$$

(3) Thus, when R is increased, bandwidth increases.

Two series resonant circuit response curves are shown in **Figure 19.13**. These curves show the effect of resistance on **bandwidth**. The curve of **Figure 19.13(b)** has high selectivity, meaning that a resonant circuit with this response curve would select a small range of frequencies. **Selectivity** is very important for radio and television tuning circuits and other electronic applications.

Example 19-7:

Calculate the resonant frequency of the series *RLC* circuit shown in **Figure 19.12**, assuming R = 1000 Ω, L = 25 mH, and C = 0.5 μF.

Solution

The resonant frequency of a series RLC circuit (f_r) is

$$f_r = \frac{1}{2\pi\sqrt{LC}}$$

$$= \frac{1}{6.28\sqrt{0.025 \times 0.5 \times 10^{-6}}}$$

$$= \frac{1}{6.28 \times 1.11 \times 10^{-4}} = \frac{1}{7.02 \times 10^{-4}} = 1424.25 \text{ Hz.}$$

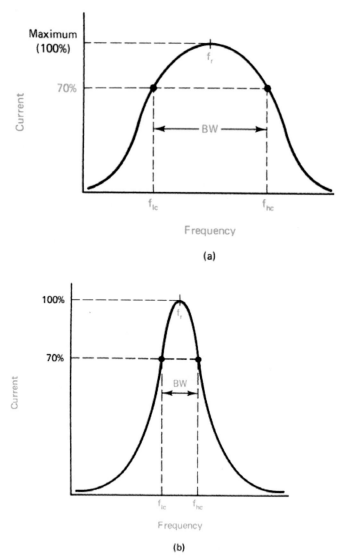

Figure 19.13   Effect of resistance on bandwidth and selectivity of resonant circuit. (a) High resistance (low selectivity). (b) Low resistance (high selectivity).

Related Problem

Calculate the resonant frequency of the series RLC circuit shown in **Figure 19.12**, assuming R = 1000 Ω, L = 50 mH, and C = 250 pF.

Parallel Resonant Circuits

Parallel resonant circuits have characteristics similar to series resonant circuits. The electrical properties of parallel resonant circuits are somewhat different, but they accomplish the same purpose as that of series circuits. This circuit is a parallel combination of L and C used to select or reject AC frequencies.

In a parallel circuit, the voltage across all components is the same. However, the current through each component may be different. With the resonant frequency (f_r) applied to a parallel resonant circuit, the following conditions occur:

(1) $X_L = X_C$
(2) $X_T = 0$
(3) $I_L = I_C$
(4) Total reactive current, $I_X = 0$
(5) $Z = R$ and is maximum
(6) Total current (I_T) is minimum

The calculations used for parallel resonant circuits are similar to those for series circuits. However, **quality factor Q** is found for parallel circuits by using the formula $Q = R/X_L$, and current values are used rather than reactive voltage drops. A sample parallel resonant circuit problem is shown in **Figure 19.14(a)**. The resonant frequency, bandwidth, lower- and upper-cutoff frequencies, quality factor, and component current values can be determined by the procedure shown in **Figure 19.14(b)**.

Example 19-8:

Calculate the resonant frequency, quality factor, and bandwidth of the parallel RLC circuit shown in **Figure 19-14**, assuming R = 1000 Ω, L = 25 mH, and C = 0.5 μF.

Solution

The resonant frequency of a series RLC circuit (f_r) is

$$f_r = \frac{1}{2\pi\sqrt{LC}}$$

$$= \frac{1}{6.28\sqrt{0.025 \times 0.5 \times 10^{-6}}}$$

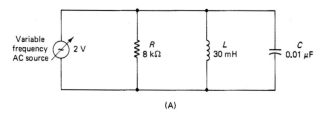

(A)

1. Find resonant frequency (f_r):

$$f_r = \frac{1}{2\pi \sqrt{L \times C}} = \frac{1}{6.28 \times \sqrt{(30 \times 10^{-3}) \times (0.01 \times 10^{-6})}}$$

$$= \frac{1}{1.0877 \times 10^{-4}} = 9194 \text{ Hz}$$

2. Find X_L and X_C at resonant frequency:
   $X_L = 2\pi \cdot f \cdot L = 6.28 \times 9194 \times (30 \times 10^{-3}) = 1732 \ \Omega$

3. Find quality factor Q:

$$Q = \frac{R}{X_L} = \frac{8000 \ \Omega}{1732 \ \Omega} = 4.62$$

4. Find bandwidth (BW):

$$BW = \frac{f_r}{Q} = \frac{9194 \text{ Hz}}{4.62} = 1990 \text{ Hz}$$

5. Find low cutoff frequency (f_{LC}):
   $f_{LC} = f_r - \frac{1}{2}BW = 9194 - 995 = 8199 \text{ Hz}$

6. Find high cutoff frequency (f_{HC}):
   $f_{HC} = f_r + \frac{1}{2}BW = 9194 + 995 = 10{,}189 \text{ Hz}$

7. Find I_R at resonant frequency:

$$I_R = \frac{E_A}{R} = \frac{2 \text{ V}}{8 \text{ k}\Omega} = 0.25 \text{ mA}$$

8. Find I_L and I_C at resonant frequency:

$$I_L = \frac{E_A}{X_L} = \frac{2 \text{ V}}{1732 \ \Omega} = 1.15 \text{ mA}$$

(B)

Figure 19.14   (a) Parallel resonant circuit. (b) Problem solving procedure.

$$= \frac{1}{6.28 \times 1.11 \times 10^{-4}} = \frac{1}{7.02 \times 10^{-4}} = 1424.25 \text{ Hz}$$

$$Q = \frac{R}{X_L} = \frac{R}{2\pi f L}$$

$$= \frac{1000}{6.28 \times 1424.25 \times 25 \times 10^{-3}}$$

$$= 4.47$$

$$BW = \frac{f_r}{Q} = \frac{1424.25}{4.47} = 318.6\,\text{Hz}.$$

Related Problem

Calculate the resonant frequency, quality factor, and bandwidth of the parallel RLC circuit shown in **Figure 19.14**, assuming R = 100 kΩ, L = 50 mH, and C = 250 pF.

Self-Examination

Answer the following questions.

11. Resonant frequency in a RLC circuit refers to a condition where

    a. $X_L > X_C$
    b. $X_L < X_C$
    c. $X_L = X_C$
    d. R = X for all frequencies

12. The impedance of a series RLC circuit at the resonant frequency is

    a. Maximum
    b. Minimum
    c. Zero
    d. Infinite

13. When the Q factor of resonant circuit is large, its bandwidth is

    a. Large
    b. Small
    c. 0
    d. Infinite

14. The impedance of a parallel RLC circuit at the resonant frequency is

    a. Maximum
    b. Minimum
    c. Zero
    d. Infinite

15. Referring to the frequency response curves shown in **Figure 19.13**, as the bandwidth increases the selectivity:

    a. Decreases

b. Increases
c. Remains constant
d. Is variable

19.4 Active Filters

An **active filter** is a device that passes electric currents at a certain frequency or frequency ranges while preventing the passage of others. An active filter is composed of a network of resistors, capacitors, or inductors that control the operation of a solid-state device, usually op-amp(s). Passive filters, discussed previously, are constructed of only a network of inductors, capacitors, and resistors.

19.5 Explain the operation of an active filter circuit.

In order to achieve objective 19.5, you should be able to:

- distinguish between active and passive components in filter circuits;
- determine the order of a given filter circuit.

Active filters have several unique advantages over passive filters. When a passive filter is inserted into a circuit, it normally causes some power and voltage losses to occur. This is called the **insertion loss**. Active filters that include amplification components such as op-amps can compensate for such insertion losses. Since an operational amplifier is capable of providing gain, the input signal is not attenuated, while the filter passes specific frequencies. Additionally, active filters typically cost far less than passive filters because inductors are expensive and are not always available in specific values. Active filters are also easy to tune (adjust) over a wide frequency range. Another advantage is that, as a result of using an operational amplifier, active filters have high input and low output impedance values providing isolation between the filter circuit and the source or load.

Some disadvantages of active filters include frequency response since it is limited by the type of operational amplifier used in the circuit. Unlike passive filters, active filters require a power supply for the operational amplifier. Several examples of active filter circuits are summarized in this chapter.

Order of a Filter Circuit

For **passive filters**, the term order describes both the number of cascaded resistor-capacitor (*R*-C) filter networks and the slope of the stop band.

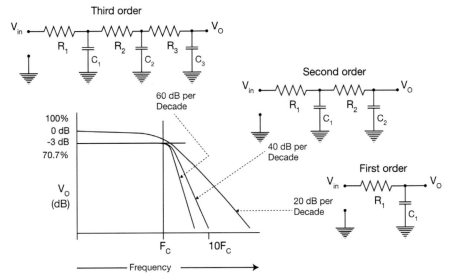

Figure 19.15 First-, second-, and third-order low-pass filters.

Figure 19.15 shows first-order, second-order, and third-order low-pass filters, along with the roll-off curve for each. The **single-stage filter** always has a roll-off rate of 20 dB per **decade**. This is true for low-pass filters and high-pass filters. Decades will be discussed in a following section.

If we **cascade** or series-connect two **first-order filter** networks, **a second-order filter**, with a roll-off rate of 40 dB per decade results. Cascading three first-order filter networks produces a **third-order filter** with a roll-off rate of 60 dB per decade. The beginning of the roll-off is still at 3 dB below the top of the curve. If the top of the curve is set at 0 dB, the start of the roll-off is -3 dB. The first-, second-, and third-order high-pass filters are also composed of cascaded first-order filter networks, and they have roll-off rates of 20, 40, and 60 dB per decade. The frequency at the -3 dB point, where the pass-band starts to roll off, is called the **cutoff frequency (f_c)**.

First-Order Active Low-Pass Filters

The **low-pass filter** allows incoming signals to be passed through the circuit with little or no attenuation up to a certain frequency. Above this frequency, the filter rejects or greatly attenuates the input signal. When working with filters, we are primarily interested in the relationship of the filter's output voltage compared to its input voltage at various frequencies. If the filter's

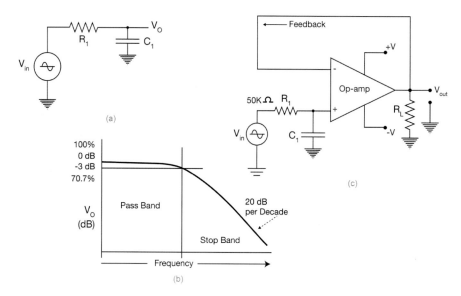

Figure 19.16   (a) Low-pass filter circuit. (b) Low-pass filter response curve. (c) First-order low-pass active filter circuit.

output voltage is larger than its input voltage, there is gain, and the dB value will be a positive number. On the other hand, if the output voltage is less than the input voltage, there is a loss, or attenuation, and the dB value will be a negative number.

A **first-order** low-pass filter's amplitude response is essentially constant up to the circuit's cutoff frequency. The range of frequencies in the region of constant amplitude (see **Figure 19.16**) is called the pass-band. As the input signal frequency exceeds the low-pass filter's cutoff frequency, the dB amplitude response decreases linearly as the frequency increases logarithmically. The frequency range above the filter's cutoff frequency is called the **stop band**. The following equation is used to calculate the cutoff frequency for all first-order filters:

$$f_c = \frac{1}{2\pi RC}.$$

The dB amplitude response in the stop band decreases linearly. The rate of decrease (slope of the line) is called the **roll-off**. This is called the "order" of the filter. Above the cutoff frequency, the roll-off of a first-order, low-pass filter is -6 dB per octave.

An **octave** is either a doubling or a halving of frequency. For example, a frequency of 500 Hz, octaves above are 1 kHz, and 2 kHz, 4 kHz and beyond. Octaves below 500 Hz are 250, 125, 62.5 Hz, and continuing. A decade is either a ten-fold increase or decrease in frequency. For example, for the 500-Hz frequency, decades above are 10,000 kHz, etc., and decades below are 25,100, 10, 1 Hz, etc.

Octaves and decades are plotted on **semi-logarithmic scales** used for the horizontal axis of frequency response curve graphs. A decade scale means that each major division on the horizontal axis is ten times the frequency of the previous division. The decade scale allows a large range of frequencies to be plotted on a sheet of graph paper.

The **octave scale** is used for audio frequency response curves because it corresponds to the musical scale. On keyboard instruments, shown in **Figure 19.17**, middle A has a frequency of 440 Hz. The A above middle A has a frequency of 880 Hz. The next-higher frequency A has a frequency of 1760 Hz. Note that the frequency doubles with each octave. The word *octave* is derived from *octal* (meaning eight) and is used because there are eight whole-note tones from A to A, from B to B, C to C, etc.

Refer back to **Figure 19.16(c)** and note that an active **low-pass filter** is nothing more than a passive low-pass filter with an added voltage-follower op-amp. The high op-amp input impedance prevents the filter from being loaded and the op-amp output can drive a significant load. Also note in **Figure 19.15** the roll-off values of 20 dB per decade for first-order low-pass filters.

A second-order, low-pass filter has a **roll-off** equal to two times a first-order filter (-12 dB per octave). A third-order filter has a roll-off of -18 dB per octave and so on for higher-order low-pass filters.

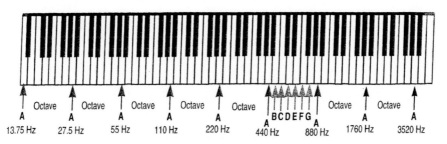

Figure 19.17   Keyboard scale – octaves.

In terms of **efficiency**, an active filter of higher than a second-order with a single operational amplifier network is not practical. A third-order filter is constructed of op-amps in series or cascade of a first-order and a second-order, low-pass section. For higher-order active filters, one or more first- and second-order sections are cascaded.

The **cutoff frequency** of a filter is the frequency at which the voltage gain drops to 0.707 times the pass-band gain. The voltage gain response is linear from the pass-band through a point that is 3 dB below the pass-band gain.

First-Order Active High-Pass Filter

An **active high-pass filter** is a passive high-pass filter with an added voltage-follower op-amp as shown in **Figure 19.18**. The high op-amp input impedance prevents the filter from being loaded, while the op-amp output can drive a significant load. The -3 dB cutoff frequency (f_c) is calculated using the same equation as low-pass filters.

Second-Order Low-Pass Filter

A **second-order** active low-pass filter uses two first-order active op-amp filters connected in cascade as shown in **Figure 19.19**. The op-amps may provide voltage gain or be unity-gain voltage followers.

Each filter section roll-off has a 20 dB per decade or 40 dB per decade total (6 dB per octave or 12 dB per octave total).

The -6 dB frequency is calculated using the following equation:

$$f_c = \frac{1}{2\pi RC}.$$

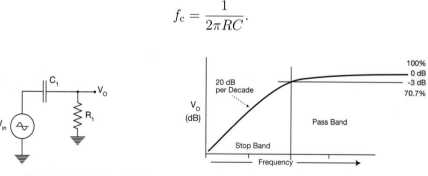

a. First-order Passive High-pass Filter Circuit

Figure 19.18   (a) First-order passive high-pass filter circuit.

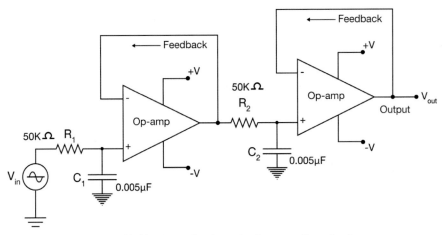

Figure 19.19   Second-order active low-pass filter circuit.

Second-Order High-Pass Filter

A **second-order** active high-pass filter uses two first-order active op-amp filters connected in cascade as shown in **Figure 19.20**. The op-amps may provide voltage gain or may be unity-gain voltage followers, just as low-pass filters.

Each filter section **roll-off** has 20 dB per decade or 40 dB per decade total (6 dB per octave or 12 dB per octave total), just as low-pass filters.

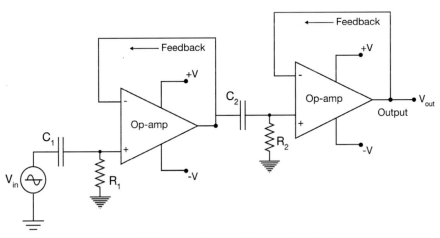

Figure 19.20   Second-order high-pass filter circuit.

The **−6 dB frequency** is calculated using the same equation as low-pass filters.

Second-Order Band-Pass Filters

Figure 19.21 is an **inverting amplifier** that uses negative feedback. The amplifier gain can be set to unity by making $R_f = R_1$. This filter is a band-pass filter, with a low-pass and a high-pass filter.

Band-pass **center frequency** is calculated as follows:

$$f_{cf} = \frac{1}{2\pi C} \sqrt{\frac{R_1 + R_2}{R_1 R_2 R_3}}. \tag{19.22}$$

Quality factor (Q) is calculated as follows:

$$Q = \pi f_{cf} C R_3. \tag{19.23}$$

Bandwidth (BW) is calculated as follows:

$$BW = f_{cf}/Q. \tag{19.24}$$

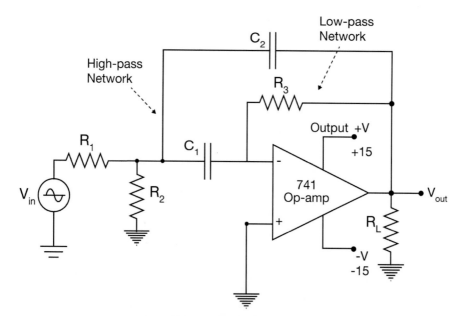

Figure 19.21   Band-pass active filter circuit.

Example 19-9:

Calculate the cutoff frequency of a low-pass active filter circuit shown in **Figure 19.16**. Assume that the value of $R_1 = 10$ kΩ and $C_1 = 1$ μF.

Solution

$$f_c = \frac{1}{2\pi R_1 C_1}$$

$$= \frac{1}{6.28 \times 10,000 \times 0.000001}$$

$$= 15.92 \, \text{Hz}.$$

Related Problem

Calculate the cutoff frequency of an active low-pass filter circuit shown in **Figure 19.16**. Assume that the value of R = 50 kΩ and C = 0.05 μF.

Self-Examination

Answer the following questions.

16. A passive filter does not have a(n):

 a. Resistor
 b. Inductor
 c. Capacitor
 d. Op-amp

17. The order of a filter circuit refers to the number of:

 a. Resistors
 b. Capacitors
 c. cascaded filters
 d. Component combinations

18. The insertion loss of an active filter is:

 a. Infinite
 b. High
 c. Medium
 d. Low

19. Referring to the first-order high-pass active filter of **Figure 19.18**, the values of the components C_1 and R_1 determine the _____.

    a. Gain

    b. Resonant frequency

    c. Cutoff frequency

    d. Inductive reactance

20. Referring to the first-order low-pass active filter of **Figure 19.16**, as the order of the filter increases, the dB loss per decade, beyond the 3 dB point will _____.

    a. Increase

    b. Decrease

    c. Remain constant

    d. Be variable

19.5 Analysis and Troubleshooting – Filter Circuits

The performance of a filter circuit is evaluated by observing its **output waveform** after applying an input signal that has a known amplitude and frequency. The output response of a filter is determined by the characteristics of the circuit and the applied frequency. In this section, a **low-pass filter** will be examined to determine its output response and possible faulty conditions.

19.6 Analyze and troubleshoot filter circuits.

In order to achieve objective 19.6, you should be able to:

• observe the input and output AC waveforms of different filter circuits on an oscilloscope;

• distinguish between normal and faulty operation of a filter circuit.

When **faults** occur in a filter circuit, there will be noticeable difference in the output response. These faults could be caused by failure of either the passive or active components used in its construction. Shorts, open-circuits, or leakage are common faults of the passive components.

Faults in the active components, such as op-amps, generally cause a change in the desired level of amplification. In most cases, repair of the active components is not possible owing to the complexity of its internal construction, and replacement is necessary.

Summary

- A low-pass filter passes low-frequency signals up to a specified cutoff frequency and blocks high-frequency signals beyond the cutoff.
- A high-pass filter passes high-frequency signals beyond a specified cutoff frequency and blocks low-frequency signals below the cutoff.
- A band-pass filter is a combination of high- and low-pass filter sections that will pass a mid-range of frequency signals and blocks other frequencies.
- A band-stop or notch filter blocks a mid-range of frequency signals and passes other frequencies.
- The performance of a filter circuit can be evaluated by using frequency response curves, which plots the power and voltage gain/loss in decibels (dB) over a range of frequencies.
- The 0-dB region of a frequency response curve signifies the portion where the output power is equal to the input power; so the signal passes through without any loss.
- The 3-dB point(s) on a frequency response curve signify the point(s) where the output power has been reduced to half the value of the input power.
- The bandwidth of a filter is the frequency range in which the output power is greater than half the value of the input power.
- The quality factor (Q) of a filter circuit refers to the ratio of the resistance and the reactance. The bandwidth of a filter and its quality factor are inversely related.
- Resonant circuits are designed to pass a certain range of certain frequencies and to block other frequencies. Resonant circuits use resistors, inductors, and capacitors in their construction, which may be connected in series or parallel.
- Active filters use amplifying components such as op-amps, in addition to passive components such as resistors, inductors, and capacitors. They are designed to pass a specific range of frequencies while blocking others.
- The order of a filter is based on the number of cascaded filter sections.
- As the order number of a filter increases, the loss in dB per decade beyond the 3-dB point increases linearly by multiples of 20 dB per decade.

Formulas

(19-1) $X_C = \frac{1}{2\pi f_c C} = R$

(19-2) $f_c = \frac{1}{2\pi RC}$

(19-3) $X_L = 2\pi f_c L = R$

(19-4) $f_c = \frac{R}{2\pi L}$

(19-5) $X_C = \frac{1}{2\pi f_c C} = R$

(19-6) $f_c = \frac{1}{2\pi RC}$

(19-7) $X_L = 2\pi f_c L = R$

(19-8) $f_c = \frac{R}{2\pi L}$

(19-9) $BW = f_{cu} - f_{cl}$

(19-10) $f_{cf} = \sqrt{f_{cu} \times f_{cl}}$

(19-11) $Q = \frac{f_{cf}}{f_{cu} - f_{cl}}$

(19-12) $BW = f_{cu} - f_{cl}$

(19-13) $f_{cf} = \sqrt{f_{cu} \times f_{cl}}$

(19-14) Power gain (dB) $= 10 \log_{10} \frac{P_{out}}{P_{in}}$; when $P_{out} > P_{in}$

(19-15) Power loss (dB) $= -10 \log_{10} \frac{P_{out}}{P_{in}}$; when $P_{out} < P_{in}$

(19-16) Voltage gain (dB) $= 20 \log_{10} \frac{V_{out}}{V_{in}}$

(19-17) $\frac{P_{out}}{P_{in}} = 10^{\frac{dB}{10}}$

(19-18) $\frac{V_{out}}{V_{in}} = 10^{\frac{dB}{20}}$

(19-19) $f_r = \frac{1}{2\pi\sqrt{LC}}$

(19-20) $Q = X_L / R$

(19-21) $BW = f_r / Q$

(19-22) $f_{cf} = \frac{1}{2\pi C} \sqrt{\frac{R_1 + R_2}{R_1 R_2 R_3}}$

(19-23) $Q = \pi f_{cf} C R_3$

(19-24) $BW = f_{cf} / Q$

Review Questions

Answer the following questions.

1. If an AC source is connected in series with a resistor and a capacitor and voltage output is taken across the capacitor, the circuit will exhibit the property of a:

    a. Low-pass filter

    b. High-pass filter

c. Band-pass filter

d. Parallel resonant circuit

2. When frequency is increased in a low-pass filter circuit, the current flow will:

   a. Increase because of greater X_L

   b. Increase because of less X_L

   c. Decrease because of greater X_L

   d. Decreases because of less X_L

3. At resonance, a series RLC circuit characteristically develops:

   a. Maximum voltage across the input terminals

   b. Minimum current though the circuit

   c. Maximum reactance of the coil and capacitor

   d. Minimum impedance between the input terminals

4. Capacitive reactance is said to be frequency sensitive. This statement means that capacitive reactance:

   a. Is independent of frequency

   b. Decreases as the frequency increases

   c. Increases with a rise of frequency

   d. Increases with a rise in capacitance

5. In a series resonant circuit, the impedance at resonance is _____ the resistance of the circuit.

   a. Always equal to

   b. Never equal to

   c. Lesser than

   d. Greater than

6. It is possible to pass a specific frequency through a resonant circuit by the proper selection of:

   a. C and R

   b. C and L

   c. R and L

   d. R and Z

7. At resonance, a parallel resonant circuit characteristically develops:

   a. Maximum voltage across the input terminals

   b. Minimum current through the circuit

   c. Maximum reactance of the coil and capacitor

   d. Minimum impedance between the input terminals

8. When frequency is increased, voltage across a capacitor in a series RC circuit will:

   a. Decrease
   b. Increase
   c. Remain constant
   d. Be infinite

9. In a series resonant circuit, when X_C and X_L equal:

   a. Line voltage leads line current
   b. Line current leads line voltage
   c. Total impedance is minimum
   d. Total impedance is maximum

10. If an AC source is connected in series with a resistor and a capacitor and a voltage output is taken across the resistor, the circuit will exhibit the property of a:

   a. Low-pass filter
   b. High-pass filter
   c. Band-pass filter
   d. Parallel resonant circuit

10. Which component would be found in an active filter but not in a passive filter:

   a. Resistor
   b. Capacitor
   c. Inductor
   d. Op-amp

11. A _____ filter passes frequencies from 0 Hz to some critical frequency and blocks all frequencies above this.

   a. Low-pass
   b. High-pass
   c. Band-pass
   d. Notch

12. A _____ filter blocks frequencies from 0 Hz to some critical frequency and passes all frequencies above this.

   a. Low-pass
   b. High-pass
   c. Band-pass
   d. Notch

13. A roll-off of a first-order low-pass filter is:

    a. -3 dB/decade
    b. -10 dB/decade
    c. -20 dB/decade
    d. -6 dB/octave

14. Cascading two first-order filters will result in a roll-off rate of:

    a. -6 dB/decade
    b. -20 dB/decade
    c. -40 dB/decade
    d. -6 dB/octave

Problems

Answer the following questions.

1. A circuit has a decibel gain of 12 dB. What is the power ratio associated with this gain:

    a. 1.58
    b. 21.2
    c. 15.8
    d. 2.12

2. A 16 H inductor is connected in series with a 1-μF capacitor. The resonant frequency is:

    a. 400 Hz
    b. 126.5 Hz
    c. 12.65 Hz
    d. 40 Hz

3. A series resonant circuit has an $f_r = 10$ kHz and a Q of 10. The bandwidth is:

    a. 10 Hz
    b. 100 Hz
    c. 1000 Hz
    d. 10,000 Hz

4. A series RLC circuit is at resonance when:

    a. $X_L = 15\ \Omega$, $X_C = 5\ \Omega$, and R = $15\ \Omega$
    b. $X_L = 10\ \Omega$, $X_C = 10\ \Omega$, and R = $100\ \Omega$

    c. $X_L = 100\ \Omega$, $X_C = 5\ \Omega$, and R = 5 Ω

    d. $X_L = 20\ \Omega$, $X_C = 5\ \Omega$, and R = 20 Ω

5. Refer to the circuit diagram of the high-pass active filter shown in **Figure 19.18(c)**. Assuming the value of $C_1 = 0.\ \mu F$ and $R_1 = $, the cutoff (or critical frequency) lies in the range:

    a. 0-0.9 kHz

    b. 1-9.9 kHz

    c. 10-99.99 kHz

    d. 100 kHz or above

Answers

Examples

19-1. $X_C = 1\ k\Omega$, C = 1.59 μF

19-2. f_c = 63.69 Hz

19-3. f_c = 3.185 kHz

19-4. f_{cf} = 2.236 kHz

19-5. BW = 100,000

19-6. V_{out}/V_{in} = 0.0224

19-7. f_r = 45.039 kHz

19-8. f_r = 45.039 kHz; Q = 7.07; BW = 6.370 kHz

19-9. f_c = 63.69 Hz

Self-Examination

19.1

    1. a. low-pass

    2. b. high-pass

    3. c. band-pass

    4. d. band-stop

    5. b. capacitive, inductive

19.2

    6. b. half

    7. b. half

    8. b. less than 1

    9. c. equal to the input

10. c. band-pass
19.3
11. c. $X_L = X_C$
12. b. minimum
13. b. small
14. a. maximum
15. a. decreases
19.4
16. d. op-amp
17. c. cascaded filters
18. d. low
19. c. cutoff frequency
20. a. increase

Terms

Amplification

An increase in value.

Attenuation

A reduction in value.

Band-pass filter

A frequency-sensitive AC circuit that allows incoming frequencies within a certain band to pass through but attenuates frequencies below or above this band.

Decibel (dB)

A unit used to express an increase or decrease in power, voltage, or current in a circuit.

Filter

A circuit used to pass certain frequencies and attenuate all other frequencies.

Frequency

The number of AC cycles per second, measured in hertz (Hz).

Frequency response

A circuit's ability to operate over a range of frequencies.

High-pass filter

A frequency-sensitive AC circuit that passes high-frequency input signals to its output and attenuates low-frequency signals.

Low-pass filter

A frequency-sensitive AC circuit that passes low-frequency input signals to its output and attenuates high-frequency signals.

Parallel resonant circuit

A circuit that has an inductor and a capacitor connected in parallel that causes it to respond to frequencies applied to the circuit.

Quality factor (Q)

The "figure of merit" or Q of frequency-sensitive circuit is the ratio of the resistance to the reactance.

Resonant circuit

A frequency-sensitive circuit in which the inductive and capacitive reactance values are used to determine the resonant frequency.

Resonant frequency

The frequency that passes through or is blocked by a resonant circuit depends on the inductive and capacitive reactance.

Selectivity

The ability a resonant or filter circuit to select a specific frequency and reject all other frequencies.

Series resonant circuit

A circuit that has an inductor and a capacitor connected in series that causes it to respond to frequencies applied to the circuit.

Cutoff frequency

The frequency at the -3 dB point, where the pass-band starts to roll off.

Order

A term that indicates the number of filter networks that are cascaded. As the order number increases, the amount of filtering increases linearly by 20 dB/decade beyond the 3-dB points.

Active filters

Filter circuits that use amplifying components such as op-amps, in addition to passive components such as resistors, inductors, and capacitors. They are designed to pass a specific range of frequencies while blocking others.

Decade

A frequency change by a factor of 10.

Octave

A frequency change by a factor of 2.

20

Oscillator Circuits

Many electronic systems employ circuits that convert the DC energy of a power supply into a useful form of AC. Oscillators and electronic clocks are examples of these circuits. In a radio receiver, for example, DC is converted into high-frequency AC to achieve signal tuning. Television receivers also have oscillators in their tuners. Oscillators are used to produce horizontal and vertical sweep signals in TVs. These signals control the electron beam of the picture tube. Calculators and computers employ an electronic clock circuit. This type of circuit produces timing pulses. Every RF transmitter employs an oscillator. This part of the system generates the signal that is radiated into the atmosphere. Oscillators are extremely important in electronics. An oscillator generally uses an amplifying device to aid in the generation of the AC output signal. Transistors and ICs can be used for this function.

There are several possible ways of describing an oscillator. This can be done according to generated frequency, operational stability, power output, signal waveforms, and components. In this chapter, it is convenient to divide oscillators according to their method of operation. This includes feedback oscillator and relaxation oscillators. As you will see, each group has a number of distinguishing features.

Objectives

After studying this chapter, you will be able to:
20.1 analyze the operation of LC, RC, and RL circuits;
20.2 analyze common feedback oscillators;
20.3 analyze common relaxation oscillators;
20.4 analyze and troubleshoot oscillator circuits.

Chapter Outline

20.1 Oscillator Fundamentals
20.2 Feedback Oscillators

20.3 Relaxation Oscillators
20.4 Analysis and Troubleshooting – Oscillator Circuits

Key Terms

astable multivibrator
capacitance ratio
continuous wave (CW)
damped sine wave
feedback oscillator
flip-flop
free-running multivibrator
monostable multivibrator
multivibrator
nonsymmetrical multivibrator
piezoelectric effect
pulse repetition rate (PRR)
regenerative feedback
relaxation oscillator
symmetrical multivibrator
capacitance ratio
one-shot multivibrator
tank circuit
sharp filter
threshold terminal
triggered multivibrator
upper trip point (UTP)
lower trip point (LTP)

20.1 Oscillator Fundamentals

An **oscillator** is an electronic circuit that generates a continuously repetitive output signal. The output signal may be alternating current or some form of varying DC. The unique feature of an oscillator is its generation function. No external input signal is applied to an oscillator circuit. An oscillator develops an output signal from the DC operating power provided by the power supply. An oscillator is, in a sense, an electronic power converter. It changes DC power into AC or some useful form of varying DC. In this section, you will learn about the two basic types of oscillators and their basic operation.

20.1 Analyze the operation of LC, RC, and RL circuits.

In order to achieve objective 20.1, you should be able to:

- determine the resonant frequency of a tank circuit;
- interpret the time constant curves of an RC circuit;
- explain the sine-wave output and damped oscillation of an LC tank circuit;
- explain the basic operation of feedback oscillator and relaxation oscillator circuits;
- describe the characteristics of feedback oscillator and relaxation oscillator circuits;
- define feedback oscillator, regenerative feedback, relaxation oscillator, tank circuit, damped sine wave, and continuous wave.

Oscillator Types

The two basic types of oscillators based on their method of operation are the feedback oscillator and the relaxation oscillator. In a **feedback oscillator**, a portion of the output signal is returned to the input circuit. This type of oscillator usually employs a tuned LC circuit. This circuit establishes the operating frequency of the oscillator. Most sine-wave oscillators are of this type. The frequency range is from a few hertz to millions of hertz. Applications of the feedback oscillator are widely used in radio and television receiver tuning circuits and in transmitters. Feedback oscillators respond well to RF generator applications.

A block diagram of a feedback oscillator is shown in **Figure 20.1**. An oscillator has an amplifier, a feedback network, frequency-determining network, and a DC power source. The amplifier is used primarily to increase the output signal to a usable level. The open-loop voltage gain provided by an amplifier is designated as A_v. Any device that has signal gain capabilities can be used for this function. The feedback network can be inductive, capacitive, or resistive. It is responsible for returning a portion of the amplifier's output to the input. The ratio of the output signal that is returned to the input with respect to the total output is called the attenuation or scaling factor and is designated as B. The feedback signal must be of the correct phase and value to cause oscillations to occur. **Regenerative feedback**, or in-phase feedback, is essential in an oscillator.

The closed-loop feedback gain (A_{cl}) of the oscillator is the product of the gain provided by the amplifying element, A_v, and the attenuation factor B.

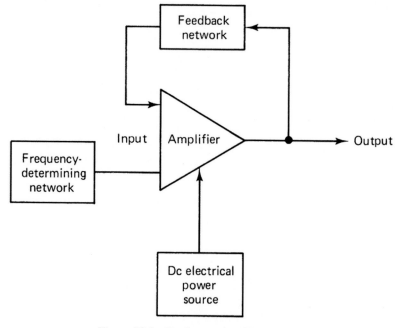

Figure 20.1   Fundamental oscillator parts.

$$A_{cl} = A_v \times B \qquad (20.1)$$

If the closed-loop feedback gain:

(a) $A_{cl} < 1$, then the amplitude output signal will rapidly decay to zero.
(b) $A_{cl} = 1$, then the amplitude of the output signal will remain constant.
(c) $A_{cl} > 1$, then the amplitude of the output signal will increase rapidly causing saturation of the device. This, in turn, will result in unacceptable distortions in the output.

For **undistorted** sustained oscillations, then the closed-loop gain should consistently be unity or one. For example, if the attenuation (B) of an oscillator circuit is 0.01, in order to continue sustained oscillations, the voltage gain (A_v) should be set to a value that will ensure a closed-loop feedback gain (A_{cl}) of 1.

$$A_{cl} = A_v \times B$$
$$1 = A_v \times 0.01$$
$$1/0.01 = A_v$$

$$100 = A_v.$$

An inductor-capacitor (LC) network determines the frequency of the oscillator. The charge and discharge actions of this network establishe the oscillating voltage. This signal is then applied to the input of the amplifier. In a sense, the LC network is energized by the feedback signal. This energy is needed to overcome the internal resistance of the LC network. With suitable feedback, a continuous AC signal can be generated. The output of a good oscillator must be uniform. It should not vary in frequency of amplitude.

A **relaxation oscillator** responds to an electronic device that goes into conduction for a certain time and then turns off for a period of time. This condition of operation repeats itself continuously. This oscillator usually responds to the charge and discharge of an RC or an RL network. Oscillators of this type usually generate square- or triangular-shaped waves. The active device of the oscillator is "triggered" into conduction by a change in voltage. Applications include the vertical and horizontal sweep generators of a television receiver and computer clock circuits. Relaxation oscillators respond extremely well to low-frequency applications.

Feedback Oscillator Operation

The operating frequency of a **feedback oscillator** is usually determined by an inductance-capacitance (LC) network. An LC network is sometimes called a **tank circuit**. The tank, in this case, has a storage capability. It stores an AC voltage that occurs at its resonant frequency. This voltage can be stored in the tank for a short period of time. In an oscillator, the **tank circuit** is responsible for the frequency of the AC voltage that is being generated.

Take a look at the **LC tank circuit** in **Figure 20.2** to see how AC is produced from DC. Note that a simple tank circuit is connected to a battery. In **Figure 20.2**, the switch is closed momentarily. This action causes the capacitor to charge to the value of the battery voltage. Note the direction of the charging current. After a short time, the switch is opened, and the accumulated charge voltage is developed on the capacitor.

To see how a tank circuit develops a sine-wave voltage, refer to **Figure 20.3**. Assume that the capacitor of **Figure 20.3(a)** has been charged to a desired voltage value by momentarily connecting it to a battery. **Figure 20.3(b)** shows the capacitor discharging through the inductor. **Discharge current** flowing through the inductor causes an electromagnetic field to expand around it. **Figure 20.3(c)** shows the capacitor charge depleted. The

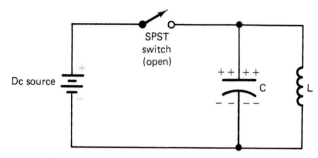

Figure 20.2   LC tank circuit being charged - charging action showing charged capacitor.

field around the inductor then collapses. This causes a continuation of current flow for a short time. **Figure 20.3(d)** shows the capacitor being charged by the induced voltage of the collapsing field.

The capacitor once again begins to discharge through the inductor. In **Figure 20.3(e)**, note that the direction of the discharge current is reversed. The electromagnetic field again expands around the inductor. Its polarity is reversed. **Figure 20.3(f)** shows the capacitor with its charge depleted. The field around the inductor collapses at this time. This causes a continuation of current flow. **Figure 20.3(g)** shows the capacitor being charged by the induced voltage of the collapsing field. This causes the capacitor to be charged to the same polarity as the beginning step. The process is then repeated. The charge and discharge of the capacitor through the inductor causes an AC voltage to occur in the circuit.

The frequency of the AC voltage generated by a tank circuit is based on the values of the inductor and capacitor. It is called the resonant frequency of a tank circuit. The formula for **resonant frequency** is

$$f_r = \frac{1}{2\pi\sqrt{LC}} \tag{20.2}$$

where f_r is in hertz, L is the inductance in henries, and C is the capacitance in farads. This formula is extremely important because it applies to all LC sine-wave oscillators. Resonance occurs when the inductive reactance (X_L) equals the capacitive reactance (X_C). A tank circuit will oscillate at this frequency for a rather long period of time without a great deal of circuit opposition. This condition is illustrated in **Figure 20.4**.

At **resonant frequency**, an LC tank circuit always has some circuit resistance. This resistance tends to oppose the AC that circulates through the circuit. The AC voltage, therefore, decreases in amplitude after a few cycles

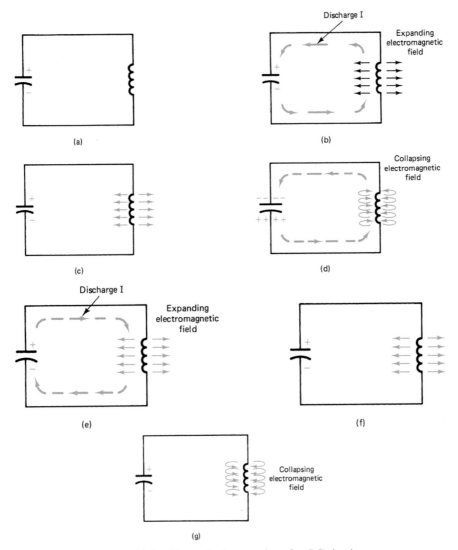

Figure 20.3   Charge-discharge action of an LC circuit.

of operation. Take a look at **Figure 20.4**. Note that **Figure 20.4(a)** shows the resulting wave of a tank circuit. Note the gradual decrease in signal amplitude. This is called a **damped sine wave**. A tank circuit produces a wave of this type.

A **tank circuit**, as we have seen, generates an AC signal when it is energized with DC. The generated wave gradually diminishes in amplitude.

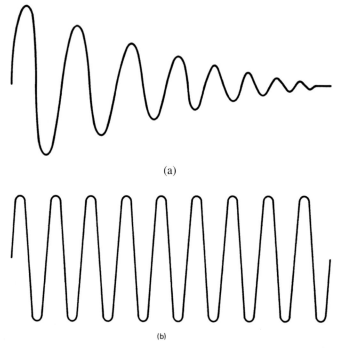

(a)

(b)

Figure 20.4   Wave types. (a) Damped oscillatory wave. (b) Continuous wave.

After a few cycles of operation, the lost energy is transformed into heat. When the lost energy is great enough, the signal stops oscillating.

The oscillations of a tank circuit could be made continuous if energy were periodically added to the circuit. The added energy would restore that lost by the circuit through heat. The added energy must be in step with the circulating tank energy. If the two are additive, the circulating signal will have a continuous amplitude. **Figure 20.4(b)** shows the **continuous wave (CW)** of an energized tank circuit.

Our original tank circuit was energized by momentarily connecting a DC source to the capacitor. This action caused the capacitor to become charged by manual switching action. To keep a tank circuit oscillating by manual switch operation is physically impossible. Switching of this type is normally achieved by an electronic device. Transistors are commonly used to perform this operation in an oscillator circuit. The output of the transistor is applied to the tank circuit in proper step with the circulating energy. When this is achieved, the oscillator generates a CW signal of a fixed frequency.

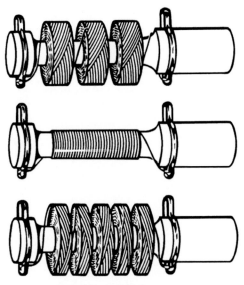

Figure 20.5 RF oscillator coils.

The **inductance** (coil) of a tank circuit varies a great deal, depending on the frequency generated. Most applications of LC oscillators are in the RF range. Some representative RF oscillator coils are shown in **Figure 20.5**. Coil inductance is usually a variable. Inductance is changed by moving the position of a ferrite core inside the coil. Adjusting the inductance permits some change in the frequency of the tank circuit.

Example 20-1:

A transistor amplifier circuit LC tank circuit has values of L = 25 mH and C = 0.1 μF. What is the resonant frequency?

Solution

$$f_r = \frac{1}{2\pi\sqrt{LC}} = \frac{1}{6.28\sqrt{25 \times 10^{-3} \times 0.1 \times 10^{-6}}}$$
$$= 3184.7\,\text{Hz} = 3.185\,\text{kHz}.$$

Related Problem

An LC tank circuit has values of L = 1 mH and C = 1 μF. What is the resonant frequency?

Relaxation Oscillator Operation

The operation of a **relaxation oscillator** usually depends on the charge and discharge of an RC or an RL network. When a source of DC is connected in series with a resistor and a capacitor, we have an RC circuit. When a source of DC is connected in series with a resistor and an inductor, we have an LC circuit.

Figure 20.6 shows an RC circuit and time constant curves. The **time constant** of an RC circuit is denoted by "t." Its value is calculated using the

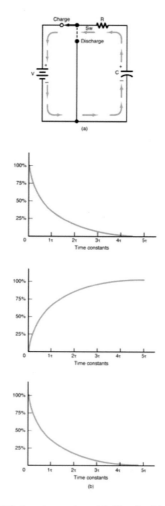

Figure 20.6   *RC* charging action. (a) Circuit. (b) Circuit values.

values of the resistance (R) and capacitance (C). It is expressed as t = R × C and is measured in seconds. The SPDT switch is used to charge and discharge the capacitor through the resistor. For **charging** the capacitor, the switch is closed, and the voltage source is applied to the RC circuit. The capacitor will charge to 63% of the applied voltage in one time constant. It takes five time constants to completely charge the capacitor. For **discharging** the capacitor, the switch is opened, thereby disconnecting the voltage source from the RC circuit. The capacitor discharges through the resistor. The capacitor gets discharged by 63% of the voltage it has been charged to in one time constant. It takes five time constants to completely discharge the capacitor.

Individual **time constant curves** are shown in **Figure 20.6(b)** for each component of the RC circuit. V_C shows the capacitor voltage with respect to time. V_R is the resistor voltage. Circuit current flow is displayed by the I curve. Note how the value of these curves change with respect to time.

The circuit switch is initially placed in the charge position. Note the direction of charge current indicated by the circuit arrows. In one time constant, the V_C curve rises to 63% of the source voltage. Beyond this point, there is a smaller change in the value of V_C. After five time constants, the capacitor is considered to be fully charged.

Note also how the circuit current and the resistor voltage change with respect to time. Initially, I and V_R rise to maximum values. After one time constant, only 37% of I flows. V_R, which is current dependent, follows the change in I. After five time constants, there is zero current flow. This shows that C has been fully charged. With C charged, there is no circuit current flow, and V_R is zero. The circuit remains in this state as long as the switch remains in the charge position.

Assume that the capacitor of the circuit of **Figure 20.7** has been fully charged. The switch is now placed in the discharge position. In this situation, the DC source is removed from the circuit. The resistor is now connected across the capacitor. C discharges through R. Current flows from the lower plate of C through R to the upper plate of C. Note the direction of the discharge current path indicated by circuit arrows.

The initial discharge current is of a maximum value. See the individual time constant curves for V_C, I, and V_R in **Figure 20.7(b)**. The discharge current shows I to be a maximum value initially. After one time constant, it drops to 37% of the maximum value. After five time constants, I drops to zero. V_C and V_R follow I in the same manner. After five time constants, C is discharged. There is no circuit current. V_C and V_R are both zero. The circuit remains in this state as long as the switch remains in the discharge position.

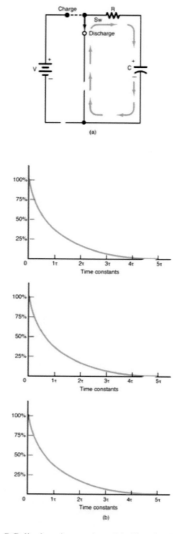

Figure 20.7   *RC* discharging action. (a) Circuit. (b) Circuit values.

Example 20-2:

Refer to the RC circuit shown in **Figure 20.6**. With the values of R = 1 kΩ, C = 0.5 μF, and an applied voltage of 10 V, determine the time it takes to charge the capacitor to 6.3 V. Assume that the capacitor is fully discharged, and the switch is set to the capacitor charge position.

Solution

t = R × C = 1000 Ω × 0.5 × 10⁻⁶ F = 5 × 10⁻⁴ = 0.5 ms.

It will, thus, take 0.5 ms to charge a capacitor to 63% of the applied voltage (10 V).

63% of 10 V is 6.3 V. Hence, the capacitor will charge from 0 to 6.3 V in 0.5 ms.

Related Problem

Refer to the RC circuit shown in **Figure 20.6**. With the values of R = 1 kΩ, C = 0.5 μF, and an applied voltage of 10 V, determine the time it takes to fully charge the capacitor. Assume that the capacitor is fully discharged, and the switch is set to the capacitor charge position.

Self-Examination

Answer the following questions.

1. A circuit that generates a continuously repetitive output signal is called a(n) _____ .

2. The four parts of a feedback oscillator circuit are _____ , _____ , _____ , and _____ .

3. A process by which a portion of a circuit's output is returned to its input is called _____ .

4. An LC network is sometimes called a(n) _____ circuit.

5. The frequency of AC voltage generated by an LC tank circuit is called _____ frequency.

6. The waveform produced by a tank circuit (without feedback) is called a(n) _____ sine wave.

7. The resonant frequency of an LC circuit with a 0.25-μF capacitor and a 6-mH inductor is _____ Hz.

8. The resonant frequency of an LC circuit (increases, decreases) as the capacitance or the inductance increases.

9. A(n) _____ oscillator goes into conduction for a certain period of time and then turns off for a period of time.

10. A(n) _____ oscillator returns a portion of the output signal back to the input.

20.2 Feedback Oscillators

Feedback, as you have learned, is a process by which a portion of the output signal of a circuit is returned to the input. This function is an important characteristic of any oscillator circuit. Feedback may be accomplished by inductance, capacitance, or resistance coupling. A variety of different circuit techniques have been developed. As a general rule, each technique has certain characteristics that distinguish it from others. This accounts for the large number of different circuit variations. This section discusses the characteristics and operation of the common **feedback oscillators**.

20.2 Analyze common feedback oscillators.

In order to achieve objective 20.2, you should be able to:

- explain the operation of common feedback oscillators;
- describe the characteristics of common feedback oscillators;
- define capacitance ratio, piezoelectric effect, and sharp filter.

Armstrong Oscillator

The **Armstrong oscillator** uses a transformer in conjunction with a capacitor to determine the oscillation frequency. A transistor is used to amplify the signal. This oscillator is also refereed to as a **tickler coil**. This type of oscillator is less common than other types because it uses an expensive transformer. The transformer is needed for providing a 180° phase shift between its primary and secondary windings, which is needed for oscillation. **Figure 20.8** is used to show the basic operation of an Armstrong oscillator. The **characteristic curve** of the transistor and its AC load line are also used to explain its operation. Note that, in the circuit, conventional bias voltages are applied to the transistor: the emitter-base junction is forward biased, and the collector-base is reverse biased. Emitter biasing is achieved by R_3. Resistors R_1 and R_2 serve as a divider network for the base voltage.

When power is first applied to the transistor, R_1 and R_2 establish the operating point (Q) near the center of the load line. The output of the transistor, at the collector, should ideally be 0-V AC. Due to the initial surge of current at turn-on time, some noise will immediately appear at the collector. This is usually of a very small value. For our discussion, let us assume that a -1 mV signal appears at the collector. The transformer inverts this voltage and steps it down by a factor of 10. The transformer (T_1) of the circuit has a 10:1 turns ratio. A $+0.1$ mV signal is, therefore, applied to C_1 in the base circuit.

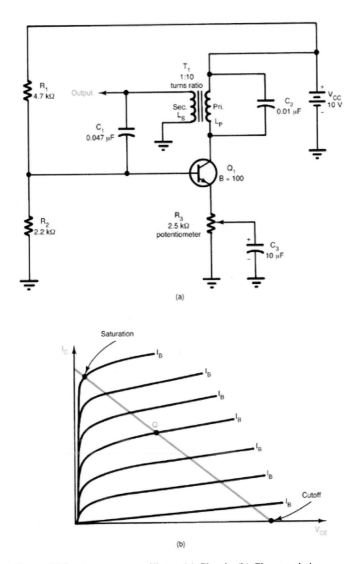

Figure 20.8 Armstrong oscillator. (a) Circuit. (b) Characteristic curves.

The transistor has a **beta** of 100. With +0.1 V applied to the base, Q_1 develops a 10-mV output signal at the collector. The polarity change from positive to negative in the output signal is due to the inverting characteristic of a common-emitter amplifier. The output signal voltage is again stepped down by the transformer and applied to the base of Q_1. A collector signal of

−10 mV causes a +1.0-mV base voltage. With transistor gain, the collector voltage immediately rises to −100 mV. The process continues, thus producing collector voltages of −1.0 V and, finally, −10 V. At this point, transistor operation is driven up the load line until saturation is reached. Note this point on the load line. The collector voltage of Q_1 will not change beyond this point.

With no change in V_C across the primary winding of T_1, the secondary voltage drops immediately to zero. The base voltage returns immediately to the Q point. This decrease of negative-going base voltage (from saturation to point Q) causes V_C to be positive going. Through transformer action, this appears as a negative-going base voltage. This action continues to drive the transistor past point Q. The process continues until the cutoff point is reached. The transformer then stops supplying input voltage to the base. Transistor operation swings immediately in the opposite direction. R_1 and R_2 cause the base voltage to again rise to point Q. The process is then repeated: Q_1 goes to saturation, to point Q, to cutoff, and returns to point Q. Operation is continuous. An AC voltage appears across the secondary of T_1.

The frequency of the **Armstrong oscillator** is based on the value of capacitor C_2 and the inductance of the primary transformer winding. The capacitor C_2 and the transformer primary form a tank circuit with the frequency of the oscillation determined by the *LC* resonant frequency formula.

The **output voltage** of the Armstrong oscillator of **Figure 20.8** can be changed by altering the value of R_3. Gain is highest when the variable arm of R_3 is at the top of its range. This adjustment may cause a great deal of signal distortion. In some cases, the output voltage may appear as a **square wave** instead of a sine wave. When the position of the variable arm is moved downward, the output wave becomes a sine wave. Adjusting R_3 to the bottom of its range may cause oscillations to stop. The gain of Q_1 may not be capable of developing enough feedback for **regeneration** at this point. Normally, 20%-30% of output must be returned to the base to assure continuous oscillation.

Hartley Oscillator

The **Hartley oscillator** of **Figure 20.9** is used rather extensively in AM and FM radio receivers. The resonant frequency of the circuit is determined by the values of L_1 and C_1. Capacitor C_2 is used to couple AC to the base of Q_1. Biasing for Q_1 is provided by R_2 and R_1. Capacitor C_4 couples AC variations in collector voltage to the lower side of L_2. The RF choke coil (L_1) prevents

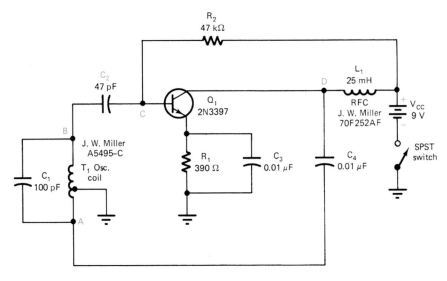

Figure 20.9   Hartley oscillator.

AC from going into the power supply. L_2 also serves as the load for the circuit. Q_1 is an NPN transistor connected in a common-emitter circuit configuration. The inductor L_1 is composed of two coils and L_a and L_b connected in series. The total inductance of the coil is thus

$$L_1 = L_a + L_b \text{ henries.}$$

When DC power is applied to the circuit, current flows from the negative side of the source through R_1 to the emitter. The collector and base are both connected to the positive side of V_{CC}. This forward biases the emitter-base junction and reverse biases the collector. I_E, I_B, and I_C flow initially through Q_1. With I_C flowing through L_2, the collector voltage drops in value. This negative-going voltage is applied to the lower side of L_1, by capacitor C_4. This causes current flow in the lower coil. An electromagnetic field expands around the coil. This, in turn, cuts across the upper part of the coil. By transformer action, the top of the upper coil swings positive. Capacitor C_1 is charged by this voltage. This same voltage is also added to the forward-bias voltage of Q_1 through C_2. Q_1 is eventually driven into saturation.

When **saturation** occurs, there is no change in the value of V_C. The field around the lower part of L_1 collapses immediately. This causes a change in the polarity of the voltage at the top of L_1. The top plate of C_1 now becomes negative and the bottom becomes positive.

The accumulated **charge** on C_1 discharges immediately through L_1 by normal tank circuit action. This negative voltage at the top of C_1 causes the base of Q_1 to swing negative. Q_1 is driven to cutoff. This, in turn, causes the V_C of Q_1 to rise very quickly. A positive-going voltage is then transferred to the lower part of L_1 by C_4, providing feedback. This adds to the voltage of C_1. C_1 will then continue to discharge.

The value change in V_C eventually stops, and no voltage is fed back through C_4. C_1 is fully discharged by this time. The field around the lower side of L_1 then collapses. C_1 charges again, with the bottom side being positive and the top negative. Q_1 again becomes conductive. The process is continuously repeated. The **tank circuit** produces a continuous wave. The circuit losses of the tank are restored by regenerative feedback.

The **frequency** of the Hartley oscillator is determined by

$$f_r = \frac{1}{2\pi\sqrt{L_1 C_1}}. \tag{20.3}$$

The **attenuation factor**, B, of the circuit is the ratio of the coil inductances used.

$$B = \frac{L_b}{L_a}. \tag{20.4}$$

In order to ensure that the oscillator will start automatically when power is initially applied to the circuit, the A_v must be greater than $1/B$. Thus,

$$A_v > \frac{1}{B}$$

$$A_v > \frac{L_b}{L_a}.$$

A distinguishing feature of the Hartley oscillator is its **tapped coil**. A number of circuit variations are possible. The coil may be placed in series with the collector. Collector current flows through the coil in normal operation. A variation of this type is called a series-fed Hartley oscillator. The circuit of **Figure 20.9** is a shunt-fed Hartley oscillator. I_C does not flow through L_1. The coil is connected in parallel or shunt with the DC voltage source. Only AC flows through the lower part of L_1. Shunt-fed Hartley oscillators tend to produce more stable output.

Example 20-3:

Refer to the Hartley oscillator circuit shown in **Figure 20.9**, which has $L_1 = 50$ mH and $C_1 = 100$ pF. Determine the frequency of oscillation.

Solution

$$f_{\mathrm{r}} = \frac{1}{2\pi\sqrt{L_1 C_1}}$$

$$= \frac{1}{6.28\sqrt{50 \times 10^{-3} \times 100 \times 10^{-12}}}$$

$$= 71212.35 \,\mathrm{Hz} = 71.212 \,\mathrm{kHz}.$$

Related Problem

Refer to the Hartley oscillator circuit shown in **Figure 20.9**, which has $L_1 = 30$ mH and $C_1 = 250$ pF. Determine the frequency of oscillation?

Colpitts Oscillator

A **Colpitts oscillator** is similar to the shunt-fed Hartley oscillator. The primary difference is in the tank circuit structure. A Colpitts oscillator uses two series-connected capacitors connected instead of the two series-connected coils used in the Hartley oscillator. An electrostatic charge across the capacitor divider network develops feedback voltage. The two capacitors in series and the inductor determine frequency of the oscillator. The formula for calculating the frequency of this oscillator is the same as that used previously.

$$f_{\mathrm{r}} = \frac{1}{2\pi\sqrt{LC}}.$$

In a **Colpitts oscillator**, however, the value of the capacitance is determined by the series combination of two capacitors C_1 and C_2. This total capacitance is called C_T and is expressed as follows:

$$C_T = \frac{C_1 \times C_2}{C_1 + C_2}. \tag{20.5}$$

The **frequency** of a Colpitts oscillator is, thus, given by

$$f_{\mathrm{r}} = \frac{1}{2\pi\sqrt{L_1 C_T}}. \tag{20.6}$$

Figure 20.10 shows a schematic of the Colpitts oscillator. Bias voltage for the base is provided by resistors R_1 and R_2. The emitter is biased by R_4. The collector is reverse biased by connection to the positive side of V_{CC} through

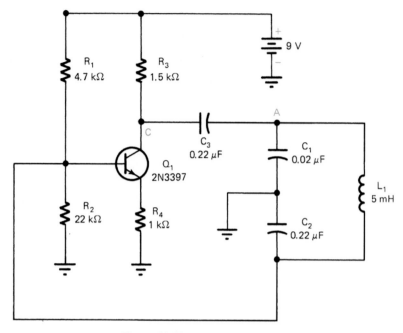

Figure 20.10   Colpitts oscillator.

R_3. This resistor also serves as the collector load. The transistor is connected in a common-emitter circuit configuration.

When DC power is applied to the circuit, current flows from the negative side of V_{CC} through R_4, Q_1, and R_3. I_C flowing through R_3 causes a drop in the positive value of V_C. This negative-going voltage is applied to the top plate of C_1 through C_3. With C_1 and C_2 connected in series, the bottom plate of C_2 takes on a positive charge. This adds to the positive voltage of the base, which, in turn, causes an increase in I_B. The conduction of Q_1 increases. The process continues until Q_1 is saturated.

When Q_1 becomes saturated, there is no further increase in I_C. The **changing** value of V_C also stops. There is no feedback to the topside of C_1. The composite charge across C_1 and C_2 discharges through L_1. Discharge current flow through L_1 is from the top plate of C_1 to the bottom plate of C_2. The positive charge on C_2 soon diminishes. The electromagnetic field around L_1 collapses. The **discharge current** continues for a short time. The bottom plate of C_2 now becomes negatively charged, and the top plate of C_1 becomes positive. The negative charge voltage of C_2 reduces the forward-bias voltage of Q_1. I_C decreases in value. V_C begins to increase in value. This change in voltage is again fed back to the top plate of C_1 through C_3. C_1 becomes more positively charged, and the bottom of C_2 becomes more negatively charged. The process continues until Q_1 is driven to cutoff.

When Q_1 reaches its **cutoff point**, no I_C flows. There is no feedback voltage supplied to C_1. The composite charge across C_1 and C_2 discharges through L_1. Discharge current flows from the bottom of C_2 to the top of C_1. The negative charge on C_2 soon diminishes. The electromagnetic field around L_1 collapses. Current flow continues. The bottom plate of C_2 now becomes positive and the top plate of C_1 becomes negative. The positive charge voltage of C_2 pulls Q_1 out of the cutoff region. I_C begins to flow again. The process then repeats from this point. Feedback energy is added to the tank circuit momentarily during each alternation. As a general rule, the Colpitts oscillator is very reliable.

The amount of **feedback** developed by the Colpitts oscillator is based on the **capacitance ratio** of C_1 and C_2. The value of C_1 in the circuit of **Figure 20.10** is much smaller than C_2. The capacitive reactance (X_C) of C_1 is, therefore, much greater than that for C_2. The voltage across C_1 is much greater than that across C_2. By making the value of C_2 smaller, the feedback voltage can be increased since a smaller C value results in increased X_C as indicated by the formula $X_{C2} = \frac{1}{2\pi f C_2}$. The feedback voltage $I \times X_{C2}$ and by making C_2 smaller, this voltage would, thus, become larger. As a general rule, large amounts of feedback may cause the generated wave to be distorted. Small values of feedback do not permit the circuit to oscillate. In practice, 10%-50% of the collector voltage is returned to the tank circuit as feedback energy.

The **attenuation factor**, B, of the circuit is the ratio of the capacitance used.

$$B = \frac{C_1}{C_2}. \tag{20.7}$$

In order to ensure that the oscillator will start automatically when power is initially applied to the circuit, the A_v must be slightly greater than 1/B in order to overcome circuit losses. The **voltage amplification** (A_v) is, thus, expressed as follows:

$$A_v > \frac{1}{B}$$

$$A_v > \frac{C_2}{C_1}.$$

Example 20-4:

Refer to the Colpitts oscillator circuit shown in **Figure 20.10**, which has $L_1 =$ 5 mH, $C_1 = 0.02$ μF, and $C_2 = 0.22$ μF. Determine the combined capacitance and the frequency of oscillation.

Solution

$$C_T = \frac{C_1 \times C_2}{C_1 + C_2} = \frac{0.02 \times 10^{-6} \times 0.22 \times 10^{-6}}{0.02 \times 10^{-6} + 0.22 \times 10^{-6}}$$
$$= 0.018 \times 10^{-6} = 0.018 \, \mu\text{F}$$

$$f_r = \frac{1}{2\pi\sqrt{L_1 C_T}} = \frac{1}{6.28\sqrt{5 \times 10^{-3} \times 0.018 \times 10^{-6}}}$$
$$= 16784.91 \text{ Hz} = 16.785 \text{ kHz}.$$

Related Problem

Refer to the Colpitts oscillator circuit shown in **Figure 20.10**, which has $L_1 =$ 15 mH, $C_1 = 0.07$ μF, and $C_2 = 0.33$ μF. Determine the combined capacitance and the frequency of oscillation.

Clapp Oscillator

A **Clapp oscillator** is simply a Colpitts oscillator circuit with a capacitor added in series with the feedback inductor of the *LC* circuit. The purpose of this capacitor is to reduce the effects of junction capacitance of the transistor on the operation of the oscillator. Junction capacitance can negatively affect the operating frequency of an oscillator circuit. In a Clapp oscillator, the value of the capacitance is determined by the series combination of three capacitors C_1, C_2, and C_2 (added as an output capacitor) around the tank circuit. This total capacitance is called C_T and is expressed as follows:

$$C_T = \frac{1}{\frac{1}{C_1} + \frac{1}{C_2} + \frac{1}{C_3}} = \frac{C_1 C_2 + C_2 C_3 + C_1 C_3}{C_1 C_2 C_3}. \qquad (20.8)$$

The **frequency** of a Clapp oscillator is given by

$$f_r = \frac{1}{2\pi\sqrt{L_1 C_T}}.$$

When the value of C_3 (output capacitor) is very small as compared to the values of C_1 and C_2, the total capacitance C_T becomes approximately equal to C_3 as shown in the following:

$$C_T = \frac{1}{\frac{1}{C_1} + \frac{1}{C_2} + \frac{1}{C_3}} \approx \frac{1}{0 + 0 + \frac{1}{C_3}} \approx \frac{1}{\frac{1}{C_3}} \approx C_3.$$

Under these conditions, the frequency of a Clapp oscillator is given by

$$f_r = \frac{1}{2\pi\sqrt{L_1 C_3}}. \tag{20.9}$$

Two of the capacitors in the tank circuits C_1 and C_2 are connected to the ground at one end. This causes the transistor junction capacitance and other stray capacitances to be in parallel with the C_1 and C_2 to ground, thereby altering their effective values. Since C_3 is not affected by the junction capacitance, its value serves as an accurate and stable factor in determining the frequency of oscillation.

Example 20-5:

Refer to the Clapp oscillator circuit shown in **Figure 20.11**, which has $L_1 = 5$ mH, $C_1 = 0.02\ \mu F$, $C_2 = 0.22\ \mu F$, and $C_3 = 200$ pF. Determine the combined capacitance and the frequency of oscillation.

Solution

$$C_T = C_T = \frac{1}{\frac{1}{C_1} + \frac{1}{C_2} + \frac{1}{C_3}} = \frac{1}{\frac{1}{0.02 \times 10^{-6}} + \frac{1}{0.22 \times 10^{-6}} + \frac{1}{200 \times 10^{-12}}}$$

$$= \frac{1}{\frac{1}{200 \times 10^{-12}}} = 200\,\text{pF}.$$

$$f_r = \frac{1}{2\pi\sqrt{L_1 C_T}} = \frac{1}{6.28\sqrt{5 \times 10^{-3} \times 200 \times 10^{-12}}}$$

$$= 159,235.67\,\text{Hz} = 159.2\,\text{kHz}.$$

Related Problem

Refer to the Clapp oscillator circuit shown in **Figure 20.11**, which has $L_1 = 5$ mH, $C_1 = 0.07\ \mu F$, $C_2 = 0.33\ \mu F$, and $C_3 = 150$ pF. Determine the combined capacitance and the frequency of oscillation.

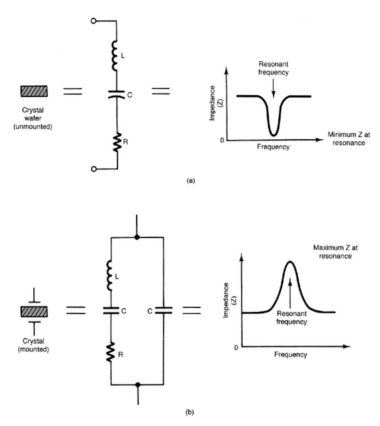

Figure 20.11 CityplaceCrystal equivalent circuits. (a) Series resonant. (b) Parallel resonant.

Crystal Oscillator

When extremely high-frequency stability is desired, **crystal oscillators** are used. The crystal of an oscillator is a finely ground wafer of quartz or Rochelle salt. It has the property to change electrical energy into vibrations of mechanical energy. It can also change mechanical vibrations into electrical energy. These two conditions of operation are called the **piezoelectric effect**.

The crystal of an oscillator is placed between two metal plates. Contact is made to each surface of the crystal by these plates. The entire assembly is then mounted in a holding case. Connection to each plate is made through terminal posts. When a crystal is placed in a circuit, it is plugged into a socket.

A crystal by itself behaves like a series-resonant circuit (see **Figure 20.11**). It has inductance (L), capacitance (C), and resistance (R). An equivalent circuit is shown in **Figure 20.11(a)**. **Inductance** is determined by the mass of the crystal. **Capacitance** deals with its ability to change mechanically. **Resistance** corresponds to the electrical equivalent of mechanical friction. The series-resonant equivalent circuit changes a great deal when the crystal is placed in a holder.

Capacitance due to the metal support plates is added in parallel with the crystal equivalent circuit. **Figure 20.11(b)** shows the equivalent circuit of a crystal placed in a holder. In a sense, a crystal can have either a series-resonant or a parallel-resonant characteristic

In an **oscillator**, the placement of a crystal in the circuit has a great deal to do with how it responds. If placed in the tank circuit, it responds as a parallel-resonant device. In some applications, the crystal serves as a tank circuit. When a crystal is placed in the feedback path, it responds as a series-resonant device and acts as a sharp filter. A **sharp filter** permits feedback only of the desired frequency. Hartley and Colpitts oscillators can be modified to accommodate a crystal of this type. The operating stability of this type of oscillator is much better than the circuit without a crystal. **Figure 20.12** shows crystal-controlled versions of the Hartley and Colpitts oscillators. Operation is essentially the same.

Pierce Oscillator

The **Pierce oscillator** of **Figure 20.13** employs a crystal as its tank circuit. In this circuit, the crystal responds as a parallel-resonant circuit. In a sense, the Pierce oscillator is a modification of the basic Colpitts oscillator. The crystal is used in place of the tank circuit inductor. A specific crystal is selected for the desired frequency to be generated. The parallel-resonant frequency of the crystal is slightly higher than its equivalent series-resonant frequency.

Operation of the **Pierce oscillator** is based on feedback from the collector to the base through C_1 and C_2. These two capacitors provide a combined 180° phase shift. The output of the common emitter amplifier is, therefore, inverted to achieve in-phase or regenerative feedback. The value ratio of C_1 and C_2 determines the level of feedback voltage. Ten percent to fifty percent of the output must be fed back to energize the crystal. When it is properly energized, the resonant frequency response of the crystal is extremely sharp. It vibrates only over a narrow range of frequency. The output at this frequency is very stable. The output of a Pierce oscillator is usually quite small. A crystal can be damaged by excessive mechanical strains and the heat caused by excessive power.

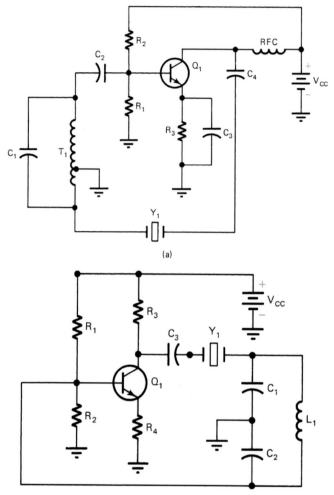

Figure 20.12   Crystal-controlled oscillators. (a) Hartley. (b) Colpitts.

Wien-Bridge Oscillator

The Wien-Bridge oscillator is used for generating low frequencies. Typically, this oscillator uses an RC bridge network to accomplish regenerative feedback when connected to an op-amp. The bridge circuit provides two feedback paths. One path provides positive feedback and is connected to the noninverting input of the op-amp. The second path provides negative feedback and is connected to the inverting input of the op-amp. **Positive feedback** is used

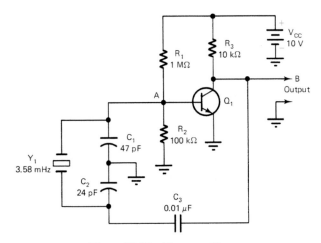

Figure 20.13 Pierce oscillator.

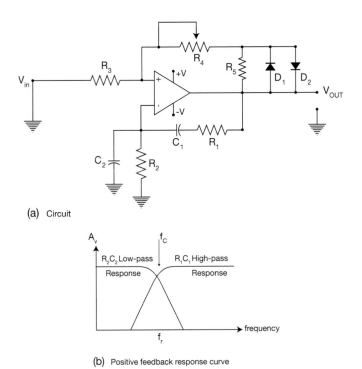

(a) Circuit

(b) Positive feedback response curve

Figure 20.14 (a) Wien-Bridge oscillator. (b) Positive frequency response curve.

for controlling the frequency of oscillation, whereas the negative feedback is used for controlling the closed-loop gain of the system.

Figure 20.14(a) shows a circuit diagram of a **Wien-Bridge oscillator**. The positive-feedback path of this circuit contains two RC networks. R_1 and C_1 are connected in series forming a high-pass filter, which has a **cutoff frequency** (f_c) of $\frac{1}{2\pi R_1 C_1}$. R_2 and C_2 are connected in parallel, forming a low-pass filter, which has a cutoff frequency (f_c) of $\frac{1}{2\pi R_2 C_2}$. These two filter circuits are then connected in series, thereby forming a band-pass filter. The resonant frequency of the band-pass filter determines the oscillating frequency of the circuit. Open simulation file Sim_20.15 which refers to **Figure 20.15**. Turn on the simulation and evaluate its operation.

Typically, $R_1 = R_2 = R$ and $C_1 = C_2 = C$, so that $R_1 \times C_1 = R_2 \times C_2 = R \times C$, so that both the low-pass and the high-pass filter sections have the same cutoff or critical frequency (f_c) of $\frac{1}{2\pi RC}$. This is shown in **Figure 20.15(b)**. The critical frequencies of the individual filters are combined to determine the resonant frequency (f_r) of the **band-pass filter** as follows:

$$f_r = \frac{1}{2\pi RC}. \tag{20.10}$$

In the operation of the Wien-Bridge circuit, **regenerative feedback** is necessary to produce oscillation. Regeneration occurs when the feedback input and output signals are in phase. When a band-pass filter operates at the resonant frequency, it does not introduce any phase shift. As a result of this, the feedback input signal appearing at noninverting input op-amp terminal is in phase with the output signal, so that the feedback is regenerative.

The **negative-feedback** portion of the circuit is used to determine the closed-loop voltage gain (A_{CL}) of the oscillator. This part of the circuit is composed of resistors R_3, R_4, and R_5, along with diodes D_1 and D_2. The **gain** of the noninverting amplifier is given as follows:

$$A_{CL} = 1 + \frac{R_f}{R_{in}}.$$

With $R_{in} = R_3$, and $R_f = R_4 + R_5$,

$$A_{CL} = 1 + \frac{R_4 + R_5}{R_3}. \tag{20.11}$$

The **potentiometer** R_4 is used to adjust the closed-loop voltage gain of the circuit. The diodes D_1 and D_2 are used to prevent the output voltage from increasing beyond the value determined by the resistors R_3 and R_4. If the output voltage increases beyond this value in the positive direction by 0.7 V, resistor R_5 is shorted out by the diode D_1 becoming forward biased. This reduces the A_{CL} to

$$A_{\text{CL}} = 1 + \frac{R_4}{R_3}.$$

The diode D_2 is used if the output voltage is exceeded in the negative direction. The combined effect of diodes D_1 and D_2 is, therefore, used to keep the amplitude from growing beyond 0.7 V of their predetermined values.

Example 20-6:

Refer to the Wien-Bridge oscillator shown in **Figure 20.15(a)**. Calculate the resonant frequency of the oscillator assuming $R_1 = R_2 = R = 25$ kΩ, and $C_1 = C_2 = C = 0.05$ μF.

Solution

$$f_{\text{r}} = \frac{1}{2\pi RC} = \frac{1}{6.28 \times 25 \times 10^3 \times 0.05 \times 10^{-6}} = 127.38 \text{ Hz}.$$

Related Problem

Refer to the Wien-Bridge oscillator shown in **Figure 20.15(a)**. Calculate the resonant frequency of the oscillator assuming $R_1 = R_2 = R = 50$ kΩ, and $C_1 = C_2 = C = 0.005$ μF.

Self-Examination

Answer the following questions.

11. The frequency of an Armstrong oscillator is based on the values of _____ and _____.
12. A distinguishing feature of a Hartley oscillator is its _____.
13. A Colpitts oscillator uses two _____ instead of a tapped coil.
14. A(n) _____ oscillator uses a crystal as its tank circuit.
15. A(n) _____ oscillator uses a RC bridge network for generating low frequencies.

20.3 Relaxation Oscillators

Relaxation oscillators are primarily responsible for the generation of nonsinusodial waveforms. Sawtooth, rectangular, and a variety of irregular-shaped waves are included in this classification. These oscillators generally depend on the charge and discharge of a capacitor-resistor network for their operation. Voltage changes from the network are used to alter the conduction of an electronic device. Transistors, unijunction transistors (UJTs), and ICs can be used to perform the control function of this oscillator.

20.3 Analyze common relaxation oscillators.

In order to achieve objective 20.3, you should be able to:

- describe the operation of astable, monostable, and bistable multivibrator circuits;
- describe the characteristics of common relaxation oscillators;
- define pulse repetition rate (PRR), multivibrator, triggered multivibrator, free-running multivibrator, astable multivibrator, symmetrical multivibrator, nonsymmetrical multivibrator, monostable multivibrator, flip-flop, threshold terminal, RC oscillator, blocking oscillator, and vertical blocking oscillator (VBO).

UJT Oscillators

The charge and discharge of a capacitor through a resistor can be used to generate a sawtooth waveform. The charge-discharge switch of **Figures 20.6** and **20.7** can be replaced with an active device. Transistors or ICs can be used to accomplish the switching action. Conduction and nonconduction of the active device regulates the charge and discharge of the RC network. Circuits connected in this manner are classified as relaxation oscillators. When the circuit device is conductive, it is active. When it is not conductive, it is relaxed. The active device of a relaxation oscillator switches states between conduction and nonconduction. This regulates the charge and discharge rates of the capacitor. A **sawtooth waveform** appears across the capacitor of this type of oscillator.

A **UJT** is used in the relaxation oscillator of **Figure 20.15**. The **RC network** is composed of R_1 and C_1. The junction of the network is connected to the emitter (E) of the UJT. The UJT will not go into conduction until a certain voltage value is reached. When conduction occurs, the emitter-base 1 (E-B$_1$) junction becomes low resistant. This provides a low-resistant discharge path for C_1. Current flows through R_3 only when the UJT is conducting. R_3 refers to the resistance of the speaker in this circuit.

Assume that power is applied to the UJT oscillator circuit. The UJT is in a nonconductive state initially. The E-B$_1$ junction is reverse biased. C_1 begins

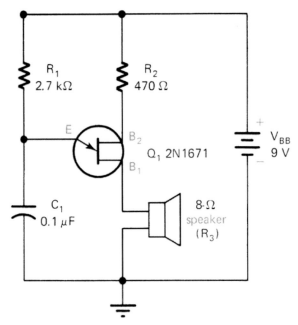

Figure 20.15 UJT oscillator.

to accumulate charge voltage after a short period of time since time constant equals R × C_1. The **intrinsic standoff ratio** (η) along with the voltage across the voltage that appears across the two base terminals of the UJT (designated as V_{BB}) determines the **peak voltage** (V_P) to which the capacitor should charge. This is expressed as follows:

$$V_P = \eta V_{BB} + 0.7. \tag{20.12}$$

Typical values of η are in the range 0.4–0.6.

Capacitor C_1 eventually charges to a voltage value that causes the E-B_1 junction to become conductive. When this point is reached, the E-B_1 junction becomes low resistant. C_1 discharges immediately through the low-resistant E-B_1 junction. This action removes the forward-bias voltage from the emitter. The UJT immediately becomes nonconductive. C_1 begins to charge again through R_1. The process is then repeated continuously as shown in **Figure 20.15**.

UJT oscillators are found in applications that require a signal with a slow rise time and a rapid fall time. The E-B_1 junction of the UJT has this type of output. Between B_1 and circuit ground, the UJT produces a spiked pulse. This type of output is frequently used in timing circuits and counting applications. The time that it takes for the waveform to repeat itself is called the **pulse repetition rate (PRR)**. This term is very similar to the hertz designation for frequency. As a general rule, UJT oscillators are very stable and are quite accurate when used with time constants of one or less.

Example 20-7:

Refer to the UJT oscillator shown in **Figure 20.15**. The capacitor $C_1 = 0.1$ μF is charged through the resistor $R_1 = 2.7$ kΩ when the UJT is off. When the UJT is triggered into conduction, capacitor C_1 discharges through the speaker that has a resistance (R_3) of 8 Ω. Calculate the charging and discharging times. Assume that $\eta = 0.56$.

Solution

The capacitor C_1 charges through the resistance R_1 until the voltage peak (V_P) is reached. When the UJT is off, the entire source voltage appears across the two base terminals, or $V_{BB} = 10$ V.

$$V_P = \eta V_{BB} + 0.7 = 0.56 \times 10 + 0.7 = 6.3 \text{V}.$$

So when the voltage at the emitter of the UJT equals or exceeds 6.3 V, it will be triggered into conduction.

The time constant of the capacitor charging circuit

$$(t_C) = R_1 \times C_1 = 2.7 \times 10^3 \times 0.1 \times 10^{-6} = 0.27 \times 10^{-3} = 0.27 \text{ms}.$$

In one time constant, the capacitor C_1 will charge to 63.3% of the applied voltage of 10 V, which is 6.3 V. Thus, in one time constant, the V_P will be reached. The capacitor will not charge beyond the value of V_P since the UJT has already been triggered into conduction.

The time constant of the capacitor discharging circuit

$$(t_D) = R_3 \times C_1 = 8 \times 0.1 \times 10^{-6} = 0.8 \ \mu\text{s}.$$

The capacitor will discharge completely in

$$T_D = 5 \times t_D = 5 \times 0.8 \ \mu\text{s} = 0.4 \ \mu\text{s}.$$

The capacitor will discharge until the valley voltage of the UJT is reached. This value is typically very small.

Related Problem

Refer to the UJT oscillator shown in **Figure 20.15**. After the UJT has been triggered into conduction, capacitor C_1 discharges through the 10-Ω speaker resistance. Calculate the time needed to discharge the capacitor fully, assuming that it has already been charged to $V_P = 6.3$ V.

Multivibrators

A **multivibrator** is a very important classification of relaxation oscillators. This type of circuit employs an RC network in its physical makeup. A rectangular-shaped wave is developed by the output. Multivibrators are considered to be either triggered devices or free running. A **triggered multivibrator** requires an input signal or timing pulse to be made operational. The output of this multivibrator is controlled, or synchronized, by an input signal. The electron beam deflection oscillators of a television receiver are triggered into operation. When the receiver is tuned to an operating channel, its oscillators are synchronized by the incoming signal. When it is tuned to an unused channel, the oscillators are free running. A **free-running multivibrator** is self-starting. It operates continuously as long as electrical power is supplied. The shape and frequency of the waveform is determined by component selection.

A multivibrator is composed of two amplifiers that are cross-coupled. The output of amplifier 1 goes to the input of amplifier 2. The output of amplifier 2 is coupled back to the input of amplifier 1. Since each amplifier inverts the polarity of the input signal, the combined effect is a positive-feedback signal. With positive feedback, an oscillator is regenerative and produces continuous output.

Astable Multivibrator

The **astable multivibrator** is a free-running multivibrator. Astable multivibrators could be used in television receivers to control the electron beam deflection of the picture tube. Computers use this type of circuit to develop timing pulses for operating.

Figure 20.16(a) shows an **astable multivibrator** circuit constructed using bipolar transistors. These amplifiers are connected in a common-emitter circuit configuration. R_2 and R_3 provide forward-bias voltage for the base of each transistor. Capacitor C_1 couples the collector of transistor Q_1 to the base of Q_2. Capacitor C_2 couples the collector of transistor Q_2 to the base of Q_1.

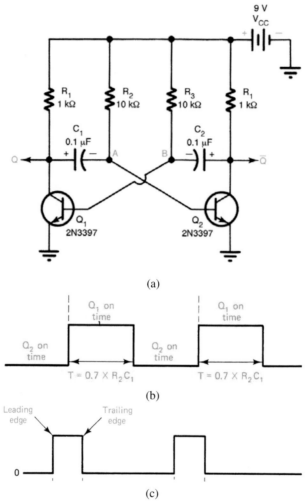

Figure 20.16   (a) Astable multivibrator circuit. (b) Symmetrical waveform. (c) Nonsymmetrical waveform.

Because of the cross-coupling, one transistor will be conductive, and one will be cut off. After a short period of time, the two transistors change states. The conducting transistor is cut off, and the off-transistor becomes conductive. The circuit oscillates between these two states. The output of the circuit is a **rectangular wave**. An output signal can be obtained from the collector of either transistor. As a rule, the output is labeled Q or $\overline{Q}$. This denotes that outputs are of an opposite polarity.

When power is applied to the multivibrator of **Figure 20.16(a)**, one transistor goes into conduction first. A slight difference in component tolerances usually accounts for this condition. For explanation purposes, assume that Q_1 conducts first. When Q_1 becomes **conductive**, there is a voltage drop across R_1. V_C becomes less than V_{CC}. This causes a negative-going voltage to be applied to C_1. The positive base voltage of Q_1 is reduced by this voltage. The conduction of Q_2 decreases. The collector voltage of Q_2 begins to rise to the value of V_{CC}. A **positive-going voltage** is applied to C_2. This voltage is added to the base voltage of Q_1. Q_1 becomes more conductive. The process continues until Q_1 becomes **saturated** and Q_2 is cut off.

When the output voltages of each transistor becomes stabilized, there is no feedback voltage. Q_2 is again forward biased by R_2. Conduction of Q_2 causes a drop in V_C. This negative-going voltage is coupled to the base of Q_1 through C_2. Q_1 becomes less conductive. The V_C of Q_1 begins to rise toward the value of V_{CC}. This is coupled to the base of Q_2 by C_1. The process continues until Q_2 is saturated and Q_1 is cut off. The output voltages then become stabilized. The process is then repeated.

The **oscillation frequency** of a multivibrator is determined by the time constants of R_2 and C_1 and R_3 and C_2. The values of R_2 and R_3 are usually selected to cause each transistor to reach saturation. C_1 and C_2 are then chosen to develop the desired operating frequency. If C_1 equals C_2 and R_2 equals R_3, the output is symmetrical (same on-time and off-time intervals). This means that each transistor is on and off for an equal amount of time as shown by the output waveform in **Figure 20.16(b)**. The output frequency of a **symmetrical multivibrator** is determined by the following formula:

$$f = \frac{1}{1.4 \times R \times C}. \tag{20.13}$$

If the resistor and capacitor values are unequal, the output is not symmetrical. One transistor could be on for a long period with the alternate transistor being on for only a short period. The output of a **nonsymmetrical multivibrator** (different on-time and off-time intervals) is described as a rectangular wave. A nonsymmetrical rectangular wave is shown in **Figure 20.16(c)**.

Monostable Multivibrator

A **monostable multivibrator** has one stable state of operation. It is often called a **one-shot multivibrator**. One trigger pulse causes the oscillator to

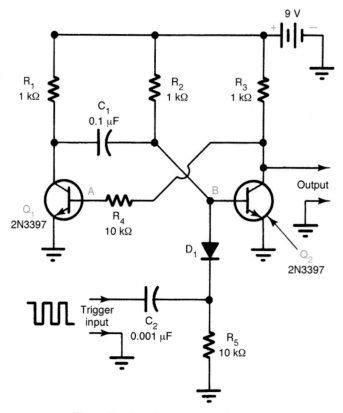

Figure 20.17   Monostable multivibrator.

change its operational state. After a short period of time, however, the oscillator returns to its original starting state. The RC time constant of this circuit determines the time period of the state change. A **monostable multivibrator** always returns to its original state. No operational change will occur until a trigger pulse is applied. A monostable multivibrator is considered to be a triggered oscillator.

Figure 20.17 shows a schematic of a **monostable multivibrator**. This circuit has two operational states. Its **stable state** is based on conduction of Q_2 with Q_1 cut off. The circuit relaxes in its stable state when no trigger pulse is supplied. The **unstable state** is initiated by a trigger pulse. When a **trigger pulse** arrives at the input, the circuit changes from its stable state to the unstable state. After the time of $0.7 \times R_2C_1$, the circuit returns to its stable state. No circuit change occurs until another trigger pulse is applied to the input.

$$\text{Pulse} - \text{width (PW)} = 0.7 \times R_2 C_1. \tag{20.14}$$

Consider the operation of a monostable multivibrator when power is first applied. No trigger pulse is applied initially. Q_2 is forward biased by a divider network consisting of R_2, D_1, and R_5. The value of R_2 is selected to cause Q_2 to reach saturation. Resistors R_1 and R_3 reverse bias each collector. With the base of Q_2 forward biased, it is driven into **saturation** immediately. The collector voltage of Q_2 drops to a very small value. This voltage coupled through R_4 is applied to the base of Q_1. V_B is not great enough to cause conduction of Q_1. The circuit, therefore, remains in this **conduction state** as long as power is applied. This represents the stable state of the circuit.

To initiate a state change in a **monostable multivibrator**, a **triggered pulse** must be applied to its input. **Figure 20.18** shows a representative trigger pulse, a wave-shaped pulse, and the resulting output of the multivibrator. C_2 and R_5 of the input circuit form a differentiator network, producing an output that corresponds to the rate of change of the input signal. The leading edge of the applied trigger pulse causes a large current flow through R_5.

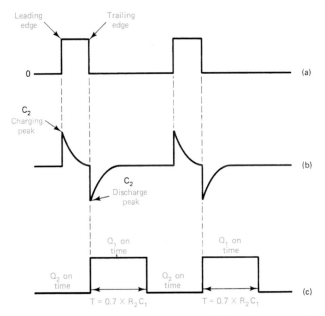

Figure 20.18 Monostable multivibrator waveforms. (a) Trigger input waveform. (b) Differentiator output waveform. (c) Monostable multivibrator output waveform.

After C_2 begins to charge, the current through R_5 begins to drop. When the trailing edge of the pulse arrives, the voltage applied to C_2 drops to zero. With no source voltage applied to C_2, the capacitor discharges through R_5. An opposite-polarity pulse, therefore, occurs at the trailing edge of the input pulse. Thus, the input pulse is changed into a positive and a negative spike that appears across R_5. D_1 conducts only during the time of the negative spike. This feeds a negative spike to the base of Q_2. With a square trigger pulse applied to the input, a single negative spike pulse is applied to the base of Q_2. This initiates a **state change** in the multivibrator.

When the base of Q_2 receives a negative spike, it is driven into **cutoff**. This causes the collector voltage of Q_2 to rise very quickly to the value of $+V_{CC}$. This, in turn, causes the base of Q_1 to become positive. Q_1, therefore, becomes conductive and Q_2 is driven into cutoff. When Q_1 conducts, the emitter-collector junction becomes very low resistant. Charging current flows through Q_1, C_1, and R_2. The bottom of R_2 immediately becomes negative due to the charging current of C_1. This drives the base of Q_2 negative. Q_2 remains in its cutoff state. The process continues until C_1 becomes charged. The charging current through R_2 then begins to slow down, and the top of R_2 eventually becomes positive. Q_2 immediately goes into conduction. This, in turn, drives Q_1 to cutoff. The circuit has, therefore, returned to its **stable state**. It will remain in this state until the next trigger pulse arrives at the input.

Bistable Multivibrator

A **bistable multivibrator** has two stable states of operation. A trigger pulse applied to the input causes the circuit to assume one stable state. A second pulse causes it to switch to the alternate, stable state. This type of multivibrator changes states only when a trigger pulse is applied. It is often called a **flip-flop**. It flips to one state when triggered and flops back to the other state when triggered. The circuit becomes stable in either state. It will not change states, or toggle, until commanded to do so by a trigger pulse. **Figure ??(a)** shows a schematic of a bistable multivibrator using bipolar transistors.

When electrical power is first applied to a multivibrator, it assumes one of its **stable states**. One transistor goes into conduction faster than the other. In the circuit of **Figure 20.19**, assume that Q_1 goes into conduction faster than Q_2. The collector voltage of Q_1, therefore, begins to drop quickly. **Direct coupling** between the collector and base causes a corresponding drop in the voltage of Q_2. A reduction in Q_2 voltage causes a decrease in I_B and I_C. The V_C of Q_2 rises to the value of $+V_{CC}$. This positive-going voltage is coupled

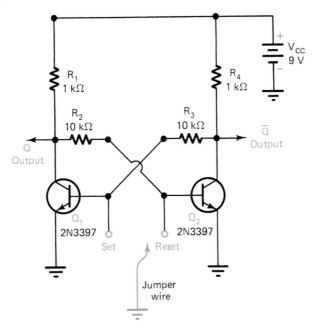

Figure 20.19 (a) Bistable multivibrator circuit. (b) Input/output relationships.

back to the base of Q_1 by R_3. This increases the conduction of Q_1, which, in turn, decreases conduction of Q_2. The process continues until Q_1 is saturated and Q_2 is **cut off**. The circuit remains in the **stable state**.

To initiate a **state change**, a trigger pulse must be applied. A negative pulse applied to the base of Q_1 causes it to go into the cutoff region. A positive pulse applied to the base of Q_2 causes it to go into conduction. This polarity applies to NPN transistors. The pulse polarity is reversed for PNP transistors.

For the circuit in **Figure 20.19**, assume that a **negative pulse** is applied to the base of Q_1 through the set input. When this occurs, the I_B and I_C of Q_1 are reduced immediately. The V_C rises toward the value of $+V_{CC}$. This positive-going voltage is coupled back to the base of Q_2. The I_B and I_C of Q_1 rise quickly. This causes a corresponding drop in the V_C of Q_2. Direct coupling of V_C through R_3 causes a decrease in the I_B and I_C of Q_1. The process continues until Q_1 is cut off and Q_2 reaches saturation. This represents the second stabilized state of operation. The circuit will remain in this state until commanded to change by a signal applied to the **Reset input** or the power is removed. The Set input is momentary pulsed low while maintaining the Reset input high. This causes the output to become high. The output remains high until it is reset. If the Reset input is momentary pulsed low while maintaining the **Set input** high, this causes the output to become low. The output now remains low until the next Set **trigger pulse** is applied.

Example 20-8:

Refer to the bistable multivibrator of **Figure 20.19(b)**. If the Set input is momentarily pulsed low at $t_1 = 10$ ms and the Reset input at $t_2 = 20$ ms, calculate the period for which the output waveform will be on (high).

Solution

The output switches on (high) when the Set input is pulsed low at $t_1 = 10$ ms.

The output switches off (low) when the Reset input is pulsed low at $t_2 = 20$ ms.

The duration of the t_2-t_1 output waveform is

$$= t_2 - t_1 = 20 - 10 \text{ ms} = 10 \text{ ms}.$$

Hence, the output remains high for 10 ms.

Related Problem

Refer to the bistable multivibrator of **Figure 20.19(b)**. Assume that the multivibrator is initially on. If the Reset input is momentarily pulsed low at $t_1 = 10$ ms and the Set input at $t_2 = 15$ ms, calculate the period for which the output waveform will be off (low).

IC Waveform Generators

The **555 IC** is a multifunction device that is widely used for applications such as **voltage-controlled oscillators**. It can be modified to respond as an astable multivibrator. This circuit can be achieved with a minimum of components and a power source. Circuit design is easy to accomplish and the operation is very reliable. This chip is available through a number of manufacturers. As a rule, the number 555 usually appears in the manufacturer's part identification number. SN72555, MC14555, SE555, LM555, XR555, and CA555 are some of the common part numbers for this chip. The internal circuitry of a 555 IC is generally viewed in functional blocks. In this regard, the chip has two comparators, a bistable flip-flop, a resistive divider, a discharge transistor, and an output stage. **Figure 20.20(a)** shows the functional blocks of a 555 IC.

The **voltage divider** of the IC consists of three 5-kΩ resistors. The network is connected internally across the $+V_{CC}$ and ground supply source. Voltage developed by the lower resistor is one-third of V_{CC}. The middle divider point is two-thirds of the value of V_{CC}. This connection is terminated at pin 5. Pin 5 is designated as the control voltage. The two comparators of the 555 IC respond as an amplifying switch circuit. A **reference voltage** is applied to one input of each comparator. A voltage value applied to the other input initiates a change in output when it differs with the reference value.

Comparator 1 is referenced at two-thirds of V_{CC} at its negative input (referred to as the upper trip point). This is where pin 5 is connected to the middle divider resistor. The other input is terminated at pin 6. This pin is called the **threshold terminal**. When the voltage at pin 6 rises above two-thirds of V_{CC}, the output of the comparator swings positive. This is then applied to the Reset input of the flip-flop. Upper trip point (UTP) is determined by the following formula:

$$\text{UTP} = \frac{2}{3}V_{CC}. \tag{20.15}$$

Comparator 2 is referenced to one-third of V_{CC} and referred to as the lower trip point. The positive input of comparator 2 is connected to the lower divider network resistor. External pin connection 2 is applied to the negative input of comparator 2. This is called the **trigger input**. If the voltage of the trigger drops below one-third of V_{CC}, the comparator output swings positive. This is applied to the set input of the flip-flop. **Lower trip point (LTP)** is determined by the following formula:

$$\text{LTP} = \frac{1}{3}V_{CC}. \tag{20.16}$$

555 Flip-Flop

The **flip-flop** of the 555 IC is a bistable multivibrator. It has **Reset** and **Set inputs** and one **output**. When the Reset input is positive, the output goes positive. A positive voltage to the Set input causes the output to go negative. The output of the flip-flop is dependent on the status of the two comparator inputs as shown in **Figure 20.20(b)**.

The **output** of the flip-flop is applied to both the output stage and the discharge transistor. The output stage is terminated at pin 3. The **discharge transistor** is connected to terminal 7. The output stage is a power amplifier and a signal inverter. A load device connected to terminal 3 will see either

$+V_{CC}$ or ground, depending on the state of the input signal. The output terminal switches between these two values. The output terminal can control load current values of up to 200 mA. A **load device** connected to $+V_{CC}$ is energized when pin 3 goes to ground. When the output goes to $+V_{CC}$, the output is off. A load device connected to ground turns on when the output goes to $+V_{CC}$. It is off when the output goes to ground. The output continuously switches between these two states, as shown in **Figure 20.20**.

Transistor Q_1 is called a **discharge transistor**. The output of the flip-flop is applied to the base of Q_1. When the flip-flop is **reset** (positive), it forward biases Q_1. Pin 7 connects to ground through Q_1 This causes pin 7 to be grounded. When the flip-flop is set (negative), it reverse biases Q_1. This causes pin 7 to be infinite or open with respect to ground. Pin 7, therefore, has two states: shorted to ground and open.

555 Astable Multivibrator

When used as an **astable multivibrator**, the 555 is an **RC oscillator**. The shape of the waveform and its frequency are determined primarily by an RC network. The astable multivibrator circuit is self-starting and operates continuously for long periods of time. **Figure 20.20** shows the LM555 connected as an astable multivibrator. A common application of this circuit is the time base generator for clock circuits and computers.

Connection of the **555 IC** as an astable multivibrator requires two resistors: a capacitor and a power source. The output of the circuit is connected to pin 3. Pin 8 is $+V_{CC}$ and pin 1 is ground. The supply voltage can be from 5-V DC to 15-V DC. Resistor R_A is connected between $+V_{CC}$ and the discharge terminal (pin 7). Resistor R_B is connected between pin 7 and the threshold terminal (pin 6). A capacitor is connected between the threshold and ground. The trigger (pin 2) and threshold (pin 6) are connected together.

When power is first applied, the capacitor charges through R_A and R_B. When the voltage at pin 6 (threshold) rises slightly above two-thirds of V_{CC}, it changes the state of comparator 1. This resets the flip-flop and causes its output to go positive. The output (pin 3) goes to ground and the base of Q_1 is forward biased. Q_1 discharges C through R_B to ground.

When the charge voltage of C drops slightly below one-third of V_{CC}, it energizes comparator 2. The trigger (pin 2) and the **threshold** (pin 6) are still connected together. Comparator 2 causes a positive voltage to go to the set input of the flip-flop. This sets the flip-flop, which causes its output to go negative. The output (pin 3) swings to $+V_{CC}$. The base of Q_1 is reverse

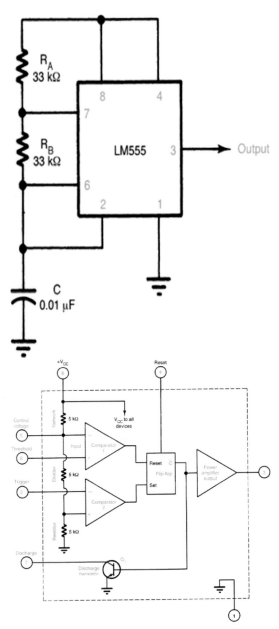

Figure 20.20 Astable multivibrator. (a) LM555 IC and external components. (b) Pin designations and internal block diagram of an astable multivibrator.

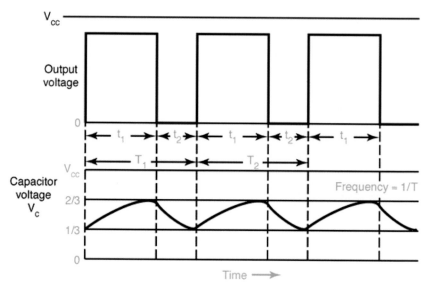

Figure 20.21   Astable multivibrator waveforms.

biased. This opens the **discharge** (pin 7). C begins to charge again to V_{CC} through R_A and R_B. The process is repeated from this point. The charge value of C varies between one-third and two-thirds of V_{CC}. See the resulting waveforms of **Figure 20.21**.

The **output frequency** of the astable multivibrator is represented as follows:

$$f_{out} = \frac{1}{T}. \tag{20.17}$$

This represents the total time needed to charge and discharge C. The **charge time** is represented by the duration t_1. In seconds, t_1 is 0.693 (R_A + R_B) C. The **discharge time** is indicated by the duration t_2. In seconds, t_2 is 0.693 $R_B C$. The combined time for one operational cycle is, therefore,

$$T = t_1 + t_2 = 0.693(R_A + R_B)C + 0.693R_B C = 0.693(R_A + 2R_B)C.$$

Expressed as frequency, this is

$$f = 1/T.$$

Combining t_1, t_2, t_3, and t_4 makes the frequency formula

$$f_{out} = \frac{1}{0.693(R_A + 2R_B)C} = \frac{1.44}{(R_A + 2R_B)C}.$$

The **resistance ratio** of R_A and R_B is quite critical in the operation of an astable multivibrator. If R_B is more than half the value of R_A, the circuit does not oscillate. Essentially, this prevents the trigger from dropping in value from two-thirds of V_{CC} to one-third of V_{CC}. This means that the IC is not capable of retriggering itself. It is, therefore, unprepared for the next operational cycle. Most IC manufacturers provide data charts that assist the user in selecting the correct R_A and R_B values with respect to C.

Some important definitions and equations for the 555 timer are listed below:

1) Cycle time (T): The time taken for one complete output cycle to occur.

$$T = 0.693(R_A + 2R_B)C. \tag{20.18}$$

2) Operating frequency (f_{out}): The frequency at which the 555 timer operates.

$$f_{out} = \frac{1}{0.693(R_A + 2R_B)C}. \tag{20.19}$$

3) Pulse-width (PW): The time duration of an active high output state.

$$PW = 0.693(R_A + R_B)C. \tag{20.20}$$

4) Duty cycle: The ratio of the pulse-width to the cycle time expressed as a percentage.

$$\frac{R_A + R_B}{R_A + 2R_B} \times 100. \tag{20.21}$$

Example 20-9:

Given the following values (see **Figure 20.20**) for a 555 timer — R_A = 1 kΩ, R_B = 2 kΩ, and C_1 = 0.02 μF, calculate 1) cycle time, 2) operating frequency, 3) pulse-width, and 4) duty cycle.

Solution

1) T = 0.693 (R_A + 2 R_B) C_1 = 0.693 (1000 + 2 $\times$ 2000) $\times$ 0.02 $\times$ 10^{-6} = 69.3 $\times$ 10^{-6} = 69.3 μs.
2) $f_{out} = \frac{1}{T} = \frac{1}{69.3 \times 10^{-6}}$ = 14, 430.0 = 14.43 kHz.
3) PW = 0.693 (R_A + R_B) C_1 = 0.693 (1000 + 2000) $\times$ 0.02 $\times$ 10^{-6} = 41.58 $\times$ 10^{-6} = 41.58 μs.
4) % duty cycle = $\frac{R_A + R_B}{R_A + 2R_B}$ $\times$ 100 = $\frac{1000 + 2000}{1000 + 2 \times 2000}$ $\times$ 100 = 60%.

Related Problem

Referring to the 555 circuit of **Figure 20.20** with $R_A = 10 \text{ k}\Omega$, $R_B = 25 \text{ k}\Omega$, and $C_1 = 0.05 \ \mu\text{F}$, calculate 1) cycle time, 2) operating frequency, 3) pulse-width, and 4) duty cycle.

Self-Examination

Answer the following questions.

16. _____ oscillators are used to generate nonsinusoidal wave-forms.
17. A unijunction transistor may be used as a(n) _____ oscillator.
18. Multivibrators are classified as either _____ or _____.
19. A(n) _____ multivibrator is a freerunning oscillator.
20. A monostable multivibrator is also called a(n) _____ multi-vibrator.
21. A bistable mutivibrator is also called a(n) _____.
22. An IC oscillator may be constructed using a(n) _____.
23. If a square wave has equal on- and off-time durations, its duty cycle is _____%.
24. When the on-time duration of a square wave is not equal to the off-time duration, the waveform is considered to be (symmetrical, unsymmetrical).
25. The time taken for one complete output cycle to occur is called the _____.

20.4 Analysis and Troubleshooting – Oscillator Circuits

In order to test **timer** and **oscillator** circuits, their operating characteristics should be understood. Familiarity with the **test equipment** used along with low- to high-frequency AC circuits, such as **function generators** and **oscillo-scopes**, is necessary for examining the operation of oscillator circuits. Some of the problems common to oscillator circuits include stray capacitances, junction capacitances, and shielding, which affect circuit operation. **Tuned oscillator** circuits are susceptible to stray capacitive effects when long cables are used for circuit interconnections, and this affects the operating frequency. Also, improper shielding wherein some of the magnetic field from one component may leak out and affect the operation of neighboring electrical components.

Oscillators are usually composed of a combination of passive and active components. Passive components such as resistors, inductors, and capacitors are externally connected to the active components such as transistors and op-amps. Passive components are subject to open-circuit, short-circuits, and leakage, which can be evaluated using test instruments. An active component failure generally requires replacement of the device due to the complexity of its internal construction.

20.4 Analyze and troubleshoot oscillator circuits.

In order to achieve objective 20.4, you should be able to:

- examine the datasheet of timer ICs to determine its operational characteristics;
- distinguish between normal and faulty operation of an oscillator circuit.

Data Sheet Analysis – Timer IC

Figure 20.22 shows the typical **data sheet** for an NE/SA/SE555/SE555C timer circuit. We will use this data sheet to analyze typical information available in integrated circuit timers.

Use this data sheet to answer the following questions:

- The number of pins in the device = _____.
- Maximum operating frequency = _____.
- Modes of operation = _____ and _____.
- Applications = _____.
- Maximum supply voltage for the NE555 = _____.
- Operating ambient temperature range = _____°C to _____°C.
- Maximum source or sink currents = _____mA.
- Maximum allowable power dissipation = _____mW.

Troubleshooting Oscillator Circuits

The active components used for amplifying the signal in an oscillator circuit generally consist of **op-amps**. By connecting external electrical components in the feedback path of the op-amp, the frequency of oscillation and the **gain** of the oscillator can be controlled. If the feedback element is shorted or open circuited during operation, the functioning of the overall circuit will be affected.

Philips Semiconductors

Product data

Timer

NE/SA/SE555/SE555C

DESCRIPTION

The 555 monolithic timing circuit is a highly stable controller capable of producing accurate time delays, or oscillation. In the time delay mode of operation, the time is precisely controlled by one external resistor and capacitor. For a stable operation as an oscillator, the free running frequency and the duty cycle are both accurately controlled with two external resistors and one capacitor. The circuit may be triggered and reset on falling waveforms, and the output structure can source or sink up to 200 mA.

FEATURES

* Turn-off time less than 2 μs
* Max. operating frequency greater than 500 kHz
* Timing from microseconds to hours
* Operates in both astable and monostable modes
* High output current
* Adjustable duty cycle
* TTL compatible
* Temperature stability of 0.005% per °C

APPLICATIONS

* Precision timing
* Pulse generation
* Sequential timing
* Time delay generation
* Pulse width modulation

PIN CONFIGURATION

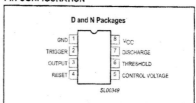

Figure 1. Pin configuration

BLOCK DIAGRAM

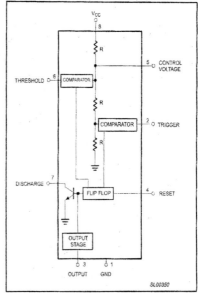

Figure 2. Block Diagram

ORDERING INFORMATION

DESCRIPTION	TEMPERATURE RANGE	ORDER CODE	DWG #
8-Pin Plastic Small Outline (SO) Package	0 to +70 °C	NE555D	SOT96-1
8-Pin Plastic Dual In-Line Package (DIP)	0 to +70 °C	NE555N	SOT97-1
8-Pin Plastic Small Outline (SO) Package	−40 °C to +85 °C	SA555D	SOT96-1
8-Pin Plastic Dual In-Line Package (DIP)	−40 °C to +85 °C	SA555N	SOT97-1
8-Pin Plastic Dual In-Line Package (DIP)	−55 °C to +125 °C	SE555CN	SOT97-1
8-Pin Plastic Dual In-Line Package (DIP)	−55 °C to +125 °C	SE555N	SOT97-1

Philips Semiconductors

Product data

Timer

NE/SA/SE555/SE555C

EQUIVALENT SCHEMATIC

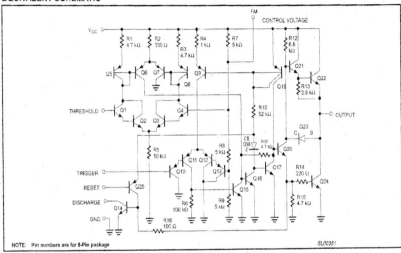

NOTE: Pin numbers are for 8-Pin package

SL00351

Figure 3. Equivalent schematic

ABSOLUTE MAXIMUM RATINGS

SYMBOL	PARAMETER	RATING	UNIT
V_{CC}	Supply voltage SE555 NE555, SE555C, SA555	+18 +16	V V
P_D	Maximum allowable power dissipation[1]	600	mW
T_{amb}	Operating ambient temperature range NE555 SA555 SE555, SE555C	0 to +70 −40 to +85 −55 to +125	°C °C °C
T_{stg}	Storage temperature range	−65 to +150	°C
T_{SOLD}	Lead soldering temperature (10 sec max)	+230	°C

NOTE:
1. The junction temperature must be kept below 125 °C for the D package and below 150°C for the N package.
   At ambient temperatures above 25 °C, where this limit would be derated by the following factors:
   D package 160 °C/W
   N package 100 °C/W

Figure 20.22 NE/SE 555C datasheet (Courtesy: Philips Semiconductors).

An op-amp circuit may be configured to operate as an **oscillator** or AC generator. It uses both positive and negative **feedback**. When the circuit is energized, it produces an AC output that drives an electrical load.

The oscillation output AC waveform can also be observed on an **oscilloscope**. If there is a problem with the positive- or negative-feedback portion of the circuit, the **frequency** and **amplitude** (gain) of the circuit may be altered.

Summary

- An oscillator generates a continuously repetitive output signal.
- Two basic types of oscillators based on their method of operation are the feedback oscillator and relaxation oscillator.
- In a feedback oscillator circuit, a portion of the output power is returned to the input circuit.
- A feedback oscillator circuit consists of an amplifier, feedback network, frequency-determining network, and a DC power source.
- A relaxation oscillator responds to an electronic device that goes into conduction for a certain time and then turns off for a period of time.
- Relaxation oscillators usually generate square- or triangular-shaped waves.
- The active device of a relaxation oscillator is triggered into conduction by a change in voltage.
- An inductive-capacitance network usually determines the operating frequency of a feedback oscillator.
- Without regenerative feedback, a tank circuit produces a damped sine wave.
- The operation of a relaxation oscillator depends on the charge and discharge of an RC or RL network.
- Feedback in a feedback oscillator can be accomplished by inductance, capacitance, or resistance coupling.
- The following oscillators are feedback oscillators: Armstrong, Hartley, Colpitts, Clapp, crystal, Pierce, and Wien-Bridge.
- An Armstrong oscillator is also referred to as a tickler coil.
- The frequency of an Armstrong oscillator is based on the value of a capacitor and the secondary of a transformer.
- The Hartley oscillator incorporates a tapped coil and is used extensively in AM and FM radio receivers.
- A Colpitts oscillator is similar to the shunt-fed Hartley oscillator except that it uses two capacitors instead of a tapped coil.
- The amount of feedback developed in a Colpitts oscillator is based on the capacitance ratio of the two capacitors in the tank circuit.

- The Clapp oscillator is similar to the Colpitts oscillator and uses an additional capacitor in series with the coil of the tank circuit.
- Crystal oscillators are used when extremely high-frequency stability is desired.
- The Hartley and Colpitts oscillators are often modified to accommodate a crystal.
- A crystal has the property to change electrical energy into mechanical energy and mechanical energy into electrical energy; this is known as the piezoelectric effect.
- The Wien-Bridge oscillator uses a RC network for providing positive feedback to an op-amp for controlling the frequency of oscillation, and negative feedback for controlling the gain.
- The Pierce oscillator is a modification of the Colpitts oscillator; it uses a crystal in place of the tank circuit inductor.
- Relaxation oscillators are typically used to generate nonsinusodial waveforms, such as the sawtooth and rectangular waveforms.
- Relaxation oscillators depend on the charge and discharge of a capacitor-resistor network.
- UJT oscillators are used in applications that require a signal with a slow rise time and a rapid fall time.
- A multivibrator is a classification of the relaxation oscillator.
- Two general types of multivibrators are triggered and free-running.
- An astable multivibrator is free-running.
- Monostable and bistable multivibrators are free-running.
- A monostable multivibrator has one stable state of operation; it is often called a one-shot multivibrator.
- A bistable multivibrator has two stable states of operation; it is often called a flip-flop.
- A 555 IC can be configured to respond as an astable multivibrator.
- The internal circuitry of a 555 IC consists of two comparators, a bistable flip-flop, a resistive divider, a discharge transistor, and an output stage.
- The shape of the waveform from an IC astable multivibrator is determined by an external RC network.

Formulas

(20-1) $A_{cl} = A_v \times B$ Closed-loop feedback gain of an oscillator

(20-2) $f_r = \frac{1}{2\pi\sqrt{LC}}$ Resonant frequency of a tank circuit.

(20-3) $f_r = \frac{1}{2\pi\sqrt{L_1 C_1}}$ Frequency of the Hartley oscillator

(20-4) $B = \frac{L_b}{L_a}$ Attenuation factor, B, of the Hartley oscillator

(20-5) $C_T = \frac{C_1 \times C_2}{C_1 + C_2}$ Total capacitance is called C_T of a Colpitts oscillator

(20-6) $f_r = \frac{1}{2\pi\sqrt{L_1 C_T}}$ Frequency of a Colpitts oscillator

(20-7) $B = \frac{C_1}{C_2}$ Attenuation factor, B, of a Colpitts oscillator

(20-8) $C_T = \frac{1}{\frac{1}{C_1} + \frac{1}{C_2} + \frac{1}{C_3}} = \frac{C_1 C_2 + C_2 C_3 + C_1 C_3}{C_1 C_2 C_3}$ Total capacitance C_T of a Clapp oscillator

(20-9) $f_r = \frac{1}{2\pi\sqrt{L_1 C_3}}$ Frequency of a Clapp oscillator when $C_3 \ll C_1$ or C_2

(20-10) $f_r = \frac{1}{2\pi RC}$ Frequency of a Wien-Bridge oscillator, when $R_1 = R_2 = R$ and $C_1 = C_2 = C$

(20-11) $A_{CL} = 1 + \frac{R_4 + R_5}{R_3}$ Closed-loop gain of a Wien-Bridge oscillator

(20-12) $V_P = \eta V_{BB} + 0.7$ Peak voltage (V_P) of UJT

(20-13) $f = \frac{1}{1.4 \times R \times C}$ Output frequency of a symmetrical multivibrator.

(20-14) $PW = 0.7 R_2 C_1$ Pulse-width of monostable multivibrator

(20-15) $UTP = \frac{2}{3} V_{CC}$ Upper trip point of a 555 IC

(20-16) $LTP = \frac{1}{3} V_{CC}$ Lower trip point of a 555 IC

(20-17) $f_{out} = \frac{1}{T}$ Output frequency of an astable multivibrator

(20-18) $T = 0.693(R_A + 2R_B)C$ Cycle time of an astable multivibrator

(20-19) $f_{out} = \frac{1}{0.693(R_A + 2R_B)C}$ Output frequency of an astable multivibrator

(20-20) $PW = 0.693(R_A + R_B)C$ Pulse width of an astable multivibrator

(20-21) % duty cycle $= \frac{On-time}{Total\ cycle\ time} = \frac{R_A + R_B}{R_A + 2R_B} \times 100$ Duty cycle of an astable multivibrator

Review Questions

Answer the following questions.

1. The amplitude of a damped oscillatory waveform (increases, decreases).
2. In a series RC circuit, if the value of the resistor is increased, the time constant of the system (increases, decreases).
3. A capacitor develops _____% of the total charge voltage in one time constant.
4. A capacitor develops _____% of the total charge voltage in five time constants.
5. The use of a _____ in the Armstrong oscillator distinguishes it from other oscillators.

6. The use of a _____ in the Hartley oscillator distinguishes it from other oscillators.

7. The capacitor divider network in a Colpitts oscillator is used for _____.

8. With reference to the series RLC circuit of **Figure 20.12(a)**, at the resonant frequency, the impedance is (maximum, minimum).

9. With reference to the parallel RLC circuit of **Figure 20.12(b)**, at the resonant frequency, the impedance is (maximum, minimum).

10. The _____ oscillator produces the most stable high-frequency output.

11. With reference to the Wien-Bridge oscillator circuit of **Figure 20.15(a)**, the components R_1 and C_1 are used for developing the _____-frequency response, while R_2 and C_2 are used for developing the _____-frequency response.

12. With reference to the Wien-Bridge oscillator circuit of **Figure 20.15(b)**, the intersection point of the low-pass frequency response and the high-pass frequency response is called the _____ frequency.

13. With reference to the UJT oscillator circuit of **Figure 20.16**, the charging time of the capacitor C_1 occurs when the UJT is (conducting, nonconducting).

14. With reference to the UJT oscillator circuit of **Figure 20.16**, the discharging time of the capacitor C_1 occurs when the UJT is (conducting, nonconducting).

15. In a square wave if the on-time is greater or lesser than the off-time, the resulting waveform is (symmetrical, nonsymmetrical).

16. In a square wave if the on-time is equal to the off-time, the resulting waveform is (symmetrical, nonsymmetrical).

17. A monostable multivibrator has _____ stable state(s), whereas an astable multivibrator has _____ stable state(s).

18. The type of multivibrator that needs an external trigger pulse to begin its operation is considered to be (monstable, astable).

19. A multivibrator that generates a continuous waveform is considered to be (bistable, monostable, astable).

20. With reference to the 555 timer circuit of **Figure 20.20**, when power is applied, the capacitor C charges through the resistor(s) _____.

Problems

Answers the following questions.

1. What is the frequency of an LC circuit using a 0.5-μF capacitor and a 100-mH inductor?

    a. 0-499 Hz
    b. 500-999 Hz
    c. 1-9.9 kHz
    d. 10 kHz or above

2. If the attenuation (B) of an amplifier circuit is 0.005, what should the voltage gain (A_v) of this amplifier be set to for oscillations to occur?

    a. 1-99
    b. 100-199
    c. 200-999
    d. 1000 or above

3. Refer to the Colpitts oscillator of **Figure 20-10**. If values of the capacitances are changed so that $C_1 = 0.05 \ \mu$F and $C_2 = 0.5 \ \mu$F, the total capacitance C_T of the circuit is:

    a. 0.001-0.009 μF
    a. 0.01-0.09 μF
    a. 0.1-0.9 μF
    a. 1 μF or above

4. Refer to the Wien-Bridge oscillator of **Figure 20-15**. If values of the components are set so that $R_1 = R_2 = R = 15$ kΩ and $C_1 = C_2 = C = 0.05$ μF, the critical or resonant frequency is:

    a. 1-49 Hz
    b. 50-99 Hz
    c. 100-199 Hz
    d. 200 Hz or above

5. If the on-time of a square wave pulse is 0.5 ms and the off-time is 0.15 ms, the duty cycle of the waveform is:

    a. 0%-24%
    a. 25%-49%
    a. 50%-74%
    a. 75% or above

Answers

Examples

20-1. 5.035 kHz

20-2. t = 0.5 ms; so T = 2.5 ms to fully charge the capacitor

20-3. 5.035 kHz

20-4. $C_T = 0.0578$ μF, $f_r = 5.410$ kHz

20-5. $C_T = 150$ pF, $f_r = 183.87$ kHz

20-6. $f_r = 636$ Hz

20-7. $t_D = 10$ μs, $T_D = 50$ μs (time taken to fully discharge the capacitor)

20-8. $t_2 - t_1 = 5$ ms

20-9. T = 2.079 ms, $f_{out} = 481$ Hz, PW = 1.213 ms, % duty cycle = 58.35%

Self-Examination

20.1

1. oscillator
2. feedback network, amplifier, frequency-determining network, and DC power source (any order)
3. feedback
4. tank
5. resonant
6. damped
7. 4.1 kHz
8. Decreases
9. relaxation
10. feedback

20.2

11. *L, C* (any order)
12. tapped coil
13. capacitors
14. crystal
15. Wien-Bridge

20.3

16. Relaxation
17. relaxation
18. triggered (or monostable or one-shot), free-running (or astable)
19. astable

20. one-shot
21. flip-flop
22. 555 IC
23. 50
24. unsymmetrical
25. cycle time

Terms

Feedback oscillator

A type of oscillator in which a portion of the output power is returned to the input circuit.

Regenerative feedback

Feedback from the output to the input that is in phase so that it is additive.

Armstrong oscillator

An oscillator that uses a transformer as a feedback element to produce a 180° phase shift which is needed for producing oscillation.

Relaxation oscillator

A nonsinusodial oscillator that has a resting or nonconductive period during its operation.

Tank circuit

A parallel-resonant LC circuit.

Damped sine wave

A wave in which successive oscillations decrease in amplitude.

Continuous wave (CW)

Uninterrupted sine waves, usually of the RF type, that are radiated into space by a transmitter.

Piezoelectric effect

The property of a crystal to change mechanical vibrations into electrical energy.

Sharp filter

A filter that permits feedback only of the desired frequency.

Pulse repetition rate (PRR)

The time that it takes for the waveform to repeat itself.

Multivibrator

A type of relaxation oscillator that employs an RC network in its physical makeup and produces a rectangular-shaped wave.

Triggered multivibrator

A type of multivibrator that uses a control technique called triggering to change its operational state.

Free-running multivibrator

A type of multivibrator that is self-starting. It operates continuously as long as electrical power is supplied. The shape and frequency of the waveform are determined by component selection.

Astable multivibrator

A free-running generator that develops a continuous square-wave output.

Symmetrical multivibrator

A circuit with an output having equal on and off times.

Nonsymmetrical multivibrator

A circuit with an output having unequal on and off times.

Monostable multivibrator

A multivibrator with one stable state. It changes to the other state momentarily and then returns to its stable state.

Flip-flop

A type of multivibrator changes states only when a trigger pulse is applied. It is called a flip-flop because it flips to one state when triggered and flops back to the other state when triggered.

Threshold terminal

The beginning or entering point of an operating condition. A terminal connection of the LM555 IC timer.

Blocking oscillator

An oscillator circuit that drives an active device into cutoff or blocks its conduction for a certain period of time.

Vertical blocking oscillator (VBO)

A TV circuit that generates the vertical sweep signal for deflection of the cathode-ray tube or picture tube.

21

Radio Frequency (RF) Communication Systems

High-level **sound amplification** makes it possible to communicate over long distances. A public-address amplifier system, for example, permits an announcer to communicate with a large number of people in a stadium or an arena. Even the most sophisticated sound system, however, has some limitations. **Sound waves** moving away from the source have a tendency to become somewhat weaker the farther they travel. An increase in signal strength does not, therefore, necessarily solve this problem. People near large speakers usually become very uncomfortable when the sound is increased to a high level. It is possible, however, to communicate with people over long distances without increasing sound levels. Electromagnetic waves in the **radio frequency (RF)** range make this type of communication possible. Radio, television, and long-distance telephone communication are achieved by this process.

This chapter investigates how sound is transmitted through the air by electromagnetic waves called **radio waves**. Radio frequency (RF) communication systems play a very important role in our daily lives. We listen to radio receivers, watch television, and talk to friends over long distances on a telephone. Therefore, it is important that we have some basic understanding of this type of communication.

A number of different **RF communication systems** are used and are covered in this chapter. As a general rule, these systems are classified according to the method by which signal information or intelligence is applied to the transmitted signal. Three very common RF communication systems are **continuous-wave (CW)**, **amplitude modulation (AM)**, and **frequency modulation (FM)**. Each system has a number of unique features that distinguish it from the others.

Objectives

After studying this chapter, you will be able to:

21.1 explain the basic operation of an RF communication system;
21.2 analyze the operation of a continuous-wave communication system;
21.3 analyze the operation of an AM communication system;
21.4 analyze the operation of an FM communication system;
21.5 troubleshoot an RF communication circuit.

Chapter Outline

21.1 RF Communication Systems
21.2 Continuous-Wave Communication
21.3 Amplitude Modulation Communication
21.4 Frequency Modulation Communication
21.5 Troubleshooting RF Communication Circuits

Key Terms

beat-frequency oscillator (BFO)
buffer amplifier
carrier wave
center frequency
channel
ganged
ground wave
heterodyning
high-level modulation
ionosphere
line-of-sight transmission
low-level modulation
modulating component
modulation
phasor
radio telegraphy
ratio detector
receiver
sidebands
sky wave

transmitter
zero beating

21.1 RF Communication Systems

Radio frequency (RF) communication systems use electromagnetic waves in the **radio frequency spectrum** for communication. These signals, like all electromagnetic waves, travel through the air at 186,000 mi/s, or 300,000,000 m/s - the speed of light. RF communication, therefore, permits sound to travel long distances instantaneously. In this section, you will learn about the fundamental parts of an RF communication system and how radio frequency waves are transmitted.

21.1 Explain the basic operation of an RF communication system.

In order to achieve objective 21.1, you should be able to:

- describe the primary function of a transmitter and receiver;
- name the basic parts of an RF communications system;
- name three common communication systems;
- define transmitter, receiver, ground wave, sky wave, ionosphere, and line-of-sight transmission.

RF Communication System Parts

Refer to **Figure 21.1** which identifies the RF portion of the electromagnetic spectrum. Note that RFs include different forms of transmissions such as AM (amplitude modulated) radio, FM (frequency modulated) radio, television, cellular communications, and microwave devices. Most of our home entertainment systems operate by processing RF signals.

An RF communication system is very similar to any other electronic system. It has an energy source, a transmission path, control, load device, and one or more indicators. These individual parts are essential to the operation of the system. The physical layout of the **RF communication system** is very important. **Figure 21.2(a)** shows a block diagram of a radio frequency (RF) communication system. The signal source of the system is an RF transmitter. The **transmitter** is the center or focal point of the system. The RF signal is sent to the remaining parts of the system through the atmosphere or air, which is the transmission path of the system shown. RF finds air to be an excellent signal path as compared to other media.

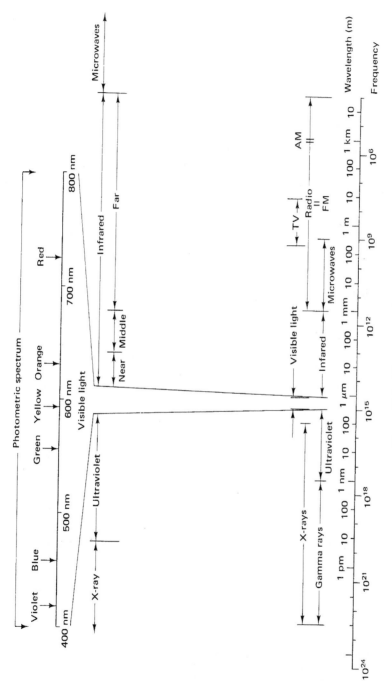

Figure 21.1   Electromagnetic spectrum.

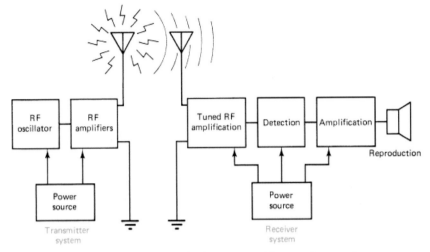

Figure 21.2 RF communication system with transmitter-receiver.

The control function of the system is directly related to the signal path. The distance that the RF signal must travel has a great deal to do with its strength. The load of the system is an infinite number of radio receivers. Each **receiver** picks the signal out of the atmosphere and uses it to do work. Any number of receivers can be used without directly influencing the output of the transmitter. System indicators may be found at a number of locations. Meters, lamps, and waveform monitoring oscilloscopes are typical indicators. These basic functions apply, in general, to all RF communication systems.

A good way to look at an RF communication system is to divide the transmitter and receiver into separate systems. The transmitter then has all its system parts in one discrete unit. Its primary function is to generate an RF signal that contains intelligence and to radiate this signal into the atmosphere. The receiver can also be viewed as an independent system. It has a complete set of system parts. Its primary function is to pick up the RF transmitted signal, extract the intelligence, and develop it into a usable output.

A one-way communication system has one transmitter and an infinite number of receivers. Commercial AM, FM, and TV communication is achieved by this method. Two-way mobile communication systems have a transmitter and a receiver at each location. In some cases, the transmitter and receiver may be combined into one unit which is called a **transceiver**. This type of system permits direct communication between each location. **Citizen's band (CB) radio** is also considered to be two-way communication.

Figure 21.2(b) shows a simplification of the transmitter and receiver as independent systems. The transmitter has an RF oscillator, amplifiers, and a power source. The antenna-ground at the output circuit serves as the load device for the system. Intelligence is applied according to the design of the system. The system may be continuous-wave (CW), amplitude modulation (AM), or frequency modulation (FM).

The **RF receiver** function is represented as an independent system. It will not be operational unless the transmitter sends out a signal. The receiver has tuned RF amplification, detection, amplification, reproduction, and a power supply. The detection function picks out the intelligence from the received signal. The reproduction unit, which is usually a speaker, serves as the load device. The receiver responds only to the RF signal sent out by the transmitter.

Radio Frequency

Radio frequency (RF) communication systems rely on the radiation of high-frequency energy from an antenna-ground network. This particular principle was discovered by Heinrich Hertz around 1885. He found that high-frequency waves produced by an electrical spark induced electrical energy in a coil of wire some distance away. He also discovered that current passing through a coil of wire produces a strong electromagnetic field. The fundamental unit of frequency is the hertz (Hz), which is the number of complete waveforms that occur in 1 s.

The basic characteristics of **radio frequency (RF) waves** are important in the operation of RF systems. Refer back to the electromagnetic spectrum of **Figure 21.1**. Note the lower right portion of the spectrum that contains radio frequencies. The **RF band** of frequencies includes those that are designated by the **Federal Communications Commission (FCC)** for AM, FM, microwave, and television signals. Electromagnetic signals are transmitted within their designated bands to distinguish one type from others. For example, standard AM signals are 535-1620 kHz, and standard FM signals are 88-108 MHz.

When direct current is applied to a coil of wire, it causes the field to remain stationary as long as current is flowing. When DC is first turned on, the field expands. When it is turned off, the field collapses and cuts across the coil. As a general rule, DC electromagnetic field development is of no significant value in radio communication. When low-frequency AC is applied to an inductor, it produces an electromagnetic field. Since AC is in a constant state of change, the resulting field is also in a state of change. It changes polarity twice during each operational cycle.

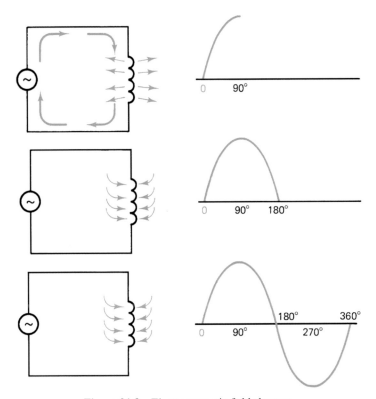

Figure 21.3   Electromagnetic field changes.

Figure 21.3 shows how the **electromagnetic field** of a coil changes when AC is applied. In **Figure 21.3(a)**, the field is expanding during the first half of the positive alternation. At the peak of the alternation, coil current begins to decrease in value. **Figure 21.3(b)** shows the field collapsing. The current value decreases to the zero reference at this time. This completes the positive alternation, or first 180°, of the sine wave. **Figure 21.3(c)** shows the field expanding for the negative alternation. The polarity of this field is now reversed. At the peak of the negative alternation, the field begins to collapse. The cycle is complete at this point. The sequence is repeated for each new sine wave.

When high-frequency AC is applied to an inductor, it also produces an electromagnetic field. The directional change of high-frequency AC, however, occurs very quickly. The change is so rapid that there is not enough time for the collapsing field to return to the coil. As a result, a portion of the field becomes detached and is forced away from the coil. In a sense, the electromagnetic field radiates out or away from the coil. The word **radio** was developed from these terms.

The **electromagnetic wave** of an RF communication system radiates from the antenna of the transmitter. This wave is the end result of an interaction between electric and magnetic fields. The resulting wave is invisible, cannot be heard, and travels at the speed of light. These waves become weaker after leaving the antenna. Stronger electromagnetic waves can be effectively used in a radio frequency communication system. These waves do not affect human beings like high-powered sound waves. RF waves are also easier to work with in the receiver circuit.

Electromagnetic wave radiation is directly related to its frequency. **Low-frequency (LF)** waves of 30-300 kHz and **medium frequency (MF)** waves of 300-3000 kHz tend to follow the curvature of the earth. This type of waves is commonly called a **ground wave**. Ground waves are usable during the day or night for distances up to approximately 100 miles.

The higher frequency portion of the electromagnetic wave radiated from an antenna is called a **sky wave**. Depending on its frequency, sky waves are reflected back to the earth by a layer of ionized particles called the **iono-sphere**. If the frequency is low, very little reflection from the ionized particles takes place. However, if the frequency is high, the reflection increases significantly. These waves permit signal reception well beyond the range of the ground wave. Sky-wave signal patterns change according to ion density and the position of the ionized layer. During the daylight hours, the ionosphere is very dense and near the surface of the earth. Signal reflection or **skip** of this type is not very suitable for long-distance communication. In the evening hours, the ionosphere is less dense and moves to higher altitudes. Hence, sky-wave reflection patterns have larger angles and travel greater distances than ground waves. **Figure 21.4** shows some sky-wave patterns and the ground wave.

The height of the **antenna**, the frequency used, and the type of antenna determine its **radiation pattern**. **Very high frequency (VHF)** signals of 30-300 MHz tend to move in straight lines because of their high frequency. VHF can, thus, be used for ionosphere-based and satellite communications owing to its straight line transmission characteristics. If the height of the antenna is increased, the radiation pattern would be altered. Since higher frequency RF signals follow a straight line, as compared to low-frequency signals, the height of an antenna extends the radiation pattern. A high-frequency RF signal, when used along with a ground wave, generates a **line-of-sight**

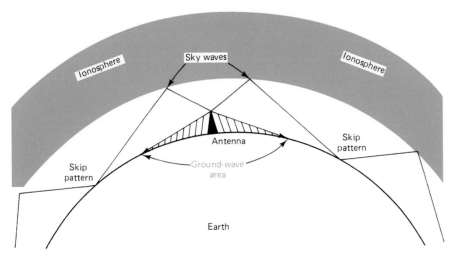

Figure 21.4 Sky-wave and ground-wave patterns.

transmission radiation pattern. FM, TV, and satellite communications are achieved by VHF signal radiation.

The physical size of the wave, or its **wavelength**, decreases with an increase in frequency. High-frequency signals have a short wavelength. These signals pass very readily through the ionosphere without being reflected or distorted. Communication between an FM or TV transmitter and a receiver on earth is limited to a few hundred miles. Communication between an earth station and a satellite can involve thousands of miles.

Self-Examination

Answer the following questions.

1. The _____ of a communications system is responsible for signal generation.
2. The _____ of a communications system intercepts the RF signal and recovers the information.
3. A (one-way, two-way) communication system has one transmitter and an infinite number of receivers.
4. A (one-way, two-way) communication system has a transmitter and a receiver at each location.
5. Three common communication systems are _____, _____, and _____.
6. _____ waves are reflected by the ionosphere.

21.2 Continuous-Wave Communication

Continuous-wave (CW) communication is the simplest of all communication systems. This type of communication system generates an electromagnetic wave of a constant frequency and amplitude. CW communication is used in radar systems and amateur radio. This section discusses CW in amateur radio - specifically, its role in the transmission of Morse code.

21.2 Analyze the operation of a continuous-wave communication system.

In order to achieve objective 21.2, you should be able to:

- determine the beat frequencies developed from the fundamental frequencies;
- describe the heterodyning process;
- explain the purpose of a beat-frequency oscillator;
- identify the parts of a continuous-wave communication system;
- define radio telegraphy, ganged, heterodyning, beat frequency, beat frequency oscillator, and zero beating.

Basic CW Parts and Operation

A **continuous-wave (CW) communication system** consists of a transmitter and a receiver. The **transmitter** employs an oscillator for the generation of an RF signal. The **oscillator** generates a continuous sine wave. In most systems, the CW signal is amplified to a desired power level by an RF power amplifier. The **output** of the final power amplifier is then connected to the antenna-ground network. The antenna radiates the signal into the atmosphere. When generating **Morse code**, information or intelligence is applied to the CW signal by turning the signal on and off. This causes the CW signal to be broken into a series of pulses, or short bursts of RF energy. The pulses conform to an intelligible code. The international Morse code is commonly used.

Figure 21.5 shows a simplification of the **Morse code**. A short burst of RF represents a dot. A burst three times longer represents a dash. Signals of this type are called keyed or coded continuous waves. The Federal Communications Commission (FCC) classified keyed CW signals as type A1. This is considered to be **radio telegraphy** with on-off keying. **Figure 21.6** shows a comparison of a CW signal and a keyed CW signal.

The **receiver** function of a CW communication system is somewhat more complex than the transmitter. It has an **antenna-ground network**, a tuning circuit with an RF amplifier, a heterodyne detector, a beat-frequency

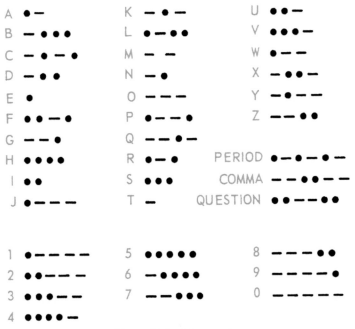

Figure 21.5 Morse code.

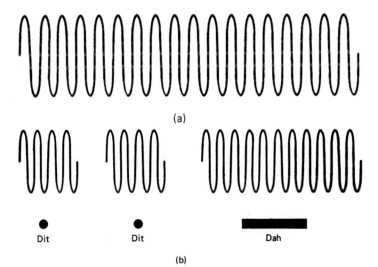

Figure 21.6 Comparison of CW signals. (a) CW signal. (b) Coded or keyed continuous wave, letter U.

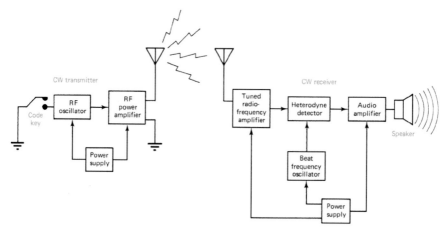

Figure 21.7   Block diagram of a CW communication system.

oscillator, an audio amplifier, and a speaker. The receiver is designed to pick up a CW signal and convert it into a sound signal that drives the speaker. **Figure 21.7** shows a block diagram of the CW communication system.

CW Transmitter

A **CW transmitter** in its simplest form has an oscillator, a power supply, and an antenna-ground network. When the CW transmitter is used to transmit Morse code, a **keying circuit** is included. The power output developed by the circuit is usually quite small. In this type of system, the active device serves jointly as an oscillator and a power amplifier. This is done to reduce the number of circuit stages in front of the antenna. An oscillator used in this manner generally has reduced frequency stability. As a rule, frequency stability is a very important operational consideration. Single-stage transmitters are rarely used because of their instability.

A **single-stage CW transmitter** is shown in **Figure 21.8**. This particular circuit is a tuned-base shunt-fed Hartley oscillator. Remember from Chapter 20 that a **Hartley oscillator** has a tapped coil that is used as part of the tuned circuit. The output frequency of the oscillator is adjustable up to 3.5 MHz. The power output is approximately 0.5 W. When the output of the oscillator is connected to a load, it usually causes a shift in frequency. The load should, therefore, be of a constant impedance value for this type of transmitter to be effective.

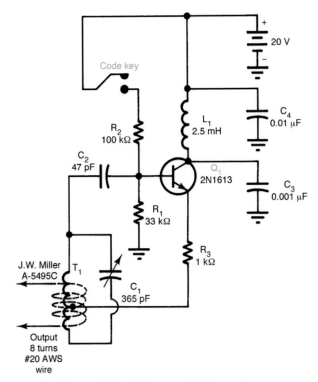

Figure 21.8 Single-stage CW transmitter.

Example 21-1:

Refer to the Hartley oscillator CW transmitter circuit of **Figure 21.8**. Assume that the T_1 value of the tank circuit = 6 μH. Calculate the resonant frequency (f_r).

Solution

$$f_r = \frac{1}{2\pi\sqrt{LC}} = \frac{1}{2\pi\sqrt{T_1 C_1}} = \frac{1}{2\pi\sqrt{6 \times 10^{-6} \times 365 \times 10^{-12}}} = 3.4\,\text{MHz}.$$

Related Problem

Change the inductance value of the circuit of **Figure 21.8** to 7.5 μH and determine the resonant frequency.

Before a **CW transmitter** can be placed into operation, it is necessary to close the code key. This operation supplies the forward bias voltage to the

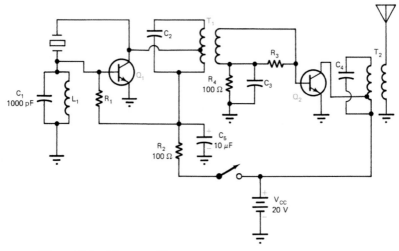

Figure 21.9   Master oscillator power amplifier (MOPA) transmitter.

base through the divider network of R_1 and R_2. With the transistor properly biased, a feedback signal is supplied to the T_1-C_1 tank circuit. This charges C_1 and causes the circuit to oscillate. Opening the key causes the oscillator to stop functioning. Intelligence can be injected into the CW signal in the form of code. The code key simply responds as a fast-acting on-off switch. The output of the transmitter is developed across a coil wound around T_1. Inductive coupling of this type is very common in RF transmitter circuits. A low-impedance output is needed to match the antenna-ground impedance. When this is accomplished, there is a maximum transfer of energy from the oscillator to the antenna.

A **single-stage CW transmitter** has a number of shortcomings. The power output is usually held to a low level. The frequency stability is rather poor. These two problems can be overcome to some extent by adding a power amplifier after the oscillator. Systems of this type are called **master oscillator power amplifier (MOPA) transmitters**.

Figure 21.9 shows the circuitry of a low-power MOPA transmitter system. This circuit is a modified Pierce oscillator. Remember from Chapter 20 that a Pierce oscillator utilizes a crystal as its tank circuit. The crystal (Y_1) provides feedback from the collector to the base of Q_1. Transformer T_1 is used to match the collector impedance of Q_1 to the base of transistor Q_2.

Q_2 is the power amplifier. It operates as a class C amplifier to provide high operational efficiency. The oscillator is, therefore, isolated from the antenna through Q_2. This provides a fixed load for the oscillator. Operational stability is improved and power output is increased with this modification.

CW Receiver

The **receiver** of a CW communication system is responsible for intercepting an RF signal and recovering the information it contains. Information of this type is in the form of a radiotelegraph code. The signal is an **RF carrier wave** that is interrupted by a coded message. A CW receiver must perform a number of functions to achieve this operation. This includes signal reception, selection, RF amplification, detection, audio frequency (AF) amplification, and sound reproduction. These functions are also used in the reception of other RF signals. In fact, CW reception is usually only one of a number of operations performed by a communications receiver. This function is generally achieved by placing the receiver in its CW mode of operation. A switch is usually needed to perform this operation.

Figure 21.10 shows the functions of a **CW communication receiver** in a block diagram. Note that electrical power is needed to energize the active components of the system. Not all blocks of the diagram are supplied operating power. This means that some of the functions can be achieved without solid-state components. The antenna, tuning circuit, and detector do not require power supply energy for operation. These functions are achieved

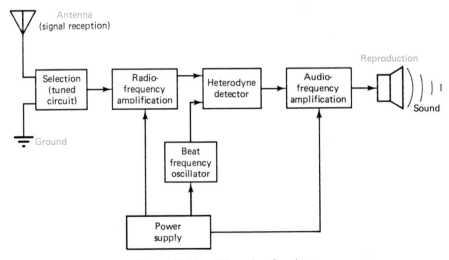

Figure 21.10 CW receiver functions.

by signal energy. A radio receiver circuit has a signal energy source and operational energy source. The power supply provides operational energy directly to the circuit. The signal source for RF energy is the transmitter. This signal must be intercepted from the air and processed by the receiver during its operation.

Antenna

The **antenna** of a CW receiver is responsible for the reception function. It intercepts a small portion of the RF signal that is sent out by the transmitter. Ordinarily, the receiving antenna only develops a few microwatts of power. In a strict sense, the antenna is a transducer. It is designed to convert electromagnetic wave energy into RF signal voltage. The signal voltage then causes a corresponding current flow in the antenna-ground network. RF antenna power is the product of signal voltage and current.

Receiving antennas come in a variety of styles and types. In two-way communication systems, the transmitter and receiving antennas are of the same unit. The antenna is switched back and forth between the transmitter and receiver. In one-way communication systems, antenna construction is not particularly critical. A long piece of wire can respond as a receiving antenna. In weak signal areas, antennas may be tuned to resonate at the particular frequency being received. Portable radio receivers use small antenna coils that are attached directly to the circuit.

When the **electromagnetic wave** of a transmitter passes over the receiving antenna, AC voltage is induced into it. This voltage causes current to flow, as shown in **Figure 21.11**. Starting at point A of the waveform, the electrons are at a standstill. The induced signal then causes the voltage to rise to its positive peak. The resulting current flow is from the antenna to the ground. It rises to a peak, and then drops to zero at point C on the curve. The polarity of the induced voltage changes at this point. Between points C and E, it rises to the peak of the negative alternation and returns again to zero. The resulting current flow is from the ground into the antenna. This induced signal is repeated for each succeeding alternation.

With radio frequency induced into the antenna, a corresponding RF is brought into the receiver by the **antenna coil**. The antenna coil serves as an impedance-matching transformer. It matches the impedance of the antenna to the impedance of the receiver input circuit. When outside antennas are used, the primary winding is an extension of the antenna. The loop antenna coil of a portable receiver is attached directly to the input circuit.

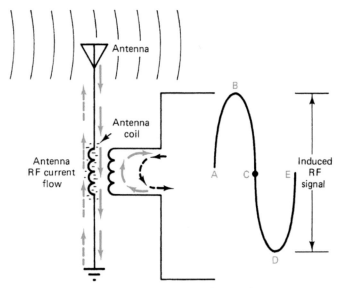

Figure 21.11 Antenna signal voltage and current.

Signal Selection

The **signal-selection** function of a radio receiver refers to its ability to pick out a desired RF signal. In receiver circuits, the antenna is generally designed to intercept a band or range of different frequencies. The receiver must then select the desired signal from all those intercepted by the antenna. Signal selection is primarily achieved by an *LC* resonant circuit. This circuit provides a low-impedance path for its resonant frequency. Nonresonant frequencies see very high impedance. In effect, the resonant frequency signal is permitted to pass through the tuner without opposition.

Figure 21.12 shows the **input tuner** of a CW receiver. In this circuit, L_2 is the secondary winding of the antenna transformer. C_1 is a variable capacitor. By changing the value of C_1, the resonant frequency of $L_2 C_1$ is tuned to the desired frequency. The selected frequency is then permitted to pass into the remainder of the receiver. Signal selection for AM, FM, and TV receivers all respond to a similar tuning circuit.

In communication receiving circuits, a number of **tuning stages** are needed for good signal selection. Each stage of tuning permits the receiver to be more selective of the desired frequency. The tuning response curve of one tuned stage is shown in **Figure 21.12**. In a crowded RF band, several stations operating near the selected frequency might be received at the same time.

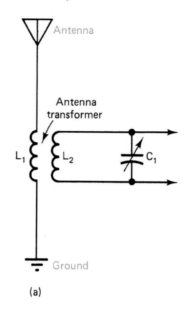

(a)

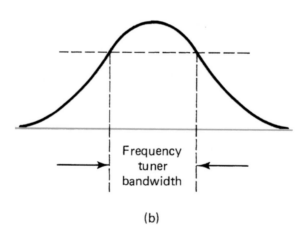

(b)

Figure 21.12   Input tuner of a CW receiver and its bandwidth. (a) Input tuning circuit. (b) Tuner frequency response curve.

Additional tuning improves the selection function by narrowing the response curve.

Figure 21.13 shows some representative **loop antenna coils**. Note that terminals 1 and 3 of each example are the end connections of the coil and

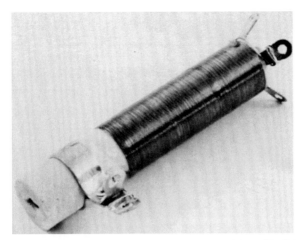

Figure 21.13   Antenna coils (Courtesy: J. W. Miller, Division of Bell Industries).

terminal 2 is the tapped connection. The full inductance (*L*) value of the coil is between terminals 1 and 3. A proportional part of the full inductance is from terminals 1 to 2 or from 2 to 3 of the coil.

Example 21-2:

Determine the bandwidth (BW) of a tuning circuit with an upper critical frequency (f_{cu}) of 20 MHz and a lower critical frequency (f_{cl}) of 6.5 MHz.

Solution

$$BW = f_{cu} - f_{cl} = 20 - 6.5 \text{ MHz} = 13.5 \text{ MHz}.$$

Related Problem

Calculate the bandwidth of a circuit with the following values of frequency: f_{cu} = 80 and 55 MHz.

RF Amplification

Most **CW receivers** employ at least one stage of RF amplification. This specific function is designed to amplify the weak RF signal intercepted by the antenna. Some degree of amplification is generally needed to boost the signal to a level where its intelligence can be recovered. RF amplification can be achieved by a variety of active devices. Vacuum tubes were used for a number of years. Solid-state circuits may now employ bipolar transistors

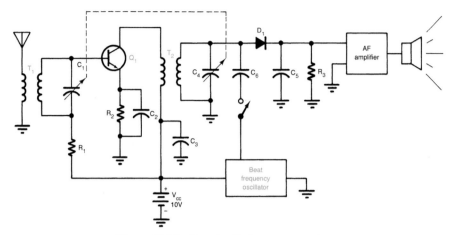

Figure 21.14   Tuned radio frequency receiver.

and MOSFETs in their design. Several manufacturers have developed IC RF amplifiers. Any of these active devices can be used to achieve RF signal amplification.

A representative **RF amplifier** circuit is shown in **Figure 21.14**. The primary function of the circuit is to control the amplification level of the RF signal applied to its input. Note that the amplifier has circuits for both input and output tuning. This circuit is called a **tuned radio-frequency (TRF) amplifier**. The broken line between the two variable capacitors indicates that they are connected together or **ganged**. When one capacitor is tuned, the second capacitor is adjusted to a corresponding value. This permits the user to make one adjustment when turning to a specific frequency. After the RF signal has been selected by the tuned amplifier, and it is then amplified by the transistor, Q_1. The amplified signal then passes through the primary of the transformer T_2. The resulting output RF signal then appears across the secondary winding of the transformer. The tuning circuit formed by the secondary and the ganged capacitor C_4 selects the specific signal frequency to be applied to diode, D_1. The RF signal is then rectified by D_1. The amplitude of the RF signal varies at the rate of the audio component it contains. At this point, the audio component is detected and then directed to the audio amplifier section by capacitor C_5, which offers a low-impedance path for the RF component. Capacitor C_7 offers a low-impedance path for the AF component, and a high-impedance path for the RF. The audio component is then processed by the AF amplifier section.

Heterodyne Detection

Heterodyne detection is an essential function of the CW receiver. The purpose of **heterodyning** is simply to mix two AC signals in a nonlinear device. In the CW receiver, the nonlinear device is a diode. The applied signals are called the **fundamental frequencies**. The resulting output of this circuit has four frequencies. Two of these are the original fundamental frequencies. The other two are beat frequencies. Beat frequencies are the addition and the difference in the fundamental frequencies. Heterodyne detectors used in other receiver circuits that are commonly called **mixers**.

Figure 21.15 shows the signal frequencies applied to a heterodyne detector. Fundamental frequency F_1 is representative of the incoming RF signal. This signal has been keyed on and off with a telegraphic code. Fundamental frequency F_2 comes from the **beat-frequency oscillator (BFO)** of the receiver as a CW signal of 1.001 MHz. This CW signal is referred to as a beat frequency. Both RF signals are applied to the receiver's diode. The resulting output of a heterodyne detector is F_1, F_2, B_1, and B_2 where

$$B_1 = F_1 + F_2 \tag{21.1}$$

and

$$B_2 = F_2 - F_1. \tag{21.2}$$

When two RF signals are applied to a diode, the resulting output is the sum and difference signals. All four signals (F_1, F_2, B_1, and B_2) appear in the output of the heterodyne detector. Typically, the difference signal will be in the AF range, thereby permitting the CW signal to be changed into an audio

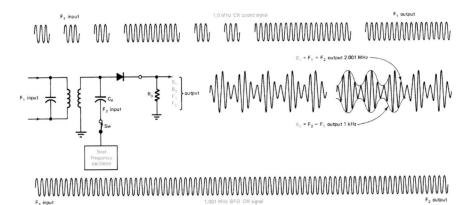

Figure 21.15   Heterodyne detector frequencies.

tone representative of the RF CW code. The other three signals will be in the RF range. If the frequency of the two input signals F_1 and F_2 are identical, the output frequency is the same as the input. This is generally called **zero beating**. There is no developed beat frequency when this occurs.

When the two RF input signals are slightly different, **beat frequencies** occur. At one instant, the signals move in the same direction. The resulting output is the sum of the two frequencies. When one signal is rising and one is falling at a different rate, there will be times when the signals cancel each other. The addition and difference of the two signals are, therefore, frequency-dependent. B_1 and B_2 are combined in a single RF wave. The amplitude of this wave varies according to frequency difference in the two signals. In a sense, the mixing process causes the amplitude of the RF wave to vary at an audio rate.

Example 21-3:

Referring to **Figure 21.15**, if F_1 is 1.0 MHz and F_2 is 1.001 MHz, determine the additive and the difference beat frequencies.

Solution:

The additive beat frequency B_1 and difference beat frequency B_2 are calculated as shown below:

$$B_1 = F_1 + F_2$$
$$= 1.0 + 1.001 \text{ MHz}$$
$$= 2.001 \text{ MHz}$$
$$B_2 = F_2 - F_1$$
$$= 1.001 - 1.0 \text{ MHz}$$
$$= 1 \text{ kHz}.$$

The output of the detector in **Figure 21.15** will contain all four frequencies, F_1, F_2, B_1, and B_2. The difference beat frequency of 1 kHz is within the range of human hearing. F_1, F_2, and F_3 are RF signals.

Related Problem

Referring to **Figure 21.15**, if F_1 is 3.0 MHz and F_2 is 3.002 MHz, determine the additive and the difference beat frequencies.

The diode of a **heterodyne detector** is also responsible for rectification of the applied signal frequencies. The rectified output of the added

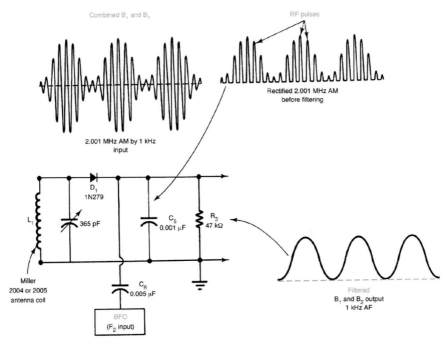

Figure 21.16   Heterodyne diode output.

beat frequency is of particular importance. It contains both RF and AF. **Figure 21.16** shows how the diode responds. Initially, the RF signals are rectified. The output of the diode is a series of RF pulses varying in amplitude at an AF rate. C_5 added to the output of the diode responds as a C-input filter. The C-input filter is used to increase the DC voltage level. Each RF pulse charges C_5 to its peak value. During the off period, C_5 discharges through R_3. The output is a low-frequency AF signal of 1 kHz. This signal occurs only when the incoming CW signal has been keyed. The AF signal is representative of the keyed information imposed on the RF signal at the transmitter.

Beat-Frequency Oscillator

Nearly any basic **oscillator** circuit could be used as the BFO of a CW receiver. **Figure 21.17** shows a Hartley oscillator. Feedback is provided by a tapped coil in the base circuit. C_1 and T_1 are the frequency-determining components. Note that this particular oscillator has variable-frequency capabilities. C_1 is usually connected to the front panel of the receiver. Adjustment of C_1 is used

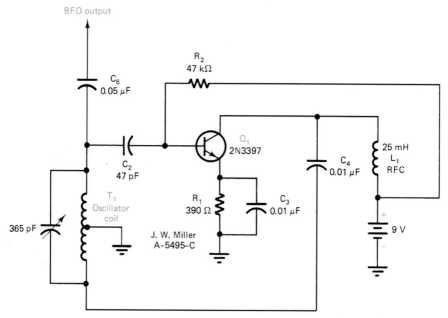

Figure 21.17   Hartley oscillator used as a beat-frequency oscillator (BFO).

to alter the **tone** of the beat-frequency signal. The relative strength of a signal determines the quality of tone. When the **BFO signal** is equal to the coded CW signal, no sound output occurs. This is where zero beating takes place. When the frequency of the BFO is slightly above or below the incoming signal frequency, a low AF tone is produced. Increasing the BFO frequency causes the pitch of the AF tone to increase. Pitch is determined by frequency. As frequency increases, pitch increases. Adjustment of the AF output is a matter of personal preference. In practice, an AF tone of 400 Hz to 1 kHz is common.

Output of the **BFO** is coupled to the diode through capacitor C_6. In a communications receiver, BFO output is controlled by a switch. With the switch on, the receiver produces an output for CW signals. With the switch is off, the receiver responds to AM and possibly FM signals. The BFO is needed only to receive coded CW signals.

AF Amplification

The **AF amplifier** of a CW receiver is responsible for increasing the level of the developed sound signal. The type and amount of signal amplification varies a great deal among different receivers. Typically, a small-signal amplifier and a power amplifier are used to provide a strong signal. The small-signal amplifier responds as a voltage amplifier. This amplifier is designed to increase the signal voltage to a level that will drive the speaker. The power output of a communications receiver rarely ever exceeds 5 W. A number of the AF amplifier circuits discussed in earlier chapters could be used in a CW receiver.

Self-Examination

Answer the following questions.

7. _____ code is used for CW transmission.
8. The ability of a receiver to pick out a desired RF signal is called _____.
9. The process of mixing two AC signals in a receiver is called _____.
10. A heterodyne detector circuit is called a(n) _____.
11. What are the basic parts of a CW transmitter?
12. How is coded information transmitted by a CW transmitter?
13. What type of signal is radiated from a CW transmitter?
14. When the beat-frequency oscillator (BFO) signal and a CW signal are at the same frequency, what is the output?
15. When the BFO signal and a CW signal are slightly above or below the incoming RF signal, what would be the range of output frequencies?

21.3 Amplitude Modulation Communication

Amplitude modulation (AM) is an extremely important form of communication. It is achieved by changing the physical size or amplitude of the RF wave by the intelligence signal. Voice, music, data, and picture intelligence can be transmitted by this method. This section discusses the process of amplitude modulation and describes the operation of AM communication system components.

21.3 Analyze the operation of an AM communication system.

In order to achieve objective 21.3, you should be able to:

- calculate percentage of modulation;
- explain the process of amplitude modulation;
- identify the components of an amplitude modulation communication system;

- define modulation, modulating component, carrier wave, sidebands, channel, high-level modulation, low-level modulation, and buffer amplifier.

Modulation

An intelligence signal must first be changed into electrical energy. This process is dependent upon **modulation**. Transducers such as microphones, tape heads, and photoelectric devices are designed to achieve this function. The developed signal is called the **modulating component**. In an AM communication system, the RF transmitted signal is much higher in frequency than the modulating component. The RF component is an uninterrupted CW wave. In practice, this part of the radiated signal is called the **carrier wave**.

An example of the signal components of an AM system is shown in **Figure 21.18**. The unmodulated RF carrier is a CW signal. This signal is generated by an oscillator. In this example, the carrier is 1000 kHz, or 1.0 MHz. A signal of this frequency is in the standard **AM broadcast band** of 535-1620 kHz. When listening to this station, a receiver would be tuned to the carrier frequency. Assume that the RF signal is modulated by the indicated

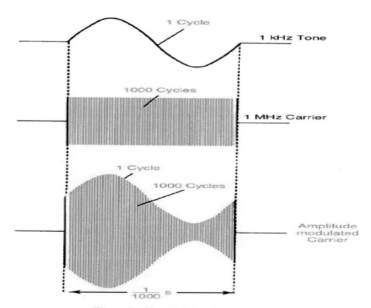

Figure 21.18   AM signal components.

1000-Hz tone. The RF component changes 1000 cycles for each AF sine wave. The amplitude of the RF signal varies according to the frequency of the modulating signal. The resulting wave is called an **amplitude-modulated RF carrier**.

To achieve amplitude modulation, the RF and AF components are applied to a nonlinear device. A solid-state device operating in its nonlinear region is used to produce modulation. In a sense, the two signals are mixed, or heterodyned, together. This operation causes two beat frequencies to be developed: B_1 and B_2. For the signals of **Figure 21.18, beat frequencies B_1 and B_2 are**

$$B_1 = F_1 + F_2$$
$$= 1000 + 1,000,000 \text{ Hz}$$
$$= 1,001,000 \text{ Hz}$$
$$B_2 = F_2 - F_1$$
$$= 1,000,000 - 1000 \text{ Hz}$$
$$= 999,000 \text{ Hz}.$$

The resulting AM signal, therefore, contains three RF signals: 1.0-MHz carrier, 0.999 MHz, and 1.001 MHz.

When the modulating component of an AM signal is music, the resulting beat frequencies become quite complex. As a rule, music involves a range or a band of frequencies. In AM systems, these are called **sidebands**. B_1 is the upper sideband, and B_2 is the lower sideband.

The space that an AM signal occupies with its frequency is called a **channel**. The bandwidth of an AM channel is twice the highest modulating frequency. For our 1-kHz modulating component, a 2-kHz bandwidth is needed. This is 1 kHz above and below the carrier frequency of 1 MHz. In commercial AM broadcasting, a station is assigned a 10-kHz channel. This limits the AM modulating component to a frequency of 5 kHz. **Figure 21.19** shows the sideband produced by a standard AM station.

It is interesting to note that the carrier wave of an AM signal contains no modulation. All the modulation appears in the **sidebands**. If the modulating component is removed, the sidebands disappear. Only the carrier is transmitted. The sidebands are directly related to the carrier and the modulating component. The carrier has a constant frequency and amplitude. The sidebands vary in frequency and amplitude according to the modulating component. In AM radio, the receiver is tuned to the carrier wave.

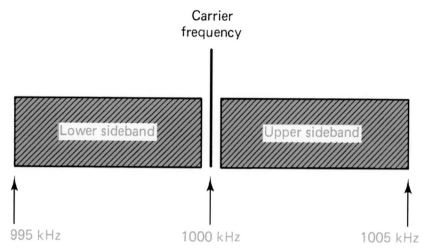

Figure 21.19   Sideband frequencies of an AM signal.

Percentage of Modulation

In AM radio, the **percentage of modulation** is not permitted by law to exceed 100%. Essentially, this means that the modulating component cannot cause the RF component to vary over 100% of its unmodulated value. **Figure 21.20** shows an AM signal with three different levels of modulation.

When the peak amplitude of modulating signal is less than the peak amplitude of the carrier, **modulation** is less than 100%. If the modulating component and carrier amplitudes are equal, 100% modulation is achieved. A modulating component which is greater than the amplitude of the carrier causes over-modulation. An **over-modulated** wave has an interrupted spot in the carrier wave. Over-modulation causes increased signal bandwidth and additional sidebands to be generated. This causes interference with adjacent channels.

Modulation percentage can be calculated or observed on an indicator. When operating voltage values are known, the percentage of modulation can be calculated. The formula is

$$\% \text{ modulation} = \frac{V_{\max} - V_{\min}}{2V_{\text{carrier}}} \times 100. \qquad (21.3)$$

For example, using the values of **Figure 21.20(a)**, the percentage of modulation is

$$\% \text{ modulation} = \frac{V_{\max} - V_{\min}}{2V_{\text{carrier}}} \times 100$$

$$= \frac{150 - 50}{2 \times 100} \times 100$$

$$= \frac{100}{200} \times 100$$

$$= 50\%.$$

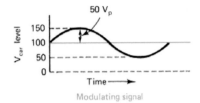

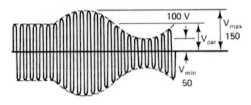

(a)

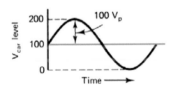

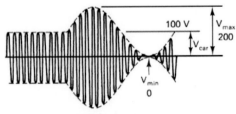

(b)

Figure 21.20 Continued.

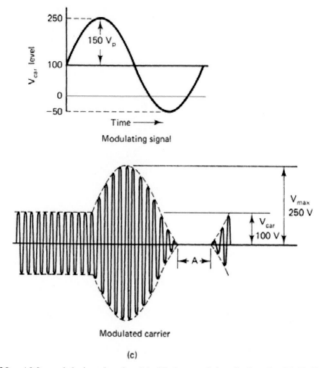

Modulated carrier

(c)

Figure 21.20   AM modulation levels. (a) Under-modulated signal. (b) Fully modulated signal. (c) Over-modulated signal.

Example 21-4:

What is the percentage of modulation for the signal in **Figure 21.20(b)**?

Solution

$$\%\text{modulation} = \frac{V_{\text{max}} - V_{\text{min}}}{2V_{\text{carrier}}} \times 100$$
$$= \frac{200 - 0}{2 \times 100} \times 100$$
$$= \frac{200}{200} \times 100$$
$$= 100\%.$$

Related Problem

What is the percentage of modulation for the signal in **Figure 21.20(c)**?

AM Communication System

A block diagram of an **AM communication system** is shown in **Figure 21.21**. The transmitter and receiver respond as independent systems. In a one-way communication system, there are one transmitter and an infinite number of receivers. Commercial AM radio is an example of one-way communication. Two-way communication systems have a transmitter and receiver

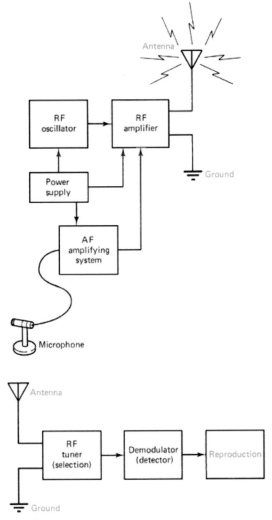

Figure 21.21 AM communication system.

at each location. CB radio systems are of the two-way type. The operating principles are basically the same for each system.

The **transmitter** of an AM system is responsible for signal generation. The RF section of the transmitter is primarily the same as that of a CW system. The modulating signal component is, however, a unique part of the transmitter. Essentially, this function is achieved by an AF amplifier. A variety of amplifier circuits can be used. Typically, one or two small-signal amplifiers and a power amplifier are suitable for low-power transmitters. The developed modulating component power must equal the power level of the RF output. Modulation of the two signals can be achieved at a number of places. **High-level modulation** occurs in the RF power amplifier. **Low-level modulation** is achieved after the RF oscillator. All amplification following the point of modulation must be linear. Only high-level modulation is discussed in this chapter.

The **receiver** of an AM system is responsible for signal interception, selection, demodulation, and reproduction. Most AM receivers employ the heterodyne principle in their operation. This type of receiver is known as a **superheterodyne** circuit. It is somewhat different from the heterodyne detector of the CW receiver. A large part of the circuit is the same as the CW receiver. This section discusses only the new functions of the AM receiver.

AM Transmitter

A wide range of **AM transmitters** are available. Toy "walkie-talkie" units with an output of less than 100 mW are very popular. **Citizen's band (CB) transmitters** are designed to operate at 27 MHz with a 5-W output. AM amateur radio transmitters are available with a power output of up to 1 kW of power. Commercial AM transmitters are assigned power-output levels from 250 W to 50 kW. These **stations** operate between **535 and 1620 kHz**. In addition to this, there are mobile communication systems, military transmitters, and public radio systems that all use AM. As a general rule, the **Federal Communications Commission** assigns frequency allocations and power levels.

An **AM transmitter** has a number of fundamental parts regardless of its operational frequency or power rating. It must have an oscillator, an AF component, an antenna, and a power supply. The **oscillator** is responsible for the RF carrier signal. The **audio** component is responsible for the intelligence being transmitted. The modulation function can be achieved in a variety of ways. The signal is radiated from the **antenna**.

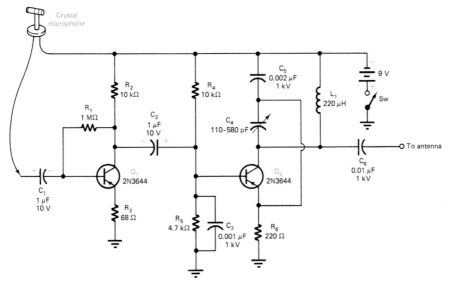

Figure 21.22 Simplified AM transmitter.

A simplified **AM transmitter** with a minimum of components is shown in **Figure 21.22**. Sound is first developed in the **crystal microphone** and then amplified by the AF signal amplifier, Q_1. The amplified signal is applied to transistor Q_2. It is then applied to transistor Q_2. This transistor is a Colpitts oscillator. L_1, C_4, and C_5 determine the RF frequency of the oscillator. The amplitude of the oscillator varies according to the AF component. Resistor R_3 is used to adjust the amplitude level of the AF signal. The developed AM output signal is applied to the transmitting antenna. With proper design of the transmitter, it can be tuned to the standard **AM broadcast band**. The frequency of the system is adjusted by capacitor C_4. The operating range of this unit is several hundred feet.

A schematic of an improved low-power AM transmitter is shown in **Figure 21.23**. Transistors Q_1, Q_2, and Q_3 of this circuit are responsible for the RF signal component. The AF component is developed by Q_4, Q_5, and Q_6. **High-level modulation** is achieved by this transmitter. The AF signal modulates the RF power amplifier Q_3.

Transistor Q_1 is the active device of a **Hartley oscillator**. This particular oscillator has a variable-frequency output of 1-3.5 MHz. The frequency-determining components are C_1 and L_{T1}. The output of the oscillator is coupled to the base of Q_2 by capacitor C_3. The **emitter-follower** output of the oscillator has very low impedance.

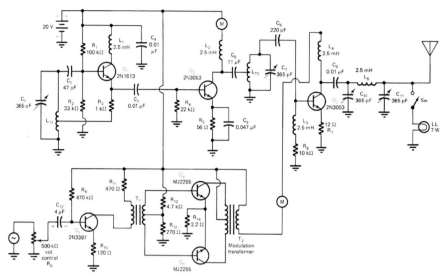

Figure 21.23   5-W AM transmitter.

Transistor Q_2 is an RF signal amplifier. This transistor is primarily responsible for increasing the signal level of the oscillator. It is also used to isolate the RF load from the oscillator. This is needed to improve oscillator frequency stability. When Q_2 is used in this regard, it is called a **buffer amplifier**.

The signal output of Q_2 must be capable of driving the **power amplifier**. In high-power transmitters, several RF signal amplifiers may be found between the oscillator and the power amplifier. Each stage is responsible for increasing the signal level to a suitable level. RF amplifiers of this type are often called **drivers**. Q_3 is an RF power amplifier. It is designed to increase the power level of the RF signal applied to its input. The output is used to drive the transmitting antenna. The load of Q_3 is a tuned circuit composed of C_{10}, L_5, and C_{11}. As you can recall from Chapter 4, this is a pi-section filter. A filter of this type is used to remove signals other than those of the resonant frequency. C_{11} is the output capacitor of the filter. It is adjusted to match the impedance of the antenna. C_{10} is the input capacitor. It is used to resonate the filter to the applied carrier frequency. **Resonance** of the tuning circuit occurs when the collector current meter dips to its lowest value. In the broadcasting field, an adjustment of this type is called dipping the final. Q_3 is operated as a class C power amplifier.

The modulating component of the transmitter is developed by an AF amplifier. Q_4 and Q_5 are **push-pull AF power amplifiers**. The developed

AF signal is applied to the modulation transformer T_2. This signal causes the collector voltage of Q_3 to vary at an AF rate instead of being DC. This causes the RF output to vary in amplitude according to the AF component. The output signal has a carrier and two sidebands. The power output is approximately 5 W. Do not connect the output of this transmitter to an outside antenna unless you hold a valid radio-telephone operator's license. A load lamp is used for operational testing. The intensity of the load lamp is a good indication of the RF power developed by the transmitter.

Simple AM Receiver

An **AM receiver** has four primary functions that must be achieved for it to be operational. No matter how complex or involved the receiver is, it must accomplish the following functions: **signal interception, selection, detection**, and **reproduction**. **Figure 21.24** shows a diagram of an AM receiver that accomplishes these functions. Note that the circuit does not employ an amplifying device. No electrical power source is needed to make this receiver operational. The signal source is intercepted by the antenna-ground network. The receiver is energized by the intercepted RF signal energy. A receiver of this type is generally called a **crystal radio**. The detector is a crystal diode.

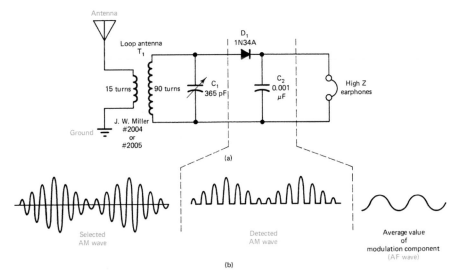

Figure 21.24 Crystal radio receiver. (a) Circuit. (b) Waveforms.

The functional operation of our crystal diode radio receiver is very similar to the CW receiver. This particular circuit does not employ an RF amplifier, a beat-frequency oscillator, or an AF amplifier. A strong AM signal is needed to make this receiver operational. Signal interception and selection are achieved in the same way as in the CW receiver. The detection, or **demodulation**, function of an AM receiver is responsible for removing the AF component from the RF signal. The detector is essentially a half-wave rectifier for the RF signal. Germanium diodes are commonly used as detectors. They are more sensitive than a silicon diode to RF signals.

The waveforms of **Figure 21.24** show how the **crystal diode** responds. The selected AM signal is applied to the diode. Detection is accomplished by rectification and filtering. The detected wave is a half-wave rectification version of the input. C_2 responds as an RF filter. It charges to the peak value of each RF pulse. It then discharges through the resistance of the earphone when the diode is reverse biased. The average value of the RF component appears across C_2. This is representative of the AF signal component. It energizes a small coil in the earphone. A thin metal disk in the earphone fluctuates according to coil energy. The earphone, therefore, changes electrical energy into sound waves. The **earphone** achieves AF signal reproduction.

Superheterodyne Receiver

Practically, all AM radio receivers in operation are of the **superheterodyne** type. This type of system accomplishes all the basic receiver functions. It has a number of circuit modifications that provide improved reception capabilities. It has excellent selectivity and is very sensitive to long-distance-signal reception. **Figure 21.25** shows a block diagram of an AM superheterodyne receiver. This particular receiver is designed for signal reception between approximately 550 and 1605 kHz, which is the AM band.

Superheterodyne operation is somewhat unusual compared with other methods of radio reception. Special RF amplifiers that are tuned to a fixed frequency are used in this circuit. These amplifiers are called **intermediate frequency (IF) amplifiers**. Each incoming station frequency is changed into the IF. The IF amplifier responds in the same manner to all incoming signals. Each receiver has a tunable CW oscillator. This CW or **local oscillator** generates an unmodulated RF signal of constant amplitude. This signal and the selected station are then mixed together. **Mixing**, or heterodyning, these two signals produces the IF signal. The IF contains all the modulation characteristics of the incoming RF signal. The IF signal is the difference beat

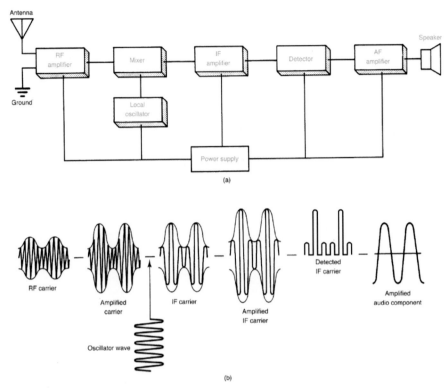

Figure 21.25 Superheterodyne receiver. (a) Block diagram. (b) Waveforms.

frequency. After suitable amplification, the IF signal is applied to the detector. The resulting AF output signal is then amplified and applied to the **speaker** for sound reproduction.

Figure 21.26 shows a diagram of the components of an **AM superheterodyne receiver**. Transistor Q_1 is a tuned RF amplifier. Adjustment of capacitor C_1 selects the desired RF signal frequency. Q_2 is the mixer stage. The selected RF signal and the local oscillator are mixed together in Q_2. Q_3 is the local oscillator. The oscillator signal is fed to the base Q_2 through C_8. Transistor Q_4 is the IF amplifier. The output of Q_2 is coupled to Q_4 by transformer T_4. This transformer is tuned to pass only an IF of 455 kHz to maintain a constant frequency. T_5 is the output **IF transformer**. The IF signal is coupled to the diode detector D_1 through this transformer. The detector rectifies the IF signal and develops the AF modulating component. C_{15} is the RF bypass filter capacitor and R_{12} is the volume control. Q_5 is the first AF signal amplifier.

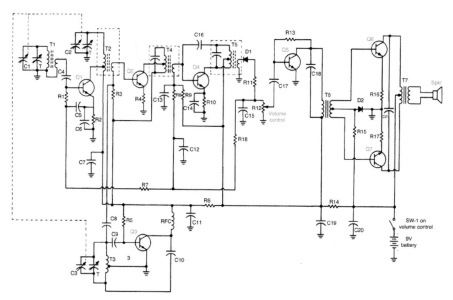

Figure 21.26   Transistor AM receiver.

Push-pull AF power amplification is achieved by transistors Q_6 and Q_7. The AF output signal is changed into sound by the speaker. The entire circuit is energized by a 9-V battery. Switch SW-1 is connected to the **volume control** and turns the circuit on and off.

Practically, all the **superheterodyne receiver** circuits have been discussed in conjunction with other system functions. The local oscillator, for example, is a variable-frequency Hartley oscillator. Mixing of the oscillator signal and selected RF signal has been described as **heterodyning**. Its output contains two input frequencies and two beat frequencies. The **detector** is essentially a crystal diode. An AF amplifying system is used to develop the audio signal component. The **speaker** was previously described as a transducer. All these operations remain the same when used in the superheterodyne receiver. Operational frequencies and circuit performance are generally somewhat different than previously described. We, therefore, direct our attention to specific circuit performance.

Using **Figures 21.25** and **21.26**, we can now see how the **superheterodyne** responds when tuned to a specific frequency. Assume that a number of AM signals are intercepted by the antenna-ground network. During the tuning process, a particular station frequency is selected from the AM band of signals. A desired **station frequency**, such as 1000 kHz, may then be selected. To do this, the **tuner** dial must be adjusted to 1000 kHz AM which then becomes the initial fundamental frequencies. The second fundamental frequency is an un-modulated CW signal that is produced by the oscillator in the receiver circuit. This frequency is designed to always be **455 kHz** above the initial fundamental frequency. For example, if the initial fundamental is 1000 kHz, the second fundamental will be 1455 kHz, which is the sum of 1000 and 455 kHz.

Adjustment of the tuning dial changes the *LC* circuit of the RF amplifier. Capacitors C_1, C_2, and C_3 are ganged together. One tuner adjustment alters the three tuned circuits at the same time. C_1 and C_2 tune the input and output of the RF amplifier to 1000 kHz. C_3 adjusts the frequency of the oscillator to 1455 kHz. The *LC* components of the oscillator are designed to generate a frequency that will be 455 kHz higher than the selected RF signal, a standard IF.

The 1000-kHz signal now passes through the input tuner and is applied to the base of Q_1. This **common-emitter amplifier** increases the voltage level of the applied signal. C_2 and T_2 are also tuned to pass the signal to the base of Q_2. The oscillator signal is coupled to the base of Q_2 through capacitor C_8. The incoming signal is AM, and the oscillator signal is CW. Through the heterodyning action, the collector of Q_2 has four signals. The fundamental frequencies are 1000-kHz AM and 1455-kHz CW. The beat frequencies are 2455-kHz AM and 455-kHz AM based on the sum and difference of the fundamental frequencies, respectively.

Example 21-5:

When an AM radio receiver is tuned to 1500 kHz, the local oscillator will produce a CW signal of _____kHz.

Solution

Local oscillator frequency = 1500-kHz AM + 455 kHz (IF frequency)
$$= 1955\text{-kHz CW}$$

Related Problem

When an AM radio receiver is tuned to 550 kHz, the local oscillator will produce a CW signal of _____kHz.

Assume that the receiver is tuned to select a station at 1340 kHz. C_1, C_2, and C_3 change to the new signal frequency. The oscillator develops a CW signal of 1795 kHz. The mixer has 1340 and 1795 kHz applied to its input.

The collector of Q_2 has these two fundamental frequencies plus 3135 and 455 kHz. The IF amplifier section processes the 455-kHz signal and applies it to the detector. The AF component is then recovered and processed for reproduction.

Suppose that the receiver is tuned to a station located at 600 kHz. This frequency is amplified and applied to the **mixer**. The oscillator now sends a CW signal of 1055 kHz to the mixer. The collector of Q_2 has signals of 600, 1055, 1655, and 455 kHz. The IF amplifier again passes only the 455-kHz signal for reproduction.

Example 21-6:

What beat frequencies are generated when an AM radio receiver is tuned to 1500 kHz, and the local oscillator produces a CW signal of 1955 kHz?

Solution
> The sum beat frequency = 1955-kHz CW + 1500-kHz AM
> = 3455-kHz AM
> The difference beat frequency = 1955-kHz CW - 1500-kHz AM
> = 455-kHz AM

Related Problem

What beat frequencies are generated when an AM radio receiver is tuned to 600 kHz, and the local oscillator produces a CW signal of 1055 kHz?

The **IF amplifier** has fixed tuning in its input and output circuits. T_4 is the input IF transformer. It is tuned to be resonant at **455 kHz**. This frequency passes into the base of Q_4 with a minimum of opposition. The other three signals encounter very high impedance. 455 kHz will be amplified by Q_4. The signal is then coupled to the detector through transformer T_5. The **detector** recovers the AF component and applies it to the AF amplifier system for sound reproduction. The AF signal component then produces sound from the speaker. The **volume control** is adjusted to a desired signal level. The receiver is operational and performing its intended function.

The basic operation of an **AM superheterodyne receiver** thus requires that the CW oscillator be designed so that it will always be higher in frequency than the incoming station frequency. In effect, its frequency is the incoming station frequency plus the IF. A standard IF for AM receivers is 455 kHz. Through this type of circuit design, the IF can be adjusted to a specific frequency.

Today, we find all the circuitry of the superhetrodyne receiver is built on a single **integrated chip (IC)**. The audio frequency drives a speaker or headphone load after amplification. Tuning is usually achieved by a switch or push button which automatically scans the available AM band for station selection.

Self-Examination

Answer the following questions.

16. The mixture of AF and an RF carrier wave is called _____.
17. The AM broadcast band is _____ kHz to _____ kHz.
18. Three levels of modulation are _____, _____, and _____.
19. AM transmitter parts consists of the _____, _____, _____, and _____.
20. Minimal AM receiver parts consists of the _____, _____, _____, and _____.
21. The detector of a crystal AM radio is a(n) _____.
22. Most AM receivers today are the _____ type.
23. Superheterodyne receivers use specially tuned RF amplifiers called _____ amplifiers.
24. If a 50 V_{p-p} carrier wave changes in amplitude from 75 to 25 V_{p-p} when modulated, what is the percent modulation?
25. How does a diode achieve demodulation of an AM signal?

21.4 Frequency Modulation Communication

Frequency modulation (FM) has a number of unique advantages over other communication systems. It uses very low audio signal levels for modulation, has excellent frequency reproduction capabilities, and is nearly immune to noise and interference. FM is commonly used in the **very high frequency (VHF)** range of 30-300 MHz and extends into the **ultra-high frequency (UHF)** range of 300-3000 MHz. The **commercial FM band** is 88-108 MHz. FM is the dominant form of communication for private two-way mobile communication services. This section discusses the process of frequency modulation and describes the operation of FM communication system components.

21.4 Analyze the operation of an FM communication system.

In order to achieve objective 21.4, you should be able to:

- explain the process of amplitude modulation;

- identify the components of an amplitude modulation communication system;
- define center frequency, ratio detector, and phasor.

Basic FM Operation

In an FM communication system, intelligence is superimposed on the carrier by variations in frequency. An FM signal does not effectively change in amplitude. The modulating component does, however, cause the carrier to shift above and below its **center frequency**. Each **FM station** has an assigned center frequency. Receivers are tuned to this frequency for signal reception. When the carrier is unmodulated, it rests at the center frequency. Frequency allocations for FM are made by the FCC.

Examine the **FM signal** of **Figure 21.27**. Note that the modulating component is low-frequency AF, and the carrier is RF. In commercial FM, the modulating component could be an audio signal of 20-15 kHz. The **carrier** would be of some RF value between **88 and 108 MHz**. As the

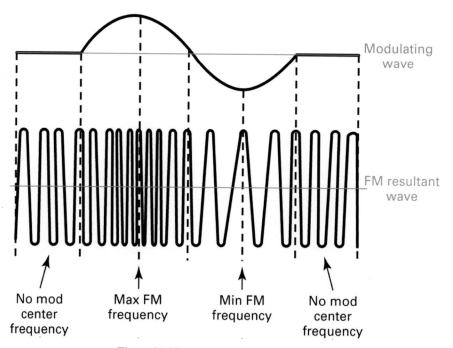

Figure 21.27   Frequency modulation waves.

modulating component changes from 0° to 90°, it causes an increase in the carrier frequency. The carrier rises above the center frequency during this time. Between 90° and 180° of the audio component, the carrier decreases in frequency. At 180°, the carrier returns to the center frequency. As the audio signal changes from 180° to 270°, it causes a decrease in carrier frequency. The carrier drops below the center frequency during this period. Between 270° and 360°, the carrier rises again to the center frequency. This shows us that without modulation applied, the carrier rests at the center frequency. With modulation, the carrier shifts above and below the center frequency.

The **amplitude** of the modulating component is also extremely important in the shifting operation. The larger the modulating signal, the greater the frequency shift. For example, if the center frequency of an FM signal is 100 MHz, a weak audio signal may cause a 1-kHz change in frequency. The carrier would change from 100.001 to 99.999 MHz. A strong audio signal could cause the carrier to change as much as 75 kHz. This action would cause a shift of 100.075-9.925 MHz. It is important to note that the amplitude of the carrier does not change. **Figure 21.28** shows how the amplitude of the audio component changes the frequency of the carrier.

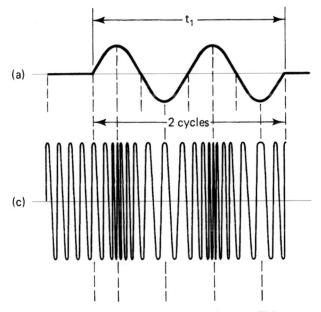

Figure 21.28 Influence of AF amplitude on FM.

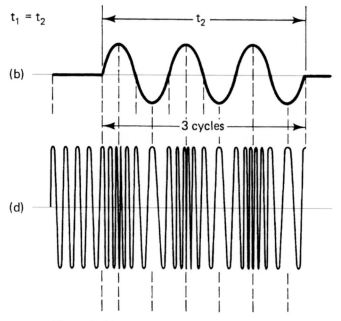

Figure 21.29   Influence of AF frequency on FM.

The frequency of the modulating component is another important consideration in FM. **Modulating frequency** determines the rate at which the frequency change takes place. With a 1-kHz modulating component, the carrier oscillates 1000 times per second. A modulating component of 10 kHz causes the carrier to shift 10,000 times per second.

Figure 21.29 shows how the **FM carrier** responds to two different frequencies. **Figures 21.29(a)** and **21.29(b)** are the **modulating frequencies** that are used to alter a high-frequency carrier signal. **Figures 21.29(c)** and **21.29(d)** are the resulting effects of frequency changes in the modulating signal. For each peak that appears in the modulating signal, the modulated carrier wave will show a corresponding dark area. This indicates that the frequency is higher as the waves get closer together. By examining an FM waveform for a given time duration, and counting the number of darker areas, we can see how the modulating frequency changes the frequency of the carrier. **Figure 21.29(c)** has two dark areas in the time duration indicated by t1, as compared to three dark areas in **Figure 21.29(d)** for the same time duration indicated by t2. Thus, the modulating frequency of the signal in **Figure 21.29(a)** is lower than that of the signal in **Figure 21.29(b)**.

FM Communication System

A block diagram of an **FM communication system** is shown in **Figure 21.30**. The transmitter and receiver respond as independent systems. In commercial FM, each station has a transmitter. There can be an infinite number of FM receivers. In two-way mobile communication systems, there is a transmitter and receiver at each location. FM mobile communication systems are classified as **narrow-band FM**. This type of system transmits only voice signals. The carrier of a narrow-band FM system deviates by only ±5 kHz. Commercial FM is considered to be **wideband FM**. The carrier of a commercial FM system deviates ±75 kHz at 100% modulation. The operational principles of narrow-band and wideband FM are primarily the same for each system.

The **transmitter** of an FM system is primarily responsible for signal generation (see **Figure 21.31(a)**). Note that it employs an audio amplifying system, RF oscillator, linear RF amplifier, linear RF power amplifier, and a power supply. The **modulating component** of an FM system is applied directly to the oscillator. Audio frequency changes in the applied signal cause the oscillator to vary in frequency. Amplifiers following the oscillator are classified as linear devices.

The **receiver** of an FM system is responsible for signal interception, selection, demodulation, and reproduction (see **Figure 21.30(b)**). Most FM

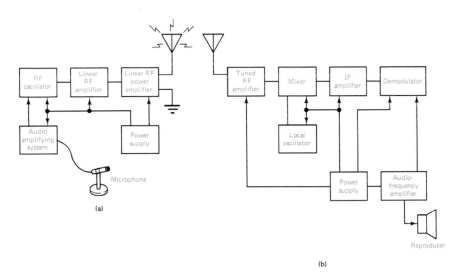

Figure 21.30 FM communication system. (a) Direct FM transmitter. (b) Superheterodyne FM receiver.

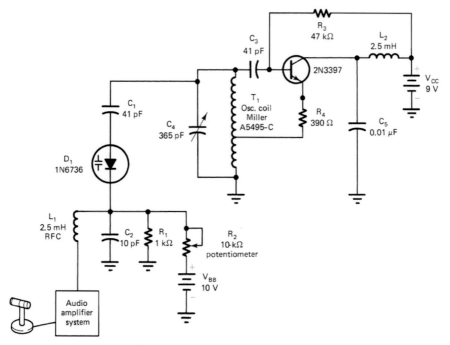

Figure 21.31   Direct FM transmitter.

receivers are of the superheterodyne type. These receivers are similar in operation to the AM superheterodyne circuit. The selection frequency is much higher (88-108 MHz) for commercial FM. A standard IF of 10.7 MHz is used in an FM receiver. The **demodulator** of an FM receiver is uniquely different from an AM detector. The FM demodulator is designed to respond to changes in the carrier frequency. The modulating component is recovered from these frequency changes. The remainder of the circuit is the same as the AM superheterodyne.

FM Transmitter

There is a rather extensive list of different FM transmitters available. **Wireless microphones** have an output of approximately 100 mW. **Maritime mobile FM communication** systems operate with a few watts of power. **Two-way land mobile communication** employs low-power transmitters. **Educational FM** transmitters can operate with a power of several hundred thousand watts. **Commercial FM** transmitters have a similar power rating. In addition to these types are public service radio and military FM. The frequency allocation and power capabilities of an FM transmitter are allocated and regulated by the FCC.

An **FM transmitter**, regardless of its power rating or operational frequency, has a number of fundamental parts. The oscillator is responsible for the development of the RF carrier. A large number of FM transmitters employ direct oscillator modulation. This means that the frequency of the oscillator is shifted by direct application of the modulating component. The RF signal is then processed by a number of amplifiers. The power amplifier ultimately feeds the FM carrier into the antenna for radiation.

A simplified **direct FM transmitter** is shown in **Figure 21.31**. A basic Hartley oscillator is used as the RF signal source. With power applied, the oscillator generates the RF carrier. The frequency of the oscillator is determined by the values of C_4 and T_1. Without modulation, the oscillator would generate a constant frequency. This is representative of the unmodulated carrier wave. **Modulation** of the oscillator is achieved by voltage changes across a varicap diode. The capacitance of this diode changes with the value of the reverse-bias voltage. Bias voltage is used to establish a DC operating level for this diode. A change in audio signal voltage causes the capacitance of the diode to change at an audio rate. Note that the series network of C_1, D_1, and C_2 is connected in parallel with capacitor C_4 of the oscillator tank circuit. A change in audio signal level causes a corresponding change in tank circuit capacitance. This, in turn, causes the oscillator frequency to change according to the modulating component. The modulating component is applied directly to the oscillator's frequency-determining components.

Direct FM can be achieved in a number of ways. For example, the varicap diode discussed in Chapter 5 could be used for transmitter equipment. This application is only one of a large number of circuit variations of the varicap diode. Circuits of this type are well suited for voice communication in **narrow-band FM**. The power output of an FM transmitter is dependent primarily on its application. In some systems, the oscillator output can be applied directly to the **antenna** for radiation. In other systems, the power level must be raised to a higher level. The output of the oscillator is applied to linear amplifiers for processing. The amplification function is primarily the same as that achieved by CW and AM systems. We do not repeat the discussion of RF power amplification for FM transmitters. The primary difference in FM power amplification is the resonant frequency of the tuned circuits. FM usually operates in the **VHF band**. This generally calls for smaller capacitance and inductance values in the resonant circuits.

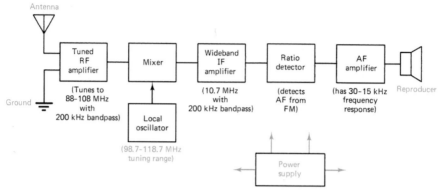

Figure 21.32   Block diagram of an FM receiver.

FM Receiver

FM signal reception is achieved in basically the same way as AM and CW. An **FM superheterodyne** circuit is designed to respond to frequencies in the VHF band. **Commercial FM** signal reception is in the range of **88-108 MHz**. Higher frequency operation generally necessitates some change in the design of antenna, RF amplifier, and mixer circuits. These differences are due primarily to the increased frequency rather than the FM signal. The RF and IF sections of an FM receiver are somewhat different. They must be capable of passing a 200-kHz bandwidth signal instead of the 10-kHz AM signal. The most significant difference in FM reception is in the **demodulator**. This part of the receiver must pick out the modulating component from a signal which changes in frequency. In general, this circuit is more complicated than the AM detector. The AF amplifier section of an FM receiver is generally better than an AM receiver. It must be capable of amplifying frequencies of 30-15 kHz. **Figure 21-32** shows a block diagram of an FM superheterodyne receiver. Some differences in FM reception are indicated by each functional block.

FM Demodulation

The **demodulator** of an FM receiver is the primary difference between AM and FM superheterodyne receivers. A number of demodulators have been used to achieve this receiver function. A **ratio detector** may be used to achieve this function. This circuit is an outgrowth of earlier FM discriminators. FM demodulation is, today, almost exclusively performed by integrated circuits making use of advanced digital techniques.

A basic ratio detector is shown in **Figure 21.33**. The operation of this detector is based on the phase relationship of an FM signal that is applied to the input. This signal comes from the IF amplifier section of the receiver. It is 10.7 MHz and deviates in frequency. The design of the circuit causes two **IF signal** components to appear in the secondary winding of the transformer. One part of this component is coupled by capacitor C_1 to the center connection of the transformer. This resulting voltage is considered to be V_1. Note the location of V_1 in the circuit and voltage line diagrams.

The second IF signal component is developed inductively by transformer action. Design of the secondary winding causes two voltage values to be developed. These voltages are labeled V_2 and V_3 in the circuit and voltage line diagrams. They are of equal amplitude and 180° out of phase with each

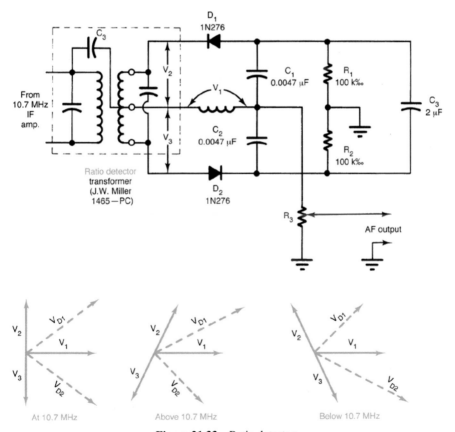

Figure 21.33 Ratio detector.

other. The **center-tap** connection of the secondary winding is used as the common reference point for these voltage values.

The resulting secondary voltage is based on the combined component voltage values of V_1, V_2, and V_3. V_{D1} and V_{D2} are the resulting secondary voltages. These two voltage values appear across diodes D_1 and D_2. The developed voltage for each diode is based on the phase relationship of the two IF components. Note the location of V_{D1} and V_{D2} in the circuit and the voltage line diagrams.

Operation of the ratio detector is based on voltage developed by the transformer for diodes D_1 and D_2. With no modulation applied to the FM carrier, the transformer has a 10.7-MHz IF applied to the transformer. The developed voltage values are shown by the center-frequency voltage line diagram. This method of voltage display is generally called a **phasor diagram** (see **Figure 21.33**). A **phasor** shows the relationship of voltage values (line length) and their phase relationship (direction). Using phasors, a sinusoidal waveform can be represented simply by a line having a certain length corresponding to its peak voltage or current amplitude. The phasor is oriented at an angle with the horizontal axes corresponding to the phase. For example, the phasor for a 10.7-MHz signal, V_1, having a certain amplitude represented by its length is shown in **Figure 21.33(b)**. It is oriented at 0° with respect to the horizontal axis, indicating that it has a 0° phase. When at the center frequency for 10.7 MHz, note that V_2 is at right angles to V_1, making it +90° out of phase with V_1. V_3 is also at right angles to V_1 but in the opposite direction, making it -90° out of phase with V_1. Thus, V_2 and V_3 are mutually 180° out of phase with each other when the system operates at the center frequency. At this frequency, note that the resulting diode voltage values (V_{D1} and V_{D2}) are of the same value as well. This means that each diode receives the same voltage value. D_1 and D_2 conduct an equal amount of current.

When the carrier swings above the center frequency, it causes the IF to swing above its resonant frequency. Note the resulting phasor for this change in frequency. V_{D1} is longer than V_{D2}. This means that D_1 conducts more current than D_2. In the same regard, note how the phasor changes when the frequency swings below resonance in **Figure 21.33(b)**. This condition causes D_1 to be less conductive and D_2 to be more conductive. Essentially, this means that the IF signal is translated into different diode voltage values.

Figure 21.34 shows how an input RF voltage (10.7-MHz IF) can be used to develop an AF signal. The two diodes of the **ratio detector** are

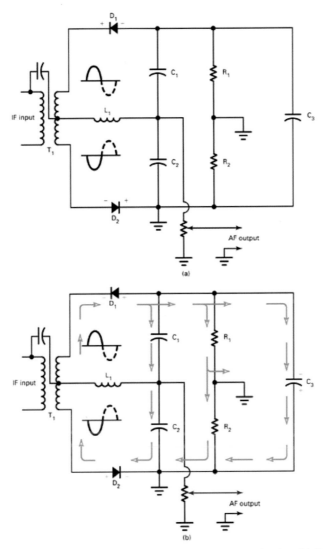

Figure 21.34 Diode conduction of a ratio detector. (a) Nonconduction. (b) Conduction.

connected in series with the secondary winding and capacitors C_1 and C_2. For one alternation of the input signal, the two diodes are reverse biased. No conduction occurs during this alternation (see **Figure 21.34(a)**). For the next alternation, both diodes are forward biased. The input signal voltage is then rectified (see **Figure 21.34(b)**). Essentially, this means that the incoming signal is changed into a pulsating waveform for one alternation.

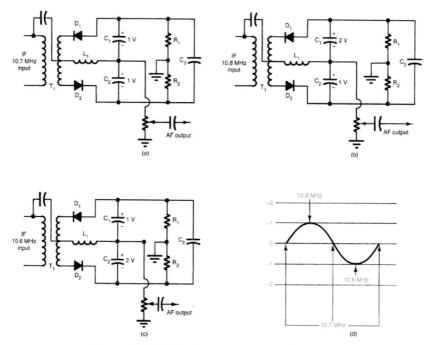

Figure 21.35   Ratio detector response to frequency.

Figure 21.35 shows how the **ratio detector** responds when the input signal deviates above and below the center frequency. Keep in mind that conduction occurs only for one alternation of the input. **Figure 21.35(a)** shows how the circuit responds when the input is at its 10.7-MHz **center frequency**. Each diode has the same input voltage value for this condition of operation. Capacitors C_1 and C_2 charge to equal voltages, as indicated. With respect to ground, the output voltages are -2 V. Note this point on the **output voltage** waveform in **Figure 21.35(d)**.

Assume that the input **IF signal** increases to 10.8 MHz as shown in **Figure 21-35(b)**. This condition causes D_1 to receive more voltage than D_2. C_1 charges to -3 V, whereas C_2 charges to -1 V. With respect to ground, the output voltage rises to -1 V. Note this point on the output voltage waveform in **Figure 21.35(d)**.

The input IF signal then decreases to 10.6 MHz as shown in **Figure 21.35(c)**. This condition causes D_2 to receive more voltage than D_1. C_2 charges to -3 V, while C_1 charges to -1 V. With respect to ground, the output voltage drops to −3 V. Note this point on the **output voltage** waveform in **Figure 21.35(d)**.

The output of the ratio detector is an **AF signal** of 2 V_{pp}. This signal corresponds to the frequency changes placed on the carrier at the transmitter. In effect, we have recovered the AF component from the FM carrier signal. This signal can then be amplified by the AF section for reproduction.

Self-Examination

Answer the following questions.

26. The FM band is from _____ MHz to _____ MHz.
27. With modulation, an FM carrier shifts above and below its _____.
28. A standard IF of _____ MHz is used in FM receivers.
29. A(n) _____ detector may be used in FM receivers.
30. What is the function of a ratio detector in an FM receiver?
31. What is the major difference between AM and FM communications?
32. In a phasor diagram, the length of a phasor signifies the _____, and the direction signifies _____.

21.5 Troubleshooting RF Communication Circuits

For **troubleshooting** radio frequency circuits, a known signal source should be available, along with instruments such as oscilloscopes and voltmeters for examining the output response. Owing to the complexity of radio frequency circuits troubleshooting is a little bit more difficult. By regarding the communication system as one composed of different blocks may make the troubleshooting process easier. An initial step in the servicing procedure would be to ensure that proper operating voltages appear at key test points, such as at the input and the output. Improper voltages, grounding, and interference problems can cause degradation of signals at the points being tested. The operation of **audio output devices** such as speakers can be verified by applying a signal to the input while examining the output response. If the output devices are functioning properly, one may examine the previous stages using the same procedure of injecting a known signal at the input and again examining the output. Successively applying a known input signal to preceding stages while observing the output permits the operation of the entire communication system to be evaluated.

21.5. Troubleshoot an RF communication circuit.

In order to achieve objective 21.5, you should be able to:

- examine the circuit diagram of a communication system to determine key signal injection points for evaluation of the system;
- distinguish between normal and faulty operation of radio frequency circuits.

Troubleshooting AM Transmitters

Examine **Figure 21.22** which shows an **AM transmitter** circuit. AM transmitters are prone to faults arising from the operation of the oscillator part of the circuit or those arising from improper adjustment of the turning circuit. The presence of such faults can be identified by examining the status of the output waveforms. For the purpose of troubleshooting, it is common practice to replace the antenna with a lamp. The lamp then serves as the load instead of the signal being radiated from the antenna. Assume that a 100-Hz audio frequency signal is applied to the input from a function generator, and power is applied to the transmitter circuit. When this occurs, an AM signal should appear across the load lamp. The brightness of the lamp should change with variations in the frequency and the amplitude of the audio signal. This can also be observed using an oscilloscope. For example, if the **audio input** into the transmitter is removed, the brightness of the lamp will remain constant. An **oscilloscope** will show that only the RF signal is present, but no modulation occurs. On the other hand, if the amplitude of the audio input is larger than the amplitude of the RF carrier, it will overdrive the transistor amplifier. This causes the RF output to be distorted. An oscilloscope connected to the load will indicate this by creating void spots and by flattening of the peaks of the output waveform.

Troubleshooting AM Receivers

Examine **Figure 21.25** which shows an AM receiver block diagram and sample waveforms that will appear at various stages of the receiver circuit. The circuit diagram shown in **Figure 21.25(a)** is fairly complex. For troubleshooting purposes, it is easier to view the circuit diagram as being composed of blocks or sub-circuits that achieve specific functions as shown in **Figure 21.25**. The proper operation of the overall receiver circuit can be determined by examining the waveforms appearing at the output of each block.

In the **troubleshooting** of this circuit, these waveforms should appear at various points of the circuit shown in **Figure 21.25**. These can be used for comparison purposes while determining which particular section of the receiver is not functioning properly. If the **waveform** at a specific test point does not appear or is distorted, this indicates that the circuit is not functioning properly. When this occurs, the waveform at the preceding test points should be evaluated to determine the problem.

AM receivers are prone to have a number of operational problems. These include **component failures** which may be caused by an excessive heating of components, power supply problems, and power surges. The **power amplifier** section is particularly vulnerable to these factors because it consumes more power. Other factors that may influence the operation of a receiver are outside interference caused by weather conditions and obstruction of the AM signal. In addition to this, internal receiver circuitry may be subjected to excessive moisture which may cause corrosion of the components and conductors. Improper **shielding** and **grounding** may also influence circuit operation. Most AM receivers do not use discrete components; however, the study of circuit functional blocks such as these are important for problem solving for electronics.

Summary

- An RF communication system consists of a transmitter and one or more receivers.
- The transmitter of an RF communication system serves as the signal source.
- The receiver of an RF communication system detects, amplifies, and reproduces the signal.
- Intelligence is applied to the transmitter according to the design of the communications system.
- Three types of communication systems are continuous wave (CW), amplitude modulation (AM), and frequency modulation (FM).
- Electromagnetic waves radiate from the antenna of the transmitter.
- Low-frequency waves (30-300 kHz) and medium frequency waves (300-3000 kHz) tend to follow the curvature of the earth; this type of radiation pattern is called a **ground wave**.
- Very high-frequency waves (30-300 MHz) tend to move in a straight line; this type of radiation pattern is called **line-of-sight transmission**.

- A continuous-wave (CW) communication system generates an electromagnetic wave of a constant frequency and amplitude.
- An amplitude modulation (AM) communication system changes the amplitude of a generated RF signal by the intelligence signal.
- A frequency modulation (FM) communication system superimposes intelligence on the carrier by variations in frequency.
- A simple CW communication system has an oscillator, power supply, and antenna; the receiver has an antenna-ground network, a tuning circuit with an RF amplifier, a heterodyne detector, a beat-frequency oscillator, an audio amplifier, and a speaker.
- To overcome the shortcomings of a single-stage CW transmitter, a power amplifier is incorporated after the oscillator.
- The signal selection function of radio receivers refers to its ability to pick out a desired RF signal.
- The purpose of heterodyning is to mix two AC signals.
- The space that an RF signal occupies along the frequency spectrum is called a channel.
- A radio receiver is responsible for signal interception, selection, demodulation, and reproduction.
- A ratio detector is used as an FM demodulator circuit.

Formulas

(21-1) $B_1 = F_1 + F_2$ Beat frequency 1.
(21-2) $B_2 = F_2 - F_1$ Beat frequency 2.
(21-3) % modulation = $\frac{V_{max} - V_{min}}{2V_{carrier}} \times 100$ Percentage of modulation in an AM communication system.

Review Questions

Answer the following questions.

1. The 535-1620 kHz frequency range is typically used for the following type of communication signals:
    a. CW
    b. AM
    c. FM
    d. AF

2. The 88-108 MHz frequency range is typically used for the following type of communication signals:

    a. CW
    b. AM
    c. FM
    d. AF

3. Which of the frequencies listed below would be associated with the VHF band?

    a. 10 kHz
    b. 100 kHz
    c. 10 MHz
    d. 100 MHz

4. What type of wave uses an antenna and causes waveforms to be reflected back to the Earth from the ionosphere for communication purposes?

    a. Ground wave
    b. Sky wave
    c. Ionosphere waves
    d. Line of sight transmission

5. What type of wave follows the curvature of the earth and is used for communications up to 100 miles during the day or night

    a. Ground wave
    b. Sky wave
    c. Ionosphere waves
    d. Satellite waves

6. In CW communication, information is added to a high-frequency constant amplitude signal by using a:

    a. Microphone
    b. Code key
    c. AM
    d. FM

7. In a CW receiver, which component is responsible for heterodyning?

    a. Audio amplifier
    b. RF amplifier
    c. Turning and station selection
    d. Beat-frequency oscillator

8. What part of the AM communication signal has the highest frequency?

    a. Modulating signal

    b. RF carrier wave

    c. AF signal

    d. BFO

9. When portions of an AM waveform are absent or distorted, it indicates:

    a. Under-modulation

    b. Over-modulation

    c. Full modulation

    d. No modulation

10. What is the distinguishing characteristic of an AM superheterodyne receiver?

    a. Local oscillator

    b. RF amplifier

    c. IF amplifier

    d. LC tuning circuit

11. A communication system in which the amplitude of the carrier remains constant, but the frequency changes, corresponding to changes in the modulating component, is a(n):

    a. AM

    b. FM

    c. CW

    d. AF

12. Which of the following components of a FM superheterodyne receiver is essential for extracting the audio signal from an FM signal?

    a. IF amplifier

    b. Antenna

    c. Turning circuit

    d. Ratio detector

13. The IF of a superheterodyne receiver is the:

    a. Frequency of the local oscillator

    b. Sum of the local oscillator and the modulated signal frequencies

    c. Difference of the local oscillator and the modulated signal frequencies

    d. Sum of the local oscillator and the audio signal frequencies

14. After the modulating component of an FM signal has been extracted, it is then processed by the:

    a. AF amplifier
    b. Detector
    c. *LC* tuner
    d. IF amplifier

15. The phase detector of an FM receiver is responsible for:

    a. Changing RF to AF
    b. Changing AF to RF
    c. Mixing the local oscillator and FM signals
    d. Maintaining a constant IF signal

Problems

Answer the following questions.

1. If an AM superheterodyne receiver is tuned to a local radio station at 680 kHz, the frequency of the local oscillator is:

    a. 680 kHz
    b. 1360 kHz
    c. 1135 kHz
    d. 225 kHz

2. If an FM superheterodyne receiver is tuned to a local radio station at 100 MHz, the frequency of the local oscillator is:

    a. 100 MHz
    b. 110.7 MHz
    c. 89.3 MHz
    d. 10.7 MHz

3. Referring to **Figure 21.19**, determine the bandwidth of an AM signal that radiated an antenna operating at a 1-MHz carrier frequency that is modulated by a 5 kHz signal:

    a. 5 kHz
    b. 10 kHz
    c. 20 kHz
    d. 1 MHz

4. Referring to **Figure 21.20**, if an AM signal has a $V_p = V_{carrier} = 100$ V and is modulated by a signal having a peak amplitude of 60 V, so that $V_{max} = 160$ V and $V_{min} = 40$ V, determine the percentage modulation:

    a. 50%
    b. 60%
    c. 100%
    d. 120%

5. Referring to **Figure 21.36**, if the frequency of the local oscillator is out of step with the incoming station frequency of 1000-kHz AM, by 100 kHz, the signal that appears at test point 5 will be:

    a. No signal
    b. Normal IF signal
    c. Only the audio component
    d. Only the unmodulated RF signal

Answers

Examples

21-1. 3.04 MHz
21-2. 25 MHz
21-3. 2 kHz
21-4. 125% (with $V_{min} = 0$ V)
21-5. 1005-kHz CW
21-6. Sum beat frequency = 1655-kHz AM, Difference beat frequency = 455-kHz AM. It may be interesting to note that the difference beat frequency is always 455-kHz AM in the IF band.

Self-Examination

21.1

1. transmitter
2. receiver
3. one-way
4. two-way
5. continuous-wave (CW), amplitude modulation (AM), frequency modulation (FM) (any order)
6. sky

21.2

7. Morse

8. signal selection

9. heterodyning

10. mixer

11. power supply, code key, RF oscillator, RF power amplifier, and antenna

12. By interrupting the RF wave produced by the oscillator

13. Constant amplitude RF signal

14. This is the 0 beating condition and no sound output will occur

15. A low tone AF is produced

21.3

16. modulation

17. 535, 1620

18. under-modulation, over-modulation, 100% modulation (any order)

19. oscillator, AF component, antenna, power supply (any order)

20. antenna, RF tuner, demodulator, sound output (any order)

21. diode *or* half-wave rectifier

22. superheterodyne

23. intermediate frequency *or* IF

24. 50%

25. The diode works as a half-wave rectifier of the incoming AM signal. This half-wave rectified voltage is used to charge a capacitor at an AF rate, thereby extracting the audio portion of the signal.

21.4

26. 88, 108

27. center frequency

28. 10.7

29. ratio

30. It detects the phase relationship of an FM signal applied to its input.

31. The manner in which the carrier signal changes with respect to the amplitude of the audio frequency signal determines the type of modulation. If the modulating component causes a change in the amplitude of the carrier, it is called amplitude modulation, and if it causes a change in the frequency of the carrier, it is called frequency modulation.

32. amplitude, phase relationship

Terms

Transmitter

The signal source of an RF communications system.

Receiver

The component of an RF communications system that picks a signal out of the atmosphere and uses it to reproduce sound.

Ground wave

Electromagnetic waves that follow the curvature of the earth.

Sky wave

An RF signal radiated from an antenna into the ionosphere.

Ionosphere

A layer of ionized particles in the atmosphere.

Line-of-sight

An RF signal transmission that radiates out in straight lines because of its short wavelength.

Radio telegraphy

The process of conveying messages by coded telegraph signals.

Ganged

Two or more components connected together by a common shaft - a three-ganged variable capacitor.

Heterodyning

The process of combining signals of independent frequencies to obtain a different frequency.

Beat frequency

A resulting frequency that develops when two frequencies are combined in a nonlinear device.

Beat-frequency oscillator (BFO)

An oscillator of a CW receiver. Its output beats with the incoming CW signal to produce an audio signal.

Zero beating

The resulting difference in frequency that occurs when two signals of the same frequency are heterodyned.

Modulation

The process of changing some characteristic of an RF carrier so that intelligence can be transmitted modulating component.

Demodulation

The process of extracting the low-frequency intelligence (usually audio) signal from the high-frequency carrier signal.

Carrier wave

An RF wave to which modulation is applied.

Sidebands

The frequencies above and below the carrier frequency that are developed because of modulation.

Amplitude modulation

The communication process in which the amplitude of the carrier wave is varied according to the changes in the amplitude or frequency of the low-frequency (usually audio) component.

Frequency modulation

The communication process in which the frequency of the carrier wave is varied according to the changes in the amplitude or frequency of the low-frequency (usually audio) component.

Channel

The space that an AM signal occupies with its frequency.

Continuous wave

An electromagnetic wave of a constant frequency and amplitude. Intelligence is injected into the continuous wave by interrupting it with a coded signal.

Channel

The space that an AM signal occupies with its frequency.

Superheterodyne

A communication receiving circuit that uses a local oscillator to produce modulated IF signal for accomplishing basic receiver functions

High-level modulation

A situation in which the modulating component is added to an RF carrier in the final power output of the transmitter.

Antenna

A component responsible for transmitting and receiving radio frequency signals.

Low-level modulation

A situation in which the modulating component is added to an RF carrier in the RF oscillator circuit.

Buffer amplifier

An RF amplifier that follows the oscillator of a transmitter. It isolates the oscillator from the load.

Center frequency

The carrier wave of an FM system without modulation applied.

Ratio detector

A circuit used as a demodulator in FM radio receivers.

Phasor

A line used to denote value by its length and phase by its position in a vector diagram.

22

Communications System Applications

A number of specialized communication systems are in common use today. These include things such as television systems, cell phones, satellite, and wireless computer communications. The primary function of a communication system is to transfer information from one point to another. The information being transmitted includes text, data, graphics, audio, and video. The value of audio and video signals can change in a continuous manner. This is considered to be an **analog signal**. Text and data, on the other hand, may change only in specific or discrete intervals. This is considered to be a **digital signal**. The trend today is to use digital signals for representing all forms of data.

Common communication methods include **CW, AM, FM**, and others. After the information is received, it must be processed for converting it into a usable format by the system. **Television** may be regarded as a specialized communication system since it uses both AM and FM transmissions. In addition to this, cell phones and computer systems represent a different type of communication technology which is predominantly digital. This chapter explains how various electronic devices and circuits are combined in different types of communications systems.

Objectives

After studying this chapter, you will be able to:

22.1. describe the basic operation of a television communication system;
22.2. identify and explain the technologies used in digital communication systems.

Chapter Outline

22.1 Television Communication Systems

22.2 Color Displays
22.3 Digital Communication Systems

Key Terms

analog-to-digital converter (ADC)
digital-to-analog converter (DAC)
access point (AP)
display unit
deflection yoke
pixel
synchronization
modulation
charge coupled device (CCD)
frame
interlaced scanning
progressive scanning
intensity
hue
cathode-ray tube (CRT)
liquid crystal display (LCD)
digital television (DTV)
high definition television (HDTV)
luminance (Y signal)
plasma display
spread spectrum
digital signal processing (DSP)
wireless local area network (WLAN)

22.1 Television Communication Systems

Nearly everyone has had an opportunity to view a television (TV) communication system in operation. This communication process plays a very important role in our lives. Very few people spend a day without watching television. It is probably the most significant application of communication electronics. We will use a television as a representative electronic system that uses many of the devices and circuits discussed in this text. The examples used in the discussions represent fundamental concepts and have evolved over the years.

22.1 Describe the operation of a television communication system.

In order to achieve objective 22.1, you should be able to:

- describe the function of a television transmitter and receiver;
- identify the block diagram components of a television transmitter and receiver;
- list the functional parts of a composite television signal;
- compare the operation of monochrome and color TV systems;
- compare the operation of various TV displays.

The **signal** of a television system is quite complex. It is made up of a number of unique parts, or components. Basically, the transmitted signal has a **picture carrier** and a **sound carrier**. These two signals are transmitted at the same time from a single antenna. The picture carrier is amplitude modulated. The sound carrier is frequency modulated. Both of these modulation methods were discussed in **Chapter 21**. A **television receiver** intercepts these two signals from the air. They are tuned, amplified, demodulated, and ultimately reproduced. Sound is reproduced by a speaker. Color and picture information are reproduced by a picture tube or display unit. Television is primarily a one-way communication system. There is a central transmitting station and a large number of receivers.

A simplification of the television communication system is shown in **Figure 22.1**. It consists of **transmitter** and **receiver** sections. The transmitter consists of an FM sound unit and an AM picture unit. A microphone is

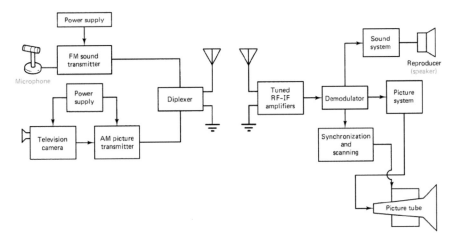

Figure 22.1 TV communication system. (a) Transmitter. (b) Receiver.

typically used for audio input, and a television camera for video input. A **diplexer** of the transmitter is a filter circuit that isolates the picture and sound carriers of the individual transmitters. The receiver consists of corresponding units for sound and picture reproduction. The receiver **demodulates** the picture and sound signals from the respective carrier waves.

The **television camera** of the system **used for this discussion** is basically a **transducer**. It changes the light energy of a televised scene into electrical signal energy. Light energy falling on a highly sensitive surface varies the conduction of current through a resistive material. The resulting current flow is proportional to the brightness of the scene. An electron beam scans horizontally and vertically across the light-sensitive surface.

Figure 22.2 shows a simplification of the type of television camera tube **used for this discussion** which is called a **Videcon**. Note that a complete circuit exists between the cathode and power supply. Electrons are emitted from the heated cathode. These are formed into a very thin beam and directed toward the back side of the photoconductive layer. Conduction through the layer is based on the intensity of light from the scene being televised. A bright or intense light area becomes low resistant. Dark areas have a higher resistance. As the electron beam scans across the back of the photoconductive layer, it sees different resistance values. Conduction through the layer is based on these resistances. A discrete area with low resistance causes a large current through the layer. Dark areas cause less current flow. Current flow is directly related to the light intensity of the televised scene. Output current

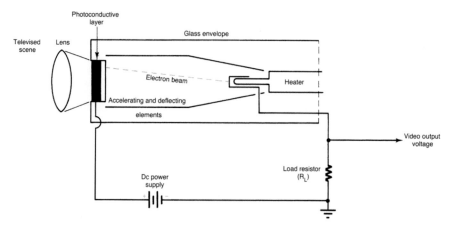

Figure 22.2   Simplification of a TV camera tube.

flow appears across the load resistor (R_L). Voltage developed across R_L is amplified and ultimately used to modulate the picture carrier. In practice, the developed camera tube voltage is called a **video signal**. **Vide** means something to see. A camera tube sees things electronically or through electronic circuitry. Camera systems have continually evolved over the years.

The scene being televised by a camera tube must be broken into very small parts called picture elements or **pixels**. For this to take place, it is necessary to scan the light-sensitive surface of a camera tube with a stream of electrons. This process is very similar to reading a printed page. Letters, words, and sentences are placed on the page by printing. We do not determine what is on a printed page at one instant. Our eyes must scan the page one line at a time, starting at the upper left-hand corner. They move left to right, drop down one line, quickly return to the left, and then scan right again for the next line. The process continues until all lines are scanned.

In a similar way, a **TV camera** that utilizes an electron beam **used for this discussion** scans the back surface of the photoconductive layer. The electron beam is deflected horizontally and vertically by an electromagnetic field. A coil fixture known as a **deflection yoke** is placed around the neck of the camera tube. This coil deflects the electron beam. Current flow in the deflection yoke is varied so that the field rises to a peak value and then drops to its starting value. **Figure 22.3** shows the deflection yoke current and the resulting electron beam scanning action. Note that each line has a trace and retrace time. During the trace period, the line is scanned from left to right. This takes a rather large portion of the complete sawtooth wave. Retrace occurs when the beam returns from right to left. Note that this takes only a small portion of the total waveform. The same condition applies to the vertical sweep waveform.

Figure 22.3 also shows another rather unique difference in the scanning lines. During the trace time, the **scanning line** is solid. This indicates that the electron beam is conducting during this time. It also shows a broken line during the retrace period. This indicates that the electron beam is nonconductive during this period. In effect, conduction occurs during the trace time and no conduction occurs during retrace. These conditions also apply to the vertical trace and retrace times.

The scanning operation of a camera tube requires two complete sets of deflection coils: one set for **horizontal deflection** and one for **vertical deflection**. The current needed to produce deflection comes from two sawtooth oscillators (Chapter 20). A vertical blocking oscillator and a horizontal multivibrator can be used for this operation.

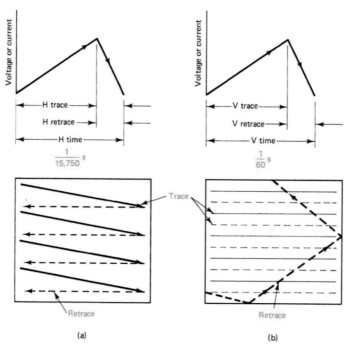

Figure 22.3   Scanning and sweep signals. (a) Horizontal. (b) Vertical.

In our example, a special image pickup device called the **charge coupled device (CCD)** is used to perform this operation. It is a solid-state device consisting of millions of photosensitive areas called cells, arranged in the form of a two-dimensional grid or **array**. The cells are also called picture elements or pixels. When photons of light energy strike the photosensitive area of a cell, it generates an analog voltage corresponding to the light intensity of the scene being detected. The photosensitive grid is continuously scanned starting from the top left-hand to the bottom right-hand area in equal intervals of time. During each scanning interval, the voltage developed on specific cells is transferred out serially for processing. A complete scan of the grid represents an electronic image of the original scene. The scanning rate of the system and the number of pixels being scanned determines the quality of the electronic image being produced. A high scan rate along with a large number of pixels will result in high-quality electronic video images.

To produce a moving television scene, there must be a least 30 complete pictures produced per second. In a TV system, a complete picture is called a

frame. For a number of years, the US television system has been regulated by the National Television System Committee (NTSC). This uses a system for transmitting and displaying 525 horizontal lines in a frame. It also uses an **aspect ratio** of 4:3, which defines the ratio of the width to the height of the display. Digital systems use a different aspect ratio.

The 525 lines are not scanned consecutively. Rather, they are divided into two fields. One field contains 262.5 odd-numbered lines, lines 1, 3, 5, 7, 9, 11, etc. The other field has 262.5 even-numbered lines, lines 2, 4, 6, 8, 10, 12, etc. A complete picture or **frame** has one odd-lined field and one even-lined field.

Picture production in a TV system of this type generally employs **interlace scanning**. To produce one frame, the odd-numbered-line field is scanned first. After scanning all the odd-numbered lines, the electron beam is deflected from the bottom position to the top. The even-numbered line field is then scanned. The odd-line field starts with a complete line and ends with a half line. The even-line field begins with a half line and ends with a complete line. The first set of lines, the odd numbered lines, is traced at 60 Hz, and the second set of lines, the even number lines is also traced at 60 Hz. This particular frequency was chosen to coincide with the AC power line frequency (60 Hz) in the USA. In some countries, the vertical frequency is 50 Hz. Since it takes two fields to make a complete picture, the picture repetition or frame rate of this television is 30 Hz. Interlaced scanning results are a perceptible flicker owing to this low scan rate.

While interlaced scanning is adequate for pictures, for displaying text clearly on monitors connected to computer systems, another type of scanning, identified as **progressive**, is often used. In this format, the image is displayed on the screen by scanning each horizontal line or row of pixels consecutively. The frame frequency in this case is at the line frequency of 60 Hz and generates a much clearer display. Television broadcasting in the USA is changing to a **digital television (DTV)** format. DTV consists of 18 different standards, which are a combination of the resolution, aspect ratio, and frame rate.

The **resolution** of a TV is defined as the process of separating or distinguishing between detailed picture elements. This is specified in terms of the number of pixels used to produce a display. For example, a standard **analog video signal** has a resolution of 704 × 480 or 0.37 mega pixels. This is also the lowest resolution of a digital TV and is used by Standard TV (SDTV). The highest resolution of a digital TV signal is generated by high definition TV (HDTV), which has a resolution of 1920 × 1080 or 2 mega pixels. Thus, the resolution of HDTV is approximately six times better than that of a standard analog TV. These values have also evolved.

The **aspect ratio** of a standard analog TV display has been 4:3 since its inception. This was done to provide a pleasing display to the human eye while viewing video. The aspect ratio of DTV ranges from 4:3 for SDTV to 16:9 for HDTV. The higher aspect ratio of HDTV is similar to the wide screen format of a movie theater.

Picture Signal

The **picture signal** of a TV transmitter contains a number of important parts. Each part of the signal plays a specific role in the operation of the system. The video signal, for example, is developed by the camera tube. It represents instantaneous variations in scene brightness. **Figure 22.4(a)** shows the video signal for one horizontal line.

The **video signal** of a television system has negative picture phase. This means that the part of the signal with highest amplitude corresponds to the darkest picture area. Bright picture areas have the lowest amplitude level. Signal levels that are 75% of the total amplitude range are considered to be in the black region. A light disappears in this region. Signal-level amplitude percentages are shown in **Figure 22.4(b)**.

In addition to the video information, the **picture signal** must also provide some way of cutting off the **electron beam** at certain times. When scanning occurs, the electron beam is driven to cut off during the retrace period. Horizontal retrace occurs at the end of one line and vertical retrace occurs at the end of each field. A **rectangular pulse** of sufficient amplitude is needed to reach cutoff. This condition permits the electron beam to retrace without producing unwanted lines. This part of the signal is called blanking. A composite signal has both horizontal and vertical blanking pulses. **Figure 22.4(a)** shows the location of a horizontal blanking pulse. Vertical blanking occurs after 262.5 horizontal lines. **Figure 22.4(b)** shows the vertical blanking time.

In television signal production, the **vertical and horizontal sweep** circuits must be properly synchronized to produce a picture. The signal sent out by the transmitter must contain synchronization, or sync, information. This signal is used to keep the oscillator of the receiver in step with the correct signal frequency. Separate generators are used to develop the sync signal, which is added to the video signal developed by the camera.

A **composite TV picture signal** also has vertical and horizontal synchronization pulses. These pulses ride on the top of the blanking pulses.

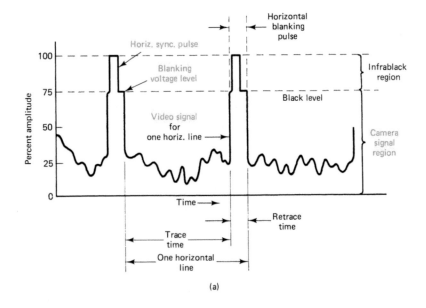

(a)

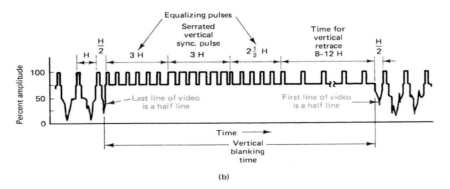

(b)

Figure 22.4 Composite TV picture signal.

Horizontal sync pulses are shown in both parts of **Figure 22.4**. The **serrated pulses** of **Figure 22.4(b)** provide continuous horizontal sync during the vertical retrace time. The vertical sync pulse is made up of six rectangular pulses near the center of the vertical blanking time. The width of a vertical sync pulse is much greater than that of the horizontal sync pulse. All these pulses, plus blanking and the video signal, are described as a composite picture signal.

Television Transmitter

The type of **television transmitter** used in this discussion is divided into two separate sections, or **divisions**, that feed outputs into a common antenna. The video section is responsible for the picture part of the signal. A crystal oscillator (**Chapter 20**) is used for carrier wave generation. As a general rule, the frequency is multiplied to bring it up to an allocated channel in the **VHF** or **UHF band**. An intermediate power amplifier and a final power amplifier (Chapters 11 and 12) follow the last multiplier. The modulating component (Chapter 21) is a composite picture signal. It contains video, blanking pulses, sync, and equalizing pulses. The composite signal is amplified and ultimately applied to the final power amplifier. The final is amplitude modulated by the composite picture signal. This section of the transmitter is essentially the same as that of a commercial AM station. There is an obvious difference in frequency and power output. **Figure 22.5** shows a block diagram of a black-and-white or monochrome TV transmitter.

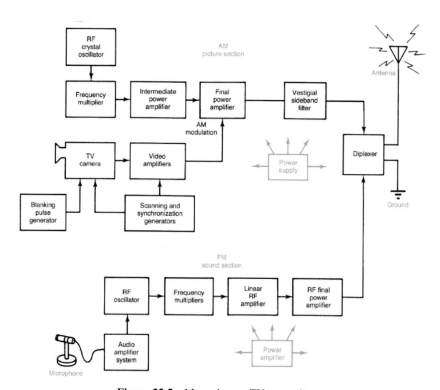

Figure 22.5   Monochrome TV transmitter.

A unique difference in TV and commercial AM transmitter circuitry appears after the final power amplifier. The TV output signal is applied to a **vestigial sideband filter**. This filter is designed to remove all sideband frequencies 1.25 MHz below the carrier frequency. The entire upper sideband of the signal is transmitted. A large portion of the lower sideband is suppressed. This is done purposely to reduce the frequency occupied by a channel. With the lower sideband suppressed, the bandwidth of a TV channel is 6 MHz.

The vestigial sideband signal is then applied to the **diplexer**. The diplexer is a filter circuit that isolates the picture and sound carriers. Essentially, the sound carrier will not pass into the picture section and the picture carrier will not pass into the sound section. This prevents undesirable interaction between the two **carrier signals**. The sound section of a TV system is primarily an FM transmitter (Chapter 21). It is very similar to a commercial FM transmitter. The center frequency of the FM carrier is always **4.5 MHz** above the picture carrier. Carrier deviation is ±25 kHz in a TV system. Modulation is normally applied to the oscillator of the FM sound system. The remainder of the sound section is similar to that of a commercial FM transmitter. See the FM sound section of the transmitter in **Figure 22.5**.

Television Receiver

The **television receiver** used in our discussion is designed to intercept the electromagnetic waves sent out by the transmitter and use them to develop sound and a picture. The received signals are in the VHF or UHF band. The **FCC (Federal Communications Commission)** has allocated a 6-MHz bandwidth for each **TV channel**. Channels 2-13 are in the **VHF band**. These frequencies are from 54 to 216 MHz. Channels 14-83 are in the **UHF band**. This ranges from 470 to 890 MHz. All channels in the immediate area induce a signal into the antenna. The desired station is selected by altering a tuning circuit. This *LC* circuit passes only the selected channel and rejects the others.

A functional block diagram of one type of **television receiver** is shown in **Figure 22.6**. Practically, all this circuitry has been used in other communication systems. The front end of a TV receiver, for example, is a **superheterodyne** circuit. It has a tuned RF amplifier, a mixer, and an oscillator. This section of the receiver is called the **tuner**. It is housed in a shielded metal container to reduce interference. The output of the tuner is an IF signal. The **IF frequency** for TV is 41.25 MHz for the sound and 45.75 MHz for the picture. The IF must pass both sound and picture carriers. This necessitates a 6-MHz bandpass for the IF amplifiers.

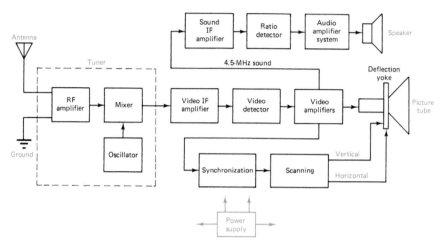

Figure 22.6   Monochrome TV receiver.

The **demodulation** function of a TV receiver is achieved by a diode. This circuit is primarily an AM detector (Chapter 21). The output of the detector has all the picture information placed on the carrier at the transmitter. After demodulation, the picture carrier is discarded. Three different kinds of signal information appear at the output of the detector. It recovers the video signal and sync signals for immediate use. The **sound carrier** passes into the sound IF section.

The **video signal** recovered by the detector is processed by the video amplifier. The output of this section is then used to control the brightness of the electron beam of the picture tube. A dark spot detected by the TV camera causes a corresponding reduction in picture tube brightness. A bright spot or an intense picture element causes the picture tube to conduct very heavily. This causes a corresponding bright spot to appear on the picture tube face. The video signal developed by the camera of the transmitter is accurately reproduced on the picture tube screen.

The **sync signal** of the video detector is used to synchronize the vertical and horizontal sweep oscillators. These oscillators develop the sawtooth waves that are used to deflect the electron beam. The electron beam must be in stepped with the transmitted signal for the picture to be usable. The transmitted sync signal is recovered and used to trigger the two receiver sweep oscillators.

An **AM video detector** does not respond effectively to frequency changes in the IF sound carrier. It does, however, **heterodyne** the two signals together.

The difference between the 45.75-MHz signal and 41.25-MHz IF signal is 4.5 MHz. This signal takes on the FM modulation characteristic of the sound carrier. 4.5 MHz is called the sound IF. The sound IF deviates ± 25 kHz above and below 4.5 MHz.

A **ratio detector** can be used to demodulate the sound IF signal (Chapter 21). The recovered audio component is then processed by an AF amplifier. It is ultimately used to drive a loudspeaker for sound reproduction. The FM sound section of a TV receiver is very similar to that of a standard FM broadcast receiver.

The **display** of a TV receiver is responsible for changing an electrical signal into light energy. A display may be a **cathode-ray tube (CRT)** or **liquid crystal display (LCD)**. We will discuss the cathode-ray tube (CRT) next in which a beam of electrons strikes a phosphor coated tube. The intensity of the electron beam changes as it scans across the face of the tube. The inside face of the tube is coated with phosphor. When the electron beam strikes tiny grains of phosphor, it produces light. A combination of different light and dark phosphor grains causes a picture to appear on the inside of the face area. The resulting picture can be observed by viewing the front of the face area.

Figure 22.7 shows a simplification of the **CRT** of a TV receiver used for this discussion. The tube is divided into three parts. The gun area is

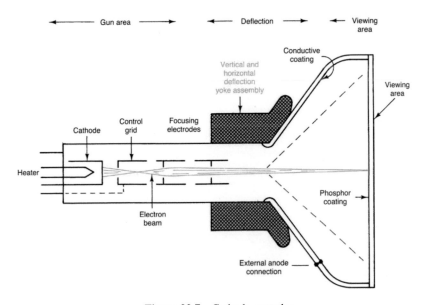

Figure 22.7   Cathode-ray tube.

responsible for electron beam production. The coil fixture attached to the neck of the tube deflects the electron beam. The viewing area changes electrical energy into light energy.

A CRT is a **thermionic emission** device. Heat applied to the cathode causes it to emit electrons. These electrons form into a fine beam and move toward the face of the tube. A high, positive charge is placed on a conductive coating on the inside of the glass housing. This coating serves as the anode. The electron beam is attracted to this area of the tube because of its positive charge. Voltage is supplied to the coating by an external anode connection terminal. The **anode voltage** of a CRT is several kilovolts (kV).

After leaving the cathode, electrons pass through the **control grid**. Voltage applied to the grid controls the flow of electrons. A strong negative voltage causes the electron beam to be cut off. Varying values of grid voltage cause the screen to be illuminated. The video signal developed by the receiver can be used to control the intensity level of the electron beam. Focusing of the electron beam is achieved by the next two electrodes. They shape the electron beam into a very fine trace. These electrodes are generally called **focusing anodes**. **Electromagnetic deflection** of the electron beam is achieved by the yoke assembly. This coil fixture produces an electromagnetic field, and the electron beam causes deflection of the beam. Both vertical and horizontal deflections are needed to produce scanning.

Color Television

The transmission of a **color television** picture is more complex than the black-and-white one used in the previous discussion. The fundamentals involved in transmission are primarily the same as those of **monochrome**, or black-and-white television. One of the problems of color TV is that it must be compatible with monochrome since some black-and-white TVs are still used. This means that programs designed for color reception are also received on a monochrome receiver. The transmitted signal must, therefore, contain color information as well as the monochrome signal. All of this must fit into the 6-MHz channel allocation.

The picture portion of the **color signal** is basically the same as the monochrome signal. The **video signal** is, however, made up of three separate color signals. Each color signal is produced by an independent pickup sensor in the camera. One sensor is used to develop a signal voltage that corresponds to the red content of the scene being televised. Blue and green sensors are used in a similar arrangement to produce the other two color signals. **Red,**

green, and blue are considered to be the **primary colors** of the video signal. White is a mixture of all three colors. Black is the absence of all three primary colors.

Color Camera

Figure 22.8 shows a simplified **color television camera**. The scene being televised is focused by a lens onto a special mirror. This mirror reflects one-third of the light and passes two-thirds of the light. The reflected image goes through a filter and is applied to the blue sensor. Sensing of colors can be achieved with a CCD. Two-thirds of the image passing through the mirror is applied to a divider mirror. One-third of it is applied to the red CCD and one-third to the green CCD. Each color CCD provides an output signal that is proportional to the light level of the primary color. The brightness or **luminance signal** is a mixture of the three primary colors. These proportions of the color signal are 59% green, 30% red, and 11% blue. The luminance signal is generally called a **Y signal**. A monochrome receiver responds only to the Y signal and produces a standard black-and-white picture.

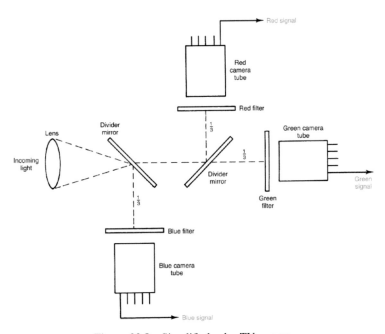

Figure 22.8   Simplified color TV camera.

The human eye needs two stimuli to perceive **color**. One of these is called **hue**. Hue refers to a specific color. Green leaves have a green hue, and a red cap has a red hue. The other consideration is called **saturation**. This refers to the amount, or level, of color present. A vivid color is highly saturated. Light or weak colors are diluted with white light. A scene being televised in color has **hue, saturation**, and **brightness**.

Color Transmitters

Figure 22.9 shows a block diagram of a **color transmitter**. The color section adds significantly to the transmitter. The remainder of the diagram has been reduced to a few blocks. This part of the transmitter is essentially the same for either a color or a black-and-white system.

In a color transmitter, the TV camera develops three separate color signals. These colors are applied to a matrix and a **mixer circuit**. The mixer develops the luminance signal. **Green, red, and blue** are mixed in correct proportions. Luminance is the equivalent of a black-and-white signal. The output of the mixer (**Y signal**) goes directly to the carrier modulator. This signal is the same as the modulating component of a black-and-white transmitter.

The **matrix** is a specialized mixing circuit. It combines the three color signals into an **I signal** and a **Q signal**. These two signals can be used to

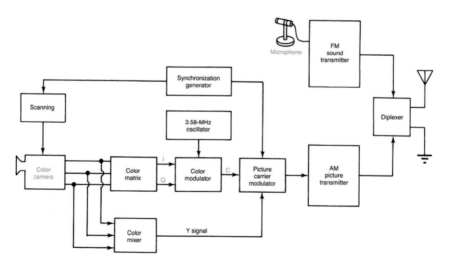

Figure 22.9   Color TV transmitter block diagram.

represent any color developed by the system. The Q signal corresponds to green or purple information. The I signal refers to orange or cyan signals. These two signals are used to amplitude modulate a **3.58-MHz subcarrier signal**. The carrier part of this signal is suppressed to prevent interference. The I signal is kept in phase with the 3.58-MHz subcarrier. The Q signal is one quadrant, or 90∘, out of phase with the subcarrier. The I and Q signals are sideband frequencies of the subcarrier. Only this sideband information is used to modulate the transmitter picture carrier.

A **synchronization signal** is also generated by the color transmitter. This signal is applied to both the modulator and the sensor scanning circuit. Scanning of the three color sensors must be synchronized with the modulated carrier output. The sync signal is an essential part of the picture carrier modulation component. The modulating component contains I, Q, Y, and sync signals.

Color Receiver

A **color television receiver** picks up the transmitted signal and uses it to develop sound and a picture. It has all the basic parts of monochrome receiver plus those needed to recover the original three color signals. **Figure 22.10** shows a block diagram of the color TV receiver. The shaded blocks denote the color section of the receiver. This is where the primary difference in color and monochrome TV receivers occurs.

Operation of the color receiver is primarily the same as that of a black-and-white receiver up to the **video demodulator**. After demodulation, the composite video signal is then divided into **chroma (C)** and **Y signals**. The Y, or luminance, signal is coupled to a delay circuit. This slows down the Y signal. It, thus, arrives at the matrix at the same time as the chroma signal. The C signal is applied to the chroma amplifier. This signal must pass through a great deal more circuitry before reaching the matrix.

Remember that the **chroma signal** contains I and Q colors and a 3.58-MHz suppressed carrier. To demodulate this signal, the 3.58-MHz signal must be reinserted. Note that the 3.58-MHz oscillator is connected to the I and Q demodulators. This is where the carrier reinsertion function takes place.

The **I demodulator** receives a 3.58-MHz signal directly from the oscillator. This is where the in-phase signal is derived. The 3.58-MHz signal fed into the **Q demodulator** is shifted 90∘. This is where the quadrature signal is derived. I and Q color signals appear at the output of the demodulator.

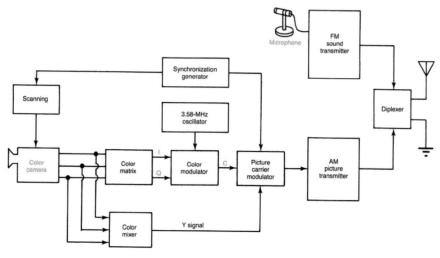

Figure 22.10   Color TV receiver block diagram.

The matrix of the receiver has three signals applied to its input. The *Y* signal contains the **luminance** information. The *I* and *Q* signals are the demodulated color signals. The matrix combines these three signals in proper proportions. Its output comprises the original red, green, and blue color signals. These signals are applied to the R, G, and B inputs of a color display. Any color can be reproduced on the face of the display tube by varying the **intensity (I)** and **hue (Q)** of the three primary colors.

Color Displays

Many of the communication technologies previously discussed in this chapter relating to video applications require an output device, and often that device is a **display**. There are several types of displays; however, the cathode-ray tube (CRT) was dominant in the electronics field for many years for systems described previously. Ultimately, the need to display color video has evolved, and, now, other types dominate the area of color displays. We need to look at some of the evolving technologies related to other color displays.

Color CRT

The **displays** used for color televisions and color computer monitors for many years were CRTs that are capable of displaying the "color" portion of the electronic signal they receive. While **monochrome** (black and white) CRTs and color CRTs share many components, there are some notable differences. The most important difference is the fact that there are three separate cathodes or guns, one for each of the red, green, and blue colors necessary to complete a color display, as shown in **Figure 22.10**. In the case of a CRT, these signals are sent to the respective color cathodes of the display tube. Second, the screen does not have one **phosphor coating**, it has three phosphors that correspond to the **red, green, and blue colors** associated with the color electron guns. Each of these phosphor colors is arranged closely to each other so that the user will perceive that one color is emitting from this small area of the screen, rather than each of the three primary colors. The electron beam of each cathode strikes closely spaced red, green, and blue phosphor spots on the face of the tube. These spots glow in differing amounts according to the signal level. The last difference is the addition of a mask on the screen with small holes that allow only the colors from this small dot or pixel of color through to the viewer. A cross-sectional view of a color CRT is shown in **Figure 22.11**.

LCD (Liquid Crystal Display)

From wristwatches to calculators, dashboards to television screens, **LCD** panels are a very popular electronic display. Rather than an image resulting from the glow of electrons striking a phosphorescent surface, a monochromatic LCD relies on light passing through a display of liquid crystals, hence the name liquid crystal display or LCD. The heart of an LCD is the liquid crystal, which is sensitive to electric current and capable of polarizing light. Current passing through the **liquid crystals** will align, but crystals with no current align parallel to each other. Two panels are found on either side of the liquid crystal, each with polarized sheets that block light in a horizontal or vertical direction. When the crystals do not have current applied and are in the parallel formation, light can pass through the vertical polarizing filter, and when current is passed through the crystals, they align, allowing light to pass through the horizontal filter.

Therefore, when no current is applied, the view of the display appears as clear, or like a mirror (the image of the rear, **vertical filter**), and when current is applied, the display appears dark (the front **horizontal filter**). A mirror to increase the amount of reflected light backs the entire "sandwich" of crystals, filters, and glass plates. The effect of this mirror can be easily seen in a large LCD. **Figure 22.12** shows each layer of the screen in order. Color LCD panels

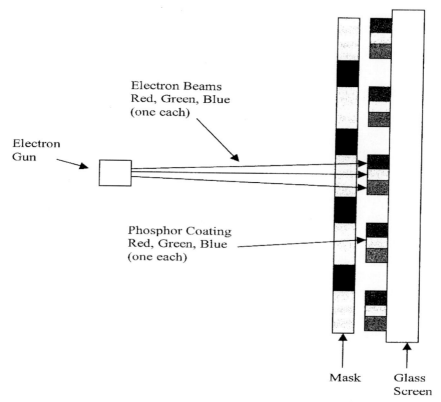

Electron Beams
Red, Green, Blue
(one each)

Electron
Gun

Phosphor Coating
Red, Green, Blue
(one each)

Mask   Glass
Screen

Figure 22.11 Cross-sectional view of a color television CRT display.

such as those found in portable computers and televisions use a different type of **liquid crystal display**. By using thin-film transistors to construct a matrix display, individual **pixels** can be switched off and on by using the transistors to act as a switch. Each pixel has three (red, blue, and green) displays that can be controlled, and brightness is directly proportional to the voltage provided to color within the pixel. By arranging these three-color pixels in a row- and column-type matrix, each pixel, the color of the pixel, and the intensity of the color in the pixel can be addressed, thereby managing the entire image.

The size and quality of the display is directly associated with the physical properties of this matrix. There are primary factors that identify the LCD, the **resolution**, and the **pixel size**. The resolution, expressed as the matrix size, gives the amount of pixels in rows and columns. The pixel size relates to the dot pitch of the actual pixels in the display.

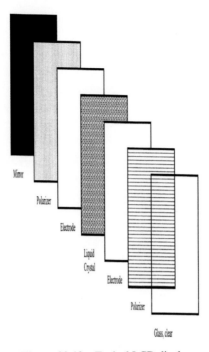

Figure 22.12 Typical LCD display.

LCDs have brought many advantages to display technologies over the CRT. These include advantages related to the lack of phosphor, such as image "burn-in," as found on CRT images that have a constant display which is on continuously, such as in ATM machines. LCD are typically more energy efficient than the CRT displays they replace and do not suffer from the "flicker" problem that affects many CRT displays, caused by low scanning rates. However, there are a few disadvantages by using the LCD technology as a display device. The most noted problem is related to the slow pixel response time. Objects moving quickly on an LCD may appear to have a shadow or ghost image trailing the active image. Another issue is that an LCD requires light be reflected through the display in order to have a viewable image. For using an LCD in darkened areas, backlighting is required.

Plasma Displays

While some **video display** devices (televisions, computer monitors, and test equipment) use a cathode-ray tube (CRT) as described previously, many of these applications are benefiting from a technology that produces a flatter visual display. One such technology is **plasma**. In a traditional CRT, the image is produced from the impact of electrons on a screen of phosphor in a vacuum. While this image is of acceptable quality, there is a drawback. The CRT is a high-current, heavy, bulky item, and not compatible with many of the needs of many pieces of consumer electronics. For applications where space or weight is an issue, manufacturers have turned their focus from the CRT toward the technology in a plasma flat screen display. One drawback is the current cost of such displays. The increased cost is due to the complexity of the structure and construction.

Plasma displays are based on the illumination of a phosphor coated surface by ionized gas or plasma. Each pixel is made up of three cells coated with red, blue, and green phosphors. **Figure 22.13** shows how these elements are arranged to comprise one **pixel** of the display. The cells are filled with Xenon or Neon gases similar to that used in fluorescent tubes. Electrodes located at the rear and front of the cell are used to ionize the gas. The charging of the electrodes is controlled using a display control sensor signal for each cell within the pixel. When **ionization** occurs, electrons are released from individual gas atoms. These electrons collide with other gas atoms causing electrons within the atom to move to a higher energy level.

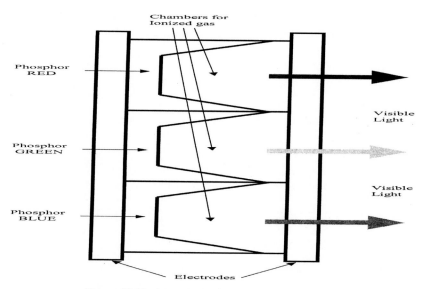

Figure 22.13   Plasma screen technology – one pixel.

When these electrons fall back to their original energy level, it causes photons of ultraviolet light to be released by the atom. The **photons** striking different color phosphors cause them to produce visible light. The display can then produce images by illuminating the color in each pixel in varying intensities and hues.

While many advantages of the **plasma display** such as the compact and lightweight design, a flatter and wider viewing area (as compared to a CRT), and other shared similarities with a CRT have already been discussed, there are some limitations to plasma display technology. Some issues found with plasma displays are increased cost, a possible degradation and loss of brightness over time, and low energy efficiency.

Self-Examination

Answer the following questions.

1. The US television system has _____ horizontal lines in a frame.
2. _____ modulation is used for the picture portion of a television signal.
3. _____ modulation is used for the audio portion of a television signal.
4. Color TVs have _____, _____, and _____ color signals.
5. The aspect ratio for an HDTV is ___:___.
6. The bandwidth of a color TV channel is _____MHz.
7. In _____ scanning, all the rows of the display are scanned sequentially.
8. The camera tube of the television system produces the _____ component of the signal.
9. A _____ TV display responds to the ionization of a gas.
10. A _____ TV display responds to the emission of electrons from a heated cathode.
11. Channels 2-13 are included in the (UHF, VHF) band.
12. The _____ signal refers to the intensity, and the _____signal to the hue in a color TV receiver.
13. A _____ is used to isolate the video and audio signals applied to a TV antenna.
14. A _____ is used to extract the audio and video components from the respective carrier signals in a TV receiver.
15. The photosensitive areas of a CCD camera are referred to as _____.

22.2 Digital Communication Systems

As stated in the beginning of the chapter, many different forms of communication systems are being utilized today. Many of these systems are categorized as simplex or one-way communication. This includes **AM/FM radio** and **television** which have been considered as analog communication systems. Now, we will be looking at duplex or two-way communication. This includes **cell phones** and **wireless computer networks** that are predominantly controlled using digital technology. This represents one of the most significant and rapidly expanding areas of communication electronics today.

22.2 Explain the operation of digital communications system.
In order to achieve objective 22.2, you should be able to:

- list the characteristics of a cordless phone;
- identify the essential functions of a cell phone communication system;
- explain how a wireless computer communicates with a network;
- list some of the important standards used in wireless network communications.

To better explain this new generation of communication systems, the concepts surrounding digital instead of analog communication systems must be defined. Central to the digital revolution in communication is **digital signal processing, or DSP**. It refers to the methodology used for processing digital information. This technology is used in cell phones, robotics, digital cable TV, digital cameras, digital audio, Internet telephones, and others. In a typical communication system, the analog signal is first converted to a digital format using a special device called an **analog-to-digital converter (ADC)**. Next, the digital signal is processed using DSP. This process may include methodologies to increase the operational speed, security, and quality of the data being transmitted. Based on variables such as speed, price, ease of use, power consumption, and flexibility, different DSP implementation technologies are available. In order to better perceive how the lines between analog and digital communication systems are blending, a few example technologies such as telephones and other important digital transmission schemes will be discussed.

Cordless Phones

An example of a device that has shared both analog and digital communication systems is the **cordless telephone**. Cordless telephones are a combination of the technologies related to a telephone and a radio transceiver.

Transceivers are built into the two major components of the cordless phone: the base and the handset. By using FCC-approved frequencies such as 900 MHz, 2.4 GHz, and 5.8 GHz, the handset and the base "talk" to each other, carrying the voice signal, data signals, and various control signals. The handset often has many features such as a keypad, battery, and LCD display for call timers, or caller ID, and a ringer. The base too has many functions, but, most importantly, it provides the interface between the cordless portion of the phone and the telephone network. The base also provides the means of charging the battery that powers the handset. As with many devices in electricity and electronics, earlier cordless phones used simple analog signals for transmission. As digital equipment has increased in reliability and decreases in price, many cordless phone manufacturers are using digital transmission of the data between the transceivers.

One concern of cordless phones has been **security**. Because the base and handset are transceivers working in **duplex**, any radio receiver, such as a scanner, can receive the same frequencies and allow a third party (perhaps unwanted) to listen to the call. Several technologies have evolved to prevent this, such as scrambling techniques in cordless phones that use analog transmissions to **digital spread spectrum (DSS)** that breaks the data over many frequencies, making reception by a listening device much more difficult.

Cellular Phones

In the simplest of context, a cellular, or "cell," phone is an extremely complex low power duplex radio. By using digital or analog transmission technologies to connect to various towers, which, in turn, are connected to traditional telephone services, **cell phone** can offer complete mobility to the user. Their name, cell phones, comes from the placement of the towers; the coverage area of the tower is considered a cell. The technologies that support cell phones have evolved rapidly in recent years. Colorful, voice-activated phones capable of phones much more than a phone conversation have replaced the early bulky tote models.

At first glance, the technology of a cellular phone does not appear complex. Most phones include an antenna, battery, speaker, microphone, a display, and a circuit board. It is on this circuit board that the complexity is found. Housed on the board are analog-to-digital and digital-to-analog converters, **digital signal processing (DSP)** circuitry, modulators and demodulators, and components related to the radio transceiver. It also includes storage capabilities and digital control circuitry.

A block diagram of a generic **cell phone** is shown in **Figure 22.14**. The block diagram is divided into two distinct sections that represent the transmission and reception functions. The output and input signals of the cells phone are radiated and received respectively by a common antenna. A **duplexer** attached to the antenna directs the input and output signals to their respective circuitry.

The **transmitter** function of a cell phone is primarily responsible for radiating a signal that has been developed by the microphone. This audio signal modulates the radio frequency component that is radiated from the antenna. Initially, the audio signal of the microphone is applied to an **analog-to-digital converter (ADC)**, which digitizes the analog component. This signal is applied to the transmitter DSP, which is responsible for improving the quality of the signal and eliminating redundant information during transmission. The DSP also adds control information to the processed audio signal. This signal is then sent to the transmitter equalizer. The equalizer is responsible for reducing defects such as phase or amplitude distortions in the audio signal being processed. This signal is then applied to the modulator, where it is mixed with the radio frequency component. The resulting output of this is an FM signal that is applied to the up-converter. The **up-converter** has another input signal applied from a synthesizer. The synthesizer generates a specific frequency such that the output frequency of the up-converter is exactly that required for transmission. This permits the cell phone to transmit multiple carrier frequencies, which is required for modern phone communications. The signal from the up-converter is now applied to a **power amplifier**, which increases the signal amplitude to an appropriate level. In a cell phone system, the amplification of a power amplifier is controlled by digital signal processing. This information is received from the base station during signal transmission. After the power amplifier, the signal is applied to the duplexer that filters the signal so that only the required frequencies are transmitted out from the antenna.

The **receiver** function of a cell phone is primarily responsible for detecting an RF signal received by the antenna and reproducing the audio component of this signal. This is functionally a reversal of the transmitting operation. In this case, the signal is first received by the antenna and then passes through the duplexer. Since the strength of the received signal is very small and may have some form of interference, the signal is then processed by a low-noise amplifier. This improves the quality of the signal. The output of the low-noise amplifier is then applied to a **down-converter**, which also accepts an input from the synthesizer. This lowers the carrier frequency of the

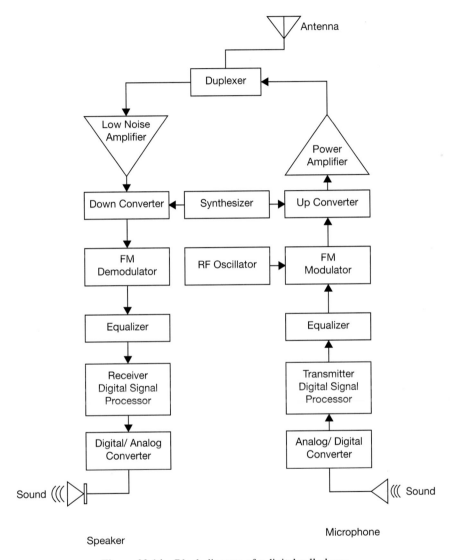

Figure 22.14 Block diagram of a digital cell phone.

received RF signal and then passes the modified signal to the demodulator. The demodulator extracts the audio component from the modulated FM carrier. Next, the receiver equalizer reduces distortions in the audio signal. This signal is then applied to the receiver DSP that corrects any further errors that may have occurred during the transmission-reception process. The output of the receiver DSP then drives a **digital-to analog-converter (DAC)**. Finally, the converter output signal drives a speaker for the reproduction of the audio component.

In addition to the circuitry, each phone carries an **electronic serial number (ESN)** that is unique to each phone, and a **system identification code (SID)** that identifies the provider the phone will use. Users recognize when the SID of the phone does not match the SID of the tower; this condition is known as roaming.

In order to communicate with a cell phone tower, various standards have been established. These include **standards** for the frequency of communication used, the type of modulation, and the security. In addition, protocols are used in computer-based communication systems which are a set of rules used by devices for exchanging information over a specific media. In the case of cell phones, the media is air.

Various analog and digital communication technologies have been developed for use with cell phones. These include:

- AMPS (advanced mobile phone system) 824-894 MHz bands, for analog transmission
- FDMA (frequency division multiple access) 824-894 MHz bands, for analog transmission
- TDMA (time division multiple access), 850-1900 MHz bands, for analog transmission, uses compression to allow more calls on one channel
- CDMA (code division multiple access) 800 and 1900 MHz bands, for digital transmission, digitally divides data over bandwidth
- GSM (global system for mobile communication), 800 or 1800 MHz bands, for digital transmission, a global standard that adds encryption to make calls more secure and allows for additional data-related services.

Wireless Computing Networks

An area for **data communications** that has grown tremendously in regard to wireless telephony extends toward computer access through computer networks. A **local area network (LAN)** is a computer network operating within a relatively small geographic area such as a home or a small office. Traditional LANs require a physical cable connection between the computer and the network. This connection can also be made using wireless technology. In order to facilitate and regulate the transmission of wireless data on these networks, the **Institute of Electrical & Electronics Engineer (IEEE)** has developed various standards for wireless **local area networks (WLANs)**. These standards include the communication frequencies, data transmission rates, data security, and others.

Radio frequencies in the 2.4-5 GHz range are commonly used in wireless computer communications. The band of frequencies used for wireless communication is called a **channel**. Specific channels must be used for data to be successfully transmitted and received using wireless devices. Wireless networks are usually implemented using either unmanaged (*ad hoc*) or managed (infrastructure) topologies. A network topology refers to the layout of the network. An *ad hoc* network topology permits devices to communicate directly with each other. It is suitable for establishing temporary connections between devices such as laptop computers. In contrast, the infrastructure topology is used for extending a wired LAN in order to include support for wireless devices. This is achieved by connecting a base station to the LAN. The base station has the capability of communicating with wireless devices using an antenna.

The **transmission** of wireless signals depends on the type of antenna used. An omni-directional antenna is used for transmitting signals in all directions, whereas a directional antenna is focused on the receiver concentrating the signal power in a certain direction. The **base station** used for transmitting and receiving signals is known as an access point (AP) or wireless access point (WAP). All data between wireless devices or between wireless devices and the LAN passes through the AP. Multiple APs can be used to extend the area of wireless coverage. Wireless devices require a configurable **Service Set Identifier (SSID)** for communicating with a given AP. The SSID of both the device and the AP has to match for communication to be established. In addition, specific channels or frequency ranges are used for communication between the device and the AP. Different methods are used for verifying the authenticity of the communicating devices prior to transmitting data. These may include the use of SSID, usernames, and passwords. Additional security is provided by encrypting the data being transferred so that it cannot be read in transit. **Encryption** is the process of altering or transforming data in order to prevent unauthorized access.

With multiple devices sharing the **wireless media**, the access method used requires devices to listen on a channel to ensure that it is not in use prior to transmitting. The manner in which data signals travel through the wireless media either altering between frequency carriers or spreading the signal over the frequency spectrum is called spread spectrum technology. This includes a method called **frequency hopping**, wherein the frequency used to send the signal is changed in a predictable pattern. Another method, the direct sequence spread spectrum, spreads the signal over the full transmission frequency spectrum. A redundant bit pattern is sent for every bit of data being transmitted, thereby enhancing reliability.

The speed of data being transferred is commonly measured in **millions of bits per second or Mbps**. Speeds of up to 54 Mbps can be achieved on a wireless networks, but the speed reduces as the number of users on a channel increases. The range of infrastructure wireless networks usually extends up to 38 m. Objects such as trees, concrete, and steel readily absorb energy from all RF signals, thus reducing the area being covered. RF interference may be caused by neighboring wireless LANs or by cordless phones and microwave devices that use the same range of frequencies in their operation. **Wireless devices** are also susceptible to electrical interference. This type of interference may be caused by the operation of motors, fluorescent lights, computers, transformers, and air conditioning units.

Table 22.1   IEEE wireless networking standards.

IEEE standard	Frequency (GHz)	Speed or data rate (bps)	Range or coverage (m)	Comment
802.11	2.5	1–2	20	One of the first wireless standards
802.11a	5	54	35	Incompatible with 802.11b and 802.11g because of frequency
802.11b	2.4	11	38	Widely used but susceptible to interference from microwave ovens
802.11g	2.4	54	38	Offers a high data rate but is also susceptible to interference from microwave ovens
802.11n	2.4 or 5	248	70	Latest standard proposed for 2009, with the highest data rates and range

A brief list of the **IEEE 802.11** standards developed for wireless networking is shown in **Table 22-1**.

With an increase in mobile computing, the popularity of **digital data transmission** for computing networks will continue to rise. As shown, new standards are being developed to meet the needs of advanced computing needs.

Self-Examination

Answer the following questions.

16. A cordless telephone has a _____and a _____.
17. Cordless phone transceivers use frequencies in the _____, _____, and _____ range for communication.
18. One primary concern in cordless phone communication is _____.
19. A form of communication in which a unit can both send and receive information signals is termed as a _____system.
20. A form of communication in which a unit can only send or receive information signals is termed as a _____system.
21. The two primary functions of a cell phone are _____ and _____.
22. A device that changes an analog signal into a digital signal is called a(n) _____.
23. A device that changes a digital signal into an analog signal is called a(n) _____.
24. The sound input of a cell phone is developed by the _____, whereas the sound output is developed by the _____.
25. When the SID of a cell phone does not match that of a cell tower, the phone is said to be _____.
26. Digital signal processing within the transmitter of a cell phone is done (before, after) the analog sound input has been converted into a digital signal.
27. Digital signal processing within the receiver of a cell phone is done (before, after) the digital input has been converted into an analog sound signal.
28. The _____block within the receiver section of a cell phone is responsible for extracting the audio component from the frequency modulated RF signal.
29. The (receiver, transmitter) function of a cell phone is responsible for detecting an RF signal received by the antenna.

30. The (receiver, transmitter) function of a cell phone is responsible for modulation of an RF signal by the audio component.
31. The frequencies that are commonly used for wireless computer communications are _____ and _____.
32. The base station used for transmitting and receiving wireless signals to or from a computer network is referred to as the _____.
33. Global system for mobile communication (GSM) operates in the _____ or _____MHz bands.
34. The range or coverage of the wireless signal in IEEE 802.11g standard is _____m.
35. Wireless networks that operate in the _____GHz range may experience RF interference due to microwave ovens and cell phones.

Summary

- Audio and video signals that can change in a continuous manner are termed as analog signals.
- Signals that carry text and data that change only in specific or discrete intervals are termed as digital signals.
- Television is a specialized type of communication that uses both AM and FM transmissions, requiring carriers for both the picture (AM) and the sound (FM).
- A microphone typically provides the audio input, and a television camera provides the video input in a TV transmitter.
- The TV receiver demodulates the picture and sound signals from the respective carrier waves, following which the video output is reproduced on a display and sound on a speaker.
- The scene being televised by a camera tube must be broken into very small parts called picture elements or pixels, which are scanned in a conventional TV camera using an electron beam, and in a CCD camera, the voltage developed on the pixels is transferred out for processing.
- Picture production in a conventional TV system generally employs interlace scanning, where, for producing one frame, the odd-numbered line of the 535 horizontal lines in a frame is scanned first, followed by the even-numbered lines.
- Digital TV consists of 18 different standards, which are a combination of the resolution, aspect ratio, and frame rate.

- The highest resolution of a digital TV signal is generated by high definition TV (HDTV), which has a resolution of 1920 × 1080, having an aspect ratio of 16:9.
- In television signal production, the vertical and horizontal sweep circuits must be properly synchronized to produce a picture.
- In a composite TV picture signal, serrated pulses provide continuous horizontal sync during the vertical retrace time, and the vertical sync pulse is made up of six rectangular pulses near the center of the vertical blanking time.
- The FCC (Federal Communications Commission) has allocated a 6-MHz bandwidth for each TV channel.
- The brightness or luminance signal, referred to as the Y signal, is a mixture of the three primary colors, in the proportion 59% green, 30% red, and 11% blue.
- Color transmitters combine the three primary color signals into an I signal and a Q signal. The I signal refers to orange or cyan signals, and the Q signal corresponds to green or purple information. Both signals are used to amplitude modulate a 3.58-MHz subcarrier signal.
- The human eye needs stimuli in the form of hue and saturation stimuli to perceive color. Hue refers to a specific color, and saturation to the amount, or level, of color present.
- Color receivers, after demodulation, divide the composite video signal into the Y signal and the chroma (C) signal which, in turn, contains the I and Q colors and a 3.58-MHz suppressed carrier signal.
- A liquid crystal display (LCD) consists of liquid crystals sensitive to electrical current and capable of polarizing light placed between transparent polarizing filters and glass plates.
- Plasma displays are based on the illumination of a phosphor coated surface by ionized gas or plasma.
- The coverage area of cellular phone towers is called a cell.
- In a typical cellular phone communication system, the transmitter contains an analog-to-digital converter (ADC), which converts the analog audio input signal into a digital format.
- Data in digital format can be suitably transformed for use in communication systems using digital signal processing (DSP) methods.
- Modulation is used for changing some property of a high-frequency carrier using the information that is to be transmitted, and demodulation extracts the information from the carrier.

- In a typical communication system, the receiver contains a digital-to-analog converter (DAC) that converts the digital input to an analog audio output.
- The transmitter function of a cell phone is responsible for radiating a signal that has been developed by the microphone, and the receiver function is primarily responsible for detecting an RF signal received by the antenna and reproducing the audio component of this signal.
- A wireless LAN (WLAN) uses RF signals for transmitting and receiving data between communicating devices.
- The base station used to extend a wireless LAN, which has the capability of communicating with wireless devices using an antenna, is called an access point (AP).
- The range of coverage in a wireless LAN is reduced due to absorption of the RF signal by objects such as walls, by RF interference due to cordless phones, and by electrical interference due to motors.

Review Questions

Answer the following questions.

1. A TV communication transmitter system consists of the following (select all that apply):
   a. FM sound transmitter
   b. AM picture transmitter
   c. Demodulator
   d. Television camera
   e. Diplexer

2. A TV communication receiver system consists of the following (select all that apply):
   a. Speaker
   b. Picture system
   c. Diplexer
   d. Demodulator
   e. Television camera

3. The function of a diplexer in a communication system is to:
   a. Amplify the TV signal
   b. Synchronize the scanning process

    c. Isolate the picture and sound carriers in the TV transmission process

    d. Change the light energy of a televised signal into electrical signals

4. The quality of the video images generated using CCD can be enhanced by (select all that apply):

    a. Increasing the number of pixels

    b. Decreasing the number of pixels

    c. Increasing the scan rate

    d. Decreasing the scan rate

5. The FCC has allocated a _____MHz bandwidth for each TV channel:

    a. 1

    b. 2

    c. 4

    d. 6

6. In a CRT display, the anode voltage can reach several:

    a. Milli-volts

    b. Volts

    c. Kilo-volts

    d. Giga-volts

7. In a color transmitter, the TV camera usually develops _____ separate color signals:

    a. 2

    b. 3

    c. 5

    d. 7

8. Some of the drawbacks associated with LCD displays as compared to CRT displays include (select all that apply):

    a. Higher weight

    b. Need for backlighting in darkened areas

    c. Slow pixel response time

    d. Noticeable flicker

9. The synthesizer unit within a cell phone:

    a. Reduces defects in the phase or amplitude of the audio signals

    b. Generates a specific frequency such that the output frequency is exactly the one required for transmission

    c. Increases the amplitude of the signal to appropriate level needed for transmission

    d. Directs the inputs and output signals to their respective processing circuitry

10. RF interference in a wireless network may be caused by (select all that apply):

    a. Cordless phones

    b. 110 V, 60 Hz AC

    c. 5 V, DC

    d. Microwave ovens

Answers

Self-Examination

22.1
1. 525
2. Amplitude
3. Frequency
4. red, blue, green
5. 16:9
6. 6
7. Progressive
8. video
9. plasma
10. CRT
11. VHF
12. I, Q
13. Diplexer
14. Demodulator
15. pixels
22.2
16. base unit, handset
17. 900 MHz, 2.4 GHz, 5.8 GHz
18. security
19. duplex
20. simplex
21. transmitting, receiving

22. analog-to-digital converter (ADC)
23. digital-to-analog converter (DAC)
24. microphone, speaker
25. roaming
26. after
27. before
28. demodulator
29. receiver
30. transmitter
31. 2.4 GHz, 5 GHz
32. access point
33. 800 MHz, 1800 MHz
34. 38 m
35. 2.4 GHz

Terms

Access point

The base station used to extend a wireless LAN, which has the capability of communicating with wireless devices using an antenna.

ADC

Analog-to-digital converter - a device that converts an analog signal into a digital format.

Ad hoc topology

An unmanaged wireless network topology, wherein devices can communicate directly with each other.

Aspect ratio

The ratio of the unit width to the unit height of a display device.

Blanking pulse

A part of the TV signal where the electron beam is turned off during the retrace period. There is both vertical and horizontal blanking.

Channel

The band of frequencies used for wireless communication.

Chroma

Short for chrominance. It refers to color in general.

Compatible

A TV system characteristic in which broadcasts in color may be received in black and white on sets not adapted for color.

DAC

Digital-to-analog converter - a device that converts a digital signal into an analog one.

Deflection

Electron beam movement of a TV system that scans the camera tube or picture tube.

Deflection yoke

A coil fixture that moves an electron beam vertically and horizontally.

DSP

Digital signal processing of information that is in digital format.

Directional antenna

An antenna that is focused at a receiver for concentrating the signal power in a certain direction.

Diplexer

A special TV transmitter coupling device that isolates the audio carrier and the picture carrier signals from each other.

Duplex

The ability of a system to transmit and receive signals from the same unit.

Encryption

The process of altering or transforming data, in order to prevent unauthorized access.

Field, even/odd lined

The even/odd-numbered scanning lines of one TV picture or frame.

Frame

A complete electronically produced TV picture of 525 horizontally scanned lines.

Hue

A color, such as red, green, or blue.

Infrastructure topology

A managed wireless network topology in which a wired LAN is extended by using a wireless base station, in order to support wireless devices.

Interlace scanning

An electronic picture production process in which the odd-numbered lines of a display are scanned first, and then the even lines are scanned next, and interleaved in order to make a complete 525-line picture.

Intermediate frequency (IF)

A single frequency that is developed by heterodyning two input signals together in a superheterodyne receiver.

I-signal (1)

A color signal of a TV system that is in phase with the 3.58-MHz color subcarrier.

LCD

A liquid crystal display that consists of liquid crystals sensitive to electrical current and capable of polarizing light placed between transparent polarizing filters and glass plates.

LAN

A local area network or computer network operating within a relatively small geographic area such as a home or a small office

Monochrome

Black-and-white television.

Negative picture phase

A video signal characteristic where the darkest part of a picture causes the greatest change in signal amplitude.

Omni-directional antenna

An antenna used for transmitting signals in all directions.

Pixel

A discrete picture element or photosensitive area of a display.

Plasma display

A display that utilizes an ionized gas to activate a picture element.

Progressive scanning

An electronic picture production process in which all the lines of a display are scanned consecutively to make a complete 525-line picture.

Protocol

In computer-based communication systems, a set of rules used by devices for exchanging information over a given media.

Q signal

A color signal of a TV system that is out of phase with the 3.58-MHz color subcarrier.

Video resolution

The process of separating or distinguishing between detailed picture elements and is specified by number of pixels used to produce a display.

Retrace

The process of returning the scanning beam of a camera or picture tube to its starting position

Saturation

The strength or intensity of a color used in a TV system.

Scanning

In a TV system, the process of moving an electron beam vertically and horizontally.

Spread spectrum

The manner in which data signals travel through the wireless media either altering between frequency carriers or spreading the signal over the frequency spectrum.

Sync

An abbreviation for synchronization.

Synchronization

A control process that keeps electronic signals in step. The sweep of a TV receiver is synchronized by the transmitted picture signal.

Topology

The layout of the computer network for transmitting and receiving data, including the physical cabling and logical flow of information.

Trace time

A period of the scanning process where picture information is reproduced or developed.

Transceiver

A communication device that is used both for transmitting and receiving signals.

WLAN

A wireless LAN that uses RF signals for transmitting and receiving data between communicating devices.

Y signal

The brightness or luminance signal of a TV system.

23

Digital Electronic Systems

Digital electronics is undoubtedly the fastest-growing area in the field of electronics technology. Personal computers, calculators, watches, clocks, video games, test instruments, and home appliances are only a few of the applications. Most of these things were unheard of only a few years ago. Digital electronics now plays an important role in our daily lives.

Remember that an **analog** device is one in which a quantity is represented on a continuous scale. Temperature, for example, is often determined by the position of a column of mercury. Voltage, current, and resistance can be determined by the movement of a coil of wire that interacts with a magnetic field. Analog devices are usually concerned with continuously changing values. An analog value could be any one of an infinite number of values. Radio, television, and sound systems process analog data.

Digital devices are considered to be counting operations. A digital watch tells time by counting generated pulses. The resulting count is then displayed by numbers representing hours, minutes, and seconds. A computer also has an electronic clock that generates pulses. These pulses are counted and, in many cases, processed as an operational control function. Digital circuits can store signal data, retrieve it when needed, and make operational decisions. Signal values are generally represented by two-state data. A pulse is either present or it is not. Data are either of high value or low value, with nothing in between.

Objectives

After studying this chapter, you will be able to:

23.1. explain the electrical differences in the response of a system that is used for analog or digital applications;

23.2. explain the differences between analog and digital systems;

23.3. change decimal numbers to an equivalent binary, binary-coded decimal (BCD), octal, or hexadecimal numbers;

23.4. change binary, BCD, or hexadecimal numbers to equivalent decimal numbers;

23.5. identify basic logic symbols of AND, OR, NOT, NOR, and NAND gates;

23.6. analyze and develop logic equations and truth tables for logic gates;

23.7. evaluate the operation of transistor logic gates;

23.8. describe the operation of *RS*, *D*, and *JK* flip-flops (triggered and clocked);

23.9. describe the operation of counting circuits.

Chapter Outline

23.1 Digital Systems
23.2 Digital Logic Circuits
23.3 Flip-Flops
23.4 Digital Counters

Key Terms

amplification
analog
base
binary
decoding
digital
digital integrated circuit
dual-in-line package (DIP)
encoding
energy
gate
hexadecimal
logic
octal
pulse
radix
semiconductor

solid state
system
transistor
AND gate
bistable
Boolean algebra
NAND gate
NOT gate
OR gate
truth table
asynchronous
bistable
clear
clock input
debounce
flip-flop
latch
memory
register
reset
toggle
BCD counter
binary counter
decade counter
down counter
incrementing
modulo
register
shift register
up counter

23.1 Digital Systems

A **digital system** is somewhat unique compared with other electrical systems. As in all electrical systems, an energy source is needed to make the digital system operational. DC electricity is primarily used as an operational energy source. This may be obtained from the AC power line and changed into a usable form of DC. Rectifier power supplies are commonly used in most systems. Portable systems are energized by batteries. Small digital systems can operate for a rather long period of time from a single battery. The DC electrical source is used primarily to energize the active components of the system. Digital systems are largely of the IC type. The amount of DC energy needed to supply this type of system is 5 V.

A digital system usually has internal **signal generation**. The digital signal is primarily a number of electrical **pulses** that occur in a given unit of time. Pulse generators and electronic clock circuits are used for this function. This part of the system is generated by an oscillator. As a rule, this type of generator produces square waves or spiked pulses. This signal is then processed through the system to achieve its counting function. In a sense, a digital system has an operational energy source and a digital signal source. Both sources are developed and applied directly to the system when it is made operational.

The other functions of a digital system are somewhat unusual. The path of electrical energy and the digital signal is achieved primarily by the metal of a printed circuit board. There is very little hand-wired electrical circuitry in a digital system. Control is achieved primarily by **logic gates**. Gates contain bipolar transistors, MOSFETs, and diodes. Most gate functions are achieved by ICs. The load of a digital system is quite unusual. The load on the operational source is all the components that are energized by the supply. The load of the digital signal is determined by the number of gates it supplies. The load in both cases does work. Indicators are included in nearly every system. **Digital displays** and sometimes cathode-ray tubes are widely used as indicators. The indicator function is ordinarily the most obvious part of the system. Operation is generally based on the response of the display.

Decimal Numbering Systems

Digital information has been used by human beings during almost all of their history. Parts of the human body were first used as a means of counting. Fingers and toes were often used to represent numbers. In fact, the word **digital** in Latin means finger or toe. This term is the basis of the word digital.

Most **counting** that we do today is based on groups of 10. This is probably an outgrowth of our dependence on fingers and toes as a counting tool. Counting with 10 as a base is called the decimal system. Ten unique symbols, or digits, are included in this system: 0, 1, 2, 3, 4, 5, 6, 7, 8, and 9. In general, the number of discrete values or symbols in a counting system is called the **base**, or **radix**. A decimal system has a base of 10.

Nearly, all numbering systems have place value. This refers to the value that a digit has with respect to its location in the number. The largest number value that can be represented at a specific location is determined by the base of the system. In the decimal system, the first position to the left of the decimal point is called the units place. Any number from 0 to 9 can be used in this place. Number values greater than 9 are expressed by using two or more places. The next location to the left of the units place is the 10-s position. Two-place numbers range from 10 through 99. Each succeeding place added to the left has a value that is 10 times as much as the preceding place. With three places, the place value of the third digit is $10 \times 10 \times 10$, or 1000. For four places, the place value is $10 \times 10 \times 10 \times 10$, or 10,000. The values continue for 100,000, 1,000,000, 10,000,000, and so on.

Any number in standard form can be expressed in expanded form by adding each weighted place value. The decimal number 2319 is expressed as $(1000 \times 2) + (100 \times 3) + (10 \times 1) + (1 \times 9)$. Note that the weight of each digit increased by 10 for each place to the left of the decimal point. In a number system, place values can also be expressed as a power of the base. For the decimal system, the place values are 10^3, 10^2, 10^1, 10^0, and so on. Each succeeding place has a value that is the next higher power of the base.

The **base 10**, or decimal, numbering system is extremely important and widely used today. Electronically, however, the decimal system is rather difficult to use. Each number would require a specific value to distinguish it from the others. Number detection would also require some unique method of distinguishing each value from the others. The electronic circuitry of a decimal system would be rather complex. In general, base 10 values are difficult to achieve and awkward to maintain.

Binary Numbering Systems

Nearly all digital electronic systems are of the binary type. This type of system uses 2 as its base, or radix. Only the numbers 0 and 1 are used in a binary system. Electronically, only two situations, such as a value or no value, are needed to express binary numbers. The number 1 is usually associated with some voltage value greater than zero. Binary systems that use voltage as 1 and no voltage as 0 are described as having positive logic. **Negative logic** uses voltage for 0 and no voltage for 1. **Positive logic** is more readily used today. Only positive logic is used in this discussion.

The two operational states or conditions of an electronic circuit can be expressed as on and off. An off-circuit usually has no voltage applied, and so represents the 0, or off, state. An on-circuit has voltage applied, and so represents 1, or on, state. With the use of electronic devices, it is possible to change states in a microsecond or less. Millions of 1s and 0s can be manipulated by a digital system in a second.

A binary digit can be expressed as either 1 or 0. The term bit is commonly used to describe a binary digit. The operational basis of a **binary system** is very similar to that of the decimal system. The base of the binary system is 2. This means that only the numbers 0 and 1 are used to denote specific numbers. The first place to the left of the binary point is the units, or 1s location. Place values to the left are expressed as powers of 2. Representative values are $2^0 = 1, 2^1 = 2, 2^2 = 4, 2^3 = 8, 2^4 = 16, 2^5 = 32, 2^6 = 64$, and so on.

When different numbering systems are used in a discussion, they usually incorporate a subscript number to identify the base of the numbering system being used. The number I 10 is a typical expression of this type. The subscript number 2 is used to denote the base of the numbering system. The binary point, which follows the 0, is usually omitted. Thus, 11010 is used to indicate when the number is expressed in decimal form. We will use other numbering systems, such as the base 8, or octal, system and the base 16, or hexadecimal, system. Numbers expressed in these bases will be identified with subscripts to avoid confusion with other numbers.

When only one numbering system is used in a discussion, the subscript notation is generally not needed. A binary number such as 101 is the equivalent of the decimal number 5. Starting at the binary point, the digit values are $1 \times 2^0 + 0 \times 2^1 + 1 \times 2^2$ or $1 + 0 + 4 = 5$. The conversion of a binary number to a decimal value is shown in **Figure 23.1**.

A shortcut version of the binary-to-decimal conversion process is shown in **Figure 23.2**. In this conversion method, first write the binary number. In this example, 1001101_2 is used. Starting at the binary point, indicate the decimal value of each binary place containing a 1. Do not indicate a value for the zero places. Add the indicated place values. Record the decimal equivalent of the binary number. Practice this procedure on several different binary numbers. With a little practice, the conversion process is very easy to achieve.

Changing a decimal number to a binary number is achieved by repetitive steps of division by the number 2. When the quotient is even with no remainder, a 0 is recorded. A remainder is also recorded. In this case, it will always be the number 1. The steps needed to convert a decimal number to a binary number are shown in **Figure 23.3**.

The conversion process is achieved by first recording the decimal value. The decimal number 30 is used in this example. Divide the recorded number

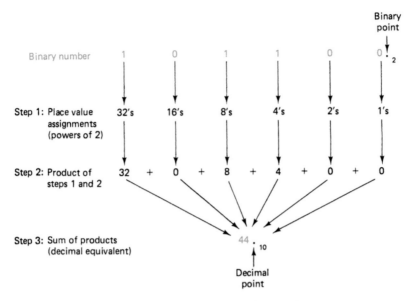

Figure 23.1 Binary-to-decimal conversion.

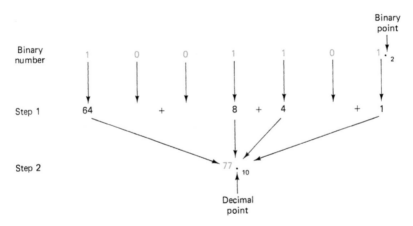

Figure 23.2 Binary-to-decimal conversion shortcut.

by 2. In this case, $30/2$ equals 15. The remainder is 0. Record the 0 as the first binary place value. Transfer 15 to position 2. Divide this value by 2. $15/2$ equals 7 with a remainder of 1. Record the remainder. Transfer 7 to position 3. $7/2$ equals 3 with a remainder of 1. Record the remainder and transfer 3 to position 4. $3/2$ equals 1 with a remainder of 1. Record the remainder and transfer 1 to position 5. $1/2$ equals 0 with a remainder of 1.

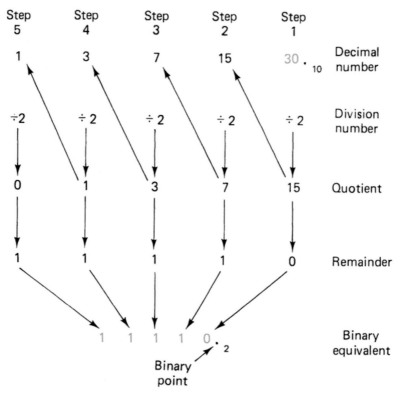

Figure 23.3   Decimal-to-binary conversion.

The conversion process is complete when the quotient has a value of 0. The binary equivalent of a decimal is indicated by the recorded remainders. For this example, the binary equivalent of 301 is 11110. Practice the decimal-to-binary conversion process with several different number values. With a little practice, the process becomes relatively easy to accomplish.

Binary Fractions

Having discussed the whole numbers of a binary system, let us now direct attention toward fractional number values. The binary point of a number decides the difference between integer and fractional values. Binary whole numbers are located to the left of the binary point. Fractional values are located to the right of the binary point. Whole numbers are identified as positive powers while fractional numbers are expressed as negative powers.

A number such as 1101.101 has both integer and fractional values. The whole number to the left of the binary point is 1101. This represents $1 \times 2^3 + 1 \times 2^2 + 0 \times 2^1 + 1 \times 2^0$ or 1310. The fractional part of this number is positioned to the right of the binary point. It is -101$_2$. This represents the sum of negative exponent values and becomes $1 \times 2^{-1} + 0 \times 2^{-2} + 1 \times 2^{-3}$ or 1/2 + 0/4 + 1/8. The decimal equivalent is 0.5 + 0.0 + 0.125 or 0.6250. The number 1101.101$_2$ equals 13.62510. The fractional part of this number is an even value of the decimal equivalent. If the binary values do not produce an even combination of the fractional number, the number will have infinite length. As a rule, only three or four fractional places are needed to define most binary numbers.

Self-Examination

1. _____ electronics is considered to be a counting operation.
2. An energy source, path, control, load, and indicator are included in the makeup of a digital _____.
3. A digital signal source produces _____ waves.
4. A decimal counting system has a base of _____.
5. In a decimal counting system, the first position to the left of the decimal point is the _____ place.
6. In a decimal counting system, any number value of 0 to _____ can be used in a place location.
7. In a decimal counting system number, values greater than 9 are expressed by using two or more _____.
8. A two-place decimal number can be used to represent numbers from 0 through _____.
9. The largest number that can be expressed at a place location in a decimal system is ____.
10. In a numbering system, place values are usually expressed as a _____ of the base.
11. In a decimal system, each place value is increased by a power of _____ as you move to the left.
12. Nearly all digital electronic systems use _____ numbers.
13. The two states or counts of a binary numbering system are _____ and _____.
14. _____ logic systems have voltage for 1 and no voltage for 0.
15. The base of a binary numbering system is _____.

16. Place values in a binary numbering system are expressed as powers of
_____.

17. The binary equivalent of the decimal number 8 is _____.

Answers

1. Digital
2. system
3. square
4. 10
5. units
6. 9
7. places
8. 99,0
9. 9
10. base
11. 10
12. binary
13. 1 and 0
14. Positive
15. 2
16. 210
17. 10002

Binary-Coded Decimal Numbers

When large numbers are to be indicated by binary numbers, they become somewhat awkward and difficult to use. For this reason, the **binary-coded decimal** method of counting was devised. In this type of system, four binary digits are used to represent each decimal digit. To illustrate this procedure, we have selected the decimal number 329, to be converted to a binary-coded decimal (BCD) number. As a binary number, $329_{10} = 101,001,001_2$. An example conversion is shown in **Figure 23.4**.

The largest decimal number to be displayed by four binary digits of a BCD number is 9_{10} or 1001_2. This means that six counts of the binary number are not being used. These are 1010, 1011, 1100, 1101, 1110, and 1111. Because of this, the **octal**, or base 8, and **hexadecimal**, or base 16, numbering systems were devised. Digital systems still process numbers in a binary form but usually display them in BCD, octal, or hexadecimal values.

Given the decimal number		105_{10}	
Step 1 Group the digits	(1)	(0)	(5)
Step 2 Convert each digit to binary group	(0001)	(0000)	(0101)
Step 3 Combine group values	0001 0000 0101 BCD		

Figure 23.4   BCD conversion example.

Octal Numbering Systems

Octal, or base 8, numbering systems are commonly used to process large numbers through digital systems. The octal system of numbers uses the same basic principles outlined with the decimal and binary systems. The octal numbering system has a base of 8. The digits 0, 1, 2, 3, 4, 5, 6, and 7 are used. The place values starting at the left of the octal point are powers of 8: $8^0 = 1$, $8^1 = 8$, $8^2 = 64$, $8^3 = 512$, $8^4 = 4096$, and so on.

The process of converting an octal number to a decimal number is the same as that used in the binary-to-decimal conversion process. In this method, however, the powers of 8 are employed instead of the powers of 2. To convert the number to an equivalent decimal number, see the procedure outlined in **Figure 23.5**.

Converting an octal number to an equivalent binary number is very similar to the BCD conversion process discussed previously. The octal number is first divided into discrete digits according to place value. Each octal digit is then converted into an equivalent binary number using only three digits. The steps of this procedure are shown in **Figure 23.6**. You may want to practice this conversion process to gain proficiency in its use.

Converting a decimal number to an octal number is a process of repetitive division by the number 8. After the quotient has been determined, the remainder is brought down as the place value. When the quotient is even with no remainder, a zero is transferred to the place position. The procedure for converting a decimal number to its octal equivalent is outlined in **Figure 23.7**.

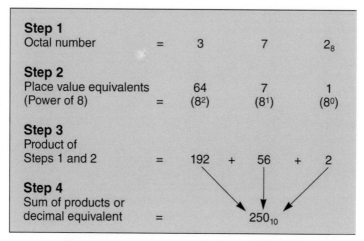

Figure 23.5   Octal-to-decimal conversion process.

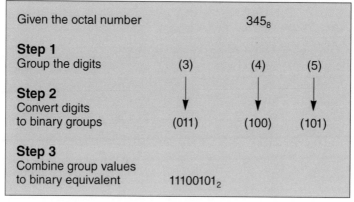

Figure 23.6   Octal-to-decimal conversion process.

Converting a binary number to an octal number is a very important conversion process found in digital systems. Binary numbers are first processed through the equipment at a very high speed. An output circuit then accepts this signal and may convert it to an octal signal that can be displayed on a readout device. Assume now that the number 10,110,101 is to be changed into an equivalent octal number. The digits must first be divided into groups of three, starting at the octal point. Each binary group is then converted into an equivalent octal number. These numbers are then combined, while remaining in their same respective places, to represent the equivalent octal number. See the conversion steps outlined in **Figure 23.8**.

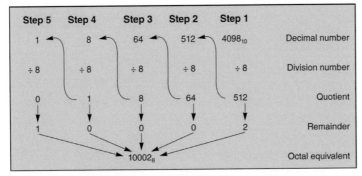

Figure 23.7 Decimal-to-octal conversion.

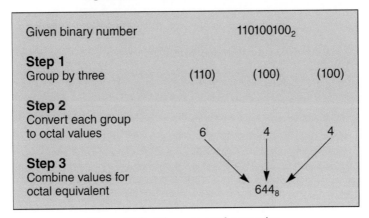

Figure 23.8 Binary-to-octal conversion.

Hexadecimal Numbering System

The hexadecimal numbering system is used to process large numbers. The base of this system is 16, which means that the largest value used in a place is 15. Digits used to display this system are the numbers 0-9 and the letters *A B, C, D, E,* and *F*. The letters *A-F* are used to denote the digits 10-15, respectively. The place values of digits to the left of the hexadecimal point are the powers of 16: $16^0 = 1$, $16^1 = 16$, $16^2 = 256$, $16^3 = 4096$, $16^4 = 65,536$, $16^5 = 1,048,576$, and so on.

The process of changing a hexadecimal number to a decimal number is achieved by the same procedure outlined for other conversions. Initially, a hexadecimal number is recorded in proper digital order as outlined in **Figure 23.9**. The powers of the base are then positioned under the respective digits. In a hexadecimal conversion, step 2 is usually added to show the values

Step 1					
Hexadecimal number	=	1	2	C	D_{16}
Step 2					
Place values (Power of 16)	=	4096s	256s	16s	1s
Step 3					
Convert letters to numbers	=	1	2	12	13
Step 4					
Product of Steps 2 and 3	=	4096 +	512 +	192 +	13
Step 5					
Sum of products or decimal equivalent	=	4813_{10}			

Figure 23.9   Hexadecimal-to-decimal conversion.

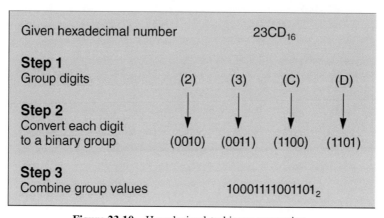

Figure 23.10   Hexadecimal-to-binary conversion.

of the letters. Each digit is then multiplied by its place value to indicate discrete place value assignments. Steps 1 and 2 are then multiplied together. In step 3, these product values are added, giving the decimal equivalent of a hexadecimal number in step 4.

The process of changing a hexadecimal number to a binary equivalent is a simple grouping operation. **Figure 23.10** shows the operational steps for making this conversion. Initially, the hexadecimal number is separated into discrete digits in step 1. Each digit is then converted to an equivalent binary number using only four digits per group. Step 3 shows the binary groups combined to form the equivalent binary number.

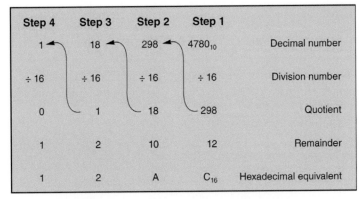

Figure 23.11   Decimal-to-hexadecimal conversion.

Given binary number		1001101101010$_2$	
Step 1			
Group by four	(0001) (0011)	(0110)	(1010)
Step 2			
Convert groups to hexadecimal values	1 3	6	10
	1 3	6	A
Step 3			
Combine values for hexadecimal equivalent	136A$_{16}$		

Figure 23.12   Binary-to-hexadecimal conversion.

The conversion of a decimal number to a hexadecimal number is achieved by the repetitive division process used with other number systems. In this procedure, however, the division factor is 16 and the remainders can be as large as 15. **Figure 23.11** shows the necessary procedural steps for achieving this conversion.

Converting a binary number to a hexadecimal equivalent is a reverse of the hexadecimal to binary process. **Figure 23.12** shows the fundamental steps of this procedure. Initially, the binary number is divided into groups of four digits, starting at the hexadecimal point. Each grouped number is then converted to a hexadecimal value and combined to form the hexadecimal equivalent.

Self-Examination

18. In a binary-coded decimal, or BCD, numbering system, _____ binary digits are used to represent each decimal digit.
19. A decimal number such as 528 is expressed in BCD form as _____.
20. The largest digit that can be expressed by any group of BCD numbers is _____.
21. An octal numbering system has a base, or radix, of _____.
22. The place values of digits of an octal numbering system are expressed as the powers of _____.
23. The decimal equivalent of the number 347_8 is _____.
24. The binary equivalent of the number 256_2 is _____.
25. The octal equivalent of the number 326_8 is _____.
26. The octal equivalent of $11,010,100_2$ is _____.
27. The radix of a hexadecimal numbering system is _____.
28. The letters A, B, C, D, E, and F are used to denote the digits _____ through _____, respectively, in the hexadecimal system.
29. The decimal equivalent of $14D9_{16}$ is _____$_{16}$.
30. The binary equivalent of $49A_{16}$ is _____$_{10}$.
31. The hexadecimal equivalent of 1235_{10} is _____$_{16}$.
32. The hexadecimal equivalent $11010,110,100_2$ is _____$_{16}$.

Answers

18. four
19. 0101-0010-100013CD
20. 9
21. 8
22. 8
23. 231
24. $010,101,110_2$
25. 506
26. 324
27. 16
28. 16 10-15
29. 16 4465
30. 16 0100,1001,1010
31. 16 4D3
32. 16 2B4

23.2 Digital Logic Circuits

Digital logic systems, no matter how complex, are composed of a small group of identical building blocks. These blocks are either decision-making circuits or memory units. A large majority of the decision-making circuits are made up of logic gates or a combination of logic gates. Logic gates respond to binary input data and produce an output that is based on the status of the input. Memory circuits are used to store binary data and release it when the need arises.

Logic gates are essentially a combination of high-speed switching circuits. These gates are the electronic equivalent of a simple switch connected in series or parallel. Digital systems combine large numbers of these gates in decision-making circuits. We investigate the simple switch type of logic circuit to explain basic logic functions.

In most digital systems, we do not use mechanical switch logic gates. For example, they respond very slowly to data. Electronic logic gates have been designed that can change states very quickly. In fact, these gates can change states so quickly that a human cannot detect the switching time. Typical switching times are less than a microsecond, or 10^{-6}s. In many microprocessor-based digital systems, switching times are in the nanosecond, or 10^{-9} s, range. This takes special circuits to detect a state change in logic gates operating at this speed. Logic gates respond in the same manner regardless of the switching speed.

Binary Logic Functions

Any bistable circuit that is used to make a series of decisions based on two-state input conditions is called a binary logic circuit. Three basic circuits of this type have been developed to make simple logic decisions: the **AND** circuit, the **OR** circuit, and the **NOT** circuit. The logic decision made by each circuit is unique and very important in digital system operations.

Electronic circuits designed to perform specific logic functions are commonly called **gates**. This term refers to the capability of a circuit to pass or block specific digital signals. A simple if-then type of statement is often used to describe the basic operation of a logic gate. For example, if the inputs applied to an AND gate are all 1, then the output will be 1. If a I is applied to any input of an OR gate, then the output will be 1. If any input is applied to a NOT gate, then the output will be reversed.

The fundamental operation of a digital system is based directly on gate applications. Technicians working with digital systems must be very familiar with each basic gate function. The input-output characteristics and operation of basic logic gates serve as the basis of this discussion.

AND Gates

An **AND gate** is designed to have two or more inputs and one output. Essentially, if all inputs are in the 1 state simultaneously, then a 1 will appear in the output. **Figure 23.13** shows a simple switch-lamp circuit of the AND gate, its symbol, and an operational table. In **Figure 23.1(a)**, a switch turned on represents a 1 condition, whereas off represents a 0. The lamp also displays this same condition by being a I when it is on and 0 when turned off. Note that the switches are labeled *A* and *B*, whereas the lamp, or output, is labeled *C*. The operational characteristics of a gate are usually simplified by describing the input-output relationship in a table. The table in **Figure 23.1(c)** shows the 1 and 0 alternatives at the input and the corresponding output that will occur as a result of the input. As a rule, such a description of a gate is called a truth table. Essentially, it shows the predictable operating conditions or a logic circuit.

Each input to an AND gate has two operational states of 1 and 0. A **two-input AND gate** would have 2, or 4, possible combinations that would influence the output. A three-input gate would have 2, or 8, combinations, and a four-input would have 21, or 16, combinations. These combinations are normally placed in the truth table in binary progression order. For a two-input gate, this would be 00, 01, 10, and 11, which shows the binary count of 0, 1, 2, and 3 in order.

Functionally, the AND gate of **Figure 23.13(a)** produces a 1 output only when switches *A* and *B* are both 1. Mathematically, this action is described as $A \times B = C$. This expression shows the multiplication operation. In a machine operation, this type of gate could be used to protect an operator from some type of physical danger. For example, it will not permit a machine to be actuated until the operator presses one button with the left hand and a second button with the right hand at the same time. This removes the hands from a dangerous operating condition.

The symbol representations of an **AND gate** shown in **Figure 23.13(b)** are very common. The symbol on the left side has been adopted by the American National Standard Institute (**ANSI**) and the Institute of Electrical and Electronic Engineers (**IEEE**). The symbol on the right side is used by the National Electrical Manufacturer's Association (**NEMA**). Both symbols are in common use today.

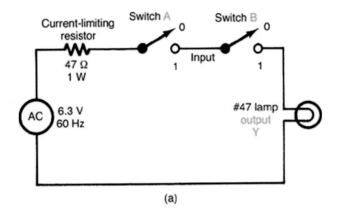

(a)

Input		Output
Switch A	Switch B	Lamp Y
0	0	0
0	1	0
1	0	0
1	1	1

(b)

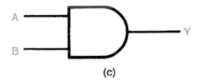

(c)

Figure 23.13   AND gates. (a) Logic circuit. (b) Truth table. (c) Symbol.

OR Gates

An **OR gate** is designed to have two or more inputs and a single output. Like the AND gate, each input to the OR gate has two possible states: I and 0. The output of this gate will produce a 1 when either or both inputs are 1. **Figure 23.14** shows a simple lamp switch analogy of the OR gate, its symbol, and a truth table. Functionally, an OR gate will produce a 1 output when both

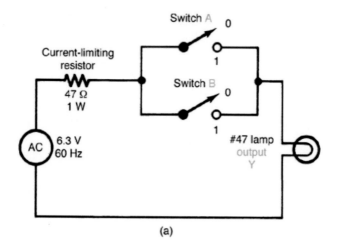

(a)

Input		Output
Switch A	Switch B	Lamp Y
0	0	0
0	1	1
1	0	1
1	1	1

(b)

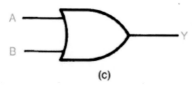

(c)

Figure 23.14   OR gates. (a) Logic circuit. (b) Truth table. (c) Symbol.

switches are 1 or when either switch *A* or *B* is a 1. Mathematically, this action is described as $A + B = C$. This expression shows OR addition. Applications of this gate are used to make logic decisions as to whether or not a 1 appears at either input. The interior light system of an automobile is controlled by an OR type of circuit. Individual door switches and the dash panel switch all control the lighting system from a different location. Essentially, when any one of the inputs is on, it will cause the interior lights to be on.

NOT Gates

A **NOT gate** has a single input and a single output, which makes it unique compared with the AND and OR gates. The output of a NOT gate is designed so that it will be opposite to that of the input state. **Figure 23.15** shows a simple switch-controlled NOT gate, its symbol, and truth table. Note that when the single-pole, single throw (SPST) switch is on, or in the 1 state, it shorts out the lamp. Likewise, placing the switch in the off condition causes the lamp to be on, or in the 1 state. NOT gates are also called inverters. Mathematically, the operation of a NOT gate is expressed as $A = \overline{A}$. The A-bar

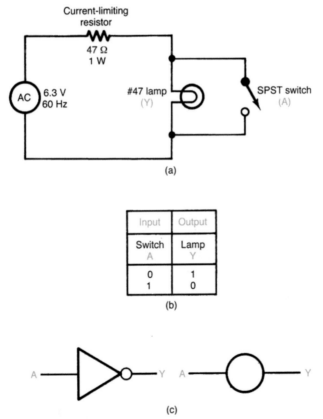

Figure 23.15 NOT gates. (a) Logic circuit. (b) Truth table. (c) Symbol.

symbol shows the inversion function. The significance of a NOT gate should be rather apparent after the following discussion of gates.

Combinational Logic Gates

When a NOT gate is combined with an AND gate or an OR gate, it is called a **combination logic function**. A NOT-AND gate is normally called a **NAND gate**. This gate is an inverted AND gate, or simply NOT and AND gate. **Figure 23.16** shows a simple switch-lamp circuit analogy of this gate, along with its symbol and truth table.

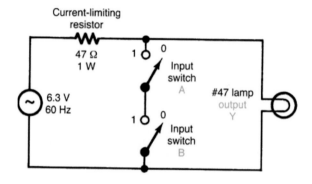

Input		Output
Switch A	Switch B	Lamp Y
0	0	1
0	1	1
1	0	1
1	1	0

Figure 23.16   NAND gates. (a) Logic circuit. (b) Truth table. (c) Symbol.

   The **NAND gate** is an inversion of the AND gate. When switches A and B are both on, or in the 1 state, the lamp C is off. When either or both switches are off, the output, or lamp C, is in the on, or 1, state. Mathematically, the operation of a NAND gate is expressed as $A \times B = e$. The bar over the C denotes the inversion, or negative function, of the gate.

   A combination of NOT-OR, or NOR, gate produces an inversion of the OR function. **Figure 23.17** shows a simple switch-lamp circuit analogy of this gate, along with its symbol and truth table. When either switch A, B or A and B are off, or 0, the output is a 1, or high. When either switch A or B is 1, the output is 0. Mathematically, the operation of a **NOR gate** is expressed as

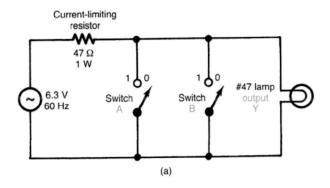

(a)

Input		Output
Switch A	Switch B	Lamp Y
0	0	1
0	1	0
1	0	0
1	1	0

(b)

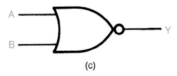

(c)

Figure 23.17   NOR gates. (a) Logic circuit. (b) Truth table. (c) Symbol.

$A + B = C$. A 1 will appear in its output only when A is 0 and B is 0. This represents a unique logic function.

Self-Examination

33. An electronic circuit designed to perform a logic function statement is often used to is called a _____.
34. A simple _____ statement describes the basic operation of a logic gate.
35. A _____ gate has one input and an inverted output.
36. In an AND gate, a 1 applied to each input will produce a _____ output.
37. A 0 and a 1 applied to the inputs of an AND gate will produce a _____ output.
38. $A \times B = C$ is the mathematical expression of a(n) _____ gate.
39. A(n) _____ will produce a I output when both inputs are 1 or when either input is 1.
40. $A + B = C$ is the mathematical expression of a(n) _____ gate.
41. A 1 applied to any one of the inputs of an OR gate will produce a(n) _____ output.
42. $A = A'$ is a mathematical expression of the _____ gate.
43. A NOT gate is also called a(n) _____.
44. A NAND gate is a combination of _____ and _____ gate functions.
45. A NOR gate is a combination of _____ and _____ gate functions.
46. Mathematically, $A \times B$ inversion is the expression of a(n) _____ gate.
47. The expression $A + B$ inversion is the mathematical function of a(n) ____ gate.
48. A 1 applied to each input on a NAND gate will produce a _____ output.
49. A 0 applied to each input of a NAND gate will produce a _____ output.
50. A 1 applied to each input of a NOR gate will produce a _____ output.
51. A 1 applied to one input and a 0 applied to the other input of a NOR gate will produce a _____ output.
52. A 1 applied to one input and a 0 applied to the other input of a NAND gate will produce a _____ output.

Answers

33. Gate
34. If-then
35. NOT
36. 1
37. 0
38. AND
39. OR
40. OR
41. 1
42. NOT
43. inverter
44. AND, NOT
45. OR, NOT
46. NAND
47. NOR
48. 0
49. 1
50. 0
51. 0
52. 1

23.3 Flip-Flops

Digital systems employ a number of devices that are not classified specifically as logic gates. These devices play a unique role in the operation of a digital system. Such things as flip-flops, counters, registers, decoders, and memory devices are included in this classification. We discuss truth tables, logic symbols, and the operational characteristics of these devices so that they may be used more effectively when the need arises. As a general rule, most of these devices are constructed entirely on IC chips. The operation is based, to a large extent, on the internal circuit construction of the IC. Very little can be done to alter the operation of these devices other than to modify the input or use its output to influence the operation of a secondary device.

Digital logic circuits are classified into two groups. We have worked with logic circuits that make up one part of this classification. **AND, OR, NOT, NAND, and NOR gates** are considered to be **combinational logic circuits**. The other part of this classification deals with sequential logic circuits. Sequential circuits involve some form of timing in their operation. The timing function permits one or several devices to be actuated at an appropriate time or in an operational sequence. Logic gates are the building blocks of combinational logic circuits. A flip-flop is the basic building block of a **sequential logic circuit**. This chapter deals with some basic flip-flop circuits.

Flip-Flop Circuits

Flip-flops are commonly used to generate signals, shape waves, and achieve division. In addition to these operations, a flip-flop may also be used as a memory device. In this capacity, it can be made to hold an output state even when the input is completely removed. It can also be made to change its output when an appropriate input signal occurs. The flip-flop is a logic device with two or more inputs and two outputs. The outputs are the complements of each other.

Figure 23.18 shows the general symbol of a flip-flop. Note that the outputs are labeled Q and Q' and this type is an *R-S* flip-flop.

Any letter designation could be used to identify the outputs. The letter Q is widely used. Q is considered to be normal, and Q' is the inverted output. The Q output can be in either the high (1) or low (0) state. Q' is always the reverse of Q. A flip-flop has two operational states: $Q = 0$, $Q' = 1$ and $Q' = 1$, $Q = 0$. A flip-flop has one or more inputs. These inputs are used to initiate a state change in the operation of a flip-flop. When an input is pulsed or triggered, it will send the output to a given state. It will remain in this state even after the input returns to normal. This is the memory characteristic of a flip-flop.

R-S Flip-Flop								
Logic Diagram		Symbol		Truth Table				
				Applied inputs		Previous outputs		Resulting outputs	
				S	R	Q	Q̄	Q	Q̄	
				0	0	1	0	1/0	0/1	Unpredictable
				0	1	1	0	1	0	
				1	0	1	0	0	1	
				1	1	1	0	1	0	
				0	0	0	1	0/1	1/0	Unpredictable
				0	1	0	1	1	0	
				1	0	0	1	0	1	
				1	1	0	1	0	1	

Figure 23.18   *R-S* flip-flop: Logic diagram, symbol, and truth table.

RS Flip-Flop

An **RS flip-flop** can be constructed from two cross-coupled NAND or NOR gates. **Figure 23.18** shows a symbol of the flip-flop, a circuit achieved with cross-coupled NAND gates, and a truth table. The outputs of the symbol and NAND gate circuit are labeled Q and Q'. The inputs are labeled R and S. R identifies the **Reset** and S is the **Set** function. The Q and Q' outputs of the flip-flop are dependent on the voltage level of the RS inputs.

In the truth table, note that the S input must be low to produce a 1 or high at the Q output. The reset must be 0 to cause Q to be 1 or high. A circle at the input identifies the state needed to produce a change. No circle at the input indicates a 1 or high-level activated device. A circle at the input indicates a low-, or 0, level activated device. A cross-connected NAND gate is an active low flip-flop and a cross-connected NOR gate circuit is an active high-level device. The presence or absence of a circle on the input of the symbol is a standard method of identifying this characteristic of a flip-flop.

Clocked Flip-Flops

Flip-flops operate as either asynchronous or synchronous devices. **Asynchronous** operation occurs when the outputs can change state any time one or more of the inputs change. An RS flip-flop is an asynchronous device. **Synchronous** operation occurs when output changes are controlled by a clock signal. Most flip-flops operate as synchronous devices. Synchronous circuits are easier to troubleshoot because the output can only change at a specific time. Everything is synchronized by the clock signal. The clock signal applied to a synchronized device is generally a rectangular series of pulses or square waves. The clock signal is distributed to all parts of a digital system. The output of specific devices can make a state change only when the clock signal occurs.

There are two basic types of **clock signals** used to trigger flip-flops. These are defined as level-triggering or edge-triggering signals. In **level clocking**, the state of a clock changes value from 0 to 1 and carries out a transfer of data or completes an action. Data cannot be changed or altered except immediately after a level change. At that time, it can be changed only once. In **edge clocking**, there is **positive-edge triggering** and negative-edge triggering. Positive-edge triggering occurs at the leading or beginning edge of a pulse. The signal makes a quick transition from 0 to 1. **Negative-edge triggering** occurs at the trailing edge or end of a pulse. This pulse changes from 1 to

0. An edge-triggered flip-flop can have its input data changed at any time. One square wave has a leading edge and a trailing edge in its makeup. A flip-flop can be triggered by only one type of pulse. Most of the flip-flops used in digital systems are classified as clocked flip-flops. The following are some principle ideas that are common to all clocked flip-flops.

1. Clocked flip-flops have an input labeled CLK, CK, or CR.
2. The symbols of clocked *RS* flip-flop have the clock input labeled CLK.
3. A circle on the symbol at the clock input indicates negative-edge triggering and no circle indicates positive-edge triggering.
4. All clocked flip-flops have one or more control inputs. Control inputs are identified by a variety of labels depending on the exact function. The control input determines the state of the output, but its effect is not realized until a clock pulse occurs. Essentially, the logic level of an input controls how the output will change, whereas the CLK signal determines when the change takes place.

The Clocked RS Flip-Flop

A **clocked RS flip-flop** has reset (R), set (S), and clock (CLK) inputs. The output is labeled Q and Q. **Figure 23.19** shows positive-edge triggered and negative-edge triggered clocked *RS* flip-flops. A positive-edge triggering device changes output states only on the leading, or positive, transition of the clock pulse. It does not respond to negative-edge transitions. Negative-edge triggering takes place on the trailing, or negative, transition of the clock pulse. This device does not respond to positive-edge transitions. The S and R inputs prepare the flip-flop for a specific state change. The time waveform diagrams show how the output responds to different combinations of the SR input. The truth table shows these possibilities. The output of a clocked flip-flop is uniquely different. The outputs are identified as $Q_.$, and $Q_.$. Q, represents the status of Q before a clock pulse is applied. The letter n of this designation refers to the status of Q(now). Qn I 1 indicates the status of Q after a clock pulse has been applied. The Qn and Q designations are widely used by IC manufacturers on data sheets.

The operational concept of a clocked *RS* flip-flop is very important. This type of flip-flop is, however, not available on an IC chip. A clocked *RS* flip-flop can be achieved by logic gates. The construction of this flip-flop is very important in the fabrication of other devices. **Figure 23.10** shows how a clocked *RS* flip-flop is built with NAND gates. This circuit employs

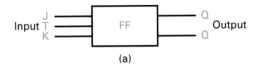

(a)

Applied inputs			Previous outputs		Resulting outputs		
J	T	K	Q	$\overline{Q}$	Q	$\overline{Q}$	
0	0	0	0	1	0	1	
0	0	1	0	1	0	1	
0	1	1	0	1	0	1	
1	0	0	0	1	0	1	
1	1	0	0	1	1	0	
1	1	1	0	1	1	0	← Toggle state
0	0	0	1	0	1	0	
0	0	1	1	0	1	0	
0	1	1	1	0	0	1	
1	0	0	1	0	1	0	
1	1	0	1	0	1	0	
1	1	1	1	0	0	1	← Toggle state

(b)

Figure 23.19 *JK* flip-flop. (a) Symbol. (b) Truth table.

NAND gates 3 and 4 in a cross-coupled *RS* flip-flop. A pulse steering circuit is formed by NAND gates 1 and 2. This part of the circuit is responsible for passing or inhibiting the Set and Reset inputs. A clock pulse is needed to enable the steering circuit so that it will pass the input on to the flip-flop. An edge detector circuit is commonly connected to one input of each steering gate. This part of the circuit is used to select the desired edge of the clock signal needed to produce triggering. The combined circuitry of an *RS* flip-flop is often used in the construction of other flip-flops. This circuit provides a convenient method of injecting a clock signal into a flip-flop.

The Clocked JK Flip-Flop

The **JK flip-flop** is frequently used. The *JK* flip-flop is often considered to be a universal device. Its operation is similar to that of other flip-flops. It can, however, be easily modified to achieve different functions. Flip-flops, in general, have two or, possibly, three input combinations. A clocked *JK* flip-flop has four possible input combinations. These are no change, enter 1, enter 0, and toggle. This makes the *JK* flip-flop a very versatile device. **Figure 23.19** shows a symbol and the truth table of a clocked *JK* flip-flop. The *J* and *K* inputs supply data to the device.

The letters *J* and *K* do not represent any particular term. They were probably chosen to distinguish this device from the Set and Reset inputs of an *RS* flip-flop. Control of a *JK* flip-flop is similar in nearly all respects to that of an *RS* flip-flop. When *J* and *K* are both 1, however, this flip-flop does not produce an unpredictable output. It causes the output to change states with each clock pulse. As a result, *Q* and -0 change states continually with the clock. This condition causes the output to respond as a toggle switch. It is generally called the **toggling mode** of operation. Toggling permits a flip-flop to achieve binary division and to shift data one bit at a time.

Flip-flops of the *JK* type have some circuit modifications. An SN7476 *JK* flip-flop is similar to the basic flip-flop but has two asynchronous inputs called **PRESET (PS)** and **CLEAR (CLR)**. The *J*, *K*, and CLK inputs are synchronous. The outputs are *Q* and *Q'*. The asynchronous inputs are designed to override the synchronous inputs. The truth table shows the response of the asynchronous inputs on the first three lines. Since the synchronous inputs are overridden by PS and CLR, an *X* is used in place of data for these entries. A prohibited state occurs when both PS and CLR are activated at the same time. This occurs when PS = 0 and CLR = 0. This means that the asynchronous inputs are activated by a negative level signal or a DC voltage value. Small circles on the PS and CLR inputs of the symbol denote negative level triggering. It is good practice to avoid using the *JK* flip-flop in its prohibited state. Note that *Q* and 0 are both 1 when this mode of operation occurs.

When both asynchronous inputs of a *JK* flip-flop are disabled, the synchronous inputs can again be used to control the output. To disable the asynchronous inputs, a 1 must be applied to PS and CLR at the same time. The bottom four lines of the truth table show the status of this operation. Note that the modes of operation are hold, reset, set, and toggle. When PS and CLR are disabled, the flip-flop is controlled by *J*, *K*, and CLK. Operation is continuous as long as power is applied, and data supplied to the appropriate inputs. Several other variations of the *JK* flip-flop are available on IC chips. Two or more *JK* flip-flops can be housed on a single chip.

Types of Flip-Flops

F-type flops are known by several names. In general, they may be called **multivibrators**, **latches**, or by a specific letter designation such as the **RS**, **RST**, **D**, or **JK** flip-flop. The letters identify the inputs. **RS** stands for reset

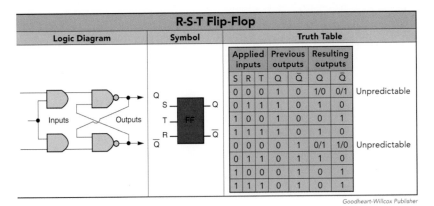

| R-S-T Flip-Flop | | |
| Logic Diagram | Symbol | Truth Table |

| Applied inputs | | | Previous outputs | | Resulting outputs | | |
S	R	T	Q	Q̄	Q	Q̄	
0	0	0	1	0	1/0	0/1	Unpredictable
0	1	1	1	0	1	0	
1	0	0	1	0	0	1	
1	1	1	1	0	1	0	
0	0	0	0	1	0/1	1/0	Unpredictable
0	1	1	0	1	1	0	
1	0	0	0	1	0	1	
1	1	1	0	1	0	1	

Goodheart-Willcox Publisher

Figure 23.20 *RST* flip-flop - logic diagram, symbol, and truth table.

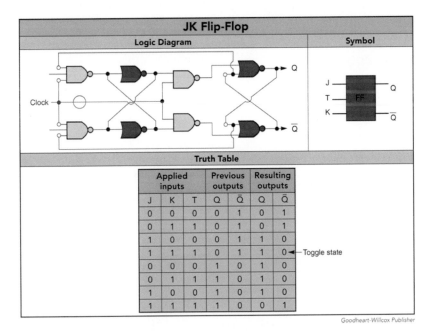

| JK Flip-Flop | |
| Logic Diagram | Symbol |

| Truth Table |

| Applied inputs | | | Previous outputs | | Resulting outputs | | |
J	K	T	Q	Q̄	Q	Q̄	
0	0	0	0	1	0	1	
0	1	1	0	1	0	1	
1	0	0	0	1	1	0	
1	1	1	0	1	1	0	—Toggle state
0	0	0	1	0	1	0	
0	1	1	1	0	1	0	
1	0	0	1	0	1	0	
1	1	1	1	0	0	1	

Goodheart-Willcox Publisher

Figure 23.21 *RST* flip-flop - logic diagram, symbol, and truth table.

and set. **RST** refers to reset, set, and trigger. **D** identifies a flip-flop that has a delay capability. A **JK** flip-flop has data inputs and a clock. Letter-designated flip-flops are more widely used than other identification means. Circuit diagrams, symbols, and truth tables of an *RST* flip-flop and a *JK* flip-flop using NAND gates are shown in **Figures 23.20** and **23.21**, respectively.

Self-Examination

53. The terms flip-flop and _____ can be used interchangeably.
54. The inputs of an *RS* flip-flop are called _____ and _____.
55. The outputs of a flip-flop are the _____ of each other.
56. NOR gates used in a flip-flop produce a _____ output when a 1 appears at either one or both of its inputs.
57. The output of a flip-flop has two operational states: $Q =$ _____ and $Q' =$ _____.
58. The *RS* flip-flop symbol in **Figure 23.18** is considered to be an active (low/high) device.
59. An active (low/high) flip-flop has no circles on the inputs of its symbol.
60. In an active low *RS* flip-flop, a 0 at the Set input and a 1 at the Reset input cause Q to be _____.
61. In an active low *RS* flip-flop, a 1 at the Set input and a 1 at the Reset input causes _____ in Q.
62. (Asynchronous, synchronous) operation occurs when the output of a flip-flop can change its state anytime one or more of the inputs change.
63. (Asynchronous, synchronous) operation of a flip-flop occurs when output changes are controlled by a clock signal.
64. A(n) _____ flip-flop is an asynchronous device.
65. _____ and _____ clock signals are used to trigger flip-flops.
66. The _____-edge or _____-edge of a clock signal can be used to trigger a flip-flop.
67. _____-edge triggering occurs at the beginning or leading edge of a clock triggering signal.
68. _____-edge triggering signals occur at the trailing edge of a clock signal.
69. A circle at the clock input symbol of a flip-flop indicates _____-edge triggering.

70. Three functional parts of a clocked *RS* flip-flop built from NAND gates are _____, _____, and _____.
71. If a flip-flop has a *D* input, it represents _____ or _____.
72. A *JK* flip-flop has (1, 2, 3, 4, 5) possible input combinations.
73. The *JK* inputs of a *JK* flip-flop supply _____ to the device.
74. When *J* = I and *K* = 1, the output _____ with each clock pulse.
75. The asynchronous inputs of a *JK* flip-flop are _____ and _____.
76. The asynchronous inputs of a *JK* flip-flop _____ the synchronous inputs.
77. When the asynchronous inputs are disabled, the _____ inputs regain control of a *JK* flip-flop.

Answers

53. latch
54. Reset and Set
55. complement
56. 0
57. $Q = 1, Q' = 1$
58. high or 1
59. low
60. 0
61. no change
62. asynchronous
63. synchronous
64. *RS*
65. Level and edge
66. positive, negative
67. Positive
68. Negative
69. negative
70. *RS* flip-flop, pulse steering circuit, edge detector
71. data or delay
72. 4
73. data
74. . toggles
75. preset (PS) clear (CLR)
76. override
77. synchronous

23.4 Digital Counters

One of the most versatile and important logic devices of a digital system is thecounter. This device, as a general rule, can be employed to count a wide variety of objects in a number of different digital system applications. While this device may be called upon to count an endless number of objects, it essentially counts only one thing - electronic pulses. These pulses may be produced electronically by a clock, electromechanically, photoelectrically, or by a number of other processes. The basic operation of the counter, however, is completely independent of the pulse generator.

Flip-Flops

A flip-flop is the basic building block of a **digital counter**. Its operation is dependent on the status of the input and in some types on the condition of control elements. A *JK* flip-flop is widely used to form counter circuits. When this flip-flop is connected so that its *JK* inputs are at a 1, or high, level, it responds in the toggle mode of operation. This means that each clock pulse applied to the input causes the output to go through a state change. Toggling causes the flip-flop to have a divide-by-two function. The output complements itself with each input clock pulse. The first pulse causes the output to go high. The next pulse causes the output to go low. This action means that the output changes one complete time with two complete input pulses. A T-connected *JK* flip-flop has a divide-by-2 capability. This is fundamental in the operation of a **binary counter**.

The polarity of the **trigger pulse** is an important consideration in the operation of a flip-flop. Some flip-flops trigger on the leading or positive edge of the trigger pulse. Other flip-flops trigger on the trailing or negative going part of the pulse. Negative-edge triggering seems to be used more frequently in counters today.

Binary Counters

A common application of the digital counter is used to count numerical information in binary form. This type of device simply employs a number of flip-flops connected so that the *Q* output of the first device drives the trigger or clock input of the next device. Each flip-flop, therefore, has a divide-by two function. **Figure 23.22** shows *JK* flip-flops connected to achieve binary counting. This counter is called a **binary ripple counter**. Each flip-flop in

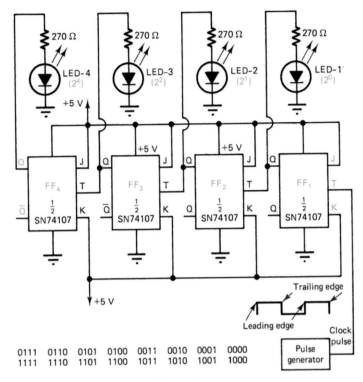

Figure 23.22 Binary counter.

this circuit has the *J* and *K* inputs held at a logic 1 level. Clock pulses applied to the input of FF-$_A$ cause a state change. This flip-flop circuit triggers only on the negative-going part of the clock pulse. The output FF-, therefore, alternates between 1 and 0 with each pulse. A 1 output appears at *Q* of FF-$_A$ for every two input pulses. This means that each flip-flop divides the input signal by a factor of two. The *Q* output of each flip-flop can then be considered as a power of two. The output of FF-$_A$ is 2^0, of FF-$_B$ is 2^1, of FF-$_C$ is 2 2, of FF-$_D$ is 2^3, and of FF-$_E$ is 2^4. Five flip-flops connected in this manner will produce a count of 2^5, or 32_{10}.

This is called a **modulo-32 counter**. The modulus of a counter is the number of different states the counter must go through to complete its counting cycle. For a modulo-32 counter, the count is 00000 to 11111. The largest count that can be achieved in this case is 1111_2. This represents 31_{10}. It occurs when all flip-flops are 1 at the same time. The next pulse applied to the input clears the counters so that 0 appears at all the *Q* outputs. The counter of **Figure 23.22**, where the output of each FF serves as the clock input for the next FF, is referred to as an **asynchronous counter**. This is done because all FFs do not change states in synchronism with the clock pulses. Only FF-$_A$ responds to the clock pulses. FF-$_B$ has to wait for FF-$_A$ to change states before it is triggered. Other FFs must also wait for a similar transition before they change. In effect, this means that there is a **delay** between the responses of each FF. The delay is usually in the range of 20-30 ns. In some applications, a counter with this amount of delay may be troublesome. Since a clock pulse applied to the input of the first FF causes a chain reaction, or ripple, to occur in the counter, this circuit is also called a **ripple counter**. The terms asynchronous counter and ripple counter are used interchangeably.

By grouping three flip-flops together, it is possible to develop the unit part of a **binary-coded-octal (BCO) counter**. Therefore, 111_2 is used to represent the seven counts, or seven units, of an octal counter. This is also called a **modulo-8 counter**. It counts from 000 to 111_2. The number of state changes that must take place is 8. Two groups of three flip-flops connected in this manner would produce a maximum count of $111\text{-}111_2$ which represents 77_8 or 63_{10}.

By placing four flip-flops together in a group, it is possible to develop the units part of a **binary-coded-hexadecimal (BCH) counter**. This could also be called a **modulo-16 counter**. Thus, 1111_2 would be used to represent F_{16} or 15_{10}. Two groups of four flip-flops can be used to produce a maximum count of $1111\text{-}1111_2$, which represents F_{16} or 15_{10}. Each succeeding group of four flip-flops is used to raise the counting possibility to the next power of 16. **Binary counters** that contain four interconnected flip-flops may be built on one IC chip. **Figure 23.23** shows the logic connections of a **4-bit binary counter**, or **modulo-16 counter**. When used as a 4-bit counter, the flip-flops will produce a maximum count of 1111_2 or 15_{10}.

Decade Counters

Since most of the mathematics that we use today is based on the decimal, or base 10, system, it is important to be able to count by this method. Digital systems are, however, designed to process information in binary form because of the ease with which a two-state signal can be manipulated. The output of a binary counter must, therefore, be changed into a decimal form before it can be used by an individual not familiar with binary numbers. The first step in this process is to change binary signals into a **binary-coded decimal** or **BCD** form. A BCD counter is shown in **Figure 23.23**.

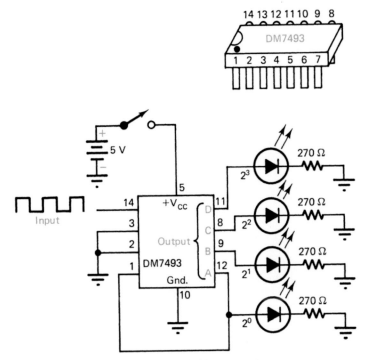

Figure 23.23 4-bit binary counter.

Self-Examination

78. When a flip-flop is connected so that the *JK* inputs are held at a 1 (high) level, it responds in its _____ mode of operation.
79. Toggling a *JK* flip-flop causes the _____ to change states with each clock pulse.
80. The first clock pulse of a *JK* flip-flop causes the output to go (high, low) and the second clock pulse causes the output to go (high, low).
81. A *T*, or toggling, *JK* flip-flop has a divide-by _____ capability.
82. (Negative, positive) edge triggering tends to be used more as counter flip-flop.
83. The output of a *JK* flip-flop occurs a short time after triggering due to _____ delay.
84. Because of the divide-by-two capability, *JK* flip-flops respond well as _____ counters.
85. A four-flip-flop counter can produce a maximum binary count of _____.

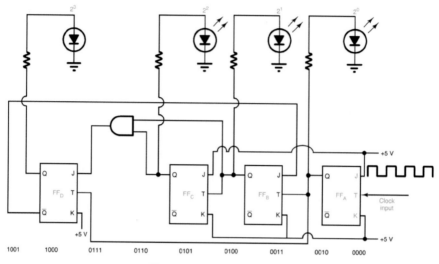

Figure 23.24 BCD counter.

86. The largest count that can be achieved by a five-flip-flop binary counter is _____.
87. After a binary counter makes its maximum count, the next clock pulse applied to the input causes the count to _____.

Answers

78. toggle
79. output
80. high, low
81. two
82. Negative
83. propagational
84. binary
85. 15
86. 111112
87. ripple

Decade Counters

Since most of the mathematics that we use today is based on the decimal, or base 10, system, it is important to be able to count by this method. Digital

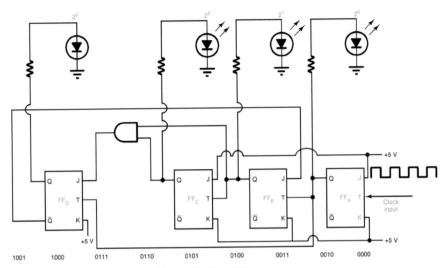

Figure 23.25 BCD counter.

systems are, however, designed to process information in binary form because of the ease with which a two-state signal can be manipulated. The output of a binary counter must, therefore, be changed into a decimal form before it can be used by an individual not familiar with binary numbers. The first step in this process is to change binary signals into a **binary-coded decimal** or **BCD** form. A BCD counter is shown in **Figure 23.25**.

Summary

- A digital system has a source of operating energy, a path, control, a load device, and an indicator.
- A decimal numbering system has 10 individual values or symbols.
- Nearly all digital systems use binary numbers in their operation.
- A binary number has 2 as its base and uses only 1s and 0s and place values are expressed as powers of 2.
- Conversion of a decimal number to a binary number is achieved by repetitive steps of division by the number 2.
- A binary-coded decimal number is used to indicate large binary numbers.
- Octal, or base 8, numbering systems are used to process large numbers through a digital system.

- Hexadecimal numbers used in digital systems have a base of 16 with digits 0-9 and the letters A, B, C, D, E, and F.
- Binary codes have been developed to interface a digital system between binary data and alphanumeric data.
- The BCD code was developed to display decimal numbers in groups of four binary bits.
- The parity code is an error-checking code.
- The odd parity code generates a 0 parity bit when the number of 1s is odd.
- The even parity code is just the opposite of the odd code.
- Three basic gates have been developed to make logic decisions - AND, OR, and NOT.
- An AND gate is designed to have two or more inputs and one output; if all inputs are 1, then the output will be 1. If any input is 0, then the output will be 0.
- Mathematically, the action of an AND gate is expressed as $A \times B = C$. This is the multiplication operation.
- An OR gate is designed to have two or more inputs and a single output; so an OR gate will produce a 1 output when a 1 appears at any input.
- Mathematically, the OR gate function is expressed as $A + B = C$, called OR addition.
- A NOT gate has a single input and a single output and is achieved by an inverter with a 1 input causing the output to be 0 and a 0 input causing the output to be 1.
- Mathematically, the operation of a NOT gate is expressed as $A = A'$ (inverse).
- When two of the basic logic gates are connected together, they form a combination logic gate, with the two most common gates - the NOT-AND and NOT-OR, referred to as NAND and NOR gates.
- A NAND gate is an inversion of the AND function; so when a 1 appears at all inputs, the output will be 0 and when a 0 appears at any input, the output will be 1.
- Mathematically, the operation of a NAND gate is expressed as $A \times B = C$.
- A NOR gate produces an inversion of the OR function; so when the inputs are all 0s, the output is 1 and when any input is 1, the corresponding output is 0.
- Mathematically, the operation of a NOR gate is expressed as $A + B = C$.
- A flip-flop is a logic device with two or more inputs and two outputs.

- A bistable latch has cross-connected NAND or NOR gates.
- An *RS* flip-flop can be designed from two cross-coupled NAND or NOR gates similar to the bistable latch.
- The clocked signal applied to a flip-flop is generally a rectangular-shaped series of waves and are defined as level triggering and edge triggering.
- In level triggering, the state of a clock changes value from 0 to 1 and carries out a transfer of data.
- Positive-edge triggering occurs at the leading edge of a pulse.
- Negative-edge triggering occurs at the trailing edge of a pulse.
- An edge-triggered flip-flop can have its data changed anytime, whereas level triggering can be changed only once at a specific level.
- A clocked *RS* flip-flop has Set, Reset, and Clock inputs with Q and Q outputs, with the inputs either positive- or negative-edge triggered.
- The outputs of an *RS* flip-flop are identified as Q and Q' (inverse).
- A *D* flip-flop is a clocked *RS* flip-flop with an inverter connected across its inputs.
- Digital counters are made up of flip-flops.
- Three flip-flops grouped together form a binary-coded octal counter called a modulo-8 counter.
- A modulo-8 counter goes from 000 to 111_2 and represents 7_{10}.
- When four flip-flops are connected together in a group, it is possible to develop the unit part of a binary-coded hexadecimal counter, called a modulo-16 counter.
- Four flip-flops are commonly connected together to form a binary-coded decimal counter.
- Synchronous, or parallel, input counters trigger each flip-flop at the same time by the clock signal.

Terms

Analog

A quantity that is continuous or has a continuous range of values.

Base

The number of symbols in a number system. A decimal system has a base of 10, a binary system has a base of 2, an octal system has a base of 8, and a hexadecimal system has a base of 16.

Binary

A system of numerical representation that uses only two symbols: 1 and 0.

Decoding

A function of a digital system that is responsible for changing coded data into alphanumeric data.

Digital

A value or quantity related to numbers of discrete values.

Digital integrated circuit

An IC that responds to two states (on-off) data.

Dual-in-line package (DIP)

A packaging method for integrated circuits.

Encoding

A function of a digital system that is responsible for changing input data, which may be in analog form, into binary data.

Energy

Something that is capable of producing work, such as heat, light, chemical, and mechanical action.

Gate

A circuit that performs special logic operations such as AND, OR, NOT, NAND, and NOR.

Hexadecimal

A base-16 numbering system that uses the symbols 0, 1, 2, 3, 4, 5, 6, 7, 8, 9, *A*, *B*, *C*, *D*, *E*, and *F*.

Integrated circuit (IC)

A circuit in which many elements are fabricated and interconnected by a single process, as opposed to a non-integrated circuit, in which individual components such as resistors, diodes, and transistors are fabricated and then assembled.

Logic

A decision-making capability of computer circuitry.

Memory

The storage capability of a device or circuit.

Octal

A base-8 numbering system that uses the symbols 0, 1, 2, 3, 4, 5, 6, and 7.

Pulse

A nonsinusodial signal or wave that occurs randomly or on a periodical basis that can be generated electronically and used to control a digital system.

Radix

The base of a numbering system.

Semiconductor

An element such as silicon or germanium that is intermediate in electrical conductivity between an insulator and a conductor.

Solid state

An area of electronics dealing with the conduction of current carriers through semiconductors.

System

A combination of functional parts (energy source, path, control, load, and indicator), which are needed to make a piece of equipment operate.

Transistor

A semiconductor device capable of transferring a signal from one circuit to another and producing amplification.

Bistable

Any device that can be set into one of two operational states or conditions such as on and off or 1 and 0.

Boolean algebra

Binary logical algebra that provides information in the form of equations and expressions.

NAND gate

A logic gate that produces a 1, or high output, for any input combination except when Is are applied to each input.

NOR gate

A logic gate that will produce a 1, or high output, only when the inputs are all 0s, or low.

NOT gate

An inverter that changes the polarity of the input in its output.

OR gate

A logic gate that provides a 1 output if there is a 1 applied to any of its inputs.

Truth table

A graph or table that displays the operation of a logic circuit with respect to its input and output data.

Asynchronous

Signals or events that can occur at any time without reference to a system clock.

Clear

An asynchronous input to a flip-flop; also called reset used to restore the Q output of a flip-flop to the logic 0 state.

Clock

A pulse generator that controls the timing of a computer through signals applied to various components.

Clock input

The terminal of a flip-flop whose condition changes with a clock signal to provide synchronized control.

Flip-flop

A digital circuit component having a stable operating condition and the capability of changing from one state to another with the application of a signal pulse.

Latch

A simple logic storage element such as a flip-flop that is used to retain data in an operational state.

Memory

A collection of latches that can store a number of different logic levels.

Register

A small group of latches used to store several bits of data.

Reset

A flip-flop input that achieves the same function as clear.

Synchronous

A digital circuit where all of the ICs are paced by a common clock signal and no activity occurs between clock pulses.

Set

An asynchronous input used to restore the Q output of a flip-flop to a logic 1 level (some flip-flops use the term **preset** for this function).

Toggle

A condition describing the output of a flip-flop where the Q and Q' output of a flip-flop changes to the complement of its previous state on each transition of the clock.

BCD counter

A circuit that counts to 10 in binary-coded numbers. A binary-coded-decimal counter.

Binary counter

A counter that progresses through a straight binary counting sequence.

Counter

A device capable of changing states in a specified sequence upon receiving appropriate input data and producing an output that is indicative of the number of input pulses.

Decade counter

A circuit that counts through 10 distinct states.

Decrementing

Decreasing the value of a counter by a known amount.

Incrementing

Increasing the value of a counter by a known amount.

Modulo

A value with respect to the modulus of a body or device such as a ***modulo-5*** counter that responds by making the equivalent of five counts.

Modulus

A value that expresses numerically the degree to which a property is possessed by a body or device.

Register

A group of flip-flops that can be used to store binary information.

Shift register

A collection of flip-flops connected so that a binary number can be shifted into or out of the flip-flops.

Up counter

A circuit that counts from its smallest numerical value to its largest value in binary progression and is incremental.

Index

About the Authors

Dale R. Patrick, Professor Emeritus in the Department of Applied Engineering and Technology at Eastern Kentucky University. His experience includes teaching and organizing laboratory classes in Electrical Engineering/Electronics Technology for many years. He completed Bachelor of Science and Graduate Degrees at Indiana State University. His experience also includes research projects, technical teacher training, and energy consulting for business and industry. He and Dr. Fardo co-authored over 30 textbooks and laboratory manuals.

Stephen W. Fardo, Currently is a Foundation Professor Emeritus now teaching part-time in the Department of Applied Engineering and Technology at Eastern Kentucky University. His experience includes teaching and organizing laboratory classes in Electrical Engineering/ Electronics Technology for many years and coordinating / advising technical teacher education undergraduate and graduate programs at EKU and for the State of Kentucky. He completed Bachelor of Science, Master of Science and Doctoral Degrees at Eastern Kentucky University and University of Kentucky. His experience also includes research projects, technical teacher education consulting, and energy consulting for business and industry. He and Dale Patrick have co-authored over 30 textbooks and laboratory manuals.

Ray E. Richardson, Currently is a Professor in the Department of Applied Engineering and Technology at Eastern Kentucky University. His experience includes teaching and organizing lecture and laboratory classes in Electrical Engineering/ Electronics Technology since 1985 as well as teaching technical seminars, technical writing, and classes in the graduate program. He completed Bachelor of Science, Master of Science Degrees at Eastern Illinois University, and doctorate level at the University of Illinois. He is certified Technology Manager at the Senior level by ATMAE. Other experience includes research projects in technical education and consulting in the manufacturing, food, and electronics areas.

Vigyan (Vigs) Chandra, Currently is a professor and coordinator of the Cyber Systems Technology-related undergraduate and graduate degree programs offered by the Departments of Computer Science and Information Technology (CSIT) and Applied Engineering and Technology (AET) at Eastern Kentucky University. He earned a doctoral degree from the University of Kentucky in Electrical and Computer Engineering and a master's degree in Career and Technical Education from Eastern. He holds certifications in various computer networking areas teaches computer systems and applications, network hardware, communication systems, and digital, analog, and machine-control electronics. He is the recipient of the 2020-21 Critical Thinking and Reading Teacher Award and the 2013 Golden Apple Award for Teaching Excellence at Eastern. His professional interests include implementing active teaching and learning strategies, metacognition, integrating open-source software/hardware with online control and computer networking technologies.

For Product Safety Concerns and Information please contact our EU
representative GPSR@taylorandfrancis.com Taylor & Francis Verlag GmbH,
Kaufingerstraße 24, 80331 München, Germany

Printed and bound by CPI Group (UK) Ltd, Croydon, CR0 4YY
01/05/2025
01858614-0001